Medical Geology

International Year of Planet Earth

The book series is dedicated to the United Nations International Year of Planet Earth. The aim of the Year is to raise worldwide public and political awareness of the vast (but often under-used) potential of Earth sciences for improving the quality of life and safeguarding the planet. Geoscientific knowledge can save lives and protect property if threatened by natural disasters. Such knowledge is also needed to sustainably satisfy the growing need for Earth's resources by more people. Earths scientists are ready to contribute to a safer, healthier and more prosperous society. IYPE aims to develop a new generation of such experts to find new resources and to develop land more sustainably.

For further volumes:
http://www.springer.com/series/8096

Olle Selinus · Robert B. Finkelman ·
Jose A. Centeno
Editors

Medical Geology

A Regional Synthesis

Springer

Editors
Olle Selinus
Geological Survey of Sweden
SE-751 28 Uppsala
Sweden
olle.selinus@gmail.com

Robert B. Finkelman
Department of Geosciences
University of Texas at Dallas
Center for Lithospheric
Studies
800 W. Campbell Road
Richardson TX 75080
USA
bobf@utdallas.edu

Jose A. Centeno
U.S. Armed Forces Institute of
Pathology (AFIP)
Department of Environmental & Toxicologic
Pathology
6825 16 Cpy NW.
Washington DC 20306
USA
tonycent@comcast.net
centeno@afip.osd.mil

ISBN 978-90-481-3429-8 e-ISBN 978-90-481-3430-4
DOI 10.1007/978-90-481-3430-4
Springer Dordrecht Heidelberg London New York

Library of Congress Control Number: 2010925026

Printed on acid-free paper

Springer is part of Springer Science+Business Media (www.springer.com)

Foreword

The International Year of Planet Earth (IYPE) was established as a means of raising worldwide public and political awareness of the vast, though frequently under-used, potential the Earth Sciences possess for improving the quality of life of the peoples of the world and safeguarding Earth's rich and diverse environments.

The International Year project was jointly initiated in 2000 by the International Union of Geological Sciences (IUGS) and the Earth Science Division of the United Nations Educational, Scientific and Cultural Organisation (UNESCO). IUGS, which is a Non-Governmental Organisation, and UNESCO, an Inter-Governmental Organisation, already shared a long record of productive cooperation in the natural sciences and their application to societal problems, including the International Geoscience Programme (IGCP) now in its fourth decade.

With its main goals of raising public awareness of, and enhancing research in the Earth sciences on a global scale in both the developed and less-developed countries of the world, two operational programmes were demanded. In 2002 and 2003, the Series Editors together with Dr. Ted Nield and Dr. Henk Schalke (all four being core members of the Management Team at that time) drew up outlines of a Science and an Outreach Programme. In 2005, following the UN proclamation of 2008 as the United Nations International Year of Planet Earth, the "Year" grew into a triennium (2007–2009).

The Outreach Programme, targeting all levels of human society from decision-makers to the general public, achieved considerable success in the hands of member states representing over 80% of the global population. The Science Programme concentrated on bringing together like-minded scientists from around the world to advance collaborative science in a number of areas of global concern. A strong emphasis on enhancing the role of the Earth sciences in building a healthier, safer and wealthier society was adopted – as declared in the Year's logo strap-line "Earth Sciences *for* Society".

The organisational approach adopted by the Science Programme involved recognition of ten global themes that embrace a broad range of problems of widespread national and international concern, as follows.

- Human health: this theme involves improving understanding of the processes by which geological materials affect human health as a means identifying and reducing a range of pathological effects.
- Climate: particularly emphasises improved detail and understanding of the non-human factor in climate change.

- Groundwater: considers the occurrence, quantity and quality of this vital resource for all living things against a background that includes potential political tension between competing neighbour-nations.
- Ocean: aims to improve understanding of the processes and environment of the ocean floors with relevance to the history of planet Earth and the potential for improved understanding of life and resources.
- Soils: this thin "skin" on Earth's surface is the vital source of nutrients that sustain life on the world's landmasses, but this living skin is vulnerable to degradation if not used wisely. This theme emphasizes greater use of soil science information in the selection, use and ensuring sustainability of agricultural soils so as to enhance production and diminish soil loss.
- Deep Earth: in view of the fundamental importance of deep the Earth in supplying basic needs, including mitigating the impact of certain natural hazards and controlling environmental degradation, this theme concentrates on developing scientific models that assist in the reconstruction of past processes and the forecasting of future processes that take place in the solid Earth.
- Megacities: this theme is concerned with means of building safer structures and expanding urban areas, including utilization of subsurface space.
- Geohazards: aims to reduce the risks posed to human communities by both natural and human-induced hazards using current knowledge and new information derived from research.
- Resources: involves advancing our knowledge of Earth's natural resources and their sustainable extraction.
- Earth and Life: it is over two and half billion years since the first effects of life began to affect Earth's atmosphere, oceans and landmasses. Earth's biological "cloak", known as the biosphere, makes our planet unique but it needs to be better known and protected. This theme aims to advance understanding of the dynamic processes of the biosphere and to use that understanding to help keep this global life-support system in good health for the benefit of all living things.

The first task of the leading Earth scientists appointed as Theme Leaders was the production of a set of theme brochures. Some 3500 of these were published, initially in English only but later translated into Portuguese, Chinese, Hungarian, Vietnamese, Italian, Spanish, Turkish, Lithuanian, Polish, Arabic, Japanese and Greek. Most of these were published in hard copy and all are listed on the IYPE website.

It is fitting that, as the International Year's triennium terminates at the end of 2009, the more than 100 scientists who participated in the ten science themes should bring together the results of their wide ranging international deliberations in a series of state-of-the-art volumes that will stand as a legacy of the International Year of Planet Earth. The book series was a direct result of interaction between the International Year and the Springer Verlag Company, a partnership which was formalised in 2008 during the acme of the triennium.

This IYPE-Springer book series contains the latest thinking on the chosen themes by a large number of Earth science professionals from around the world. The books are written at the advanced level demanded by a potential readership consisting of Earth science professionals and students. Thus, the series is a legacy of the Science Programme, but it is also a counterweight to the Earth science information in several media formats already delivered by the numerous National Committees of the International Year in their pursuit of world-wide popularization under the Outreach Programme.

The discerning reader will recognise that the books in this series provide not only a comprehensive account of the individual themes but also share much common ground that makes the series greater than the sum of the individual volumes. It is to be hoped that the scientific perspective thus provided will enhance the reader's appreciation of the nature and scale of Earth science as well as the guidance it can offer to governments, decision-makers and others seeking solutions to national and global problems, thereby improving everyday life for present and future residents of Planet Earth.

Eduardo F.J. de Mulder
Executive Director International Secretariat
International Year of Planet Earth

Edward Derbyshire
Goodwill Ambassador
International Year of Planet Earth

Preface

This book series is one of the many important results of the International Year of Planet Earth (IYPE), a joint initiative of UNESCO and the International Union of Geological Sciences (IUGS), launched with the aim of ensuring greater and more effective use by society of the knowledge and skills provided by the Earth Sciences.

It was originally intended that the IYPE would run from the beginning of 2007 until the end of 2009, with the core year of the triennium (2008) being proclaimed as a UN Year by the United Nations General Assembly. During all three years, a series of activities included in the IYPE's science and outreach programmes had a strong mobilizing effect around the globe, not only among Earth Scientists but also within the general public and, especially, among children and young people.

The Outreach Programme has served to enhance cooperation among earth scientists, administrators, politicians and civil society and to generate public awareness of the wide ranging importance of the geosciences for human life and prosperity. It has also helped to develop a better understanding of Planet Earth and the importance of this knowledge in the building of a safer, healthier and wealthier society.

The Scientific Programme, focused upon ten themes of relevance to society, has successfully raised geoscientists' awareness of the need to develop further the international coordination of their activities. The Programme has also led to some important updating of the main challenges the geosciences are, and will be confronting within an agenda closely focused on societal benefit.

An important outcome of the work of the IYPE's scientific themes includes this thematic book as one of the volumes making up the IYPE-Springer Series, which was designed to provide an important element of the legacy of the International Year of Planet Earth. Many prestigious scientists, drawn from different disciplines and with a wide range of nationalities, are warmly thanked for their contributions to a series of books that epitomize the most advanced, up-to-date and useful information on evolution and life, water resources, soils, changing climate, deep earth, oceans, non-renewable resources, earth and health, natural hazards, megacities.

This legacy opens a bridge to the future. It is published in the hope that the core message and the concerted actions of the International Year of Planet Earth throughout the triennium will continue and, ultimately, go some way towards helping to establish an improved equilibrium between human society and its home planet. As

Contributors

José Ângelo S.A. Dos Anjos University of Salvador, Salvador, BA, Brazil

Rômulo S. Angélica Federal University of Para, Belem, PA, Brazil

M. Aurora Armienta Universidad Nacional Autónoma de México, Instituto de Geofísica, México, DF 04510, México, victoria@geofisica.unam.mx

Zheng Baoshan The State Key Laboratory of Environmental Geochemistry Institute of Geochemistry, Chinese Academy of Sciences, Guiyang 55002, P. R. China, zhengbaoshan@vip.skleg.cn

Natalia Baranovskaya Polytechnic University, Lenin Avenue 634050, Tomsk, Russia, dgazn@narod.ru

Ospan B. Beiseyev Satpaev Kazakh National Technic University, Republic of Kazakhstan 050013, Almaty, Kazakhstan, beiseyev@mail.ru

Almas O. Beiseyev Al-Faraby Kazakh National University, Republic of Kazakhstan 050065, Almaty, Kazakhstan, beiseyev@mail.ru

Wang Binbin The State Key Laboratory of Environmental Geochemistry Institute of Geochemistry, Chinese Academy of Sciences, Guiyang 55002, P. R. China, wangbinbin@mails.gyig.ac.cn

Maxim A. Bogdasarov State Brest University named after A.S. Pushkin, Republic of Belarus 224016, Brest, bogdasarov73@mail.ru

Mark Cave British Geological Survey, Keyworth, Nottingham NG12 5GG, UK

Jose A. Centeno U.S. Armed Forces Institute of Pathology, Washington, DC 20306, USA, tonycent@comcast.net; centeno@afip.osd.mil

Rohana Chandrajith Department of Geology, University of Peradeniya, Peradeniya, Sri Lanka, rohanac@pdn.ac.lk

Georgy E. Chernogoryuk Siberian State Medical University (SibGMU), Russian Federation 634050, Tomsk, Russia, chernogoryuk@yandex.ru

Angus Cook School of Population Health, University of Western Australia, Perth, WA, Australia

Fernanda G. Cunha Geological Survey of Brazil – CPRM, Rio de Janeiro, RJ, Brazil

T.C. Davies Department of Mining and Environmental Geology, University of Venda, Thohoyandou 0950, Limpopo Province, Republic of South Africa, daviestheo@hotmail.com

Eduardo M. De Capitani University of Campinas, Campinas, SP, Brazil

Alecos Demetriades Institute of Geology and Mineral Exploration, Acharnae, Hellas Gr-136 77, Greece, ademetriades@igme.gr

Olga A. Denisova Siberian State Medical University (SibGMU), Russian Federation 634050, Tomsk, Russia, oadeni@sibmail.com

Gabriela M. Di Giulio Environmental Studies Center, University of Campinas, Campinas, SP, Brazil

C.B. Dissanayake Institute of Fundamental Studies (IFS), Kandy, Sri Lanka, cbdissa@hotmail.com

Anastassia L. Dorozhko Sergeev Institute of Environmental Geoscience Russian Academy of Sciences (IEG RAS), Moscow 101000, Russia, a_dorozhko@mail.ru

Evgeny G. Farrakhov Russian Geological Society, Moscow 115191, Russia, geo@rosgeo.org

Bernardino R. Figueiredo Institute of Geosciences, University of Campinas, Campinas, SP, Brazil, berna@ige.unicamp.br

Robert B. Finkelman University of Texas at Dallas, Richardson, TX 75080, USA; China University of Geosciences, Beijing 100083, P. R. China, bobf@utdallas.edu

Olga V. Frank-Kamenetskaya Department of Geology, St. Petersburg State University, Russian Federation 199034, St. Petersburg, Russia, ofrank-kam@mail.ru

Kunio Furuno Research Institute of Environmental Geology, Chiba. 3-5-1, Image-kaigan, Mihama-ku, Chiba City, 261-0005, Japan, kuniofurunojp@gmail-com

Cristina Garzón INGEOMINAS, Bogotá, Colombia

Heather Gingerich University of Queensland, School of Population Health Brisbane, Australia

Murat Z. Kajtukov Community Scientific Organization of Veterans and Invalids, Vladikavkaz 362013, Russia, alania@rosnedra.com

Alexey E. Khitrov Fedorovsky VIMS, Russian Federation 119017, Moscow, Russia, altitel@mail.ru

Anne Kousa Geological Survey of Finland, GTK, FI-70211 Kuopio, Finland

Gary Krieger NewFields, Denver, CO 80202, USA

Takashi Kusuda Research Institute of Environmental Geology, Chiba. 3-5-1, Image-kaigan, Mihama-ku, Chiba City, 261-0005, Japan, uy4t-ksd@earth.email.ne.jp

Gerald Lalor International Centre for Environmental and Nuclear Sciences, 2 Anguilla Close, UWI, Mona, Kingston 7, Jamaica

Otávio A. Licht MINEROPAR, Curitiba, PR, Brazil

Marta I. Litter Atomic Energy National Commission, Consejo Nacional de Investigaciones Científicas y Técnicas and University of San Martín, San Martín, Buenos Aires, Argentina

Karin Ljung School of Population Health, University of Western Australia, Perth, WA, Australia; Institute of Environmental Medicine, Karolinska Institutet, Stockholm, Sweden, Karin.Calluna@gmail.com

Sandra C. Londono National University of Colombia, Bogotá, Colombia

Nelly Mañay University of the Republic of Uruguay, Montevideo, Uruguay

Olga V. Menchinskaya Institute of Mineralogy, Geochemistry and Crystal Chemistry of Rare Elements, Moscow 121357, Russia, monir@imgre.ru

Humam Misconi Consultant, Baghdad, Iraq, humammisconi@yahoo.com

Michele Monteil University of West Indies, St. Augustine, Trinidad and Tobago, Jamaica, mmonteil@tstt.net.tt

Maria Celeste Morita University of Londrina, Londrina, PR, Brazil

Maryam Navi National Geoscience Database of Iran, Geological Survey of Iran, Tehran, Iran, maryamnavi@gmail.com

Hisashi Nirei Japan Branch of the International Medical Geology Association, 1277, Kamauchiya, Motoyahagi, Katori City, Chiba Prefecture, 287-0025 Japan, nireihisashi@msn.com

Alla V. Oderova Fedorovsky VIMS, Russian Federation 119017, Moscow, Russia, inter-freska@yandex.ru

Mônica M.B. Paoliello University of Londrina, Londrina, PR, Brazil

Igor G. Pechenkin Fedorovsky VIMS, Russian Federation 119017, Moscow, Russia, vims-pechenkin@mail.ru

Igor M. Petrov Ltd. Infomine Research Group, Moscow 119049, Russia, ipetrov@infomine.ru

Oxana L. Pikhur St. Petersburg Medical Academy of Postgraduate Education, Russian Federation 191015, St. Petersburg, Russia, pol0012@mail.ru

Yulia V. Plotkina Institute of Precambrian Geology and Geochronology RAS, Russian Federation 199034, St. Petersburg, Russia, jplotkina@ya.ru

Goeffry Plumlee U.S. Geological Survey, Denver, CO 80225, USA

Anatoly P. Pronin Geotech VIMS, Fedorovsky VIMS, Moscow 119017, Russia, vims@df.ru

C.R.M. Rao Geological Survey of India, Bandlaguda, Hyderabad 500068, India, chebrolumohan@hotmail.com

Robin Rattray International Centre for Environmental and Nuclear Sciences, 2 Anguilla Close, UWI, Mona, Kingston 7, Jamaica

Leonid P. Rikhvanov Department of Geology, Polytechnic University, Lenin Avenue 634050, Tomsk, Russia, rikhvanov@tpu.ru

Ramiro Rodríguez Universidad Nacional Autónoma de México, Instituto de Geofísica, México, DF 04510, México, ramiro@geofisica.unam.mx

Ana Maria Rojas INGEOMINAS, Bogotá, Colombia

Elena V. Rosseeva Department of Geology, St. Petersburg State University, Russian Federation 199034, St. Petersburg, Russia, rosseev@mail.ru

Armen K. Saghatelyan Center for Ecological-Noosphere Studies of NAS RA (Armenia), Republic of Armenia 0025, Yerevan, ecocentr@sci.am; eco-centr@rambler.ru

Lilit V. Sahakyan Center for Ecological-Noosphere Studies of NAS RA (Armenia), Republic of Armenia 0025, Yerevan, ecocentr@sci.am; eco-centr@rambler.ru

Alice M. Sakuma Adolfo Lutz Institute, Sao Paulo, SP, Brazil

Nuria Segovia Universidad Nacional Autónoma de México, Instituto de Geofísica, México, DF 04510, México, nurina@terra.com.mx

Olle Selinus Geological Survey of Sweden, Uppsala SE-751 28, Sweden

Cássio R. Silva Geological Survey of Brazil – CPRM, Rio de Janeiro, RJ, Brazil

Eduardo Ferreira Da Silva Universidade de Aveiro, Aveiro, Portugal

Eiliv Steinnes Department of Chemistry, Norwegian University of Science and Technology, Trondheim NO-7491, Norway

Tommaso Tosiani Central University of Venezuela, Caracas, Venezuela

Zukhra H. Uzdenova Department of Medicine, Kabardino-Balkaria State University, Nalchik, KBR 360004 Russia, bsk@kbsu.ru

Jaques Varet Geological Survey of France, BRGM, Orléans cedex 02, France

Iosif F. Volfson Russian Geological Society, Moscow 115191, Russia, iosif_volfson@mail.ru

Annemarie de Vos School of Population Health, University of Western Australia, Perth, WA, Australia

Philip Weinstein School of Population Health, University of Queensland, Brisbane, QLD, Australia

Paul R.D. Wright International Centre for Environmental and Nuclear Sciences, 2 Anguilla Close, UWI, Mona, Kingston 7, Jamaica, paul.wright@uwimona.edu.jm

Tamara D. Zangiyeva Institute of Mineralogy, Geochemistry and Crystal Chemistry of Rare Elements, Moscow 121357, Russia, monir@imgre.ru

Reviewers

Kaj Lax Geological Survey of Sweden, Uppsala, Sweden

Reijo Salminen Geological Survey of Finland, Espoo, Finland

Pauline Smedley British Geological Survey, Keyworth, Nottingham, UK

Philip Weinstein University of Queensland, Brisbane, QLD, Australia

Bob Finkelman University of Dallas at Texas, Texas, USA

Jose Centeno US Armed Forces Institute of Pathology, Washington, DC, USA

Eduardo Ferreira Da Silva Universidade de Aveiro, Aveiro, Portugal

Monica Nordberg Karolinska Institute, Stockholm, Sweden

Ulf Lindh Uppsala University, Uppsala, Sweden

Erland Johansson Uppsala University, Uppsala, Sweden

Introduction

Medical Geology – A Regional Synthesis

This book, part of a series that will be a legacy to the International Year of Planet Earth, focuses on earth and health, or, as this subject is commonly referred to, medical geology. Medical geology is the science dealing with the relationship between natural geological factors and health in man and animals. It does not deal with pure anthropogenic factors. There is, however, a gray zone. The arsenic catastrophe in Bangladesh is caused by geology factors (naturally occurring arsenic in the groundwater); however, the health problem was triggered by boring millions of tube wells to bring this water to the surface. Mining brings ores and minerals from depth to the surface environment. Unfortunately, there are often serious health consequences caused by oxidation of the ores, liberation of the minerals, etc. Organic pollutants are generally not an issue for medical geology because they are usually anthropogenic; however, there are serious health problems when these compounds are transported by groundwater and deposited in soils. Geoscientists and the information they generate can have important roles to play even in these 'gray zones.'

Medical geology is a rapidly expanding field concerned with the relationship between natural geological factors and human and animal health, including understanding the influence of environmental factors on the geographical distribution of health problems. Medical geology brings together geoscientists and medical/public health researchers to address health problems caused or exacerbated by geological materials; ultimately, it is only with multidisciplinary collaborations that interventions can be devised to reduce morbidity and mortality from such problems. Medical geology also deals with the many health benefits of geologic materials and processes.

Rocks are the source of most chemical elements found on the earth. Many elements in the right quantities are essential for plant, animal, and human health. Most of these elements enter the human body via food and water in the diet and through the air that we breathe. Through weathering processes, rocks break down to form soils on which crops and animals that constitute the food supply are raised. Drinking water moves through rocks and soils as part of the hydrological cycle. Much of the natural dust and some of the gases present in the atmosphere are the result of geological processes. Elements that are essential for our well-being and non-elements, some potentially toxic, exist side-by-side in bedrock or soils and may become a direct risk for human and animal health if present in low quantities (deficiency) or if present in excessive quantities (toxicity). The inability of the environment to provide the correct chemical balance can lead to serious health problems. The links between environment and health are particularly important for subsistence populations that are heavily dependent on the local environment for their food supply. Trace element deficiencies in crops and animals are commonplace over large areas of the world and mineral supplementation programs are widely practiced in agriculture.

Infectious diseases in humans are also dramatically affected by the geological environment, albeit indirectly. Geological forces shape the environments in which microbes thrive, sometimes creating opportunities for the emergence of infectious diseases as major public health problems.

Because of the emergence of these health problems, there has been a growing awareness of the interaction between the natural environment and animal and human health for the past several decades – medical geology. More and more people in developed and developing countries are becoming aware of the potential health impacts caused by geologic processes along with human activities of all kinds that redistribute elements and minerals from sites where they are harmless to places where they adversely impact animal and human health.

Medical geology issues transcend political boundaries, and many issues are found in countries around the globe putting at risk the health of billions of people. A few examples will be mentioned here.

Arsenic in drinking water is one example of international concern. Millions of people in Bangladesh and West Bengal, India, suffer from exposure to high arsenic levels in drinking water. Also millions of people in many countries on other continents are suffering from arsenic exposure or are at risk of arsenic poisoning, a potentially fatal health problem.

High **fluorine** content in drinking water is another medical geology issue causing problems with teeth and limbs all over the world. Fluorides are ubiquitous in nature and are present in rocks, soil, water, plants, foods, and even air. Excessive ingestion of fluoride through water, food, or dust causes acute toxicity or a debilitating disease called 'fluorosis.' Chronic fluoride poisoning is more common and can affect animals as well as humans. Excessive intake during pre-eruptive stage of teeth leads to dental fluorosis and further continued ingestion over years and decades causes bony or skeletal fluorosis.

Iodine deficiency resulting in health problems, such as goiter, affects many millions of people and is a consequence of the local geology, deficiency of the element iodine in bedrock and soils and thus a deficiency of iodine in the diet.

Selenium is an essential trace element. However, selenium deficiency (due to soils low in selenium) has been shown to cause severe physiological impairment and organ damage. Several areas in the world have been demonstrated to have soils deficient in selenium. Selenium deficiency is particularly prevalent in China.

Geology is the most important factor controlling the source and distribution of **radon**. Relatively high levels of radon emissions are associated with particular types of bedrock and unconsolidated deposits, for example, some, but not all, granites, phosphatic rocks, and shales rich in organic materials. Inhalation of radon was a severe occupational hazard for uranium miners, commonly resulting in fatal lung cancer.

Balkan endemic nephropathy (BEN), an irreversible kidney disease associated with renal pelvic cancer, was thought to be confined to several rural regions of the Balkans, but recently it has been discovered in other parts of the world by medical geology scientists.

Atmospheric dust is a global phenomenon. Dust storms from Africa regularly reach the European Alps and the Western Hemisphere. Asian dust can reach

California in less than a week, some of the dust ultimately crossing the Atlantic and reaching Europe. The ways in which mineral dust impacts upon life and health are wide ranging. These include changes in the planet's radiative balance, transport of disease bacteria to densely populated regions, dumping of wind-blown sediment on pristine coral reefs, general reduction of air quality, provision of essential nutrients to tropical rainforests, and transport of toxic substances. Mobilization of dust is both a natural and an anthropogenically triggered process.

This book will cover all these aspects and many more, on a global and regional scale. Many specialists from all around the globe have contributed with their expertise. The book gives many examples of environmental health issues from different continents and also an overview of what is going on now in the field of medical geology. *A truly international book like this also reflects the different scientific cultures from all around the world with different terminologies and different ways of thinking. The readers must bear this in mind when reading the book.*

Uppsala, Sweden — Selinus, Olle
Washington, DC — Centeno, J.A.
Richardson, TX — Finkelman, R.B.

Medical Geology Issues in North America

Robert B. Finkelman, Heather Gingerich, Jose A. Centeno, and Gary Krieger

Abstract To a larger degree than most others, North Americans are shielded from the natural environment. Nevertheless, health problems caused by geologic materials and geologic processes do occur in North America. In contrast to the acute health problems caused by the geologic environment in developing countries, in North America these health concerns are more likely to be chronic, caused by long-term, low-level exposures. Among the potential health concerns that have received public health attention are exposure to trace elements such as fluorine, arsenic, and radon; exposure to natural mineral dusts; occupational and community exposures to trace elements; and ingestion of naturally occurring organic compounds in drinking water. This chapter provides North American examples of each of these environmental health problems and suggestions how the earth sciences can be an integral part of multi-disciplinary teams working to mitigate these problems.

Keywords United States · Canada · Fluorosis · Cancer · Asbestos · Black lung disease · BEN · Organic compounds · Radon · Arsenic · Diabetes · Fluorine

Introduction

To a larger degree than most other people, North Americans are shielded from the natural environment. North Americans commonly live, work, and travel in air-conditioned environments; in supermarkets they purchase foods grown all over the world; most drink municipal water that has been purified; and many take daily vitamins to supplement dietary deficiencies. Despite this shielding from the natural environment, health problems caused by geologic material and geologic process do occur in North America. In contrast to the acute health problems caused by the geologic environment in developing countries, in North America these problems are generally chronic, caused by long-term, low-level exposures.

There is a rich medical geology history in North America dating back to 1792 when John Rouelle described the medicinal properties of mineral water in Virginia (Rouelle, 1792). The 1970s enjoyed a renaissance marked by the appearance of a series of publications focused on the impacts of trace elements on human health. These publications included Geological Society of America (GSA) Memoir 123 (Cannon and Hopps, 1971); GSA Special Paper 140 (Cannon and Hopps, 1972); Annals of the New York Academy of Sciences Volume 199 (Hopps and Cannon, 1972); U.S. Geological Survey Professional Paper 574-C (Shacklette et al., 1970); GSA Special Paper 155 (Freedman, 1975); and National Research Council (1979).

Changes in staffing and priorities resulted in a hiatus in North American medical geology activities from the 1980s until the mid-1990s. The past decade has seen a resurgence of interest in medical geology evidenced by the inclusion of human health issues in the USGS Strategic Plan (U.S. Geological Survey, 2007), the formation of GSA's Geology and Health Division, and strong North American leadership and representation in the International Medical Geology Association.

R.B. Finkelman (✉)
University of Texas at Dallas, Richardson, TX, 75080, USA
e-mail: bobf@utdallas.edu

With contributions by Geoff Plumlee on Evolving Concerns about Asbestos and A Growing Role for Earth Sciences in Environmental Disaster Response and Planning.

O. Selinus et al. (eds.), *Medical Geology*, International Year of Planet Earth,
DOI 10.1007/978-90-481-3430-4_1, © Springer Science+Business Media B.V. 2010

This chapter is not intended to be a compendium of geology and health issues in North America. Rather, our objective is to highlight some of the more prevalent issues such as exposure to selected trace elements, dust, naturally occurring organics, and occupational health issues providing both historical and current examples of heath issues caused by geologic materials.

With renewed interest in the links between geology and health, natural elemental hazards are an important consideration in the rural areas as well as the densely populated urban centers of North America where cumulative, low-dose, and long-term exposures can lead to some of the chronic illness that consumes moderate percentages of the national incomes (as GDP) of the United States (15.9%), Canada (9.7%), and Mexico (6.7%), respectively (*The Economist*, July 19, 2007).

Trace Element Exposure

One of the major themes of the geology and health story involves acquiring an understanding of the consequences of both natural and perturbed cycling of elements between and within the lithosphere, the atmosphere, the hydrosphere, and the biosphere. In some cases, human settlement on our increasingly crowded landscape has expanded to include environments that are "naturally impaired" by either an over-abundance or a deficiency of elements relative to human and animal biological requirements.

Where elemental exposures create toxicity, safeguarding health becomes a matter of finding the "point(s) of intervention" in the Source–Path–Trap relationship, whether it be in air, water, earth materials or the food chain, and adapting appropriately to the natural environment in sustainable ways. In other instances, the challenge is to distinguish natural elemental occurrences, which we generally can only avoid through awareness, from "anthropogenic overprinting." Naturally occurring elements such as fluorine, radon, and arsenic have had variable impacts on different communities in North America as a function of route of exposure and dose. For example, US EPA considers that radon, "... is the leading cause of lung cancer among non-smokers. Radon is the second leading cause of lung cancer in America and claims about 20,000 lives annually." (http://www.epa.gov/radon/). Similarly, naturally occurring arsenic levels in ground water sources can be a significant issue in many parts of the United States especially in the western mountain regions where ground water levels often exceed regulatory standards (http://www.epa.gov/safewater/arsenic/index.html).

Fluorine

Owing to its extensive use in preventive dentistry in the post-World War II era, fluorine (F), which is commonly referred to as its ionic form, *fluoride*, is one of the most familiar and controversial elements of the Periodic Table. Safe, responsible, and sustainable use of fluorides is dependent on decision makers (whether they be politicians or parents) having a firm grasp on three key principles: (i) fluorine is not so much "essential" as it is "everywhere," (ii) recent human activities have significantly increased fluorine exposures to the biosphere, and (iii) fluorine has biogeochemical effects beyond bones and teeth.

Some of the fluorine that is ingested, imbibed, inhaled, or absorbed through the skin is excreted via the kidneys after having spent time in the circulatory system (WHO, 1997), whereas the balance is integrated into the body's mineralized tissues of teeth, bone (Ledbetter et al., 1960), the pineal gland (Luke, 2001), and sometimes as constituents of calculi or "stones" in the kidney, gallbladder, and tonsils. The easily identified and irreversible cosmetic, and sometimes structural, damage of *dental fluorosis* (Fig. 1) is caused by ingestion of excessive fluoride prior to the eruption of the tooth through the gum-line in childhood (Ruan et al., 2007). *Skeletal fluorosis* mimics a host of osteological disorders, including osteoarthritis and osteoporosis, and primarily affects adults in middle age although earlier incidence can occur in severe cases (Skinner, 2005). The full range in dysfunction associated with *systemic fluorosis* is still not well understood, although research has found correlations with thyroid disorders (National Research Council, 2006), certain cancers (Bassin et al., 2006), and deleterious effects on the brain (Mullenix et al., 1995).

Fluorine is ubiquitous in the natural environment. At the atomic level, fluorine is not only the 14th most abundant element in the Earth's crust –more abundant than any other halogen and even the "Basic Building Block of Life", carbon (C) –but also supremely reactive and oxidative, with the highest Pauling value for electronegativity of 4.0 and an unusually low dissociation energy, which means that it has the tendency to "steal" electrons from most other elements or at least "share" them by forming strong bonds. The fluoride ion (F^-) has a charge of (–1) and a similar ionic radius (1.33 Å) to that of the anions of oxygen (O^{2-}, 1.40 Å) and the hydroxyl group (OH^-, 1.32 Å), but being so much more electronegative, it often substitutes for the hydroxyl group in mineral complexes and increases the stability of the crystal lattice structure. The resulting resistance to dissolution in acid of fluorapatite as compared with hydroxylapatite is the basis of fluoride's use in preventive dentistry.

Earth materials that are characteristically rich in fluorine are organic clays and shales, carbonatites, phosphates, hydrothermal ores, and silicic igneous rocks like rhyolites, dacites, and granites –especially the Rapakivi and alkali type (Boyle, 1976). In short, every geologic environment in the lithosphere –igneous, metamorphic, sedimentary –contains fluorine, though the amount that is bioavailable can vary considerably.

By volume, the greatest amount of naturally occurring fluorine in the atmosphere occurs as HF gas related to volcanic activity. Fluoride also regularly makes its way into the atmosphere as marine aerosols and by the diffusion from the surface of the water-soluble fluoride mineral *fluorite* (CaF_2), which is common to the non-volcanic subsurface environment of the Mississippi Valley of the United States and some parts of Ontario and Québec in Canada (Boyle, 1976).

Within the biosphere, plants appear to be relatively tolerant to high fluoride in groundwater and soils (Kabata-Pendias, 2001) and respond by accumulating this element in the leaves (which is why tea is a fluoride-rich foodstuff). While not obviously toxic to the plant itself, fluorine-laden particulates associated with volcanic activity can accumulate on forage materials that are ingested by herbivores and then concentrated along the food chain (Fleming et al., 1987).

Aside from the rare instances when one of North America's many active volcanoes like Mount St. Helen's or Popocatepetl erupts, drinking water is the primary route of fluoride exposure for individuals and communities. The average concentration of fluoride in seawater is 1.35 mg/l, but with the exception of alkali lakes, other surface waters and precipitation are generally naturally low in dissolved fluoride ion (0.01–0.3 mg/l), depending on atmospheric inputs and the geochemistry of the earth materials contacted (Edmunds and Smedley, 2005). Groundwater, on the other hand, can vary tremendously with respect to fluoride content.

Dental Fluorosis: A Century of "Colorado Brown Stain"

Nestled at the foot of the Pikes Peak at the eastern end of the Rocky Mountain Chain, Colorado Springs at the turn of the last century was a breathtakingly beautiful, albeit challenging, first posting for the young dentist, Dr. Frederick McKay. Fresh out of dental school and far from his native Massachusetts, McKay was puzzled by the mottled and sometimes pitted appearance of many of his new patients' teeth, locally referred to as "Colorado Brown Stain" that reminded him of a phenomenon that was prevalent in the local population in St. Louis, Missouri, in the heart of the Mississippi River Valley, where he completed his training in orthodontia. Upon further investigation, it was found that 87.5% of surveyed school children that were born in the Pikes Peak region had some degree of what is now known as *dental fluorosis* that residents attributed to "something in the water." It was the Spring of 1909. (Paraphrased from the Pierre Fauchard Academy International Hall of Fame of Dentistry induction speech.) (Fig .1).

In 1931, US Public Health Service researcher Henry Trendley Dean concluded that the "something" that caused Colorado Brown Stain was naturally elevated levels of fluoride (F^-) in the local water supply, which

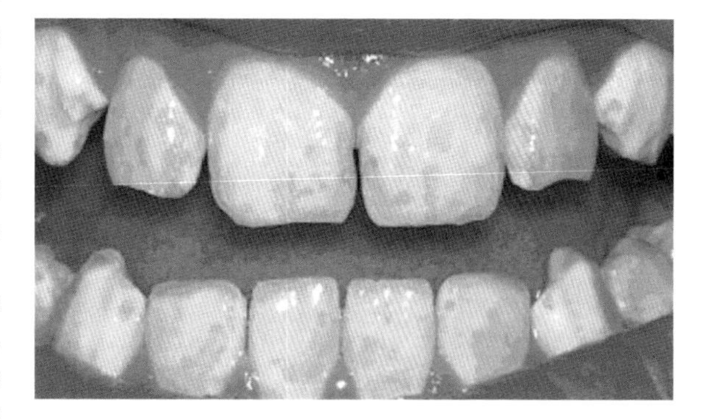

Fig. 1 *Colorado Brown Stain.* Dental fluorosis that would be classified as "moderate" on Dean's Index. All enamel surfaces of the teeth are affected, and the surfaces subject to attrition show wear. Brown stain is frequently a disfiguring feature that lasts for life. Source: Photo courtesy of Hardy Limeback

substituted for hydroxyl (OH^-) ions of similar size and charge in the apatite [$Ca_{10}(PO_4)_6(OH,F,Cl,Br)_2$] crystals of teeth. Looking at the geology of the Colorado Springs area, this is hardly surprising, as there are no less than three significant geogenic sources of fluoride in the environment. With meteoric waters of low ionic strength and slightly acidic pH cascading off the 1.1 billion year-old Rapakivi granites of the Pikes Peak Batholith, and at least two major mineralized faults (the Ute Pass and Rampart Range Faults) that are associated with the Laramide Orogeny (or mountain building event) 65 million years ago, fluoride from easily dissolved minerals would be picked up and concentrated all along the flow path. Being a black (organic-rich), and therefore also likely fluorine enriched, Cretaceous sedimentary unit deposited sometime over the last 65–145.5 million years, the Pierre Shale that underlies Colorado Springs would have contributed even more fluoride to the local water supply once it bubbled up through the network of fractures and fissures in the bedrock of the plateau (Fig. 2).

In the decades that followed, dental data have been instrumental in identifying fluoride-rich environments, as children are especially susceptible to developing the tell-tale sign of mottled teeth that they carry into adulthood due to high fluoride exposure relative to their low body weight during critical phases in enamel formation. H. Trendley Dean became the first director of the National Institute of Dental Research in 1948 and was able to identify several other geochemical environments that are associated with high naturally occurring fluoride resulting in adverse biological effects within the continental United States. Notable among them are the petroleum-rich Lone Star State which is home to "Texas Teeth" (a southern cousin of "Colorado Brown Stain"), the fault- and fracture-riddled bedrock that hosts the metal-sulfide ore deposits of the Upper Mississippi River Valley (noted by Frederick McKay during his time in St. Louis), the highly metamorphosed coal-bearing hills of Virginia, Maryland, and Pennsylvania along the eastern seaboard, and the phosphate-rich Florida Panhandle (Fig. 3).

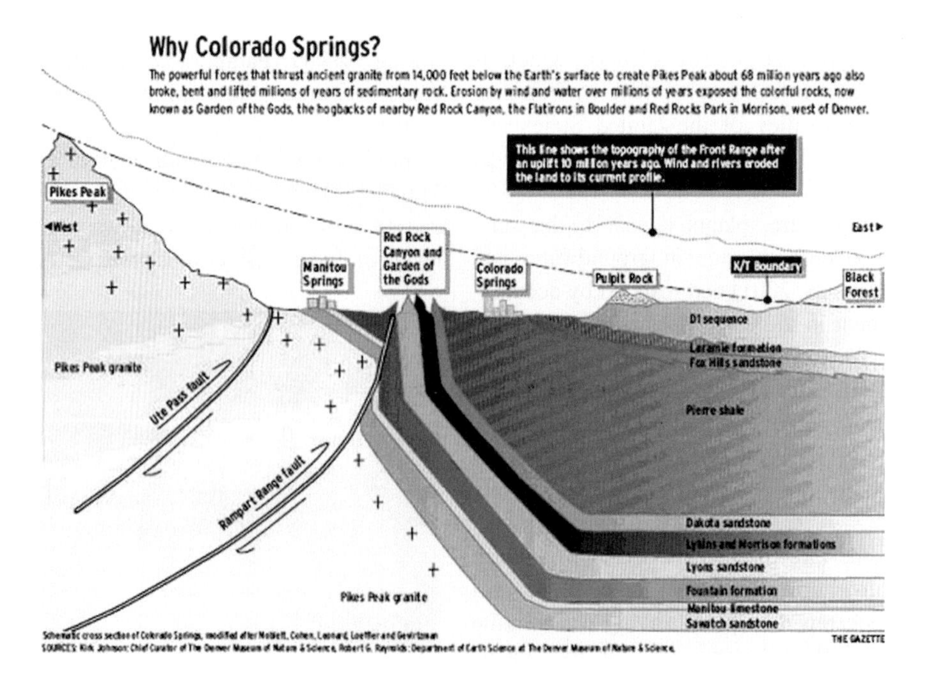

Fig. 2 *Geologic cross-section of the Pikes Peak area.* Clearly an environment that would be prone to the development of fluorosis in children with three natural sources of fluoride. Naturally occurring fluoride in the water supply of nearby Manitou Springs is reportedly in the 3.0–3.6 mg/L range. Source: From http://gazetteoutthere.blogspot. com/2007/07/colorado-springs-geology-rocks.html

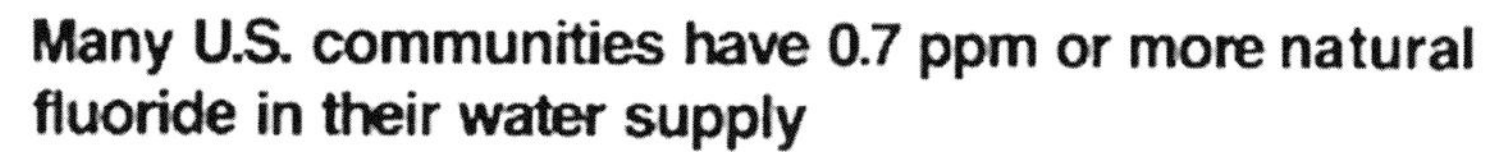

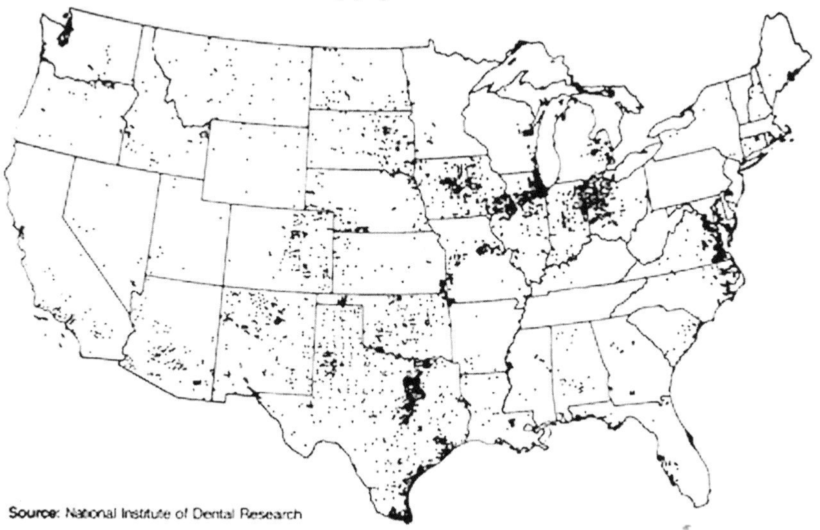

Fig. 3 Many US communities have 0.7 ppm or more natural fluoride in their water supply

Recent Human Activities Have Perturbed Natural Fluorine Cycling

While volcanic eruptions do not happen every day, and few people live in close proximity to an economically significant mineral deposit, the fluorine exposures of modern North American residents have increased dramatically since the end of World War II when industrial fluorine use became commonplace. Today, the primary sources of anthropogenic fluorine emissions include phosphate fertilizer production, aluminum and magnesium smelting, coal burning, oil refining, steel production, chemical production, primary copper and nickel production, clay production, lead and zinc smelting, glass and enamel making, brick and ceramic manufacturing, glues and adhesives production, fluoridation of drinking water, waste from sewage sludge, and the production of uranium trifluoride (UF_3) and uranium hexafluoride (UF_6) for the nuclear industry (Environment Canada, 2001). Most of these new sources of fluorine to the atmosphere and hydrosphere are the result of processing Earth materials that had previously been largely biologically unavailable as little as 60 years ago.

Taken together with the advent of fluorinated agrichemicals and phosphate fertilizers introduced to the food chain via intensive agricultural practices (Hudlicky and Pavlath, 1995), the development and marketing of an FDA-approved fluoridated toothpaste by Proctor and Gamble Company, and the application of deep well drilling technology to groundwater resources (Bailey, 2006, personal communication) that also coincided with the post-war era, it is perhaps not too much of an exaggeration to say that consumer choices (i.e., what you eat, where you live, where you get your drinking water) now play a bigger role in determining overall fluorine exposures than Nature. And whether water in particular is considered to be "high" or "low" in fluoride is relative only to established benchmarks that seem to vary considerably depending on its intended use but not its eventual fate.

In broad terms, it can be said that groundwater supplies in arid climates in North America are more fluoride rich than those in temperate climates; that aquifers influenced by certain igneous and metamorphic earth materials are generally more fluoride-enriched than aquifers pumping from sedimentary rock (although sedimentary aquifers are more widespread); that deep-source bedrock water wells produce more fluoriferous water as compared to shallow overburden wells; and that bedrock aquifers that are highly fractured due to industrial blasting, hydraulic fracturing, meteorite impacts, seismicity, and/or glacial isostasy are higher in fluoride than undisturbed formations (Boyle, 1976).

Skeletal Fluorosis: The Danger of Drilling Deeper

In the 1970s, when the shallow dug wells of a small town in Canada's Gaspé Peninsula were no longer able to meet their needs, the residents of Maria, Québec, did what most rural residents would do –they drilled deeper. This has become standard practice in most parts of Canada, where 33% of the population currently rely on groundwater resources, especially with mounting concerns over the vulnerability of shallow wells to dropping water table levels associated with climate change and susceptibility to anthropogenic surface contaminants like road salt run-off, landfill and septic system leachate, fertilizers, pesticides, floodwaters, and manure. What the residents did not anticipate when they drilled through the 10–30 m (30–100 feet) of Quaternary glacial tills, whose mean fluoride concentration was 0.1 mg/l, was that they had been exposing themselves to 100 times more fluoride from "fossil waters" extracted from the highly mineralized and naturally softened Carboniferous sedimentary bedrock aquifer (Fig. 4).

Although it took several years, eventually the dramatically increased incidence of osteoarthritis-like symptoms among Maria residents caught the attention of local public health authorities who, with the help of a geoscientist well versed in fluoride geochemistry, diagnosed *skeletal fluorosis* resulting from the consumption of fluoride-enriched drinking water for as little as 6 years (Boyle and Chagnon, 1995). X-rays of the affected individuals would have been familiar to Dr. Kaj Roholm, who had observed similar effects in the skeletons of Danish cryolite (Na_3AlF_6) workers in Greenland in the 1930s, documenting his findings in the landmark publication, *Fluorine Intoxication: A Clinical-Hygienic Study* in 1937 (Roholm, 1937: source of photos below) (Fig. 5).

Dental fluorosis was not prevalent in either the cryolite workers in Greenland, whose primary route of exposure was via inhaled dust particles, or the mostly adult population of 2,500 in the town of Maria because the critical period of exposure in the biomineralization of enamel occurs before a tooth erupts through the gum-line during early childhood. The human skeleton is continuously being re-modeled through the action of osteoclasts and osteoblasts, replacing itself completely three times over the course of an average lifespan. Therefore, skeletal fluorosis is the most obvious sign of pathology in adults exposed to high levels of fluorine but there is some hope of recovery once fluoride inputs have been significantly reduced, whereas dental fluorosis is irreversible.

Fluorine from any source has biological effects beyond human bones and teeth that are consistent with its unique chemical characteristics. From a geology and health perspective, it is important to realize that

Fig. 4 British postage stamp showing the glaciated terrain of the Gaspé Peninsula and its location on the brackish waters of the St. Lawrence Seaway. The overburden is under the influence of meteoric waters, whereas the presence of sodium increases the solubility of fluoride minerals in the cement of the Carboniferous bedrock sediments

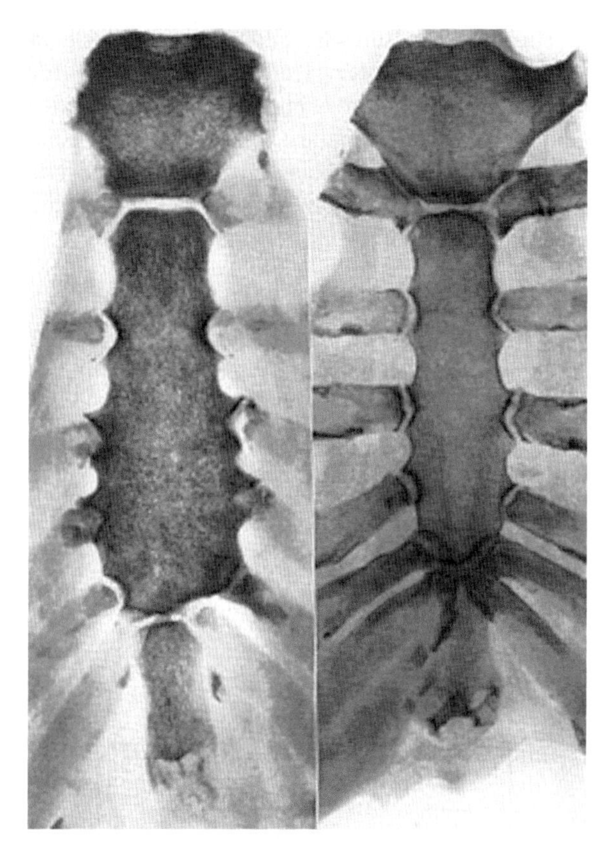

Fig. 5 Chest X-rays of a 51-year-old Danish cryolite worker (*left*) and a 50-year-old non-worker (*right*). The sternum is thickened and misshapen and the articular surfaces of the costal (rib) bones are sclerosed, making inhalation difficult

fluorine's function in nature is to facilitate chemical reactions. It lowers the energy required to both bring certain elements together and break molecules apart – its behavior remains the same across the lithosphere, atmosphere, biosphere, and hydrosphere. This raises questions regarding the extent of the effects of anthropogenic fluorine exposures in the human system as well as the ecosystem, as municipal and industrial wastewaters are discharged into surface water bodies at levels that are greater than 10 times what sensitive aquatic species can tolerate (Environment Canada, 2001).

The combination of natural and man-made fluorine compounds in the upper atmosphere is resulting in "greenhouse powerhouse" gases like trifluoromethyl sulfur pentafluoride (SF_5CF_3) (Sturges et al., 2000) and fluoride-facilitated mineralization in the pineal gland early in life is thought to be contributing to the premature sexual maturation (Luke, 1997) with all of its associated "social system" implications related to early parenting. Infancy appears to be a critical time with regard to fluorine exposures and the development of symptoms of systemic fluorosis in adulthood, an aspect that was not captured in the early water fluoridation studies that lasted only 10 years (Bast et al., 1950).

One of fluorine's often over-looked effects in the hydrosphere is that it makes lead-, copper-, and cadmium-containing compounds in plumbing pipes and fixtures and cookware more soluble in water, particularly at the higher temperatures that might be experienced from the water heater to the tap or while cooking with copper pots (Boyle, 1976). Geochemistry tells us that the corrosivity of a low pH solution (i.e., surface water) is further enhanced when Na^+ and Cl^- ions are also present (Barnes, 1979), as is the case when certain disinfecting agents like sodium hypochlorite or chlorine bleach (NaClO) and chloramine (NH_2Cl) are added to naturally or artificially softened water. Many naturally low-fluoride North American municipal water supplies are now artificially fluoridated to levels of between 0.6 and 1.2 mg/l using fluoride salts (NaF, villiaumite also known as sodium fluoride) or, more commonly, with fluorine-rich by-products of phosphate fertilizer processing like hydrofluosilicic acid (H_2SiF_6) that also contain other elemental constituents like lead, arsenic, and natural radionuclides because phosphate mineral deposits are never 100% pure (Wedepohl, 1978).

Systemic Fluorosis: The Catalyst of the Universe

Lead is among the handful of elements that are known to have adverse neurological effects to which unborn and young children are the most vulnerable due to their low body weight and susceptible stage of brain development (Gavaghan, 2002). Aside from updating old infrastructure, recommending to homeowners that lead plumbing be replaced, and incorporating the use of lead-free materials into the building code, managers of municipal water supplies can also control corrosion in the system by using some additives that adjust pH and alkalinity and others that are meant to inhibit leaching through the formation of an inorganic film on the inside of the pipes. The effectiveness of this last measure will depend heavily on the purity of the corrosion inhibitors (typically orthophosphate, polyphosphate and sodium silicate), as phosphate rock contains between 10,400 and 42,000 mg/kg of fluorine (Boyle, 1976) and some sodium silicate minerals contain up to 5% fluorine by weight (Wedepohl, 1978), possibly doing more harm than good in the cases where artificial fluoridation is the cause of the corrosion (Fig. 6).

High lead exposures in childhood can have significant long-term neurological effects that can range from extremely subtle to gross impacts on motor function. Some research data indicate that high early childhood exposures are associated with significant long-term effects including lower intellectual performance

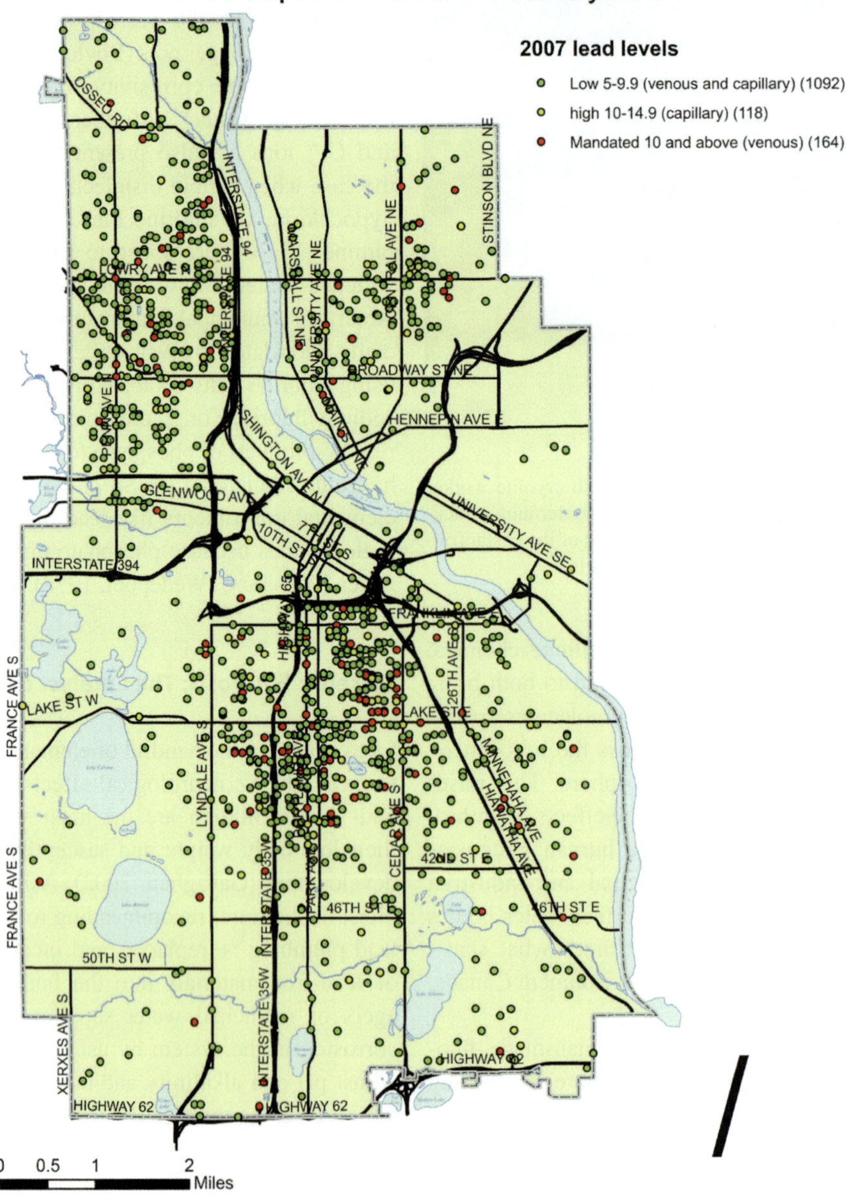

Fig. 6 Blood levels in children in Minneapolis collected by the Public Health Unit

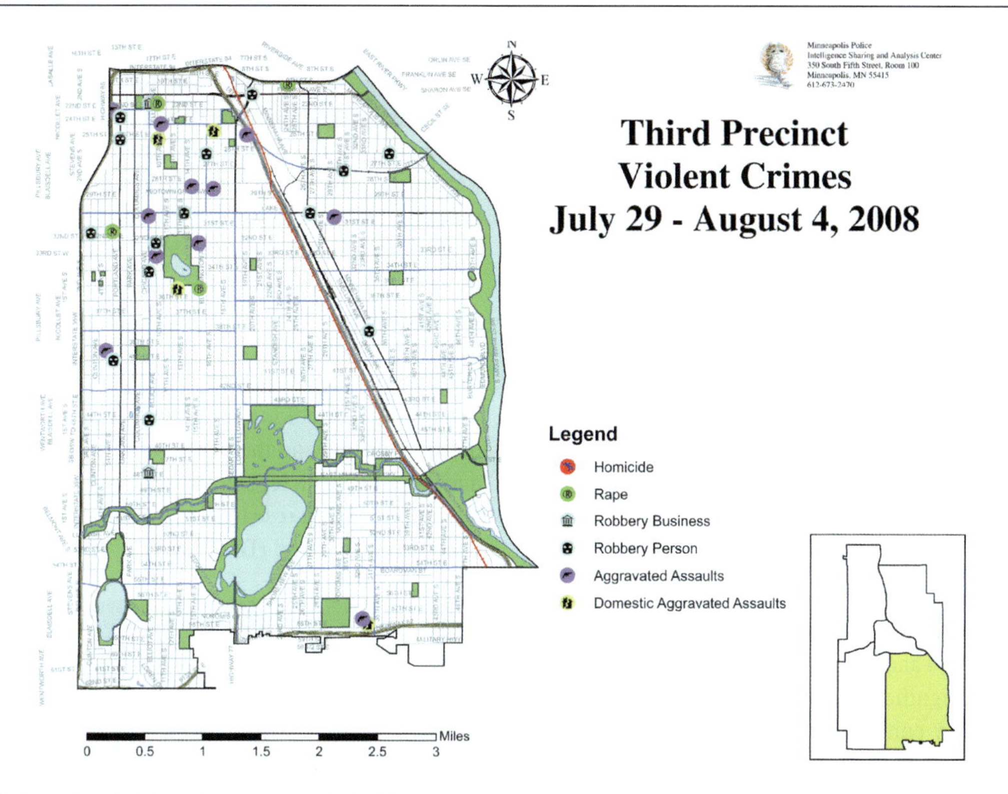

Fig. 7 Violent crime incidence for the week of July 29 – August 4, 2008 in the Third Precinct of Minneapolis, a US city that artificially fluoridates its drinking water. Adding fluorosilicates to weakly-mineralized and low-pH surface water from lakes and rivers increases community lead exposure, particularly in older and economically-depressed neighborhoods where lead plumbing has not been updated. Fluoride in water also increases the leaching of copper, cadmium and zinc from infrastructure, contributing to metal toxicity in consumers

and higher rates of delinquency. Large and long-established communities like Minneapolis, Minnesota (incorporated in 1867), are particularly at risk because of the metals in the pipes of the old infrastructure that is common to the more densely populated and poorer parts of town where nutritional status is also low. Residences closest to the treatment and distribution centers might also be receiving higher doses of chlorine and fluorine (up to 1.2 mg/l) added to Mississippi River water as compared to those at "the end of the line." Though there are more questions than answers at this point, aided by GIS technology, a multidisciplinary team of researchers is currently investigating different aspects of the relationship between fluoride-enhanced lead exposures via drinking water and violent crime rates in major US cities (Fig. 7).

Prior to the Industrial Revolution, overly fluorine-rich "provinces" were generally restricted to specific natural environments like coastal, volcanic, and arid areas, or to places with deposits of soluble fluoride minerals like fluorite (CaF_2), villiaumite (NaF), and cryolite (Na_3AlF_6) and/or rifted, faulted, and fractured terrain (Edmunds and Smedley, 2005). Given its chemical properties, it becomes clear that sufficient fluorine exposure to meet the biological requirements of a particular organism is easily obtained from a variety of sources –even outside of these "fluoriferous" natural environments –without any effort, and so attention needs to be focused on avoiding toxicity.

Fluoride's geochemical associations with some "bad company" in the public health world, together with the enhanced solubility of certain metals found in plumbing and fixtures in the presence of fluoridated water (particularly with a low pH and low calcium–magnesium content as is common to surface water sources), make it a potentially useful tool for identifying communities in environments that are naturally prone to arsenicosis, heavy metal toxicity, and radon gas exposure from the radioactive decay of uranium in earth materials.

Radon

The harmful isotope-222 of the noble gas, *radon,* is produced from the natural radioactive decay of radium-226 and uranium-238, but this was not known when Czech silver miners were suffering from Bergkrankheit or "mountain sickness" in 1550 (Witschi, 2001). The ultimate cause of this pulmonary illness that was the cause of death among miners 75% of the time, now diagnosed as lung cancer, was not identified until 1924 –almost 3 decades after Becquerel had discovered radioactivity (in 1896) and the husband and wife team of Pierre and Marie Curie had isolated radium (in 1898), and well over a century after Martin Klaproth extracted uranium from the mineral pitchblende (in 1789) or uraninite (UO_2) (Porter and Ogilvie, 2000). The carcinogenicity of radon gas is associated with its decay products, primarily polonium-218 and polonium-214, which deliver a radiologically significant dose of alpha and gamma rays to pulmonary epithelium as inhaled aerosols of short-lived progeny on the way to eventually becoming the stable isotope of lead-206 –a process that takes about 22 years (National Research Council, 1999).

Between the mid-1940s and 1990 in North America, several cohorts of more than 3,500 fluorspar and uranium miners receiving high occupational exposures of radon-222 were followed. But before these studies had concluded, public health researchers were already looking at the carcinogenicity of radon gas exposures in North American residential settings (National Research Council, 1999). By the mid-1990s it was concluded that 10–15% of lung cancers in the United States and Canada were caused by radon gas, 90% of which were also smokers.

Radiation in the Navajo Nation

Being primarily engaged in agriculture and other "open-air" activities and originally having low smoking rates, lung cancer among Native American Indians in the southwest was once a rare occurrence (Gottlieb and Luverne, 1982). But when vast uranium deposits were discovered beneath their territory in the years leading up to World War II, many male members of the Navajo Nation were eager for work in the mines so close to their homes, despite the fact that they would be paid less than minimum wage (between $0.81 and 1.00) for their back-breaking labor. Although they were doing most of the work by hand in an enclosed space, they were not provided with any protective equipment, as ventilation measures developed from the European experience with "mountain sickness" were purported to have drastically reduced radon exposures. As is common, the greatest public health failure seems to have been related to poor communication and lack of education, as there is no word for "radiation" in the Navajo language, and few Navajo miners spoke much English. Adding to the constellation of factors were the high rates of smoking among all miners, particularly during wartime and the period immediately following, and lack of clear legislation that required employers and public health researchers to inform workers of the occupational hazard (Brugge et al., 2006) (Fig. 8).

Fig. 8 Uranium mines and mills in the southwest United States employed many local Native Americans who suffered from high rates of pulmonary illness, including lung cancer. Source: http://www.ehponline.org/docs/2007/115-12/ mining.jpg

The radon story parallels that of fluorine. The earth materials that may produce radon are also similar to those that are known to be high in fluorine, namely silicic igneous rocks, organic-rich clays and shales, phosphates, carbonatites, and hydrothermal or metasomatic ore deposits in "structural disconformities" such as are found in Canada's Athabasca Basin and various locations in the United States (Fig. 9).

The atmosphere and the hydrosphere provide subsurface pathways for radon into the biosphere, where human exposures occur in the home. Depending on its location and construction, radon may be inhaled as it

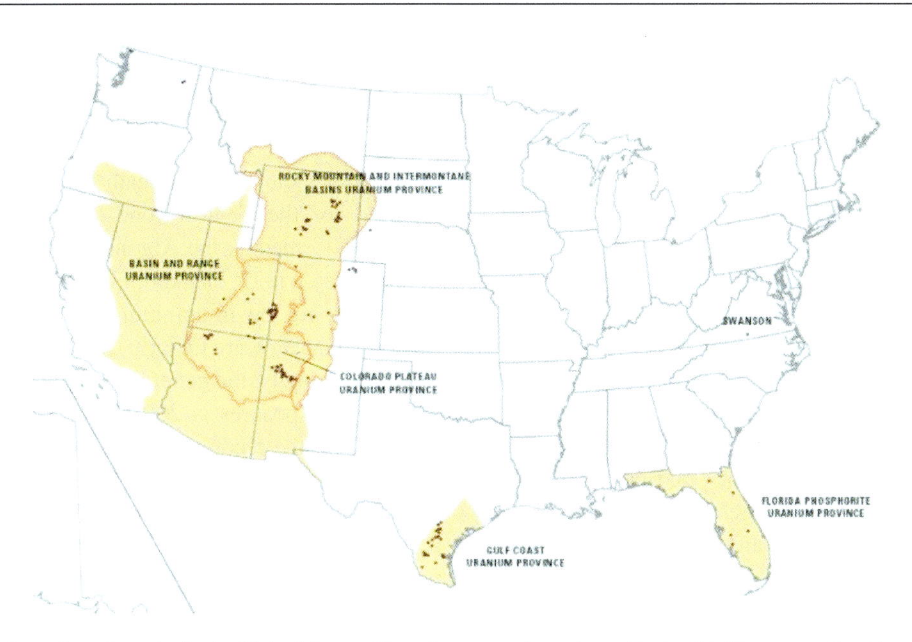

Fig. 9 Map of uranium deposits in the United States. Note the areas of overlap with the drinking water fluoride map in Box 1. Source: USGS (http://www.energy-net.org/01NUKE/u-mining/us-uranium-usgs.jpg)

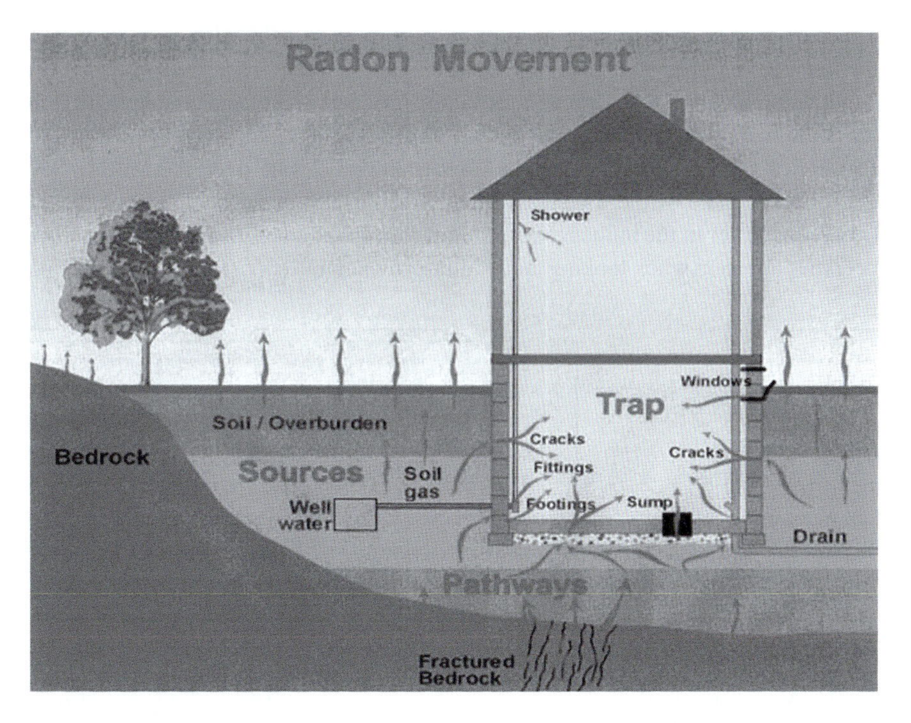

Fig. 10 Diagram of the many pathways radon gas can employ in migrating from its source in uranium-rich earth materials to its trap within the home. Source: Zielinski (2008)

vaporizes from running water and drains, or as it seeps through ground-level windows, foundation cracks, and gaps created by fittings and footings. Like fluoride, uranium (and therefore, dissolved radon) concentrations tend to be higher in groundwater as opposed to surface water (Wedepohl, 1978) but unlike fluoride, where bioaccumulation and the "halo effect" of processing must always be borne in mind, there is little evidence that food chain exposures of radon are much of a concern (Kabata-Pendias, 2001) (Fig. 10).

The pathways that allow the migration of radon gas into homes with foundations are primarily controlled by rock mechanics and climate, which determine the number and "connectedness" of potential conduits, mineral surface area, atmospheric pressure, and permeability of the channels due to saturation with water. In this way, radon gas hazards may present intermittently in a wide variety of geochemical environments, making geologic hazard identification difficult, especially when considering the effects of global climate change. It is conceivable that dropping water table levels and the "drying up" of clay- and organic-rich soils will create new pathways for the liberation of significant amounts of radon gas into existing homes or that soils relocated by extreme weather events like hurricanes or airborne dusts from dried-up lakebeds and deforestation could lay dangerous foundations for new construction in areas that are not currently considered to be radon gas prone.

Lung Cancer Hot Spots

Thick Palaeozoic sedimentary bedrock units underlie the most densely populated parts of Ontario, and so it is not surprising that lung cancer rates in this province lag significantly behind those of Nova Scotia, New Brunswick, and Québec (Canadian Cancer Society, 2008) where more of the Canadian Shield is exposed (Fig. 11). However, the influence of uranium-rich clays becomes apparent when looking at the intra-provincial lung cancer statistics. Both lung cancer morbidity and mortality rates for the province are highest in areas underlain by heavy clay (Fig. 12, zone 1) and Canadian Shield (zones 10, 12, 13, and 14). Predictably, the lowest rates of incidence and death from lung cancer are reported for the Central zone (8) which covers the highlands formed by the Algonquin Arch but which is underlain by a thick blanket of coarse Quaternary sediments that are low in uranium content and Palaeozoic bedrock (Cancer Care Ontario, 2006 2004 data).

Internationally, the generally accepted guideline for residential exposures is 200 Bq/m^3. If this threshold is exceeded, the recommended remedial measures are inexpensive, simple to implement, and can generally be completed in a short period of time. In so doing, it is estimated that hundreds of cases of lung cancer per year could be prevented in North America, thus significantly reducing the burden on the health-care system.

Arsenic

Arsenic contamination through natural (geogenic) and anthropogenic sources is a serious threat to humans all over the world. Several epidemiological studies have documented the global impact of arsenic contamination and the characterization of the sources of exposure. The health effects of chronic exposure to

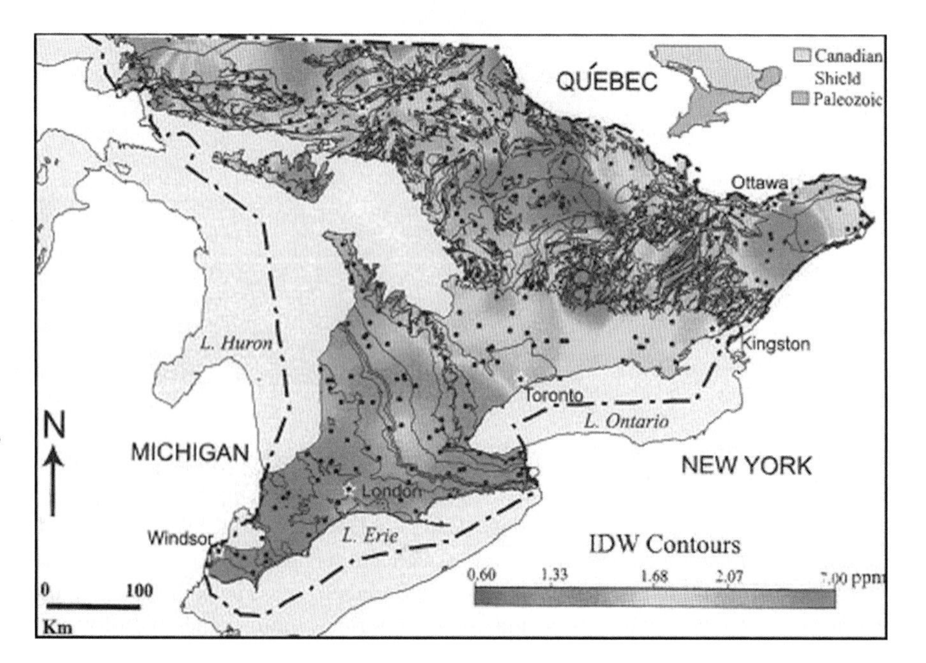

Fig. 11 Map of uranium (U) in the A horizon of southern Ontario. Dark gray "hot spots" of higher U concentrations in soil are indicative of a potential radon gas hazard. Source: Map compiled by R. Klassen from unpublished Geological Survey of Canada data collected by D. Sharpe

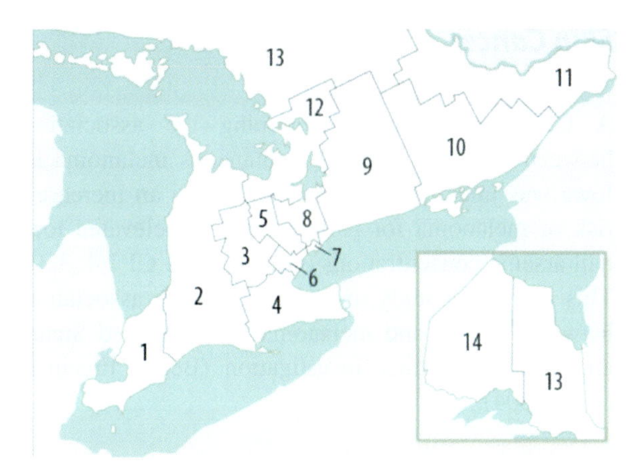

Fig. 12 Map of local health integration networks in the province of Ontario. 1 – Erie St. Clair, 2 – South West, 3 – Waterloo Wellington, 4 – Hamilton Niagara Haldimand Brant, 5 – Central West, 6 – Mississauga Halton, 7 – Toronto Central, 8 – Central, 9 – Central East, 10 – South East, 11 – Champlain, 12 – North Simcoe Muskoka, 13 – North East, 14 – North West. Source: http://www.lhins.on.ca/FindYourLHIN.aspx

arsenic are well established in countries with high arsenic in drinking water; however, such evidence is not so readily available in countries with lower levels of environmental arsenic, or treated drinking water systems. Of more relevance to developed and industrialized countries, such as the United States, are the potential health consequences of long–term, low-level exposure via drinking water or through occupation or medical use although the relationship between health effects and exposure to drinking water arsenic is not well established in US populations (Lewis et al., 1999). We provide here an overview of the state of arsenic studies in the United States and a discussion of the available epidemiological and human health literature in the United States.

Epidemiological and Environmental Health Studies on Arsenic Exposure

Areas of the United States have been affected by arsenic in drinking water, especially in areas close to mining sites, for example, Twisp, Okanogan County, Washington (Peplow and Edmonds 2004) and New Hampshire (Karagas et al., 2002). Recently, the US Geological Survey (USGS) published an updated version of a map illustrating "Arsenic in Ground Water of the United States." The map was generated from the most recent arsenic measurements from 31,350 wells and springs showing national level patterns of naturally occurring arsenic in ground water resources of the United States and Puerto Rico. The data set displayed on this map is a moving 75th percentile, which is the maximum arsenic concentration found in 75% of samples within a moving 50 km radius (the median size of a US county; USGS, 2005). The USGS map shows that there are parts of many states affected by high levels of arsenic in groundwater, following the new USEPA MCL for 10 μg/l (http://water.usgs.gov/nawqa/trace/pubs/geo_v46n11/fig2.html).

Several recent studies have been published concerning the distribution of arsenic levels in the US drinking water supplies. A study by Frost et al. (2003) identified 33 counties in 11 states with an estimated mean drinking water arsenic concentration of 10 μg/l or greater. A total of 11 of these counties have an estimated mean arsenic concentration of 20 μg/l or more, and two have an estimated mean arsenic concentration of 50 μg/l or more. Domestic wells are a particular issue, for example, in New Hampshire, domestic wells serve roughly 40% of the population, and about 10% of these contain arsenic concentrations in the controversial range of 10–50 μg/l (Karagas et al., 2002). Based on census data, between 1950 and 1999 there were approximately 51.1 million person-years of exposure to drinking water arsenic at levels of 10 μg/l or more, 8.2 million at levels of 20 μg/l or more, and 0.9 million at levels of 50 μg/l or more (Frost et al., 2003).

Although several studies have described the distribution of arsenic within the US drinking water supplies, there is a lack of epidemiologic data to indicate disease associations within the United States (Brown and Ross, 2002). It has been suggested that the mortality and incidence of diseases known to be associated with arsenic exposure should be examined in other high exposure State/Counties as part of an assessment of arsenic health effect in US populations (Frost et al., 2003).

Cardiovascular Effects

The association of drinking water arsenic and mortality outcome was investigated in a cohort of residents from Millard County, Utah. Median drinking water arsenic concentrations for selected towns ranged from 14 to 166 μg/l and were from public and private samples. Standard mortality ratios (SMRs) were calculated. Statistically significant findings included increased mortality from hypertensive heart disease among males (SMR 2.20, 95% CI (confidence interval) 1.36–3.36), increased mortality for hypertensive heart disease among females (SMR 1.73, 95% CI 1.11–2.58), and for the category of all other heart disease (SMR 1.43, 95% CI 1.11–1.80) (Lewis et al., 1999).

A case–control study in the United States investigated the association between chemicals in maternal drinking water consumed during pregnancy and congenital heart disease in the offspring. Two hundred and seventy affected children and 665 healthy children took part in the study. Data included information on contaminant levels in maternal drinking water and on health, pregnancy management, and demographic characteristics. Nine inorganic metals were analyzed for detection of an association with congenital heart disease. Arsenic exposure at any detectable level was associated with a threefold increase in occurrence of contraction of the aorta (OR (odd ratios) 3.4, 95% CI 1.3–8.9) (Zierler et al., 1988).

A recent US study analyzed water samples and used self-report format for 1,185 people who reported drinking arsenic-contaminated water for more than 20 years. They found that respondents with arsenic levels of >2 μg/l were statistically more likely to report a history of depression, high blood pressure, circulatory problems, and heart bypass surgery (Zierold et al., 2004).

Diabetes Mellitus

Diabetes is a major source of morbidity and mortality in the United States and recent studies have reported an increase in the incidence and prevalence of this disease. For example, the overall prevalence rose from 4.9% in 1990 to 6.5% in 1998 (Mokdad et al., 2001). As discussed, diabetes has recently been found to be associated with arsenic exposure in some epidemiologic studies, predominantly on studies conducted in Taiwan (Lai et al., 1994; Tseng et al., 2002), However, no work to date has been undertaken examining the relationship between diabetes mellitus with lower chronic doses of arsenic in drinking water in the United States.

Skin Cancer

A case–control study examining the association between drinking water and cutaneous melanoma in Iowa was undertaken. This study found an increased risk of melanoma for participants with elevated toenail arsenic concentrations (OR 2.1, 95% CI 1.4, 3.3). This is the first study to find a significant association between arsenic and melanoma in the United States and warrants further investigation (Beane Freeman et al., 2004).

Lung, Stomach, and Colon Cancer

Increased risk for lung cancer with arsenic exposure has been consistently observed in ecological, case–control, and cohort studies in Taiwan, Japan, Chile, and Argentina, but not yet in the United States (WHO, 2004). Significant excess mortality from cancers of the digestive tract has been observed among copper smelter workers in Anaconda, Montana, with a standardized mortality ratio of 1.3 (Enterline et al., 1987) only a slight excess in mortality from digestive tract cancer was observed among smelter workers in Tacoma, Washington. Colon cancer mortality has also been significantly associated with chronic exposures to inorganic arsenic among copper smelter workers in Tacoma, Washington, with a significant standardized mortality ratio of 2.1 for those who were employed before respirators were implemented in the smelter exposure areas (Enterline et al., 1987).

Breast Cancer

An association between arsenic exposure and breast cancer has been investigated but not confirmed. The associations between toenail levels of five trace elements and breast cancer risk were studied among a cohort of 62,641 US women who were free from diagnosed breast cancer in 1982. Among 433 cases of breast cancer identified during 4 years of follow-up and their matched controls, the OR comparing the highest with the lowest quintiles was 1.12 (95% CI 0.66–1.91). Even though breast cancer is examined as part of this epidemiological exercise, results to date do not provide evidence for an effect of arsenic on breast cancer risk (Garland et al., 1996).

Bladder Cancer

New Hampshire is known to have high arsenic in groundwater and along with other States in New England has among the highest bladder cancer mortality rates in the country. These facts prompted a large case–control study to be undertaken in this region examining toenail arsenic levels and their association with skin and bladder cancer (Karagas et al., 1998). The OR for squamous cell carcinoma (SCC) and basal cell carcinoma (BCC) were close to unity in all but the highest arsenic percentile category. Among individuals with toenail arsenic concentrations above the 97th percentile, the adjusted ORs were 2.07 (95% CI 0.92–4.66) for SCC and 1.44 (95% CI 0.74–2.81) for BCC, compared with those with concentrations at or below the median (Karagas et al., 2002). Among smokers, an elevated OR for bladder cancer was observed for the uppermost arsenic category (OR 2.17, 95% CI 0.92–5.11) (Karagas et al., 2004). However, because the 95% CI includes the value 1.0, arsenic levels cannot be considered a useful predictor variable, thus these results merely suggest an association between arsenic and the cancers examined and indicate that smoking may act as a co-carcinogen.

Ayotte et al. (2006) found a statistically significant positive correlation between residential bladder cancer mortality rates and private water supply use in New England. Previous studies (Montgomery et al., 2003) found elevated levels of arsenic in the well water.

Prostate Cancer

Epidemiologic studies have suggested a possible association between exposure to inorganic arsenic and prostate cancer (Wu et al., 1989; Chen and Wang, 1990), including a recent study of populations residing in the United States, in which SMR analysis by low-, medium-, and high-arsenic exposure groups indicated a dose relationship for prostate cancer (Lewis et al., 1999). Prostate cancer SMR was 1.45 (95% CI 1.07–1.91). A study by Achanzar et al. (2002) which found human prostate epithelial cells are directly susceptible to the transforming effects of inorganic arsenite. Another study on the effect of chronic oral exposure to arsenic on male mouse testicular and accessory sex organ weights, sperm parameters, testicular marker enzymes, and distribution of arsenic in reproductive organs, found a significant accumulation of arsenic in testes, epididymis, seminal vesicle, and prostate gland in treated animals (Pant et al., 2004). These laboratory studies show biological plausibility in the relationship between arsenic and prostate cancer.

The health effects of chronic exposure to arsenic are well established in countries with high levels of drinking water arsenic; however, such evidence is not so readily available in countries with lower levels of arsenic, such as the United States. The research that has been undertaken in the United States is limited and much of what has been undertaken is not definitive.

From the studies that have been carried out in the United States, some have found a statistically significant association between disease and arsenic levels, most consistently the cardiovascular studies (Lewis et al., 1999, Zierler, et al., 1988, Zierold et al., 2004). Prostate cancer also showed a significant association with arsenic levels (Lewis et al., 1999), which has been found to be biologically plausible in laboratory studies (Achanzar et al., 2002, Pant et al., 2004). Finally, a study on the risk of melanoma found a significant association with elevated toenail arsenic concentrations (Beane Freeman et al., 2004). In the United States the arsenic-related environmental etiology of several chronic diseases warrants further investigation.

Naturally Occurring Organic Compounds

Long-term exposure to low levels of organic substances leached from low-rank coal aquifers has been hypothesized to be linked to kidney disease and urinary system cancers (Orem et al., 2007). This link has been most firmly established in the case of Balkan Endemic Nephropathy (BEN), a fatal kidney disease with associated renal/pelvic cancer that occurs in restricted areas of the Balkans where Pliocene lignites deposits are present (Tatu et al., 1998). In northwestern Louisiana, where lignites underlie a region having an elevated incidence of renal pelvic cancers, recent studies have shown that a suite of coal-derived organic substances occur in drinking water supplies, resembling those observed in drinking water supplies from areas impacted by BEN (Bunnell et al., 2006). Recently (Peterson et al., 2009) collected water samples from two counties in East Texas reported to have relatively high incidences of renal/pelvic cancer. Dichloromethane extractable organic compounds in the water samples were identified by gas chromatography/mass spectrometry. Compounds present in the water samples included some substituted polycyclic aromatic hydrocarbons (mostly lower molecular weight), heterocyclic and substituted aromatics, phenols, terpenoids, and aliphatic hydrocarbons at sub-microgram/liter levels. This suite of organic compounds is similar to those observed in other coal containing aquifers. These results suggest that a BEN-like health problem may exist in the Gulf Coast region of the United States.

Evolving Concerns About Asbestos

Asbestos is a term applied to a group of fibrous silicate minerals that were mined and used commercially for centuries, primarily because of their flexibility, weavability, high tensile strength, and resistance to heat, chemicals, and electricity. The potential health effects of exposure to asbestos have been recognized for decades and include diseases such as asbestosis, mesothelioma cancer, lung cancer, pleural plaques, pleural thickening, and others (Roggli et al., 2004; Dodson and Hammar, 2006). These diseases are known to result from occupational exposures to high levels of asbestos dusts. As a result, mining and processing of asbestos ceased in the United States in 2002, and US commercial use of asbestos has plummeted since the 1970s; however, some asbestos is still imported for use in some commercial applications such as roofing (Virta, 2009).

In spite of decades of research into the health effects of asbestos, substantial debate continues as to the exact mechanisms of toxicity, the potential health effects of low-level exposures, potential health effects of elongated mineral particles that do not fit compositional or morphological definitions of asbestos, and other aspects (NIOSH, 2009; Plumlee and Ziegler, 2007; Roggli et al., 2004; Dodson and Hammar, 2006). Further, the last decade has seen a substantial increase in public health concern about potential health effects resulting from occupational and environmental exposures to dusts generated by mining of mineral deposits that contain asbestos as a natural contaminant, or by natural weathering or human disturbance of rock types that contain naturally occurring asbestos (Van Gosen, 2009; Harper, 2008; Pan et al., 2005).

Vermiculite deposits mined at Libby, Montana, from 1923 to 1990 provided the bulk of the Nation's vermiculite used commercially during that time period. High rates of asbestos-related disease are noted in Libby vermiculite miners and workers exposed occupationally, in their family members, and in Libby townspeople with no occupational connection to the mine (Peipins et al., 2003; Sullivan, 2007). Elevated rates of asbestos-related disease are also noted in communities elsewhere in the United States that received and processed vermiculite ore from Libby (Horton et al., 2008). It is generally agreed that the source of these diseases was exposure to asbestiform and fibrous amphiboles intergrown with the Libby vermiculite that were released into the air as the vermiculite was mined, processed, or otherwise disturbed.

Rocks that contain natural asbestos are found in 35 of the 50 United States (Van Gosen, 2009). As these rocks are weathered naturally, mined for aggregate or industrial mineral commodities, or exposed by earth moving for road and building construction, there is the potential for release of asbestos-containing dusts into the environment. In the last decade, substantial debate has developed as to whether these dusts pose a health risk to residents and workers alike. The debate has been particularly intense in the area around El Dorado Hills, California, an area that is underlain by serpentinites and other ultramafic rocks that contain chrysotile and tremolite–actinolite asbestos (Harper, 2008; Van Gosen, 2007; USEPA, 2006). Asbestos occurrences in or released from their original geologic source rocks, or soils derived naturally from those rocks have been termed "naturally occurring asbestos," or "NOA" (Harper, 2008).

How Earth Sciences Can Help

Earth sciences investigations are providing important contributions on a wide variety of asbestos issues, ranging from policy-making to fundamental health research. On a local to regional scale, fundamental geologic mapping, remote sensing, geophysical, and related characterization studies are helping map the distribution and mineralogical characteristics of asbestos-containing rocks (Churchill and Hill, 2000; Clinkenbeard et al., 2002; Swayze et al., 2009). In some jurisdictions such as California and Fairfax County, Virginia, geologic mapping forms the basis of policies requiring enhanced dust control measures and other practices in areas underlain by asbestos-containing rocks.

Studies are underway to inventory asbestos localities and map the distribution of geologically favorable rock types containing asbestos across the United States (Van Gosen, 2007, 2009). These studies are needed to aid in the interpretation of regional to national scale epidemiological data on asbestos-related diseases, with the ultimate goal of understanding better the health risks associated with living on or rocks containing NOA (Pan et al., 2005).

Mineralogical characterization studies are essential to help understand the nature of the particles to which affected populations have been exposed. For example, mineralogical characterization has shown that the Libby amphiboles occur in a range of morphologies and chemical compositions, only a subset of which has been traditionally regulated as "asbestos" (Meeker et. al, 2003). In El Dorado Hills, California, mineralogical studies also documented a wide range of elongated particle types in air samples that would all be counted as potentially hazardous under current regulations, but only a subset of these particles is truly asbestiform (Meeker et al., 2006).

Other areas in which earth scientists work in collaboration with health scientists can contribute include (Plumlee et al., 2006; Plumlee and Ziegler, 2007; NIOSH, 2009) characterization of dosing materials used in asbestos-related toxicity testing; characterizing particles to which various worker cohorts having elevated asbestos-related disease rates were exposed; examining dissolution rates of asbestos and other mineral particles in body fluids encountered along inhalation and ingestion exposure pathways; examining other chemical interactions of particles with the body, and how these interactions (such as oxidation–reduction reactions that generate free radicals and oxidative stress) may influence toxicity in the body.

A Growing Role for Earth Sciences in Environmental Disaster Response and Planning

Natural and anthropogenic disasters (e.g., earthquakes, volcanic eruptions, wildfires, urban fires, landslides, hurricanes, tsunamis, floods, industrial spills, terrorist attacks) can produce copious solid, gaseous, or liquid materials of potential environmental and public health concern. Examples include contaminated and/or pathogen-bearing waters, dusts, soils, and sediments; liquids; gases; smoke; ash; and debris. Many of these materials are derived from the earth, geochemical processes influence how they evolve in the environment, and their geochemical characteristics can strongly influence their potential impacts on the environment and health (Plumlee and Ziegler, 2007). As a result, there is a growing role for process-focused earth science expertise and methods applied to environmental disaster response and planning.

Volcanologists have had a long history of helping assess impacts of volcanic ash and gases on the environment and health, and Mt. St. Helens and Kilauea volcanoes in the United States have played prominently in these studies. Water leach studies of ash have documented that rain waters falling onto fresh ash could become quite acidic, due to the liberation of acidic gas species that had condensed onto the ash particles in the eruption cloud (Hinkley and Smith, 1987; Witham et al., 2005). Studies of other eruptions from around the world have regularly demonstrated that fluoride can be leached in sufficient quantities from ash to trigger fluorosis in humans and other animals that consume water affected by ash, or animals that eat vegetation coated by ash (Weinstein and Cook, 2005). Volcanology collaborations with the public health community have helped elucidate and communicate to the public the health hazards associated with exposure to volcanic fog (vog) emanating from volcanoes, and laze (lava haze) produced when lava flows into seawater (USGS, 2008). Recent articles provide excellent summaries of the potential health hazards associated with respiratory exposure to volcanic ash (Hansell et al., 2006; Horwell and Baxter, 2006; Weinstein and Cook, 2005), and toxicologically important reactions involving ash (Horwell et al., 2007). Other information on volcanic ash can be found through web sites of the US Geological Survey (http://volcanoes.usgs.gov/ash/health/) and International Volcanic Health Hazards Network (http://www.ivhhn.org).

The outbreak of the fungal disease Valley Fever in humans and sea otters immediately following the 1994 Northridge, California, earthquake is often cited as an example of the links between earth processes, geophysical disasters, and health. Landslide specialists worked directly with public health experts to help examine the timing and spatial distribution of sickened people and were able to determine that the outbreak originated through exposures to dust clouds generated during earthquake-triggered landslides (Jibson et al., 1998; Schneider et al., 1997). The landslides had disturbed soils that contained spores of the pathogenic soil fungus *Coccidioides immitis*, the etiological agent of Valley Fever that is endemic to soils in much of the US desert southwest. Since 1994, substantial research has focused on determining potential geological and other controls on the habitat of *C. immitis* in soils (Bultman et al., 2005; Fisher et al., 2007).

In the last decade, interdisciplinary earth science studies have helped provide insights into the potential environmental- and health-hazard characteristics of materials produced by a number of natural and anthropogenic disasters in addition to volcanic eruptions, including

- dust and debris produced by the attacks on and collapse of the World Trade Center towers (Clark et al., 2001, 2005; Plumlee et al., 2005; Meeker et al., 2005; Swayze et al., 2005);
- flood waters and flood sediments left in the greater New Orleans area by 2005 hurricanes Katrina and Rita (e.g., Pardue et al., 2005; Plumlee et al., 2006; Presley et al., 2006; Van Metre et al., 2006; Cobb et al., 2006; Plumlee et al., 2007a; Farris et al., 2007; USEPA, 2005–2008; Griffin et al., 2009);
- ash and burned soils from wildfires (Plumlee et al., 2007b; Wolf et al., 2008);
- mud from the ongoing LUSI, East Java mud volcano eruption (Plumlee et al., 2008).

How Earth Sciences Can Help

These studies collectively demonstrate that important contributions can be made in disaster response from across the earth science disciplines, including geology, geophysics, geochemistry, hydrology, remote sensing, geomicrobiology, and others. For example, during a disaster, earth scientists can, in collaboration with emergency managers, public health experts, microbiologists, and ecologists, help to

- characterize the physical, chemical, and microbial makeup of materials generated by the disaster;
- identify source(s) of the materials;
- monitor, map, and/or model dispersal and evolution of materials in the environment;
- understand how the materials are modified by environmental processes;
- identify key characteristics and processes that influence the materials' toxicity to exposed humans and ecosystems; and
- estimate shifts away from pre-disaster environmental baseline conditions.

The key goal of all these studies is to help define the magnitude and nature of the environmental and health risks posed by a specific disaster, both spatially and over time.

In addition, results from past responses can be used to anticipate and plan for future disasters. As part of disaster preparedness, earth scientists can

- measure and compile data on pre-disaster environmental baseline conditions;
- develop models for the environmental behavior and effects of materials generated by similar types of disasters;
- develop a resource of appropriate data and methods that can be used to enhance and guide responses to future environmental disasters; and
- work with emergency planners to help mitigate effects of and improve resiliency to future disasters.

Occupational Health Issues

Occupational health issues associated with metals exposure are both industry and metal specific. While there are a large number of metals that are important in a workplace setting, there are six "heavy metals" that are considered to be the most significant in terms of potential toxicity to exposed workers. These are arsenic, beryllium, cadmium, hexavalent chromium, lead, and mercury. The other metals of significance include Al, Sb, Co, Cu, Fe, Mn, Mo, Ni, Se, Ag, Sn, V, and Zn. Each of these metals can also produce potentially significant dose-related health effects.

Unlike the situation associated with community exposures where either the food pathway or direct soil ingestion routes dominate, in an industrial setting, the dominant pathway is air. Occupational exposure standards are primarily focused on the air pathway, as neither the dermal nor direct ingestion routes of exposure are typically as significant in comparison to the air route. The exceptions to this observation are beryllium and hexavalent chromium where dermal exposures are important potential routes of exposure and health effects.

In general, there are several important questions that should be considered when analyzing the potential for adverse effects from workplace exposures to metals:

1. What industrial operations are typically associated with workplace exposures to the metal under consideration?
2. What health effects are associated with exposure dose to the specific metal?
3. How are exposure and/or dose evaluated?
4. What are the relevant regulatory exposure standards?
5. What are some of the typical strategies utilized to minimize or prevent exposure?

The first four questions tend to be substance specific and will be briefly summarized for each of the most important "heavy metals." However, regulatory standards are country specific and the reader should always consult the most current available guidance. Health effects are metal and dose specific. Simply listing observed generic clinical effects, e.g., rash, cough, fatigue, provides little or no useful information as health effects are strongly dose related. Exposure minimization strategies apply to all of the metals. Excellent sources of information for both toxicology and occupational exposures, including regulatory considerations are (i) Agency for Toxic Substance and Disease Registry (ATSDR) site, http://www.atsdr.cdc.gov/toxpro2.html; (ii) Occupational Health and Safety Administration (OSHA) site, http://www.osha.gov/SLTC/metalsheavy/index.html; and (iii) Canadian Centre for Occupational health and Safety (CCOHS) site, http://www.ccohs.ca/

Arsenic: Potential exposures to inorganic arsenic and its compounds include coke oven emissions, smelting, wood preservation, glass production, nonferrous metal alloys, and electronic semiconductor manufacturing industries.

Chronic long-term exposure can lead to specific forms of dermatitis, mild pigmentation keratosis of the skin, vasospasticity, gross pigmentation with hyperkeratinization of exposed areas, and wart formation. These health effects are not due to direct dermal exposure but instead can be observed in individuals exposed by inhalation or ingestion, particularly via contaminated drinking water. Arsenic-related skin lesions have been well described when exposures were due to high levels, either anthropogenic or naturally occurring, in drinking water. Arsenic is considered to be a known human carcinogen via chronic high-dose inhalation exposure.

Exposures are determined by a combination of (i) airborne industrial hygiene sampling, e.g., personal and/or area and/or (ii) biological monitoring which can include blood and urine sampling.

Beryllium: Occupational exposure can occur in mining, extraction, and in the processing of alloy metals containing beryllium. There are significant applications in the aerospace, nuclear, and manufacturing industries. Beryllium is extremely light weight and stiff; hence it has wide usage in dental appliances, golf clubs, non-sparking tools, wheel chairs, and electronic devices.

The adverse health effects of beryllium exposure are due to an immunological reaction that produces an allergic-type response. According to OSHA, approximately 1–15% of all people occupationally exposed to beryllium in air become sensitive to beryllium and may develop chronic beryllium disease (CBD), an irreversible and sometimes fatal scarring of the lungs.

Air, wipe, and bulk sampling techniques can be used to measure occupational exposures. A specific blood test, the beryllium lymphocyte proliferation test (BeLPT), is an important screening tool that has been developed in order to determine which workers have become sensitized to beryllium.

Cadmium: Most cadmium exposures are associated with workplaces where ores are processed or smelted. Exposures in electroplating, and welding operations have also been observed. Cadmium is also a component of certain industrial paints and is utilized in the manufacture of nickel–cadmium batteries.

Cadmium toxicology is quite complex with target organs primarily centered on the lung and kidney. Cadmium is considered to be a potential lung human carcinogen via inhalation exposures. The kidney is an important target that exhibits threshold effects, i.e., an overall "critical concentration" is observed prior to development of observable adverse effects.

Both industrial hygiene (air) and biological monitoring (urine) are used for determining exposures and as regulatory compliance monitors.

Chromium-hexavalent (Cr+6): Toxicity to chromium is directly related to valence state, i.e., the more common naturally occurring trivalent form (Cr+3) has an extremely low toxicity profile. According to OSHA, occupational exposures are associated with workers who handle pigments containing dry chromate, spray paints and coatings containing chromate, operate chrome-plating baths, and weld or cut metals containing chromium, such as stainless steel. Hexavalent chromium compounds include chromate pigments in dyes, paints, inks, and plastics; chromates added as anticorrosive agents to paints, primers, and other surface coatings; and chromic acid electroplated onto metal parts to provide a decorative or protective coating. Hexavalent chromium can be created during "hot work" i.e., welding on stainless steel or melting chromium metal. In these situations the high temperatures involved in the process oxidize the Cr+3 to the hexavalent state.

Exposures can occur via inhalation, dermal exposure, and ingestion. The most significant routes are inhalation and dermal (skin and eye) and the upper respiratory tract and lung are the most important target organs for toxicity with increased lung cancer rates observed in a dose-related exposure pattern.

Both industrial hygiene (air) and biological monitoring (urine) are used for determining exposures. Air exposures are used for regulatory compliance monitors. Biological exposure indices can be performed; however, pre- and post-shift monitoring of urine is necessary as the urine test measures Cr+3 a normal dietary intake component.

Lead: Lead exposures are one of the most common in an industrial setting. OSHA considers lead exposure as a leading cause of chemical-related workplace illness. Aside from mining and smelting, lead is found in numerous industrial settings. In addition, workers can be exposed to lead while working at certain hazardous waste sites or during remediation of older housing stock where lead-based paints were commonly used. According to OSHA, standard particulate sampling techniques are used to evaluate lead exposures. Wipe sampling can indicate potential for lead ingestion.

Toxicity in workers can be acute or chronic and is typically dose related. Symptoms of lead toxicity include loss of appetite, nausea, vomiting, stomach cramps, constipation, difficulty in sleeping, fatigue, moodiness, headache, joint or muscle aches, anemia, and decreased sexual drive. However, these symptoms are non-specific and can be commonly elicited during a history and physical examination. Hence the use of biological monitoring is extremely important, i.e., measuring the blood lead level.

Target organs include blood-forming, nervous, renal, and reproductive systems. Both air sampling and biological monitoring are mainstays for exposure determination and are also used for regulatory compliance.

Mercury: Exposure can occur during mining and refining of gold and silver ores. Unlike commercial gold miners who use cyanide-based processes, "artisanal" miners commonly use mercury in their small-scale extraction efforts. Mercury can be found in thermometers, manometers, barometers, gauges, valves, switches, batteries, and high-intensity discharge (HID) lamps. Mercury is also used in amalgams for dentistry, preservatives, heat transfer technology, pigments, catalysts, and lubricating oils.

In the workplace, mercury can be found in three forms (i) elemental (metallic); (ii) inorganic, i.e., typically combined with chlorine, sulfur, and oxygen; and (iii) organic, i.e., combined with carbon or carbon-containing substances into straight (alkyl) or aromatic (aryl) ring forms. In general the alkyl forms are the most toxic although all forms exhibit a dose-related toxicity pattern.

Both air sampling and biological monitoring are mainstays for exposure determination. Regulatory agencies have issued specific standards for the major mercury compounds; hence, it is critical to determine which category a compound belongs to before comparing it with a standard or determining its relative toxicity.

Regardless of the specific chemical, workplace exposure control strategies are based on engineering controls, administrative controls, (i.e., good work practices and training), and the use of personal protective equipment (PPE), e.g., clothing and respirators, where required. According to OSHA, engineering controls include material substitution, isolation, process/equipment modification, and local ventilation. Administrative controls include good housekeeping, appropriate personal hygiene practices, periodic inspection and maintenance of process and control equipment, proper procedures to perform a task, and appropriate supervision to ensure that the proper procedures are followed.

North America, United States, and Canada have a 100+ year history of mining activity. While overall base metal mining and smelting has declined, particularly in the United States, mining activity continues to be an extremely important economic driver in many discrete geographic areas. In Canada, as opposed to the United States, mining is far more important economically as a percent of the overall Gross Domestic product (GDP). In the United States, mining is extremely geographically focused and often centered on gold mining as opposed to more "traditional" base metals, e.g., lead, copper, zinc.

As discussed in this chapter, the medical geology of mining activity is minimally a function of the (i) specific suite of metals in the ore and their chemical speciation and mineralogy; (ii) the type of extractive procedures used on the ore, i.e., smelting, flotation, leaching, or closed circuit cyanide extraction processes; (iii) background environmental characteristics of the mine site; (iv) the fate-transport (chemodynamics) characteristics of the individual metals (e.g., existing relationship between surface and ground water, air patterns,); and (v) demographic characteristics of the potentially exposed populations, e.g., socio-economics, age distributions, housing stock ages. All of these elements influence the degree in which source→ exposure→dose→potential health effects. Various complex conceptual models have been developed by the regulatory agencies and risk assessment professionals to illustrate this fundamental sequence. Each historic and active site has a unique set of circumstances that should be considered.

Despite these site-specific difference, over the last 30 years there has been a substantial regulatory history of evaluating and managing the environmental and human health effects from mining activity, particularly legacy sites that are no longer actively operating. Many of these "legacy sites" are extremely complex and are often referred to as "mining megasites." In 2005, the US National Academy of Sciences (NAS) published a large monograph, entitled "Superfund and Mining Megasites: Lessons form the Coeur d'Alene River Basin" (http://www.nap.edu/catalog/11359.html). While there are a large number of US-centric features about the monograph these are entirely related to the legal context of managing sites as opposed to the medical geology of a complex megasite. From a scientific perspective, the NAS analysis is an extremely useful and detailed road map of the twists and turns of analyzing large legacy mining sites. While focused on inactive or legacy issues, the technical analysis is easily applicable to active locations.

One of the critical insights of the report is that approaches to evaluating sites, including their ultimate remediation, are constantly evolving. The disposition and effects of contaminants within a geographical location have evolved significantly over the last 30+ years; hence, decisions are always made based on limited technical information. The information database grows and evolves in sophistication and complexity. In addition, the medical/health effects database does not remain static. For example, background levels of blood lead in US children under age 7 have dramatically fallen almost sevenfold over the last 35 years. Blood lead levels that were considered "normal" in the 1970s and 1980s now trigger intense regulatory and medical investigation. The medical toxicology has clearly shifted from a consideration of gross effects to subtle often sub-clinical considerations, i.e., findings not observable on physical examination but instead requiring sophisticated longitudinal testing procedures. At many of the US megasites, e.g., Bunker Hill lead site in the Coeur d'Alene basin, this has been the situation. Medically site investigations have evolved from gross exposure in emergency rooms to long-term medical monitoring protocols for both previous workers and exposed community residents. Similarly, the environmental assessment has transitioned from "hot spot" analysis of gross contamination to concerns over low-level exceedances of background or calculated "safe" soil levels. For example, at the Tacoma smelter in Washington State, United States, arsenic levels in soil greater than 20 ppm are considered a "public health concern." ("Dirt Alert: Arsenic and lead in Soils" Washington State Department of Ecology, www.ecy.wa.gov/programs/tcp/sites/tacoma_smelter/ts_hp.htm). Similarly, lead in soils above 250 ppm is also considered a "public health concern." As previously noted in the chapter, background arsenic and lead soil levels often significantly exceed these administrative levels of "public health concern."

The history of megasite remediation indicates that decades of effort and hundreds of millions of dollars are spent on remediation. Critics often ask whether this concerted effort is "worth it." As pointed out by the NAS, this question goes beyond matters that science alone can address. Complex economic and societal values tend to dominate the megasites and while "science" is necessary it rarely seems "sufficient" based on the historic record. Thus, medical geology practitioners need to realize that the process may be more art than science and framed by a heavy dose of legal and political considerations. Megasites, whether Bunker Hill Idaho, Leadville Colorado, Tacoma smelter site, Washington, copper sites in the Gaspé Region of Quebec Canada, all illustrate these observations.

Black Lung Disease

Although the incidence of Coal Worker's Pneumoconiosis (CWP: Black Lung Disease), a progressive, debilitating respiratory problem caused by inhalation of coal dust has decreased dramatically in the United States, it still takes a heavy toll on former miners coal miners. For more than 100 years it has been assumed that black lung disease was caused by the inhalation of the black pulverized particles of coal. Recent research (Huang et al., 2004), however, has shown that CWP may be initiated not by the coal particles but by inhalation of pulverized pyrite, a common coal mineral. The pyrite dissolves in the lung fluids releasing strong acids that irritate the lung tissues. Particles (coal dust, quartz, clay, pyrite, etc.) that then contact the irritated tissues will cause the fibrosis leading to decreased oxygen exchange capacity. Thus, knowledge of the mineral composition of the coal may be a key parameter in anticipating the incidence of CWP. Information on the mineralogy of the coal being mined may provide essential data needed to protect the health of the miners in cost-effective ways thus reducing the enormous financial burden of health care and lost productivity.

Community Effects

Community level exposures to metals generically occur in two forms: (i) anthropogenic, i.e., man-made activity and (ii) natural, i.e., low-level concentrations found in soils. The anthropogenic exposures to metals are generally related to industrial activity, either legacy (e.g., past mining and smelting) or current operations with releases to air, water, and soils. For example, coal combustion associated with electrical power generation is a common source of atmospheric metals emissions, e.g., mercury and selenium. Depending upon the specific form, naturally occurring soil levels of virtually all of the metals is easily demonstrated. These "background" levels exhibit significant heterogeneity as a function of geography. Hence, background is invariably presented as a wide range as variation in natural mineralization is extremely common. In many situations it is important to fully understand the local and regional background levels of a specific metal as the variability can easily span a factor of 10, i.e., an order of magnitude. This variability becomes important at the two ends of the concentration distribution profile and both critical micronutrient deficiencies, e.g., iodine and selenium, and over exposures, e.g., lead, arsenic, and cadmium are well documented in the literature. A convenient source of high-quality information describing the most important metals can be found at the Agency for Toxic Substance and Disease Registry (ATSDR) site, http://www.atsdr.cdc.gov/toxpro2.html. ATSDR has published extensive "Toxicological Profiles" for a wide variety of chemicals, including the medically most significant chemicals.

From a medical toxicology perspective, community exposures are typically focused on four common "heavy metals": (i) arsenic, (ii) cadmium, (iii) lead, and (iv) mercury. Each of these metals has a distinctive fate-transport (how it moves in the environment in air, water, and soil) profile that influences how and where the metal accumulates once released from an anthropogenic source. For example, all of these metals exist in multiple forms that preferentially accumulate in soils, sediments, and biota. The geochemistry of each metal is the key factor and must be understood in order to anticipate whether a potential human health problem actually exists. Many metals are present in forms that have very low bioavailability; hence, while an individual is exposed to the metal, the actual internal dose is quite low and may have little or no discernable clinical effects. Therefore, it is extremely important that specific chemical forms of the metal under consideration are evaluated and characterized.

For example, legacy issues (arsenic, cadmium, lead, and mercury) associated with mining and smelting have been a major focus of the regulatory agencies in the United States with development of extensive and multi-million dollar remediation effects. Much of this effort is driven by the observation that some metals, e.g. lead, do not have a known physiological function in humans. In contrast, certain forms of arsenic do have an essential biochemical function in humans.

In general, exposure to the heavy metals is dominated by two routes: inhalation and ingestion. The third pathway, dermal (skin), is usually considered to be extremely minor. Unless there is an ongoing industrial operation, e.g., smelting, incineration, power plant emissions, the inhalation or air pathway is usually quite small relative to ingestion. Ingestion is dominated by food, water, and direct contact with soils. In a non-legacy situation, i.e., significant concentrations above naturally occurring background are not present, overall human exposures are primarily driven by dietary intake. Overall diet intake is a function of the natural level found in soils, the geochemistry of the soils, e.g., pH, carbon content, cationic exchange capacity, and the specific types of crops and/or livestock that are in local or regional production. In addition, some personal choice habits, e.g., smoking, dramatically influence the daily exposure level to specific metals, e.g., cadmium. Smokers have twice the body burden of cadmium versus a non-smoker. In situations associated with daily intake of cadmium from food, the margin of safety becomes significantly narrower for an individual who smokes.

In the northern hemisphere, the legacy problems associated with heavy metals have often dominated the discussion surrounding medical geology. However, the micronutrient deficiency burden of disease at a community level is significantly larger in terms of overall attributable morbidity (illness) and mortality (death). A low daily intake (usually less than 100 μg/day) of a wide variety of metals, e.g., Fe, I, Se, Cu, Zn, Co, Mn, and Mo, is essential for normal health and wellness. The normal daily intake of these metals comes from diet. In areas that have naturally occurring deficiencies in these metals a variety of gross and subtle clinical effects can be found. Iron and iodine deficiency are two of the most common micronutrient deficiencies, particularly. In rural sub-Saharan Africa, geographically distributed pockets of micronutrient deficiencies can be found where there is a largely rural subsistence economy and soils are deficient in one or more of the essential metals.

For example, selenium serves an extremely important role in number of important physiologically important processes, particularly enzymes involved in antioxidant defense mechanisms, thyroid hormone metabolism, and redox control of intracellular reactions (ATSDR, Toxicological Profile for Selenium, 2003). Both excessive intake and deficiency scenarios are possible and have been clinically documented. The creation of an excess or deficiency risk is largely a function of background soil concentration, specific soil geochemistry, and daily diet. For humans, meat products generally contain the highest concentration of selenium, while vegetables and fruits contain the lowest. In some areas of the world, e.g., foothills of the Andes Mountains, soils contain elevated levels of selenium and local agricultural products, such as Brazil nuts, can have extremely high levels of selenium (ATSDR, 2003). In the United States, the highest levels of selenium in soil are found in certain areas of the West and Midwest.

Acknowledgments Parts of this chapter has been partially published in "*Metal Contaminants in New Zealand –Sources, Treatments and Effects on Ecology and Human Health*" (Moore TA, Black A, Centeno JA, et al., 2005, Editors; resolutionz press, Christchurch, New Zealand. ISBN: 0-476-01619-3). JAC is grateful to Dr. Tim Moore and resolution press for allowing the use of this material. This chapter has been reviewed in accordance with the policy and guidelines of the Armed Forces Institute of Pathology and the Department of Defense, and approved for publication. Approval should not be construed to reflect the views and policies of the Department of the Army, the Department of the Navy, the Department of Defense, or the United States Government, nor does mention of trade names or commercial products constitute endorsement or recommendation for use.

References

Achanzar WE, Brambila EM, Diwan BA, Webber MM, Waalkes MP (2002) Inorganic arsenite-induced malignant transformation of human prostate epithelial cells. J Natl Cancer Inst 94:1888–1891.

ATSDR (Agency for Toxic Substances and Disease Registry) (2003) Toxicological profile for selenium. http://www.atsdr.cdc.gov/toxprofiles/tp92.html

Ayotte JD, Baris D, Cantor KP, Colt J, Robinson Jr, GR, Lubin JH, Karagas M, Hoover RN, Fraumeni Jr, JF, Silverman DT

(2006) Bladder cancer mortality and private well use in New England: an ecological study. J Epidemiol Comm Health 60:168–172.

Barnes HL (1979) Geochemistry of Hydrothermal Ore Deposits. New York: John Wiley & Sons.

Bast D, Finn SB, McCaffrey I (1950) The Newburg-Kingston Caries Fluorine Study. Am J Public Health 40:716–724.

Beane Freeman LE, Dennis LK, Lynch CF, Thorne PS, Just CL (2004) Toenail arsenic content and cutaneous melanoma in Iowa. Am J Epidemiol 160:679–687.

Boyle DR (1976) The Geochemistry of Fluorine and Its Application in Mineral Exploration. Ph.D. Thesis. London: University of London.

Boyle DR, Chagnon M (1995) An incidence of skeletal fluorosis associated with groundwaters of the Maritime Carboniferous Basin, Gaspé region, Quebec, Canada. Environ Geochem Health17:5–12.

Brown KG, Ross GL (2002) American council on science and health. Arsenic, drinking water, and health: a position paper of the American council on science and health. Regul Toxicol Pharmacol 36:162–174.

Brugge D, Benally T, Yazzie-Lewis E (2006) The Navajo People and Uranium Mining. Albuquerque, NM: University of New Mexico Press.

Bultman MW, Fisher FS, Pappagianis D (2005) The ecology of soil-borne human pathogens. In O Selinus, B Alloway, J Centeno, R Finkelman, R Fuge, U Lindh, P Smedley (Eds), Essentials of Medical Geology (pp. 481–512). Amsterdam: Elsevier Inc.

Bunnell JE, Tatu CA, Bushon RN, Stoeckel DM, Brady AMG, Beck M, Lerch, HE, McGee B, Hanson BC, Shi R, Orem WH (2006) Possible Linkages Between Lignite Aquifers, Pathogenic Microbes, and Renal Pelvic Cancer in Northwestern Louisiana, USA. Environ Geochem Health 28:577–587.

Canadian Cancer Society/National Cancer Institute of Canada (2008) Canadian Cancer Statistics 2008. Toronto, Canada.

Cancer Care Ontario 2006. New cases and deaths for all cancers (2004) Geographic patterns of cancer in Ontario. http://www.cancercare.on.ca/ocs/snapshot/cancerstats/

Cannon HL, Hopps (1971) Environmental Geochemistry in Health and disease. Ottawa, Canada: Geological Society of America Memoir 123.

Cannon HL, Hopps HC (1972) Geochemical Environment in Relation to Health and Disease. Ottawa, Canada: Geological Society of America Special Paper 140.

Chen CJ, Wang CJ (1990) Ecological correlation between arsenic level in well water and age-adjusted mortality from malignant neoplasms. Cancer Res 50:5470–5474.

Churchill RK, Hill RL (2000) A general location guide for ultramafic rocks in California –Areas more likely to contain naturally occurring asbestos: Sacramento, Calif., California Department of Conservation, Division of Mines and Geology, DMG Open-File Report 2000–19, http://www.consrv.ca.gov/

Clark RN, Green RO, Swayze GA, Meeker GP, Sutley S, Hoefen TM, Livo KE, Plumlee G, Pavri B, Sarture C, Wilson S, Hageman PL, Lamothe PJ, Vance JS, Boardman J, Brownfield I, Gent C, Morath LC, Taggart J, Theodorakos PM, Adams M (2001) Environmental Studies of the World Trade Center area after the September 11, 2001 attack: U.S. Geological Survey Open File Report 01-0429, http://pubs.usgs.gov/of/2001/ofr-01-0429/

Clark RN, Swayze GA, Hoefen TM, Green RO, Livo KE, Meeker G, Sutley S, Plumlee G, Pavri B, Sarture C, Boardman J, Brownfield I, Morath LC (2005) Environmental mapping of the World Trade Center area with imaging spectroscopy after the September 11, 2001 attack. In JS Gaffney, NA Marley (eds) Urban Aerosols and Their Impacts: Lessons Learned from the World Trade Center Tragedy (pp. 66–83). American Chemical Society Series 919.

Clinkenbeard JP, Churchill RK, Lee, Kiyoung (eds) (2002) Guidelines for geologic investigations of naturally occurring asbestos in California: Sacramento, Calif., California Department of Conservation, California Geological Survey Special Publication 124, p. 70.

Cobb GP, Abel MT, Rainwater TR, Austin GP, Cox SB, Kendall RJ, Marsland EJ, Anderson TA, Leftwich BD, Zak JC, Presley SM (2006) Metal distributions in New Orleans following hurricanes Katrina and Rita: A continuation study. Environ Sci Technol 40:4571–4577.

Dodson RF, Hammar SP (eds) (2006) Asbestos: Risk Assessment, Epidemiology, and Health Effects. Boca Raton, FL: Taylor and Francis.

Edmunds M, Smedley P (2005) Fluoride in natural waters. In Selinus et al. (eds) Essentials of Medical Geology (pp. 301–329). Amsterdam: Elsevier.

Enterline PE, Henderson VL, Marsh GM (1987) Exposure to arsenic and respiratory cancer: A reanalysis. Am J Epidemiol 25:929–938.

Environment Canada (2001) Canadian Water Quality Guidelines for the Protection of Aquatic Life: Inorganic Fluorides. National Standards Office, Ottawa.

Farris GS, Smith GJ, Crane MP, Demas CR, Robbins LL, Lavoie DL, (eds) (2007) Science and the storms: The USGS response to the hurricanes of 2005: U.S. Geological Survey Circular 1306.

Fisher FS, Bultman MW, Johnson SM, Pappagianis D, Zaborsky E (2007) *Coccidioides* niches and habitat parameters in the southwestern United States-A matter of scale. Ann New York Acad Sci 1111:47–72.

Fleming WJ, Grue CE, Schuler CA, Bunck CM (1987) Effects of oral doses of fluoride on nestling European starlings. Arch Environ Contam Toxicol 16:483–489.

Freedman J (ed) (1975) Trace element geochemistry in health and disease. Geolog Soc Am, Special Paper 155.

Frost FJ, Muller T, Petersen HV, Thomson B, Tollestrup K (2003) Identifying US populations for the study of health effects related to drinking water arsenic. J Expo Anal Environ Epidemiol 13:231–239.

Garland M, Morris JS, Colditz GA, Stampfer MJ, Spate VL, Baskett CK, Rosner B, Speizer FE, Willett WC, Hunter DJ (1996) Toenail trace element levels and breast cancer: a prospective study. Am J Epidemiol 144:653–660.

Gavaghan H 2002 Lead, unsafe at any level. Bull World Health Organ [online] 80(1): 82–82.

Gottlieb LS, Luverne AH (1982) Lung cancer among Navajo uranium miners. Chest, 81:449–452.

Griffin DW, Petrosky T, Morman SA, Luna VA (2009) A survey of the occurrence of *Bacillus anthracis* in North American soils over two long-range transects and within post-Katrina New Orleans: Applied Geochemistry 24:1464–1471.

Hansell A, Horwell CJ, Oppenheimer C (2006) The health hazards of volcanoes and geothermal areas. Occup Environ Med, Continuing Professional Development Series, 63:149–156.

Huang X, Li W, Attfield MD, Nadas A, Frenkel K, Finkelman RB (2004) Mapping and prediction of Coal Workers' Pneumoconiosis with bioavailable iron content in bituminous coals. Environ Health Persp 113(8):964–968.

Harper M (2008) 10th anniversary critical review: Naturally-occurring asbestos. J Environ Monit 10:1394–1408.

Hinkley TK, Smith KS (1987) Leachate chemistry of ash from the May 18, 1980 eruption of Mount St. Helens, with a section on laboratory methods developed for Mount St. Helens water and acid leaching studies, by SA Wilson and CA Gent. In T.K. Hinkley (ed) Chemistry of Ash and Leachates from the May 18, 1980 Eruption of Mount St. Helens. Washington: U.S. Geological Survey Professional Paper 1397.

Hopps HC, Cannon HL (1972) Geochemical environment in relation to health and disease. Annal New York Acad Sci 199.

Horton DK, Bove F, Kapil V (2008) Select mortality and cancer incidence among residents in various U.S. communities that received asbestos-contaminated vermiculite ore from Libby, Montana. Inhal Toxicol 20:767–775.

Horwell CJ, Baxter PJ (2006) The respiratory health hazards of volcanic ash: a review for volcanic risk mitigation. Bull Volcanol 69:1–24.

Horwell CJ, Fenoglio I, Fubini B (2007) Iron-induced hydroxyl radical generation from basaltic volcanic ash. Earth Plan Sci Lett 261:662–669.

Hudlicky M, Pavlath AE. (1995) Chemistry of organic fluorine compounds II: A critical review. (ACS Monograph Series 187). Washington, DC: American Chemical Society.

Jibson, RW, Harp EL, Schneider E, Hajjeh RA, Spiegel RA (1998) An outbreak of coccidioidomycosis (valley fever) caused by landslides triggered by the 1994 Northridge, California, earthquake. Geol Soc Rev Eng Geol 12:53–61.

Kabata-Pendias A (2001) Trace elements in soil and plants (3rd ed.). Boca Raton, FL: CRC Press.

Karagas MR, Tosteson TD, Blum J, Morris JS, Baron JA, Klaue B (1998) Design of an epidemiologic study of drinking water arsenic exposure and skin and bladder cancer risk in a U.S. population. Environ Health Perspect 4:1047–1050.

Karagas MR, Stukel TA, Tosteson TD (2002) Assessment of cancer risk and environmental levels of arsenic in New Hampshire. Int J Hyg Environ Health 205:85–94.

Karagas MR, Tosteson TD, Morris JS, Demidenko E, Mott LA, Heaney J, Schned A (2004) Incidence of transitional cell carcinoma of the bladder and arsenic exposure in New Hampshire. Cancer Causes Cont 15:465–472.

Lai MS, Hsueh YM, Chen CJ, Shyu MP, Chen SY, Kuo TL, Wu MM, Tai TY (1994) Ingested inorganic arsenic and prevalence of diabetes mellitus. Am J Epidemiol 139:484–492.

Ledbetter MC, Mavridubneau R, Weiss AJ (1960) Distribution studies of radioactive fluoride^{18} and stable fluoride^{19} in tomato plants. Contrib Boyce Thompson Inst 20:331–348.

Lewis DR, Southwick JW, Ouellet-Hellstrom R, Rench J, Calderon RL (1999) Drinking water arsenic in Utah: A cohort mortality study. Environ Health Perspect 107: 359–365.

Luke J (1997) The effect of fluoride on the physiology of the Pineal Gland. Ph.D. Thesis. University of Surrey, Guildford

Luke J (2001) Fluoride deposition in the aged human pineal gland. Caries Res 35:125–128.

Meeker GP, Bern AM, Brownfield IK, Lowers HA, Sutley SJ, Hoefen TM, Vance JS (2003) The composition and morphology of amphiboles from the Rainy Creek complex, near Libby, Montana: Am Mineral 88:1955–1969.

Meeker GP, Sutley SJ, Brownfield IK, Lowers HA, Bern AM, Swayze GA, Hoefen TM, Plumlee GS, Clark RN, Gent CA (2005) Materials characterization of dusts generated by the collapse of the World Trade Center. In J.S. Gaffney, N.A. Marley (eds) Urban aerosols and their impacts: Lessons learned from the World Trade Center Tragedy (pp. 84–102). American Chemical Society Series 919.

Meeker GP, Lowers HA, Swayze GA, Van Gosen BS, Sutley SJ, Brownfield IK (2006) Mineralogy and Morphology of Amphiboles Observed in Soils and Rocks in El Dorado Hills, California: U.S. Geological Survey Open File Report 2006-1362, http://pubs.usgs.gov/of/2006/1362/downloads/pdf/OF06-1362_508.pdf

Mokdad AH, Bowman BA, Ford ES, Vinicor F, Marks JS, Koplan JP (2001) The continuing epidemics of obesity and diabetes in the United States. JAMA 286: 1195–1200.

Montgomery DL, Ayotte JD, Carroll PR, Hamlin P (2003) Arsenic Concentrations In Private Bedrock Wells In Southeastern New Hampshire. U. S. Geological Survey Fact Sheet 051-03.

Mullenix PJ, Denbesten PK, Schunior A, Kernan WJ (1995) Neurotoxicity of sodium fluoride in rats. Neurotoxicol Teratol 17(2):169–177.

National Research Council (1979) Geochemistry of water in relation to cardiovascular disease / Panel on the geochemistry of water in relation to cardiovascular disease, Subcommittee on the Geochemical Environment in Relation to Health and Disease, U.S. National Committee for Geochemistry. Washington, DC.

National Research Council (1999) Health Effects of Radon: BEIR VI. Washington, DC: National Academy Press.

National Research Council (2006) Fluoride in drinking water: A scientific review of EPA's standards. Washington, DC: National Academies Press.

NIOSH (2009) Asbestos Fibers and Other Elongated Mineral Particles: State of the Science and Roadmap for Research, revised draft: National Institutes of Occupational Safety and Health. Accessed June, 2009. http://www.cdc.gov/niosh/review/public/099/

Orem W, Tatu C, Pavlovic N, Bunnell J, Lerch H, Paunescu V, Ordodi V, Flores D, Corum M, Bates A (2007) Health effects of toxic organic substances from coal: Toward "Pandmic" nephropathy. Ambio 36(1):98–102.

Pan X, Day HW, Wang W, Beckett LA, Schenker MB (2005) Residential proximity to naturally occurring asbestos and mesothelioma risk in California. Am J Res Crit Care Med 72:1019–1025.

Pant N, Murthy RC, Srivastava SP (2004) Male reproductive toxicity of sodium arsenite in mice. Hum Exp Toxicol 23:399–403.

Pardue JH, Moe WM, McInnis D, Thibodeaux LJ, Valsaraj KT, Maciasz E, Van Heerden I, Korevec N, Yuan QZ (2005)

Chemical and microbiological parameters in New Orleans floodwater following hurricane Katrina. Environ Sci Tech 39:8591–8599.

Peipins LA, Lewin M, Campolucci S, Lybarger JA, Miller A, Middleton D, Weis C, Spence M, Black B, Kapil V (2003) Radiographic abnormalities and exposure to asbestos-contaminated vermiculite in the community of Libby, Montana, U.S.A.. Environ Health Perspect 111:1753–1759.

Pardue JH, Moe WM, McInnis D, Thibodeaux LJ, Valsaraj KT, Maciasz E, Van Heerden I, Korevec N, Yuan QZ (2005) Chemical and microbiological parameters in New Orleans floodwater following hurricane Katrina. Environ Sci Tech 39:8591–8599.

Peplow D, Edmonds R (2004) Health risks associated with contamination of groundwater by abandoned mines near Twisp in Okanogan County, Washington, USA. Environ Geochem Health 26:69–79.

Peterson M, Finkelman RB, Orem W, Lerch H, Bunnell J, Tatu C (2009) Organic compounds in the Carrizo-Wilcox aquifer of East Texas and possible health implications. Geolog Soc Am Abs Prog 41(2):13.

Plumlee GS, Casadevall TJ, Wibowo HT, Rosenbauer RJ, Johnson CA, Breit GN, Lowers HA, Wolf RE, Hageman PL, Goldstein H, Berry CJ, Fey DL, Meeker GP, Morman SA (2008) Preliminary analytical results for a mud sample collected from the LUSI mud volcano, Sidoarjo, East Java, Indonesia. U.S. Geological Survey Open-File Report 2008-1019.

Plumlee GS, Foreman WT, Griffin DW, Lovelace JK, Meeker GP, Demas CR (2007a) Characterization of flood sediments from Hurricane Katrina and Rita and potential implications for human health and the environment. In GS Farris, GJ Smith, MP Crane, CR Demas, LL Robbins, DL. Lavoie, , (eds) Science and the storms: The USGS response to the hurricanes of 2005: U.S. Geological Survey Circular 1306, pp. 246–257.

Plumlee GS, Hageman PL, Lamothe PJ, Ziegler TL, Meeker GP, Theodorakos P, Brownfield I, Adams M, Swayze GA, Hoefen T, Taggart JE, Clark RN, Wilson S, Sutley S (2005) Inorganic chemical composition and chemical reactivity of settled dust generated by the World Trade Center building collapse. In JS Gaffney, NA Marley (eds) Urban Aerosols and Their Impacts: Lessons Learned from the World Trade Center Tragedy (pp. 238–276). Washington, DC: American Chemical Society Symposium Series 919.

Plumlee GS, Martin DA, Hoefen T, Kokaly R, Hageman P, Eckberg A, Meeker GP, Adams M, Anthony M, Lamothe PJ (2007b) Preliminary analytical results for ash and burned soils from the October 2007 Southern California Wildfires: U.S. Geological Survey Open-File Report 2007-1407 http://pubs.usgs.gov/of/2007/1407

Plumlee GS, Meeker GP, Lovelace JK, Rosenbauer R, Lamothe PJ, Furlong ET, Demas CR (2006) USGS environmental characterization of flood sediments left in the New Orleans area after Hurricanes Katrina and Rita, 2005 –progress report: U.S. Geological Survey Open-File Report 2006-1023. http://pubs.usgs.gov/of/2006/1023

Plumlee GS, Morman SA, Ziegler TL (2006) The toxicological geochemistry of earth materials: an overview of processes and the interdisciplinary methods used to study them, In N Sahai, M Schoonen (eds) Reviews in Mineralogy and Geochemistry, 64, Medical Mineralogy and Geochemistry (pp. 5–58).

Plumlee GS, Ziegler TL (2007) The medical geochemistry of dusts, soils, and other earth materials –2007 web update. In BS Lollar (ed) Treatise on Geochemistry, online version, Volume 9, Chapter 7, Elsevier, 1–61. http://www.sciencedirect.com/science/referenceworks/9780080437514

Porter R, Ogilvie M (2000) The Biographical Dictionary of Scientists (3rd ed). New York: Oxford University Press.

Presley SM, Rainwater TR, Austin GP, Platt SG, Zak JC, Cobb GP, Marsland EJ, Tian K, Zhang B, Anderson TA, Cox SB, Abel MT, Leftwich BD, Huddleston JR, Jeter RM, Kendall RJ (2006) Assessment of pathogens and toxicants in New Orleans, LA following Hurricane Katrina. Environ Sci Tech 40:468–474.

Roggli VL, Oury TD, Sporn TA (eds) (2004) Pathology of Asbestos-Associated Diseases (2nd Ed, pp. 1–16). New York: Springer.

Roholm K (1937) Fluorine Intoxication: A Clinical-Hygienic Study. Copenhagen: NYT Nordisk Forlag.

Rouelle J (1792) A complete treatise on the mineral waters of Virginia containing a description of their situation, their natural history, their analysis, contents, and their use in medicine (68 p). Philadelphia, PA: Charles Cist Publisher.

Ruan JP, Bårdsen A, Astrøm AN, Huang RZ, Wang ZL, Bjorvatn K (2007) Dental fluorosis in children in areas with fluoride-polluted air, high-fluoride water, and low-fluoride water as well as low-fluoride air: A study of deciduous and permanent teeth in the Shaanxi province, China. Acta Odontol Scand 65(2):65–71.

Schneider E, Spiegel RA, Jibson RW, Harp EL, Marshall GA, Gunn RA, McNeil MM, Pinner RW, Baron RC, Burger RC, Hutwagner LC, Crump C, Kaufman L, Reef SE, Feldman GM, Pappagianis D, Werner B (1997) A coccidioidomycosis outbreak following the Northridge, Calif, earthquake. J Am Med Assoc 277:904–908.

Shacklette HT, Saur HI, Miesch AT (1970) Geochemical environmental and cardiovascular mortality rates in Georgia. U.S. Geological Survey Professional Paper 574-C.

Skinner HCW (2005) Mineralogy of bone. Contrib. Essentials of Medical Geology (pp. 667–693). Amsterdam: Elsevier Press.

Sturges WT, Wallington TJ, Hurley MD, Shrine KP, Sihra K, Engel A, Oram DE, Penkett SA, Mulvaney R and Brenninkmeijer CAM (2000) A potent greenhouse gass identified in the atmosphere: SF5CF3. Science, 289(5479): 611–613.

Sullivan PA (2007) Vermiculite, respiratory disease, and asbestos exposure in Libby, Montana –Update of a cohort mortality study. Environ Health Perspect 115:579–585.

Swayze GA, Hoefen TM, Sutley SJ, Clark RN, Livo KE, Meeker GP, Plumlee GS, Morath LC, Brownfield IK (2005) Spectroscopic and x-ray diffraction analyses of asbestos in the World Trade Center dust. In JS Gaffney, NA Marley (eds) Urban Aerosols and Their Impacts: Lessons Learned from the World Trade Center Tragedy (40–65). American Chemical Society Series 919.

Swayze GA, Kokaly RF, Higgins CT, Clinkenbeard JP, Clark RN, Lowers HA, Sutley SJ (2009) Mapping potentially asbestos-bearing rocks using imaging spectroscopy. Geology 37:763–766.

Tatu CA, Orem WH, Finkelman RB, Feder GL (1998) The etiology of Balkan Endemic Nephropathy: Still more questions than answers. Environ Health Persp 106(11): 689–700.

Tseng CH, Tseng CP, Chiou HY, Hsueh YM, Chong CK, Chen CJ (2002) Epidemiologic evidence of diabetogenic effect of arsenic. Toxicol Lett 133:69–76.

USEPA (U.S. Environmental Protection Agency) (2006) Response to the November 2005 National Stone, Sand & Gravel Association Report Prepared by the R.J. Lee Group, Inc. 'Evaluation of EPA's Analytical Data from the El Dorado Hills Asbestos Evaluation Project', United States Environmental Protection Agency, Region IX, San Francisco, CA, www.epa.gov/region09/toxic/noa/ eldorado/index.html

USEPA (U.S. Environmental Protection Agency) (2005–2008) Response to 2005 hurricanes: U.S. Environmental Protection Agency, http://www.epa.gov/katrina/, accessed 6/2009.

USGS (U.S. Geological Survey) (2008) Volcanic gases and their effects: U.S. Geological Survey Web Site, http://volcanoes.usgs.gov/hazards/gas/index.php, accessed 6-09.

USGS (U.S. Geological Survey) (2007) Facing Tomorrow's Challenges –U.S. Geological Survey Science in the Decade 2007–2017. U.S. Geological Survey Circular 1309.

USGS (U.S. Geological Survey) (2005) National Atlas of the United States. The Arsenic in Ground Water of the United States. http://nationalatlas.gov/natlas/natlasstart.asp

Van Gosen BS (2007) The geology of asbestos in the United States and its practical applications: Environ Eng Geosci 13(1):55–68.

Van Gosen BS (2009) The geologic relationships of industrial mineral deposits and asbestos in the Western United States: Preprints of the 2009 SME Annual Meeting, February 22–25, 2009, Denver, Colorado: Littleton, Colo., Society for Mining, Metallurgy, and Exploration, Inc. (SME) Preprint 09-061, 6 p.

Van Metre PC, Horowitz AJ, Mahler BJ, Foreman WT, Fuller CC, Burkhardt MR, Elrick KA, Furlong ET, Skrobialowski SC, Smith JJ, Wilson JT, Zaugg SD (2006) Effects of hurricanes Katrina and Rita on the chemistry of bottom sediments in Lake Pontchartrain, Louisiana, USA. Environ Sci Tech 40:6894–6902.

Virta RL (2009) Asbestos: Mineral Commodity Summaries, 2009, U.S. Geological Survey, 24–25.

Wedepohl KH (1978) Handbook of Geochemistry. Volume II/1. New York: Springer-Verlag.

Weinstein P, Cook A (2005) Volcanic emissions and health. In O Selinus, B Alloway, J Centeno, R Finkelman, R Fuge, U Lindh P Smedley (eds) Essentials of Medical Geology (pp. 203–226). Amsterdam: Elsevier Inc.

Witschi H (2001) A short history of lung cancer. Toxicol Sci 64:4–6.

Witham CS, Oppenheimer C, Horwell CJ (2005) Volcanic ash-leachates: A review and recommendations for sampling methods. J Volcanol Geother Res 141:299–326.

Wolf RE, Morman SA, Plumlee GS, Hageman PL, Adams M (2008) Release of hexavalent chromium by ash and soils in wildfire-impacted areas: U.S. Geological Survey Open-File Report 2008–1345 http://pubs.usgs.gov/of/2008/1345/

World Health Organization (WHO) (1997) Environmental health criteria for fluorides and fluorosis (Second ed). Internal Technical Report. International Program on Safety. Geneva, Switzerland.

World Health Organization (WHO) International Agency for Research on Cancer. (2004) IARC Monographs on the Evaluation of Carcinogenic Risks to Humans. Volume 84. Some Drinking-water Disinfectants and Contaminants, including Arsenic. Lyon, France.

Wu MM, Kuo TL, Hwang YH, Chen CJ (1989) Dose-response relation between arsenic concentration in well water and mortality from cancers and vascular diseases. Am J Epidemiol 130:1123–1132.

Zielinski J (2008) Presentation "Residential radon and lung cancer: a case study in risk assessment and risk management" and "RC-7 epidemiological methods: residential radon and cancer risk". Bureau of Environmental Health Science, Health Canada. Department of Epidemiological and Community Medicine, University of Ottawa.

Zierler S, Theodore M, Cohen A, Rothman KJ (1988) Chemical quality of maternal drinking water and congenital heart disease. Int J Epidemiol 17:589–594.

Zierold KM, Knobeloch L, Anderson H (2004) Prevalence of chronic diseases in adults exposed to arsenic-contaminated drinking water. Am J Public Health 94: 1936–1937.

A Regional Perspective of Medical Geology – Cadmium in Jamaica

Paul R. D. Wright, Robin Rattray, and Gerald Lalor

Abstract High levels of geogenic cadmium (Cd) have been found in some soils of Jamaica, particularly in the central part of the island. The potential health impact on residents who consumed foods grown on these soils was determined. The foods which have shown the greatest affinity for Cd uptake are mainly green leafy vegetables, tubers and root vegetables. Diet studies showed that some residents are at risk of ingesting Cd in excess of the 7 μg/kg body weight PTWI established by the WHO. Cd exposure and proximal tubular dysfunction were measured as urine Cd (U-Cd) and β_2-microglobulin (β_2-MG) in urine, respectively. Mean U-Cd and β_2-MG concentrations of 4.6 μg Cd/g_cr and 290 μg/g_cr confirm that the population is being exposed to elevated Cd levels and there is evidence of very mild tubular proteinuria. The proteinuria detected in the population was related to Cd exposure, evidenced by the relationship between U-Cd and β_2-MG. While positive results were obtained for the identification of Cd-related renal biomarkers in the study population and there was a clear association between U-Cd and β_2-MG, the absolute concentrations obtained were well below critical limits for the onset of acute or chronic renal effects. Women typically ingested less Cd than men but absorbed greater amounts than males in the same cohort groups.

Keywords Jamaica · Cadmium · Diet · Uptake · Proteinuria · Biomarkers

P.R.D. Wright (✉)
International Centre for Environmental and Nuclear Sciences, UWI, Mona, Kingston 7, Jamaica
e-mail: paul.wright@uwimona.edu.jm

Introduction

Medical geology focuses on the relationship between natural geology and health problems in humans, animals and plants. The discipline integrates the fields of geology, geochemistry, medicine, nutrition and public health in order to solve complex problems of trace element interactions with the human body. The uniqueness of the geochemistry of Jamaican soils, particularly their trace element content, may have associated potential health impacts. The occurrences of high natural levels of arsenic, lead and cadmium or deficiencies in iodine or selenium may bring associated morbidities to resident populations. In Jamaica, specific geochemical anomalies exist that may impact human and animal health. Some soils have geogenic cadmium concentrations that are orders of magnitude higher (Lalor, 1995) than regulated worldwide levels. These levels were established to limit Cd exposure and minimise the potential nephrotoxic effects of its bioavailability. Given the fact that the population has an elevated rate of chronic renal failure (Barton et al., 2004), it was important to establish cadmium's contribution to the incidence of renal disease. To this end, multidisciplinary studies were carried out in areas with significantly elevated soil cadmium concentrations in order to identify its contribution to any related morbidity in the population.

Aspects of Environmental Cadmium Toxicity

Health effects from occupational exposure to cadmium were first identified in nickel–cadmium battery workers (Friberg 1950), who breathed toxic vapours.

O. Selinus et al. (eds.), *Medical Geology*, International Year of Planet Earth,
DOI 10.1007/978-90-481-3430-4_2, © Springer Science+Business Media B.V. 2010

The resulting metal fume fever rapidly progressed in many cases to chemical pneumonitis, pulmonary edema and even death as a result of respiratory failure (Beton et al., 1966; Seidal et al., 1993). As an environmental hazard, cadmium was responsible for the occurrence of itai-itai disease in Japan in the 1950s. Itai-itai, which is Japanese for "ouch-ouch", is a form of acquired Fanconi's syndrome and is characterised by chronic calcium loss and the development of brittle, easily broken bones. Following World War II, farming populations living along the banks of the Jinzu River were exposed to food and water contaminated by mine wastes (Nogawa and Kido, 1996). The cadmium accumulated in rice crops grown downstream, which were consumed by residents of nearby villages, resulting in renal abnormalities, including osteomalacia, proteinuria and glucosuria (Jarup et al., 1998; Nordberg et al., 1997). Older post-menopausal, multiparous, malnourished women were most affected (Shigematsu, 1984).

In non-occupational settings the general population is exposed to cadmium through tobacco smoking and ingestion from dietary sources. Smoking carries a greater risk of Cd exposure and it has been estimated that about 10% of the cadmium content of a cigarette is inhaled (Morrow, 2001; NTP, 2005). The lungs have a very high absorption efficiency compared to the gut and as much as 50% of the cadmium inhaled may be absorbed (Friberg, 1983). On average, smokers are reported to have four to five times higher blood cadmium concentrations and two to three times higher kidney cadmium concentrations than non-smokers (Lewis et al., 1972, Vahter et al., 2002).

The dietary route provides the most significant pathway for non-smoking populations (Jarup et al., 1998). Cadmium accumulates in leafy vegetables such as lettuce and spinach and staples such as root tubers and grains (ATSDR, 2008). The exposure risk is high for persons who consume large amounts of these foods and also some shellfish and organ meats (liver and kidney). The estimated daily intakes of cadmium in non-smoking adult males and females living in the United States are 0.35 and 0.30 μg Cd/kg/day, respectively. Females generally absorb greater amounts of cadmium in the gastrointestinal tract than do males (ATSDR, 2008). Cadmium exposure in humans is typically measured by concentration increases in the urine (U-Cd), an indication of body burden and/or the blood (B-Cd), a measure of more recent exposure (ATSDR, 2008; Olsson et al., 2002).

Etiology of Toxicity

The primary target organs for Cd are the kidney and liver but it can be also found in bone, gall bladder, pancreas, prostate, testes and the thyroid. It is generally considered to be one of the more toxic elements and consequently its effects have been thoroughly reviewed (Agency for Toxic Substances and Disease Registry (ATSDR), 2008). As a result of the significant body of work following the itai-itai incident in Japan (Aoshima, 1997; Nogawa, 1981; Fan et al., 1998), many countries have placed maximum limits on the content of Cd in food. The World Health Organization maintains a provisional tolerable weekly intake (PTWI) for Cd of 7 μg/kg body weight (JECFA, 2000). This value was recently revised by the EU to give a TWI of 2.5 μg/kg body weight (EFSA, 2009). There have been, however, populations such as those in Shipham in the United Kingdom that despite high exposures to Cd have been spared related health problems (Baker et al., 1977; Ewers et al., 1985; McKenzie-Parnell and Eynon, 1987; Kido et al., 1988; Strehlow and Barlthrop, 1988; Sarasua et al., 1995).

The etiology of Cd toxicity has been widely studied and the mechanisms of uptake and concentration in the proximal renal tubules are well understood (Nordberg, 1984; Jarup et al., 1998; Bernard, 2008). An illustration of the mechanism is shown in Fig. 1 (Bernard, 2008). Following exposure, Cd is bound by metallothionein (MT) in the liver forming a Cd-metallothionein complex (Cd-MT). The low molecular weight of Cd-MT allows free passage through the glomerular membrane of the kidney into the tubular fluid. The Cd-MT is transported by pinocytosis into the proximal tubular cells of the kidney where it is degraded by lysosomes, releasing free Cd into the cytoplasm. This free Cd induces de novo synthesis of MT in the kidney tubuli, once again forming protective Cd-MT. Tubular damage can occur if the influx of Cd-MT into the tubuli is very high and the kidney cannot produce sufficient MT to bind the Cd released by the lysosomes (Nordberg et al., 2000; Friberg, 1984). At this stage the non-MT-bound Cd becomes toxic, resulting in degradation of the proximal tubules. The resulting renal dysfunction is characterised by a reduced filtration capacity and is measured by determining the concentrations of low molecular weight proteins such as α_1-microglobulin (α_1-MG), β_2-microglobulin (β_2-MG) or retinol-binding protein in urine (Friberg et al., 1986). These are known to

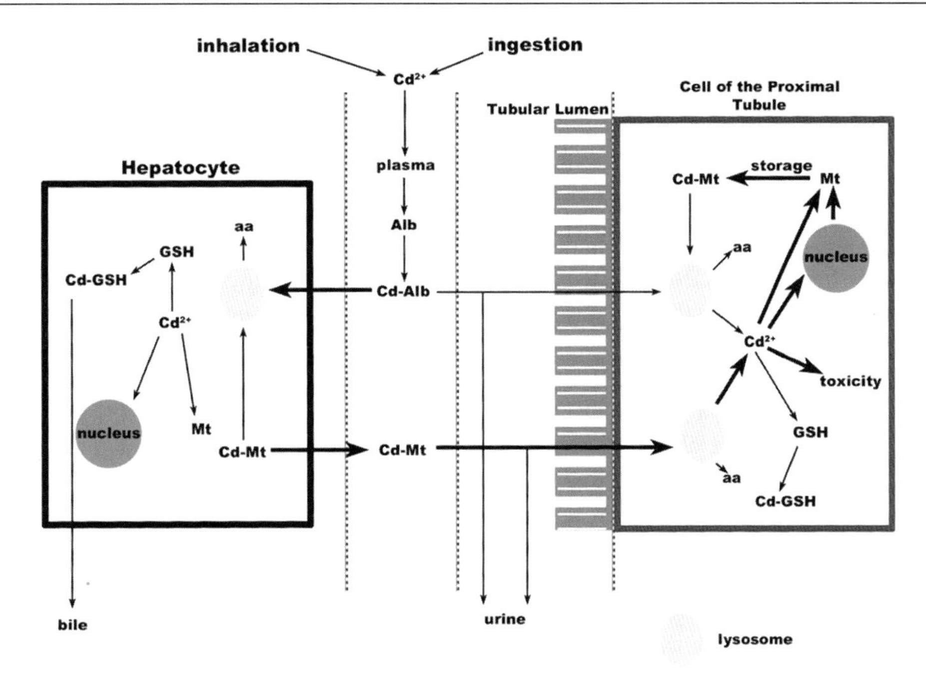

Fig. 1 Illustration of the mechanism of Cd accumulation in proximal tubular cells of the kidney (Reprinted by permission from Bernard, 2008), copyright, the IJMR

be early indicators of the onset of proximal tubular dysfunction (Kjellstrom et al., 1977; Friberg, 1984; Lauwerys et al., 1984; Buchet et al., 1990; Piscator, 1991). Japanese studies have shown that there is a non-linear relationship between Cd and β_2-MG in urine, with the β_2-MG increasing sharply once a threshold value of U-Cd is exceeded (Ikeda et al., 2005).

Micronutrient Amelioration of Cadmium Toxicity

Cadmium has the capacity to interact with other micronutrients (chromium, magnesium, zinc, selenium, and iron) having similar physical and chemical properties. This interaction may change the rate of absorption of this metal by the body, resulting in either increased or reduced toxicity. For example, exposure to Cd coupled with a low-iron diet may contribute to the increase in the toxicity of this metal. These interactions may happen in different phases of the absorption, distribution and excretion of Cd and its elements (Brzóska and Moniuszko-Jakoniuk, 1997). The absorption of Cd via the gastrointestinal tract may be affected by several factors, such as age, gender and nutritional status. Among these, being younger, having iron deficiency in the organism and especially being a woman are characteristics that may accelerate the absorption of the metal (Olsson et al., 2002; Satarug and Moore, 2004; Horiguchi et al., 2004). Women are generally more affected by the toxic effects of Cd due to the fact that they have less iron burden than men. In some cases, one of the symptoms associated with cadmium intoxication is the development of anaemia on exposed individuals, as a result of an inhibiting effect of Cd in the metabolism and absorption of iron (Vahter et al., 2002). It was found that in a study of urine and blood Cd in 105 individuals in Sweden, there were 1.6 times higher Cd levels in women compared to men (Olsson et al., 2002). According to the authors, the fact may be attributed to the lower iron levels in women as there were no statistically significant differences between the Cd intakes.

Local Health Concerns

Chronic non-communicable diseases continue to be a drain on the national budget. According to the available records which reflect discharges from public hospitals, the most common occurrences are conditions related to the circulatory system, malignant neoplasms, diabetes mellitus and renal failure in that order. In 2002, circulatory system, malignant neoplasms and diabetes mellitus ranked among the top five, renal failure ranked ninth among the listed disease conditions. All these have been associated in one way or other with Cd but the records do not allow links to be made and show no evident geographic links. It has been suggested by members of the medical fraternity in Jamaica that the prevalence of chronic renal failure (CRF) in the population is extremely high. The Diabetes Association of Jamaica has estimated that about 400–600 new cases of CRF per million occur in Jamaica each year. A study carried out on a group of persons with CRF from across the island estimated prevalence in 1999 to be 327 cases of CRF per million (Barton et al., 2004). However, there were concerns of serious underreporting making the potential impact even greater. Of great interest is the possible influence of the nephrotoxicity of Cd on the prevalence of renal impairment in Jamaica.

A population autopsy survey carried out in Jamaica with persons not resident in high cadmium areas has suggested that the concentrations of Cd in the kidneys are elevated when compared to other parts of the world (Lalor et al., 2004). We were unable to differentiate between smoking and non-smoking subgroups, probably because of elevations in the background levels by other sources. In central Jamaica, where the cadmium levels in the soils are highest, indicative data collected suggest a doubling in the concentration of Cd in kidneys of local autopsy subjects when compared to the population (unpublished data). This could imply correlations between Cd concentrations in soil and urine and possibly also any associated renal dysfunction.

Given worldwide concerns of the toxicity of cadmium and efforts taken to minimise exposure to the general population, a thorough examination of the local situation was required. Elevated Cd concentrations in some farm soils in Jamaica result in accumulations in some crops grown for local consumption (Howe et al., 2005). Having confirmed elevated concentrations in the kidneys of the general population it was necessary to examine what health risk was posed by exposure to diet-sourced cadmium.

Geologic Origins of the Caribbean and Jamaica

Jamaica is located in the Caribbean Sea (Fig. 2) at approximately latitude 18°N and longitude 77°W, about 4.5° south of the Tropic of Cancer or midway between the southern tip of Florida and the Panama Canal. The island has an area of approximately 11,244 km^2 (or about 4,411 square miles) (Fig. 3).

The islands of both the Greater Antilles (except Cuba) and Lesser Antilles are located on the Caribbean Plate, an oceanic tectonic plate underlying Central America and the Caribbean Sea off the north coast of

Fig. 2 Map of the Greater Antilles

Fig. 3 Parish map of Jamaica

South America. The plate is approximately 3.2 million square kilometres (1.2 million square miles) in area and borders the North and South American Nazca and Cocos Plates. These bordering plates are intensely seismic, with frequent earthquakes, occasional tsunamis and volcanic eruptions. The eastern boundary is a subduction zone, resulting from convergence of the Caribbean Plate and either the North or South American Plates or both. The subduction activity formed the volcanic islands of the Lesser Antilles which are home to 17 active volcanoes including Montserrat's Soufriere Hills, Mount Pelée on Martinique, Guadeloupe's La Grande Soufrière and Soufrière Saint Vincent on Saint Vincent. The southern boundary of the Caribbean Plate is the result of a combination of transform and thrust faulting along with some subduction. Its interaction with the South American Plate formed Barbados and Trinidad (both on the South American Plate) and Tobago (on the Caribbean Plate) and islands off the coast of Venezuela (including the Leeward Antilles) and Colombia.

Two models of the tectonic evolution of the Caribbean Plate and genesis of the Cretaceous submarine flood basalts have been postulated. Generally called the Pacific and Atlantic models, both propose that the submarine flood basalts originated from a hotspot/mantle plume, but differ on the location. The Pacific model suggests that the thick, Caribbean oceanic crust and its flood basalts formed in the Pacific Ocean from the Galapagos hotspot in the Farallon Plate and migrated to its present location (Burke et al., 1978; Ross and Scotese, 1988; Pindell and Barrett, 1990). The Atlantic model suggests that the Caribbean crust was formed in-place on a mantle plume between North and South America (Donnelly, 1985; Meschede and Frisch, 1998).

Jamaica is situated on the northwest rim of the Caribbean Plate (Green, 1977) and is seen as the emergent, uplifted, easterly tip of the Nicaragua rise. The Caribbean Plate has moved east–north–east by about 1,400 km with respect to North America since the late Eocene period. The east–west trending, seismically active Cayman Trench separates Jamaica from the North American Plate. The island shares many of its geological features with Cuba, Hispaniola and Puerto Rico. The earliest Caribbean seafloor was formed during the Jurassic period and the oldest dated rocks in Jamaica are consistent with the late Jurassic/early Cretaceous ages of ophiolitic rocks typically found in the Greater Antilles (Wadge et al., 1982). The island can be described as a Cretaceous volcano-plutonic arc, 70% of which is now covered unconformably by Cenozoic sediments. Approximately 10% of the present-day land surface is covered by the major Cretaceous inliers. Calc-alkaline igneous rocks and volcaniclastics occur in the Blue Mountains inlier (Grippi and Burke, 1980). During the mid-Eocene, yellow limestones unconformably overlay tectonically

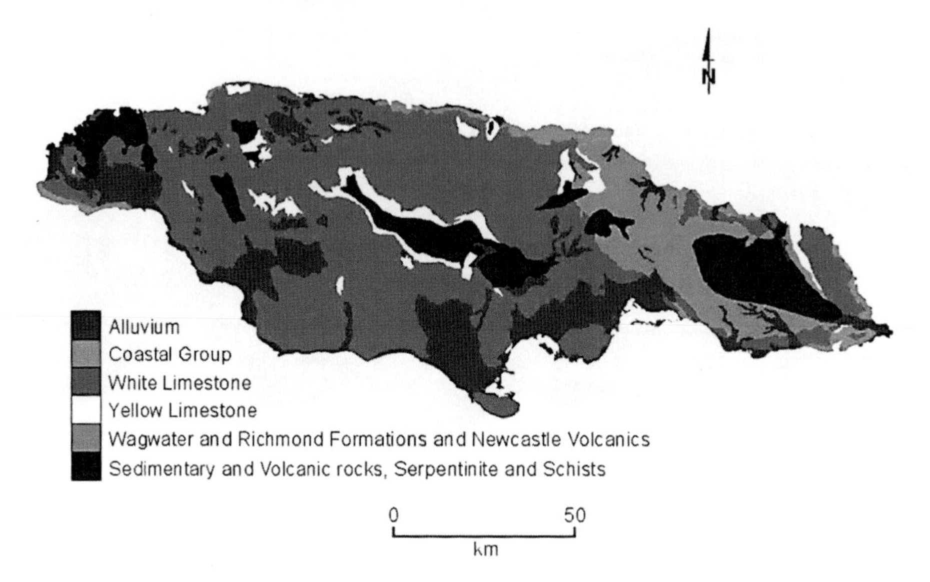

Fig. 4 Simplified geology map of Jamaica highlighting major formations

stable blocks, while a conformable cover extended over the subsiding terranes of the Wagwater and Montpelier-Newmarket belts. Approximately 60% of the surface formations on the island are deposits of the White Limestone Group that accumulated during the late Eocene/early Miocene periods. A simplified geology map of Jamaica is shown in Fig. 4.

During the Miocene period the northern end of the Blue Mountains block saw extensive uplift and volcanic activity. During this time, the limestone cover was removed from numerous tectonically stable localities of the island interior. The late Miocene sandstone and limestone of the present-day Coastal Group were produced by continued sedimentation around the coastal margins. The uplift that created the Blue Mountain range occurred during the Pliocene. That uplift continued over the whole island during the Pleistocene, resulting in the southward tilting and down-faulting of the southern coastal plains.

Soils

Topologically, Jamaica can be divided into two regions: a highland interior and a flatter coastal periphery. Approximately 15% of the soils have slopes of 5% or less and about 50% have slopes above 20%. Jamaica shows diverse variation in soil types as a result of differing parent materials, topography and rainfall. About two-thirds of the land is covered with soils underlain by limestone with the remainder weathered from conglomerates, tuffs, shales, igneous and metamorphic rocks. These soils are classified into three broad categories, which reflect the underlying geology and altitude (JRA, 1982).

Highland Soils

These occur at altitudes greater than 1,000 m in the Blue Mountains and in sections of the central and western mountainous axis. They are mainly derived from the weathering of Cretaceous volcano-sedimentary rocks. Two soil types can be distinguished:

1. Lithosols: Immature, highly porous soils which have been leached to acidity. Deforestation has resulted in the erosion and subsequent truncation of the soil profile resulting in low-nutrient, infertile soils. Exceptions exist in areas of forest cover, which allows the development of an upper layer that is organic rich and fertile.
2. Less acidic clays, produced by the weathering of shales in areas with reduced drainage.

Upland Plateau Soils

These soils account for about 60% of Jamaica's land surface and occur over limestone between 700 and 1,000 m. Two types exist, terra rossa (red limestone) and rendzina (black limestone). Terra rossa soils occur mainly in the parishes of Manchester and St. Ann and are leached, well oxidised and dehydrated residual bauxitic soils. They vary greatly in depth and have no distinctive profile. The fine-grained rendzina soils are weathered over the exposed stone and are poorly drained. Their high Ca content and wetness limit ferrous to ferric oxidation resulting in a darker colour and an absence of the typical red colour associated with bauxites.

Bauxite

The bauxite deposits are found mainly in areas underlain by Cenozoic white limestones deposited in shallow water. They are mainly of blanket type, although the deposits can concentrate in depressions such as those that occur in the extensively karstified Cockpit Country. The genesis of bauxites and their relationship to the limestone bedrock have been the subject of much discussion and are described by three principal models:

1. Bauxites were formed by the concentration of impurities following dissolution of the underlying limestone bedrock.
2. Alumina-rich residues, derived from the weathering of the volcanic rocks of the Cretaceous inliers, were transported onto and into the limestone terrain, with bauxites resulting from in situ weathering (Zans, 1952).
3. The deposition of ash falls derived from external volcanic activity that occurred close to the island (Comer, 1974).

Alluvial Plain and Valley Soils

These are deposited by rivers and are most extensive on the southern slopes of the island. Narrow strips also occur in the central valleys and along the north coast. The soils are highly fertile and consist of loam, sand and gravel. Other soil variations include riverine soils covered by heavy clay soils of marine origin and swamp soils, notably in the Black River and Negril areas.

Background Trace Element Research in Jamaica

Jamaica has been the focus of significant geological research, resulting in extensive bibliographies (Draper and Lewis, 1990; Wright and Robinson, 1993). Early geochemical work was focused on defining the elemental content of the economically important bauxite deposits. Detailed mapping of the island-wide soil geochemistry was undertaken during the 1980 s following the launch of the Centre for Nuclear Sciences (now the International Centre for Environmental and Nuclear Sciences, ICENS) and the commissioning of the SLOWPOKE nuclear reactor used for Instrumental Neutron Activation analysis (INAA). ICENS was established to conduct integrated research programmes based on environmental geochemistry.

Orientation Survey

A detailed geochemical orientation survey was conducted as a precursor to more extensive mapping and established the sampling and sample preparation methodologies used for subsequent work (Simpson et al., 1991). Six sample sites were chosen that covered the main lithologic and climatic zones of Jamaica. Samples were analysed for 35 elements, using neutron activation analysis, X-ray fluorescence analysis and direct reading optical emission spectrometry.

Jamaican soils were found to contain a high proportion of fine fraction <150 μm (<100 mesh) material. This fraction was found to provide better geochemical differentiation than higher fractions and showed minimal variation of element concentrations with particle size. This fine fraction was chosen as the sampling medium for the regional geochemical survey of Jamaica.

Radiation Survey

Approximately 6,700 km of roads and tracks were traversed by a Land Rover, with a rigidly mounted 6" × 4" sodium iodide scintillation detector (Fig. 5). The

Fig. 5 Radiometric monitoring equipment used during the survey

lowest dose rates recorded in the island (<2 μR/h) were in the Blue Mountain inlier. The limestones of the Coastal Group, which form a narrow discontinuous belt 2–5 km in width along the northern and southwestern coasts of the island, had gamma dose rates below 4 μR/h, and similar levels were detected on adjacent formations.

The pattern of background gamma activity in Jamaica reflected many of the features of the bedrock geology. The data showed average or below average gamma activities across the Cretaceous inliers with small outcrops having higher values. Several slightly radioactive mineral springs occur in Jamaica, and most of the activity appears to be derived from deep underground sources (Vincenz, 1959). Only the Black River Spring in St. Elizabeth seems to be associated with surface activity. Dose rates of up to 20 μR/h were found over the Above Rocks Cretaceous inlier, probably a result of uranium. The highest activities appeared to be associated with iron-stained shear structures transecting the Border Volcanics (Lalor et al., 1989).

Higher activities matched bauxite distribution almost identically, despite the low radioactivity of the underlying limestone. The resemblance of the gamma distribution map with that of aluminium was so remarkable that it was suggested that gamma-ray mapping might well be an aid to bauxite exploration in Jamaica (Lalor et al., 1989).

Geochemical Mapping

Many naturally occurring elements have biological effects. Some are necessary in quite large amounts, others are essential at low concentrations but deficiencies or small excesses can lead to disease or death. Many are toxic even at low concentrations and exposure to these, and to all radioactive elements, is restricted to the lowest possible levels achievable. The Jamaican regional survey provided an overall picture of the levels of occurrence and the distribution of the elements has provided the background for more detailed studies. This work resulted in the publication of a geochemical atlas of Jamaica (Lalor, 1995).

The survey density was one sample in 64 km^2 using surface samples collected from between 5 and 30 cm depth. Data are now available for 54 elements on this scale island-wide and at much higher sample density for much of central Jamaica. The trace element geochemistry of Jamaican soils was found to be extremely variable due to their diverse parent materials, varying maturity and the variety of soil forming processes. Leaching has led to depletion of silica and base cations, Ca^{2+}, Mg^{2+}, Na^{+} and K^{+}, and thereby to elevated Fe and Al levels typical of the many tropical soils. Many of the naturally occurring elements are biologically active, and interactions between some pairs have both antagonistic and synergistic effects.

Toxic heavy metals, arsenic, cadmium and to a lesser extent lead, were identified at elevated concentrations in some soils. Arsenic was shown to have a low risk to human health (Lalor et al., 1999), while lead was shown to be a significant risk to children exposed to contamination from a former mine (Anglin-Brown et al., 1996; Lalor et al., 2001). Cadmium, a nephrotoxin, has been the focus of more recent investigations given the massively high concentrations present in the soils (Lalor et al., 1998, 2001, 2004).

Identification of a Potential Medical Geology Problem in Jamaica

Arsenic in Jamaican Soils

The soil arsenic anomaly discovered during the island-wide geochemical survey of Jamaica was investigated and concentrations of up to 400 μg As/g were found near Maggotty in the parish of St. Elizabeth (the dark spot in Fig. 6) (Lalor et al., 1999). The area exceeded the 95th percentile (>65 μg As/g) of the island-wide concentrations and covered an area of at least 10 km^2. The anomalous values may have been sourced from an ancient hot spring environment which was responsible for the introduction and deposition of Fe–As–S as pyrite and arsenopyrite in the limestone bedrock. These were subsequently oxidised and weathered to yield arsenic-rich soils. The soils were also enhanced in elements such as Sb, Fe and Co. Despite the high arsenic content, the arsenic concentration in the surface water was low and health risk to the residents at that time was assessed to be low. The area, however, could present a potential hazard with changing land use.

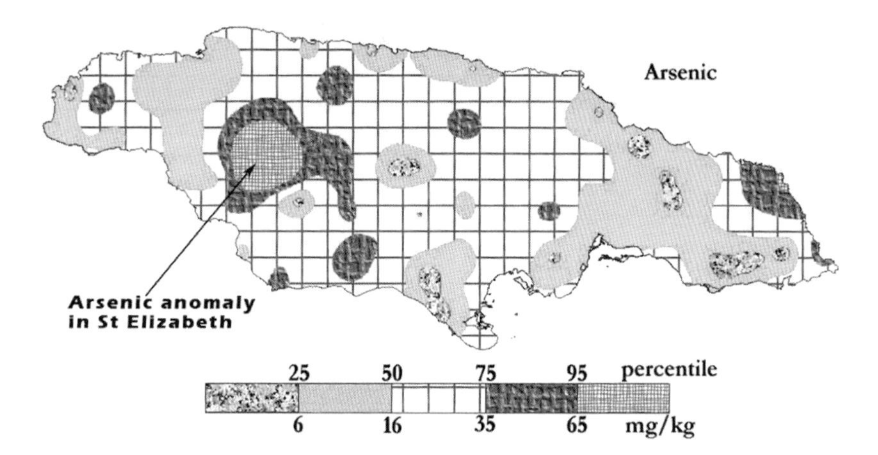

Fig. 6 Arsenic distribution in Jamaican soils

Cadmium in Jamaican Soils

Origins of Cadmium

The higher levels of cadmium in Jamaican soils are associated with the white limestone geology and the overlying aluminium-rich bauxite deposits in the central parishes (Lalor, 1995) (Fig. 7). The soils containing the highest soil cadmium concentrations typically occur at elevations greater than 200 m.

The two theories of the origin of cadmium in Jamaica's soils are the same as those postulated for the genesis of bauxite (Zans, 1952; Comer, 1974). The competing theories of weathering from parent limestone (Zans) or deposition of Miocene volcanic ash (Comer) have to be examined in light of the identification of a phosphatic band at the bauxite–limestone interface rich in Cd and other elements such as Be and Y. This band, seen in some soils at varying depths, forms the interface between the limestone and overlying bauxite soils. It was first identified by Eyles (1958) and examined in greater detail more recently (Garrett et al., 2008), which suggests a relationship to late Miocene or Pliocene phosphorites derived from the weathering of fossilised sea bird guano deposits from rookeries that spanned the central part of the island. The material contains fossilised fish bones and teeth suggesting a relation to the marine environment. The accurate dating of this material would be useful

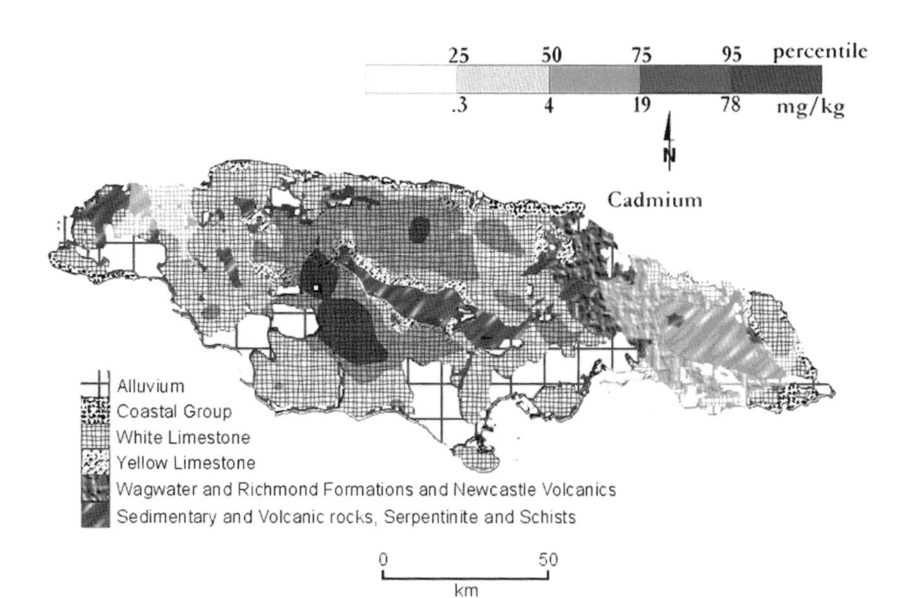

Fig. 7 Cadmium association with underlying geology

in determining whether the Cd is a result of concentration in limestone solutions as suggested by Zans or is the result of leachate from overlying volcanic parent minerals, which was postulated by Comer, weathered to form the bauxites.

Concentration of Cd in Soils

The Jamaican soil Cd concentrations vary widely, with a mean value of approximately 20 mg/kg with maximum values exceeding 900 mg/kg in some areas (Lalor et al., 1998). These concentrations are in stark contrast to world averages of approximately 0.1–0.5 mg/kg (Adriano, 1986; Waldron, 1980). The extraordinary levels in Jamaican soils are compared with world values in Table 1.

Table 1 Comparison of Cd concentrations in selected Jamaican and world soils

	Manchester range	Manchester mean	World range
Cd (mg/kg)	0.56–931	54	0.005–2.9

These, with the exception of some values from the Dominican Republic, appear to be the highest geogenic levels observed in the literature, often orders of magnitude larger than for example the Canadian quality guideline for Cd of 1.4 mg/kg (CCME, 1999) in agricultural soils. The mapping in Fig. 8 illustrates the concentration distribution.

As a result of the perceived toxicity of Cd, several countries have placed maximum limits on soil Cd concentrations which affect usage. Critical limits above which the soils are considered inappropriate for any human use are shown in Table 2 (DeVries et al., 1998) and are significantly lower than the average concentrations in Jamaica.

Table 2 Some critical limits for Cd and Zn in soils (mg/kg DW)

Country	Cd	Zn
Canada	0.5	50
Denmark	0.3	100
Finland	0.3	90
The Netherlands	0.8	140
Germany[a]	0.4–1.5	60–200
Switzerland	0.8	200
Czech Republic	0.4	150
Eastern Europe[b]	2	100
Ireland	1.0	150

[a]The first value is for sandy soils and second value for clay soils
[b]Eastern Europe includes Russia, Ukraine, Moldavia and Belarus

Cadmium in Food

Some plant species have been shown to have an affinity for the uptake of cadmium. The soils of central Jamaica support major crop growing areas and investigations were undertaken to identify crops sensitive to Cd uptake as shown in Table 3 for samples of fruit, legumes, leafy and root vegetables and root crops. Significant variation in Cd uptake was found between crop types. Soil Cd content is the main factor other than species type in controlling uptake (Howe et al., 2005).

There is significant variability in the concentrations of Cd in food crops, which extends to over three orders of magnitude in the analysed samples. Root crops (excluding root vegetables) had concentrations

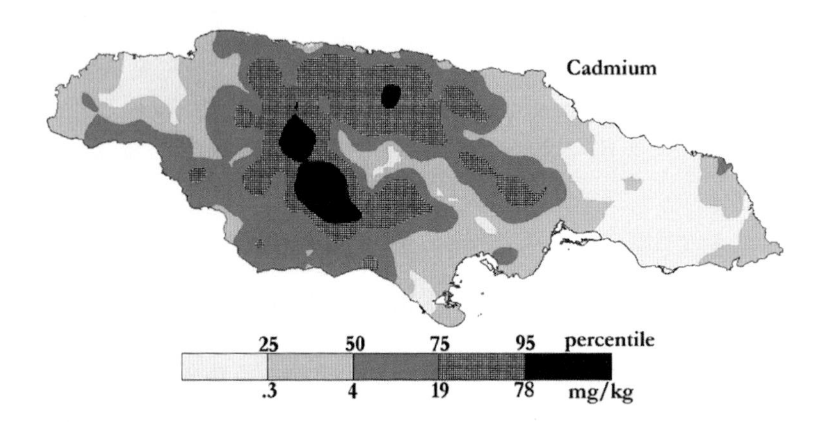

Fig. 8 Cadmium distribution in Jamaican soils

Table 3 Comparisons of the Cd compositions of Jamaican foods by category

		Cd concentrations (mg/kg fresh weight)		
Food category	Number of samples	Range	Mean	EU limit
Fruit	18	0.005–0.14	0.03	0.005
Legumes	4	0.026–0.132	0.17	0.2
Vegetables: leafy	12	0.02–1.71	0.383	0.2
Vegetables: root	9	0.43–0.94	0.49	0.1
Other root crops	35	0.026–1.04	0.18	0.1
Food category	**Types of samples**			
Fruit	Ackee, banana, breadfruit, corn, cucumber, orange, ortanique, plantain, tomato, pumpkin, squash, sweet pepper, zucchini			
Legumes	Cow peas, French bean (string bean), gungo peas, red kidney beans			
Vegetables: leafy	Broccoli, cabbage, callaloo, cauliflower, lettuce, pak choi, thyme, scallion			
Vegetables: root	Beet, carrot, onion, turnip			
Other root crops	Cassava, coco, dasheen, ginger, potato, sweet potato, yam			

Table 4 Comparisons of mean Cd concentration of food for three countries (mg/kg fresh weight)

Food category	Jamaica	United Kingdom	United States
Fruit	0.029	0.002	0.004
Legumes	0.33	0.059	0.054
Leafy vegetables	0.4	0.023	0.058
Root vegetables	0.4	0.011	0.014
Root crops	0.2	0.026	0.018

between 0.004 and 6.5 mg/kg, root vegetables ranged from 0.001 to 1.8 mg/kg and leafy vegetables covered a 0.002–1.7 mg/kg range. Jamaican foods have Cd concentrations 10–20 times higher than those reported in the literature (Table 4), and the overall average is greatly influenced by the values from central Jamaica.

There was significantly less variability in the concentrations of other elements in the food crops. The difference observed for Cd may be attributable to the absence of a mechanism of homeostatic control as that which exists for the other elements. The absence of this uptake regulating mechanism is illustrated very well using yams and carrots, both of which are able to concentrate significant amounts of Cd (Fig. 9).

Comparing the concentration range for Cd in the samples with Fe, Zn, Cu, K, Ca, Mg and P, it is seen that there is a much wider spread of data for Cd, up to three orders of magnitude. The remaining elements of interest, Se (no data for carrots), As, Co, Mn, Cr and Sc, show increased spread but not as much as for Cd. It is apparent from the data that the variability of Cd in soils is also reflected in the plants. Also, despite varying levels of trace elements in the soils, both yam and carrot regulate their uptake of bioessential elements ensuring the maintenance of constant internal levels to support growth.

The cadmium concentration in some foods is directly related to the soil Cd, illustrated by yams in Fig. 10. The cadmium concentrations of yams show a strong correlation with the soil Cd up to a point, followed by a decrease at very high soil Cd concentrations, suggesting possible toxicity.

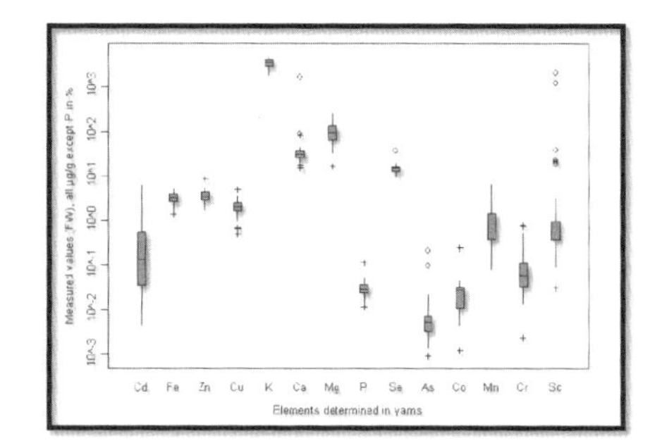

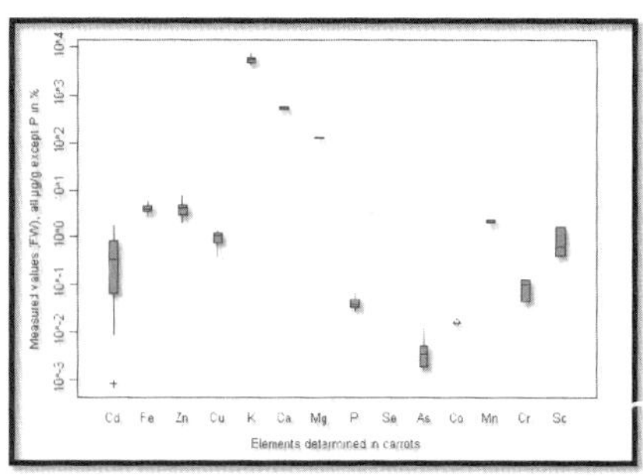

Fig. 9 Distribution of Cd and other trace elements in yams and carrots

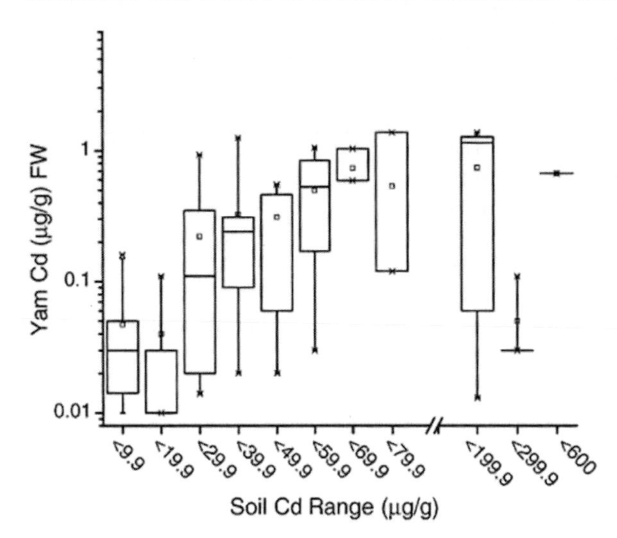

Fig. 10 Variation of the median concentrations of Cd in soils and yams

Extractability of Cd in Gastric Juices

Tests of the extractability of Cd using simulated gastric juices carried out on a set of raw and cooked foods show that (a) cooking does not reduce the Cd content and (b) all the Cd is extracted (Hoo-Fung et al., 2005–2006). For those crops examined (callaloo, carrot, pak choi, sweet potato and yam) there is no loss of Cd during the usual cooking procedures of steaming and boiling and all the Cd present is effectively extracted by simulated gastric juices. This means that all cadmium ingested is available for absorption across the gut. Typically, absorption of Cd from the gut is reported as between 5 and 10% and varies depending on factors such as the dietary intake of iron, where higher levels have been linked to reduced absorption.

Assessment of Cd Exposure in a Jamaican Population

The confirmation of cadmium accumulation in some foods and its availability following gastric dissolution suggested a possible exposure risk to populations consuming foods grown in high-cadmium areas. Accumulation of cadmium in the kidneys was confirmed in the general population following examination of cortical cadmium concentrations using data from autopsy subjects (Lalor et al., 2004). Cortical Cd concentrations were elevated when compared to other countries, except Japan.

In order to better determine what contribution food consumption had on this observation, a population in the highest cadmium section of the island was examined. Dietary assessments and the determination of food consumption frequencies were used to determine the intake of cadmium and also examine variations throughout the population by age group. The exposure was further confirmed by the examination of cadmium in the urine of persons in these areas and also examinations of renal dysfunction biomarkers as an indication of impact of cadmium intake.

Description of Study Area

The area chosen for studies on cadmium exposure is enclosed by the box shown in Fig. 11 and consists of contiguous sections of the central parishes of Clarendon, St. Ann, St. Elizabeth, Manchester and Trelawny. Cadmium concentrations vary from very low to extremely high. In the main, however, the study area parishes represent areas of highest cadmium concentrations island-wide.

The geology of the area is dominated by white limestone and contains some areas of yellow limestone, sedimentary and volcanic rocks (Serpentinite and Schists). The study area encloses an intensively farmed area of 22 km by 34 km, about 74,800 ha (Fig. 12), and is one of Jamaica's "bread basket" regions.

Using soil data collected prior to the study, the area was further divided into four strata according to Cd status shown in Table 5. The lowest soil Cd concentration strata had an upper limit of 4 mg/kg, already considered very high for uncontaminated soils in most countries. The lowest concentration of Cd obtained within the study area was 1.3 mg/kg, a figure, as has been shown, considered elevated in most other countries (CCME, 1999).

Population Demographics

The total population of the study area was 47,266 according to the 2001 census (Statistical Institute of Jamaica) distributed relatively evenly between males and females (52/48%). The number of persons over

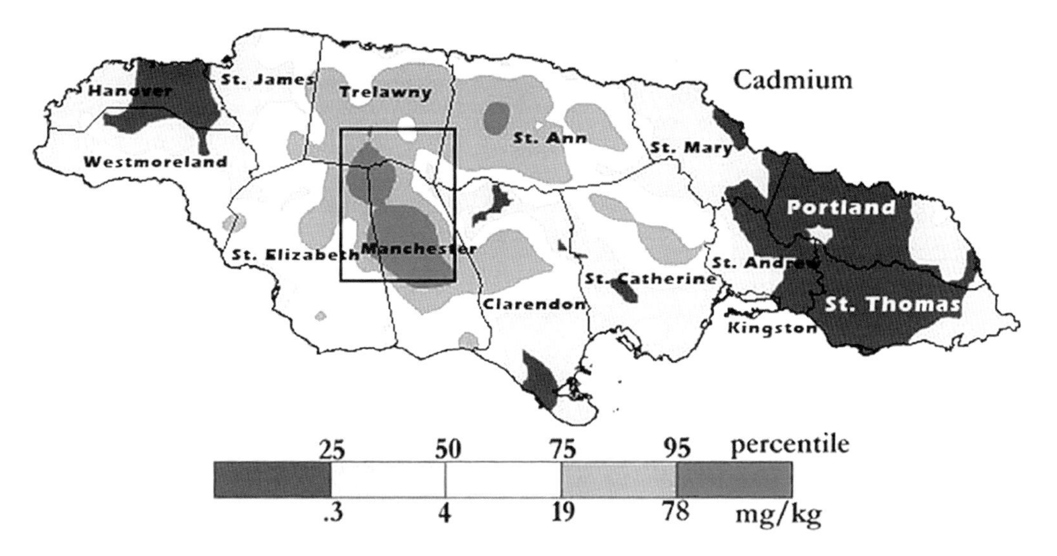

Fig. 11 Island-wide distribution of cadmium highlighting the study area

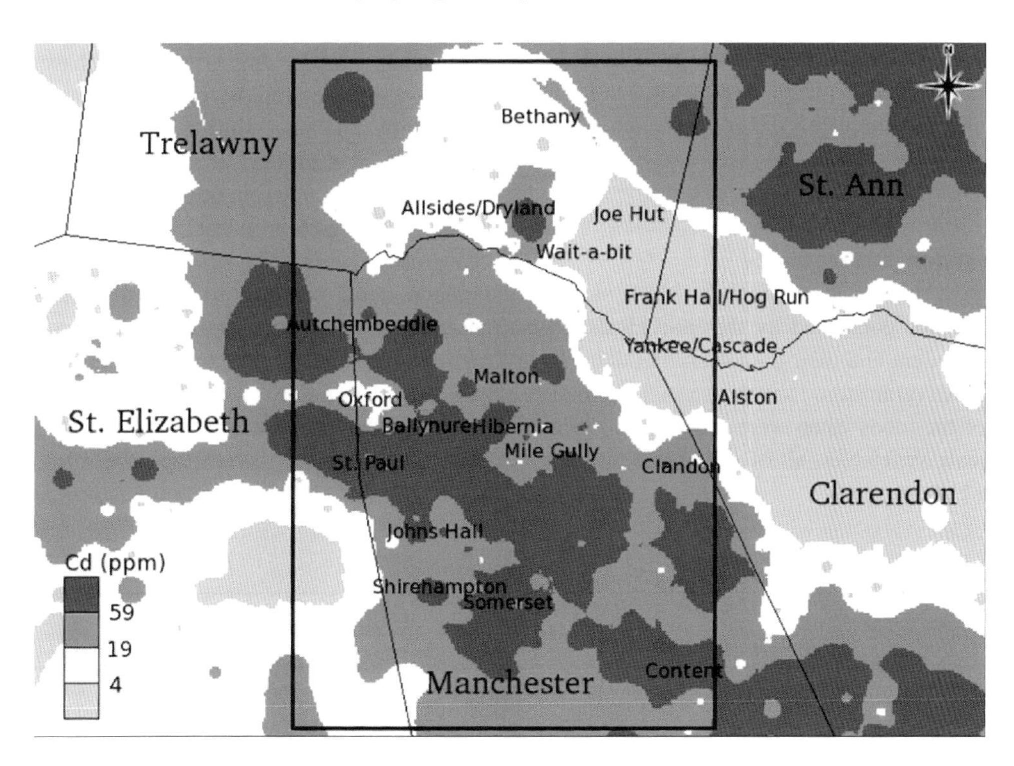

Fig. 12 Study area showing local Cd distribution

the age of 40 based on the census data is approximately 13,800 persons again with 52% of those being male. A total of 678 individuals were interviewed, with age from 18 to 85 years (mean 41.9 years). In this total, 64.3% were female, with the greatest number of individuals in the 25- to 34-year-old range.

Dietary Patterns

Cadmium intake was determined from the dietary patterns of persons in the study population. The daily intakes were compared to the WHO provisional tolerable weekly intake (PTWI) of 7 μg/kg bodyweight/

Table 5 Definition of stratified soil Cd areas and the population distribution

Strata	Soil Cd status (mg/kg)	Percentage of study population in stratum
Low	<4	38
Medium	5–19	5
High	20–59	28
Very high	>60	27

week (JECFA, 2000) and an estimate of risk for excessive dietary intake ascertained. A sub-sample of persons >40 years of age was surveyed for renal biomarkers of cadmium exposure to estimate what effects, if any, dietary cadmium was having on the health of the population. The study included the occurrence of Cd and other elements in soils, food and urine. The dietary patterns of selected populations and consequent intakes from food have been evaluated. Also investigated were blood pressure, body mass index and urine dipstick measurements of blood, protein, pH and sugars.

Food Consumption Frequency

The major foods consumed by the individuals, their related Cd content and their corresponding daily consumption rates (grams/day) are shown in Table 6. The majority of the foods eaten were tubers, root vegetables and leafy vegetables, all of which can have high uptakes of Cd.

Cd Intake and Micronutrients Consumption

The mean Cd intake was 9.3 μg/kg bodyweight/week; median: 5.4 μg/kg bodyweight/week. In the population, 39.2% of persons had Cd intake above the WHO PTWI of 7 μg/kg bodyweight/week. Cd intake means were higher and statistically significant ($p \leq 0.05$) for those individuals that lived in medium and low Cd strata, males, farmers, individuals with ages 18–34 years, current smokers and alcohol consumers. The intake and calculated risk factors are shown in Table 7.

Health Effects of Dietary Exposure to Cadmium

Other than the cases of itai-itai disease in Asia, especially in Japan, no reports of significant morbidity associated with diet-sourced cadmium have been published in the literature record. In Jamaica, a unique opportunity has been afforded to study the extent to which the nephrotoxic effects of cadmium can be identified in a population of persons residing and consuming foods grown on enriched soils. The presence of significant morbidity associated with cadmium in Jamaica would serve to confirm what has been suggested about the toxicity of cadmium and the need to regulate its intake in the human body. On the other hand, if identifiable morbidities in Jamaica remain low and cadmium intakes are significantly higher than the PTWI, the real risk to the population will need to be assessed. It will be important to examine and

Table 6 Rankings of foods eaten and food Cd contents

		Cd concentration (mg/kg) by percentile			
Food	Daily consumption (g)	25th percentile	50th percentile	75th percentile	95th percentile
Total carrot juice	164	0.072	0.073	0.076	0.079
Total bush tea	117	0.0002	0.0005	0.0009	0.003
Total yam	69	0.04	0.158	0.548	1.231
Total ackee	46	0.089	0.178	0.305	0.654
Total pak choi	29	0.34	0.57	0.81	1
Total callaloo	29	0.06	0.34	0.89	1.54
Total peanuts	27	2.063	2.442	2.82	3.124
Total chicken liver	15	No value	No value	No value	No value
Total carrot	15	0.07	0.33	0.84	1.6
Total cabbage	11	0.006	0.01	0.012	0.02
Total cow liver	11	0.15	0.27	1.02	1.56
Total lettuce	9	0.06	0.07	0.083	0.09
Total cow kidney	7	1.14	3.13	6.14	12.1

Table 7 Cd intake and calculated risk factors

			≤7 μg Cd/kg[a]		>7 μg Cd/kg[a]		
Variables	Cd mean	*p*-Value	*N*	(%)	*N*	(%)	*p*-Value*
Gender							
Male	13.4	0.000 *	116	28.6	125	46.3	0.000
Female	7.1		290	71.4	145	53.7	
Age groups							
18–24	12.2	0.000*	61	15.1	62	23	0.003
25–34	11.7		75	18.6	67	24.8	
35–44	9.3		81	20	54	20	
45—54	7.1		72	17.8	32	11.9	
55–64	7.6		61	15.1	35	13	
64+	5.5		54	13.4	20	7.4	
Occupation							
Farmers, labourers	11.5	0.000*	112	27.6	105	38.9	0.002
Others	7.0		294	72.4	165	61.1	
Area							
Low/median	10.8	0.009*	114/37	28.1/9.1	96/27	35.6/10	0.14
High/very high	8.2		144/111	35.5/27.3	88/59	32.6/21.9	
Smoker							
Non-smoker	7.9	0.000*	348	85.7	197	73	
Past smoker	11.2		13	3.2	7	2.6	
Current smoker	13.6		30	7.4	31	11.5	0.000
Smoke other substances	17.7		15	3.7	35	13	

* *t*-Test; significant $p < 0.05$

[a] PTWI/WHO – Provisional tolerable weekly intake recommended by World Health Organization. Men had a higher average ingestion of Cd than women (13.4 and 7.1 μg/kg bodyweight/week) as they generally tend to consume higher amounts of food than women and may have as a consequence a higher intake of the metal (Watanabe et al., 2004)

quantify the intake of potential ameliorating micronutrients and assess if any genetic adaptations exist that may provide additional protection. The co-ordination of multi-disciplinary studies must be furthered in order to fully identify and assess any Cd-related morbid conditions in the population and to help better understand their prevalences of lack thereof. Some of this work has already commenced and it is hoped that further studies will be conducted with collaborators to better understand the local etiology of Cd toxicity.

Cd Exposure and Renal Biomarker Survey

Urine was sampled from persons living in areas of differing soil Cd concentrations within the study area. The group was a subset of the larger sample of persons that took part in the diet survey that quantified their intake of Cd. The urine samples allowed the determination of a quantitative relationship between urine cadmium (U-Cd) and β_2-microglobulin (β_2-MG) and also facilitated the identification of the threshold U-Cd that causes a significant increase in β_2-MG. The link between increased U-Cd and geographic location and the associated soil Cd concentrations was also investigated.

Cd (Creatinine Adjusted – U-Cd/g_cr) and β_2-Microglobulin (Creatinine Adjusted – β_2-MG/g_cr) in Urine

Results for Cd (Table 8) and β_2-MG (Table 9) in urine were obtained for 143 participants in the urine survey. The range of U-Cd in the total sample was 0–35.9 with a mean of 4.6 and a median of 3.1. The range of β_2-MG in the samples was 2–10,440 μg Cd/g_cr with a mean

Table 8 Cadmium in urine

		U-Cd (μg/g_cr)	
Gender	*N*	Range	Median
Male	51	0.3–22.3	2.46
Female	92	0.01–35.9	3.42

Table 9 β_2-*MG* in urine

		β_2-MG (μg/g_cr)	
Gender	*N*	Range	Median
Male	51	9–10,440	78
Female	92	2–8897	74

of 290 and a median of 74. About 17% of the subjects exceed 200 μg/g creatinine, the level considered by the University Hospital of the West Indies (UHWI) Chemical Pathology Department to be normal. Urine data were adjusted using creatinine content to correct for dilution between samples.

Age

The median concentrations of U-Cd and β_2-MG plus median Cd intakes for a series of increasing age groups with ranges of 10 years, divided into total population, male and female groups, are shown in Table 10.

Figures 13, 14 and 15 are box plots of U-Cd, β_2-MG and Cd intake, respectively, vs. age groups for the total population, male and female subgroups. A very gradual increase in U-Cd concentrations is observed for the 70–79 age group, after which a significant decrease is observed (Fig. 13). No discernible pattern is observed in the median values measured for β_2-MG concentrations between the subject age groupings (Fig. 14). Cd intake gradually decreases with age in the total population (Fig. 15) and similarly in the male population. The intake of Cd by females is more consistent across age groups, decreasing significantly in the 80–89 age group; however, the number of persons in that group was very low (n=2).

Table 10 Median U-Cd (μg/g_cr), β_2-*MG* (μg/g_cr) and Cd intake (μg Cd/kg bodyweight/week) by age range

	Total (n = 143)			Male (n = 51)			Female (n = 92)		
Age range	U-Cd	β_2-MG	Cd intake	U-Cd	β_2-MG	Cd intake	U-Cd	β_2-MG	Cd intake
40–49	3 (41)	67	3.4	3.14 (17)	81	4.02	2.96 (25)	58	2.41
50–59	2.7 (36)	83	2.77	2.5 (11)	140	3.16	3.5 (25)	81	2.40
60–69	3.5 (37)	87	2.93	2.04 (12)	84	1.93	3.6 (25)	117	3.31
70–79	4.5 (23)	59	2.26	6.94 (8)	54.5	1.22	3.68 (15)	88	2.33
80–89	0.8 (5)	51	1.37	0.8 (3)	48	5.93	3 (2)	215	0.80

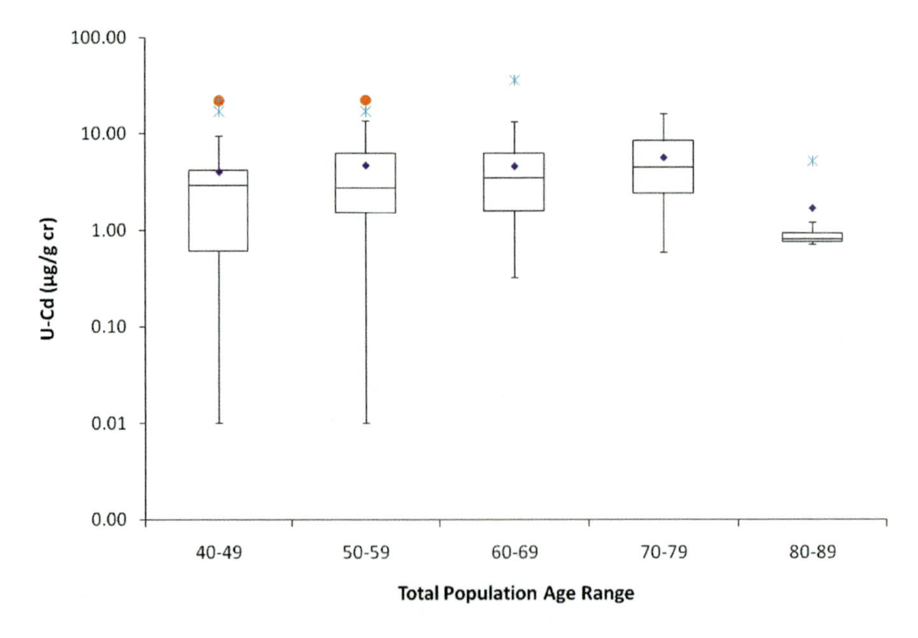

Fig. 13 Total population urine Cd by age range

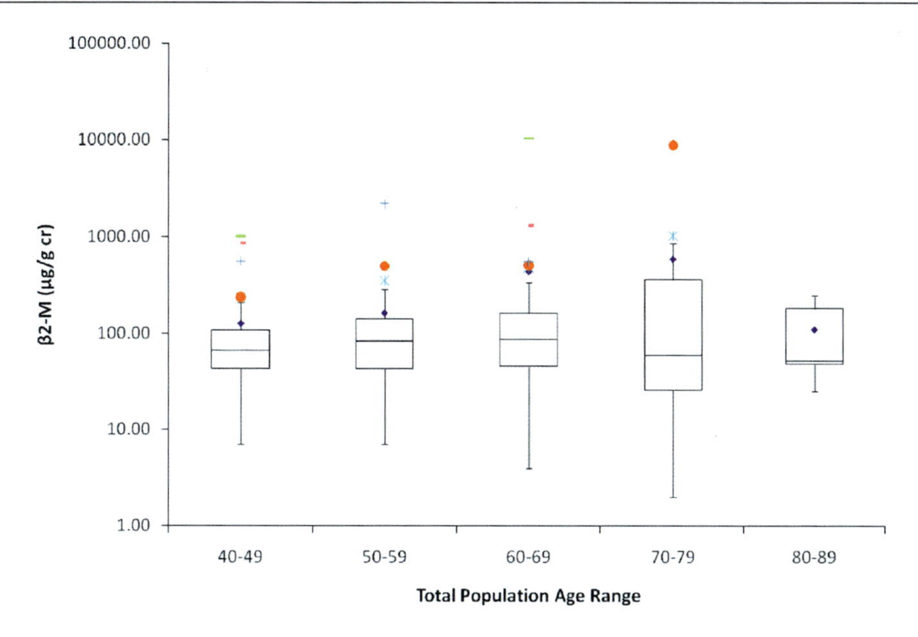

Fig. 14 Total population β_2-MG in urine by age range

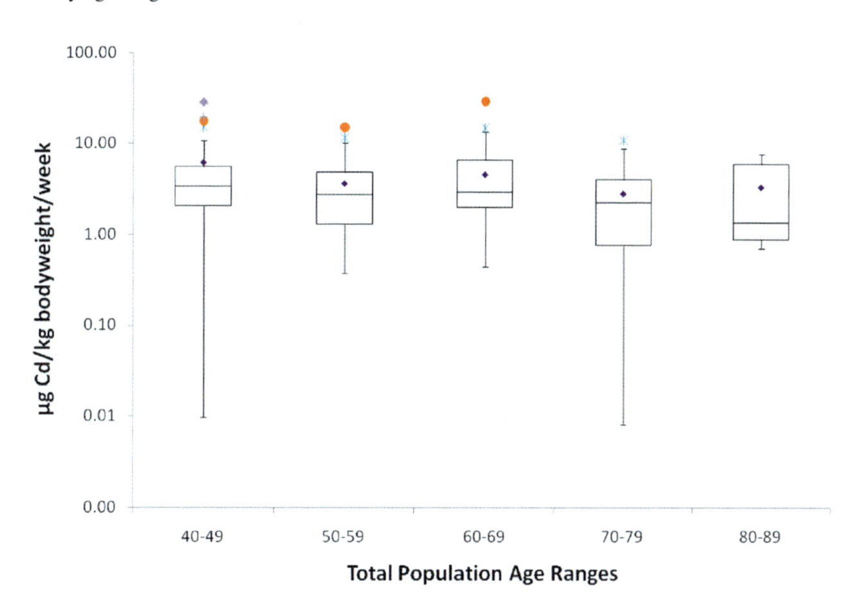

Fig. 15 Total population Cd intake by age range

Gender

The median concentrations of U-Cd and β_2-MG plus median Cd intakes for survey subjects aged 40 years and greater are shown in Table 11. Box plots showing U-Cd and β_2-MG concentrations and Cd intake, respectively, for both gender groups have been plotted in Figs. 16, 17 and 18. The median concentration of cadmium in the urine of females is slightly higher than that for the males (3.4 vs. 2.5 μg Cd/g_cr), while the β_2-MG concentrations are almost the same (78 vs. 74 μg/g_cr). The relative Cd intake is higher for males than females (3.4 vs. 2.6 μg Cd/kg bodyweight/week).

Table 11 Median U-Cd (μg/g_cr), β_2-MG (μg/g_cr) and Cd intake (μg Cd/kg bodyweight/week) by gender

Male ($n = 51$)			Female ($n = 92$)		
U-Cd	β_2-*MG*	Cd intake	U-Cd	β_2-MG	Cd intake
2.5	78	3.4	3.4	74	2.6

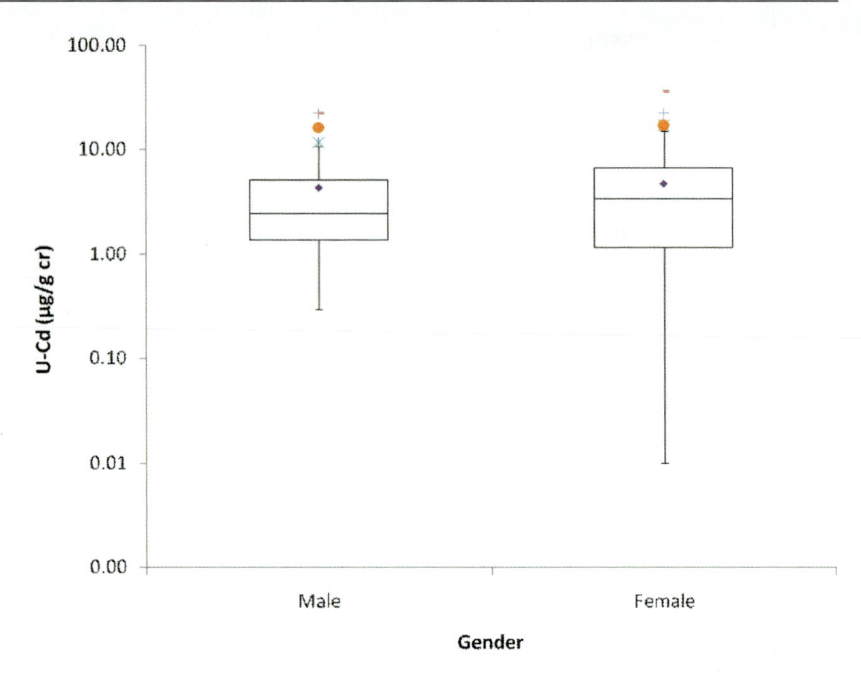

Fig. 16 Box plots of U-Cd by gender

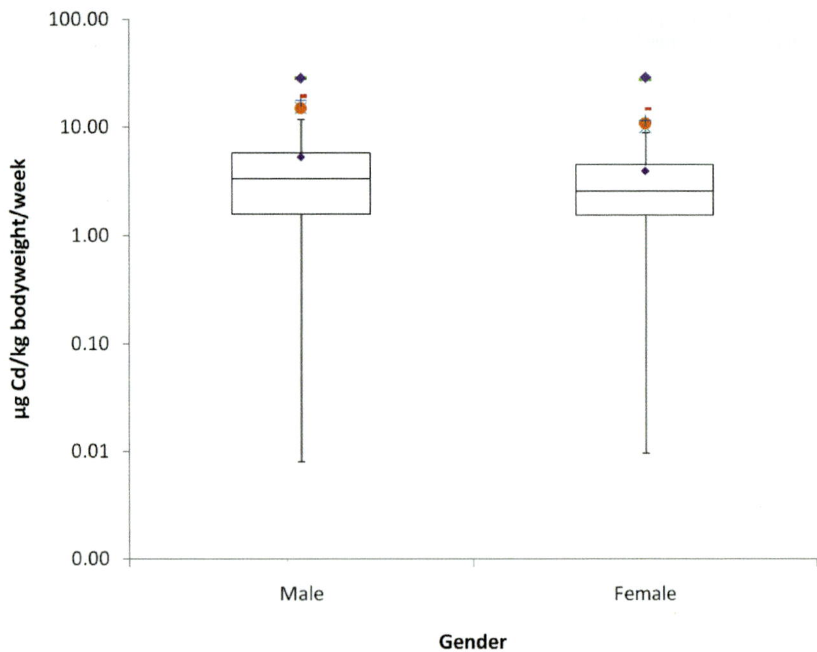

Fig. 17 Box plots of Cd intake by gender in >40-year-old population

The Cd intake variations with age were assessed and the median concentrations are shown in Table 12. Relative values of U-Cd by gender group are shown in Fig. 19. In the population below 40 years of age, males have more than double (6.8 vs. 3.0 µg Cd/g_cr) the Cd intake of females with the highest intakes recorded for the 20- to 29-year-old males. Compared to their respective >40 age groups, the Cd intake was significantly higher for younger males while younger females were only slightly higher.

Contribution of Smoking to Cd Exposure

The median concentrations of U-Cd, β_2-MG and median Cd intakes for smoking status cohort groups

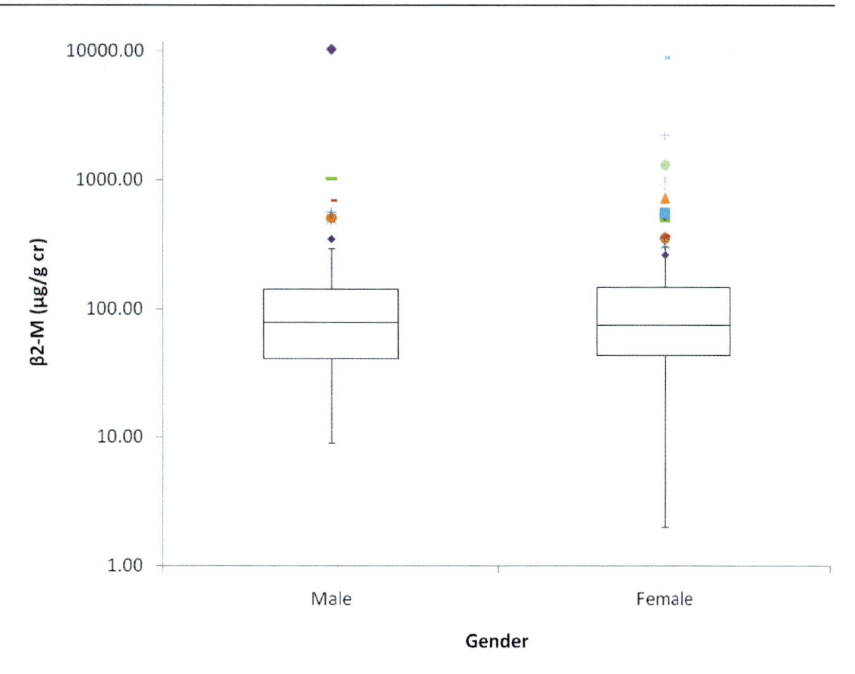

Fig. 18 Box plots of β_2-MG by gender

Table 12 Cd intake in <40-year-old population by gender

Male < 40 years ($n = 124$) Cd intake	Female < 40 years ($n = 219$) Cd intake
6.8	3.0

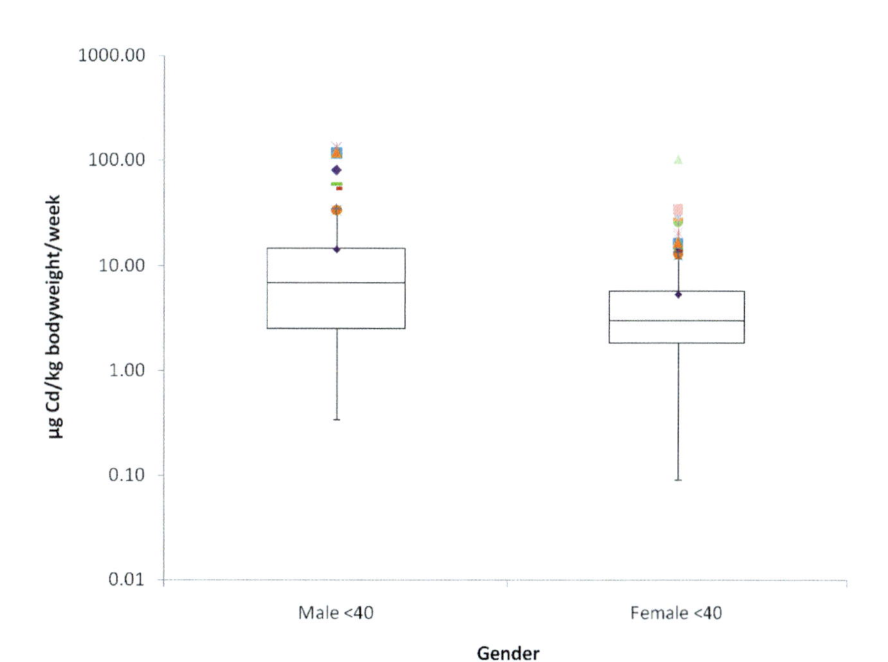

Fig. 19 Plot of Cd intake in <40-year-old population

Table 13 Median U-Cd (μg/g_cr), β_2-*MG* (μg/g_cr) and Cd intake (μg Cd/kg bodyweight/week) by smoking status and gender

Smoking status	Median total (*n*)			Median male (*n*)			Median female (*n*)		
Cohort	U-Cd	β_2-MG	Cd intake	U-Cd	β_2-MG	Cd intake	U-Cd	β_2-MG	Cd intake
Smoker	4.7 (13)	68	2.7	4.6 (9)	68	3.2	9.2 (4)	291.5	1.7
Non-smoker	3.1 (110)	72.5	2.8	2.4 (26)	72.5	3.3	3.3 (84)	74	2.6
Past smoker	1.8 (9)	82	2.6	1.5 (6)	61	3.6	3 (3)	89	2.6
Other material	4.2 (11)	86	3.6	4 (10)	105	3.7	5.4 (1)	26	0.1

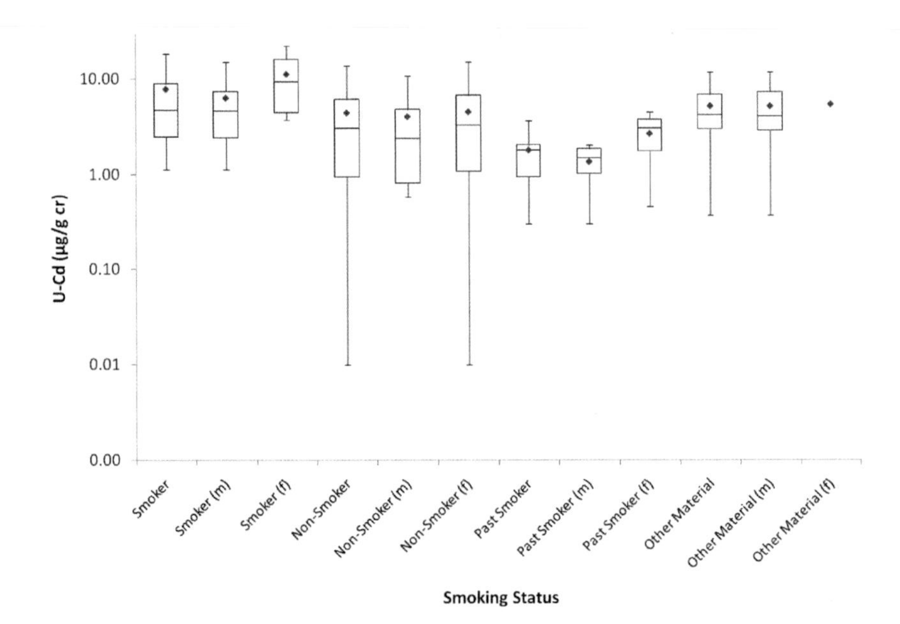

Fig. 20 Box plots of U-Cd by smoking status and gender

are shown in Table 13. The data for U-Cd by smoking cohort group are shown graphically as a box plot (Fig. 20). Smokers of all substances comprised 16.8% of the total population sampled. About 40.4% of all males sampled smoked cigarettes, unfiltered tobacco or marijuana, while only 5.2% of the female population did the same. For both males and females, cigarettes were the major smoking materials at 9.1 and 4.2% of their total respective populations. The U-Cd for all female smoking status cohorts was higher than corresponding males. The female smokers had the highest U-Cd concentration, twice that of the males. Past female smokers and the only female marijuana smoker had U-Cd higher than corresponding males. The non-smoking females had slightly higher U-Cd than non-smoking males, suggesting that even without the smoking contribution women had a higher Cd body burden than men.

Cigarette smoking seemed to have no impact on β_2-MG concentrations in males. The men who smoked marijuana and unfiltered cigar smokers had the highest β_2-MG concentrations. The β_2-MG concentrations of female smokers were the highest seen in any subgroup at 291.5, significantly higher than corresponding males. Male smokers ingested almost twice the amount of Cd than females (3.2 vs. 1.7 μg Cd/g_cr); however, women had significantly higher U-Cd.

Effect of Diabetes Status on Renal Biomarkers

Diabetes is a potential confounder for increased β_2-MG in urine. The median concentrations of U-Cd, β_2-MG and median Cd intakes for diabetic and non-diabetic cohort groups are shown in Table 14. For the total population as well as male and female subgroups the median concentrations of β_2-MG were lower for the diabetics. The same held true for the U-Cd concentrations and Cd intakes. Female diabetics had twice the U-Cd concentrations of males and also higher Cd intake. The U-Cd concentration of the female non-diabetics was slightly higher than their

Table 14 Median U-Cd (μg/g_cr), β_2-*MG* (μg/g_cr) and Cd intake (μg Cd/kg bodyweight/week) by diabetes status and gender

Diabetes status	Median total (*n*)			Median male (*n*)			Median female (*n*)		
Cohort	U-Cd	β_2-MG	Cd intake	U-Cd	β_2-MG	Cd intake	U-Cd	β_2-MG	Cd intake
Diabetes	1.5 (19)	56	1.7	1.1 (8)	52	1.3	3.0 (11)	57	2.4
No diabetes	3.3 (124)	81	2.9	2.8 (43)	82	3.7	3.5 (83)	80	2.6

male counterparts whose Cd intake was slightly elevated by comparison.

Effect of Hypertension Status on Renal Biomarkers

Hypertension is another potential confounder for elevations in β_2-MG concentrations. The median concentrations of β_2-MG, U-Cd and median Cd intakes for hypertensive and non-hypertensive cohort groups are shown in Table 15. The β_2-MG is virtually the same in both total population cohorts and showed very little variation between male and female cohorts. Both female cohorts have higher U-Cd concentrations than males but have either the same Cd intake (hypertensive) or significantly lower Cd intake (non-hypertensive).

Soil Cd Status

The median concentrations of U-Cd and β_2-MG for survey subjects located in areas stratified by soil Cd concentration are shown in Table 16. U-Cd is directly related to the soil Cd status of the residence. The β_2-MG was not related to the soil Cd status, with concentrations being relatively consistent across soil Cd strata groups (Fig. 21).

Relationship Between Urine Cadmium and β_2-Microglobulin

The total population, male and female median β_2-MG concentrations for a series of increasing U-Cd concentration groups with ranges of 4 μg/g_cr are shown in Table 17.

The data were plotted using bands of 4 μg Cd/g_cr (Fig. 22), and the medians outline the central J-shaped trend in the data with a U-Cd flexure point between 8 and 12 μg/g_cr. The females show a similar trend, with a flexure range of approximately 8–12 μg Cd/g_cr with the 16–20 U-Cd range showing a much higher value for the β_2-MG concentration (602.5 vs. 351 μg/g_cr). Median concentrations of β_2-MG in the male subgroup are similar to those obtained for the females with a similar flexure range of 8–12 μg Cd/g_cr. The development of the exponential increase is, however, limited

Table 15 Median U-Cd (μg/g_cr), β_2-*MG* (μg/g_cr) and Cd intake (μg Cd/kg bodyweight/week) by hypertension status and gender

Hypertension status	Median total (*n*)			Median male (*n*)			Median female (*n*)		
Cohort	U-Cd	β_2-MG	Cd intake	U-Cd	β_2-MG	Cd intake	U-Cd	β_2-MG	Cd intake
Hypertension	4.1 (46)	74	2.5	3.1 (41)	49.5	2.5	3.9 (39)	88	2.5
No hypertension	3.1 (78)	76	3.2	2.5 (37)	82	3.9	3.3 (51)	84	2.6

Table 16 Median U-Cd (μg/g_cr), β_2-*MG* (μg/g_cr) and Cd intake (μg Cd/kg bodyweight/week) by soil Cd status and gender

Soil Cd status	Total population median (*n*)		Male population median (*n*)		Female population median (*n*)	
Strata	U-Cd	β_2-MG	U-Cd	β_2-MG	U-Cd	β_2-MG
Low	1.2 (41)	70	1.4 (16)	84.5	1.0 (25)	64
Medium	1.8 (11)	147	1.8 (1)	196	2.3 (10)	147
High	3.3 (51)	71	2.8 (21)	71	4.8 (30)	70.5
Very High	5.1 (40)	66.5	5.2 (13)	82	5.0 (27)	59

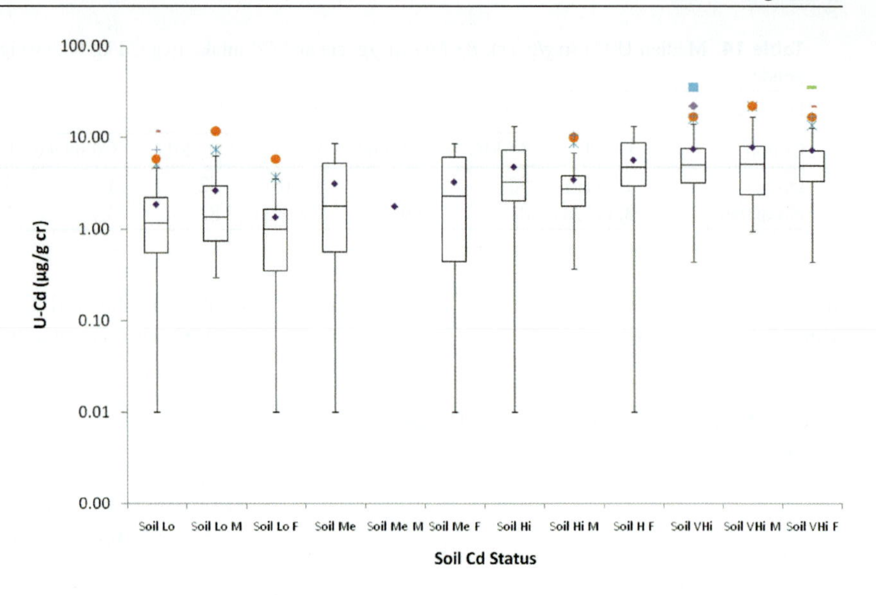

Fig. 21 Box plots of U-Cd by soil Cd status and gender

Table 17 Median U-Cd (μg/g_cr) (in 4 μg ranges) and β_2-*MG* (μg/g_cr) for total population and by gender

U-Cd	Total population median (*n*) β_2-*MG*	Female population median (*n*) β_2-*MG*	Male population median (*n*) β_2-*MG*
<4	68.5 (90)	59 (56)	76 (34)
4–8	70.5 (30)	72 (21)	67 (9)
8–12	120 (13)	108.5 (8)	142 (5)
12–16	131 (3)	131 (3)	
16–20	351 (3)	602.5 (2)	28 (1)
20–24	557 (3)	2,205 (1)	342 (2)
>24	551 (1)	551(1)	

by a lack of males with U-Cd values in excess of 12 μg Cd/g_cr.

Discussion

Much of the concern about environmental cadmium toxicity is based on the Japanese experience of the horrible itai-itai disease. The symptoms were first noted in the Toyama Prefecture, Japan, in 1912 (Imamura, 2007), and after World War II, an outbreak in the nearby Jinzu basin attracted worldwide attention (Aoshima, 1997; Nogawa, 1981; Fan et al., 1998). The

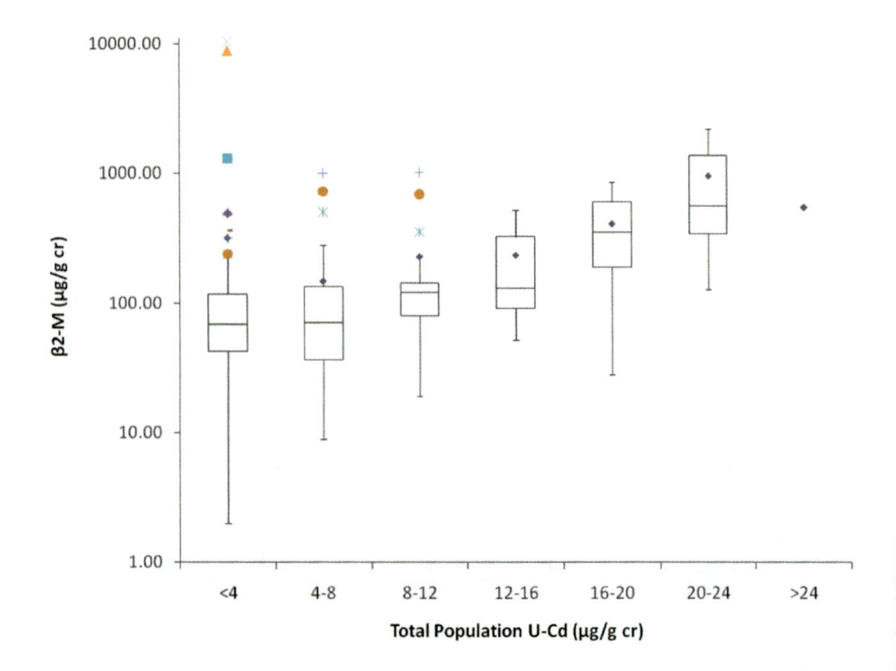

Fig. 22 Plot of creatinine adjusted β_2-MG vs. creatinine adjusted U-Cd as box plots for each 4 μg Cd/g_cr interval

sufferers were elderly, multiparous women (Nogawa, 1981; Aoshima, 1997) who had lived for more than 30 years near a Cd-contaminated stream and whose main diet was rice grown in cadmium-contaminated paddies (Tsukahara et al., 2003a; Tsukahara et al., 2003b; Ikeda et al., 2006; Chaney et al., 2000; Simmons et al., 2003).

Itai-itai like diseases seem to have occurred only in Asia, linked to the staple diet of rice grown in some circumstances on soils with as little as 2 mg/kg Cd (Cai et al., 1990). The high cadmium uptake of rice was accompanied by low levels of calcium, iron and especially zinc, likely a strong contributing factor influencing the absorption of Cd and its distribution in organs and tissues (Berglund et al., 1994; Brzóska and Moniuszko-Jakoniuk, 1997; Chaney et al., 2000; Chaney et al., 2001; Flanagan et al., 1978; Fox et al., 1979; Reeves and Vanderpool, 1998). Such deficiencies exacerbate Cd-related health risks (Hallberg et al., 1977; Pedersen and Eggum, 1983). Higher Zn/Cd ratios are reported as particularly important in ameliorating the absorption and possible ill effects of Cd on humans (Chaney et al., 2000).

Itai-itai affected a population with specific characteristics and the confluence of factors was different from what is obtained in the West and indeed elsewhere in Japan. Several studies have indicated an increased cadmium intake in women with reduced body iron stores (Berglund et al., 1994; Flanagan et al., 1978; Fox et al., 1979; Pedersen and Eggum, 1983). Zn and Fe deficiencies can increase Cd retention and risk and would very much apply to an essential element-deficient population. There may be other confounding factors including Cd bioavailability differences among foods, the percentage of absorption of dietary Cd in a population, genetics, lifestyle and antagonistic or synergistic exposures to other elements. The disease has not to the best of our knowledge been reported outside of Asia and certainly not in Jamaica where some of the soil Cd are so high that a helpful anonymous referee of our first paper which contained the values (Lalor et al., 1998) queried the analytic accuracy, suggesting that with levels like these there should be many dead people around. He was only convinced, though still surprised, by confirmatory results using neutron activation analysis. There is also no question about cadmium's bioavailability and accumulation in segments of the Jamaican population.

In numerous examples in the literature, cadmium is linked with one of several afflictions including renal disease, osteomalacia, prostate cancer and diabetes. Incidences of hypertension, diabetes, cardiac disease, prostate cancer and end-stage renal disease are considered high in Jamaica. There, however, seems to be no relationship between the observed incidence of end-stage renal failure and soil Cd concentrations (Barton et al., 2004). There is still no evidence of increased morbidity or mortality that can be attributed to Cd in Jamaica. The life expectancy of Jamaicans at birth is 73.59 years and ranks #100 in the world (CIA, 2008), comparing well with Caribbean countries with different soils. Also, it is observed that in the >40-year-old population, the highest median age of death is found in the parish with the highest soil Cd concentrations.

At the levels being found, the Jamaican situation may be somewhat puzzling but is not entirely unique as there are at least five other examples in the literature with reports of high cadmium concentrations that too have led to no obvious effects. They are France (de Burure et al., 2003), Germany (Ewers et al., 1993), United Kingdom (Strehlow and Barlthrop, 1988), United States (Sarasua et al., 1995) and New Zealand (McKenzie-Parnell and Eynon, 1987). Where comparisons can presently be made, the general situation in Jamaica appears very similar to that in the UK sites. Aside from the itai-itai cases there seems to be no other significant morbidity recorded in literature as a result of cadmium in the diet. However, there are several dose–response studies at various levels of cadmium exposure and clearly a renal response exists and this dose–response is also observed between urinary cadmium and β_2-MG in the Jamaican population. But the levels being observed are simply not high enough to be of concern for the onset of significant renal impairment.

Studies of Cd exposure in occupational settings and the Asian itai-itai experiences have defined the physiological effects of Cd toxicity. Such information was the base of many regulations and guidelines and informed the World Health Organization's provisional tolerable weekly intake (PTWI) value for Cd (7 μg Cd/kg bodyweight/week) (JECFA, 2000) and the recently revised EU TWI of 2.5 μg Cd/kg bodyweight/week (EFSA, 2009). There has been a great deal of work reporting links of cadmium with numerous diseases so the reported soil concentrations in Jamaica, the observed transfers to foods and to people raised justifiable concerns about the potential public health of cadmium on the population.

U-Cd may be influenced by smoking and age, while the renal impacts of diabetes and hypertension may increase the β_2-MG concentrations observed. U-Cd typically increases with age (Satarug et al., 2002; Satarug and Moore, 2004; Benedetti et al., 1999), and in the Jamaican population it was found that an increase in U-Cd did occur with age and was mainly attributed to the women, who showed a distinct, gradual increase. However, the median age of the population studied was 58 years and there was no bias in the sample towards persons with higher accumulations of Cd (70–79 years). The influence of smoking was evident in both males and females, where U-Cd concentrations were significantly higher than respective non-smokers. The females demonstrated enhanced susceptibility to Cd absorption also identified by other workers (Nishijo et al., 2004; Jarup et al., 1998). Regardless of the fact that male smokers ingested twice the diet-sourced Cd of females, the U-Cd concentration for females was significantly larger. The difference between female smokers and non-smokers was significantly larger than that between the male smoking cohorts. Though the smoking cohort showed significantly higher U-Cd than non-smokers, the group comprised only 16.8% of persons surveyed. The majority of persons were therefore influenced only by diet-sourced cadmium. Chronic illnesses such as diabetes and hypertension can increase β_2-MG concentrations (Hong, 2000). There have been some studies that suggest diabetics may be more susceptible to Cd toxicity (Akesson et al., 2005; Buchet et al., 1990). For this work, diabetes and hypertension showed no confounding influence on the β_2-MG concentrations.

The concentrations of cadmium in tubers, green leafy vegetables and root vegetables are directly influenced by the soil Cd concentrations. The plants affinity for Cd is exacerbated by what seems to be a lack of the homeostatic controls seen for essential elements. The result is that cadmium uptake continues, while essential elements are regulated and have a very narrow concentration range across soils of varying concentrations. This lack of homeostatic control presents a larger problem, as the ratios between Cd and ameliorating elements such as Zn, Se, Ca or Fe increase, potentially increasing the toxicity of Cd.

The Asian experience has shown that rice is an excellent vector for Cd and the potential for micronutrient amelioration is reduced given low levels of calcium, iron and zinc (Chaney et al., 2000; Simmons et al., 2003). It has been argued that other elements in foods, zinc in particular, may influence the absorption of Cd and its distribution in organs and tissues (Berglund et al., 1994; Brzóska and Moniuszko-Jakoniuk, 1997; Chaney et al., 2001; Flanagan et al., 1978; Fox et al., 1979; Reeves and Vanderpool, 1998) and low plant concentrations may exacerbate Cd-related health risks (Hallberg et al., 1977; Pedersen and Eggum, 1983).

The Zn/Cd ratios in soils are usually >100 (Holmgren et al., 1993) but low values appear to be an island-wide characteristic (Lalor et al., 2001). In Jamaica, the spatial distribution of zinc is similar to Cd, but the Zn/Cd ratios are very low in the soils of central Jamaica. These low Zn/Cd ratios in the soils are reflected in the crops grown therein, resulting in ratios much lower than those from the United States or United Kingdom (Table 18). In Manchester, mean plant Zn/Cd ratio is as low as 8, and in the area with the highest Cd concentration the ratio falls to 0.3. The low ratios may deserve attention in terms of potential health hazards (Chaney et al., 1999) as it may reduce the potential of zinc to ameliorate the effects of Cd following ingestion. However, the soils also show

Table 18 Comparison of Cd concentrations with EU limits, Zn concentrations and Zn/Cd ratios (Jamaica, United Kingdom and United States)

Food category		Cd	Zn	Cd limit (EU)	Zn/Cd (JA)	Zn/Cd (UK)	Zn/Cd (US)
Fruit	*n*	65	69	0.05			
	Mean	0.057	2.47		43	425	475
Legumes	*n*	18	18	0.2			
	Mean	0.638	21.27		33	508	389
Leafy vegetables	*n*	84	88	0.2			
	Mean	0.163	4.43		27	170	43
Root vegetables	*n*	90	94	0.1			
	Mean	0.350	3.64		10	218	200
Other root crops	*n*	542	424	0.1			
	Mean	0.200	3.81		19	127	167

enhancements in other elements, such as calcium and iron which may also restrict the uptake and effects of cadmium.

For the total population, it is observed that both men and women have mean Cd intakes that are significantly higher than the WHO PTWI. Given that Cd exposure has a cumulative impact, the current intake of Cd through the diet is not a suitable measure of lifetime Cd exposure. To get a better proxy of the fluctuation of individual Cd intake with age, intakes were examined across increasing 10-year age groups in the study population. Younger men ingested significantly more cadmium than older males and the same was true for women, however, with a smaller difference between the groups. Suffice to say, the Jamaican population is being exposed to an elevated amount of cadmium over time.

The mean and median values of U-Cd in the Jamaican study population were 4.6 and 3.1 μg Cd/g_cr, respectively. The results suggest an elevated exposure to Cd with the potential for an increase in excretion of tubular enzymes (Table 19). Japanese studies have reported itai-itai disease patients with U-Cd as high as 20–30 μg Cd/g_cr (Nogawa and Kido, 1993) and values of approximately 6–10 μg Cd/g_cr in the areas where Cd pollution was identified in the environment and tubular dysfunction (no itai-itai disease) was detected in the population (Watanabe et al., 2000).

Table 19 Renal effects associated with increasing urine cadmium concentrations

Urine Cd μg/g_creatinine	Exposure status (non-occupational)	Effects/response
≤2	Normal value	Between 0.5 and 2.0 increased urinary excretion of NAG-B may occur
2–5	Overexposure	Potential increase in urinary excretion of tubular enzymes or antigens (total NAG, AAP)
5–10	Significant overexposure	Potential renal tubular damage and associated low and high molecular weight proteinuria (increase in β_2-MG). Onset of osteomalacia, glucosuria. Primary as well as secondary prevention intervention

Table 20 Interpretation of elevated values of urinary β_2-microglobulin (β_2-MG) and retinol-binding protein (RBP) induced by occupational or environmental exposure to Cd

β_2-MG in urine (μg/g_cr)	Renal significance
<300	Normal
300–1,000	Incipient Cd tubulopathy (potential reversal of condition if exposure is removed and urinary Cd is not too elevated, i.e. below 20 μg/g_cr)
1000–10,000	Irreversible tubular proteinuria that may accelerate the decline of glomerular filtration rate with age. At this stage glomerular filtration rate is normal or slightly impaired
>10,000	Overt Cd nephropathy usually associated with a decreased glomerular filtration rate

Modified from Bernard (2004)

In Jamaica, the values are well below both sets of values. The mean and median β_2-MG concentrations were 290 and 74 μg/g_cr, respectively, suggesting no significant renal tubular dysfunction (Table 20) (Bernard, 2004). There were only five subjects with β_2-MG concentrations greater than 1,000 μg/g_cr, the level where there should be serious consideration of irreversible tubular proteinuria with retention of normal glomerular filtration rates.

The data illustrate the tendency for females to absorb greater amounts of Cd than males. Females ingested less Cd than males but their U-Cd excretion was similar or greater, suggesting that they absorbed more. Increased Cd absorption in females has been covered in the literature (Nishijo et al., 2004; Jarup et al., 1998) and it was postulated that as a result of reduced iron stores in women (menstruation blood loss) their absorption of Cd across the intestines was greater than that of males (Akesson et al., 2002; Berglund et al., 1994; Vahter et al., 1996). For women, this increases any risk associated with exposure to diet-sourced cadmium compared to males.

The data also show that as the soil Cd increased in concentration, so did the median U-Cd. A significant number of the persons surveyed were subsistence farmers and their dietary Cd intake was most likely related to the ingestion of foods whose Cd content was related to the local farm soils. The relationship between U-Cd and soil Cd was not evident between β_2-MG and soil, but is not surprising as the extent of absorption depends on a host of factors including amounts of protective metallothionein available to bind Cd and the presence of ameliorating elements that limit absorption in the gut (e.g. iron). The variability of these factors in any population makes direct linkages of β_2-MG to Cd sources difficult.

While a direct link could not be established between renal dysfunction and soil Cd (and by extension food) there was clearly a dose–response between U-Cd and β_2-MG. For this study, median β_2-MG was 1.5 times higher in subjects with U-Cd/g_cr >5 μg Cd/g_cr compared with those <5 μg Cd/g_cr, increasing for values in excess of 9 μg Cd/g_cr. The dose–response was characterised by an exponential increase in β_2-MG concentrations once a threshold level of U-Cd between 9 and 12 μg Cd/g_cr was exceeded. This threshold level is similar to values of 10 μg Cd/g_cr postulated in the literature (Nordberg et al., 2002; Ikeda et al., 2003). The majority of persons with U-Cd concentrations in excess of 9 and 12 μg Cd/g_cr were females. It was also shown that β_2-MG exceeds the "normal" value of 300 μg/g_cr at a threshold U-Cd concentration between 11 and 12 μg Cd/g_cr, similar to the value of 11.5 μg Cd/g_cr suggested in the literature (Lauwerys et al., 1994). These factors notwithstanding, there was no group of persons whose median β_2-MG exceeded 1,000 μg/g_cr (irreversible renal damage) and there were only four persons or 2.8% of the total population where U-Cd exceeded the 20 μg Cd/g_cr level, identified in Japanese itai-itai patients (Nogawa and Kido, 1993).

The link between the soil Cd and the U-Cd in Jamaica is clear as is the Cd dose–response with β_2-MG concentrations once the threshold of 9–12 μg Cd/g_cr for U-Cd is exceeded. However, the U-Cd concentrations are not at levels that would suggest possible onset of renal impairment. Similarly the concentrations of β_2-MG measured are below levels that would suggest irreversible renal damage. The population though exposed to an elevated amount of Cd is not displaying significant renal impairment, most likely a result of limited absorption.

Summary

Geogenic levels of Cd in the soils of central Jamaica are high and have been shown to be available to some foods, mainly green leafy vegetables, tubers and root vegetables. The ingestion of these foods exposes the persons consuming them to an elevated dose of Cd. Cd exposure has been confirmed in the urine of residents of high-cadmium areas. Mean urine cadmium and β_2-MG concentrations of 4.6 μg Cd/g_cr and 290 μg/g_cr, respectively, confirm that the (a) population is being exposed to elevated Cd levels and (b) there is evidence of very mild tubular proteinuria. There was clear evidence of an association between U-Cd and soil Cd (and by extension food Cd) grouped by strata. The proteinuria detected in the population was related to Cd exposure, evidenced by the relationship between U-Cd and β_2-MG. While positive results were obtained for the identification of Cd-related renal biomarkers in the study population and there was a clear association between U-Cd and β_2-MG, the absolute concentrations obtained were well below critical limits for the onset of acute or chronic renal effects. The study identified that women absorbed greater amounts of Cd than men in the same cohort groups. Even with a lower dietary intake of Cd than men, women clearly retained more of what they ingested, evidenced by U-Cd concentrations that were either slightly or significantly more than males, depending on cohort group.

The reasons why the effects of Cd in the Jamaican environment are seemingly so slight may be attributed to dietary balance, nutrient amelioration or perhaps genetic adaptation. A thorough assessment of the toxicity of Cd, taking into account ingestion of ameliorating nutrients, is warranted and the real risk to the population must be quantified. It may be that the reputation of cadmium as an environmental toxin may be overrated in normal conditions, with the absence of the specific population characteristics associated with the Asian itai-itai occurrence.

Acknowledgements Thanks are due to the entire staff of ICENS who contributed in many ways to this work. Kameaka Duncan and Richard Hanson, who coordinated the field component of the urine sampling programme. Richard also contributed significantly to the analyses, the results of which are also being used in the pursuit of his masters of sciences degree. The International Development Research Centre (IDRC) who provided the major funding for this project. We would also acknowledge the following agencies and persons:

Dr. Pauline Samuda, Dr. Beverley Lawrence and Pauline Johnson, Caribbean Food and Nutrition Institute;

Dr. Peta-Anne Baker and Horace Levy, Social Work Unit;

Dr. Leslie Simpson, Caribbean Agricultural Research and Development Institute;

Dr. Nadia Williams, Department of Pathology, University Hospital of the West Indies;

Dr. Blossom Anglin-Brown, Director, University Health Centre;

Professor Everard Barton, Head and Consultant Nephrologist, Department of Medicine (UWI);

Dr. Robert G. Garrett, Geological Survey of Canada;

Dr. Edna Massae Yokoo, Federal University of Mato Grosso, Brazil;

The Environmental Foundation of Jamaica;

The Government of Jamaica;

The Organisation of American States;

The University of the West Indies, Mona.

References

Adriano DC (1986) Trace Elements in the Terrestrial Environment. New York: Springer-Verlag.

Agency for Toxic Substances and Disease Registry (ATSDR) (2008) Draft Toxicological Profile for Cadmium. Atlanta: ATSDR.

Akesson A, Berglund M, Schutz A, Bjellerup P, Bremme K, Vahter M (2002) Cadmium exposure in pregnancy and lactation in relation to iron status. Am J Public Health 92(2):284–287.

Akesson A, Lundh T, Vahter M, Bjellerup P, Lidfeldt J, Nerbrand C, et al. (2005) Tubular and glomerular kidney effects in swedish women with low environmental cadmium exposure. Env Health Perspect 113:1627–1631.

Anglin-Brown B, Armour-Brown A, Lalor GC, Preston J, Vutchkov MK (1996) Lead in a residential environment in Jamaica. Environ. Geochem. Health, 18:129–133.

Aoshima K (1997) Recent advances in studies of Itai-itai disease. Jpn J Toxicol Environ Health 43(6): 317–330.

Baker EL, Hayes GG, Landrigan PJ, handke JL, Leger RT, Housworth WJ, et al. (1977) A nationwide survey of heavy metal absorption in children living near primay copper, lead and zinc smelters. Am. J. Epidemiol., 106, 261–273.

Barton EN, Sareant LA, Samuels D, Smith R, James J, Wilson R, et al. (2004) A Survey of Chronic Renal failure in Jamaica. West Indian Med J 53(2):81–84.

Benedetti J-L, Samuel O, Dewailly E, Gingras S, Lefebvre M (1999) Levels of cadmium in kidney and liver tissues among a Canadian population (Province of Quebec). J Toxicol Environ Health 56:146–163.

Berglund M, Akesson A, Nermell B, Vahter M (1994) Intestinal absorption of dietary cadmium in women depends on body-iron stores and fiber intake. Environ Health Perspect, 102:1058–1066.

Bernard A (2008) Cadmium and its adverse effects on human health. Indian J Med Res 128:557–564.

Bernard A (2004) Renal dysfunction induced by cadmium: biomarkers of critical effects. Biometals 17:519–523.

Beton DC, Andrews GS, Davies HJ, Howells L, Smith GF (1966) Acute cadmium fume poisoning; five cases with one death from renal necrosis. Br J Ind Med 23(4): 292–301.

Brzóska MM, Moniuszko-Jakoniuk J (1997) The influence of calcium content in diet on the accumulation and toxicology of cadmium in the organism. Arch Toxicol 72:63–73.

Buchet JP, Lauwerys R, Roels H (1990) Renal effects of cadmium body burden in the general population. Lancet 336:699–702.

Burke K, Fox PJ, Sengor AM (1978) Buoyant ocean floor and the evolution of the Caribbean. J Geophys Res 83 3949–3954.

Cai SW, Yue L, Hue ZN, Zhong XZ, Ye ZL, Xu HD (1990) Cadmium exposure and health effects among residents in an irrigation area with ore dressing wastewater. Sci Tot Environ 90:67–73.

CCME (1999) Canadian Soil Quality Guidelines for the Protection of Environmental and Human Health: Cadmium. Canadian Environmental Quality Guidelines.

Chaney RL, Reeves PG, Angle JS (2001) Rice plant nutritional and human nutritional characteristics role in human Cd toxicity. In WJ Horst (ed) Plant nutrition: Food security and sustainability of agro-systems (pp. 288–289). Netherlands: Kluwer Academic Publishers.

Chaney RL, Ryan JA, Li YM, Angle JS (2000) Transfer of cadmium through plants to the food chain. In JK Skyres, M Gochfeld (ed) Proceedings of the SCOPE Workshop Environmental cadmium in the Food Chain: Sources, Pathways and Risks. Brussels: Belgian Academy of Sciences. 13–16 September.

Chaney RL, Ryan JA, Li YM, Brown SL (1999) Soil cadmium as a threat to human health. In MJ McLaughlin, BR Singh (eds) Cadmium in Soil and Plants (pp. 219–256). Netherlands: Kluwer Publishers.

CIA (2008) The 2008 World Factbook. Central Intelligence Agency.

Comer JB (1974) Genesis of Jamaican Bauxite. Economic Geology 69(8):1251–1264.

de Burure C, Buchet JP, Bernard A, Leroyer A, Nisse C, Haguenoer JM, et al. (2003) Biomarkers of renal effects in children and adults with low environmental exposure to heavy metals. J Toxicol Environ Health A66: 783–798.

DeVries W, Bakker DJ (1998) Manual for calculating critical loads of heavy metals for terrestrial ecosystem: Guidelines for cirtical limits, calculation methods and input data. TNO Institute of Environmental Sciences, Energy Research and Process Innovation. The Netherlands: Den Helder; 144 pp.

Donnelly TW (1985) Mesozoic and Cenzoic plate evolution of the caribbean region. In FG Stehli, SD Webb (eds) The Great American Biotic Interchange (pp. 89–121). New York: Plenum Press.

Draper G, Lewis JF (1990) Geology and tectonic evolution of the northern Caribbean margin. In G Dengo JE Case (eds) The Caribbean Region (Vols. Geology of North America v. H, 120). Boulder, CO: Geological Society of America.

EFSA (2009) Scientific opinion of the Panel on Contaminats in the food chain on a request from the European Commission on cadmium in food. EFSA J 980:1–139.

Ewers U, Brockhous A, Dolgner R, Freier I, Jermann E, Bernard A, et al. (1985) Environmental exposure to cadmium and renal function in elderly women living in cadmium-polluted areas of the Federal Republic of Germany. Int Arch Occup Environ Health 55:217–239.

Ewers U, Freier I, Turfeld M, Brockhaus A, Hofsetter L, Konig W, et al. (1993) Heavymetals in garden soil and vegetables from private gardens located in lead/zinc smelter area and exposure of gardeners to lead and cadmium (in German). Gesundheitswesen 55: 318–325.

Eyles VA (1958) A phosphatic band underlying bauxite deposits in Jamaica. Nature 182:1367–1368.

Fan J, Aoshima K, Katoh T, Teranishi H, Kasuya M (1998) A follow-up study on renal tubular dysfunction in women living in the cadmium-polluted Jinzu River basin in Toyama, Japan. Nippon Eiseigaku Zasshi 53(3):545–557.

Flanagan PR, McLellan JS, Haist J, Cherian G, Chamberlain MJ, Valberg LS (1978) Increased dietary cadmium absorption in mice and human subjects with iron deficiency. Gastroenterology 74:841–846.

Fox MR, Jacobs RM, Jones AO, Fry BE (1979) Effects of nutritional factors on metabolism of dietary cadmium at levels similar to those of man. Env Health Perspect 28:107–114.

Friberg L (1983) Cadmium. Ann Rev Public Health 4:367–373.

Friberg L (1984) Cadmium and the Kidney. Environ Health Perspect 554:1–11.

Friberg L (1950) Health hazards in the manufacture of alkaline accumulators with special reference to chronic cadmium poisoning. Acta Med Scand 133(Supple 240):1–124.

Friberg L, Kjellstrom T, Nordberg GF (1986) Cadmium. In L Friberg GF Nordberg VB Vouk (eds) Handbook on the Toxicology of Metals (2nd ed., pp. 130–184). Amsterdam: Elsevier.

Garrett RG, Porter AR, Hunt PA, Lalor GC (2008) The prescence of anomalous trace element levels in Jamaican soils and the geochistry of late-miocene or pliocene phosphorites. Appl Geochem 23:822–834.

Green GW (1977) Structure and stratigraphy of the Wagwater belt. Overseas Geol Miner Resour 48:21.

Grippi J, Burke K (1980) Ubmarine canyon comples among Cretaceous island arc sediments, Western Jamaica. Geol Soc Am Bull 91:179.

Hallberg L, Bjorn-Rasmussen E, Rossander L, Suwanik R (1977) Iron absorption from Southeast Asian diets. II. Role of various factors that might explain low absorption. Am J Clin Nutr 30:539–548.

Holmgren GG, Meyer MW, Chaney RL, Daniels RB (1993) Cadmium, lead, copper and nickel in agricultural soils of the United States of America. J Environ Qual 22:335–348.

Hong C (2000) Urinary protein excretion in Type 2 diabetes with complications. J Diabet Complic 14(5):259–265.

Hoo-Fung L, Lalor G, Rattray R (2005–2006) Effect of cooking and extraction with simulated gastric juice on the cadmium content of selected foods. Jamaica J Sci Tech 16&17:26–29.

Horiguchi H, Oguma E, Sasaki S, Miyamoto K, Ikeda Y, Machida M, et al. (2004) Dietary exposure to cadmium at close to the curent provisional tolerable weekly intake does not affect renal function among female japanese farmers. Environ Res 95:20–31.

Howe A, Hoofung L, Lalor GC, Rattray R, Vutchkov M (2005) Elemental composition of Jamaican foods 1: A survey of five food crop categories. Environ Geochem Health 27: 19–30.

Ikeda M, Ezaki T, Moriguchi J, Fukui Y, Ukai H, Okamoto S, et al. (2005) The threshold cadmium level that causes a substantial increase in B-2 Microglobulin in urine of general populations. Tohoku J Exp Med 205:247–261.

Ikeda M, Ezaki T, Tsukahara T, Moriguchi J, Furuki K, Fukui Y, et al. (2003) Threshold levels of urinary cadmium in relation to increases in urinary B2-microglobulin among general Japanese populations. Toxicol Lett 137:135–141.

Ikeda M, Shimbo S, Watanabe T, Yamagami T (2006) Correlation among cadmum levels in river sediment, in rice, in daily foods and in urine of residents in 11 prefectures in Japan. Int Arch Occup Environ Health 79: 365–370.

Imamura T (2007) History of public health crises in Japan. J Public Health Policy 2(8):221–237.

Jarup L, Berglund M, Elinder CG (1998) Health effects of cadmium exposure – a review of the literature and risk estimate. Scand J Work Environ Health 24:1–52.

Jarup L, Berglund M, Elinder CG, Nordberg G, Vahter M (1998) Health effects of cadmium exposure– a review of the literature and a risk estimate. Scand J Work Environ Health 24:11–51.

JECFA (2000) Summary and conclusions of the fifty-fifty meeting, Geneva, 6–15 June 2000. Geneva: World Health Organisation, Joint FAO/WHO Expert Comittee on Food Additives.

JRA (1982) Comprehensive Resource Inventory and Evaluation System (CRIES) Project. Jamaica Resource Assessment. Jamaica: Ministry of Agriculture.

Kido T, Honda R, Tsuritani I, Yamaya H, Ishizaki M, Yamada Y, et al. (1988) Progress of renal dysfunction in inhabitants environmentally exposed to cadmium. Arch Env Health 43:213–217.

Kjellstrom T, Ervin PE, Rahnster B (1977) Dose-response relationship of cadmium induced tubular proteinuria. Environ Res 13:303–317.

Lalor GC (1995) A Geochemical Atlas of Jamaica. Jamaica: Canoe Press, Universty of the West Indies.

Lalor GC, Davies BE, Vutchkov M (2001) Cadmium in Jamaican soils. Environmental Cadmium in the Food Chain: Sources, pathways and risks. Proceedings of the SCOPE Workshop, Belgian Academy of Sciences, Brussels, Belgium, 13–16 September 2000. Scientific Commitee on Problems of the Environment (SCOPE) of the international Council for Science (ICSU).

Lalor GC, Miller JM, Robotham H, Simpson PR (1989) Gamma radiometric survey of Jamaica. Tran Inst Min Metall 98: 34–37.

Lalor GC, Rattray R, Simpson P, Vutchkov M (1998) Heavy metals in Jamaica part 3: The distribution of cadmium in Jamaican soils. Rev Int Cont Ambient, 14, 7–12.

Lalor GC, Rattray R, Williams N, Wright P (2004) Cadmium levels in kidney and liver of Jamaicans at autopsy. West Ind Med J 53(2):76–80.

Lalor G, Rattray R, Simpson P, Vutchkov MK (1999) Geochemistry of an arsenic anomaly in St. Elizabeth Jamaica. Environ Geochem Health 21:3–11.

Lalor G, Rattray R, Vutchkov M, Campbell B, Lewis-Bell K (2001) Blood lead levels in Jamaican school children. Sci Tot Environ 269:171–181.

Lauwerys RR, Bernard A, Roels HA, al e (1984) Characterisation of cadmium proteinuria in man and rat. Environ Health Perspect 54:147–152.

Lauwerys R, Bernard A, Roels H, Buchet JP (1994) Cadmium: Exposure markers as predictors of nephrotoxic effects. Clin Chem 40(7):1391–1394.

Lewis GP, Coughlin LL, Jusko WJ, Hartz S (1972) Contribution of cigarette smoking to cadmium accumulation in man. Lancet 1(7745):291–292.

McKenzie-Parnell JM, Eynon G (1987) Effect on new Zealand adults consuming large amounts of cadmium in oysters. In DD Hemphill (ed) Trace Substances in Environmental Health – XXI (pp. 420–430). Columbia, MO: University of Missouri Press.

Meschede M, Frisch W (1998) A plate-tectonic model for the Mesozoic and early Cenzoic history of the Caribbean plate. Tectonophysics 296:269–291.

Morrow H (2001) Cadmium and cadmium alloys. In Kirk-Othmer encyclopaedia of chemical technology (pp. 471–507). John Wiley and Sons.

Nishijo M, Satarug S, Honda R, Tsuritani I, Aoshima K (2004) The gender differences in health effects of environmental cadmium exposure and potential mechanisms. Mol Cell Biochem 255(1–2):87–92.

Nogawa K (1981) Itai-itai disease and follow-up studies. In JO Nriagu (ed) Cadmium in the Environment (Vol. 2, pp. 1–37). New York: Wiley.

Nogawa K, Kido T (1993) Biological monitoring of cadmium exposure in itai-itai disease epidemiology. Int Arch Occup environ health 65:S43–S46.

Nogawa K, Kido T (1996) Itai-itai disease and health effects of cadmium. In LW Chang (ed) Toxicology of Metals (pp. 353–369). New York: CRC.

Nordberg GF, Jin T, Bernard A, al, e (1997) Biological monitoring of cadmium exposure and renal effects. Sci Total Environ 199:111–114.

Nordberg G, Jin T, Bernard A (2002) Low bone density and renal dysfunction following environmental cadmium exposure in China. Ambio 31(6):478–481.

Nordberg G, Jin T, Leffler P, Svensson M, Zhou T, Nordberg M (2000) Metallothioneins and diseases with special reference to cadmium poisoning. Analusis 28(5):396–400.

Nordberg M (1984) General aspects of cadmium:Transport, uptake and metabolism by the kidney. Environ Health Perspect 5:13–20.

NTP (2005) Report on carcinogens. 11th ed. Research Triangle Park, NC: U.S. Department of Health and Human Services, Public Health Service, National Toxicology Program.

Olsson I, Bensryd I, Lundh T, Ottosson H, Skerfving S, Oskarsson A (2002) Cadmium in blood and urine – impact of sex, age, dietary intake, iron status and former smoking – association of renal effects. Environ Health Perspect 110(12):1185–1190.

Pedersen B, Eggum BO (1983) The influence of milling on the nutritive value of flour from cereal grains: 4. Rice. Plant Foods Hum Nutr 33:267–278.

Pindell JL, Barrett SF (1990) Geological evolution of the Caribbean region: A plate tectonic perspective. In G Dengo, JE Case (eds) The Caribbean Region, Geology of North America (Vol. H, pp. 405–432).Washington, DC:Geological Society of America.

Piscator M (1991) Early detection of tubular dysfunction. Kidney Int 40(Suppl 34):15–17.

Reeves PG, Vanderpool RA (1998) Organ content and fecal excretion of cadmium in male and female rats consuming variable amounts of naturally occurring cadmium in confectionery sunflower kernels (*Helianthus annuus L.*). J Nutr Biochem 9:636–644.

Ross MI, Scotese CR (1988) A hierarchal tectonic model of the Gulf of Mexico and Caribbean region. In CR Scotese, WW Sager (eds) Tectonophysics 155:139–168.

Sarasua SM, McGeehin MA, Stallings FL, Terracciano G, Amler RW, Logue JN, et al. (1995). Technical assistance to the Pennsylvania department of health. Biologic indicators of exposure to cadmium and lead. Final Report. Palmerton, PA.

Satarug S, Moore MR (2004) Adverse health effects of chronic exposure to low-level cadmium in foodstuffs and cigarette smoke. Environ Health Persp 112(10):1099–1103.

Satarug S, Baker JR, Reilly PE, Moore MR, Williams DJ (2002) Cadmium levels in the lung, liver, kidney cortex and urine samples from Australians without occupational exposure to metals. Arch Environ Health 57:69–77.

Seidal K, Jorgensen N, Elinder C (1993) Fatal cadmium induced pneumonitis. Scand J Work Environ Health 19: 429–431.

Shigematsu I (1984) The epidemiological approach to cadmium pollution in Japan. Ann Acad Med, 13:231–236.

Simmons RW, Pongsakul P, Chaney RL, Saiyasitpanich D, Klinphoklap S, Nobuntou W (2003) The relative exclusion of zinc and iron from rice grain in relation to rice grain cadmium as compared to soybean: Implications for human health. Plant Soil 257:163–170.

Simpson PR, Hurdley J, Lalor GC, Plant JA, Robotham H, Thompson C (1991) Orientation studies in Jamaica for multipurpose geochemical mapping of the Caribbean region. Transactions of the Institution of Mining and Metallurgy 100:B98–B110.

Strehlow CD, Barlthrop D (1988) The Shipham report – An investigation into cadmium concentrations and its implications for human health: 6. Health Studies. Sci Tot Environ 75:101–133.

Tsukahara T, Ezaki T, Moriguchi J, Furuki K, Fukui Y, Ukai H, et al. (2003b). No significant effect of iron deficiency on cadmium body burden or kidney dysfunction among women in the general population in Japan. Int Arch Occup Environ Health 76:275–281.

Tsukahara T, Ezaki T, Moriguchi J, Furuki K, Shimbo S, Matsuda-Inoguchi N, et al. (2003a) Rice as the most influential source of cadmium intake among general Japanese population. Sci Tot Environ 305:41–51.

Vahter M, Berglund M, Akesson A, Liden C (2002) Metals and womens health. Environ Res 88(3):145–155.

Vahter M, Berglund M, Nermell B, Akesson A (1996) Bioavailability of cadmium from shellfish and mixed diet in women. Toxicol Appl Pharmacol 136(2):332–341.

Vincenz SA (1959) Some observations of radiation emitted from a mineral spring in Jamaica. Geophys Prospect 7(4): 422–434.

Wadge G, Jackson TA, Isaacs MC, Smith TE (1982) Ophiolitic bath dunrobin formation, Jamaica: Significance for the cretaceous plate margin evolution in the north-western Caribbean. J Geol Soc London 139:321.

Waldron HA (ed) (1980) Metals in the Environment. London: Academic Press Inc.

Watanabe K, Kobayashi E, Suwazono Y, Okubo Y, Kido T, Nogawa K (2004) Tolearable life-time cadmium intake calculated from the inhabitants living in the Jinzu River Basin. Jpn Bull Environ Contamin Toxicol 72:1091–1097.

Watanabe T, Zhang ZW, Moon CS, Shimbo S, Nakatsuka H, Matsuda-Inoguchi N, et al. (2000) Cadmium exposure of women in general populations in Japan during 1991–1997 compared with 1977–1981. Int Arch Occup Environ Health 73:26–34.

Wright RM, Robinson E (1993) Biostratigraphy of Jamaica. Washington, DC: Geological Society of America, Memoir.

Zans VA (1952) Bauxite resources of Jamaica and their development. Colon Geol Mine Resour 3:307–333.

Medical Geology in Mexico, Central America and the Caribbean

M. Aurora Armienta, Ramiro Rodríguez, Nuria Segovia, and Michele Monteil

Abstract An overview of the occurrence, concentrations, and possible sources of toxic elements released by geogenic processes that may threat the health of millions of people of Mexico, Central America, and the Caribbean is presented. The geology and tectonic characteristics of Mexico and Central America constitute an appropriate environment for the presence of arsenic and fluoride in groundwater of many zones of the area. Health problems linked with As-tainted water consumption have been documented in Mexico and Nicaragua where epidemiological and toxicological studies have been developed. Fluorosis has been recognized mainly in the central and northern part of Mexico and also in Antigua, Puerto Rico, and Trinidad. Specific health effects resulting from exposure to natural dust transported from Africa have been identified in the Caribbean. Radon exposure may also affect the population living in volcanic and active tectonic environments. However, this problem has only been studied by some researchers, mainly in Mexico and Nicaragua. Collaboration among research groups and authorities has been scarce. The review presented here, although not exhaustive, shows the urgency for increasing that collaboration specially to identify polluted areas, sources, and health effects, of the routine collection and analysis of arsenic and fluoride in all potable water sources of the area, and of developing short-term measures to decrease their concentrations to safe levels.

M.A. Armienta (✉)
Universidad Nacional Autónoma de Mexico, Instituto de Geofísica, Mexico, DF 04510, Mexico
e-mail: victoria@geofisica.unam.mx

Keywords Mexico · Central America · Nicaragua · El Salvador · Cuba · Caribbean · Arsenic · Fluorine · Radon

Introduction

The geological framework of Mexico and Central America is defined by its volcanic origin and active tectonic setting. The Trans-Mexican Volcanic Belt, TVB, crosses Mexico, 900 km from west to east. The Central American Volcanic Arc confers a similar environment to the rest of the Central American countries.

Mexico is almost 2,000,000 km^2 in extent. The Global Volcanism Program (GVP, 2009) lists about 30 Holocene volcanoes and 4 Pleistocene volcanoes with geothermal activity, many of which are located along the TVB. The southern part of its Pacific Coast is one of the most seismically active areas in the world due to the subduction of the Cocos and Rivera Plates under the North American Plate (Fig. 1). The recurrence period for large earthquakes (up to Ms 8.2) in this area has been estimated to be in the range from 30 to 70 years (Singh and Mortera, 1991).

The Mexican Highlands are surrounded by two large mountain ranges. The Sierra Madre Occidental in Northwestern Mexico hosts some of the largest fluorite deposits in the world (Ruiz et al., 1985). El Bajio, in Central Mexico, is characterized by felsic volcanic rocks with Sn, Au, Ag, and F mineralization (Mérida et al., 1998).

The countries located in Central America, Belize, Costa Rica, El Salvador, Guatemala, Honduras, Nicaragua, Panamá, about 520,000 km^2 in extent have a common volcanic axis, the Central American

O. Selinus et al. (eds.), *Medical Geology*, International Year of Planet Earth,
DOI 10.1007/978-90-481-3430-4_3, © Springer Science+Business Media B.V. 2010

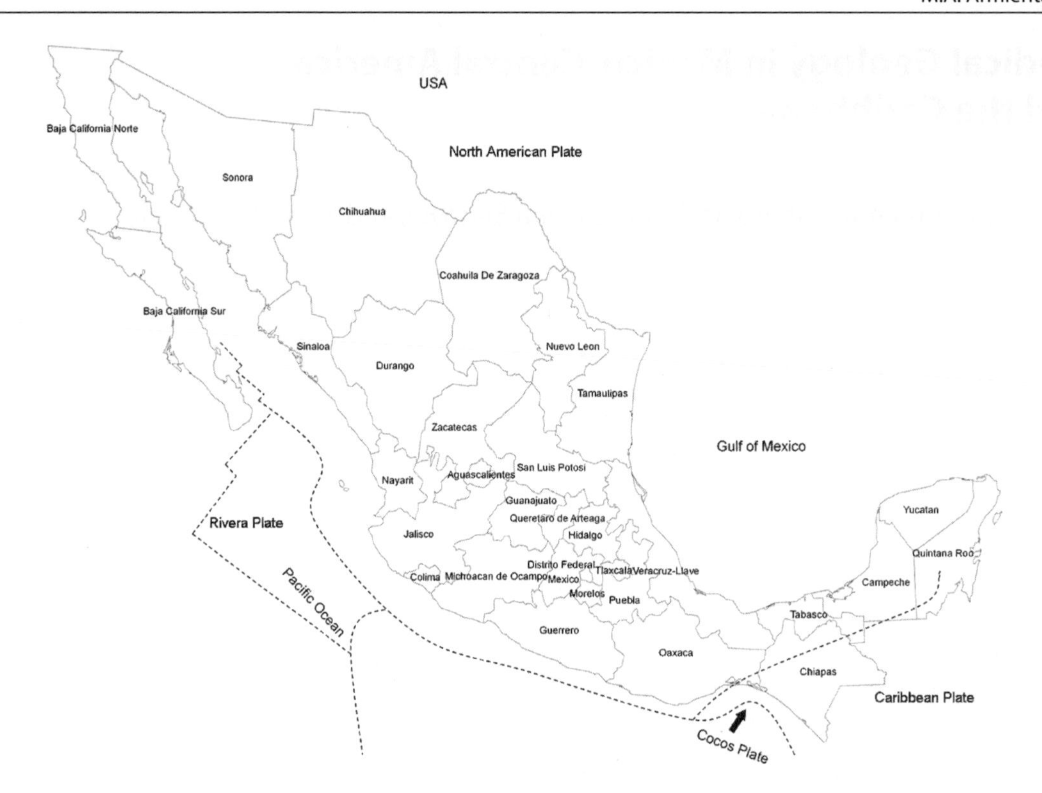

Fig. 1 Tectonic setting and political division of Mexico

Fig. 2 Tectonic setting and countries of Central America

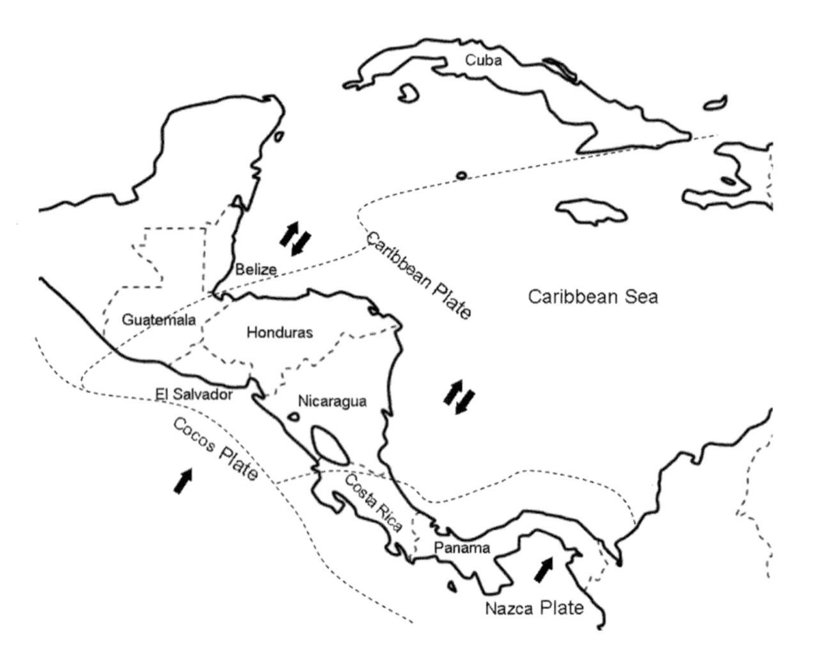

Volcanic Arc. The subduction of the Cocos and Nazca Plates under the North American and Caribbean Plates (Fig. 2) generates large earthquakes in the zone that affect recurrently the populated areas of large cities.

The regionally prevalent rocks may contain As-bearing minerals. Arsenic minerals are also found in intrusive bodies in calcareous rocks in mining areas. Arsenic can be incorporated in groundwater flow. The high dependence of aquifer abstraction for urban and rural water supply increases the risk for the population to be exposed to arsenic or other risky elements.

In the highlands and in some large intermountain valleys, conditions for the occurrence of lacustrine systems are present. In such environments fluorine, F, is associated with evaporitic layers and with saline shallow groundwater (Levy et al., 1998). Apatite and fluorite are frequently found in silica-saturated and silica-undersaturated igneous rocks and sediments (Mattheß, 1994). Metamorphic rocks such as hornblende gneiss also contribute to the presence of fluorine in groundwater (Ravishankar, 1987).

The Caribbean Islands and the Antilles (Anguilla, Antigua and Barbuda, Araba, Bahamas, Barbados, British Virgin Islands, Cayman Islands, Cuba, Dominica, Dominican Republic, Grenada, Guadeloupe, Haiti, Jamaica, Martinique, Montserrat, Netherlands Antilles, Puerto Rico, St. Barthelemy, St. Kitts and Nevis, St. Lucia, St. Martin, St. Vincent and the Grenadines, Trinidad and Tobago, Turks and Caicos, US Virgin Islands) have evolved from an arc of very old volcanoes. Different periods of submersion formed layers of limestone over the igneous and metamorphic rocks. The Caribbean Islands, from Cuba to Venezuela, are characterized by absence of volcanic activity, and some common geological characteristics.

Thermal waters facilitate the leaching of fluorine into groundwater. In most of the thermal areas of the region there is a high correlation between temperature and high fluorine concentrations.

Groundwater is the main source of drinking water for the entire population of the whole geographic area, so that the local geology may affect directly, through water ingestion, the health of the inhabitants. Being low-income countries, it is worth mentioning that the precarious state of dwellings in the region is one of the serious problems related to health for most people. Dwellings are subject to damage or destruction from earthquakes, volcanic eruptions, ground slides, floods, and tropical hurricanes. About 40% of the people cannot accede adequate lodging and 32% of the urban population live in precarious conditions, where the risk factors and health impact maximize (Peña, 2005). Water quality, ground slides, tropical diseases, and non-controlled natural and anthropogenic pollution in rivers, air at urban areas, food products, and catastrophic natural events occur systematically in the area.

Arsenic in Mexico

Natural As-enriched groundwater occurs in many parts of Mexico, mainly within a mineralized belt that crosses from the northwest to the south of the country. In Chihuahua State, north of Mexico, the "Junta Central de Agua y Saneamiento" (local water supplier) informed in 2006 that 35 out of 104 sampled locations presented As levels from 0.006 to 0.474 mg/L. Two As sources have been identified in distinct parts of the state: groundwater interaction with volcanic rocks and As release from lacustrine sediments (Reyes-Cortés et al., 2006a, 2006b; Mahlknecht et al., 2008). A mineralogical study identified leaching of arsenogoyazite $(SrAl_3[(OH)_5(AsO_4)_2])H_2O$, contained in the Cenozoic volcanic tuff as one of the sources of As to groundwater at El Mimbre area, which is one of the zones that supplies potable water to Chihuahua City (Rodrigu et al., 2007). Authorities have partially solved the population exposure to As in drinking water by installing small reverse osmosis treatment plants, partially paid by the consumers (Calderón-Fernández, 2006).

Arsenic has also been detected in one of the Mexico–USA border states in northwestern Mexico (Sonora). A study developed in early 1990s including sampling of 173 wells and storage tanks showed an As range from 0.002 to 0.305 mg/L and revealed three areas with As levels above 0.05 mg/L. However, a decrease in As content was detected at some of the polluted sites re-sampled in 1995 (Wyatt et al., 1998a). The lack of an anthropogenic source and the basin geology indicates that the As originates from a natural source. Another study was carried out in the Yaqui Valley to determine arsenic exposure via drinking water and to characterize urinary arsenic excretion. The town of Esperanza had the highest daily mean As intake through drinking water, with a mean value of 65.5 mg/day. A geometric mean concentration (65.1 mg/L) of arsenic in urine was determined for that population. The authors also found a positive correlation between total arsenic intake from drinking water/day and total arsenic concentration in urine (Meza et al., 2004). Similarly, Wyatt et al. (1998b) determined that urine As was also correlated with water contents at Hermosillo City.

La Comarca Lagunera (Durango and Coahuila states of northern Mexico) was the first place where health problems related to As intake were detected in Mexico in 1958 (Cebrián et al., 1994). Periferical vascular diseases including black foot disease, keratho-sis, skin pigmentation, and gastrointestinal sickness were observed in people drinking As-polluted water. A population of approximately 400,000 persons was considered to be exposed. Concentrations from 0.008 to 0.624 mg/L were measured in groundwater from 128 wells sampled in 1990 (Del Razo et al., 1990). The highest concentrations have been measured in the towns of Francisco I. Madero (up to 0.740 mg/L in 1992), Tlahualilo (up to 0.590 mg/L in 1992 and 0.718 mg/L in 2000), and San Pedro (0.490 mg/L in 1992 and 0.26 mg/L in 2000) (Rosas et al., 1999; Molina, 2004). Various sources have been hypothesized as origin of the arsenic: extinct hydrothermal activity, evaporation processes, oxidation of sulfides, desorption, and dissolution of Fe and Mn oxides (González-Hita et al., 1991; Ortega-Guerrero, 2003; Molina, 2004; Gutiérrez-Ojeda, 2009). Investigations on As-related health problems in this area have shown relationships between As speciation in urine and signs of dermatological affectations (Del Razo et al., 1997). In addition, a significant increase in the frequency of chromatide deletions, isochromatides in lymphocytes, and in micronuclei of urinary and oral epithelial cells were observed in people exposed to As (Gonsebatt et al., 1997). Higher risk of having diabetes was determined in people with higher concentrations of total As in urine, in a case–control study developed in 2003 in this area (Coronado-González et al., 2007).

In Aguascalientes City, Central Mexico, an average of 0.0145 ± 0.0077 mg/L of As was determined in one-third of the potable water sources. Minimum (1.6×10^{-4}) and maximum (1.81×10^{-3}) exposure doses were calculated based on As concentrations in water. Estimation of cancer risk yielded 9.5 cases per 100,000 inhabitants for the lower water As content and 1.63 cases per 1,000 inhabitants for the highest As content, suggesting a health risk for the population (Trejo-Vázquez and Bonilla-Petriciolet, 2002).

In Guanajuato State, within the Mexican Highlands, many studies trying to explain the occurrence and health affectations of arsenic have been carried out. In the Independence aquifer, As (up to 0.12 mg/L) contents were found in groundwater (CODEREG, 2000). Cuerámaro, near Cuitzeo Lake, presented As in the main supply well. A thermal well located in Cerrito Agua Caliente (east of Cuerámaro) had an As content of 0.22 mg/L. High temperatures and As concentrations are associated in Abasolo with concentrations ranging between 0.10 and 0.17 mg/L (Martínez and García, 2007). Other localities and concentrations of As in groundwater of Guanajuato State are shown in Table 1.

Table 1 Arsenic contents in localities in Guanajuato State

Locality	As concentration (mg/L)	Reference
Acámbaro	0.05–0.08 (in 3 wells)	SSP (2009)
San Miguel Allende	0.03 (1 well)	ESF (2006)
Silao-Romita	0.031–0.059 (agricultural wells)	Horst (2006)
El Copal	0.30	Rodríguez et al. (2006)
Salamanca	0.02–0.18	Mejía et al. (2007)

Salamanca is the only region where an anthropogenic origin was identified for arsenic in Guanajuato. Leachates from fallout of power plant particulates rich in As result in variations of groundwater As content (0.02–0.18 mg/L). However, As coming from rocks has also been recognized (Mejía et al., 2007).

Many arsenic minerals occur in the Mexican territory such as arsenopyrite, mimetite, adamite, scorodite, and tennantite. At the La Ojuela mine in Durango State, about 29 distinct As minerals have been recognized such as conichalcite, adamite, and legrandite. In addition, arsenic may occur associated with silver, lead, zinc, and iron sulfides. Mining zones may be polluted by arsenic from natural sources but also by centuries of ore extraction and processing activities. Studies regarding concentrations and geochemical processes releasing As to groundwater have been developed only in a few of the mineralized areas. One of the most studied areas has been Zimapán, in Hidalgo State, Central Mexico. Mineralization comprises the complete spectrum of high-temperature Mexican minerals in carbonates. More than eight As minerals have been reported in the area, arsenopyrite being the most widespread. In 1993 up to 1.1 mg/L of arsenic was measured in one of the deep wells used as drinking water supply (Simons and Mapes-Vazquez, 1956; Megaw et al., 1988; Armienta et al., 1997a). The aquifer system is composed of deep fractured limestone and granular perched aquifers derived from the calcareous rocks. Mining wastes polluted shallow wells within the granular aquifer. The deep aquifer presents the highest arsenic contents released by a natural source. Arsenic enrichment is produced by dissolution of arsenic-bearing minerals, mainly arsenopyrite and scorodite in the mineralized zones within the limestones (Armienta and Rodríguez, 1996; Armienta et al., 2001). Arsenic is carried through fractures and faults in a complex regional flow system influenced by seasonal processes (Rodríguez et al., 2004). However, until about 5 years ago, the limestone aquifer provided most of the potable water. Currently, good quality water is pumped from a non-polluted well out of the valley and mixed with tainted deep water. This mixing results on varying As concentrations along time (0.10 mg/L in November 2008).

Toxicity and effects of groundwater As intake on the inhabitants' health have also been investigated in this area. Experimental studies showed a concentration–response relationship between SCE (sister chromatid exchanges) in root tip meristems of the broad bean *Vicia faba* and As contents of Zimapán wells (Gómez-Arroyo et al., 1997). Hyperkeratosis, hypopigmentation, and hyperpigmentation were associated with As concentrations in hair and drinking water in a sample of 120 Zimapán inhabitants (Armienta et al., 1997b). Reséndiz and Zúñiga (2003) carried out a similar study including another 71 individuals; an average of 9.74 mg/kg As was determined in the hair of that group. More than half (56.1%) of the studied population showed dermal damage. The highest As content in hair (average of 29.25 mg/kg) was measured in people ranging from 50.1 to 60 years old. One-factor ANOVA test showed correlation between dermal health effects and As concentrations in hair. Valenzuela et al. (2007) observed at least one sign of arsenicism in about half of a high As (0.104 mg/L on average) exposed group of 72 women. Besides, association of arsenic presence with the increment of transforming growth factor alpha (TGF-α) levels in bladder urothelial cells was also shown. TGF-α levels in urothelial cells were significantly linked with As in urine in those individuals presenting skin lesions.

At the historical mining zone of Villa de la Paz, San Luis Potosí State, high As concentrations were measured in water, soil, sediments, and maize leaves, as well as elevated levels of Pb in soils and sediments (Castro-Larragoitia et al., 1997; Razo et al., 2004). Increased DNA damage was observed in children exposed to As and Pb in this area (Yáñez et al., 2003). Gamiño and Monroy (2009) also determined high soil As and Pb bioaccesibility and demonstrated that children showed concentrations higher than the maximum national and international criteria for children health risk prevention, resulting from exposure to naturally enriched soils and mining wastes. Similar results would be expected as well at other mining sites of the country processing skarn ore deposits.

Arsenic is also present in aquifers of geothermal areas like Los Humeros and Acoculco located east of the TVB. Concentrations as high as 100 mg/L occur at a depth from 600 to 3,000 m below sea level in geothermal reservoirs along the TVB. Physical and chemical reactions at elevated temperatures increase As contents derived from dissolution of volcanic host rock in geothermal fluids (Armienta and Segovia 2008; Birkle and Bundschuh, 2009). High As concentrations (up to 0.2629 mg/L) were measured also in wells used as potable water sources at Los Altos de Jalisco region, west of the TVB. In 2005, 34% of 129 sampled wells contained As above the national drinking water standard of 0.025 mg/L. Estimated exposure doses were 1.1–7.6 μg/kg/day for babies, 0.7–5.1 μg/kg/day for children, and 14.7–101.9 μg/kg/day for adults. Two groups of water samples were identified in this zone; in one of them As was correlated with temperature (Hurtado-Jiménez and Gardea-Torresdey, 2009).

Concentrations above the Mexican drinking water standard have also been determined in wells of Zacatecas and San Luis Potosí, Central Mexico. The town of San Ramón was considered to represent the greatest health problem in Zacatecas based on water As levels (about 0.500 mg/L), groundwater abstracted volumes, and exposed population (Leal-Ascencio and Gelover-Santiago, 2006). In the Rio Verde locality, in San Luis Potosí State, the presence of As in groundwater (up to 0.05 mg/L) was associated with its release from lacustrine sediments (Planer-Friedrich et al., 2001).

Fluorine in Mexico

Fluoride is one of the main contaminants in groundwater in Mexico. Its presence has been reported in many areas, often in the same areas with high As levels. It is estimated that a population of 4 million people live in areas naturally enriched in fluorine (Estupiñán-Day et al., 2005). To evaluate the information about fluoride use (mainly from fluoridated salt) and fluorosis in Mexico and to provide recommendations to cope with fluoride excess in some areas of the country, a meeting took place in 2004 with the participation of the National Commission of Water and the Secretary of Health (Comisión Nacional del Agua, Secretaría de Salud) of Mexico, the Pan American Health Organization, and the Center for Disease Control and Prevention of the USA. The final report points out that the states of Aguascalientes, Sonora, Zacatecas, San Luis Potosí, Baja California, and Durango have groundwater concentrations equal or higher than 0.7 mg/L, 24% overall prevalence of dental fluorosis, and represent 67% of the national cases. Distribution of fluoridated salt was prohibited by law in those states. However, fluoride groundwater contents above 0.7 mg/L occur at other locations such as Comarca Lagunera and Monclova, in Coahuila State, several localities in Chuihuahua, and Acapetahua in Chiapas State. About 1,750,000 children and teenagers living in Aguascalientes, Chihuahua, Durango, and Jalisco states were expected to be at risk of developing dental fluorosis (Estupiñán-Day et al., 2005). Although measures have been taken to reduce the exposure level consisting mainly of dilution with good quality water, exposure to fluoride concentrations above the drinking water standard is still a health problem in some areas of the country. A critical review of the publications concerning dental fluorosis in Mexico was written by Soto-Rojas et al. (2004). Specific studies regarding natural fluoride occurrence in water and fluorosis are presented next.

At Comarca Lagunera, concentrations from less than 0.5 to 3.7 mg/L fluoride in groundwater were reported in 1993. Higher fluoride contents were determined in wells located in As-polluted areas (Del Razo et al., 1993). At Hermosillo City, arsenic concentrations also correlated with those of fluoride (Wyatt et al., 1998b). In Durango State, fluoride concentrations above drinking water standards (1.5–5.67 mg/L) were measured in groundwater of Durango City (Alarcón-Herrera et al., 2001a; Ortiz et al., 1998). About 95% of the population was estimated to be exposed to fluoride levels higher than 2.0 mg/L (Ortiz et al., 1998) and showed dental fluorosis and increased bone fractures (Alarcón-Herrera et al., 2001a). A positive relationship between water fluoride intake and dental fluorosis in children was also determined (Alarcón-Herrera et al., 2001a). In addition, about half of the 74 sampled wells used as potable water sources for Durango City exceeded 0.05 mg/L As (Alarcón-Herrera et al., 2001b). Concentrations of fluoride at Los Altos de Jalisco, an area also with high As groundwater contents, ranged from 0.1 to 17.7 mg/L. About half of the 105 wells sampled in 2002 contained fluoride concentrations above the Mexican drinking water standard (1.5 mg/L). The potential risk of dental, skeletal fluorosis, and bone fracture for the inhabitants was estimated based on calculation of exposure doses (Hurtado-Jiménez and Gardea-Torresdey, 2005).

San Luis Potosí is one of the most studied zones regarding fluoride occurrence and health impact. The deep fractured volcanic aquifer is the main source of fluoride in the groundwater (Carrillo-Rivera et al., 1996). However, increased pumping contaminated also shallower waters. Mineralogical and geochemical evidence indicate that water interaction with rhyolitic rocks has contaminated the aquifer. The higher the water temperature, the higher the fluoride content (Cardona et al., 1993, Carrillo-Rivera et al., 2002). High fluoride concentrations in drinking water produced dental fluorosis in children of San Luis Potosí and also in Aguascalientes State. Concentrations of fluoride in urine and severity of dental fluorosis correlated with drinking water contents of fluorine (Grimaldo et al., 1995; Loyola-Rodríguez et al., 2000; Hernández-Montoya et al., 2003). Neurotoxicological effects due to tap water fluoride intake were also found in children living in San Luis Potosí. Three tests were used to evaluate the neuropsychological development: (1) Wechsler Intelligence Scale for Children Revisited version for Mexico (WISC-RM), (2) Rey Osterrelth-Complex Figure test, and (3) Continuous Performance Test (CPT). Urinary fluoride

(1.6–10.8 mg/g creatinine) correlated positively with reaction time and inversely with the scores in visuospatial organization (Calderón et al., 2000). Another study carried out in children from San Luis Potosí and Durango states showed that arsenic and fluoride intake result in increased risk of reduced IQ scores (Rocha-Amador et al., 2007).

Within the Mexican Highlands, high water fluoride contents and associated health effects have been documented for some cities. Rodríguez et al. (1997) determined F concentrations of up to 4 mg/L in urban wells supplying Aguascalientes City, whose origin was associated with local igneous rocks. In Irapuato and Salamanca, within Guanajuato State, groundwater concentrations ranged from 1 to 3 mg/L fluoride (Rodríguez et al., 2000, 2006). Cases of dental fluorosis were observed in this area (Ovalle, 1996) and also in students at Leon City (Garcia and Ovalle, 1994). Wells with 4.0 mg/L F were identified in San Miguel de Allende, Guanajuato. However, health affectations have not been reported in this city (ESF, 2006). Up to 16 mg/L of fluoride in groundwater were attributed to weathering of acid volcanic rocks in the Independence aquifer in Queretaro State (Mahlknecht et al., 2004). Dental fluorosis linked with water consumption was recognized in Tequisquiapan, in the same state (Sanchez-Garcia et al., 2004).

Although official reports of groundwater fluoride contents are not available in many regions, occurrence of health effects indicates its presence in some areas. In El Bajio Guanajuatense, Central Mexico, e.g., the presence of fluorine is deduced from cases of dental fluorosis in the population. Fragoso et al. (1997) mentioned that inhabitants of San Francisco del Rincon also suffer from that illness. Abasolo and Juventino Rosas cities and Valtierrilla, a small settlement near Salamanca City, are supplied with thermal water that probably contains fluoride since dwellers show evidence of dental fluorosis. Occasionally, the local water supply organizations recognize the presence of F and As in groundwater. JUMAPA (local water supplier) in Celaya City reported concentrations of those elements over the national drinking water standard (1.5 and 0.025 mg/L, respectively) in some urban wells (A. M., 2008).

Irigoyen and col. (1995) informed the presence of F in concentrations of up to 2.8 mg/L in supplied water affecting teeth of children in Xochimilco, Basin of Mexico.

Fluoride has also been detected in sedimentary rock environments, such as Campeche in southern Mexico, where children groups showed injured teeth (Vallejo-Sanchez et al., 1998).

Arsenic in Central America

Arsenic occurrence and health problems related to the ingestion of polluted water have been discussed by researchers and authorities in various Central American countries. This information is mainly available from Nicaragua and El Salvador.

Nicaragua

Hydroarsenicism cases have been observed in some locations in Nicaragua. In 1996, arsenic-related health problems were reported in a community of El Zapote town in Sébaco Valley caused by people drinking water from a well with an arsenic concentration of 1.32 mg/L during a 6-month period (Altamirano, 2005). The INAA (Instituto Nicaragüense de Acueductos y Alcantarillados) detected 11 wells containing As above the drinking water standard of Nicaragua (10 μg/L), with a maximum concentration of 0.289 mg/L As in that valley. The areas of El Zapote-San Isidro, Santa Rosa del Peñón, La Cruz de la India, Susucayán-Ciudad Sandino, and Mina de Agua-Rincón de García near to thermal zones parallel the Nicaragua depression and some locations in Boaco, Chontales, Estelí, Jinotega, Terrabona, and Matagalpa have been found to contain As above 0.010 mg/L in groundwater (Estrada, 2003; Armienta et al., 2008). Some correlation was found between sites with high As contents and the fault system. Respiratory and skin problems were detected in dwellers consuming water with the highest As concentrations (from 0.080 to 0.380 mg/L) (González, 1998; Armienta et al., 2008).

A national study on water quality developed in 2004 identified the northern zone of the country as having the greatest As contents (Barragne, 2004). UNICEF in 2005 developed a sampling program in Llano de la Tejera and results showed 87% of wells with As concentrations over the drinking water standards (Altamirano-Espinoza and Bundschuh, 2009).

A hydrogeochemical study carried out during 2002–2003 in the southwestern part of Sébaco Valley revealed groundwater As contents above 0.010 mg/L in 21 out of 57 samples. Consolidated volcanic rocks, hydrothermally altered Tertiary volcanic rocks, and alluvial sediments were identified as the As source. Arsenic is mobilized from the solid matrix by weathering and changing redox conditions (Altamirano-Espinoza and Bundschuh, 2009).

At San Juan de Limay, Estelí Department, the presence of geogenic arsenic (up to 0.115 mg/L) was revealed by a 2005 study in that rural area. Concentrations showed significant variability and As(V) was the dominant species. A volcanic ash layer appeared to be the main groundwater As source (Morales et al., 2009).

El Salvador

El Salvador is located on the western part of the Caribbean Plate. Subduction, oblique to the plate boundary of the Cocos Plate, beneath the North American and Caribbean Plates produced a prominent magmatic arc, the modern Central American arc, in the western active margin. This tectonic setting has resulted also in the existence of many active volcanoes and geothermal zones (Agostini et al., 2006). Arsenic occurrence was revealed by studies in some of these areas.

Although it is known that several institutions have analyzed arsenic in waters at El Salvador, only part of the information is available to the public. Studies carried out in volcanic lakes showed As presence. Coatepeque Lake had concentrations from 0.09 up to 3.09 mg/L, Ilopango Lake from 0.15 to 0.78 mg/L, and one sample from Olomega Lake had an As content of 4.21 mg/L. Those lakes are the largest water reservoirs of the country. Furthermore, in spite of high As and boron levels, people living near their shores employ the water for domestic use and irrigation (López et al., 2009; Armienta et al., 2008). Fluxes of gases seem to occur south of the Ilopango Caldera, where higher concentrations of As and boron were also measured (López et al., 2009).

Other sites with As presence in Nicaragua are the Ahuachapán and Berlín geothermal fields with concentrations from 0.01 to 0.21 mg/L and 0.002 to 0.285 mg/L, respectively. Arsenic also occurs in Las Burras (0.164 mg/L) and Obrajuelo (0.016–0.330 mg/L). The polluted water is not directly used by dwellers; however, thermal discharges may contaminate superficial waters (Armienta et al., 2008).

Fluorine in Central America

Less information than that regarding arsenic is available about fluorine in waters and soils of Central American countries. Folleti and Paz (2001) reported cases of dental fluorosis in 39 small communities in Sula Valley, Honduras. The highest concentration measured was 7.5 mg/L. Other sites with fluoride presence are Kimuna, in La Libertad Municipality. In six villages more than 1,200 persons were affected by fluoride ingestion. In Nicaragua, the maximum groundwater fluorine content in a survey carried out by UNICEF during 2002 was 1 mg/L (Barragne, 2002). In Honduras and Nicaragua F is associated with groundwater flow through volcanic rocks.

Strock et al. (2008) reported high F concentrations in groundwater of Belize. They did not specify places, concentrations, or origin.

Fluorosis and natural fluoride-enriched water were identified in Chorotega and in central Costa Rica. However, high prevalence of fluorosis in Huetar Norte and Huetar Atlántica regions and low prevalence in Central Pacific Area need to be explained by more studies (Salas-Pereira et al., 2008).

Aspects of Medical Geology in the Caribbean

Introduction

The Caribbean archipelago borders the Caribbean Sea and extends from the Bahamas in the north (situated in between 24°15′N latitude and 76°00′W longitude) to the twin island republic of Trinidad and Tobago in the south (situated between 10°2′ and 11°12′N latitude and 60° 30′ and 61° 56′ W longitude) (Fig. 3). Also known as the West Indies, these islands are noted for their physical beauty, wide variety of flora and fauna, and cultural diversity. Table 2 lists the Caribbean Islands and summarizes important demographic and geologic information.

Fig. 3 The Caribbean region (taken from the all free map website – www.english.freemap.jp)

There is comparatively little in the published English medical literature about medical geological disorders in the Caribbean. They were searched to identify relevant articles. The information presented in this overview was gleaned from the articles, abstracts, and conference proceedings that were identified from PubMed, Google Scholar, and MedCArib databases.

Heavy Metal Toxicity

The most widely reported heavy metal poisoning in the Caribbean is lead poisoning. In 1997, a report from the Lead Research Group of the Pan American Health Organization (PAHO) expressed concern about the potential for increased lead exposure following rapid industrialization in Latin America and the Caribbean. There was particular mention about lead exposure as a result of widespread use of leaded gasoline, lead in industrial emissions, battery recycling, use of lead paints and varnishes, and ingestion of contaminated food and water (Romieu et al., 1997).

Since the Pan American Health Organization (PAHO) Lead Research Group report, most Caribbean Islands have implemented the use of lead-free gasoline but recent national surveys show that lead toxicity remains a problem in some territories. Nationwide surveys of blood lead levels in young children have been conducted in Jamaica (Lalor et al., 2007) and Trinidad and Tobago (Rajkumar et al., 2006). Lalor and colleagues surveyed 1,081 children, aged 2–6 years old, and found that blood lead levels ranged from 1.4 to 202 micrograms (mcg)/dL with a geometric mean of 4.35 mcg/dL. Of these, 230 children (21.3%) had blood lead levels above 10 mcg/dL and of these 80 children received medical attention, 11 (1%) had chelation therapy, and 6 had acute lead poisoning and required repeated sessions of chelation. The situation was less severe in Trinidad and Tobago. Rajkumar and colleagues surveyed 1761 children aged 5–7 years and found a smaller range of blood lead levels from <1 to 28.6 mcg/dL. Fifteen children (0.9%) had blood lead levels above 10 mcg/dL and only three children (0.2%) met the criteria for lead poisoning. The higher blood lead levels recorded in some Jamaican children were linked to lead-contaminated areas.

Previous studies of lead toxicity in other Caribbean Islands have focused on children and adults with increased risk for lead poisoning. In Barbados, Koplan and colleagues (1977) described high blood levels in potters and their families where there has been traditional use of lead glazes. High blood levels were also noted in members of a community living near a previous auto battery recycling plant in the Dominican

Table 2 Summary of the demography and geography of the Caribbean Islands

Country	Population	Total area (km^2)++	Rock type++	Max. elevation (m)++
Anguilla	13,477	102	L	65
Antigua and Barbuda	85,632	442.6	L/V	402
Aruba	104,494	193		188
Bahamas	323,000	13,940	L	63
Barbados	279,000	431	L/S	336
British Virgin Islands	22,000	153	S/M/V	521
Cayman Islands	48,000	262	L	43
Cuba	11,382,000	110,860	L/S/M	2,005
Dominica	72,660	754	V	1,447
Dominican Republic	9,183,394	48,730	V/L	3,175
Grenada	103,000	344	V	840
Guadeloupe	448,000	1,628		1,467
Haiti	8,528,000	27,750		2,680
Jamaica	2,651,000	10,991	L/S/M	2,556
Martinique	396,000	1,128	V	1,397
Montserrat	4,000	102	V	930
Netherland Antilles	183,000	960		862
Puerto Rico	3,955,000	13,790	L/S/M	1,339
St. Barthelemy	8,450	21		286
St. Kitts and Nevis	43,000	261	V	1,156
St. Lucia	161,000	616	V	950
St. Martin	35,000	54.4		424
St. Vincent and Grenadines	119,000	389	V	1,234
Trinidad and Tobago	1,305,000	5,128	L/S/M- T'dad M/V – T'go	940- T'dad 572 – T'go
Turks and Caicos	26,000	430	L	49
US Virgin Islands	112,000	1,910	S/M/V	475

Rock Type: L, limestone; M, metamorphic; S, sedimentary; V, volcanic; data from Wikipedia-List of Caribbean Island Countries by Population
(www.en.wikipedia.org/wiki/List_of_Caribbean_island_ countries_by_population). www.geographic.org and www. montrosetravel.com//cs_cgeography.html

Republic (Kaul and Mukerjee, 1999) and in children living near backyard smelters of lead in Jamaica (Lalor et al., 2006). Sanchez-Nazario et al. (2003) reported on lead levels in the house dust of homes of children living in a former landfill in Puerto Rico and found that lead on window sills and toys, and geophagia were significant predictors of higher blood lead levels in the study population.

There is little information about other forms of heavy metal poisoning in the Caribbean area. The Caribbean Environmental Program (CEP) web site lists several sources of mercury exposure in the region including high volcanic activity, "gold mining, non-metal extraction, dental laboratories and hospitals, fish canning, municipal, mercury lamps, batteries and electrical components" (CEP website, 2009). The frequent consumption of fish by island populations has also raised concern about possible mercury poisoning as a significant public health problem. The UN Atlas of the Ocean has noted high levels of mercury in sharks, swordfish, marlins, and tunas which are among the fish that are widely consumed by Caribbean people and hence lends support for possible chronic mercury poisoning in groups with heavy fish consumption. Additionally, a study in Trinidad and Venezuela of heavy metal contamination of sediments, mussels, and oysters showed that mercury levels in the sediments tested exceeded internationally recognized guidelines (Astudillo et al., 2005) and lend support to the concern of high mercury exposure through regular consumption of these mollusks.

High cadmium levels have been found in food crops in Jamaica (Howe et al., 2005) and high levels of cadmium, chromium, and zinc have been found in sediments from multiple sites in the Caribbean and Latin America. There is concern that marine activities could further increase the concentrations of these metals which could be taken up by marine organisms and enter the food chain (CEP website, 2009).

Arsenic has been reported in soil in bauxite occurrence areas in the parish of St. Elizabeth in Jamaica (Lalor et al., 1999); in gold deposit areas in the Dominican Republic (Nelson, 2000); and in ground water in the volcanic island of Nevis (USACE, 2004). There have been no reports of arsenic-associated health problems from any of these sites. However, a study of arsenic and other heavy metals in the gold mining district of Delita area of Isla de la Juventud in Cuba showed that there was leaching of arsenic into the well water in this area and that children were exposed to high levels of this metal (Toujague et al., 2003).

Fluorosis

There have been a few reports of dental fluorosis in Antigua (Vignarajah, 1993), Puerto Rico (Elias-Boneta et al., 2006), and Trinidad (Naidu et al., 2006). Further, a study of the fluoride content of different water sources in and around Port-au-Prince, Haiti, has shown that people, who obtained water from areas with sedimentary formations, were at increased risk of dental fluorosis though it was not shown, in that paper, that there was an increased prevalence of dental fluorosis in the susceptible population (Angerville et al., 2000).

Cuellar-Luna and Garcia-Melian (2003) found high natural fluoride concentrations in eastern Cuba where the water was obtained from wells located in Cretaceous volcanic arc rocks but this report did not assess any associated increased prevalence in dental fluorosis in these areas. In all territories, more severe forms of fluorosis were not reported.

Elemental Deficiencies

Iodine deficiency was previously a significant public health problem in some parts of the Caribbean such as mountainous areas of Cuba (Rodríguez-Ojea et al., 1998). However, widespread use of iodized salt has ameliorated this situation.

Selenium deficiency has been linked with cases of epidemic neuropathy in parts of Cuba (Beck et al., 2003) and with cardiomegaly and hepatomegaly in malnourished children from limestone areas of Jamaica (Bennett et al., 1983). An outbreak of coxsackievirus infection in parts of Cuba led to the development of neuropathy in patients who smoked and had various nutritional deficiencies. Affected persons were noted to have significantly lower levels of selenium compared to unaffected persons. It was postulated that selenium deficiency led to a state of oxidative stress in the host, which in turn, facilitated more virulent infection (Beck et al., 2003). Malnourished children with selenium deficiency were more likely to have enlarged hearts on chest radiographs, large livers, skin ulceration, and edema. There was also a tendency for selenium-deficient patients to gain weight more slowly during recovery (Bennett et al., 1983).

Long-Distance Natural Dust

Every year an estimated 70×10^6 tons of dust from the Saharan desert reaches the Caribbean Islands (Kaufman et al., 2005). These clouds of "Saharan dust" impart a characteristic reddish hue to prevailing cloud cover because of its relatively high concentration of iron, primarily in the ferric form (Zhu et al., 1997). Chemical contaminants (Garrison et al., 2006), fungi, and microbes (Prospero et al., 2005) have also been identified in these clouds.

The dust clouds that reach the Caribbean originate from two major sources: the Bodele Depression in Chad and an area extending from eastern Mauritania, western Mali, and Algeria (Goudie and Middleton, 2001). Movement of the dust clouds westward leads to the deposition of heavier particles over the west coast of Africa and the Atlantic Ocean. The clouds reaching the Caribbean region are enriched in particles of 2.5 μm or less aerodynamic diameter. Particles of this size can enter small airways in the lungs and cause respiratory disease. Particles of <1 μm diameter, which are also present in the dust clouds, have been implicated in acute cardiovascular disorders. The possibility that Saharan dust clouds could cause or exacerbate disease in Caribbean populations has led to ongoing research in several territories.

Gyan and colleagues (Gyan et al., 2005) conducted a retrospective assessment of Saharan dust cover and pediatric asthma admissions to a single hospital in Trinidad. They found a weak positive association between dust cover and asthma admissions. In contrast, Prospero et al. (2008) did not find any association with emergency room asthma admissions in Barbados and short-term surges in dust levels when they examined pediatric emergency room asthma admissions to the major public hospital in Barbados in relation to Saharan dust cover from 1996 to 1997. To date, as far as the authors are aware, there have been no published reports of acute cardiovascular disease in relation to Saharan dust cover in the region.

Radon

The potential hazards posed by exposure to alpha radiation from air radon have been of concern worldwide, especially associated with increased lung cancer risk.

A tiny bit of radon is incorporated into our body in each breath of air, due essentially to the geology of our residence region. It is also found in groundwater, but soil-borne radon provides the main source of exposure for the general population. Radon becomes a problem when it collects in our homes at high concentrations (Segovia et al., 2007). Domestic radon has been identified as the most important environmental risk factor for lung cancer. UNSCEAR (2000) estimation indicates that radon and its radioactive daughters contribute 75% to the effective dose due to natural terrestrial sources.

In Central America, the Caribbean, and Mexico, several radon projects were performed during the last quarter of the 20th century. Projects related to geology and geophysics such as the study of soil and groundwater radon variations in seismic zones and around active volcanoes were especially welcomed in this part of the world where long chains of active volcanoes are present and where very large earthquakes occur systematically, due to the subduction of the Pacific Plates under the continent.

The methodologies to evaluate radon concentration levels in air and water in the region consist mainly in the use of passive detectors (solid-state nuclear track detectors, electrets, and activated charcoal) (Espinosa and Gammage, 2003; Canoba et al., 2002), active detectors (Armienta et al., 2002), and the liquid scintillation method for determination of radon in water samples (Segovia et al., 2003).

The interest in the effective radiation dose received by the general population due to inhalation of radon and its radioactive daughters, that adhere to suspended atmospheric particulate matter, was also demonstrated in the evaluation of radon in closed atmospheres such as dwellings, caves, or spa buildings, together with the amount of ^{226}Ra, ^{222}Rn, and daughters in groundwater used mainly for drinking water supply in the whole region. Recent studies to relate atmospheric pollutants and lung cancer cases have already started.

Radon in Mexico

In Mexico radon in the soil and groundwater studies started in the 1980s around active volcanoes such as Popocatepetl and Colima (Segovia et al., 2005), and along the seismic zone of the Pacific Coast (Segovia et al., 1999).

Many radon-in-soil surveys have been conducted to obtain an estimate of the radon concentration levels in Mexico. Those values obtained from stations deployed at more than 150 sites were presented in a map covering approximately one-third of the territory of Mexico. Radon-in-soil levels were relatively low; the mean value was approximately 10 kBq/m^3. The highest values appear in the north of the country where a radioactive-rich mineralization zone exists, with maximum values up to 500 kBq/m^3 related to a uranium-rich zone in the state of Sonora. In the southern part of the country, moderate radon levels of 18 kBq/m^3 were associated with intrusive rocks and with enhanced hydrothermal activity. In Central Mexico and Mexico City maximum values of 8 and 15 kBq/m^3, respectively, were reported. The lowest values within the studied sites were found in active volcanoes (Colima, El Chichon, and Popocatepetl volcanoes).

Soil radon determination has also been systematically performed along the Pacific Coast. During 1997, the seismicity was extremely intense along the coastal zone of the state of Guerrero and some radon peaks (between 20 and 30 kBq/m^3) were observed at Acapulco. The variations observed corresponded with a period of particularly intense meteorological perturbations produced as a consequence of *El Niño,* culminating with the formation of a tropical hurricane that struck the Guerrero coast (Segovia et al., 1999).

The first measurements of radon in groundwater were made in some samples from alluvial and volcanic aquifers. The water samples were taken from wells and springs in San Luis Potosí, Michoacán, and Mexico states. Radon content varied depending on the sample source, reaching a maximum value of 11.3 kBq/m^3 at Toluca City. However, the highest radon concentrations values (34.2 and 27.3 kBq/m^3) in groundwater and drinking water supply system, respectively, were measured in Aldama, Chihuahua, located in the northern part of the country (Colmenero et al., 2004).

Outdoor radon concentrations in different geological zones were quite comparable; average values were between 13 and 23 Bq/m^3. The small fluctuations observed can be attributed to air convection during the sampling time.

Indoor radon concentrations were measured in family houses located at several towns within the Mexican Volcanic Belt, and in other family houses from towns

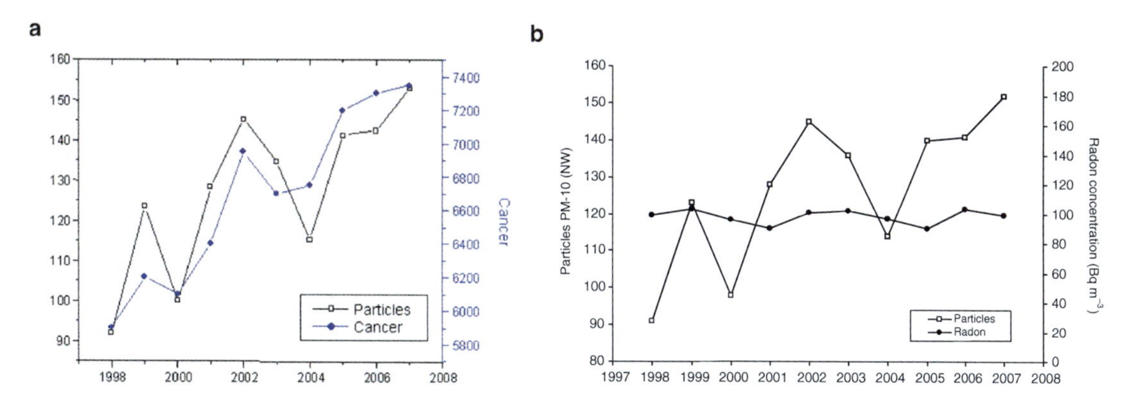

Fig. 4 Lung cancer case trends along 10 years vs. (**a**) SPM (PM-10) and (**b**) average indoor radon

in the north part of the country. The Mexican data median value for dwellings (72 Bq/m^3) is higher by a factor of 1.7 than the worldwide radon concentration median value (46 Bq/m^3) reported by UNSCEAR (2000). The average effective dose derived by use of the equation

$$H_E = C_{\mathrm{Rn}}\mathrm{Eq}T\mathrm{CF}_{\mathrm{Rn}}10^{-6},$$

where H_{E} is the annual effective dose by inhalation of radon (mSv/y), C_{Rn} is the radon concentration in air (Bq/m^3), Eq is the equilibrium factor (0.4 and 0.6 for indoor and outdoor environments, respectively), T is the occupancy factor (7000 and 1760 h/y for the indoor and outdoor environments to the general public, respectively), $\mathrm{CF}_{\mathrm{Rn}}$ is a dose conversion factor for radon (9 and 6 nSv/Bq/h m^3 in terms of the equilibrium equivalent radon concentration and equilibrium factors of 0.4 and 0.6 for indoor and outdoor environments, respectively), results in the calculation that the average effective dose (2.1 ± 1.1 mSv/y) was twice the world average effective dose (1.0 mSv/y) for indoor environments (UNSCEAR, 2000). Chihuahua State in northern Mexico was the zone showing highest values (Colmenero et al., 2004).

In Mexicali City near the Mexican–USA border, indoor radon concentration was measured in 95 houses. Reyna-Carranza and López-Badilla (2002) determined that the number of deaths was higher in neighborhoods without pavement with respect to those paved; Rn levels were also higher at homes registering a death. More female than male deaths were ascribed to their longer permanence periods at home.

In Mexico City, the calculated average indoor radon concentrations and range taking into consideration all the values reported were 73.8 and 4300 Bq/m^3, respectively. In certain areas of Mexico City, many sources of suspended particulate matter (SPM), from motor vehicles, factories, or deforested soil, among others, induce high concentration of particles that remain suspended in the atmosphere practically all year long (Espinosa et al., 2008). Radon daughters can adhere easily to the SPM. These often-radioactive carriers are dangerous due to their high mobility to enter through the nasal fosses into the respiratory system and reaching the alveolus. Lung cancer is the first cause of death by malignancies in Mexico; in recent decades an increasing frequency of lung cancer has been reported. The reasons for this increase have not been adequately elucidated. Data on mortality of the last 10 years, data routinely collected, and tendencies published from the Health Ministry (SSA, 2003) indicate that yearly PM-10 average concentrations at the center-north part of Mexico City, between 1998 and 2007, follow the same tendency as lung cancer cases reported in the same area of the city (Fig. 4). Both curves show the same shape and an increase with time. On the other hand, the average indoor radon behavior at the same region shows stability with time (Espinosa et al., 2009). However, the role of radon as compared with SPM was not conclusive in this case.

Radon in Central America

The active volcanic range along Central America is characterized by the presence of settlements where thousands of people live around volcanoes. Studies of gaseous emissions from the craters, fumaroles, and ground fractures are rarely applied to determine the effects of the emissions on people or the environment. However, results of the degassing monitoring can be analyzed in the future to estimate the impact on health of the local population (Baxter, 2005).

Costa Rica

In Costa Rica, radon-in-soil studies started in 1981 related to volcanic eruptions. Several volcanoes such as Poas and Arenal have been continuously monitored for soil radon (Barquero et al., 2005). The determination of the effective dose received by construction workers of a tunnel due to radon concentration levels of up to 1.5 kBq/m^3 in the underground air showed that the maximum effective dose was 13.5 mSv/y, for people working in the tunnel 8 h a day (Garcia Vindas and Mora, 2004). Recently, regional cooperation to promote nuclear science and technology (ARCAL-IAEA, 2005) initiated proposals for regional projects. Among them, a project to evaluate radon as a possible precursor of seismic events and volcanic eruptions together with the role of radon in environmental pollution was proposed by Costa Rica for the 2007–2008 ARCAL schedule. Several Latin American countries (Ecuador, Chile, Colombia, Argentina, and Guatemala) were interested in the Costa Rica proposal.

Nicaragua

Radon studies related to fault location were reported since 1982 in active volcanoes from Nicaragua (Crenshaw et al., 1982, Connor et al., 1996). These studies continue within cooperation programs with universities from abroad.

El Salvador

The same interest for volcanic radon gas related to the volcanic monitoring for risk assessment occurs in El Salvador, where studies on volcanoes such as San Miguel have included radon and other volcanic gases (Cartagena et al., 2004). Radon was also measured in several studies developed at Ilopango Lake. Soil Rn concentrations were higher than 100 pCi/L in several locations of the caldera. The strongest anomalies of Rn and carbon dioxide were observed in a region located between the south and southwest of Ilopango Lake, close to a region of frequent seismicity (López et al., 2009).

The Caribbean

Cuba

In Cuba, projects in association with international organizations have shown interesting results performed at Elguea Spa, in the north-central part of the island, the most important natural thermal resource of radioactive groundwater. The hot spring is associated with a system of geological faults where anomalous values of uranium, thorium, and potassium have been found. The gamma anomaly is the most intense in Cuba and is associated with Cretaceous limestones. ^{226}Ra and ^{222}Rn concentration values in the spring water were up to 32.5 and 220.5 kBq/m^3, respectively. During a 10-year period dosimetry studies of the effective dose control of the spa workers were performed. Indoor radon concentrations were up to 52 kBq/m^3(Manchado et al., 2005). As reported by the authors, the systematic medical observation of professional workers exposed at this high-dose radiation showed very low indicators of negative health effects, concluding that an evidence of hormesis effect for ^{222}Rn occurred. Studies of radioactive isotopes in sediments from La Habana Bay also revealed no effects related to health on the population (Gela et al., 2001).

Final Remarks

Studies of As and fluoride including concentration levels, distribution, and their impact on health have been performed in various zones within Mexico, Central America, and the Caribbean. However, most of them were conducted by isolated research groups from universities or the government. International organizations have also been involved or have given support for specific projects. One of those joint efforts was developed during the years 2003–2005 in Nicaragua, when UNICEF supported ENACAL-GAR (Empresa Nicaragüense de Acueductos y Alcantarillados-Gerencia de Acueductos Rurales) through the project "Determinación de la Calidad del Agua de Consumo Humano, con énfasis en Arsénico" (Determination of drinking water quality with emphasis on arsenic), to determine arsenic contents in drinking water sources at the departments of Boaco, Chontales, Estelí, Jinotega, and Matagalpa (Altamirano-Espinoza and Bundschuh, 2009). On the other hand, the large area (about 2,000,000 km^2) of Mexico is one of the causes of scarce cooperation among research groups within the country. Nevertheless, more joint efforts have been started recently. In 2006, the creation of the Iberoarsen network (supported by the Spanish Agency CYTED) promoted a networking enhancement on arsenic studies within Latin America and the Iberian countries. This group focused particularly on information exchange, joint projects development, and formation of experts on the themes of distribution, analytical determination, and remediation. On the other hand, As sources have been revealed only in certain regions, and people are still consuming As- and/or fluoride-tainted water at many sites. There is not a consensus on the number of people exposed to As in the area, but about 4 million people exposed have been estimated (Bundschuh et al., 2008). However, the problem is exacerbated in rural areas with a lack of access to unpolluted water.

More research studies are needed in the whole area to identify arsenic sources, environmental fate, and risks to human, animals, and vegetation. There is also a need to develop innovative and affordable methods for the removal of arsenic and fluoride from water. At the same time, water authorities should be willing to install those new methodologies on a routine basis. The influence of seasonal changes on As and fluoride on their concentrations and fate in waters should also be investigated.

Mexico was the seventh country in the world to adopt salt fluoridation as a "mass measure" adding 250 mg of potassium fluoride to the salt for human consumption since 1994 (Pazos, 1991). Toothpastes contain about 1,000 ppm F, although in regions with high contents of F, there are no Mexican toothpastes without fluorine available. Since 1997 the US Food and Drug Administration, USFAD, put a warning label on toothpastes and restricts the maximum F content per package to 276 mg (USFDA, 2003). In Central American countries there are no similar restrictions.

The overview of aspects of medical geological disorders in the Caribbean suggests that there is a paucity of published reports on a range of geologically related medical disorders in the Caribbean region. However, it must be noted that only English medical databases were searched and that relevant articles may exist in the Spanish, French, and Dutch medical literature, since these are also languages of the Caribbean region. Nonetheless, this region offers opportunities for unique and important research on a diverse range of topics such as the effects of long-range dust on humans, animals, and marine life; and the impact of rapid industrial development in small island states.

The review presented here, although being non-exhaustive, shows the urgency of establishing the determination of arsenic and fluoride routinely in all potable water sources of the area, and to adopt measures immediately to decrease their concentrations to safe levels in all drinking water sources.

References

Agostini S, Corti G, Doglioni C, Carminati E, Innocenti F, Tonarini S, Manetti P, Di Vincenzo G, Montanari D (2006) Tectonic and magmatic evolution of the active volcanic front in El Salvador: Insight into the Berlín and Ahuachapán geothermal areas. Geothermics 35:368–408.

Alarcón-Herrera MT, Martín-Domínguez IR, Trejo-Vázquez R, Rodríguez-Dosal S (2001a) Well water fluoride, dental fluorosis, bone fractures in the Guadiana Valley of Mexico. Fluoride 34:139–149.

Alarcón-Herrera MT, Flores-Montenegro I, Romero-Navar P et al. (2001b) Contenido de arsénico en el agua potable del valle del Guadiana, Mexico. Ing Hidraul Mex XVI: 63–70.

Altamirano M (2005) Distribution of arsenic pollution in the groundwater of the South West basin of Sebaco valley-Matagalpa, Nicaragua. Proceedings Hydrogeology and Water Resources Management: Working together for the Future. March 2005, Managua.

Altamirano-Espinoza M, Bundschuh J (2009) Natural arsenic groundwater contamination of the sedimentary aquifers of southwestern Sébaco valley, Nicaragua. In J Bundschuh, MA Armienta, P Birkle, P Bhattacharya, J Matschullat, AB Mukherjee (eds) Natural arsenic in groundwaters of Latin America – Occurrence, health impact and remediation. London: Taylor and Francis Group.

A. M. (2008) Consumen dos colonias agua con flúor y arsénico. A. M., Regional newspaper. Feb 23, 2008, Guanajuato.

Angerville R, Emmanuel E, Nelson J, Saint-Hilaire P (2000) Assessment of the fluoride concentration in the water resources in the hydrographic region of central south Haiti. Case Studies on Water and Envrionment Issues in Haiti.UNESCO Regional Office for Science & Technology for Latin America and the Caribbean (Uruguay) Document Code, UY/2000/SC/PHI/PI/1.

Armienta MA, Rodríguez R (1996) Arsénico en el Valle de Zimapán, Mexico: Problemática Ambiental. Revista MAPFRE Segur 63:33–43.

Armienta MA, Rodríguez R, Aguayo A, Ceniceros N, Villaseñor G, Cruz O (1997a) Arsenic contamination of groundwater at Zimapán, Mexico. Hydrogeol J 5:39–46.

Armienta MA, Rodríguez R, Cruz O (1997b) Arsenic content in hair of people exposed to natural arsenic polluted groundwater at Zimapán, Mexico. Bull Environ Contam Toxicol 59:583–589.

Armienta MA, Villaseñor G, Rodríguez R et al. (2001) The role of arsenic bearing rocks in groundwater pollution at Zimapán Valley, Mexico. Environ Geol 40:571–581

Armienta MA, Varley N, Ramos E (2002) Radon and hydrogeochemical monitoring at Popocatepetl volcano, Mexico. Geofis Int 41:271–276.

Armienta MA, Amat PD, Larios T, López DL (2008) América Central y Mexico. In J Bundschuh, M Litter, A Pérez-Carrera (eds) IBEROARSEN, Distribución del arsénico en las regiones Ibérica e Iberoamericana. CYTED, Buenos Aires.

Armienta MA, Segovia, N (2008) Arsenic and fluoride in the groundwater of Mexico. Environ Geochem Health 30: 345–353.

Astudillo LR, Yen IC, Bekele I (2005) Heavy metals in sediments, mussels and oysters from Trinidad and Venezuela. Rev Biol Trop 53:41–53.

Barragne P (2002) Evaluación rápida de la contaminación por arsénico y metales pesados de las aguas subterráneas de Nicaragua. PIDMA-UNICEF Managua.

Barragne P (2004) Contribución al estudio de cinco zonas contaminadas naturalmente por arsénico en Nicaragua. Managua, Nicaragua: UNICEF.

Barquero J, Fernandez E, Monnin M et al. (2005) Water chemistry and radon survey at the Poas volcano (Costa Rica). An Geophys 48:33–42.

Baxter P (2005) Human impacts of volcanoes. In J Marti, GGJ Ernst (eds) Volcanoes and the Environment. Cambridge: Cambridge University Press.

Beck MA, Levander OA, Handy J (2003) Selenium deficiency and viral infection. J Nutr 138:1643S–1467S.

Bennett FI, Golden MH, Golden BE (1983) Red cell gluthathione peroxidase concentration in Jamaican children with malnutrition. Proceedings 28th Scientific meeting of the Commonwealth Caribbean Medical Research Council, April 20–23 Kingston.

Birkle P, Bundschuh J (2009) The abundance of natural arsenic in deep thermal fluids of geothermal and petroleum reservoirs in Mexico. In J Bundschuh, MA Armienta, P Birkle, P Bhattacharya, J Matschullat, AB Mukherjee (eds) Natural Arsenic in Groundwaters of Latin America – Occurrence, Health Impact and Remediation. London: Taylor and Francis Group.

Bundschuh J, Perez-Carrera A, Litter M (2008) Introducción: Distribución del arsénico en las regions Ibérica e Iberoamericana. In Bundschuh J, Litter M, Pérez-Carrera A (eds) IBEROARSEN, Distribución del arsénico en las regiones Ibérica e Iberoamericana. Buenos Aires: CYTED.

Calderón J, Machado B, Navarro M et al. (2000) Influence of fluoride exposure on reaction time and visuospatial organization in children. Epidemiol 11 (4) Supplement S153.

Calderón-Fernández ML (2006) Alternativas de depuración de agua para consumo humano en el estado de Chihuahua. Junta Central de Agua y Saneamiento de Chihuahua. Documentos Foro Mundial del Agua. Mexico, DF.

Canoba A, Lopez FO, Arnaud MI et al. (2002) Indoor radon measurements in six Latin American countries. Geofis Int 41:453–457.

Cardona BA, Carrillo RJJ, Armienta HMA (1993) Elemento traza: Contaminación y valores de fondo en aguas subterráneas de San Luis Potosí, SLP, Mexico. Geofis Int 32:277–286.

Carrillo-Rivera JJ, Cardona A, Moss D (1996) Importance of the vertical component of groundwater flow: a hydrogeochemical approach in the valley of San Luis Potosí, Mexico. J Hydrol 23:23–44.

Carrillo-Rivera JJ, Cardona A, Edmundo WM (2002) Use of abstraction regime and knowledge of hydrogeological conditions to control high-fluoride concentration in abstracted groundwater: San Luis Potosí Basin, Mexico. J Hydrol 261:24–47.

Cartagena R, Olmos R, Lopez D et al. (2004) Diffuse soil degassing of carbon dioxide, radon and mercury at San Miguel volcano, El Salvador. In W Rose, J Bommer, D Lopez, M Carr, J Major (eds) Natural Hazards in El Salvador. Boulder: Geological Survey of America.

Castro-Larragoitia J, Kramar U, Puchelt H (1997) 200 years of mining activities at La Paz/San Luis Potosi/Mexico – Consequences for environment and geochemical exploration. J Geochem Explor 58:81–91.

Cebrián ME, Albores A, García-Vargas G, Del Razo LM (1994) Chronic arsenic poisoning in humans: The case of Mexico. In JO Nriagu (ed) Arsenic in the Environment, Part II. New York: John Wiley & Sons Inc.

CEP website (2009) Caribbean Environmental Programme website, http://www.cep.unep.org/publications-and-resources/marine-and-coastal-issues-links/heavy-metals. Accesed 20 March 2009.

CODEREG (2000) Acuífero de la Independencia, municipios de San José Iturbide, Doctor Mora, San Luis de la Paz, Dolores Hidalgo, San Felipe, San Diego de la Unión y San Miguel de Allende, Guanajuato. Technical Report inedit. Consejo para el Desarrollo Regional Noreste y Norte, Guanajuato.

Colmenero SL, Montero CME, Villalba L et al. (2004) Uranium-238 and thorium-232 series concentrations in soil, radon-222 indoor and drinking water concentrations and dose assessment in the city of Aldama, Chihuahua, Mexico. J Environ Radioact 77:205–219.

Connor CB, Hill BE, La Femina PC et al. (1996) Soil radon pulse during the 1995 phreatic eruption of Cerro Negro, Nicaragua. J Volcanol Geotherm Res 73:119–127.

Coronado-González JA, Del Razo LM, García-Vargas G et al. (2007) Inorganic arsenic exposure and type 2 diabetes mellitus in Mexico. Environ Res 104:383–389.

Crenshaw WB, William SN, Stoiber RE (1982) Fault location by radon and mercury at an active volcano in Nicaragua. Nat 300:345–346.

Cuellar-Luna L, Garcia-Melian M (2003) Fluoride in drinking water in Cuba and its association with geological and geographical variables. Pan Am J Public Health 14: 341–349.

Del Razo LM, Arellano MA, Cebrián ME (1990) The oxidation states of arsenic in well-water from a chronic arsenicism area of northern Mexico. Environ Pollut 64:143–153.

Del Razo LM, Corona JC, García-Vargas G et al. (1993) Fluoride levels in well-water from a chronic arsenicism area of northern Mexico. Environ Pollut 80:91–94.

Del Razo LM, Garcia-Vargas GG, Vargas H et al. (1997) Altered profile of urinary arsenic metabolites in adults with chronic arsenicism. A pilot study. Arch Toxicol 71:211–217.

Elias-Boneta AR, Psoter W, Elias-Viera AE et al. (2006) Relationship between dental caries experience (DMFS) and dental fluorosis in 12-year-old Puerto Ricans. Community Dent Health 23:244–250.

ESF (2006) Well water quality in San Miguel de Allende. Phase I: Results and Conclusions. Ecosystem Sciences Foundation, ESF, San Miguel de Allende Municipality. Technical Report inedit, Guanuajuato.

Espinosa G, Gammage RB (2003) A representative survey of indoor radon in the sixteen regions in Mexico City. Radiat Prot Dosim 103:73–76.

Espinosa G, Golzarri JI, Bogard J et al.(2008) Indoor radon measurements in Mexico City. Radiat Meas 43: S431–S434.

Espinosa G, Golzarri JI, Ponciano RG et al. (2009) Population vulnerability due to the exposure to radon and airborne particulate matter (PM10), in Mexico City. Submitted to Radiat Meas.

Estrada F (2003) Estudio preliminar de la incidencia del arsénico en aguas subterráneas con relación al medio físico natural en la región noroeste y sureste de Nicaragua: Periodo 2001–2002. Universidad Nacional de Ingeniería UNI. Programa de Investigación y Docencia en Medio Ambiente, Managua.

Estupiñán-Day S, Vera H, Duchon K et al. (2005) Final Report "Task-Force Meeting" Defluoridation systems in Latin America and the Caribbean, Washington DC, 22–24 October, 2004. Comisión Nacional del Agua, Secretaría de Salud de Mexico, Pan American Health Organization, WHO, Washington DC.

Folleti C, Paz G (2001) Diagnóstico de fluorosis dental en 39 comunidades del valle de Sula, Honduras. En Superación sanitaria y ambiental: el reto. AIDIS, Tegucigalpa.

Fragoso R, Jackson G, Cuairan V, Gaitan L (1997) Efectividad del acido clorhídrico como blanqueador dental en piezas con fluorosis dental. Rev ADM LIV:219–222.

Gamiño SP, Monroy M (2009) Evaluation of children exposition and effect biomarkers in a mining site with high concentration of arsenic and lead bioaccesibility as a case study for abandoned sites associated to Pb-Zn-Cu-Ag skarn deposits in Central Mexico. Geological Society of America Abstracts with Programs, 16–17 March Dallas, TX 41(2):13, Dallas.

Garcia S, Ovalle J (1994) Grado de fluorosis dental en pacientes de la Univ. Del Bajío. Rev ADM 51:162–168.

Garcia Vindas JR, Mora P (2004) Radon concentration and dose assessment in a tunnel under construction in Costa Rica. Radioisot53:517–522.

Garrison VH, Foreman WT, Genualdi S et al. (2006) Saharan dust- a carrier of the persistent organic pollutants, metals and microbes to the Caribbean? Rev Biol Trop 54(supl.3): 9–21.

Gela A, Solo J, Diaz, Simon MJ et al. (2001) Radiological evaluation of sediments from the Havana Bay. VII Workshop on Nuclear Physics, La Habana.

Gómez A (2002) Monitoreo y atención de intoxicados con arsénico en El Zapote-San Isidro, Departamento de Matagalpa. Nicaragua. Reporte Técnico. Estudio realizado con apoyo financiero de UNICEF por MINSA, Managua.

Gómez-Arroyo S, Armienta MA, Cortés-Eslava J, Villalobos-Pietrini R (1997) Sister chromatid exchanges in Vicia faba induced by arsenic-contaminated drinking water from Zimapan, Hidalgo, Mexico. Mutat Res 394:1–7.

González-Hita L, Sánchez L, Mata I (1991) Estudio hidrogeoquímico e isotópico del acuífero granular de la Comarca Lagunera. Instituto Mexicano de Tecnología del Agua. Technical Report inedit, Jiutepec.

Gonsebatt ME, Vega L, Salazar AM et al. (1997) Cytogenetic effects in human exposure to arsenic. Mutat Res 386: 219–228.

González M (1998) Exposición al arsénico en comunidades rurales de San Isidro, Matagalpa. OPS/OMS, Managua.

Goudie AS, Middleton NJ (2001) Saharan dust storms: Nature and consequences. Earth Sci Rev 56:179–204.

Grimaldo M, Borja-Aburto VH, Ramírez AL et al. (1995) Endemic fluorosis in San Luis Potosí, Mexico. Environ Res 68:25–30.

Gutiérrez-Ojeda C (2009) Determining the origin of arsenic in the Lagunera region aquifer, Mexico using geochemical modeling. In J Bundschuh, MA Armienta, P Birkle, P Bhattacharya, J Matschullat, AB Mukherjee (eds) Natural arsenic in groundwaters of Latin America – Occurrence, health impact and remediation. London: Taylor and Francis group.

GVP (2009) Mexico, Volcanoes of Mexico and Central America, Global Volcanism Program, Smithsonian Institution, Washington DC, http://www.volcano.si.edu/world/region.cfm?rnum=1401, Accessed 16 April 2009.

Gyan K, Henry W, Lacaille S et al. (2005) African dust clouds are associated with increased paediatric asthma accident and emergency admissions on the Caribbean island of Trinidad. Int J Biometeorol 49:371–376.

Hernández-Montoya V, Bueno-López JI, Sánchez-Ruelas AM et al. (2003) Fluorosis y caries dental en niños de 9 a 11 años del estado de Aguascalientes, Mexico. Rev Int Contam Ambient 19:197–204.

Horst A (2006) Use of Stable and Radioactive Isotopes and Gaseous Tracers for Estimating Groundwater Recharge, Time of Residence, Mixing of the Different Types of Groundwater and Origin in the Silao Romita Aquifer, Guanajuato, Central Mexico. FOG, Freiberg Online Geol 17, Freiberg.

Howe A, Fung LH, Lalor G et al. (2005) Elemental composition of Jamaican foods 1: A survey of five food crop categories. Environ Geochem Health 27:19–30.

Hurtado-Jiménez R, Gardea-Torresdey J (2005) Estimación de la exposición a fluoruros en Los Altos de Jalisco, Mexico. Rev Panam Salud Publica 20:236–247.

Hurtado-Jiménez R, Gardea-Torresdey J (2009) Contamination of drinking water supply with geothermal arsenic. In J Bundschuh, MA Armienta, P Birkle, P Bhattacharya, J Matschullat, AB Mukherjee (eds) Natural arsenic in groundwaters of Latin America – Occurrence, health impact and remediation. London: Taylor and Francis Group.

Irigoyen ME, Molina N, Luengas I (1995) Prevalence and severity of dental fluorosis in a Mexican community with above optimal fluoride concentrations in drinking water. Community Dent Oral Epidemiol 23(4):243–245.

Kaufman YJ, Koren I, Remer LA et al. (2005) Dust transport and deposition observed from the Terra-Moderate Resolution Imaging Spectroradiometer (MODIS) spacecraft over the Atlantic Ocean. J Geophys Res 110:D10S12.

Kaul B, Mukerjee H (1999) Elevated blood lead and erythrocyte protopophyrin levels of children near a battery-recycling plant in Haina, Dominican Republic. Int J Occup Environ Health 5:307–312.

Koplan JP, Wells AV, Diggory HJ et al. (1997) Lead absorption in a community of potters in Barbados. Int J Epidemiol 6: 225–229.

Lalor GC, Rattray R, Simpson P, Vutchkov M (1999) Geochemistry of an arsenic anomaly in St. Elizabeth Jamaica. Environ Geochem Health 21:3–11.

Lalor GC, Vutchkov MK, Bryan ST et al. (2006) Acute lead poisoning associated with backyard lead smelting in Jamaica. West Indian Med J 55:394–398.

Lalor GC, Vutchkov MK, Bryan ST (2007) Blood lead levels of Jamaican children island-wide. Sci Total Environ 374: 235–241.

Leal-Ascencio MT, Gelover-Santiago S (2006) Evaluación de acuíferos de la mesa del norte, V Congreso Internacional y XI Congreso Nacional de Ciencias Ambientales, Memorias Oaxtepec, Morelos, 7–9 de junio, 2006, Oaxtepec.

Levy DB, Schramke JA, Esposito KJ et al. (1998) The shallow ground water chemistry of arsenic, fluorine, and major elements: Eastern Owens Lake, California. Appl Geochem 14:53–65.

López DL, Ransom L, Monterrosa J et al. (2009) Volcanic arsenic and boron pollution of Ilopango lake, El Salvador. In J Bundschuh, MA Armienta, P Birkle, P Bhattacharya, J Matschullat, AB Mukherjee (eds) Natural arsenic in groundwaters of Latin America – Occurrence, health impact and remediation. London: Taylor and Francis Group.

Loyola-Rodríguez JP, Pozos-Guillén, CD, Hernández-Guerrero JC et al. (2000) Fluorosis en dentición temporal en un área con hidrofluorosis endémica. Salud Publ Mex 42: 194–200.

Manchado A, Cervantes P, Lantigua L (2005) Evidencias sobre la hormesis por 222Rn en el balneario Elguea, Cuba. Proceedings of the First Convention on Earth Sciences, La Habana.

Mahlknecht J, Steinich B, Navarro I (2004) Groundwater chemistry and mass transfers in the Independence aquifer, Central Mexico, by using multivariate statics and mass-balance models. Environ Geol 45:781–795.

Mahlknecht J, Horst A, Hernández-Limón G, Aravena R (2008) Groundwater geochemistry of the Chihuahua City region in the Rio Conchos Basin (Northern Mexico) and implications for water resources Management. Hydrol Proces 22: 4736–4751.

Martínez P, García M (2007) Distribución de iones mayores y metales en el agua subterránea de la subcuenca del Río Turbio, estados de Guanajuato y Jalisco. Rev Geocienc 1:37–54.

Mattheß G (1994) Groundwater Properties. Hydrogeologie, Band 2. Berlin, Stuttgart: Gebriider Borntraeger.

Megaw PK, Ruiz JR, Titley SR (1988) High temperature, carbonate.hosted Ag-Pb-Zn(Cu) deposits of northern Mexico. Econ Geol 83:131–140.

Mejía JA, Rodriguez R, Armienta A et al. (2007) Aquifer vulnerability zoning, an indicator of atmospheric pollutants input? Vanadium in the Salamanca aquifer, Mexico. Water, Air, Soil Pollut 185:95–100.

Mérida R, Reyes A, Hernández I (1998) Texto Guía Carta Magnética Guanajuato. Consejo Recursos Minerales, SECOFI, Mexico, Mexico DF.

Meza MM, Kopplin M, Burges JL, Gandolfi J (2004) Arsenic drinking water exposure and urinary excretion among adults in the Yaqui Valley, Sonora, Mexico. Environ Res 96: 119–126.

Molina MA (2004) Estudio hidrogeoquímico en la Comarca Lagunera, Mexico. M.Sc. Thesis, Posgrado en Ciencias de la Tierra. Universidad Nacional Autónoma de Mexico, Mexico DF.

Morales L, Puigdomènech C, Puntí A et al. (2009) Arsenic and water quality of rural community wells in San Juan de Limay, Nicaragua. In J Bundschuh, MA Armienta, P Birkle, P Bhattacharya, J Matschullat, AB Mukherjee (eds) Natural arsenic in groundwaters of Latin America – Occurrence, health impact and remediation. London: Taylor and Francis group.

Naidu R, Prevatt I, Simeon D (2006) The oral health and treatment needs of schoolchildren in Trinidad and Tobago: Findings of a national survey. Int J Paediatr Dent 16: 412–418.

Nelson C (2000) Volcanic domes and gold mineralization in the Pueblo Viejo district Dominican Republic. Miner Depos 35:511–525.

Ortega-Guerrero A (2003) Origin and geochemical evolution of groundwater in a closed-basin clayey aquitard, Northern Mexico. J Hydrol 284:26–44.

Ortiz D, Castro L, Turrubiartes F et al. (1998) Assessment of the exposure to fluoride from drinking water in Durango, Mexico, using a geographic information system. Fluoride 31:183–187.

Ovalle J (1996) Fluorosis dental de la población escolar de Salamanca Guanajuato. Rev ADM 53:289–294.

Pazos L (1991) Salt Fluoridation in Mexico. National Coordination of the Salt Fluoridation Program in Mexico. On May 27, 1991. Mexico City.

Peña M (2005) Reto del milenio en los asentamientos precarios de America Latina y el Caribe. Invited Conference. Regional Symposium on Healthy Houses, San Jose.

Planer-Friedrich B, Armienta MA, Merkel BJ (2001) Origin of arsenic in the groundwater of the Rioverde Basin, Mexico. Environ Geol 40:1290–1298.

Prospero JM, Blades E, Mathison G, Naidu, R (2005) Interhemispheric transport of viable fungi and bacteria from Africa to the Caribbean with soil dust. Aerobiol 21: 1–19.

Prospero JM, Blades E, Naidu et al. (2008) Relationship between African dust carried in the Atlantic trade winds and surges in paediatric asthma attendances in the Caribbean. Int J Biometeorol 52:823–832.

Rajkumar WS, Manohar J, Doon R et al. (2006) Blood lead levels in primary school children in Trinidad and Tobago. Sci Total Environ 361:81–87.

Ravishankar S (1987) Status of Geothermal Exploration in Maharashtra and Madhya Pradesh (C.R.). GSI Rec 115(6): 7–29.

Razo I, Carrizales L, Díaz-Barriga F, Monroy M (2004) Arsenic and heavy metal pollution of soil, water and sediments in a semi-arid climate mining area in Mexico. Water, Air, Soil Pollut 152:129–152.

Reséndiz MRI, Zúñiga LJC (2003) Evaluación de la Exposición al arsénico en pobladores del municipio de Zimapán, Hidalgo, Bachelors' Thesis Eng. Chem., Universidad Tecnológica de Mexico, Mexico, DF.

Reyes-Cortés IA, Reyes-Cortés M, Villalba L et al. (2006a) Origen del As en las cuencas endorreicas, Chihuahua, Mexico. Geos 26:39.

Reyes-Cortés IA, Vázquez-Balderas JF, Ledesma-Ruiz R (2006b) As en el sistema hidrogeológico del valle de Delicias, Chihuahua, Mexico. Geos 26:40.

Reyna-Carranza MA, López-Badilla G (2002) Estudio del efecto del radón en los casos de muerte por cáncer pulmonar en la población de Mexicali, Baja California, Mexico. Rev Mex Ing Biomed XXIII:68–73.

Rocha-Amador D, Navarro ME, Carrizales L, Morales R, Calderón J (2007) Decreased intelligence in children and exposure to fluoride and arsenic in drinking water. Cad Saúde Pública Rio de Janeiro 23suppl. 4:5579–5587.

Rodrigu A, Ren M, Goodell P (2007) Potential bedrock source of groundwater arsenic anomaly in northeastern Chihuahua City, Chihuahua, Mexico. American Geophysical Union, Fall Meeting 2007, San Francisco Ca, 10–14 Dec, 2007, abstract #H11E-0824.

Rodriguez-Ojea A, Menendez R, Terry B et al. (1998) Low levels of urinary iodine excretion in schoolchildren of rural areas in Cuba. Eur J Clin Nutr 52(5):372–375.

Rodríguez R, Hernández G, González T, Cortes A (1997) Definición del flujo regional de agua subterránea, su potencialidad y uso en la zona de la Cd. De Aguascalientes, Ags. Technical Report inedit, IGF-UNAM, Mexico, DF.

Rodríguez R, Mejía JA, Berlín J, Armienta A, González T (2000) Estudio para la determinación del grado de alteración de la calidad del agua subterránea por compuestos orgánicos en Salamanca, Gto.. Technical Report inedit, CEASG, IGF-UNAM, Mexico, DF.

Rodríguez R, Ramos JA, Armienta MA (2004) Groundwater arsenic variations: The role of local geology and rainfall. Appl Geochem 19:245–250.

Rodríguez R, Armienta MA, Morales P, Silva T, Hernández H (2006) Evaluación de Vulnerabilidad Acuífera del valle de Irapuato Gto. Technical Report inedit, JAPAMI, CONCyTEG, IGF UNAM. Mexico, DF.

Romieu I, Lacasana M, McConnell R (1997) Lead exposure in Latin America and the Caribbean. Lead research group of the Pan-American health organization. Environ Health Perspect 105:398–405.

Rosas I, Belmont R, Armienta A, Baez A (1999) Arsenic concentrations in water, soil, milk and forage in Comarca Lagunera, Mexico. Water, Air, Soil Pollut 112:133–149.

Ruiz J, Kesler S, Jones L (1985) Strontium isotope geochemistry of fluorite mineralization associated with fluorine-rich igneous rocks from The Sierra Occidental, Mexico; possible exploration significance. Econo Geol 80:33–42.

Salas-Pereira MT, Beltrán-Aguilar ED, Chavarría P et al. (2008) Enamel fluorosis in 12- and 15-year-old school children in Costa Rica. Results of a national survey, 1999. Commun Dent Health 25:178–184.

Sanchez-Garcia S, Pontigo-Loyola A, Heredia-Ponce E, Ugalde-Arellano J (2004) Fluorosis dental en adolescentes de tres comunidades del estado de Querétaro. Rev Mex Pediatr 17:5–9.

Sanchez-Nazario EE, Mansilla-Rivera I, Derieux-Cortes JC et al. (2003) The association of lead-contaminated house dust and blood lead levels of children living on a former landfill in Puerto Rico. P R Health Sci 22:153–159.

Segovia N, Armienta MA, Valdes C et al. (2003) Volcanic monitoring for radon and chemical species in the soil and in spring water samples. Radiat Meas 36:379–383.

Segovia N, Gaso MI, Armienta MA (2007) Environmental radon studies in Mexico. Environ Geochem Health 29: 143–153.

Segovia N, Mena M, Peña P, et al. (1999) Soil radon time series: Surveys in seismic and volcanic areas. Radiat Meas 31: 307–312.

Segovia N, Peña P, Valdes C, et al. (2005) Radon, water chemistry and pollution check by volatile organic compounds in springs around Popocatepetl volcano, Mexico. Ann Geophys 48:85–91.

Simons SF, Mapes-Vazquez VE (1956) Geology and ore deposits of the Zimapán mining district, State of Hidalgo, Mexico. US Geological Survey Professional Paper 284, Washington, DC.

Singh SK, Mortera F (1991) Source time functions of large Mexican subduction earthquakes, morphology of the Benioff Zone, age of the plate, and their tectonic implications. J Geophys Res 96:21487–21502.

Soto-Rojas AE, Ureña-Cirett JL, Martínez-Mier EA (2004) A review of the prevalence of dental fluorosis in Mexico. Rev Panam Salud Publ 15:9–18.

SSA (2003) Mortalidad observada por cáncer pulmonar según sexo y entidad federativa. Reporte Interno. Dirección General de Epidemiología. Secretaría de Salud. Mexico. Mexico, DF.

SSP (2009) http://proteccioncivil.guanajuato.gob.mx/atlas/ sanitario/acambaro.php. Accessed 15 March 2009.

Strock C, Songer A, Fiori Ch (2008) Appropriate technologies for sustainable access to safe drinking water: A case study in Belize. Proceedings Construction in Developing Countries International Symposium, Trinidad & Tobago, WI, Trinidad and Tobago.

Toujague RT, Leonarte A, Reyes Verdecia BL, Miravet RM (2003) Arsénico y metales pesados en aguas del área Delita, Isla de la Juventud. Ciencias de la Tierra y el Espacio, Cuba 4:5–8.

Trejo-Vázquez R, Bonilla-Petriciolet A (2002) Cuantificación de arsénico en el agua subterránea de la ciudad de Aguascalientes, Mexico, y evaluación de riesgos entre la población. Rev Ing Hidraul Mex 17:79–88.

UNSCEAR (2000) United Nations Scientific Committee on the Effects of Atomic Radiation, Sources and effects of ionizing radiation. Annex A and B, United Nations, E.00.IX.3, New York.

USACE (2004) Water resources assessment of Dominica, Antigua, Barbuda, St. Kitts and Nevis. US Army Corps of Engineers Report.
USFDA (2003) Oral health care drug products for over the counter human use. 21 CFR Part 356. Department of Health and Human Services, Food and Drug Administration.
Valenzuela OL, Germolec DR, Borja-Aburto VH et al. (2007) Chronic arsenic exposure increases TGFalpha concentration in bladder urothelial cells of Mexican populations environmentally exposed to inorganic arsenic. Toxicol Appl Pharmacol 222:264–270.
Vallejo-Sanchez A, Perez-Olivares S, Casanova-Rosales A, Gutierrez Salazar M (1998) Prevalencia, severidad de fluorosis y caries dental en una población escolar de 6 a 12 años de edad en la Cd. De Campeche, 1997–98. Rev ADM 55:266–271.
Vignarajah S (1993) Dental caries experience and enamel opacities in children residing in urban and rural areas of Antigua with different levels of natural fluoride in drinking water. Commun Dent Health 10:159–166.
Wyatt, CJ, Fimbres C, Romo L et al. (1998a) Incidence of heavy metal contamination in water supplies in Northern Mexico. Environ Res A 76:114–119.
Wyatt CJ, Lopez Quiroga V, Olivas-Acosta RT, Méndez RO (1998b) Excretion of arsenic (As) in urine of children, 7–11 years, exposed to elevated levels of as in the city water supply in Hermosillo, Sonora, Mexico. Environ Res A 78: 19–24.
Yáñez L, García-Nieto E, Rojas E et al. (2003) DNA damage in blood cells from children exposed to arsenic and lead in a mining area. Environ Res 93:231–240.
Zhu XR, Prospero JM, Millero FJ (1997) Diel variability of soluble Fe (II) and soluble total Fe in North African dust in the trade winds at Barbados. J Geophys Res 102(D17): 21297–21305.

Medical Geology Studies in South America

Bernardino R. Figueiredo, Marta I. Litter, Cássio R. Silva, Nelly Mañay, Sandra C. Londono, Ana Maria Rojas, Cristina Garzón, Tommaso Tosiani, Gabriela M. Di Giulio, Eduardo M. De Capitani, José Ângelo S. A. Dos Anjos, Rômulo S. Angélica, Maria Celeste Morita, Mônica M.B. Paoliello, Fernanda G. Cunha, Alice M. Sakuma, and Otávio A. Licht

Abstract "Earth and Health" or medical geology has been promoted worldwide as one of the fundamental themes of the International Year of Planet Earth (2007–2009). This was in response to relevant achievements noted in this new field of applied science from the time of the IGCP 454 project which led to foundation of the International Medical Geology Association (IMGA) in 2004. In association with international movements, several academic, professional, and student groups in South America began to study medical geology which started with scientific meetings held in Chile, Brazil, and Uruguay in 2002 and 2003. In this chapter, an attempt is made to describe South American scientists' relevant contributions to various subjects such as arsenic, lead, mercury, and selenium as well as fluorine and environmental problems affecting different parts of the continent. Some societal issues arising from medical geology studies are also highlighted from the point of view of the international risk communication and risk governance debate and the pioneering ethnographic descriptions of geophagy in the Andean and Amazonian countries. Finally, some ongoing medical geology projects in South America are identified as inspiring initiatives that may encourage future educational and research activities in this science field.

Keywords South America · Brazil · Colombia · Uruguay · Argentina · Metals · Arsenic · Lead · Selenium · Mercury · Risk communication · Geophagy

B.R. Figueiredo (✉)
Institute of Geosciences, University of Campinas, Campinas, São Paulo, Brazil
e-mail: berna@ige.unicamp.br

Historical Background and Scope of Medical Geology Studies

Medical geology has been recognized worldwide as a promising scientific field in which geosciences, environmental medicine, and other disciplines contribute together in attempting to test relationships between natural geological factors and the health of humans and of other living beings. Not coincidently, medical geology was included as one of the ten main themes of the International Year of Planet Earth under the title "Earth and Health – for a safer environment." Under the coordination of IUGS and UNESCO, the International Year was conceived to show the general public how earth sciences are contributing to society's well-being by making the world safer and healthier. Many initiatives and an extraordinary number of scientific activities were undertaken in the current decade to disseminate concepts and methods adapted to this new area of applied science. In South America the first short course on medical geology was held in Santiago in 2002 followed by the first workshops dedicated to the subject in Brazil, Uruguay, and Argentina from 2003 to 2007. Since then, South American professionals have made a modest but significant scientific contribution to this field. As an example, the first book on medical geology (in Portuguese) was published in 2006, only a year after "Essentials of Medical Geology" (Selinus et al., 2005) appeared. It is well known that other textbooks and an impressive number of papers were published in all continents during this period, showing that medical geology is already a consolidated international science.

Taking this historical background into account, it is easy to see that everything has happened very quickly for medical geology all around the world.

O. Selinus et al. (eds.), *Medical Geology*, International Year of Planet Earth,
DOI 10.1007/978-90-481-3430-4_4, © Springer Science+Business Media B.V. 2010

South America is not an exception in this regard and, as in other parts of world, compiling medical geology accomplishments in the continent is not an easy task.

This chapter presents the first attempt to report recent and ongoing medical geology studies in South America and is far from being complete and should therefore be regarded as a partial inventory. Its aim is to motivate reflection among academics, professionals, and students on what is going on in some countries. The present monograph may stimulate updating efforts in the near future and hopefully it will inspire other researchers to further pursue the scientific lines described here. In South America there are around 400 million people interested in how medical geology can contribute to improve their standard of living without compromising the environment and save resources by pointing out the beneficial properties of rocks and minerals; by preventing adverse health effects for humans, animals, and plants; and by protecting those in need of more protection.

The presence of metals and other substances in South America is addressed and available information on human exposure and health effects is provided in this chapter. Additionally, some societal issues arising from medical geology studies are also highlighted from the point of view of the international debate on risk governance as well as from the pioneering ethnographic descriptions of geophagy in the Andean and Amazonian countries. Finally, some ongoing medical geology projects in South America are identified as inspiring initiatives that may encourage future educational and research activities in this science field.

In the following sections, the distribution of arsenic, lead, and mercury in South America is identified according to recently published reports. For a long time these substances have been considered the most dangerous toxic substances for humans (ATSDR – CERCLA, 2003). They are widely dispersed in the environment and may be as much a threat to human health as some agrochemicals and radioactive substances. These reported case studies were carried out in Argentina, Chile, Brazil, and Uruguay.

Arsenic in South America

Water contamination from arsenic is a worldwide problem when considering the number of regions where arsenic-contaminated surface water and groundwater are being consumed by the people. This problem is estimated to affect more than 100 million people worldwide.

Long-term ingestion of water with high concentrations of arsenic can lead to a disease known as chronic endemic regional hydro-arsenicism, highly prevalent in Asia and Latin America. Keratosis, hyperkeratosis, damage to the central nervous system and lever, loss of hair, incidence of different types of cancer such as skin cancer and cancer of internal organs (lung, kidney, and bladder) are epidemiological evidence of exposure to inorganic arsenic. There is not yet a medical treatment for hydro-arsenicism; hence, prevention and attenuation of human exposure to arsenic is the only way to solve the problem.

The presence of arsenic in water and soil comes from anthropogenic causes (mining, metal smelting and refining, use of pesticides, etc.) as well as natural processes. Arsenic can be mobilized from rocks and soils toward water bodies through biogeochemical processes, changes in pH, volcanism, and microbiological action (Mansilla and Cornejo, 2002).

The World Health Organization recommends the limit of 10 μg/L As for drinking water (WHO, 2004). Considering this limit, around 14 million people in Latin America are exposed to prolonged consumption of arsenic-contaminated water and threatened by serious health problems. Potentially dangerous situations in South America have been known for decades, especially in Argentina, Chile, and Peru. However, governmental intervention intended to solve this problem has not been effective and does not exist in many isolated regions where communities do not have access to pre-treated potable water.

The most critical areas of Latin America concerning arsenic contamination of surface water and groundwater are depicted in Fig. 1. Estimated population exposure for different countries in relation to the former limit of 50 μg/L As in drinking water is quoted, according to Bundschuh et al. (2008a) and Bundschuh et al. (2008b). In these countries arsenic is found mainly in groundwater as geogenic arsenic associated with Andean volcanism.

Considering a maximum limit of 50 μg/L As in drinking water, the exposed population in Mexico, Argentina, and Chile in relation to the total population is estimated to be 0.4, 3.0–5.1, and 3.1%, respectively. Only recently in Argentina a new regulation was introduced lowering the maximum arsenic content in drinking water to 10 μg/L (Argentina, 2007) in accordance to WHO regulations.

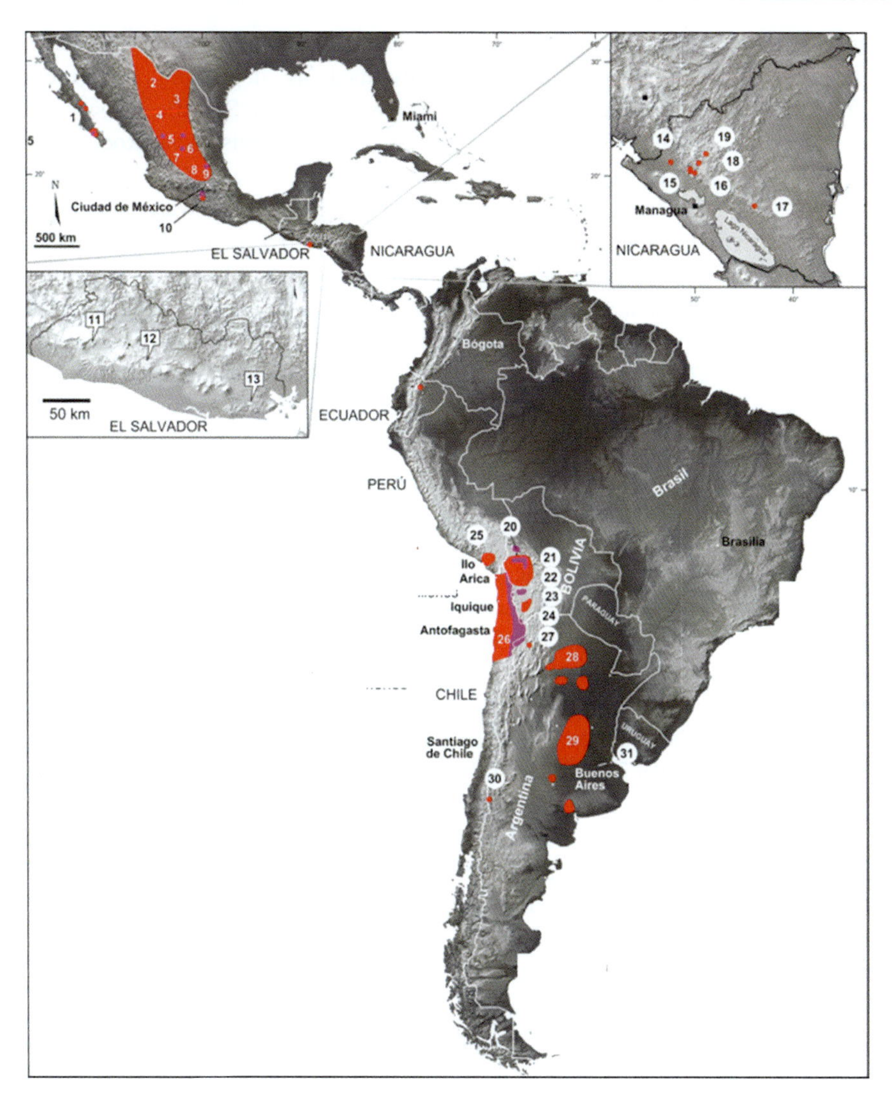

Fig. 1 Areas with high arsenic contents in surface water and groundwater in Latin America. 1 South Baja California, 2–9 Mexico, 10 Guerrero, 11–19 Central America, Mexico, 20–24 Bolivia, 25 Peru, 26 Atacama, Chile, 27–31 Argentina (Modified from Bundschuh et al., 2008a, b)

Argentina

In Argentina, the highest arsenic contents in groundwater are found in the Chacopampean plain, Puna, and Cuyo. However, the Patagonia region still needs to be studied in more detail. Updated information about these regions is presented below.

Chacopampean Plain

This is the largest and most populated geographic region in Argentina, covering more than 1 million square kilometers. Comprehensive information about arsenic sources, mobility, and concentration factors in shallow aquifers is not available due to the vast size of the area. However, a number of local studies point to several risk zones located in the north of La Pampa province, south and southeast of Cordoba province, Buenos Aires province, Santa Fe province, Santiago del Estero, Chaco and Salta provinces, and in the eastern portion of Tucumán province. A total population of 2 million people is estimated to be at risk of hydro-arsenicism. Besuschio et al. (1980) and Hopenhayn-Rich et al. (1996) have described several arsenic-related health problems affecting people in some parts of this region and drawn attention to an increasing prevalence of certain types of skin, bladder, digestive tract, and lung neoplasm.

Groundwater composition in the central north and southeast of La Pampa province varies. Arsenic contents range from less than 4 to 5,300 μg/L, and 99% of water samples exceed the reference value of 10 μg/L (Nicolli et al., 1997; Smedley et al., 1998, 2002, 2006).

For most sites located in southern Cordoba province, arsenic contents in groundwater fall in the narrow interval of 0–100 μg/L mainly consisting of As (V); however, at Alejo Ledesma groundwater was found to contain up to 300 μg/L As. In the province's southeast plain, arsenic contents in groundwater vary from less than 10 to 3,810 μg/L with 46% of samples in the interval 100–316 μg/L (Nicolli et al., 1989). The highest anomalies were found in the San Justo, Marcos Juárez, Unión, Río Cuarto, Río Primero, and General San Martín departments where arsenic contents in groundwater range from 10 to 4,550 μg/L (Nicolli et al., 1985, 1989; Pinedo and Zigarán, 1998; Pérez Carrera et al., 2005).

Arsenic in water in Buenos Aires province is particularly but not exclusively related to the Pampean aquifer beneath a large area of the Pampean Plain. This aquifer is the major groundwater source for the whole region (south of the Santa Fe province: 0.13 mg/L As; Atlantic coast: 0.1–0.3 mg/L As; La Pampa province: 0.04–0.5 mg/L As). Fortunately, the northern part of Buenos Aires province is served by the Puelche aquifer with better water quality (As < 0.01 mg/L; F < 1.5 mg/L), although excess of nitrate may occasionally be caused by domestic sewers, waste disposal, or agrochemicals.

This Puelche aquifer in the western region of Santa Fe Province contains high-saline water, inappropriate for consumption. In the shallows, which contain bicarbonate-sodic water, arsenic is present at high concentrations, up to 0.78 mg/L As, along with some fluorine content (Nicolli et al., 2007a).

In Santiago del Estero province, Bhattacharya et al. (2006) determined that groundwater originated from a shallow aquifer, 12 m deep, had a mean arsenic content of 53 μg/L, and a maximum value of 14,969 μg/L As. The authors found that As (III) content in the water was 1.9–45% of total arsenic with an average of 125 μg/L As. Some phenomena such as mobilization of organic matter by excessive irrigation may alter the local physical–chemical conditions, facilitating arsenic mobility in the area.

In the Chaco province, people are affected by endemic hydro-arsenicism caused by ingestion of drinking water with high As content, with cases described as depending on genetic factors and exposure length. Specific damage was identified in patients from Resistencia, Roque Sáenz Peña, Santa Sylvina, Santa Iglesia, Charata, and other places (Web Odontológica, 2007). Water with high arsenic content (0.01–0.8 μg/L), mainly As (V), and high fluorine content led to a risk of hydro-arsenicism along with dental and skeletal fluorosis among rural and urban populations from Comandante Fernández, Independencia, Quitilipi, Maipú, Almirante Brown, General Belgrano, 9 de Julio, 25 de Mayo, 12 de Octubre, Mayor J. Fontana, and San Lorenzo (Osicka et al., 2007). According to Concha et al. (1998), in Taco Pozo, some individuals with 9.1 and 11 μg/L As detected in blood presented hydro-arsenicism from consumption of water with 200 μg/L As.

Groundwater from the Salí River basin is the most important arsenic reservoir in the eastern part of the Tucumán province, as identified by Nicolli et al. (1989), Smedley et al. (2002), and Tineo et al. (1998). Positive correlations between arsenic, fluorine, and vanadium contents in water were identified by Nicolli et al. (2001a). Arsenic and related elements (F, B, and other elements) are present in rather high amounts in quaternary loess deposit volcanic materials. High pH values (up to 9.24) in groundwater favor dissolution of volcanic glass and leaching of pyroclastic rocks, as stated by Nicolli et al. (1989). As (V) was shown to be the dominant species by Nicolli et al. (2007b), who noted as well an increase of the arsenic content predominantly in shallow waters caused by desorption phenomena. Similar arsenic concentrations were also found in this part of Tucumán province in the small basin of Burruyacú as identified by Nicolli et al. (2001, 2006).

Puna Province

The geological province of Puna, at the southern part of the Bolivian and Peruvian Altiplano, enters Argentinean territory from the Bolivia border (21°45′) down to San Buenaventura Cordillera (26°15′). Puna Alto has median altitudes exceeding 3,500 m above sea level. Prevalence of arid climate, oxidizing conditions, and high water salinity yield anomalous concentrations of certain anions such as arsenic complexes and fluoride in water. In the Pompeya and Antuco spring fields, elevated As contents were found in thermal water. Elevated arsenic concentrations in water from La Puna

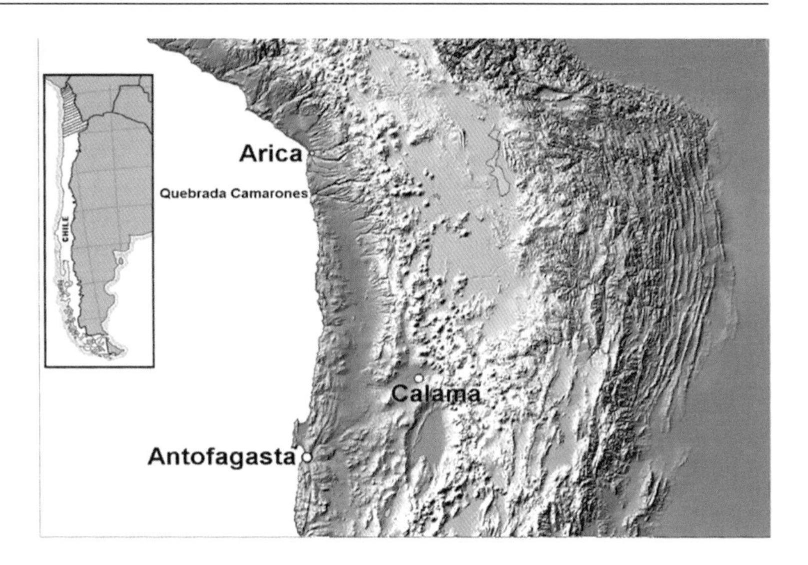

Fig. 2 Geographic map of northern Chile (Modified from Bundschuh et al., 2008b)

region were found by Farías et al. (2006). The highest As content in drinking water (up to 2,030 μg/L) was found in the locality called San Antonio de los Cobres.

Cuyo Region

Studies carried out in the southeastern part of the San Juan Province by Gómez et al. (2004) showed As concentrations greater than 150 μg/L in groundwater consumed by the 1000 rural and urban inhabitants of El Encón. In the study, arsenic contents in biological samples (hair, nail, and urine) and food samples were analyzed.

In the Mendoza province, elevated arsenic levels in water were found in the northern district of Lavalle. In the province of San Luis, arsenic contents in water exceeding 50 μg/L were recorded in the districts of Coronel Pringles, General Pedernera, and Gobernador Dupuy. In the Buena Esperanza district, arsenic contents ranging from 43 to 170 μg/L were found in water from 13 wells and from 90 to 97 μg/L As in treated water (González et al., 2003, 2004).

In the Patagonia region, data are scarce up to the present. Sandali and Diez (2004) monitored arsenic in drinking water from 27 localities in the Chubut province. In 25 of them, arsenic concentrations were below 20 μg/L, whereas contents of 30 and 50 μg/L were found at Garayalde and Camarones, respectively. Additional studies are needed in the region because the geological context suggests that higher As concentrations in water are to be expected.

Chile

In Chile, arsenic-related problems have been solved for most parts of the country but the arsenic levels found in drinking water in the north are still of great concern. This region is located between parallels 17°30′ and 28°30′S (Fig. 2), 250,000 km^2 in the middle of the Atacama Desert and extending from Arica to Antofagasta cities. In this region where 420 volcanoes can be seen, some still active, 35,000 km^2 is made up of quaternary rocks (Cabello et al., 2007).

Total arsenic contents are very high in Chilean water sources such as rivers, springs, and brines, exceeding national and international standards by as much as 6–300 times. In some places, communities are at this very moment consuming water with As contents above the limit of 10 μg/L established by Chilean Regulation 409/1 in 2005 (Chile, 2005). These communities live in the middle of the largest desert in the world, where precipitation is zero and water is scarce. Surface water in this zone may contain 1,000–5,100 μg/L As and arsenic can also be found in bedrock, soil, and plants on riverbanks.

Arica and Parinacota Region

People from the rural zone of Arica have been affected by hydro-arsenicism for more than 4,500 years (Mansilla et al., 2003; Cornejo et al., 2006). Arsenic is derived from natural processes related to volcanic activity in Andean Cordillera. In Arica Province, arsenic is very common in water, soil and plants originating from minerals, and rocks transported from Cordilleran peaks by drainage.

Lluta and Azapa are two alluvium valleys that are very important for agriculture in the Arica and Parinacota region. The Lluta Valley is located only 10 km from the Peruvian border. This valley is formed by the 150 km long Lluta River that extends from the Tacora volcano to the sea, carrying water with around 200 μg/L As. In the Azapa Valley, only low arsenic concentrations in water have been observed (Cabello et al., 2007).

The 4,500-ha Camarones Valley is located 100 km south of Arica city and is crossed by the Camarones River, historically the most important water source in the region (Cornejo et al., 2006). Small local communities such as Esquiña, Illapata, Camarones, Taltape, and Huancarane with around 60 inhabitants each (Mansilla and Cornejo, 2002) were shown to be exposed to arsenic through water consumption in several different ways. People at Esquiña consume low-arsenic water from springs (Yañez et al., 2005) while people from the other communities are consuming surface water with high arsenic contents, in some cases greater than 1 mg/L As (Lara et al., 2006).

Antofagasta Region

Antofagasta is known as the mining capital of Chile and is one of the most important copper producers in the world. The region is extremely arid and Andean volcanism has been particularly intense considering the frequent volcanic eruptions and the number of geysers and thermal springs. The waters in the region altered by these phenomena have variable chemical compositions. Surface water originating in the Andean Precordillera is the main water source for human consumption and irrigation for around 3,000 people.

The Antofagasta Rivers are different in size and present variable arsenic concentrations in water ranging from 10 to 3,000 μg/L (Queirolo et al., 2000). In the Loa province (22°12′ to 23°45′S; 68°20′W), however, high arsenic concentrations (100–1,900 μg/L) were found in water used in the area for human consumption and irrigation. In some rivers very high arsenic concentrations in excess of 3,000 μg/L were found by Oyarzun et al. (2004).

The Elqui Valley in the Coquimbo Region

This 9,800 km^2 valley is also located in northern Chile. The river originates in the Andean peaks and in its course drains important hydrothermal alteration zones and epithermal ore deposits that contain copper, gold, and arsenic in the context of the famous El Indio mineral district. According to several recent studies (Oyarzun et al., 2004), river sediments and ancient lake sediments in the area are highly enriched with arsenic that comes not only from modern mining activity but mainly from long-term erosion processes that affected natural deposits of arsenic-bearing minerals and ores.

Brazil

Integrated studies on environmental and anthropogenic arsenic contamination have been carried out in only three areas of Brazil as seen in Fig. 3. They are (1) the Iron Quadrangle in the state of Minas Gerais, where large amounts of As have been released into drainage, soil, and the atmosphere as a result of gold mining over the last 300 years; (2) the Ribeira Valley, where As was dispersed as a byproduct of Pb–Zn mining over the last century and also as a result of weathering of natural Au-sulfide deposits located downstream from the mining area; (3) the Santana district in the Amazon region where As occurs in association with manganese ore processed locally in the last 50 years. To date, Brazil has had no reports of diffuse pollution sources (geological formations, rivers, and aquifers which extend across the region) such as those described in Argentina and Chile.

The Iron Quadrangle

The Iron Quadrangle (in the state of Minas Gerais) has been the most famous gold-producing area in Brazil since colonial times. Arsenopyrite is commonly associated with gold ores hosted in metamorphosed banded iron formations, schist, meta-basalts, and sedimentary rocks. These terrains of the Archean and Paleoproterozoic ages are an important geochemical As anomaly in the southern portion of the San Francisco craton. During the last 300 years, most of the As-rich waste has been discarded into drainage or

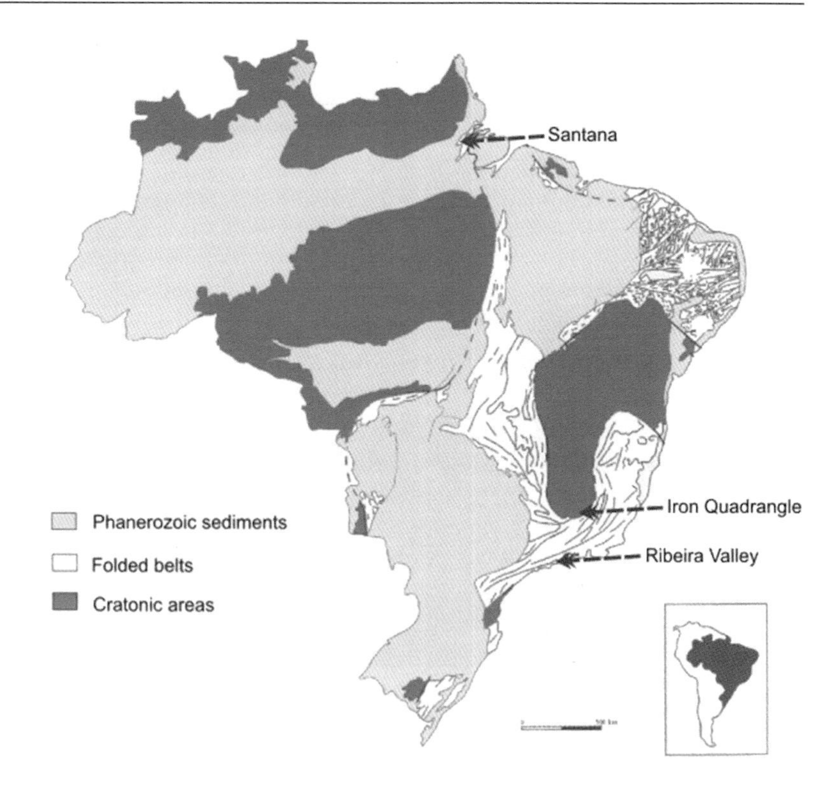

Fig. 3 Contaminated areas of arsenic in Brazil with geologic-tectonic units indicated (After Figueiredo et al., 2007)

stored in tailing piles along river banks and until the seventies was also used for arsenic oxide production. Oliveira et al. (1979), Deschamps et al. (2002), and Deschamps and Mello (2007) have shown that soils around Iron Quadrangle gold deposits are enriched with As. The release of natural As originating from the oxidation of arsenopyrite into water was examined in situ by Borba and Figueiredo (2004) in several underground mines.

As previously mentioned, there is significant anthropogenic contribution to As pollution in the Iron Quadrangle. Arsenic contents in stream sediments (<63 μm) are at very high levels throughout the entire region and concentrations up to 4,000 mg/kg are common. On the other hand, high arsenic contents of up to 350 μg/L in surface water were only found near mines and tailing piles whereas exceptionally high As contents of up to 2,980 μg/L were observed for runoff water from some old gold mines in the region (Borba et al., 2003). In general, arsenic contents in surface water rarely exceeded the threshold of 50 μg/L established by former Brazilian regulations for non-treated water. Low As values, rarely exceeding the limit of 10 μg/L, were also found in spring water and tap water.

In 1998, human screening was carried out among school children (7–12 years) in two municipalities in the Iron Quadrangle which used arsenic content in urine as a bioindicator (Matschullat et al., 2000). The mean value of inorganic As content in urine for a population sample of 126 children was 25.7 μg/L with 20% of samples above 40 μg/L As, for which adverse health effects cannot be excluded on a long-term basis. The probable route of exposure was contact with contaminated soil and dust since As content in the domestic water supply did not exceed 10 μg/L. During the following monitoring campaigns the percentage of individuals in this class (>40 μg/L As) decreased consistently down to 3% in 2003 (Couto et al., 2007).

The Ribeira Valley

The Ribeira Valley is located in the southeastern region of Brazil (Fig. 3) extending for about 500 km in territories of the states of Parana and São Paulo.

Lead and arsenic contamination in the Ribeira River as a result of Pb–Zn ore production and smelting operations in the Upper Valley during the last century has long been demonstrated. In addition, in the

Fig. 4 Lead-contaminated site at La Teja in Montevideo, Uruguay

Middle Valley, a number of noneconomic gold-sulfide deposits occur, forming a northeasterly geochemical anomaly, locally known as the Piririca belt (CPRM, 1982; Perrota, 1996). Most of Pb–Zn ore in the Upper Valley originated from lode ore deposits found in metamorphosed carbonate rocks and schist whereas the Piririca gold deposits are associated with metapelites and basic intrusions.

In the period 1999–2003, human exposure to arsenic was evaluated for the populations of all municipalities affected by mining and metal refining activities in the Upper Ribeira Valley and also for several resident communities in the Middle Valley (De Capitani et al., 2006). The population of Cerro Azul village, which is located far from the mining district, was chosen as a reference group. Different communities' arsenic contents in urine as a bioindicator of recent human exposure are shown in Table 1.

As expected, the lowest values were obtained in control area Cerro Azul (3.60 μg/L for children, $n = 73$; 3.87 μg/L for adults, $n = 83$). The highest As contents in urine for the Upper Valley were found in Serra district, municipality of Iporanga (8.94 μg/L for children, $n = 89$; 8.54 μg/L for adults, $n = 86$). The difference in As levels between these groups was proven to be statistically significant, a finding that may be explained by the fact that the population of the Serra district is exposed to an environment that has been affected by Pb–Zn–Ag mines located in the vicinity.

Table 1 Arsenic contents in urine from various communities (children and adults) of Ribeira Valley, Brazil

Locality	No. of samples	Mean (μg/L)	Range
Cerro Azul	156	3.86	1–34.12
Serra district	175	8.90	1–62.54
Iporanga town	112	8.14	1–33.49
Pilões	49	4.63	1–68.92
Castelhanos	54	9.48	1–60.32
São Pedro	51	11.35	1–76.19
Ivaporunduva	30	10.02	1–34.57
Maria Rosa	26	2.24	1–24.34
Nhungara	22	6.98	1–36.55

Source: Sakuma (2004) and De Capitani et al. (2006)

In the Middle Valley, weathering of mineralized rocks gave rise to soils with high arsenic contents of up to 764 mg/kg As (Abreu and Figueiredo, 2004). In this area, stream sediments with as much as 355 mg/kg As (Toujague, 1999) can be found in contrast with low arsenic concentrations in surface water that do not exceed 10 μg/L according to Takamori and Figueiredo (2002).

Arsenic contents in urine were also low for this area (Table 1). However, the communities with highest As levels in urine are those living near Piririca belt. The difference between As levels in urine obtained at the Castelhanos, Ivaporunduva, and São Pedro localities and that from the control area in Cerro Azul is statistically significant and indicates that the quality of the environment is an important determining factor for human As exposure level. For Castelhanos and São Pedro only 3.4 and 11.8% of urine samples yielded arsenic concentrations greater than 40 μg/L, respectively (De Capitani et al., 2006).

The Santana District

The Santana district is located on the margin of Amazon River, in the state of Amapa. Arsenic dispersion originated from beneficiation of arsenopyrite-bearing manganese ore of the Precambrian Serra do Navio deposit that has been mined for more than 50 years. At Santana wastes contain up to 1,700 mg/kg As and some wells close to the facility were shown to contain extremely high As contents, as much as 2,000 μg/L.

Several studies were carried out in the area when arsenic hazards were revealed and these included surface and groundwater, stream sediment, soil, ore, and waste (Lima, 2003; Santos et al., 2003). Arsenic concentration in surface water was found to range from 5 to 231 μg/L, but most of the As values fell below 50 or even below 10 μg/L and they did not exceed 0.5 μg/L in residential tap water. Stream sediments and suspended particulate were particularly rich for arsenic, yielding maximum values of 1,600 and 696 mg/kg As, respectively.

Arsenic content in hair was determined in a sampling population of 512 people of a total population of around 2,000 residents at Santana. According to Santos et al. (2003), As mean value in hair was 0.2 μg/g with maximum concentrations lower than 2 μg/g As. According to Choucair and Ajax (1988) and Franzblau and Lilis (1989), As concentrations in hair and nails of 1 μg/g or less should be considered to be normal, which was also defined by ATSDR (2000).

These results suggest low human exposure to arsenic at Santana, which is rather consistent with ingestion of low-arsenic drinking water and with restricted contact of humans with poisoned soil and sediments.

These recurrent observations of low exposure levels for arsenic and low arsenic contents in natural water in different places in Brazil were interpreted by Figueiredo et al. (2007) in terms of relative immobility of arsenic in soil and derived sediments under the influence of strong, chemical weathering processes of tropical and sub-tropical regions. In addition, they observed that the communities investigated were less dependent on groundwater consumption and mostly made domestic use of treated surface water. Nevertheless, periodic health monitoring of these communities and periodic assessment of the environment were recommended to assure that the risk remains under control.

Additional Arsenic Cases in Brazil

Other point-source contaminated areas for arsenic are known in Brazil. Also related to gold mining, arsenic occurs in several places such as northern Bahia, Goiás, and northern Minas Gerais states, but no human monitoring has been reported for these places. Coal formations in southern Brazil are known to contain significant concentrations of sulfur and arsenic. A very preliminary hydrogeochemical study carried out in 21 wells yielded very low arsenic contents in groundwater from the Rio Bonito aquifer in the state of Paraná. This aquifer is located in coal-bearing formations that extend southward to the states of Santa Catarina and Rio Grande do Sul.

In Brazil 50% of municipalities are known to have some degree of dependence on groundwater usage for agriculture, industry, and domestic use. However, analytical data for arsenic are still very scarce for groundwater in Brazil. To date the arsenic content of major Brazilian aquifers remains unknown and any probable non-point-source contaminated areas for arsenic such as those previously identified in Argentina and Chile have never been revealed in Brazil.

Lead in South America

Lead has long been listed among the three most hazardous metals worldwide. Beyond recycled lead, which accounts for 50% of consumption, a significant

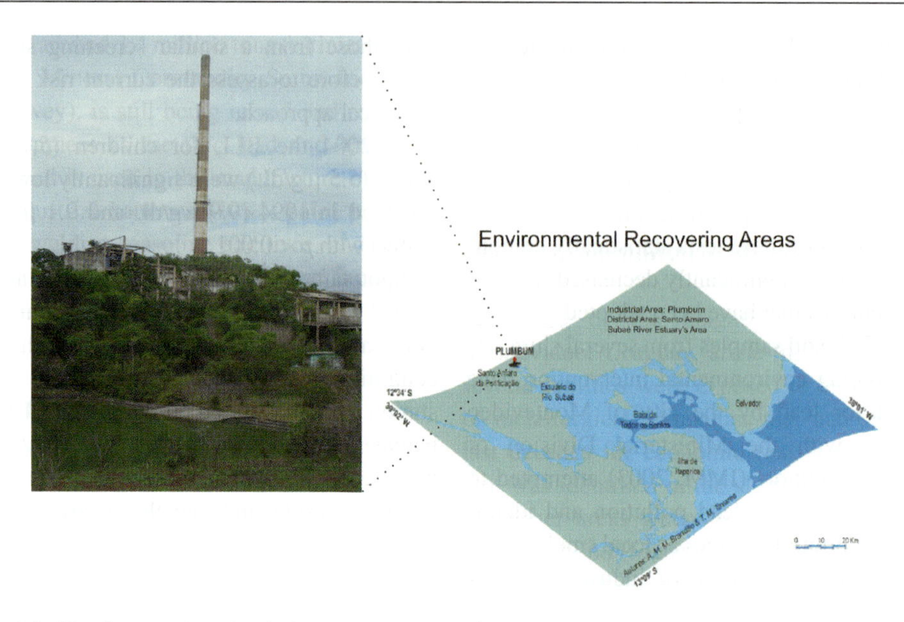

Fig. 5 Ruins of the Plumbum smelter plant in Santo Amaro da Purificação, state of Bahia, Brazil (location map of Todos os Santos Bay and Salvador city)

well as to be used as landfill in the backyards of some houses in past decades.

Several studies carried out since 1980 have shown elevated lead contents in surface water, edible mollusks, cattle, and greens. Because of the size of population, the extension of lead and cadmium dispersion in the city, and the very high levels of human exposure at the time the facility was active, Santo Amaro is considered the most serious case of lead contamination ever described in Brazil.

Three human monitoring campaigns carried out among children in the period 1980–1998 yielded the mean BLL shown in Table 3. Regardless of sample population size, the decreasing lead exposure levels may reflect the shutdown of the facility and the effects of several environmental and sanitary actions implemented in the area since the results of initial studies in Santo Amaro were disclosed.

Table 3 Blood lead concentrations in children from Santo Amaro da Purificação, state of Bahia, Brazil

	1980	1985	1998
Age	1–9	1–4	1–4
N	555	53	47
Mean BLL (μg/dL)	59.1 ± 25.0	36.9 ± 22.9	17.1 ± 7.3
BLL > 20 μg/dL	32%		

Source: Carvalho et al. (2003)

Lead Contamination in Adrianópolis

Adrianópolis is located in the state of Paraná and until recently was the most important site of lead–zinc mining in the Ribeira Valley, southeastern Brazil. During much of the last century, several mines were in production in the region leading to widespread metal contamination in the Ribeira River basin, as identified in previous studies. More recently, additional geochemical and toxicological studies were carried out in the Upper Ribeira Valley and reported by Paoliello et al. (2002, 2003), Paoliello and De Capitani (2005), and Cunha et al. (2005). They clearly showed that sites with the most impact were the villages Vila Mota and Capelinha, in the vicinity of the Plumbum smelter and the Panelas mine both located not far from Adrianópolis center (Fig. 6). The Plumbum industry was in production from 1945 to 1995 when it was shut down probably for financial and technological reasons.

Among all screened groups, the highest mean blood lead levels among children were found in these two villages and the results are shown in Table 4. Human monitoring among adults yielded quite similar results with a mean BLL of 8.80 μg/dL ($n = 101$).

High lead concentrations in soil were found in the surrounding industrial areas by Cunha et al. (2005)

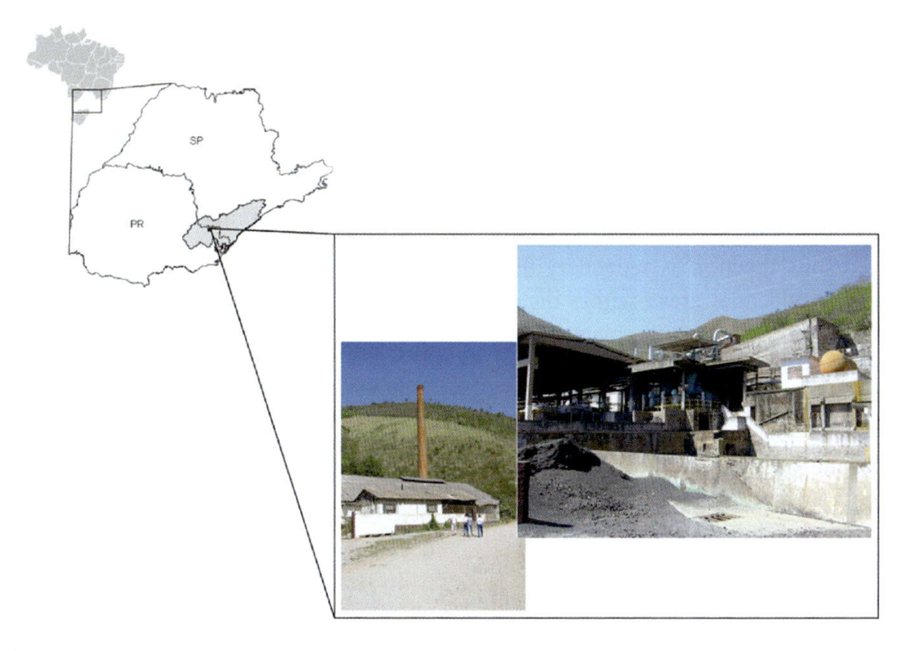

Fig. 6 Ruins of the Plumbum smelter in Adrianópolis and location map of the Ribeira Valley

Table 4 Blood lead levels of children from Adrianópolis and from Bauru city, Brazil

	Adrianópolis (2001)	Bauru (2002)
Age	7–14	0–12
n	94	850
Mean BLL (μg/dL)	11.89	7.3
BLL > 10 μg/dL	59.6%	36.6%
BLL > 20 μg/dL	12.8%	8.05%

Source: Paoliello et al. (2002, 2003), Cunha et al. (2005), Freitas et al. (2007)

with a maximum of 916 μg/g Pb in surface and vegetable garden soils from Vila Mota. Subsequently, Lamoglia et al. (2006) found lead contents in the interval from 100 to 1,500 μg/g in soil samples collected in residential areas. Lead concentrations in edibles from Vila Mota far exceeded the regulated limits in Brazil. These results led to the conclusion that the main routes for human contamination were thus exposure to soil and dust as well as consumption of contaminated food.

Lead Contamination in Bauru City

Environmental and human contamination for lead was brought to light in the city of Bauru (around 360,000 inhabitants) in 2002 when the State Agency for Environmental Control (CETESB) suspended the activities of Ajax company, a battery recycling plant that had been in production since 1974. Studies carried out by the agency and universities were then oriented to assess the quality of the environment and human contamination in the surrounding areas. The results of a very comprehensive study among children are shown in Table 4. The highest mean BLL was found in the 3- to 6-year-old age group. Blood lead values decreased with the distance of the residential area to the industry.

Soil sampling carried out 5 months after the facility was shut down yielded a maximum lead content of 1,071 μg/g inside the industrial facility. However, only low Pb concentrations were found in the surrounding area with maximum at 92 μg/g Pb. Soil samples were collected from 0 to 20 cm depth and slightly higher lead contents on the surface (0 to 2 cm depth) were noted.

Groundwater from one well close to the facility had 60 μg/L Pb, far exceeding the regulated limit of 10 μg/L Pb for drinking water. Several samples of edibles were analyzed in the period from September 2002 to August 2003 but excessive lead concentrations were found only for those collected inside the industrial area.

Following these studies some environmental intervention actions were implemented such as removal of topsoil from the area surrounding the facility, cleaning of houses, and restoration of water reservoirs.

The results summarized above for Brazil deserve some comments. In Adrianópolis as well as in Santo Amaro, poisoned soils function as secondary diffuse sources of lead in residential areas within distances of 500–1,500 m from the industrial plants. In both localities, elevated BLL levels reveal that people are coexisting with lead pollution for several years after the plant shut down. In the case of the Ajax battery facility, despite being located in an urban area (Bauru city), both environmental impact and human exposure were found to be less severe than in Santo Amaro and Adrianópolis. Additionally, environmental and societal actions were less difficult to implement in cosmopolitan Bauru than in other areas.

Few decades ago the area dominated by the Cubatão petrochemical industrial complex near the coast of the state of São Paulo was known as the most polluted site in South America. After a period when several environmental campaigns were undertaken, some data on human exposure for lead have been reported by Santos Filho et al. (1993). Mean BLL of 17.8 ± 5.8 μg/dL was found for a group of 250 children 1–10 years old from Cubatão. This case has not received much comment in recent times.

Also in the Ribeira Valley, human monitoring for lead was carried out among a group of 43 children from the Serra district located not far from some Pb–Zn mines. Cunha et al. (2005) reported a mean blood lead value of 5.36 μg/dL Pb with only 9.3% of samples exceeding the normal limit 10 μg/dL for children. These figures lie well below the exposure levels found in Adrianópolis where the Panelas mine and the Plumbum refinery were simultaneously in operation for decades. Although both sites could be considered industrial point sources, the dispersion of lead into the environment was more widespread in Adrianópolis than in the Serra district.

Additional point-source lead pollution associated with mining may be found at the Morro Agudo zinc mine (in the state of Minas Gerais) and around past lead-ore producers such as the Boquira mine (in the state of Bahia) and several middle-sized mines in the Ribeira Valley (in the states of Paraná and São Paulo).

Mercury in the Amazon

Human exposure to mercury is mainly by ingestion of contaminated food and can lead to adverse health effects such as fever, lung edema, pneumonitis, anorexia, irritability, emotional disturbances, photophobia, disorders of memory and cognitive functions.

Since the Minamata tragedy in the 1950s, mercury dispersion into the environment has become a subject of concern worldwide. Mercury is being investigated in a number of countries especially in places where it is frequently used for gold recovery in low-technology based mines. Hence, several studies have been carried out for mercury in South America in the past decades.

In Brazil most studies concentrated in the Amazon where rudimentary gold mining, locally called "garimpo," made use of mercury for gold amalgam for decades at an annual rate of 80–100 tons according to reliable estimations (Veiga et al., 1999). The Brazilian Amazon covers 58% of the country and has a population of 20 million inhabitants (12% of Brazilian population).

It is well known that water and river sediments in "garimpo" areas often have higher Hg contents than those from remote areas. A compilation of mercury data for Amazon fishes from three major rivers (Madeira, Tapajós, and Negro) and from two water reservoirs (Tucuruí and Balbina) from various authors made by Malm (1998) indicate an average value less than 0.2 μg/g Hg (wet wt) for virgin areas and mercury contents of 2–6 μg/g Hg or even higher for contaminated areas, far exceeding the maximum tolerable limit of 0.5 μg/g established by Brazilian regulation in 1975.

Several determinations of mercury content in human hair from residents in the Tapajós River basin were compiled from various authors by Lima de Sá et al. (2006). Mercury contents from five exposed groups (n = 1,287) fell in the interval 11.8–25.3 μg/g whereas Hg contents from seven less exposed communities (n = 1,644) ranged from 4.0 to 10.8 μg/g. The latter were very consistent with the maximum tolerable level of 10 μg/g Hg established by WHO (1990).

Malm (1998) reported an even more comprehensive data set on mercury content in human hair from

Table 5 Total mercury contents in human hair from various Amazonian areas

Locality	No. of samples	Average (μg/g)
Madeira River	169	8.98
Madeira River	242	17.20
Tapajós River	432	16.76
Tapajós River	96	13.20
Negro River	154	75.50
Tucuruí Reservoir	125	35.00
Balbina Reservoir	58	5.78

Source: Malm (1998) compilation from various authors

different areas and human groups in the Amazon as shown in Table 5. These data revealed that communities exposed to mercury in Amazon are not restricted to "garimpo" areas but may be found around water reservoirs, newly deforested regions, and even in protected indigenous peoples' areas. This find was subsequently confirmed by Jardim and Fadini (2001) who reported elevated mercury contents in soil, water, and air in the Rio Negro River basin where no "garimpo" or any kind of factory has ever existed. The authors argued that the mercury levels in the environment were nevertheless comparable to other known industrial centers in developed countries.

After decades of research work on mercury in the Amazon, scientists are coming to the conclusion that mercury distribution in the region is much more widespread than was thought before. According to Malm (1998), concentration of methyl-mercury in sediment, water, and fish is a function of total mercury concentration, microbiological activity, concentration of organic matter, presence of methyl-group donors, pH, Eh, oxygen activity, and other factors. High mercury methylation rates were found in tropical aqueous systems such as seasonal inundated forest and at root zones of floating meadows formed by aquatic vegetation (Guimarães et al., 1997).

The physical, chemical, and biological functions of the Amazon rainforest to retain and recycle mercury of natural origin probably related to Andean volcanism still require future studies to be understood.

An interesting matter concerning mercury studies in the Brazilian Amazon is the lack of information on human health effects caused by mercury exposure. It is widely thought that the Amazonian population is protected against mercury intoxication by a selenium-rich diet that is widespread in the region.

Selenium in Venezuela

Since the 1970s some pioneering works on human and animal exposure to selenium have been carried out in Venezuela. Jaffé et al. (1972) described some symptoms such as hair loss and anomalies in the skin and nails among children from Villa Bruzual (in the state of Portuguesa, Fig. 7) and pointed out that those children presented a mean blood selenium concentration (813 mg/L) much higher than children from the capital, Caracas (0.355 mg/L). Mondragón and Jaffé (1976) found selenium concentrations in milk, egg, cheese, pork, and chicken produced in Venezuela exceeding selenium contents in similar products imported from eastern USA by 5 to 10 times. The authors believed that the probable cause of contamination is the Se-rich sesame cultivated in the Turén district, produced for animal feed.

Furthermore, it was noted that nursing children in the region ingest 10 times more selenium than children from Finland (Jaffé, 1992). This fact was anticipated by Otaiza et al. (1977) who investigated the blood Se levels in cows from three regions in Venezuela. They found an average Se level of 0.21 mg/L in the central region whereas in Portuguesa this average was 0.67 mg/L, reaching 1.02–3.24 mg/L for cows in the farms located in Turén.

More recently, studies on selenium were extended to the state of Merida, where an attempt was made by Burguera et al. (1990) to correlate blood Se levels with cancer prevalence in humans. Selenium contents fell in the interval 58–115 mg/L, well correlated with soil selenium. Cancer was less common in population

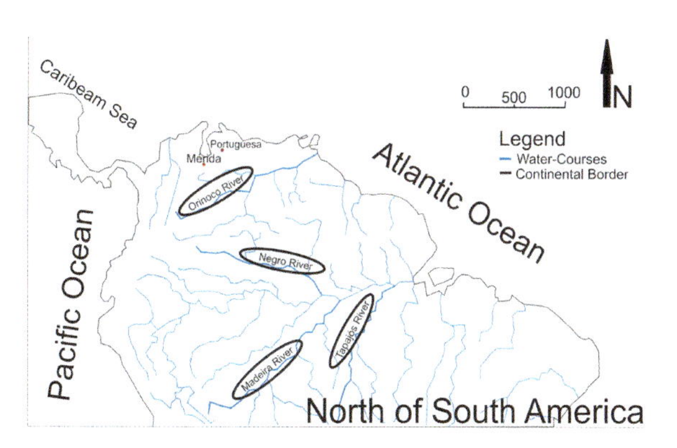

Fig. 7 Map of the Amazon region with indication of locations cited in the text

nurtures the animals from the forest in these sites as if it were mother's milk. They also believe that ingesting soil and water from natural licks is constrained to animals, but that they can benefit from these sites by hunting. Natural licks are important sites for the survival of these societies. That is why they can be listed as keystone resources (Montenegro, 2004; Primack, 1993), resources critical or limiting in particular habitats but crucial for many species of a community (Primack, 1993).

Natural licks are a result of particular hydrologic, ecologic, sedimentary, and mineralogical conditions. In some areas they may be controlled by the tectonic setting (Montenegro, 1998). Thus they are related to the environmental and geological evolution of a region. They also seem to be dynamic: they can dry out and another lick may appear somewhere else. Geologists can contribute to the study of these key resources. This knowledge is important for conservation plans, for natural resource management, and to understand the natural history and the world that has created us.

Geophagy in Humans: Positive Feedbacks

The Otomac tribe from South America was well known for their custom of eating large amounts of clay. The Otomacs used to live in the Neotropics, in the lower llanos of Apure, and on the banks of the Orinoco River (Venezuela, Fig. 7). Today they have disappeared. Among their representatives were the Guamos, Taparitos, Otomacos, and Yaruros. They all shared the taste for soil consumption.

The Otomacs caught the attention of naturalists such as Humboldt (1985) and different missionaries such as Father Gumilla (published in 1944) and Father Bueno (published in 1933). They recorded the Otomacs' habit of eating clay in copious quantities. Clay was extracted from the alluvial beds from the Orinoco River, during high water levels, a period where the food seemed to be scarce, although geophagy was practiced all year long. Otomacs enjoyed eating clay, they cooked it and made small balls to be carried and consumed as a snack or as a small meal, and they also had recipes such as "clay bread": They mixed clay with corn and turtle or caiman fat, left the mix for fermentation, and then ate it (Gumilla cited in Rosenblat, 1964).

In the Northern Andes, on the coast of Ecuador, certain "clay" artifacts were recovered from the Japotó and Acatames archeological sites by Guinea (2006). The objects were described as being composed of calcite-rich soils with a significant content of iron and other minerals such as feldspars, quartz, and sometimes very small amounts of smectite clay. They had prints of leaves and had elongated or spherical shapes (Fig. 9).

They are reminiscent of *tamales, bollos,* and other traditional foods in which dough, paste, or batter is wrapped in leaves which add a natural flavor. They were found to have been cooked, and because of

Fig. 9 Geological materials found in the Atacames and Japoto archeological sites. Note the leaf printing and the shape. (Reproduced with permission of M. Guinea, 2006)

heating they could be preserved. Guinea (2006) assigns the earth to the *taku,* or *pasa*, or even *poya* type.

The pre-conquest cultures had vast knowledge about their environment. Their medical system was completely fused with their culture and religious views, as an emanation from the cultural frame (Gutierrez, 1985), making efforts to explain geophagy considering the ecological and cultural framework. Although traditional explanations of diseases could include supernatural forces or curses, some therapies have proven to be efficient, and some of them have been observed in very different cultures such as the ingestion of clays to treat digestive upset.

The use of clays as detoxifying agents is well known in traditional cultures. Johns (1986) reports the handling of clay material composed of kaolinite, interstratified illite-smectite, and chlorite by indigenous groups from Peru, Bolivia, and Arizona. Poisonous species of wild potato were covered with clay and then consumed without harm. Clays eliminate the bitter taste and absorbed the glycoalkaloid tomatine, preventing stomach aches and vomiting. A study conducted in the Central Andes (Bolivia and Argentina) by Browman (2004), reports 24 types of earths used by the inhabitants either as a food supplement or as drugs; the samples are composed of phyllosilicates, sodic and calcium earths, sulfur minerals, and another group made up of iron and copper salts. The inhabitants of the Central Andes may have learned the geophageal behavior from the Camelids.

This knowledge is part of their cultural heritage, as there were some samples recovered in archeological contexts and these raw materials can currently be found in local markets. The handling of these earthy materials dates back at least five millennia. Browman (2004) reports the traditional names as *Pásа* and *C̆hago* for raw materials composed of mixed clays (smectites, kaolinite, chlorite, illite) being highly appreciated earths for their properties to alleviate gastrointestinal upsets related to phytotoxins present in domesticated plants (such as solanine in potatoes or saponine in quinoa).

Another case comes from the Aymaras that used to live in the high planes of Lake Titicaca. Women ate a plastic soil called Chaco as candy (Patiño, 1984). Half a century ago, Rowe (1946) stated that in the Inca Empire "a quantity of edible clay was collected and exchanged with some frequency in the southern mountains."

During the evangelization period in South America, some Franciscan friars were prominent and well versed in medicine. A complete *vademecum* found in the provincial files was published in 2002 by the Colombian Academy of Exact, Physical and Natural Sciences. It was written by an anonymous Franciscan friar in the 18th century (Díaz and Mantilla, 2002). It is a result of mixing the legacy of occidental medicine and the traditional indigenous medicine.

Numerous reports accounted for geophagy as a form of pica since the period of the conquest. In the Franciscan *vademecum* mentioned in previous section, children with the habit of eating earth were treated with cow milk mixed with a little lead carbonate in the mornings, until they detoxify by vomiting (Díaz and Mantilla, 2002). In Colombia, reported remnants of that habit are found in Magdalena and in Naré; the disease caused by eating dirt was called "jipatera." Colombian writer Gabriel García Márquez captured geophagy in his novel "One hundred years of solitude": Rebecca was an orphan girl, with the custom of eating dirt and lime from the walls when she faced emotional crisis.

In the northwest Amazon basin, in Colombia, Londoño (2007) reports the ingestion of clay minerals in solution by the Uitotos indigenous society as a therapy for alleviating digestive upsets, detoxification of the liver, and counteracting poisonous compounds. The raw materials were dominated by clay and the composition was the same as that reported by Johns and Browman. In the Uitoto classification they belong to the *Nógoras* group (Fig. 10).

Fig. 10 Medicine man and woman from the Uitoto indigenous society collecting clay for medicine purposes in the Amazon rainforest. (After Londoño, 2007)

These examples reflect gaps that exist in our knowledge; the intricate relation between culture, ecology, and geological elements; and the interesting challenge that geophagy represents.

Soil Living Helminthes (Geo-Helminthes)

Soils and minerals may contain toxic compounds and pathogenic organisms. In fact, the excessive consumption of an essential mineral may lead to health problems. Geo-helminthiasis (caused by parasites that live in soil) is one of the most widespread parasites in the world, affecting almost 2,000 million people all over the world (WHO, 2002). The pathogenic agents (*Ascaris lumbricoides, Trichuris trichiura, Uncinaria* sp.) depend on many environmental variables during their early stages of life to become infectious for humans. Some of these variables include soil temperature, weather, air humidity, and soil properties. A first approach to assess this issue was attempted in 2005 with the study "Influence of geological factors on the prevalence of soil transmitted helminthes in Colombia" (Valencia et al., 2008). The main objective was to identify the most relevant environmental variables in an epidemiological sense.

The study was performed with epidemiological data in the form of prevalence of helminthiasis statistically correlated with environmental and geological data. A positive relationship was found between temperature, annual precipitation, and index of basic unsatisfied needs.

This compiled information leads to a conclusion that geophagia is a part of natural history. It is a behavior now restricted to certain cultural patterns and wild animals, its causes and consequences cannot be generalized. Medical geology can be an important contributor to the study of geophagia in trying to explain this behavior by integrating a variety of perspectives and disciplines.

Risk Communication and Risk Governance

In recent years, societies have had to face complex environmental and health risks, which are characterized by controversial values, high stakes, and urgent decisions. These problems, associated with chemical substances, unknown effects of hazards, and uncertainty within the scientific community demand a new approach to deal with these risks, called risk governance.

This approach considers that risk is more than a situation or an event during which something of human value, including humans themselves, is at stake and where the outcome is uncertain. Since risks can only be seen and measured within a social context, it must be understood as a social construction. This means that hazards interact with psychological, social, institutional, and cultural processes in ways that greatly affects public reaction to risk (Hannigan, 1995; Slovic, 1987).

This approach also takes into account that risks are part of our routine. This means that all actors involved in a risk situation have the right to participate in the definition of and solution to the problems that they face. This participation is facilitated when dialogue is encouraged. This dialogue is known as risk communication.

Conceptual Framework of Risk Communication and Risk Governance

Risk communication practices should not be limited to a knowledge deficit model, which considers that experts have to communicate their scientific knowledge to the lay public in order to stop them from continuing to live in a state of ignorance. The idea that lay knowledge is not irrational characterizes the dialogue among social actors involved within the risk management arena. It includes considering that value judgments and subjective influences are present in every phase of risk management, also dividing the experts opinions (Guivant, 2004). In this approach, risk communication must be a process that provides guidelines and strategic tools for scientists and authorities to create a confident atmosphere with all social actors.

The risk communication concept includes strategies that ensure information is supplied in a clear and explanatory manner, so that local populations are informed about and aided to understand pertinent information and its subsequent implications as well as being encouraged to actively participate in mitigation of risk (Horlick-Jones, Sime, and Pidgeon, 2003;

Renn, 2003; Lundgren and McMakin, 2004; Wynne, 1989).

The concept of risk communication has undergone significant advancements and an increase in popularity since the Chernobyl nuclear accident that occurred in the Ukraine in 1986. The accident illustrated the difficulties that researchers were faced with in trying to adequately disseminate information to the public regarding risk assessments and their uncertainties (Wynne, 1989). Since then, scientists, communicators, and public administrators have highlighted the need to put an interactive communication process into practice, which can accommodate the necessary exchange of opinions between the various social actors involved during a risk management process. Today, more than 20 years after the Chernobyl incident, risk communication has gained ground on the public agenda. This change could be seen as better addressing the wide range of situation of risks that we live with, such as natural disasters, environmental changes, outbreaks of infectious or chronic diseases, as well as terrorism.

These events also help to promote a discussion about risk governance. The concept of governance is related to a new institutional arrangement where the decision-making process is collective and involves governmental and non-governmental actors. In this practice, the power of society is recognized and respected. Experts and authorities recognize that complex problems, such as those that are faced in situation of risks, demand more than a technical solution. The choice of these solutions is not a technical decision, but also has political, social, cultural, and economic aspects.

According to Jasanoff and Martello (2004), the concept of governance means rules, processes, and behaviors that affect how power is practiced, particularly regarding participation, accountability, efficiency, and responsibility in the decision-making process. These authors agree that in environmental situations, when the limits of science are so evident, the social dimensions of knowledge production must be recognized and more attention must be given to local perspectives and traditional knowledge.

The great interest in this new approach to dealing with risks (involving risk communication and governance) is a result of the debate about justice, trust, public participation, and democracy that has been happening in societies in recent years. These subjects have had a central role in the development of research and political agendas.

Interest is also related to the idea that it is possible to deal more efficiently with public answers to risk if people that really live these hazard situations are involved in the decision-making process. This means that the risk management must be an analytical and deliberative process in which the effects of social amplification of risk are included as an important element in the decision.

Social amplification of risk denotes the phenomenon by which information processes, institutional structures, social group behavior, and individual responses shape the social experience of risk, thereby contributing to consequences of risk (Kasperson et al., 2005). Therefore acceptance or non-acceptance of existence of a risk is determined by a number of elements such as an individual's familiarity with the problem, ability to solve the problem, exposure to the media coverage, beliefs, and personal feelings (Duncam, 2004; Smith, 1992; Weyman and Kelly, 1999; Sturloni, 2006). The legitimacy of the institutions involved in risk management as well as the lack of direct participation during the decision-making processes of local people involved in a situation of risk influences the level of concern felt by a given population.

Many different experiences with environmental and health risks have shown that when people are involved in the decision-making process, they know and preserve better the place where they live and act individually and collectively to reduce risks. Besides, the chances to avoid a community or place being stigmatized by the risks that they are facing decrease when people are integrated in the solution of their problems.

Public participation in assessment and management of risks also has direct implications in the development of democratic principles. Even if not all desires and aspirations of the communities are satisfied, the simple fact that people are involved in associative models because they have a political project in common that could improve their life quality is a significant advance. It is the chance to develop those communities' personal capacities of analyses and argument, tolerance, and solidarity which is so important in democratic societies.

Environmental Contamination and Health Risks

The debate and practices about risk communication and governance is more advanced in the United States and Europe. In the United States risk communication is often conducted as a result of a law, regulation, or other government inducement (Lundgren and McMakin, 2004), but it is a regular procedure during risk management. One example of this is the act known as CERCLA or Superfund (Comprehensive Environmental Response Compensation, and Liability Act). It requires that specific procedures be implemented to assess the release of hazardous substances at inactive waste sites. Those procedures involve the inclusion of "community relations" in the evaluation process. According to Lundgren and McMakin (2004) "community relations" refers to developing a working relationship with the public to determine acceptable ways to clean up the site so that the community is included in the decision-making process and participates effectively in risk management.

In Europe, although each country has a unique way to deal with risks related to contamination, there is a common idea about promoting an open debate with the public and giving more attention to local and traditional knowledge to face problems. The United Kingdom and France, for instance, have legislation about getting input from the public before some actions are taken by authorities in uncertain areas.

Unfortunately, the debate about risk communication and governance is still reduced in developing countries. The paradigm of assessment and management of risks includes numerical data about the intensity of pollution and measures to reduce threats, but it does not take into account – or perhaps barely takes into account – how people perceive and live with these risks.

Nonetheless, environmental and health researchers (and in some circumstances government) are convinced about the necessity to promote a dialogue between experts that assess and manage risks and the people that really live these risks. The latter's experiences, especially those that involve environmental contamination and human exposure to dangerous substances, show the necessity to include social dimensions and all kinds of knowledge (technical, lay, traditional, and local) to deal with hazard situations (Di Giulio et al., 2008a, b, c).

This necessity is evident in a study focusing on three cases in Brazil (Adrianopolis, Santo Amaro, and Bauru) and one in Uruguay (La Teja) in which resident communities had to deal with environmental contamination and human exposure to lead.

Based on bibliographic research, analysis of journalistic articles, and interviews conducted with people who played different roles in the events, a lack of adequate planning for information release to local people was identified in Adrianópolis. This circumstance seriously undermined the relationships between researchers and community and contributed to a misperception of risk among people. There, the absence of a community involvement plan produced a feeling of exclusion from the decision-making process. In Santo Amaro a concern about risk communication was noted among researchers, although local people had always demanded more information about the problem. There is a local association of contamination victims that presses the government for solutions but the level of public involvement is probably not as extensive as it could be. In Bauru, the relatively good relationship between researchers, authorities, and local people was facilitated by the existence of a risk communication plan. The strategies of risk communication aimed at encouraging public participation in the solution of the problem, although this participation was limited to only public pressure. Local people insisted on actions directed to pollutant control, searched for clarification regarding the health effects of these pollutants and for assistance from the public health service. Finally, in La Teja (Montevideo) local people were from the beginning very engaged in handling the risks, although the risk communication strategies were not enough to open dialogue and to empower community representatives to actively participate in risk management. People's engagement in La Teja resulted from the potential existence of social mobilization represented by a neighborhood association that played an important role in guaranteeing information to local people and to press governmental institutions to action.

These experiences show that even when there is a risk communication plan, the strategies formulated are still limited to a knowledge deficit model and not to contributing to create a trustful and confident relationship between all social actors involved in the risk arena. Without trust it becomes difficult to promote social interaction to handle the risks. This difficulty is more evident when experts and authorities adopt a risk management approach that underestimates potential input from the public. However, these experiences also highlighted that being aware of public values and character as well as taking into account social organizations' input might be the easiest way to mobilize people and promote risk governance.

Other Medical Geology Studies in South America

As mentioned earlier in this chapter the list of environmental problems discussed above is far from exhaustive. Other hazardous substances such as chromium, cadmium, siliceous particulate, metallic or asbestos-bearing dust, agrochemicals, radioactive elements, and many other substances may cause harmful effects to exposed populations. These themes are focused in a number of ongoing projects and publications in South America. Additional information on Se, As, and Cu in Uruguay; on chemical composition of drinking water in several metropolitan areas in Brazil; on human exposure to atmospheric particulates in urban and rural areas; on natural gamma radiation in the environment as well as at U–Th mine sites; and on the presence of agro toxins in edibles are available in a number of papers, technical reports, dissertations, doctoral theses, and presentations in diverse congresses.

These environmental and human health studies cover a wide spectrum of scientific themes in response to real problems identified in South American countries. They benefit from a laboratory infrastructure that could be considered good especially in the more developed regions. In the past decades getting high-quality analytical data, for example, for lead in blood, arsenic speciation, or any trace-element concentration at ppb or ppt levels was an obstacle to scientific research in South America. However, substantial advances in laboratory facilities have occurred in recent years in many places that can largely benefit scientists, professionals, and students from less developed regions. Participation in various web lists and research networks that are active in the continent may be the best way to improve scientific research on medical geology in this part of world.

Future of Medical Geology in South America

In Brazil, the most promising medical geological initiative undertaken so far is the Geomedicine Project of the State of Paraná. Since 2006 a comprehensive study on surface geochemistry and public health has been conducted by the Pele Pequeno Principe Research Institute in cooperation with the local geological company Mineropar. Interest in this study has arisen from the fact that in the state there is an elevated incidence of TP33 genetic alteration generally associated with adrenal cortex cancer among children and mammal cancer. The project was designed to verify its existence and indentify probable relationships between health problems and environmental parameters. The research team takes advantage of the excellent geochemical database generated by Mineropar since the 1990s. A substantial amount of public health, demographic, social-economic, and surface geochemical data are being stored in a web mapping system from which different user friendly maps are produced for system users, according to Pedrini et al. (2010).

Current and future achievements of the Parana Geomedicine Project illustrate very well the thesis that significant advances in future medical geology will occur in those regions where geological and geochemical mapping efforts succeed. That brings attention to the strategic plans of geological surveys and other geological institutions in the continent. Incidentally, the Geological Survey of Brazil (CPRM) recently announced additional investments in low-density geochemical mapping of the country (Silva, 2008). These geochemical studies include the chemical compositions of surface water, stream sediments, and soil as well as trace-element analyses of treated water for domestic use in some metropolitan areas.

In connection with regional, geological, and geochemical mapping, a great number of new research initiatives may be undertaken by South American universities and research centers. They will also to a great extent be based on the results from current research work, which covers a wide thematic spectrum from investigations of air quality in urban and rural areas and probable correlations with lung and heart diseases and mortality rates; remediation experiments at laboratory and pilot scales on arsenic removal from acid mine drainage and chromium removal from contaminated water; use of phosphate or bio-accumulators for rehabilitation of contaminated soil for lead and other metals, etc.

Despite the difficulty of being able to create multidisciplinary research teams and interinstitutional collaboration between Earth, life and social scientific institutions, this seems to be the best model to follow and the most effective way for future medical geology studies in the continent to succeed in response to societal needs and scientific challenges.

References

Abreu MC, Figueiredo BR (2004) Mapeamento geoquímico de arsênio e metais pesados em solo da Unidade Piririca, Vale do Ribeira (SP). In Brazilian Geological Congress, 41, Proocedings, João Pessoa, Paraíba.

Acosta H, Cavelier J, Londoño S (1996) Aportes al conocimiento de la biología de la danta de montaña *Tapirus pinchaque* en los Andes Centrales de Colombia. Biotropica 28:258–265.

ATSDR (Agency for Toxic Substances and Disease Registry) (2000) Toxicological profile for Arsenic. Atlanta, GA: U.S. Department of Health and Human Services, Public Health Service.

Alvarez C, Piastra C, Cousillas A, Mañay N (2003a) Importancia del dato analítico en la contaminación por plomo durante 2001–2002 en Uruguay. In Congresso Brasileiro de Toxicologia, XIII, Londrina, Brasil.

Alvarez C, Piastra C, Queirolo E, Pereira A, Mañay N (2003b) Evolución de plumbemias en niños de Montevideo-Uruguay. In Congresso Brasileiro de Toxicologia, XIII, Londrina, Brasil.

Andreazzini MJ, Figueiredo BR, Licht OAB (2006) Comportamento geoquímico do flúor nas águas e sedimentos fluviais da Região de Cerro Azul, Estado do Paraná. Rev Bras Geoc 36:74–94.

Argentina (2007) Código Alimentario Argentino, modification of articles 982 and 983, May 22, 2007, Buenos Aires, Argentina. http://www.anmat.gov.ar/normativa/normativa/Alimentos/Resolucion_Conj_68-2007_96-2007.pdf.

ATSDR-CERCLA (2003) Top 20 Hazardous Substances from the CERCLA Priority List of Hazardous Substances, Comprehensive Environmental Response, Compensation, and Liability Act, CERCLA/Superfund, 2002.

Besuschio SC, Desanzo AC, Pérez A, Croci M (1980) Epidemiological associations between arsenic and cancer in Argentina. Biol Trace Elem. Res 2:41–55.

Bhattacharya P, Claesson M, Bundschuh J, Sracek O, Fagerberg J, Jacks G, Martin RA, Storniolo A. del R, Thir JM (2006) Distribution and mobility of arsenic in the Río Dulce alluvial aquifers in Santiago del Estero Province, Argentina. Sci Total Environ 358:97–120.

Borba RP, Figueiredo BF (2004) A influência das condições geoquímicas na oxidação da arsenopirita e na mobilidade do arsênio em ambientes superficiais tropicais. Revista Brasileira de Geociências 34(3):489–500.

Borba RP, Figueiredo BR, Matschullat J (2003) Geochemical distribution of arsenic in waters, sediments and weathered gold mineralizes rocks from iron quadrangle, Brazil. Environ Geol 44(1):39–52.

Browman DL (2004) Tierras comestibles de la Cuenca del Titicaca: Geofagia en la prehistoria boliviana. Estudios Atacameños 28:133–141.

Bueno R (1933) Apuntes sobre la provincia misionera 1785–1804. Caracas, Tipografía Americana.

Bundschuh J, García ME, Birkle P, Cumbal LH, Bhattacharya P, Matschullat J (2008a) Occurrence, health effects and remediation of arsenic in groundwater of Latin America. In J Bundschuh, MA Armienta, P Bhattacharya, J Matschullat, P Birkle, AB Mukherjee (eds) Natural Arsenic in Groundwater of Latin America — Occurrence, Health Impact and Remediation. Leiden, Holland: A.A. Balkema Publishers.

Bundschuh J, Pérez Carrera A, Litter MI (2008b) Distribución del Arsénico en la Región Ibérica e Iberoamericana", Argentina, CYTED, octubre de 2008. ISBN 13 978-84-96023-61-1.

Burguera JL, Burguera M, Gallignani M, Alarcón OM, Burguera JA (1990) Blood serum selenium in the province of Mérida, Venezuela, related to sex, cancer incidence and soil selenium content. J Trace Elem Electrolytes Health Dis 4: 73–77.

Cabello G, Cornejo L, Arriaza B, Santero C (2007) Comprendamos el desierto para que seamos protagonistas del desarrollo de nuestra región. In G Cabello Fernández (ed) Proyecto Explora-Conicyt, ISBN N 978-956-8649-00-5.

Cardoso L, Morita MC, Alves JC, Licht OAB (2001) Anomalia hidrogeoquímica e ocorrência de fluorose dentária em Itambaracá-Pr. In Congresso Brasileiro de Geoquímica, VIII, SBGq, Curitiba. Resumos (CD Rom).

Carvalho FM, Silvany Neto AM, Tavares TM, Costa ACA, Chaves CR, Nascimento LD, Reis MA (2003) Chumbo no sangue de crianças e passivo ambiental de uma fundição de chumbo no Brasil. Revista Panamericana de Salud Pública 13(1):1–10.

Castilho LS, Ferreira EF, Jorge WV, Menegasse LN, Fantinel LM (2004) Geologia, odontologia e saúde: instrução de comunidades rurais sobre aspectos geoambientais e epidemiológicos da fluorose dentária em São Francisco, MG. In Encontro de Extensão da Universidade Federal de Minas Gerais, 7, Belo Horizonte, Anais.

CDC (1991) United States Center of Disease Control. Preventing Lead Poisoning in Young Children. A Statement by the Center of Disease Control US. Atlanta, Georgia: Department of Health and Human Services.

CETESB (2005) Estabelecimento de Valores Orientadores para Solos e Águas Subterrâneas no Estado de São Paulo. Companhia de Tecnologia de Saneamento Ambiental. Decisão Diretoria 195-05, São Paulo, 4 p.

Chile (2005) Instituto Nacional de Normalización (INN), Chile, Norma Chilena 409/1 of 2005, Agua Potable Parte 1: Requisitos.

Choucair AK, Ajax ET (1988) Hair and nails in arsenical neuropathy. Ann Neurol 23(6):628–629.

CPRM (Companhia de Pesquisa de Recursos Minerais) (1982) Projeto Eldorado, Relatório Final Integrado de Pesquisa (Final Report), CPRM, São Paulo.

CONAMA (2005) Resolution 357, March 17, 2005 CONAMA (National Council for the Environment), Brazil

Concha G, Nemel B, Vahter M (1998) Metabolism of inorganic arsenic in children with chronic high as exposure in Northern Argentina. Environ Health Perspect 6:355–359.

Cornejo L, Mansilla H, Arenas J, Flores M, Flores V, Figueroa L, Yánez J (2006) Remoción de arsénico en aguas del río Camarones, Arica, Chile, utilizando la Tecnología RAOS modificada. In M Litter, A Jiménez (eds) Avances en tecnologías económicas solares para la desinfección, descontaminación y remoción de arsénico en aguas de comunidades rurales de América Latina (métodos FH y RAOS). Proyecto OEA AE141/2001, ISBN N 978-95081-9-X, Cap. 4:pp. 85–92.

Couto N, Mattos S, Matschullat J (2007) Biomonitoramento humano. In E Deschamps, J Matschullat (org), Arsênio antropogênico e natural, Fundação Estadual do Meio Ambiente, Belo Horizonte, pp. 241–269.

Cunha FG, Figueiredo BR, Paoliello MMB, De Capitani EM, Sakuma AM (2005) Human and environmental lead contamination in the upper Ribeira valley, southeastern Brazil. Terrae 2(1–2):28–36.

De Capitani EM, Sakuma AM, Figueiredo BR, Paoliello MMB, Okada IA, Duran MC, Okura RI (2006) Exposição humana ao arsênio no Médio Vale do Ribeira, São Paulo, Brasil. In CR Da Silva et al. (eds) Geologia Médica no Brasil, CPRM-Serviço Geológico do Brasil, Rio de Janeiro, pp. 82–87.

Deschamps E, Mello J (2007) Solos e sedimentos. In: E. Deschamps, J Matschullat (org), Arsênio Antropogênico e Natural, Fundação Estadual do Meio Ambiente, Belo Horizonte, pp. 200–215.

Deschamps E, Ciminelli VST, Lange FT, Matschullat J, Raue B, Schmidt H (2002) Soil and sediment geochemistry of the Iron Quadrangle, Brazil: The case of arsenic. J Soils Sediments 2(4):216–222.

Díaz S, Mantilla LC (2002) La terapéutica en el Nuevo Reino de Granadaun recetario franciscano del siglo XVIII. Academia Colombiana de Ciencias Exactas, Físicas y Naturales, II. Bogotá, Colombia, 207 p.

Di Giulio GM, Pereira NM, Figueiredo BR (2008a) Lead contamination, the media and risk communication: A case study from the Ribeira Valley, Brazil. In DGE Liverman, C Pereira, B Marker (Org.) Communicating Environmental Geoscience. London: Geological Society, London, Special Publications, 305, pp. 63–74.

Di Giulio GM, Pereira NM, Figueiredo BR (2008b) O papel da mídia na construção social do risco: o caso Adrianópolis, no Vale do Ribeira. História, Ciências, Saúde-Manguinhos, 15, pp. 293–311.

Di Giulio GM, Figueiredo BR, Ferreira LC (2008c) Communicating environmental risks as a long-term policy. In Berlin Conference on the Human Dimensions of Global Environmental Change, Berlin, Germany.

Dos Anjos JASA (2003) Avaliação da eficiência de uma zona alagadiça (wetland) no controle da poluição por metais pesados: o caso da Plumbum em santo Amaro da Purificação, BA. Doctoral Thesis, Escola Politécnica, USP, 326 p.

Downer CC (1996) The mountain tapir, endangered "flagship" species of the high Andes. Oryx 30:45–58.

Duncam B (2004) Percepción pública y comunicación eficaz del riesgo. http://www.jrc.es/home/report/spanish/articles/vol82/-welcome.html.

Emmons LH, Stark NM (1979) Elemental composition of a natural mineral lick in Amazonia. Biotrópica 11(4):311–313.

Farías SS, Bianco de Salas G, Servant RE, Bovi Mitre G, Escalante J, Ponce RI, Ávila Carrera ME (2006) Survey of arsenic in drinking water and assessment of water intake of arsenic in La Puna, Argentina. In J Bundschuh, MA Armienta, P Bhattacharya, J Matschullat, AB Mukherjee (eds) Natural Arsenic in Groundwaters of Latin America – Occurrence, Health Impact and Remediation, Leiden, Balkema.

Figueiredo BR, Borba RP, Angélica RS (2007) Arsenic occurrence in Brazil and human exposure. Environ Geochem Health, Springer, Netherlands 29:109–118.

Franzblau A, Lilis R (1989) Acute arsenic intoxication from environmental arsenic exposure. Archiv Environ Health 44(6):385–390.

Freitas CU, De Capitani EM, Gouveia N, Simonetti MH, Silva MRP, Kira CS, Sakuma AM, Carvalho MFH, Duran MC, Tiglea P, Abreu MH, (2007) Lead exposure in an urban community: Investigation of risk factors and assessment of the impact of lead abatement measures. Environ Res 103(3):338–344.

Gilardi JD, Duffey SS, Munn CA, Tell LA (1999) Biochemical functions of geophagy in parrots: Detoxification of dietary toxins and cytoprotective effects. J Chem Ecol 25:897–922.

Gómez D, Molina S, Naranjo A, Cabrera I (2004) Contaminación del agua de consumo con arsénico en la población de El Encón. In Taller de Evaluación y Manejo de Riesgos por Exposición a Arsénico en Agua de Consumo, 10, S. M. de Tucumán, Argentina, (unpublished).

González DM, Ferrúa NH, Cid J, Sansone MG, Jiménez I (2003) Arsénico en aguas de San Luis (Argentina). Uso de un equipo alternativo Al De Gutzeit modificado. Acta Toxicológica Argentina. 11 (1), p. 3–6.

González DM, Ferrúa NH, Sansone MG, Ferrari S, Cid JA (2004) Arsénico en agua de consumo humano en poblaciones de las provincias de San Luis y Buenos Aires. Acta Toxicológica Argentina. 12:7.

Guimarães JRD, Meili M, Hylander LD, Castro e Silva E, Roulete M, Mauroa JBN, Lemos RA (1997) Mercury net methylation in five tropical flood plain regions of Brazil: high in the root zone of floating macrophyte mats but low in surface sediments and flooded soils. The Science of the Total Environment, 261:99–107.

Guinea M (2006) El uso de tierras comestibles por los pueblos costeros del Periodo de Integración en los Andes septentrionales. Bulletin de l'Institut Français d'Études Andines 35(3):321–334

Guivant JS (2004) A governança dos riscos e os desafios para a redefinição da arena pública do Brasil. In Ciência, Tecnologia + Sociedade. Novos Modelos de Governança. Brasiília, dezembro, 2004. http://www.nisra.ufsc.br/pdf/A%20governa%5B1%5D.pdf

Gumilla, J. (1944) El Orinoco ilustrado. Tomo I. Bilbioteca popular de cultura colombiana. 360 p.

Gutiérrez V (1985) Medicina Tradicional de Colombia. Magia, religión y Curanderismo. V II, Universidad Nacional de Colombia, Bogotá, Colombia.

Hannigan J (1995) Sociologia Ambiental – a formação de uma perspectiva social. Lisboa: Instituto Piaget.

Horlick-Jones T, Sime J, Pidgeon N (2003) The social dynamics of environmental risk perception: implications for risk communication and practice. In N Pidgeon, RE Kasperson, P Slovic, The Social Amplification of Risk (pp. 262–285). Cambridge: Cambridge University Press.

Hopenhayn-Rich C, Biggs ML, Fuchs A, Bergoglio R, Tello E, Nicolli H, Smith AH (1996) Bladder cancer mortality associated with arsenic in drinking water in Córdoba,

Argentina. Epidemiology 7:117–124. http://wonder.cdc.gov/wonder/prevguid/p0000029/p0000029.asp
http://www.ufmg.br/proex/arquivos/7Encontro/Saude95.pdf (May 26, 2009)

Humboldt A (1985) [1799] Viaje a las Regiones Equinocciales del Nuevo Continente. Caracas: Monte Ávila Editores, T. 4, 601 p. (translated by Lisandro Alvarado).

IMM (2003) Contaminación por Metales en suelo. In Informe Ambiental de Montevideo. Intendencia Municipal de Montevideo, Documentos de Desarrollo Ambiental. http://www.montevideo.gub.uy/ambiente/documentos/infoamb03c.pdf.

Jaffé WG (1992) Selenium, an essential and toxic element. Latin American data. Arch Latinoam Nutr 42:90–93.

Jaffé WG, Ruphael MD, Mondragón MC, Cuevas MA (1972) Clinical and biomedical studies on school children from a seleniferous zone. Arch Latinoam Nutr 22: 595–611.

Jardim WF, Fadini PS (2001) A origem do mercúrio nas águas do rio Negro. Ciência Hoje 30:62–64.

Jasanoff S, Martello ML (2004) Earthy Politics: Local and Global in Environmental Governance. EUA: MIT Press.

Johns T (1986) Detoxification functions of geophagy and domestication of the potato. J Chem Ecol 12: 635–646.

Kasperson RE, Renn O, Slovic P, Brown HS, Emel J, Goble R, Kasperson JX, Ratick S (2005) The social amplification of risk: a conceptual framework. In J Kasperson, R Kasperson, (eds) The Social Contours of Risk: Publics, Risk Communication and the Social Amplification of Risk (pp. 99–114). London: Earthscan.

Lamoglia T, Figueiredo BR, Sakuma AM, Buzzo ML, Okada IA, Kira CS (2006) Lead in Food and Soil from a Mining Area in Brazil and Human Exposure. Chinese Journal of Geochemistry, Supplementary Issue dedicated to the 7th International Symposium on Environmental Geochemistry, Beijing, September, p. 66.

Lara F, Cornejo L, Yáñez J, Freer J, Mansilla HD (2006) Solar-light assisted arsenic removal from natural waters: the effect of iron and citrate concentrations. J Chem Technol Biotechnol 81:1282–1287.

Lima de Sá A, Herculano AM, Pinheiro MC, Silveiras LCL, Do Nascimento JLM, Crespo-López ME (2006) Exposição humana ao mercúrio na região Oeste do Estado do Pará, Revista Paraense de Medicina, 20(1), Belém.

Lima MO (2003) Caracterização geoquímica de arsênio total em águas e sedimentos em áreas de rejeitos de minérios de manganês no Município de Santana, Estado do Amapá. MSc Dissertation, Universidade Federal do Pará.

Lips JM, Duivenvoorden JF (1991) Morphological and chemical features of mineral salt licks in the middle Caqueta basin, Amazonas, Colombia, Colombia Amazónica, 5(1), pp. 119–130.

Lizcano DJ, Cavelier J (2004) Características químicas de salados y hábitos alimenticios de la Danta de montaña (Tapirus pinchaque Roulin, 1829) en los Andes Centrales de Colombia. Mastozool Neotrop 11(2):193–201.

Londoño SC (2007) Caracterización Geoquímica Preliminar De Las Arcillas Con Potencial De Uso Medicinal Presentes En Araracuara, Caquetá, Colombia. BSc Monograph, Geology Program, Universidad Nacional, Bogotá, Colombia.

Lundgren R, McMakin A (2004) Risk Communication: A Handbook for Communicating Environmental, Safety and Health Risks. Ohio: Battelle Press.

Malm O (1998) Gold mining as a source of mercury exposure in the Brazilian Amazon. Environ Res Section A 77:73–78.

Mañay N, Alonzo C, Dol I (2003) Contaminación por plomo en el barrio La Teja, Montevideo-Uruguay. In Suplemento "Experiencia Latinoamericana" Salud Publica de México 45:268–275.

Mañay N, Alvarez C, Cousillas A, Pereira L, Baranano R, Heller T (2006) Changes in blood lead levels in Uruguayan populations. In MC Alpoim, PV Morais, MA Santos, A Cristovao J Centeno P Collery (eds) Metals Ions in Biology and Medicine, 9 (pp. 530–534). Paris: John Libbey Eurotext.

Mañay N, Cousillas AZ, Alvarez C, Heller T (2008) Lead contamination in Uruguay: The "La Teja" neighborhood case. In DM Whitacre (ed) Reviews of Environmental Contamination and Toxicology, 93, Springer, 195, pp. 93–115.

Mansilla H, Cornejo L (2002) Chile, relevamiento de comunidades rurales de América Latina para la aplicación de tecnologías económicas para la potabilización de aguas. In M Litter (ed) Proyecto OEA AE 141/2001, ISBN N 987-43-54127, 2002, pp. 43–58.

Mansilla H, Cornejo L, Lara F, Yánez J, Lizama C, Figueroa L (2003) Remoción de arsénico de aguas del río Camarones, Arica, Chile. In M Litter, H Mansilla (eds) Remoción de Arsénico Asistida por Luz Solar en Comunidades Rurales de Latinoamérica. ISBN N 987-43-6943-4. Cap. 2, pp. 35–46.

Matschullat J, Borba RP, Deschamps E, Figueiredo BR, Gabrio T, Schwenk M (2000) Human and environmental contamination in the Iron Quadrangle, Brazil. Appl Geochem 15: 181–190.

MINEROPAR (2001) *Atlas Geoquímico do Estado do Paraná.* Curitiba, 80 p.

Mondragón MC, Jaffé WG (1976) Ingestion of selenium in Caracas, compared with some other cities. Arch Latinoam Nutr 26:343–352.

Montenegro O (1998) The Behavior of Lowland Tapir (*Tapirus terrestris*) at a Natural Mineral Lick in the Peruvian Amazon. Master Dissertation, University of Florida.

Montenegro O (2004) Natural Licks as Keystone Resources for Wildlife and People in Amazonia. PhD Thesis, University of Florida.

Nicolli HB, O'Connor TE, Suriano JM, Koukharsky MML, Gómez Peral MA, Bertini A, Cohen LM, Corradi LI, Baleani OA, Abril EG (1985) Geoquímica del arsénico y de otros oligoelementos en aguas subterráneas de la llanura sudoriental de la provincia de Córdoba. Miscelánea No. 71, Acad. Nac. Ciencias. Córdoba, Argentina.

Nicolli HB, Smedley PL, Tullio JO (1997) Aguas subterráneas con altos contenidos de flúor, arsénico y otros oligoelementos en el norte de la provincia de La Pampa: estudio preliminar. In Congreso Internacional sobre Aguas y Workshop sobre Química Ambiental y Salud, Abstracts, III-40, Buenos Aires, Argentina.

Nicolli HB, Suriano JM, Gómez Peral MA, Ferpozzi LH, Baleani OM (1989) Groundwater contamination with arsenic and other trace elements in an area of the Pampa, Province of Córdoba. Argentina Environ Geol Water Sci 14(1):3–16.

Nicolli HB, Tineo A, Falcón CM, García JW, Merino MH, Etchichury MC, Alonso MS, Tofalo OR (2006)

Hydrogeochemistry of arsenic in groundwaters from Burruyacú basin, Tucumán Province, Argentina. In J Bundschuh, MA Armienta, P Bhattacharya, J Matschullat, AB Mukherjee (eds) Natural Arsenic in Groundwaters of Latin America – Occurrence, Health Impact and Remediation. Leiden: Balkema.

Nicolli HB, Tineo A, Falcón CM, Merino MH (2001a) Movilidad del arsénico y de otros oligoelementos asociados en aguas subterráneas de la cuenca de Burruyacú, provincia de Tucumán, República Argentina. In A Medina, J Carrera y L Vives (eds) Congreso Las Caras del Agua Subterránea I (pp. 27–33). Madrid: Instituto Geológico y Minero de España.

Nicolli HB, Tineo A, García JW, Falcón CM, Merino MH (2001) Trace-element quality problems in groundwater from Tucumán, Argentina. In R Cidu (ed) Water-Rock Interaction 2 (pp. 993–996). Lisse: Balkema.

Nicolli HB, Tineo A, García JW, Falcón CM, Merino MH, Etchichury MC, Alonso MS, Tofalo OR (2007b) Arsenic-contamination source of groundwater from Salí basin, Argentina. In TD Bullen, Y Wang (eds) Water-Rock Interaction. Leiden: Balkema.

Nicolli HB, Tujchneider OC, Paris M del C, Barros AJ (2007a) Fuentes y movilidad del arsénico en aguas subterráneas del área centro-norte de la provincia de Santa Fe (unpublished).

Oliveira JJC, Ribeiro JH, Souza Oki S, Barros JRR (1979) Projeto Geoquímica do Quadrilátero Ferrífero: Levantamento orientativo e regional. CPRM (Geological Survey of Brazil), Final Report (Vol. I).

Osicka RM, Agulló NS, Herrera Aguad CE, Jiménez MC (2007) Evaluación de las concentraciones de fluoruro y arsénico en las aguas subterráneas del Domo Central de la Província del Chaco. Website: www.msal.gov.ar/htm/site/pdf/FyAsChaco.pdf, 2002 (visited in 2007).

Otaiza ER, Valeri H, Cumare V (1977) Selenium content in the blood cattle from Venezuela. I. Central and Portugueste zone. Arch Latinoam Nutr 27:233–246.

Oyarzun R, Lillo J, Higueras P, Oyarzú J, Maturana H (2004) Strong arsenic enrichment in sediments from the Elqui watershed, Northern Chile: industrial (gold mining at El Indio–Tambo district) vs. geologic processes. J Geochem Explor 84:53–64.

Paoliello MMB, De Capitani EM (2005) Environment contamination and human exposure to lead in Brazil. Rev Environ Contam Toxicol 184:59–96.

Paoliello MMB, Capitani EM, Cunha FG, Carvalho MF, Matsuo T, Sakuma A, Figueiredo BR (2003) Determinants of blood lead levels in an adult population from a mining area in Brazil. Journal de Physique IV 107:127–130.

Paoliello MMB, Capitani EM, Cunha FG, Matsuo T, Carvalho MF, Sakuma A, Figueiredo BR (2002) Exposure of children to lead and cadmium from a mining area of Brazil. Environ. Res., Section A 88:120–128.

Passwater RA (1996) Selenium against cancer and AIDS. New Canaan CT: Keats, 1996, pp. 47–48.

Patiño VM (1984) La alimentación en Colombia y en los países Vecinos. Tomo I de Historia de la cultura material en la América Equinoccial, pp. 23–24.

Pedrini H, Ibáñez HC, Figueiredo BC (2010) Sistema Web Mapping. In: B.C. Figueiredo and H.C. Ibáñez, GeoMedicina no Paraná, Instituto de pesquisa Pelé Pequeño Príncipe, Secretaria de Estado de Ciencia, Tecnología e Ensino Superior, Curitiba, PR (in forthcoming).

Pérez Carrera A, Moscuzza C, Fernández Cirelli A (2005) Aporte de macrominerales del agua de bebida a la dieta de bovinos de leche (Córdoba, Argentina). Revista Argentina de Producción Animal 25(3–4):115–121.

Perrota MM (1996) Potencial aurífero de uma região no Vale do Ribeira, São Paulo, estimado por modelagem de dados geológicos, geoquímicos, geofísicos e de sensores remotos num sistema de informações geográficas. PhD Thesis, University of São Paulo.

Pinedo M, Zigarán A (1998) Hidroarsenicismo en la Provincia de Córdoba, actualización del mapa de riesgo e incidencia. In Congreso Interamericano de Ingeniería Sanitaria y Ambiental, 26, Lima, Perú.

Primack RB (1993) Essentials of Conservation Biology. Sinauer Associates Inc: Sunderland, Mass.

Queirolo F, Stegena S, Mondaca J, Cortés R, Rojas R, Contreras C, Muñoz L, Schwuger MJ, Ostapczuk P (2000) Total arsenic, lead, cadmium, copper, and zinc in some salt rivers in the northern Andes of Antofagasta, Chile. The Sci of the Total Environ 255:85–95.

Renn O (2003) Social amplification of risk in participation: Two case studies. In N Pidgeon, RE Kasperson, P Slovic (eds) The Social Amplification of Risk (pp. 374–401). Cambridge: Cambridge University Press.

Rosenblat A (1964) Los otomacos y taparitas de los llanos de Venezuela. Estudio etnográfico y lingüístico. Tomo I. Instituto de antropología e historia.

Rowe JH (1946) Inca culture at the time of the Spanish conquest. In J Steward (ed) The Andean Civilizations, Bureau of American Ethnology, Washington DC, Bulletin 143(2): 183–330.

Sakuma AM (2004) Avaliação da Exposição Humana ao Arsênio no Alto Vale do Ribeira, Brasil, Doctoral Thesis, Faculdade de Ciencias Médicas, University of Campinas, 196 p.

Sandali G, Diez E (2004) Determinación del contenido de arsénico en agua de consumo humano en la provincia del Chubut. Acta Toxicológica Argentina Supl 12:13–14.

Santos Filho E, Silva RS, Barreto HHC, Inomata ONK, Lemes VRR, Sakuma AM, Scorsafava MA (1993) Concentrações sanguíneas de metais pesados e praguicidas organoclorados em crianças de 1 a 10 anos. Rev Saúde Pública 27(1): 59–67.

Santos ECO, Jesus IM, Brabo ES, Fayal KF, Lima MO (2003) Exposição ao mercúrio e ao arsênio em estados da Amazônia: síntese dos estudos do Instituto Evandro Chagas/FUNASA. Revista Brasileira de Epidemiologia 6(2):171–185.

Schrauzer GN (1977) Cancer mortality correlation studies–III: Statistical associations with dietary selenium intakes. Bioinorg Chem 7(1):23–31.

Selinus O, Alloway B, Centeno JA, Finkelman RB, Fuge R, Lindh U, Smedley P (2005). Essentials of Medical Geology. Amsterdam: Elsevier, 812 pp.

Shamberger R, Frost D (1969) Possible protective effect of selenium against human cancer. Can Med Assoc J 100:682.

Silva CR (2008) Programa Nacional de Pesquisa em Geoquímica Ambiental e Geologia Médica – PGAGEM. In Congresso Brasileiro de Geologia, 44, Curitiba, SBG, Anais, p. 374.

Slovic P (1987) Perception of risk. Science 236:280–285.

Smedley PL, Nicolli HB, Barros AJ, Tullio JO (1998) Origin and mobility of arsenic in groundwater from the Pampean Plain, Argentina. In EB Arehart, JR Hulston (eds) Water-Rock Interaction (pp. 275–278). Rotterdam: Balkema.

Smedley PL, Nicolli HB, Macdonald DMJ, Barros AJ, Tullio JO (2002) Hydrogeochemistry of arsenic and other inorganic constituents in groundwater from La Pampa, Argentina. Appl Geochem 17(3):259–284.

Smedley PL, Nicolli HB, Macdonald DMJ, Kinniburgh DG (2006) Arsenic in groundwater and sediments from La Pampa Province, Argentina. In J Bundschuh, MA Armienta, P Bhattacharya, J Matschullat, AB Mukherjee (eds) Natural Arsenic in Groundwaters of Latin America – Occurrence, Health Impact and Remediation. Leiden: Balkema.

Smith KS, Huyck LO (1999) An overview of the abundance, relative mobility, bioavailability, and human toxicity of metals. In GS Plumlee, MJ Logsdon, The Environmental Geochemistry of Mineral Deposits, Part A: Processes, Techniques, and Health Issues. Reviews in Economic Geology, 6A, pp. 29–70.

Smith K (1992) Environmental Hazards – Assessing Risk and Reducing Disaster. London: Routledge.

Sturloni G (2006) Le mele de Chernobyl sono buone: mezo secolo di rischio tecnológico. Editore Sironi, Milano. Italy.

Takamori AY, Figueiredo BR (2002) Monitoramento da qualidade de água do rio Ribeira de Iguape para arsênio e metais pesados. Brazilian Geological Congress, 41, Proocedings, João Pessoa Paraíba, p. 255.

Tineo A, Falcón C, García J, D'Urso C, Rodríguez G (1998) Hidrogeología. In Geología de Tucumán, (2nd ed.). Colegio de Geólogos de Tucumán (publicación especial), S. M. de Tucumán, Argentina, pp. 259–274.

Toujague RDR (1999) Arsênio e metais associados na região aurífera do Piririca, Vale do Ribeira, São Paulo, Brasil. MSc Dissertation, University of Campinas, Brazil.

Valencia HC, Fernández NJ, Londoño SC, Cucunubá Z, Reyes HP, Lopez PM (2008) Geological Influences on the Soil, Contributions to Infections Transmitted by Helminths in Colombia. In International Geological Congress, 33, Oslo, 2008, Proceedings.

Veiga ATC, Salomão EP, Veiga MM, Barros JGC (1999) Tapajós Project – A Proposal for a Clean "Garimpagem" in the Brazilian Amazon. Mercury as a Global Pollutant. In 5th International Conference, Rio de Janeiro, CETEM/UFF, Abstracts, p. 291.

Velásquez LNM, Fantinel LM, Uhlein A, Iglésias MM, Ferreira EF, Castilho LS, Vargas AMD, Rodriges PCH (2008) Investigação hidrogeológica do flúor em aqüíferos carbonáticos do Médio São Francisco-MG e epidemiologia da fluorose dentária associada. In Congresso Brasileiro Geologia, 44, Curitiba, SBG, Anais, p. 371.

Web Odontológica (2007) Website: http://webodontologica.com/index.asp (visited in 2007).

Weyman AK, Kelly CJ (1999) Risk Perception and Risk Communication – A Review Literature. Broad Lane, Sheffield: HSE Books,.

WHO (1990) Methylmercury. IPCS Environmental Health Criteria Document 101. Geneve: World Health Organization.

WHO (2002) Prevention and Control of Schistosomiasis and Soil Transmitted Infections Report of EHO Expert Committee. Geneve: World Heath Organization.

WHO (2004) Guidelines for Drinking Water Quality (Vol. 1, 3rd ed.). Recommendations. Geneve: World Health Organization.

Wynne B (1989) Sheep farming after Chernobyl – A case study in communicating scientific information. Environ Mag 31:10–15.

Yañez J, Mansilla H, Figueroa L, Cornejo L (2005) Arsenic speciation in human hair: A new perspective for epidemiological assessment in chronic arsenicism. J Environ Monit 7:1335–1341.

An Overview of Medical Geology Issues in Australia and Oceania

Karin Ljung, Annemarie de Vos, Angus Cook, and Philip Weinstein

Abstract Australia and Oceania together make up some of the oldest and youngest geologic formations on the planet, ranging from rocks dating back to 4,400 million years to newly formed volcanic isles in the Pacific. The health issues related to these diverse geological materials range from those derived from exposure to metals and minerals to volcanic emissions including gas and ash, bushfires, dust storms, as well as health threats posed by natural hazards. With the position of a large part of this region within the Ring of Fire, many Australian and Oceanian lives are impacted upon by the forces related to tectonic movement, including earthquakes, tsunamis, and volcanic eruptions. There are also a number of medical geology issues related to the soil in this area. These include geophagy, melioidosis – an infectious disease caused by soil bacteria, and the impacts from ecosystem transformations caused by the disturbance of acid sulfate soils. An example of this is the increase of vector-borne mosquitoes carrying the Ross River virus with the formation of acidic ponds through acid sulfate soil oxidation. The potential adverse health outcomes from disturbing some parts of the land have long been acknowledged by traditional Aboriginal landowners of Australia, who refer to an area particularly rich in uranium and other metals as Sickness Country.

Keywords Australia · Oceania · metals · Volcanic eruptions · Acid sulfate soils · Geophagy · Vector-borne diseases · Sickness country

K. Ljung (✉)
School of Population Health, University of Western Australia, Perth, WA, Australia; Institute of Environmental Medicine, Karolinska Institutet, Stockholm, Sweden
e-mail: Karin.Calluna@gmail.com

General Background

Australia and Oceania together consist of a few large islands, including Australia, New Zealand, Papua New Guinea, and Fiji, and thousands of small coral atolls and volcanic islands within Melanesia, Micronesia, and Polynesia. These islands are all defined as "small island states" as they cover less than 10,000 km^2 and are inhabited by less than 500,000 people (McMichael et al., 2003) (Fig. 1).

The total population of the region is around 36 million, with nearly 22 million in continental Australia, 6 million in Papua New Guinea, and 4 million in New Zealand and the Oceanian part of Indonesia. Hawaii, although politically part of the United States, is with its population of just over 1 million generally included in Oceania, but is not specifically included in this chapter. The remaining 3 million inhabit just over 20 islands or groups of islands, with Fiji and the Solomon Islands being the largest with 800,000 and 500,000 inhabitants, respectively, and the Pitcairn Islands having the smallest population of only 47 (CIA, 2009).

It has been suggested that altered weather patterns, including gradual warming secondary to greenhouse gas emissions, may have resulted in increasing average global temperatures over the last 30 years. Island habitats are particularly vulnerable to probable consequences of global warming, such as rising sea levels. The Intergovernmental Panel on Climate Change (IPCC) has indicated that small islands are vulnerable to increasing amplitudes and frequencies of high tides, greater wave damage, and intrusion of salt water into the islands underground reserves of freshwater. Atolls, such as the Kiribati, are at particular risk from even small changes in sea level (Nunn, 1991). Many low-lying states may partially disappear, eroding

O. Selinus et al. (eds.), *Medical Geology*, International Year of Planet Earth,
DOI 10.1007/978-90-481-3430-4_5, © Springer Science+Business Media B.V. 2010

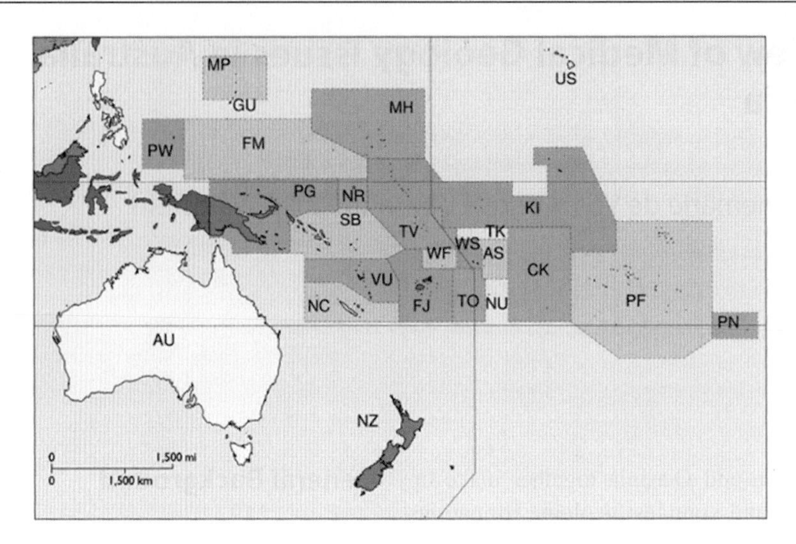

Fig. 1 Australia and Oceania. Courtesy of Reis (2006) *Australasia*: AU = Australia; NZ = New Zealand; *Melanesia*: FJ = Fiji; NC = New Caledonia; PG = Papua New Guinea; SB = Solomon Islands; VU = Vanuatu Micronesia; FM = Federated States of Micronesia; GU = Guam (USA); KI = Kiribati; MH = Marshall Islands; NR = Nauru; MP = Northern Mariana Islands; PW = Palau; *Polynesia*: AS = American Samoa; CK = Cook Islands (NZ); PF = French Polynesia (France); NU = Niue (NZ); PN = Pitcairn Islands (UK); WS = Samoa; TK = Tokelau (NZ); TO = Tonga; TV = Tuvalu; WF = Wallis and Futuna (France)

terrestrial habitats and triggering further competition for use of the remaining land. However, this chapter mainly relates to medical geology issues of concern in the areas which house the largest populations. Although geologically mediated natural hazards such as earthquakes, tsunamis, and volcanoes are included, other threats more related to weather as well as the threat of rising sea levels with climate change are not considered issues of medical geology and are only covered in brief.

Geology of Australia and Oceania

Australia did not fall out of the sun, as was proposed by Rev. W.B. Clarke in 1861 (Johnson, 2004). Instead of being a new continent as was initially believed, it is one of the oldest land masses on Earth, containing almost all known rock types from all geological time periods. The continental crust of Australia has been dated at older than 3,500 million years, with metamorphosed sedimentary rocks in some regions containing zircons that are even older, dated at 4,400 million years, making it the oldest material known on Earth (Johnson, 2004). In contrast, much of the geology of New Zealand is relatively recent and reflects its location between the tectonic plates of Australia and the Pacific. Its largest city, Auckland, sits on top of an active volcano field, while Wellington, its capital, is built on the active Wellington fault, with approximately one earthquake per week registered below the city. On the South Island, a 600 km active fault cuts across the landscape, slipping at a rate of approximately 25 mm/year. Similarly, many of the islands in the Pacific are characterized by tectonic and volcanic activities resulting from the long collision of the Australian, Eurasian, and Pacific plates and comprise some of the youngest land formations on Earth (Te Ara, 2007).

Geological Features Related to Health

The Australian continent encompasses a range of climatic zones from arid to humid temperate to wet tropical as well as areas of seasonal freezing and snow. The lowest temperature on record was –23°C (–9.4°F) measured in New South Wales in 1994, while the highest temperature of 53°C (127.4°F) was recorded in Queensland about a hundred years earlier, in 1889. The small town of Marble Bar in Western Australia boasts the record as the hottest town in Australia after its record spell of 160 days in a row with temperatures recorded above 37.5°C (99.5°F). Annual rainfalls also vary, with a record precipitation set in 1979 when over 11 m of rain fell over Mount Bellenden Ker in Queensland. The same state holds the wettest annual average of 4,048 mm. At the other end of the spectrum, Lake Eyre in south Australia has experienced the lowest rainfall on record – an annual precipitation of a mere 125 mm (Johnson, 2004).

Although many areas experience frequent floods, the Australian continent is one of the driest, with approximately one-third being desert and another third semiarid (Johnson, 2004). As a result, Australia is dusty in large parts of the inland areas with subsequent health issues. Lake Eyre, in spite of its name, only filled three times in the past century and is together with the surrounding deserts a significant source of dust during its usual dry condition. Residents of Sydney recently experienced this as they woke up to a "Mars-like" city on September 23, 2009, when dust from inland Australia was uplifted and transported some 4,000 km to the coast, filling the air with red dust and decreasing visibility to 300 m. It has been suggested that the Lake Eyre area is the largest source of airborne dust in the Southern Hemisphere, although its contribution to global dust levels is small compared to, for example, the Sahara desert (Earth Observatory, 2006). Prolonged periods of drought coupled with high winds and the flammable nature of eucalyptus also provide excellent conditions for bushfires. In 2009, swathes of the state of Victoria were consumed by raging February bushfires, which caused the death of 210 people and destroyed over 2,000 homes.

In addition to particular natural events affecting human health, there are a number of geological features of the Australian landscape that cause chronic exposures and sometimes more subtle or long-term health effects. Examples of potential hazards include the occurrence of natural asbestos as well as a range of minerals, metals, and radioactive elements. Some areas particularly rich in uranium are referred to as Sickness Country by the Aboriginal landowners, who warn that disturbing the land will cause disease.

With New Zealand's position between two moving tectonic plates, the geological issues of concern with regard to health mainly include those related to this activity. Earthquakes and volcanic eruptions cause direct physical injury, and volcanic emissions and geothermal springs may greatly increase exposure to potentially toxic elements. Other natural hazards of relevance in New Zealand include tsunamis and landslides. Papua New Guinea and the numerous islands of the Pacific are characterized by their position in the Ring of Fire, and similar to New Zealand, these are frequented by geological events that can threaten local communities (Box 1).

This chapter has been divided into four main parts, each discussing area-specific geological features and their relation to human health. The first section covers minerals and metals, the second discusses natural hazards, the third covers geological emissions, and the fourth covers issues related to soil.

Box 1 Island Life

Among the most vulnerable areas of inhabited land are coral atolls and volcanic islands of small size (Fig. 2). With severe constraints on available space, people must live and function in relatively close proximity to any industrial process, such as mining. Many islands have limited capacity to produce and store food resources and also suffer freshwater shortages needed for both human and ecosystem health. Among the most vulnerable sites are small atolls, which may be fully dependent on rainwater or imported water supplies. In such delimited locations, reserves are depleted from two directions: from the "bottom-up" by salinization of the underground lens of freshwater and from the "top-down" contamination

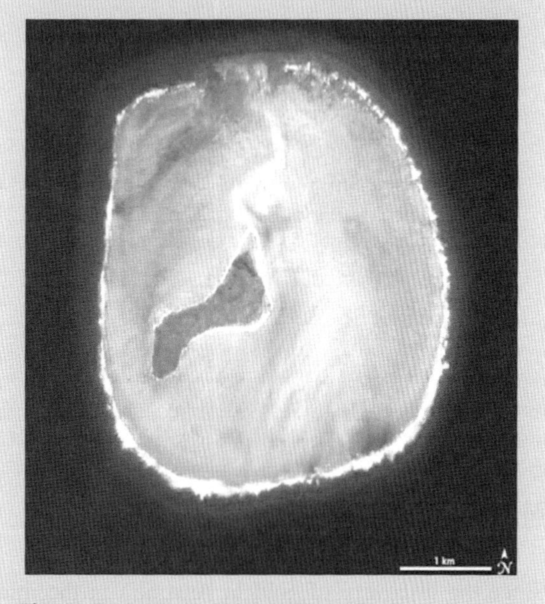

Fig. 2 Oeno Island, 280 km NW of Pitcairn Islands in the Pacific Ocean (Ong, 2006) covers 0.5 km^2 and has a maximum elevation of 5 m

of freshwater by agricultural and human wastes. On occasion, groundwater supplies totally dry up, as occurred on Vanuatu's Nguna Island in 1992: the island population was forced to consume coconut milk to maintain hydration (Thistlethwaite and Davis, 1996).

Islands are also vulnerable to catastrophic events that create sudden imbalances between utilizable land and population. In the 1800s, both the Cook Islands and Mangareva in French Polynesia underwent cycles of drought and severe damage of crops by hurricanes, resulting in recurring famine; the Micronesian Island of Pingelap experienced a similar fate over the last century, spurring migration to neighboring centers. Climatic-driven changes may limit the capacity for islanders to utilize their aquatic resources, such as by increasing the risk of diarrheal illness, particularly on islands where water reserves are already scarce (Singh et al., 2001). Marine food sources may also be placed in jeopardy, such as through the threat of ciguatera (fish poisoning). Ciguatera, occurring from the consumption of fish which have ingested toxins produced by algal microflagellates, affects Pacific Island populations with greater frequency where local warming of the oceanic surface occurs (Hales et al., 1999). It is predicted that the acceleration of global warming will result in an increased incidence of ciguatera in populations dependent on reef fish as a food supply.

A common problem that compromises the safety of crops and waterways of many island communities is soil erosion. Removal of forests for firewood and agriculture accelerates this process and precipitates flash flooding. Even islands with a diverse range of habitats may be severely affected. In Papua New Guinea and other Melanesian Islands, the combination of steep topography, volcanic soils, and torrential rainfall may often trigger lethal mudslides.

Minerals and Metals

Since the discovery of gold in Australia in the mid-1800s, the mining industry has been a significant contributor to the Australian economy (Australian Bureau of Statistics, 2008a). Australia currently mines, or has unworked deposits of, almost all mineral commodities. It is one of the world's leading miners of coal, bauxite, diamonds, gold, iron ore, lead, manganese, nickel, titanium, uranium, zinc, and zircon (Australian Bureau of Statistics, 2006). It is ranked as one of the world's leading mining nations with significant mineral and fuel resources close to the surface. In the 1960s, Australia emerged as a major trader in black coal, bauxite, iron ore, nickel, manganese, titanium, and zirconium, and by the 1980s, new technologies expanded gold production and the processing of bauxite into alumina and aluminum. At the end of the 20th century, Australia was a world producer and exporter of alumina, diamonds, gold, copper, silver, manganese ore, uranium, and tantalum (Australian Bureau of Statistics, 2006), the world's largest producer of bauxite, mineral sands, and tantalum and one of the top producers of uranium, iron ore, lead, zinc, and nickel (Australian Bureau of Statistics, 2008a) (Box 4).

Other areas of Oceania also have rich deposits of metals, with gold mining in New Zealand, Papua New Guinea, and Fiji as well as in a number of smaller states (CIA, 2009). Other mineral deposits in Oceania include copper, lead, nickel, zinc, manganese, chromium, molybdenum, cobalt, and phosphate (Advameg, 2009). Although there are occupational health issues as a result of mining, environmental exposure may also exert negative effects through elevated exposures from drinking water and crops as well as from soil and dust (Box 2). An area particularly rich in mineral deposits in northern Australia has been dubbed Sickness Country as it is believed by the Aboriginal landowners that sickness will fall upon those that enter there. Geochemical analyses have found that this area coincides with unusually high natural levels of thorium, uranium, arsenic, mercury, fluorine, and radon in groundwater and drinking water (McKay et al., 2008) (Box 3). In this section, we will look at how the minerals and metals present in the land affect human health.

Box 2 Silent Spring in Esperance

In December 2006, birds started falling from the sky in the small coastal town of Esperance in Western Australia. By the end of January 2007, over 20 incidents had been reported with hundreds of birds found dead. At least 500 birds were confirmed dead over the 2 months, with extrapolation using local bird density data suggesting around 4,000 birds may have died in the town during this time (Department of Environment and Conservation, 2007a). By the end of March, this estimate had reached 9,500 birds (Western Australia Legislative Committee, 2007). At first, a viral or bacterial cause was suspected (Department of Environment and Conservation, 2007a), but subsequent testing revealed the cause of death to be lead poisoning (Department of Environment and Conservation, 2007b). Isotope testing showed the source to be lead carbonate from a mining company which used the Esperance Port for shipping out their product (Western Australia Legislative Committee, 2007).

Export of lead carbonate through the Esperance port started in June 2005 and continued up until the suspicion of its environmental contamination in March 2007. The lead is mined near Wiluna, which was expected to provide 2% of the world's lead at target production. The lead carbonate was processed here and transported around 900 km south to Esperance where it was shipped to China, mainly for the manufacture of car batteries (Esperance Port Authority, 2008). As a result of the finding of lead contamination causing the bird deaths, the Port Authority put an immediate stop to lead carbonate shipments through the port in mid-March 2007. A few days later, the Department of Environment and Conservation (DEC) issued a prevention notice for further transport into and out of the port, which left 9,000 tons of lead carbonate stored on port grounds (Government of Western Australia, 2007).

Concern was raised about the lead exposure to the children in the town, and free blood lead testing was offered. Out of the 600 children tested, 81, or 13.5% had blood lead levels above or equal to the internationally acceptable level of 5 μg/l (WACL, 2007). The contribution of lead from the Wiluna mine to the blood of the children tested varied between 30 and 87%, and of the children with blood lead levels above 3 μg/l, 84% had at least 50% Wiluna lead (WACL, 2007). An inquiry by the Legislative Committee into the lead contamination of Esperance revealed that the incidence was both foreseeable and foreseen and in spite of a detailed examination it remained unclear how the events leading up to the contamination could happen anyway. Sadly, it became evident that without the death of the birds and the persistence of local community members, the exposure would have continued.

Box 3 Sickness Country

Our land was first created by Bula, who came from saltwater country to the north. With his two wives, the Ngallenjilenji, he hunted across the land and in doing so transformed the landscape through his actions. Bula finally went under the ground at a number of locations north of Katherine in an area known to us as 'Sickness Country.' It is called this because the area is very dangerous and should not be disturbed for fear that earthquakes and fire will destroy the world.

Bolung the rainbow serpent inhabits the deep green pools found in the Second Gorge. We do not fish in the pools where Bolung sits. When fishing close to these pools, we can take only a small portion of the fish caught and throw back the rest in order to appease Bolung. Drinking water must not be taken from these deep pools but rather from the shallow associated waters. Pregnant women and new initiates may not swim in the Katherine River for fear of disturbing Bolung, who must not be spoken to and must be left undisturbed. (Dreamtime of Jawoyn people)

Sickness Country described above is located in the northern inland part of the Northern Territory of Australia (Fig. 3). As the Dreamtime explains, the Jawoyn people who are the traditional Aboriginal landowners have long been aware of potentially adverse health effects from disturbing this land. Interestingly, the location of Sickness Country coincides with some of the largest uranium deposits in Australia (Fig. 3) (McKay et al., 2008). Although many areas within Sickness Country are sacred to avoid disturbing Bula, in modern times they are also very attractive to mining companies. In the early 1990s, the Jawoyn people opposed the proposal to mine Coronation Hill, or Guratba, which has been referred to as "one of the great unmined

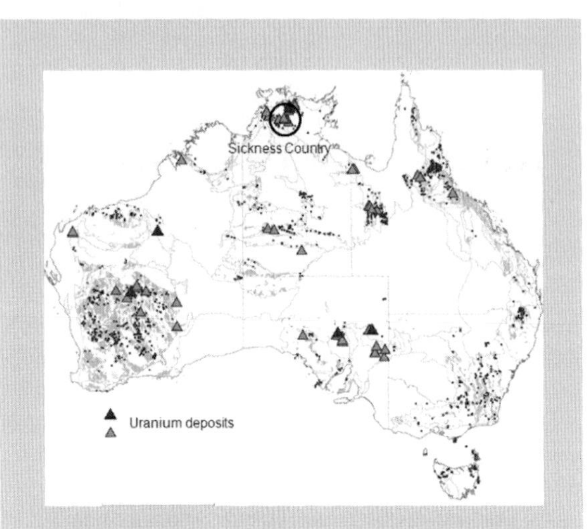

Fig. 3 Location of uranium deposits in Australia and Sickness Country

and barely studied uranium ore deposits on the planet" (Davidson, 2009). After great political debate, the area was designated a conservation zone of the Kakadu National Park, one of four Australian sites included on the World Heritage List for outstanding cultural and natural universal values.

Naturally Occurring Asbestos

Asbestos is a term applied to a group of silicate minerals that easily separate into thin flexible fibers. There are six types of asbestos minerals recognized by the International Agency for Research on Cancer (IARC) for their carcinogenicity and these include chrysotile (white asbestos), amosite (brown asbestos), crocidolite (white asbestos), tremolite, anthophyllite, and actinolite (IARC, 1987). Asbestos minerals can exist as homogenous mineral deposits but more commonly in substantial amounts together with other minerals. Because of their flexibility and heat resistance, asbestos minerals have been widely used in building materials and in fabrics with over 3,000 asbestos products having been identified. Many Australian homes contain some form of asbestos product, including insulation, fencing, pipes, paints, and textiles. Asbestos was also widely used in clutches and brake linings of cars (Asbestos Diseases Society of Australia Inc., 2004). Asbestos deposits have been mined in northern New South Wales (chrysotile or white asbestos) and in Wittenoom of Western Australia (crocidolite or blue asbestos, see Box 4).

Box 4 Whispers of Wittenoom

Around 150,000 tons of crocidolite, or blue asbestos, were extracted from three mines in the Pilbara region of Western Australia between 1937 and 1966, when the mines were closed due to health concerns of asbestos exposure. The town of Wittenoom was established in 1950 to cater to around 7,000 mine workers and their families, and its population quickly reached around 20,000. The waste ore containing about 5% asbestos was used in the construction of the town and was used to pave roads, school yards, and the airstrip. It is estimated that natural weathering has now spread the asbestos waste over an approximately 10 km^2 large area (Twohig, 2006). Since the late 1970s, the government has phased down Wittenoom as a town site to about eight people and in October 2006, the decision to close the town altogether was made. Mail deliveries, police patrols, and the power supply were all halted and the town was degazetted by the state government in 2007. Visits to the town are discouraged and a brochure on the risks involved is made available by the state.

Most of the people who moved to Wittenoom to work in the mine were young post-war emigrants, mainly from Italy. Because of their young age, many of them have lived long enough to experience asbestos-related diseases with long latency periods (Musk et al., 2008). A study on the mortality rate of almost 7,000 former mine workers found that the overall increase was 13–90% higher than the general Australian population, with much of the increase resulting from asbestos exposure (Musk et al., 2008). Non-occupational or environmental exposures have also resulted in increased risks of mesothelioma, and an exposure–response relationship was confirmed in a study on 4,659 former Wittenoom residents who were not directly working in the mining industry (Fig. 4). The relationship was confirmed at much lower exposure times (2 months) and intensities (measured as 0.53 f/ml: presence of fibers <0.53 μm long per milliliter of air) than had previously been

Fig. 4 (a) Road sign at the entrance of Wittenoom. (b) Children playing with asbestos in a Wittenoom backyard. Photographs courtesy of Asbestos Diseases Society of Australia Inc

reported (Hansen et al., 1998). Increased cancer mortality rates were also found for women exposed in their environment or homes when compared to the Western Australian female population (Reid et al., 2008). In a study on the disease incidence in an Aboriginal population living in the Pilbara, one of the highest population-based rates of environmental exposure was recorded at 250 per million at age 15 and over. The exposure had mainly resulted from asbestos transport from the Wittenoom operation (Musk et al., 1995). In 1999, the Western Australian mesothelioma incidence rate was the highest in the country at 47.7 cases per million, while the national annual incidence rate of 31.8 cases per million was the highest national rate reported in the world (Leigh and Driscoll, 2003).

Because of the fiber-like structure of asbestos minerals, they can penetrate deep into lungs when inhaled and cause significant damage to affected tissues. The six types of asbestos minerals listed above are listed as Class 1 carcinogens by world health authorities (IARC, 1987) and are known to cause four main illnesses, including asbestosis (chronic chest disease leading to fibrosis), benign pleural plaques (thickening of pleural membranes), mesothelioma (cancer of the pleura), and lung cancer of the bronchial tubes, lung, and alveoli. In addition, exposure to asbestos has been linked to laryngeal cancer, diffuse pleural thickening as well as cancers in the gastrointestinal tract. The toxicity of asbestos fibers is affected by three main factors: fiber dimension, biopersistence, and dose. The fiber dimension, i.e., the relationship between fiber length and fiber diameter, determines the place of deposition in the lung. Smaller fibers or particles can reach deeper down in the lung and those of a diameter <0.5 μm are generally more likely to cross membranes and enter lymphatic and circulatory systems. Normally these particles would be engulfed by macrophages and transported out of the lungs, but because of the length of these fibers, this function is inhibited. As a result, enzymes, inflammatory mediators, oxidants, and other contents of the macrophages are released into the lung where they can be toxic. With regard to fiber dimension, lung cancer has primarily been associated with fibers >10 μm in length and >0.15 μm in diameter, mesothelioma with fibers >5 μm in length and <0.1 μm in diameter while fibrosis has been associated with particles >5 μm in length (National Institute of Occupational Health and Safety, 2007). Biopersistence of asbestos fibers in the lung is determined by a number of factors including the site and rate of deposition, where a high rate can overwhelm macrophages and affect toxicity, rate of clearance, particle solubility, rate and pattern of breakage as well as translocation across membranes. Chrysotile mineral fibers make up around 90% of the asbestos present today. The chrysotile fibers are curly or wavy (serpentine asbestos) while the other asbestos forms have more needle-like straight and long fibers and are referred to as amphibole asbestos. Amphibole asbestos minerals are usually dominating in mesothelioma cases, while the forms amosite and crocidolite are considered the most hazardous and can persist in the lungs for a long time, even decades (National Institute of Occupational Health and Safety, 2007).

The use of amosite and crocidolite has been banned in Australia since the mid-1980s. Chrysotile was banned by the end of 2003, and it is at present illegal to import, export, use, re-use, or sell any product containing any form of asbestos in Australia, but this ban does not apply to products already in use. Since the closure of Australia's asbestos mines, the largest exposure source today is from renovation of houses built from asbestos-containing material. Australia is ranked 11th in the world for mesothelioma mortality rates because of the widespread use of asbestos-containing products and the high rate of occupational exposure. Around 10% of the cases have no known exposure source (Leigh and Driscoll, 2003) and it has been suggested that these cases may be due to naturally occurring asbestos (NOA) rather than from commercial products (Hendrickx, 2008). This hypothesis is supported by a study from California that found significant correlations between mesothelioma and residential proximity to NOA (Pan et al., 2005).

In eastern Australia, around 0.2% of the land area has potential asbestos-bearing rocks, and this proportion increases to around 2.5% of areas in southern Australia (Hendrickx, 2008). It should be noted that NOA does not constitute a health risk unless disturbed, but those working or living near new urban developments as well as farmers, construction workers, and quarry workers whose activities routinely disturb soil and rock may be at risk of excessive exposures (Hendrickx, 2008). For natural weathering of asbestos minerals, Hendrickx (2008) calculated an annual release rate of about 72.5 tons of asbestos across eastern Australia, although it is unknown to which degree the weathering products result in forms that constitute a risk for human health. Asbestos-bearing minerals have been identified in Western Australia, Tasmania, South Australia, New South Wales, and Queensland.

A high incidence of malignant pleural mesothelioma has also been observed in New Caledonia. In a case–control study including all cases diagnosed with respiratory cancers during 1993–1995, a significant association was found between mesothelioma and the use of a whitewash produced from asbestos-bearing minerals (Luce et al., 2000). A subsequent study including all mesothelioma cases between 1984 and 2002 demonstrated a significant relationship between mesothelioma and chrysotile in soil, which is in turn a likely constituent of the whitewash (Baumann et al., 2007).

Fig. 5 Map of Australian manganese mines, 2008. Adapted from Geoscience Australia: © Commonwealth of Australia (Geoscience Australia) [2009] http://www.ga.gov.au/servlet/BigObjFileManager?bigobjid=GA3030

Metals and Mental Health

Australia is the world's third largest producer of manganese ore, holding 12% of the world's economically demonstrated resources of manganese (Geoscience Australia, 2003; McKay et al., 2008). Manganese is mined at three main sites, of which one is located in Western Australia and two in the Northern Territory (Fig. 5). The largest supplier of manganese ore is the mine in Groote Eylandt, which produces 10% of the world's high-grade manganese ore (BHP Billiton, 2008).

Manganese, albeit an essential element, is a known neurotoxin. A Parkinson-like disease, manganism, was first noted in 1837 in five pyrolusite mill workers and has since been reported in hundreds of occupationally exposed workers (Mergler, 1999; World Health Organization, 2004). Excessive manganese exposure has also been linked to learning disabilities in children (Collipp et al., 1983; Takser et al., 2003; Wasserman et al., 2006; Ericson et al., 2007), manganism in elderly exposed from drinking water (Kondakis et al., 1989), and to a neurological disease among an Aboriginal community of Groote Eylandt in the Northern Territory of Australia (Purdey, 2004), known locally as "bird disease" with features similar to the neurodegenerative disease amyotrophic lateral sclerosis, ALS (Donaldson, 2002). The disease has also been locally called the Groote Eylandt syndrome, but its proper name is Machado Joseph disease (MJD), or spinocerebellar ataxia type 3 (SCA3) a rare incurable hereditary neurodegenerative disease that affects a disproportionately large population of the island.

Machado Joseph disease was first described in the Azores Islands 1972, when two families of Portuguese/Azorean descent, Machado and Joseph, were described with the unique symptoms of the disease. The Azorean Island of Flores holds the highest prevalence of the disease in the world, with 1 in 140 affected, and the prevalence is generally highest among people of Portuguese/Azorean descent. For this reason it has been hypothesized that the disease was introduced into Groote Eylandt and the area by visiting fishermen from Sulawesi, many of whom have Portuguese ancestry (Burt et al., 1993). However, MJD has been identified among a diverse array of ethnic populations around the world, also in families of isolated areas with no apparent Portuguese descent (Purdey, 2004; National Institute of Neurological Disorders and Stroke, 2008).

The familial motor symptom disorders resembling those later attributed to MJD have been reported in Aboriginal communities of the east Arnhem land in the Northern Territory of Australia since the 1960s (Kiloh et al., 1980; Burt et al., 1993; Burt et al., 1996; Purdey, 2004). In 1980, Kiloh and coworkers reported 13 cases of neurological disorder with up to 28 years duration in a population of approximately 1,100 Aborigines living on Groote Eylandt and its adjacent mainland. The authors hypothesized the causative mechanisms to be both genetic and environmental, recognizing the large manganese deposits on the island, which had been mined by open-cut methods since 1962; the main staple food of the toxic cycad nuts; and the ingestion of clay cakes for medicinal and other purposes likely holding high concentrations of manganese (Kiloh et al., 1980).

Purdey strongly supported the hypothesis of an environmental cause of the disease and suggested that the primary cause for the MJD expansion mutation is initiated by external environmental and dietary factors which bring about an excess of manganese and deficiency of magnesium in the human body. He tested this hypothesis by carrying out an epidemiological and environmental study on affected Aboriginal communities on Groote Eylandt and Azorean communities. The study included a survey on dietary, occupational, and lifestyle status as well as analysis of soil, water, and staple foods for a range of trace elements. It was suggested that the onset of the disease on Groote Eylandt in the 1960s was attributed to the 1942 relocation of the community from low- to high-manganese areas and also to the development of the area into an open-cut manganese mine in 1962, resulting in a majority of the population residing in close proximity to a dusty dump of manganese ore (Kiloh et al., 1980; Purdey, 2004). The vegetables grown locally held a manganese concentration of around 1,000 mg/kg, to be compared with yams from non-MJD areas having a mean concentration of around 30 mg/kg. In addition, the customary cooking practice involved covering the food with soil, further increasing manganese exposure (Purdey, 2004). In contrast, the magnesium intake was generally low for the populations due to low concentrations in soil, food, and water, and body stores were likely further depleted from excessive use of culinary salt.

It is generally recognized that manganese can readily substitute for magnesium binding sites involved in the catalysis of certain key enzymes and vice versa. Within normal cells, the concentrations of magnesium ions greatly exceed that of manganese ions. Whenever this balance is upset, the resulting inactivation of enzyme groups dependent on this balance can have devastating consequences. Endonuclease-1 is one such enzyme whose active centers require a combination of magnesium and manganese for catalysis to proceed. This enzyme protects against the expansion and contraction of the same CAG trinucleotide repeats that are implicated in MJD pathogenesis. Since two magnesium ions are needed for the successful catalysis of endonuclease-1, Purdey proposed that the manganese-loaded/magnesium-depleted individual fails to activate this enzyme, thereby leading to the development of the expansion mutation of the ataxin-3 protein that has been heralded as the central core of MJD pathogenesis (Purdey, 2004).

The extensive study by Purdey suggests that while MJD is indeed a genetic disorder, its onset and clinical manifestations are exacerbated by a geologic environment which simultaneously provides excessive manganese and deficient magnesium. Indeed, the Groote Eylandt is the site of one of Australia's three main manganese mines, and the open-cut mine is located in the village of Angurugu, which is also the residence of most of those on Groote Eylandt who suffer from MJD (Cowan, 2006).

In a review on "manganese madness" by Donaldson (2002), it is suggested that several clusters of neurological disorders in the Australia and Oceania region may be derived from a geologic anomaly connected to Groote Eylandt by a line running through approximately 140° longitude. This anomaly has high manganese and/or aluminum concentration and low magnesium and/or calcium. Studies have revealed a correlation between a high incidence of motor neuron disorders in Western Pacific regions with low concentrations of calcium and magnesium in soil and water. The role of calcium is likely its disruption of critical absorptive processes which results in the otherwise regulated uptake from food and water becoming toxic (see also Section "Lytico-Bodig – A Medical Geology Disease?").

Natural Hazards

The Pacific Ring of Fire encompasses a horseshoe-shaped area in the Pacific Ocean in which frequent earthquakes and volcanic eruptions occur (see Fig. 6). It is estimated that 75% of the world's active and dormant volcanoes are situated within the Ring of Fire, which also experiences 80% of the world's earthquakes, including 90% of the world's largest ones. Many of the countries of Oceania are situated in the Ring of Fire and are subsequently affected by some of the 452 volcanoes within this area. The natural activities which impact on human health include earthquakes, volcanic eruptions and landslides, as well as tsunamis, caused by movements generated from these activities.

Earthquakes and Tsunamis

In New Zealand, around 14,000 earthquakes shake the ground every year, although only around 150–200 are large enough to be felt at the surface (Te Ara, 2007). In the last 150 years, eight major earthquakes have caused considerable damage and killed around 300 people. Most deaths occurred as a result of collapsing buildings and falling debris, but collapsing mines, landslides, and mudflows have also caused fatal injuries. The earthquake that caused the most fatalities in New Zealand occurred in 1931 with a magnitude of 7.8, when 256 people died in the Hawke's Bay region and thousands more were injured as the cities of Napier and Hastings were devastated (Fig. 7). As a result of

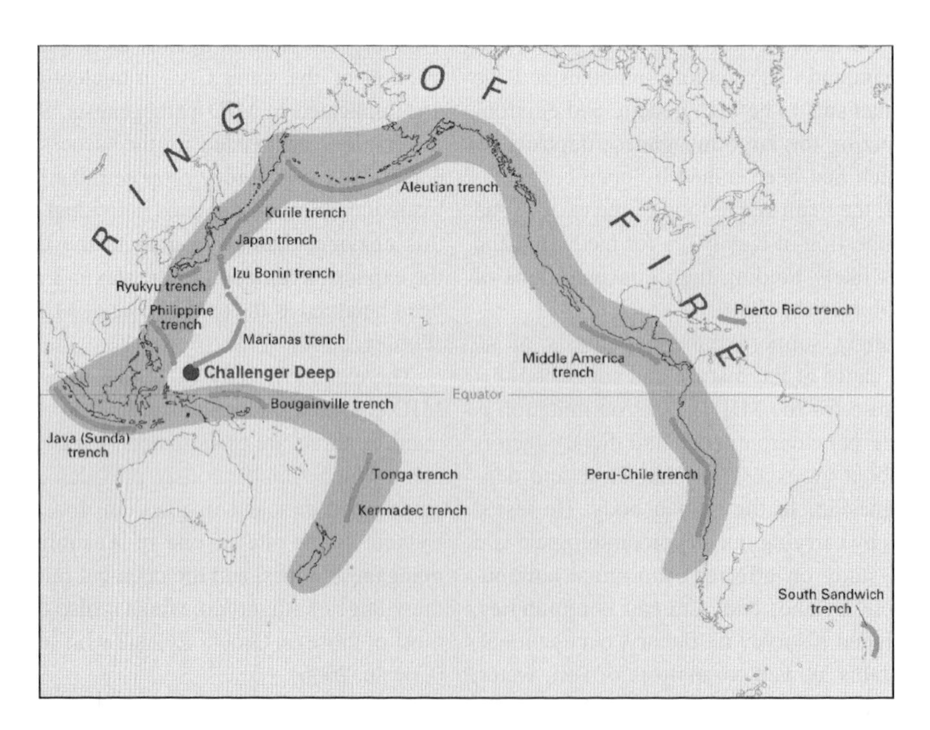

Fig. 6 The Ring of Fire – earthquakes and volcanic activities in the Pacific Ocean. Image courtesy of the US Geological Survey

Fig. 7 Post office damaged by the 1931 earthquake in Hastings, New Zealand Courtesy of Alexander Turnbull Library

the devastation caused by collapsing buildings, guidelines were developed for safer buildings which were the forerunners of the building codes used throughout New Zealand today (Te Ara, 2007).

Earthquakes also commonly occur in other regions of Oceania. The World Data Center for Seismology has recorded numerous earthquakes in Vanuatu, Solomon Islands, Guam, Australia, Tonga, and other islands in the Pacific although few resulted in serious disasters. In Papua New Guinea, 18 earthquakes were reported between 1998 and 2007. Six of them (in 1987, 1993, 1995, 1998, 2000, and 2002) were reported to the UN Office for the Coordination of Humanitarian Affairs (OCHA), which gathers information on humanitarian emergencies and disasters. An earthquake in 1998, which struck 25 km off the Papua New Guinea coast with a magnitude of 7.0, caused an undersea landslide which in turn caused a tsunami. When the tsunami hit the coast, over 2,000 people were killed, thousands were injured and almost 10,000 were made homeless (US Geological Survey, 1998). According to the Intergovernmental Oceanographic Commission (IOC), around 60% of all tsunamis take place in the Pacific Ocean and are mainly caused by undersea earthquakes (Intergovernmental Oceanographic Commission, 2008). More recently, on September 29, 2009, an earthquake of magnitude 8.0 occurred off the coast of Samoa, triggering a tsunami which resulted in at least 110 people killed in Samoa, 22 in American Samoa, and 7 in Tonga and was felt as far away as Wallis and Futuna Islands (US Geological Survey, 2009).

Volcanic Eruptions

It has been estimated that around 300,000 people have been killed in the past 400 years as a result of volcanic eruptions. At present, more than 500 million people, or 10% of the world's population, live within 100 km of an historically active volcano (Feldman and Tilling, 2007). More than half of the world's active volcanoes above sea level are found in the Ring of Fire. However, the definition of an "active" volcano varies with the time frame being considered: according to the Smithsonian Institute, around 20 volcanoes are erupting now, 50–70 each year, 150 each decade, and around 550 eruptions have occurred historically, which implies that the event was documented (Smithsonian Institute, 2009). Death and disease from volcanic eruptions have resulted from physical injury from ejecta, lava, lahars, and tephra and also from indirect physical injuries similar to those of earthquakes, including collapsing buildings. The health hazards related to volcanic eruptions are not only the result of the actual eruption but also due to emissions of potentially toxic substances in the periods between eruptions (Hansell et al., 2006). According to Feldman and Tilling (2007), approximately 70% of fatalities associated with volcanoes are due to direct impacts of eruptions, such as pyroclastic flows, lahars, tsunamis, and toxic gases, while the remainder historically has been caused by post-eruption famine and disease epidemics. Health effects of a non-physical character, such as increased exposure to metals and volcanic ashfalls, will be discussed in "Geologic Emissions" below.

During the 20th century, 490 volcanic events around the world resulted in impacts on 4–6 million people, including the destruction of homes and evacuation. Around half of the events had fatal outcomes, resulting in the deaths of 80,000–100,000 people (Hansell et al., 2006). In the Oceania region, the Relief Web of the United Nations registered 10 volcano-related emergencies in Papua New Guinea between 1984 and 2006 and 5 in Vanuatu between 1995 and 2008. Several of these reported cases of respiratory and eye irritation complaints and a risk for contamination of water supplies and crops from ashfalls (Global Identifier Number, 2009). In New Zealand, the deadliest eruption in recent history was the Tarawera eruption of 1886. At the time, it was not realized that the mountain was a volcano, and tourists were drawn to the many hot springs in the area, especially the Pink and White Terraces on the shore of Lake Rotomahana. In an eruption lasting less than 6 h, a 17 km rift spewed steam, mud, and ash covering hundreds of square kilometers (Te Ara, 2007). Where the terraces had once been, a 100-m deep crater was now opened up. Steam eruptions continued for many months and within 15 years, the crater had filled up with water, and a new and larger Lake Rotomahana took form (Te Ara, 2007).

The Ring of Fire coincides with some of the most populated areas in the world, and as the global population increases, the risk for catastrophic results from future eruptions is considerable. Lava flows only account for a small part of the fatalities of volcano eruptions, given that these are often predictable and allow time for evacuation. Larger numbers of people are affected by destroyed crops and animal food supplies, contaminated water supplies, and air polluted by volcanic ash (Hansell et al., 2006). Another threat is from lahars, an Indonesian term for volcanic mudflows, which form when water mixes with hot or cold volcanic debris to a consistency similar to wet concrete. They can travel very rapidly down the flank of a volcano and are extremely dangerous as they can form not only before, during, or after an eruption but also by non-eruptive forces such as heavy rainfall and earthquakes (Box 5). Any settlements in the path of these mudflows can be rapidly buried (Feldman and Tilling, 2007).

Box 5 Lashing Lahars

Lahars present an ongoing threat to communities and travelers near Mount Ruapehu in New Zealand. In 1953, a railway bridge was smashed from a lahar flowing down the side of the mountain just minutes before a passenger train approached, resulting in 151 deaths. Smaller lahars reportedly traveled down the mountainside in 1969, 1971, 1995, and 1996. Thanks to an early warning system, no one was injured in 2007 when a lahar again flowed down the mountainside to form a river of mud 30–40 m across, which rose 6–8 m above an access bridge (Earth Observatory, 2008). While no injuries were reported during the increased activity of Ruapehu in 1995/1996, it had large impacts on the local economy as the popular ski areas had to be closed, power supply was frequently interrupted by ashfalls, and traffic had to take detours. Ashfalls also intermittently disrupted air travel in the North Island of New Zealand, with financial losses of NZ$2.4 million (Te Ara, 2007).

Geologic Emissions

Dust

The issue of dust and associated disease in the Australia and Oceania region is mainly limited to the Australian continent. Approximately 40% of the land area is in the arid zone, covered by dunes, playa lakes, and other desert landforms which all represent major sources of dust (Greene et al., 2001). Dust storms occur frequently in Australia and are for the most part restricted to the inland areas where significant amounts of topsoil are uplifted by the wind, with devastating consequences in arable areas (Fig. 8). With respect to

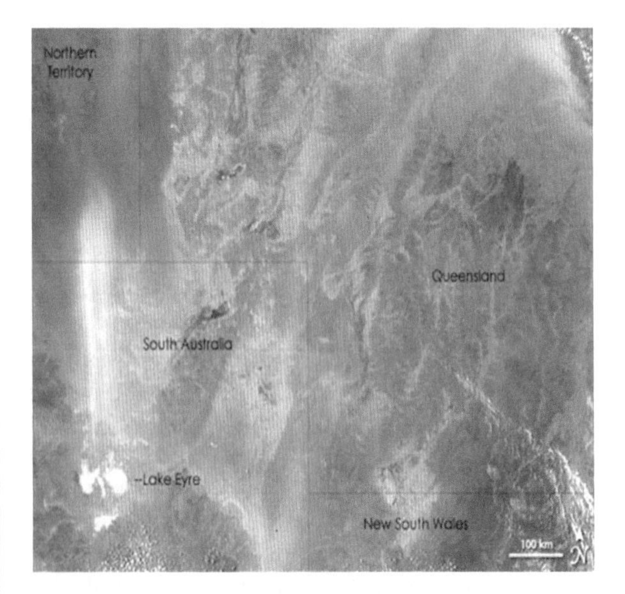

Fig. 8 Dust storm over eastern Australia, February 2, 2005. An estimated 5 million tons of topsoil was blown away in Queensland and plumes of windblown dust can be seen in the bottom left corner, escaping from the dry salt pan lake bed of Lake Eyre (Earth Observatory, 2005a)

human exposure, the effects of dust storms are mainly observed in drought years when huge amounts of dust can be transported by wind to the more densely populated coastal areas. Many dust storms were generated during the extended drought of the 1930s and 1940s (Bureau of Meteorology, 2009a, b). In Sydney in 1944, the dust falls were sufficiently severe for city lights to be turned on in the afternoon, and railway lines were blocked by sand. In Adelaide, several dust storms in December of the same year disrupted train traffic and also required lights to be turned on, in some cases early in the morning. In January of the following year, a number of dust storms struck both New South Wales and Victoria, resulting in power cuts and four children losing their way home from school, only to be found after the storm had subsided. After a continued drought inland, the sun over Melbourne appeared blue through the dust in late March of 1945 (Bureau of Meteorology, 2009c). Box 6 describes one of the worst dust storm events in modern time, which descended on the city of Melbourne in 1983 (Fig. 9).

Box 6 Dusty Dawn – Melbourne, February 1983

Late in the morning of February 8, 1983, a strong, but dry, cold front began crossing the state of Victoria, preceded by hot, gusty northerly winds. It was an El Nino year and the normally dry sandy areas of Mallee and Wimmera located west of Melbourne were exceptionally dry. The loose topsoil was quickly uplifted by the wind and eventually grew into a massive cloud of dust as the front moved east. By 11 am, the dust cloud was visible at Horsham, in western Victoria; an hour later it had obscured the sky. In Melbourne, the temperature rose quickly as the north wind strengthened and had reached a February maximum of 43.2°C (109.8°F) by mid-afternoon. Half an hour later, at 3 o'clock, visibility was reduced to 100 m as a spectacular reddish-brown cloud reached Melbourne, accompanied by a rapid temperature drop and a wind strong enough to uproot trees and unroof about 50 houses (Fig. 9). The worst of the dust storm was over by 4 o'clock when the wind speed dropped rapidly (Bureau of Meteorology, 2009c).

At its height, the dust storm extended across the entire width of Victoria. The dust cloud was up to 320 m deep when it struck Melbourne, but in other areas extended thousands of meters into the atmosphere. It was estimated that about 50,000 tons of topsoil was stripped from the Mallee – about a fifth was dumped on the city – leaving the ground bare and clogging open water channels with sand and dirt (Australian Bureau of Statistics, 2008b).

The enormous volumes of dust transported by wind and deposited around Australia increase the load of inhalable dust in the air, which may cause ill-health (Greene et al., 2001). Few studies have been conducted on the associations between naturally occurring particulate exposures and respiratory health outcomes, as most studies on respiratory disease have focused on urban emissions. However, a number of dust events near Brisbane between 1992 and 1994 were significantly associated with changes in asthma severity although general relationships were not determined (Rutherford et al., 1999). In a study on respiratory health of two remote Aboriginal communities in Western Australia, the community located in the central desert had greater prevalence of respiratory morbidity than the one from a coastal community in the tropical north. The differences were not explained by doctor-diagnosed asthma, allergy, or increased airway responsiveness and were present despite the lower prevalence of cigarette smoking in the desert community adults. It was concluded that other environmental

Fig. 9 Dust storm over Melbourne 1983. Photograph courtesy of Trevor Farrar (Bureau of Meteorology, 2009c)

factors, such as infection, may be responsible for the differences observed between the communities, but their residence in a dusty environment has also likely had an impact (Verheijden et al., 2002).

The impacts of dust on health have been recognized by a number of Australian state governments. Fact sheets on the potential health effects of dust exposure have been issued by the New South Wales Department of Health and the Victoria Department of Human Services, and a guide on reducing the health impacts of dust has been included in the National Indigenous Housing Guide (Commonwealth of Australia, 2008). In addition to the direct health effects of dust, such as trachoma, respiratory disease, and skin infections, the impact on "health hardware" is recognized, which shortens the life span of washing machines and water heaters, which in turn may cause disease through decreased hygiene. It is also recognized that dust can transport pathogens and contaminate water supplies. Almost 60% of the 3,662 houses across Australia included in surveys conducted on health and indigenous housing by the Australian Government were located in a hot, dry, and dusty environment, with summer temperatures regularly reaching above 40°C (104°F) (Commonwealth of Australia, 2008). Moreover, safety issues on the road are recognized as visibility may deteriorate very quickly during a dust storm, increasing the risk of accidents (New South Wales Health, 2003; Department of Human Services, 2008).

Because of the large mining industry in Australia, and the often arid location of mining sites, dust generated from mining activities may impact on residents nearby. It is primarily the nuisance of the dust itself that has received most attention. In Port Hedland, Western Australia, iron ore is exported through the port and a study on the health impacts of this activity was prompted by concerns from local residents. It was found that the residents of the town have a small but increased risk of hospitalization due to respiratory conditions. However, it remains uncertain whether this health effect could be attributed to the dust levels. A similar increase in rates of cardiovascular and gastrointestinal diseases was also observed (Department of Health, 2006). An additional problem arising from the movement of dust across Australia is the spread of salt from dry lakes, which can cause soil salinity as well as respiratory problems.

Deposition of dust may also have beneficial effects. A wind-blown clay deposited in southeastern Australia during the Quaternary is referred to as "parna" (Summerell et al., 2000), and its high clay content has positive effects on the physical and chemical characteristics of soil (Greene et al., 2001). It seems parna was primarily deposited in vegetated areas of high rainfall, as the vegetation slowed the winds transporting the dust and also reduced any subsequent remobilization once deposited. Further, higher rainfall resulted in taller vegetation and a higher rate of deposition (Summerell et al., 2000). It has been estimated that, over a specific area, 13 million cubic meters of dust has been deposited over 500,000 years, while 26 billion cubic meters was lost (Greene et al., 2001), indicating a significant movement of soil and dust across the continent.

In addition to the exposure to dust particles, exposures to toxic agents within the particulates may result in respiratory disease. Exposures to particulates containing elevated concentrations of silica (Gillette, 1997) and asbestos, or exposure to particulates containing these agents at lower concentrations but under high dust loading rates, as from occupational exposures can cause lung cancer (Baris et al., 1987; Attfield and Costello, 2004). Exposure to metals in particulates is known to cause interstitial pulmonary fibrosis (Balmes, 1987), asthma (Costa and Dreher, 1997), and chronic obstructive lung disease (Nemery, 1990). Two primary sources of toxic agents in dust are volcanic emissions and bushfires, which will be examined in more detail below.

Volcanic Emissions

Volcanic activity results in a number of hazards to human health. In addition to the more obvious impacts from the actual explosion, such as direct physical injury caused by ejecta (the products of eruption, including gas, lava, tephra, and lahars) or from indirect effects of eruption (such as building collapse, population displacement, and fire), volcanic activity also results in dispersal of a great number of metals, sometimes in great quantities (Cook and Weinstein, 2005). Volcanic activity may also release fatal gases, such as sulfur dioxide, carbon dioxide and carbonic acid, and hydrogen sulfide. For example, three people died of suffocation in 1990 as carbon dioxide was released from a vent of the side of the volcano Tavurvur in Papua New Guinea. Three more people died when trying to retrieve the bodies (Global Volcanism Program, 1990). In spite of the potentially large impact on human health from volcanic gases, a recent systematic review found few primary studies (Hansell and Oppenheimer, 2004). However, with regard to the geological activity in the Australia and Oceania region, it can be assumed that health effects described elsewhere from volcanic emissions are likely to also affect the Pacific population. For example, the Ambrym volcano on Vanuatu was the site of the most significant point source of sulfur dioxide on the planet in the first months of 2005. The emissions were not due to a large eruption but rather to leaking from passive or non-eruptive emissions. The resulting acid rain killed crops and contaminated water supplies, leaving the population dependent on food aid (Earth Observatory, 2005b).

Gaseous Emissions

Carbon dioxide and hydrogen sulfide are both highly toxic gases that are released from volcanic eruptions and because they are denser than air they can collect in low-lying areas. Carbon dioxide can cause unconsciousness and death if inhaled at high concentrations (>20–30%) while hydrogen sulfide can cause death after a single breath if concentrations are high (Feldman and Tilling, 2007). Asphyxiation has been reported from accumulations of hydrogen sulfide and carbon dioxide from volcanic and geothermal sources while chronic exposure to hydrogen sulfide has been associated with increases in nervous system and respiratory diseases (Hansell and Oppenheimer, 2004). Sulfur dioxide and acid aerosols from eruptions and degassing events have been associated with respiratory morbidity and mortality in those with pre-existing disease. In a recent epidemiological study on exposure to volcanic sulfur dioxide emissions, 35% of the 335 participants perceived health effects from a volcanic eruption, and increased risk estimates were associated with a number of both self-reported and diagnosed cardiorespiratory diseases (Longo, 2009). Hydrogen fluoride is also emitted from volcanic activity and although there have been no fatalities reported as a direct result of its exposure, it can cause crop failure and death of livestock, with subsequent famine (Feldman and Tilling, 2007).

In Rotorua, New Zealand, the population is continuously exposed to geothermal emissions. A health assessment conducted in the town found exposure–response trends, particularly for nervous system, respiratory, and cardiovascular diseases (adjusted for age, ethnicity, and gender) (Bates et al., 2002). These results are supported by the findings of a spatial analysis of respiratory disease in the same area, where clusters of increased relative risk for obstructive pulmonary disease coincided with geothermal fields (Durand and Wilson, 2006). A study conducted in Hawaii also found significant associations between cardiorespiratory diseases and chronic exposure to sulfur-containing volcanic emissions (Longo et al., 2008). Both of the two latter studies concluded that there is an urgent need for further community studies and for greater collaboration between epidemiologists and volcanologists.

Volcanic Ashfalls

Ash from volcanoes does not refer to the burnt substances often regarded as ash, but instead to ejected particles, including very fine material that may be inhaled. Although inhalation of ash may be a relatively benign health hazard to healthy individuals as the particles are often too large to inhale, and any high-dust concentrations settle relatively quickly, the fine ash fraction can be dangerous to individuals with asthma or other respiratory diseases (Feldman and Tilling, 2007). For example, the 1995/1996 Ruapehu eruption in New Zealand resulted in the deposition of 7 million tons of ash over North Island and the principal health outcome was a 44% increase in hospital admissions for bronchitis (Cook and Weinstein, 2005). Another issue with regard to ashfall is the risk of death or injury from house collapse since as little as 10 cm on a flat roof may cause its collapse, especially if it gets wet (Fig. 10). The 1991 eruption of Mt. Pinatubo in the Philippines unluckily coincided with a typhoon which resulted in the killing of at least 300 people from collapsing roofs due to ashfall. Ash may also increase injuries from road accidents as visibility decreases and roads become slippery (Feldman and Tilling, 2007).

One of the most severe consequences of ashfalls is the destruction of crops and other vegetation and its contamination of waterways. According to Feldman and Tilling (2007) famine from destroyed crops was the number one cause of death associated with volcanic

Fig. 10 Vehicles buried in ash in the 1994 Rabaul eruption, Papua New Guinea. Courtesy of Cascades Volcano Observatory/US Geological Survey

activity before the 20th century. For example, while 10,000 people died from direct injuries during the largest historical eruption on Earth, in Indonesia in 1815, over 82,000 died from disease and starvation as a result of ruined crops. In 1865, the Maori told of how the eruption of Mt. Ngauruhoe in New Zealand resulted in several inches of black dust covering the ground and how the fish living in a small lake in the area had been poisoned by the ash (Te Ara, 2007). Nearby, ashfall from the eruption of Mt. Ruapehu in 1945 gave rise to "Ruapehu throat" (throat irritation) and eye irritation. Ash also killed crops, contaminated water supplies, and affected stock as pastures were rendered inedible by the ash cover. Dry ash was stirred up from the ground and in addition to being a nuisance, it damaged machinery. As a result, the New Zealand army had to move 700 vehicles from its base in the Ruapehu area (Te Ara, 2007).

Volcanic eruptions may also contribute to the mobilization of harmful metal and halide ions. In a review on water contamination from ashfalls and subsequent health issues, it was concluded that fluoride was the main element of concern together with an increased risk for infectious disease outbreaks due to inhibited drinking water disinfection (Stewart et al., 2006). In a simple model of the impact from the 1995/1996 Ruapehu eruptions on water supplies, it was found that there was a risk of exceeding World Health Organization (WHO) drinking water guidelines for iron, aluminum, and manganese. Whereas guideline values for iron and aluminum are set as aesthetic limits, the value for manganese in drinking water is set with regard to the increased risk for neurological damage (World Health Organization, 2004). In a study on White Island in New Zealand, an active offshore volcano serving as research station and tourist attraction, 10 volunteers measured increased urinary aluminum after only 20 min exposure (Durand et al., 2004). In a separate study of volcanic ash effects on public health in Vanuatu, it was found that drinking water supplies exceeded guidelines for fluoride, even in times of low volcanic activity, increasing the risk for chronic dental and skeletal fluorosis (Cronin and Sharp, 2002). In general, acute health effects from volcanic emissions have been the focus of past epidemiological research because they are easier to detect. In contrast, little work has been conducted on the long-term effects of living close to a volcano. Metal exposure from such events has been suggested as an under-evaluated health risk related to volcanoes (Cook and Weinstein, 2005).

Bushfire

Particulates released from wildland fires can include a mix of smoke particles derived from vegetation combustion and soil particles brought into the plume through convective updrafts: combustion-derived organic compounds such as polycyclic aromatic hydrocarbons may thus mix with mercury and other metals, as well as crystalline silica or asbestos if such minerals are components of the underlying soils. Wildfire ash and burned soils can be the source of dust with elevated levels of some metals. The types and abundances of the metals or metalloids are a function of the underlying bedrock geology, soil type, vegetation burned, and intensity of the burn. Dust from areas recently burned by wildfires may contain elevated levels of wildfire ash, which is a well-known source of caustic alkalinity that can lead to similar irritation of the eyes, respiratory tract, and skin (Plumlee and Ziegler, 2007).

Bushfires are naturally occurring events across the Australian continent at any time of the year, with the greatest extent of bushfires occurring in the savannas of northern Australia and in the southeastern corner of the continent, south of a line from Sydney to Adelaide. Due to climatic variation across the nation's vast land mass, the bushfire seasons vary in different parts of Australia (Fig. 11). In southeastern Australia,

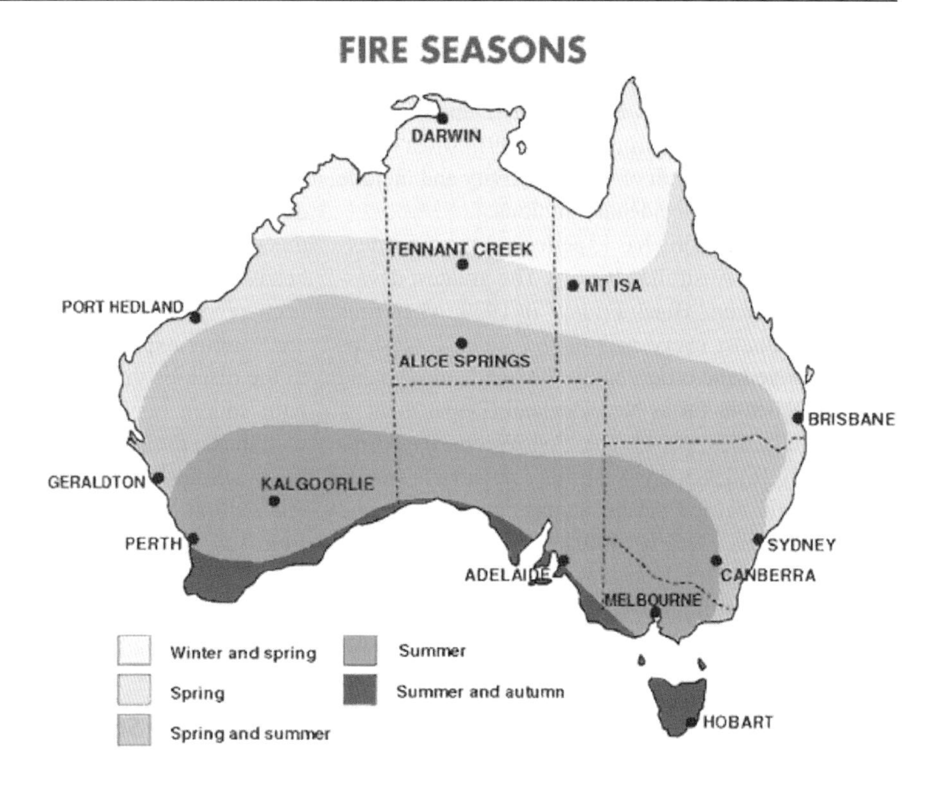

Fig. 11 Bushfire seasons across Australia (Bureau of Meteorology, 2006)

including Tasmania, the major fire season is summer and early autumn; in coastal New South Wales and southern Queensland, spring and early summer; and in much of northern Australia, winter and spring. In the arid zones of Australia large fires most commonly occur in the months following an abnormally wet season, when there is enough vegetation to provide fuel (Bureau of Meteorology, 2006).

For more than 40,000 years, Aboriginal people have used fire for a variety of purposes, including signaling, game hunting, cooking, tree felling, illumination, and clearing the land to stimulate re-growth of native grasses (Singh and Geissler, 1985). The journals of early navigators contain many references to the sighting of smoke along the Australian coastline. In the Geelvinck, a ship's log from Willem De Vlamingh's Dutch expedition to Terra Australis (1696–1697), Rottnest Island outside Perth in Western Australia is referred to as Mist Eiland (Fog Island). It is suggested that the island came into view through a haze of smoke blown from the mainland, where many bushfires, lit by the Aborigines, were then burning (Playford, 1998). The consistent observation of the first explorers was of open forests and woodlands with grassy understorey, with an impression of an annual conflagration during the fire season. In May 1770, Captain James Cook described the vegetation and country on Botany Bay's southern shore as follows:

> The moors looked like our moors in England and as no trees grew upon it but everything was covered with a thin brush of plants about as high as the knees (Banks, 1770).

Cook's description fits with vegetation which is burned annually or, in the case of heathlands, every 2 or 3 years (Cheney, 1995). These observations were repeated by other explorers, who referred frequently to fires smoldering or raging, according to the nature of the fuel conditions. During the 19th century, Australia underwent a process of sparse settlement, largely through use of axes and fire sticks to carve small farms out of the bush. Despite the general view that European settlement increased the frequency of bushfires in Australia, it is clear from historical evidence that as settlement increased, fire frequency decreased. However, as a result, the fuels built up and, consequently, fires that did occur were more intense than those under the Aboriginal annual burning regime (Luke and McArthur, 1978).

Bushfires are caused by natural processes, human activities, or a combination of both. Lightning strikes are the cause of virtually all naturally occurring bushfires and account for approximately 26% of all bushfires on public land in Victoria (Department of Sustainability and Environment, 2009). Most other bushfires start as a result of human activity and include both deliberate and accidental ignitions.

The state of Victoria has experienced the most devastating fires in Australian history. The greatest devastation occurred on "Black Friday" in 1939, when fires burned in almost every part of the state. Townships were destroyed and others badly damaged, and ash and smoke reached as far as New Zealand (Department of Primary Industries, 2004). The "Ash Wednesday" fires of February 16, 1983, are Australia's most well-known bushfires and caused severe damage in Victoria and South Australia. Nearly 400,000 ha of land were burned, and property damage was estimated at approximately $200 million (Department of Sustainability and Environment, 2006). Most recent is the "Black Saturday" bushfires in Victoria in February 2009. With a death toll of 210, and an area of 450,000 ha burned, this is the worst disaster in recorded history in 110 years in Australia.

Many epidemiological studies worldwide have demonstrated a consistent association between particle air pollution from bushfire smoke and respiratory disease. Yet, only four of six Australian studies have shown an association between bushfire smoke and respiratory health outcomes, such as asthma (Churches and Corbett, 1991; Cooper et al., 1994; Smith et al., 1996; Jalaludin et al., 2000; Johnston et al., 2002; Chen et al., 2006). Given the high frequency and severity of bushfires in Australia, and the impact on both the environment and the population, it is evident that bushfires play a key role in the lives of many Australians. Whether it is through the destruction of land and structures, the devastating effects on flora and fauna, or the injuries and loss of human life, the impacts of bushfires are dramatic, extensive, and may have long-term consequences.

Soil

Soil quality can have profound effects on human health. It influences the degree of human exposure to both potential contaminants and beneficial elements through its role as substrate for plant growth; it acts as a natural filter for drinking water and a contributor to the dust load in ambient air. The main exposure pathway to soil is through ingestion, which primarily occurs involuntarily and can give rise to diseases linked to bacteria in the soil as in the case of melioidosis discussed below, or through toxic exposure to metals. Soil may also be ingested deliberately and this behavior is referred to as geophagia or geophagy. It has been reported from all parts of the world and there are various explanations for its practice. In addition to the diseases that result from the natural interaction with soil in our everyday lives, human interference with the landscape may interrupt ecosystems and result in sometimes unexpected effects on human health. In this section, both the natural impact on human health from soil and impacts resulting from human interference will be explored in the settings of Australia and Oceania.

Acid Sulfate Soils

Acid sulfate soils are naturally occurring soils containing sulfidic minerals which, when in contact with air, oxidize to produce sulfuric acid. The acid produced causes a significant drop in soil pH, an increase in soil metal mobility, and a decrease in soil nutrient availability (Dent, 1986). For these reasons, their impact on vegetation as well as water quality is severe, with reported soil scalds (areas lacking vegetation) and massive fish kills (Brown et al., 1983; Callinan et al., 1993; Russell and Helmke, 2002; Rosicky et al., 2004; Fältmarsch, 2006; Stephens and Ingram, 2006; White et al., 2006). The main impact of acid sulfate soils once disturbed constitutes the increased mobility of acid and metals, which may pollute drinking water sources with potentially toxic elements and result in food shortages through ruined crops and killing of aquatic organisms. Acid sulfate soil scalds are relatively minor on a global scale, but may have impacts on air quality on a local scale through increased dust generation and thereby respiratory health. In Australia, an increase in vector-borne diseases including the Ross River and Barma viruses has been reported because of the preferred breeding patterns of disease-carrying mosquitoes in acidic pools (Broom et al., 2000; Macdonald et al., 2006). Figure 12 shows the range of direct and indirect impacts of improper

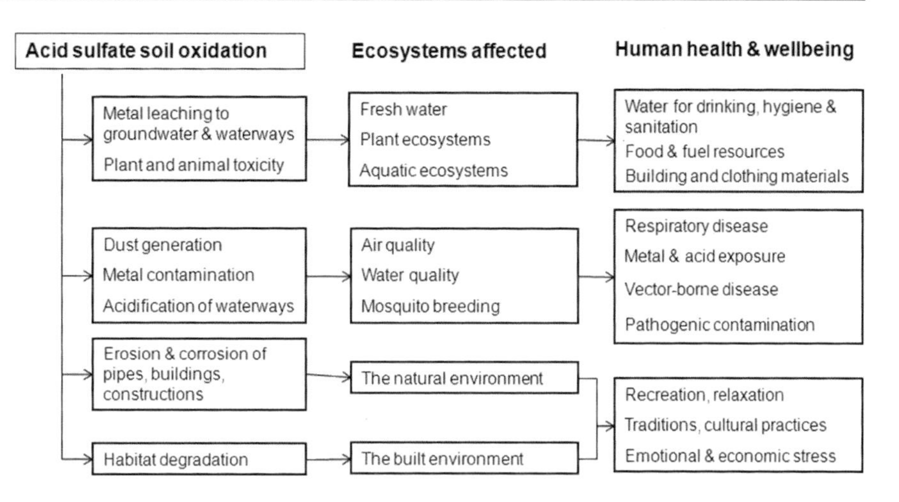

Fig. 12 Environmental effects of acid sulfate soil disturbance, corresponding ecosystem services that may be affected, and adverse impacts on human well-being (adapted from Ljung et al., 2009)

management of acid sulfate soils on ecosystems and how these impacts in turn affect human health and well-being.

Acid sulfate soils have been identified as a major problem in coastal areas of Australia, as almost all of its coastal land is underlain by sediments containing iron sulfide minerals. These floodplains are under pressure from both urban and agricultural development. About 80% of Australians currently live in the coastal zone, with 25–30% of the coastline being subjected to increasing development expected to continue over the next 5 decades (White et al., 1999). The coastal floodplains on the eastern coast of Australia were the first to be developed for agriculture following European settlement because of their favorable climate and environmental conditions (White et al., 2006). The government actively encouraged development and flood mitigation through drainage and flood gates which successfully controlled excess water, but have resulted in severe water quality and ecological problems (White et al., 1999). In Queensland, drainage for sugarcane production has resulted in the release of 72,000 tons of sulfuric acid since 1976, acidifying 110 ha of land and causing an increase in aluminum, zinc, iron, and arsenic concentrations in nearby waters (Hicks et al., 1999). Concern has been raised about indirect effects on the Great Barrier Reef Marine Park, with the location of important spawning and nursery areas for reef fish in recipient areas of acid sulfate soil discharge (Cook et al., 2000). Fish kills and run-off from acid sulfate soils have affected the livelihoods of commercial fishers, tourism operators, oyster farmers, and sugarcane producers (White et al., 2006). In Queensland, costs attributed to acid sulfate soils have been estimated at AUS$189 million (Powell and Martens, 2005), and in New South Wales, the overall annual costs to fishery resources have ranged between AUS$ 2.2 and 23 million (New South Wales Government, 2008). Because of the alterations to natural drainage caused by past mitigation schemes, episodic ecological emergencies can continue for decades. Floodplains continue to discharge acid water for centuries once the oxidation processes have been initiated, and floodgates form acid reservoirs that leak acidic water to the surrounding environment for months (White et al., 1997).

Disturbance of acid sulfate soil underlying a suburb to the Western Australian capital of Perth resulted in a severe decrease in groundwater quality in 2002. The contamination was due to the excavation of peat-based wetlands in the area for housing developments and subsequent disposal of the material in stockpiles which underwent acidification as the peat oxidized. In addition, a combination of prolonged dry seasons and groundwater abstraction in the area contributed to lowering the groundwater table by as much as 6 m, causing a subsequent oxidation of subsurface acid sulfate soils (Angeloni et al., 2004). The issue came to light after a resident reported that their vegetable gardens were dying as a result of bore water irrigation (Water and Rivers Commission, 2002). Investigations showed that the ground water in the area had pH values as low as 2.25, arsenic concentrations up to 1,000 times the WHO guideline value for drinking water, and elevated concentrations of a number of metals.

Box 7 Soil Salinity – The White Death

The area of salt-affected land in Western Australia is increasing at a rate of one football field per hour, and soil salinity costs both New South Wales and South Australia millions of dollars every year in road and building maintenance (National Action Plan, 2001). Salinization of soil in dry areas, dryland salinity, is the result of forest clearance, often for agricultural production, which replaces deep-rooted plants with shallow-rooted ones. As a result, natural salts previously bound in the unsaturated soil above the groundwater table now dissolve and are brought to the surface as the groundwater table rises (National Action Plan, 2001).

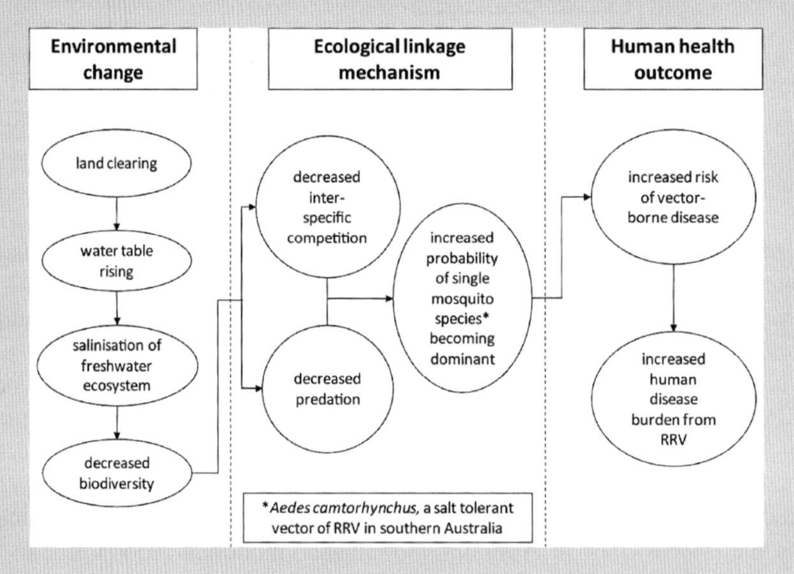

Fig. 13 Linkages between deforestation, mosquitoes, and human health. Modified from Jardine et al. (2007)

In Australia, an unexpected indirect health effect of salinization has been identified. A study on population dynamics of mosquitoes in Western Australia found that the development of dryland salinity and water logging has led to an increase in the mosquito specie primarily responsible for carrying the Ross River virus (Fig. 13) (Jardine et al., 2008). In addition to increased risk of Ross River virus, a recent review on the health implications of dryland salinity in Western Australia concluded that the salinity-induced environmental degradation may affect respiratory health through increased wind-borne dust and also mental health through adverse effects on livelihood and recreational areas (Jardine et al., 2007).

Melioidosis

Melioidosis is an infectious disease caused by soil bacteria (*Burkholderia pseudomallei*). Exposure pathways include ingestion of contaminated soil and/or water, inhalation of dust, or exposure through skin wounds. The disease is manifested by abscesses in skin and internal organs, pneumonia-like symptoms, and septicemia (presence of pathogenic organisms in the bloodstream), but it is not uncommon for the infected person to remain free of symptoms. However, in those who develop clinical disease, the outcome may be fatal even with correct treatment. The bacterium is found in wet soils, accounting for the greater frequency of the melioidosis in tropical areas. It has commonly been referred to as "Nightcliff gardener's disease," referring to the high incidence among home gardeners in Darwin's northern suburbs, including Nightcliff. Other common names for the disease include paddy-field disease, due to the susceptibility of rice paddy farmers who work in close contact with muddy waters, and Vietnamese tuberculosis, after the tuberculosis-like symptoms of the chronic forms of the disease which affected many Americans after the Vietnam War (White, 2003). The risk of mortality greatly increases in patients with predisposing conditions such as lung or kidney disease, cancer, alcohol abuse, and those on long-term immunosuppressive drugs. It has also been reported that up to 50% of patients with symptomatic melioidosis have pre-existing type II diabetes (White, 2003). Relapse is rare in children but adult patients require follow-ups throughout their lives (White, 2003).

The disease is endemic in parts of Southeast Asia, in northern Australia, and in the Torres Strait Islands. The incidence rate is particularly high in the Northern Territory of Australia with an annual disease rate of 16.5 per 100,000 and it has been recognized as the most common cause of fatal community-acquired bacterial pneumonia at the Royal Darwin Hospital (Currie et al., 2000). During a 10-year period starting in 1989, 252 cases were confirmed in this area, most during the wet season (85%) with 49 deaths (19%). The mean age of the affected patients was 47 years, but ranged from 16 months to 91 years. Most patients were men (75%) and just over half were Aboriginal (Currie et al., 2000). Between October 2004 and February 2005, 24 cases were confirmed, a significant increase compared with previous years (Healthy Living Northern Territory, 2005). The mortality rate has decreased in recent years due to earlier case detection and intensive care treatment (White, 2003). Three main features distinguish the presentation of the disease in Australia compared to that of Thailand: the proportion of children affected is lower (4% compared to 15%); infection of the urinary system and reproductive organs is more common (15% compared to 2%); and brainstem encephalitis is more common (4% compared to <0.2%) (White, 2003).

Geophagy

Geophagy, or the eating of soil, occurs both due to neuropsychiatry and developmental disorders and as a normal practice in cultures all over the world. In the extensive review by Laufer published in 1930, the applications of soil are described as varied as the cultures in which geophagy is practiced: as substitute for food in times of scarcity, as condiment, as medical ailment for a range of disorders, and as a taste enhancer (Laufer, 1930). Rowland (2002) points out that the ingestion of soil during famine may arise from its detoxifying properties when people are confined to eating less nutritious and often toxic "famine foods." Laufer stresses that the behavior is individual and should never be ascribed to an entire group/tribe/nation only because it has been observed by some of its members. It occurs everywhere, regardless of climate, race, culture, affluence, or modernity. With these caveats in mind, we will present the practice of geophagy in the areas of Australia and Oceania.

Geophagy was reportedly practiced by the Maori of New Zealand in times of scarcity. The village of Kaiuku, for example, means "eat clay" and got its name from the ingestion of uku, a white soapy clay, as food supplies ran out during a long siege in 1832 (Laufer, 1930). Clay eating in times of scarcity was also reported from the Rotorua and Lake Taupo regions on the North Island. The British settlers referred to a fine gray yellow ooze ejected by the volcanoes in the area as "native porridge," and while the same mud springs are described elsewhere, the use of them was reportedly limited to cooking and not eating. Geophagy is not widely reported from any other locations in Polynesia, where instead soil was mainly used as a base for paint on buildings (Laufer, 1930).

Many travelers reported geophagy in New Guinea, where red, gray, and yellow clays were eaten. The soils of the different regions were distinguishable in taste and were used for barter. For example, a white clay found inland was used as a relish and a face-paint and was brought to the coast for exchange. A different type of clay was found in coastal areas, a gray marine clay, which was described as being helpful for digestion. It was particularly popular with pregnant women who stated that it was beneficial for the fetus, with a recommendation of daily ingestion until delivery. A fatty (heavy) clay of ochre color was reportedly used on an island off the coast of Papua New Guinea for the treatment of digestive problems (Laufer, 1930). Soil ingestion was also observed in New Caledonia, in some areas by men and in others only by women, reportedly eaten to fill the stomach or in response to craving sensations, similar to the use of tobacco or opium. Different clays were eaten in different areas, including steatite containing a significant amount of copper eaten in Baladea; a friable grayish earth in Baaba; weathered yellow marl crushed and eaten in Mare; and red earth eaten after it had been burnt. An aluminous soil full of organic debris was reportedly eaten in the Loyalty Islands, while a tough dark brown clay mixed with organic materials was ingested by pregnant women in North Santo and Malekula. In Santo, the clay was flattened out like a biscuit and dried in smoke, while it was shaped into small balls, dried, and sucked on in Malekula (Laufer, 1930).

In Australia, geophagy has been reported in some Aboriginal tribes but not others. Laufer (1930) included only two cases from Australia in his extensive review on geophagy: In Queensland, clay or mud pills were reportedly prescribed for diarrhea and in Western Australia, a certain type of clay was pounded and ingested together with the root of the mene (*Haemodorum coccineum* or scarlet bloodroot). Rowland (2002) suggested that the detoxifying characteristics of clay may have enabled the first arrivals to the continent some 40,000 years ago to adapt to some of the native toxic plants. In contrast to Laufer's two examples, Rowland includes numerous examples of geophagy in different Aboriginal tribes across Australia for a range of reasons. For example, the Anbarra in the Northern Territory reportedly ingested mineral matter as a normal part of a meal, and at times collected white clay and material called galamba from ant-bed as food in its own right. This material has in fact been found to contain some protein (Rowland, 2002). As reported for Queensland, soil derived from ant-hills or termite mounds was also used to treat diarrhea in some Northern Territory tribes, while a fatty red clay was used for the same purpose in parts of Western Australia, in some places mixed with seeds and roots and baked in hot ashes prior to ingestion (Rowland, 2002) (Fig. 14).

Fig. 14 Termite mounds near Pine Creek, Northern Territory. Courtesy of Kobold (2006)

In some tribes of Queensland, diarrhea was treated by a roasted special clay, and tuberculosis and cough in a tribe living on Cape York were treated by ingestion of a white clay (Rowland, 2002). In the Queensland rainforest, the Kuku Yalanji tribe ate a clay called gumburu during famine. In another rainforest tribe, clay was allegedly ingested for abortive and contraceptive purposes. On Groote Eylandt, a variety of clays were ingested for different purposes, but reportedly only by women: a reddish or yellow clay as well as material from termite mounds reportedly cured mineral deficiencies; white clay was used for diarrhea, to "make the stomach cool," and if fish was craved because of the clay's similarity in taste. On Mornington Island, white clay was used to cure internal pains, headaches, joint pains, eye complaints, and snakebite wounds and to increase the flow of breast milk. On Merr Island, pregnant women ate a chocolate-like clay which was rolled up in banana leaves and roasted to ensure the birth of a fair skinned baby, similar to the practice of New Guinean women. It was also eaten by children in the belief that it would make them braver, stronger, and sturdier (Rowland, 2002).

In a psychiatric review on geophagy among Australian Aboriginals by Eastwell (1979), the practice of eating soil was linked with social insecurity. He describes an "epidemic" in 1958–1959 among middle-aged women, allegedly brought on by the introduction of a cash economy system which eliminated the role of these women in the provision of food for the community. Another epidemic was reported in 1974 with the introduction of higher wages and social welfare payments to Aboriginals from the Australian government, with the result that working mothers returned to their domestic duties but the elderly women who had previously cared for the children became increasingly marginalized in the local society. According to the author, these events caused depression among some and brought on the ingestion of soil, suggesting it to be a "medicine for life's dissatisfactions" (Robbins, 1980).

Lytico-Bodig – A Medical Geology Disease?

Research on the cause of the unusually high incidence of neurodegenerative disorders on the Pacific Island of Guam has been ongoing for the past 60 years. During this time, various causes have been hypothesized and dismissed, ranging from genetics to the endemic cycads to fruit bats. Researchers of many disciplines have wanted to solve the puzzle of the Guam disorder in the hope that by doing so it may also provide some clues to other neurological diseases such as Parkinson, amyotrophic lateral sclerosis (ALS), and Alzheimer's. The following review will focus on the environmental exposures, although both genetic and environmental risk factors have been recognized but not yet fully elucidated (Steele and McGeer, 2008).

The first record of a paralyzing neurodegenerative disease on Guam is from 1904, although the island was described as early as 1565 by a Portuguese explorer. Two years earlier, the Chamorro people of Guam had been described as "remarkably free of disease" and reportedly lived long. The number of cases of neurological disease grew quickly until it became the leading cause of death in adult Chamorro between 1940 and 1965. The symptoms of the disease vary from ALS to Parkinson's with or without dementia, and it was originally considered as two separate entities (Hof and Perl, 2002). Locally it was called lytico-bodig, where lytico referred to a progressive paralysis resembling ALS and bodig to the Parkinsonian symptoms (van den Enden, 2009). One of the first studies in 1954 found that the Chamorro people were 100 times more likely to suffer the symptoms of ALS than the general population (Steele and McGeer, 2008). The high frequency in some families and uneven distribution on the island initially suggested a genetic link. This may have been secondary to the drastic population decrease from around 100,000 people in the 17th century down to 1,000 by the end of the 1800s due to disease, famine, and slavery after colonial exploitation. The recovery from near extinction may have enabled the inheritance of the genetic polymorphism and an exceptionally high disease frequency in the remaining population (Ince and Codd, 2005). However, genetic influences could not alone explain its distribution, and it was concluded that an unidentified environmental exposure also played an important role. In the 1980s, the incidence of lytico-bodig on Guam seemed to decrease but has of today not disappeared. However, over time, its clinical characteristics have shifted from ALS to Parkinsonism and now to dementia. The age of its onset has also increased (Steele and McGeer, 2008).

Nutritional deficiencies of calcium and magnesium were suggested in the 1980s, but proved inaccurate when symptoms persisted in spite of elimination of the deficiencies. The finding of a high incidence of an unusual retinal disease in patients suffering from the Guam syndrome, similar to the pathology of botfly infections, suggested a transmissible agent of the disease. However, follow-up studies to identify a vector of the disease have not been reported (Steele and McGeer, 2008). A popular hypothesis throughout the years has been that lytico-bodig is caused by the toxic effects from cycads, a palm-like plant used locally to make flour for tortillas and dumplings. This possible causative agent was first suggested in the 1950s but has repeatedly failed to find supporting evidence. It was first dismissed in the 1970s after intense investigations continuously failed. The toxic agent was identified in the 1980s as being beta-methylamino-alanine (BMAA), and the cycad hypothesis gained renewed focus (Caparros-Lefebvre and Lees, 2005). By 2002, it was suggested that the toxin found in cycads is bioaccumulated in fruit bats, which then are eaten by the Chamorro, causing a toxic insult decades after consumption (Cox and Sacks, 2002). This proposal corresponded with the previous finding by Reed et al. (1987) that traditional Chamorro diet is the only factor known to correlate with disease incidence (Banack et al., 2006). Elevated concentrations of the toxin BMAA were found in various foodstuffs of the Chamorro diet, as well as in brains of Chamorro with the disease, but not in a comparison population (Banack et al., 2006). The role of BMAA in the Guam disease and the cycad hypothesis is still controversial and a number of issues still need to be resolved (Ince and Codd, 2005). For example, a subsequent study failed to confirm the findings of BMAA in human brain (Montine et al., 2005), although it was later argued that the reason was a different methodology for BMAA detection and failure to measure protein-bound BMAA (Cox et al., 2006). Another hypothesis regarding toxicity from food plant sources originates from Guadeloupe in the Caribbean where a similar type of Parkinson as noted on Guam was reported. It was suggested that a significant environmental factor was the exposure to plants containing alkaloid toxins, agents that were simultaneously toxic for dopaminergic neurons and inhibited mitochondrial respiration. These toxins were derived from herbal medicine, high intake of tropical fruits, and herbal tea (Caparros-Lefebvre and Lees, 2005).

Cox and coworkers (2003) found that BMAA is in fact produced by cyanobacteria living inside the roots of the cycads. The bacteria provide the plant with nitrogen and the cycad relocate the BMAA to its seeds, possibly as protection against herbivores (van den Enden, 2009). This may explain why the disease has occurred in places where cycads or fruit bats are not consumed. The most recent study (Pablo et al., 2009) suggests that cyanobacteria and its related neurotoxins are an environmental factor for certain sporadic neurodegenerative diseases in vulnerable individuals. The exact cause of the disease is thus still debated, although it seems generally accepted that both genetic and environmental factors play significant roles. This disease is an exceptional example of the need for multidisciplinary collaboration and an area where the importance of medical geology is highlighted.

References

Advameg (2009) Encyclopedia of the Nations, Vol 2009. Illinois: Advameg Inc.

Angeloni J, Peek A, Appleyard S, Wong S, Watkins R (2004) Acid sulfate soils: distribution, impacts and regulation (a Western Australian perspective). Corros Prev 106:1–13.

Asbestos Diseases Society of Australia Inc. (2004) Asbestos Products. In Asbestos Information. Perth: Asbestos Diseases Society of Australia Inc.

Attfield MD, Costello J (2004) Quantitative exposure-response for silica dust and lung cancer in Vermont granite workers. Am J Ind Med 45:129–138.

Australian Bureau of Statistics (2006) Special article – a century of mining in Australia. In Coal and Mineral Industries Division (ed) Australian Mining Industry 1998–1999. Adelaide: Department of Industry Science and Resources.

Australian Bureau of Statistics 2008a: Mining. In B Pink (ed) Yearbook of Australia 2008 (Vol. 1301). Canberra: Australian Bureau of Statistics.

Australian Bureau of Statistics 2008b: Feature article 4: Natural disasters in Australia. In: Pink B (ed) Yearbook of Australia 2008, vol 1301. Australian Bureau of Statistics, Canberra.

Balmes JR (1987) Respiratory effects of hard-metal dust exposure. Occup Med 2:327–244.

Banack SA, Murch SJ, Cox PA (2006) Neurotoxic flying foxes as dietary items for the Chamorro people, Marianas Islands. J Ethnopharmacol 106:97–104.

Banks (1770) Banks Journal – Daily Entries. In National Library of Australia Southseas Collection. Canberra: National Library of Australia,

Baris I, Simonato L, Artvinli M, Pooley F, Saracci R, Skidmore J, Wagner C (1987) Epidemiological and environmental evidence of the health effects of exposure to erionite fibres: A four-year study in the Cappadocian region of Turkey. Int J Cancer 39:10–17.

Bates MN, Garrett N, Shoemack P (2002) Investigation of health effects of hydrogen sulfide from a geothermal source. Arch Environ Health 57:405–411.

Baumann F, Rougier Y, Ambrosi JP, Robineau BP (2007) Pleural mesothelioma in New Caledonia: an acute environmental concern. Cancer Detect Prev 31:70–76.

BHP Billiton (2008) About BHP Billiton and Manganese. Melbourne: BHP Billiton.

Broom A, Lindsay M, Oliveira N, Jasinska E, Heuzen Bv, Sturrock K, Vetten S, Maley F, Durling S, Dodsley N, Voss S, Ypelaar I, Smith D, Shellam G (2000) The University of Western Australia Arbovirus Surveillance and Research Laboratory Annual Report: 1998–2000. In The University of Western Australia Arbovirus Surveillance and Research Laboratory Annual Report, Perth.

Brown T, Morley A, Sandersen N, Tait R (1983) Report of a large fish kill resulting from natural acid water. J Fish Biology 22:335–350.

Bureau of Meteorology (2006) Fire Seasons. In Bureau of Meteorology, Australian Government (eds). Melbourne: Commonwealth of Australia.

Bureau of Meteorology 2009a: "Darkness at noon" – the summer of 1944/1945. In Bureau of Meteorology, Australian Government (eds). Melbourne: Commonwealth of Australia.

Bureau of Meteorology 2009b: Dust-storms. In Bureau of Meteorology, Australian Government (eds). Melbourne: Commonwealth of Australia.

Bureau of Meteorology 2009c: The Melbourne dust-storm of February 1983. In Bureau of Meteorology, Australian Government (eds). Melbourne: Commonwealth of Australia.

Burt T, Blumbergs P, Currie B (1993) A dominant hereditary ataxia resembling Machado-Joseph disease in Arnhem Land, Australia. Neurology 43:1750–1752.

Burt T, Currie B, Kilburn C, Lethlean AK, Dempsey K, Blair I, Cohen A, Nicholson G (1996) Machado-Joseph disease in east Arnhem Land, Australia: Chromosome 14q32.1 expanded repeat confirmed in four families. Neurology 46:1118–1122.

Callinan R, Fraser G, Melville M (1993) Seasonally recurrent fish mortalities and ulcerative disease outbreaks associated with acid sulfate soils in Australian estuaries. In D Dent, M van Mensvoort (eds) Selected Papers of the Ho Chi Minh City Symposium on Acid Sulphate Soils, Vol Publication 53. Wageningen: International Institute for Land Reclamation and Improvement.

Caparros-Lefebvre D, Lees AJ (2005) Atypical unclassifiable parkinsonism on Guadeloupe: an environmental toxic hypothesis. Mov Disord 20(Suppl 12):S114–S118.

Chen L, Verrall K, Tong S (2006) Air particulate pollution due to bushfires and respiratory hospital admissions in Brisbane, Australia. Int J Environ Health Res 16:181–191.

Cheney NP (1995) Bushfires – An Integral Part of Australia's Environment. In Yearbook of Australia 1995, Vol. 1301.

Churches T, Corbett S (1991) Asthma and air pollution in Sydney. NSW Public Health Bull 2:72–73.

CIA (2009) The 2008 World Factbook. US Government. https://www.cia.gov/library/publications/the-world-factbook/index.html. Accessed 31 March 2009.

Collipp PJ, Chen SY, Maitinsky S (1983) Manganese in infant formulas and learning disability. Ann Nutr Metab 27: 488–494.

Commonwealth of Australia (2008) National Indigenous Housing Guide. Department of Families Housing, Community Services and Indigenous Affairs (ed). Canberra: Commonwealth of Australia.

Cook A, Weinstein P (2005) Volcanic emissions and health risks of metal contaminants in New Zealand. In T Moore, A Black, J Centeno, J Harding, D Trumm (eds) Metal contaminants: sources, effects and integration in New Zealand. Christchurch: University of Canterbury Press.

Cook FJ, Hicks W, Gardner EA, Carlin GD, Froggatt DW (2000) Export of acidity in drainage water from acid sulphate soils. Marine Pollut Bull 41:319–326.

Cooper CW, Mira M, Danforth M, Abraham K, Fasher B, Bolton P (1994) Acute exacerbations of asthma and bushfires. Lancet 343:1509.

Costa D, Dreher K (1997) Bioavailable transition metals in particulate matter mediate cardiopulmonary injury in healthy and compromised animal models. Environ Health Perspect 105:1053–1060.

Cowan J (2006) Angurugu Community Grapples with Machado Joseph Disease. ABC Online – Stateline Northern Territory. http://www.abc.net.au/stateline/nt/content/2006/s1606126.htm. Accessed 31 March 2009.

Cox P, Banack SA, Murch S (2003) Biomagnification of cyanobacterial neurotoxins and neurodegenerative disease among the Chamorro people of Guam. PNAS 100: 13380–13383.

Cox PA, Banack S, Murch S, Sacks O (2006) Commentary on: Return of the cycad hypothesis does the amyotrophic lateral sclerosis/Parkinsonism dementia complex (ALS/PDC) of Guam have new implications for global health? Neuropathol Appl Neurobiol 32:679–682.

Cox PA, Sacks OW (2002) Cycad neurotoxins, consumption of flying foxes, and ALS-PDC disease in Guam. Neurology 58:956–959.

Cronin SJ, Sharp DS (2002) Environmental impacts on health from continuous volcanic activity at Yasur (Tanna) and Ambrym, Vanuatu. Int J Environ Health Res 12:109–123.

Currie B, Fisher D, Howard D, Burrom J, Lo D, Selva-nayagam S, Anstey N, Huffam S, Snelling P, Marks P, Stephens D, Lum G, Jacups S, Krause V (2000) Endemic melioidosis in tropical Northern Territory: a 10-year prospective study and review of the literature. Clin Infect Dis 31:981–986.

Davidson G (2009) Uranium- and platinum-rich unconformity-style ore deposit genesis in the South Alligator Valley, Northern Territory. In School of Earth Sciences (ed) ARC Centre of Excellence in Ore Deposits, Vol. 2009. Hobart: University of Tasmania.

Dent D (1986) Acid Sulphate Soils: A Baseline for Research and Development (Vol. 39). Wageningen: ILRI Publications.

Department of Environment and Conservation 2007a: DEC investigates further bird deaths. Media statement 17 January 2007.

Department of Environment and Conservation 2007b: Esperance bird death test results. Media statement 9 March 2007.

Department of Health (2006) Port Hedland dust study results released. Department of Health, Media Release 13 July 2006. Government of Western Australia. http://www.health.wa.gov.au/press/view_press.cfm?id=606. Accessed 31 March 2009.

Department of Human Services (2008) Community factsheet – dust storms and health. Department of Human Services (ed). Public Health Emergency Management Community Fact Sheet. www.health.vic.gov.au/environment/downloads/dust_storms_and_health.pdf. Accessed 31 March 2009.

Department of Primary Industries (2004) The 1939 Fires. Knowledge Resource Centre (ed). The Virtual Exhibition. Victoria: Department of Primary Industries. http://www.dpi.vic.gov.au/virtualexhibition/39fires/index.htm. Accessed 31 March 2009.

Department of Sustainability and Environment (2006) Ash Wednesday – Significant Fire Years (ed). Victoria: Department of Sustainability and Environment. http://www.dse.vic.gov.au/DSE/nrenfoe.nsf/childdocs/-D79E4FB0C437E1B6CA256DA60008B9EF-7157D5E68CDC2002CA256DAB0027ECA3?open. Accessed 31 March 2009.

Department of Sustainability and Environment (2009) Bushfire Statistics. Victoria: Department of Sustainability and Environment. http://www.dse.vic.gov.au/DSE/nrenfoe.nsf/childdocs/-D79E4FB0C437E1B6CA256DA60008B9EF-3D569492B50F5A78CA256DA600095F56?open. Accessed 31 March 2009.

Donaldson J (2002) "Manganese madness" Clues to the aetiology of human brain disease emerges from a geological anomaly. Med Geol Newsl 4:8–10.

Durand M, Florkowski C, George P, Walmsley T, Weinstein P, Cole J (2004) Elevated trace element output in urine following acute volcanic gas exposure. J Volcanol Geother Res 134:139–148.

Durand M, Wilson JG (2006) Spatial analysis of respiratory disease on an urbanized geothermal field. Environ Res 101:238–245.

Earth Observatory 2005a: Dust storm in Australia. NASA (ed) Earth Observatory, 2009. EOS Project Science Office, NASA Goddard Space Flight Center. http://earthobservatory.nasa.gov/NaturalHazards/view.php?id=14580. Accessed 31 March 2009.

Earth Observatory 2005b: Sulfur dioxide leaks from the Ambrym volcano. NASA (ed) Earth Observatory, 2009. EOS Project Science Office, NASA Goddard Space Flight Center. http://earthobservatory.nasa.gov/NaturalHazards/view.php?id=14796. Accessed 31 March 2009.

Earth Observatory (2006) Ghostly face in South Australian desert. NASA (ed) Earth Observatory, 2009. EOS Project Science Office, NASA Goddard Space Flight Center. http://earthobservatory.nasa.gov/IOTD/view.php?id=7076. Accessed 31 March 2009.

Earth Observatory (2008) Lahar on Mount Ruapehu, New Zealand. NASA (ed) Earth Observatory, 2009. EOS Project Science Office, NASA Goddard Space Flight Center.

http://earthobservatory.nasa.gov/NaturalHazards/view.php?id=18166. Accessed 31 March 2009.

Eastwell H (1979) A pica epidemic: a price for sedentarism among Australian ex-hunter-gatherers. Psychiatry 42:264–273.

Ericson JE, Crinella FM, Clarke-Stewart KA, Allhusen VD, Chan T, Robertson RT (2007) Prenatal manganese levels linked to childhood behavioral disinhibition. Neurotoxicol Teratol 29:181–187.

Esperance Port Authority (2008) Lead Removal Plan. C/2300. Retrieved 5 October 2009 from http://www.esperanceport.com.au/downloads/report/LRP_Final.pdf.

Fältmarsch R (2006) Effect of toxic metals mobilised from Finnish acid sulphate soils on terrestrial and aquatic biota and human health: A literature review. In 18th World Congress of Soil Science (Vol. 174–4). Philadelphia.

Feldman J, Tilling RI (2007) Danger Lurks Deep: The Human Impact of Volcanoes. Feature Article. Geotimes November 2007.

Geoscience Australia (2003) Australian Mines Atlas: Copper. Department of Industry, Tourism and Resources (ed) AIMR Report, 2005. Canberra: Commonwealth of Australia.

Gillette D (1997) Soil derived dust as a source of silica: Aerosol properties, emissions, deposition, and transport. J Expo Anal Environ Epidemiol 7:303–311.

Global Identifier Number (2009). Asian Disaster Reduction Centre (ACDR). http://www.glidenumber.net/glide/public/search/search.jsp. Accessed 4 March 2009.

Global Volcanism Program (1990) BGVN 15:06. Rabaul. CO_2 kills six at Tarvurvur, seismicity remains low. In Global Volcanism Program (ed) Bulletin of the Global Volcanism Network, vol 2009. Smithsonian National Museum for Natural History.

Government of Western Australia (2007) Response of the Western Australian government to the Western Australian Legislative Assembly, Education and Health Standing Committee in relation to the cause and extent of lead pollution in the Esperance area.

Greene R, Gatehouse R, Scott K, Chen X (2001) Symposium report: Aeolian dust – implications for Australian mineral exploration and environmental management. Austr J of Soil Res 39:1–6.

Hales S, Weinstein P, Woodward A (1999) Ciguatera (Fish Poisoning), El Nino, and Pacific sea surface temperatures. Ecosyst Health 5:20–25.

Hansell A, Horwell C, Oppenheimer C (2006) The health hazards of volcanoes and geothermal areas. Occup Environ Med 63:149–156.

Hansell A, Oppenheimer C (2004) Health hazards from volcanic gases: a systematic literature review. Arch Environ Health 59:628–639.

Hansen J, de Klerk NH, Musk AW, Hobbs MS (1998) Environmental exposure to crocidolite and mesothelioma: exposure-response relationships. Am J Respir Crit Care Med 157:69–75.

Healthy Living Northern Territory (2005) Melioidosis – information sheet. Diabetes Association of the Northern Territory (ed). www.healthylivingnt.org.au/content/?action=getfile&id=45-. Accessed 31 March 2009.

Hendrickx M (2008) Naturally occurring asbestos in eastern Australia: A reveiw of geological occurrence, disturbance and mesothelioma risk. Environ Geol. Doi 10.1007/s00254-008-1370-5. Accessed 31 March 2009.

Hicks W, Bowman G, Fitzpatrick R (1999) Environmental impacts of acid sulfate soils near Cairns, QLD. Technical Report 15/99. CSIRO Land and Water. http://www.clw.csiro.au/publications/technical99/tr15-99.pdf. Accessed 31 March 2009.

Hof PR, Perl DP (2002) Neurofibrillary tangles in the primary motor cortex in Guamanian amyotrophic lateral sclerosis/parkinsonism-dementia complex. Neurosci Lett 328:294–298.

IARC (1987) Asbestos. In IARC Monographs on the Evaluation of Carcinogenic Risks to Humans, Overall Evaluations of Carcinogenicity: An Updating (Vol. 1, Suppl 7). Lyon: Who Health Organization, International Agency for Research on Cancer.

Ince PG, Codd GA (2005) Return of the cycad hypothesis – does the amyotrophic lateral sclerosis/parkinsonism dementia complex (ALS/PDC) of Guam have new implications for global health? Neuropathol Appl Neurobiol 31: 345–353.

Intergovernmental Oceanographic Commission (2008) Tsunami, The Great Waves, Revised edition. In: Intergovernmental Oceanographic Commission (ed) IOC Brochure 2008-1. Paris: UNESCO, .

Jalaludin B, Smith M, O'Toole B, Leeder S (2000) Acute effects of bushfires on peak expiratory flow rates in children with wheeze: A time series analysis. Austr NZ J Public Health 24:174–177.

Jardine A, Lindsay M, Johansen C, Cook A, Weinstein P (2008) Impact of dryland salinity on population dynamics of vector mosquitoes (diptera: culicidae) of Ross River virus in areas of southwestern Western Australia. J Med Entomol 45: 1011–1022.

Jardine A, Speldewinde P, Carver S, Weinstein P (2007) Dryland salinity and ecosystem distress syndrome: Human health implications. Eco Health 4:10–17.

Johnson D (2004) The Geology of Australia, Illustrated edn. Cambridge: Cambridge University Press, .

Johnston FH, Kavanagh AM, Bowman DMJS, Scott RK (2002) Exposure to bushfire smoke and asthma: An ecological study. Med J Austr 76:535–538.

Kiloh LG, Lethlean AK, Morgan G, Cawte JE, Harris M (1980) An endemic neurological disorder in tribal Australian aborigines. J Neurol Neurosurg Psychiatry 43: 661–668.

Kobold K (2006) Termite Mounds NT. Image Licensed Under the Creative Commons Attribution SharAlike 2.0: http://creativecommons.org/licenses/by-sa/2.0/

Kondakis XG, Makris N, Leotsinidis M, Prinou M, Papapetropoulos T (1989) Possible health effects of high manganese concentration in drinking water. Arch Environ Health 44:175–178.

Laufer B (1930) Geophagy (Vol. XVIII). Chicago: Field Museum Press.

Leigh J, Driscoll T (2003) Malignant mesothelioma in Australia, 1945–2002. Int J Occup Environ Health 9: 206–217.

Ljung K, Maley F, Cook A, Weinstein P (2009) Acid sulfate soils and human health – a millenium ecosystem assessment. Environ Int 35:1234–1242.

Longo BM (2009) The Kilauea Volcano adult health study. Nurs Res 58:23–31.

Longo BM, Rossignol A, Green JB (2008) Cardiorespiratory health effects associated with sulphurous volcanic air pollution. Public Health 122:809–820.

Luce D, Bugel I, Goldberg P, Goldberg M, Salomon C, Billon-Galland MA, Nicolau J, Quenel P, Fevotte J, Brochard P (2000) Environmental exposure to tremolite and respiratory cancer in New Caledonia: a case-control study. Am J Epidemiol 151:259–265.

Luke RH, McArthur AG (1978) Bushfires in Australia. Canberra: Australian Government Publishing Service.

Macdonald B, White I, Heath L, Smith J, Keene A, Tunks M, Kinsela A (2006) Tracing the outputs from drained acid sulphate flood plains to minimize threats to coastal lakes. In C Hoanh, T Tuong, J Gowing, B Hardy (eds) Environment and livelihoods in tropical coastal zones. Wallingford, Oxfordshire: CAB International.

McKay AD, Miezitis Y, Jaireth S (2008) Australian Uranium Resources, May 2008 edn. 1:10,000,000 scale map. Australian Government (ed), May 2008 edn. Canberra. www.ga.gov.au/servlet/BigObjFileManager?bigobjid=GA 11404 -. Accessed 31 March 2009.

McMichael A, Woodruff R, Whetton P, Hennessy K, Nicholls N, Hales S, Woodward A, Kjellstrom T (2003) Human health and climate change in Oceania: A risk assessment 2002. Commonwealth Department of Health and Ageing (ed). Canberra: Commonwealth of Australia.

Mergler D (1999) Neurotoxic effects of low level exposure to manganese in human populations. Environ Res 80: 99–102.

Montine TJ, Li K, Perl DP, Galasko D (2005) Lack of beta-methylamino-l-alanine in brain from controls, AD, or Chamorros with PDC. Neurology 65:768–769.

Musk AW, de Klerk NH, Eccles JL, Hansen J, Shilkin KB (1995) Malignant mesothelioma in Pilbara Aborigines. Aust J Public Health 19:520–522.

Musk AW, de Klerk NH, Reid A, Ambrosini GL, Fritschi L, Olsen NJ, Merler E, Hobbs MS, Berry G (2008) Mortality of former crocidolite (blue asbestos) miners and millers at Wittenoom. Occup Environ Med 65:541–543.

National Action Plan (2001) Australia's Salinity Problem. Department of Agriculture, Fisheries and Forestry Department of the Environment and Water Resources (eds). Canberra: National Action Plan, Government of Australia.

National Institute of Neurological Disorders and Stroke (2008) Machado Joseph Disease Fact Sheet. Bethesda: National Institute of Health. Publication No. 02-2716.

National Institute for Occupational Safety and Health (2007). Asbestos and Other Mineral Fibers: A Roadmap for Scientific Research. Retrieved 5 October 2008 from http://www.cdc.gov/niosh/review/public/099/pdfs/NIOSHAsbestosRoadmap.pdf.

Nemery B (1990) Metal toxicity and the respiratory tract. Eur Respirol J 3:202–219.

New South Wales Health (2003) Dust Storms. In Department of Health (ed). New South Wales Health Factsheet. Sydney: New South Wales Government.

New South Wales Government (2008) Acid Sulfate Soils, What are the Effects of Acid Sulfate Soils? http://www.naturalresources.nsw.gov.au/soils/what_effects.shtml. Accessed 23 June 2008.

Nunn P (1991) Causes of environmental changes on Pacific Islands in the last millennium: implications for decision makers. In T Johnson, J Flenley (eds) Aspects of Environmental Change. Palmerston North: Massey University.

Ong L (2006) Island Evolution Part 2: Oeno Island. In NASA (ed) EO-1 Science Team, Earth Observatory. http://earthobservatory.nasa.gov/IOTD/view.php?id=6786. Accessed 31 March 2009.

Pablo J, Banack SA, Cox PA, Johnson TE, Papapetropoulos S, Bradley WG, Buck A, Mash DC (2009) Cyanobacterial neurotoxin BMAA in ALS and Alzheimer's disease. Acta Neurol Scand 4:216–225.

Pan XL, Day HW, Wang W, Beckett LA, Schenker MB (2005) Residential proximity to naturally occurring asbestos and mesothelioma risk in California. Am J Respir Crit Care Med 172:1019–1025.

Playford PE (1998) Voyage of discovery to Terra Australis, by Willem De Vlamingh, 1696–1697. Perth: Western Australian Museum.

Plumlee G, Ziegler T (2007) The medical geochemistry of dusts, soils and other earth materials. In B Lollar (ed) Treatise on Geochemistry (Vol. 9.07). Amsterdam: Elsevier Press.

Powell B, Martens M (2005) A review of acid sulfate soil impacts, actions and policies that impact on water quality in Great Barrier Reef catchments, including a case study on remediation at East Trinity. Marine Pollut Bull 51: 149–164.

Purdey M (2004) The pathogenesis of Machado Joseph Disease: a high manganese/low magnesium initiated CAG expansion mutation in susceptible genotypes? J Am Coll Nutr 23: 715S–729S.

Reed D, Labarthe D, Chen KM, Stallones R (1987) A cohort study of amyotrophic lateral sclerosis and parkinsonism-dementia on Guam and Rota. Am J Epidemiol 125:92–100.

Reid A, Heyworth J, de Klerk N, Musk AW (2008) The mortality of women exposed environmentally and domestically to blue asbestos at Wittenoom, Western Australia. Occup Environ Med 65:743–749.

Reis J (2006) Oceania_ISO_3166-1. Image licensed under the Creative Commons Attribution ShareAlike 2.5 License: http://creativecommons.org/licenses/by-sa/2.5/.

Robbins K (1980) Abstracts and Reviews: 4 Australia: A Pica Epidemic: A price for sedentarism among Australian ex-hunter-gatherers by Harry D. Eastwell. Psychiatry 42 (1979): 264–273. Transcult Psychiatry 17:168–170.

Rosicky M, Sullivan L, Slavich P, Hughes M (2004) Factors contributing to the acid sulfate soil scalding process in the coastal floodplains of New South Wales, Australia. Aust J Soil Sci 42:587–594.

Rowland M (2002) Geophagy: An assessment of implications for the development of Australian Indigenous plant processing technologies. Aust Aborig Stud 1:51–66.

Russell D, Helmke S (2002) Impacts of acid leachate on water quality and fisheries resources of a coastal creek in northern Australia. Marine Freshw Res 53:19–33.

Rutherford S, Clark E, McTainsh G, Simpson R, Mitchell C (1999) Characteristics of rural dust events shown to impact on asthma severity in Brisbane, Australia. Int J Biometeorol 42:217–225.

Singh G, Geissler EA (1985) Late Cainozoic history of vegetation, fire, lake levels and climate, at Lake George, New South Wales, Australia. Philos Trans Royal Soc Lond Series B 311:379–447.
Singh R, Hales S, de Wet N, Raj R, Hearnden M, Weinstein P (2001) The influence of climate variation and change on diarrhoeal disease in the Pacific Islands. Environ Health Perspect 109:155–159.
Smith M, Jalaludin B, Byles J, Lim L, Leeder S (1996) Asthma presentations to emergency departments in western Sydney during the January 1994 Bushfires. Int J Epidemiology 25:1227–1236.
Smithsonian Institute (2009) How Many Active Volcanoes are There in the World? Program GV (ed) 2009. http://www.volcano.si.edu/faq/index.cfm?faq=03. Accessed 31 March 2009.
Steele JC, McGeer PL (2008) The ALS/PDC syndrome of Guam and the cycad hypothesis. Neurology 70:1984–1990.
Stephens F, Ingram M (2006) Two cases of fish mortality in low pH, aluminium rich water. J Fish Dis 29:765–770.
Stewart C, Johnston D, Leonard G, Horwell C, Thordarson T, SJC (2006) Contamination of water supplies by volcanic ashfall: A literature review and simple impact modeling. J Volcanol Geotherm Res 158:296–306.
Summerell G, Dowling T, Richardson D, Walker J, Lees B (2000) Modelling current parna distribution in a local area. Aust J Soil Res 38:867–878.
Takser L, Mergler D, Hellier G, Sahuquillo J, Huel G (2003) Manganese, monoamine metabolite levels at birth, and child psychomotor development. Neurotoxicology 24: 667–674.
Te Ara (2007) Life on the Edge: New Zealand's Natural Hazards and Disasters. Auckland: David Bateman Ltd.
Thistlethwaite B, Davis D (1996) Pacific 2010: A sustainable future for Melanesia? In National Centre for Development Studies (ed), Canberra.
Twohig J (2006) Management of asbestos contamination in Wittenoom. In Department of Industry and Resources, Department of Local Government and Regional Development (eds). Canberra: GHD Pty Ltd.
U.S. Geological Survey (1998) Magnitude 7.0 near north coast of Papua New Guinea, 1998 July 17. In Preliminary Earthquake Report. Denver: National Earthquake Information Center, World Data Center for Seismology.
U.S. Geological Survey (2009) Magnitude 8.0 – Samoa islands region. Earthquakes Hazards Program. In: Preliminary Earthquake Report. Denver: National Earthquake Information Center, World Data Center for Seismology.
van den Enden E (2009) Lytico-Bodig. In Medical Problems Caused by Plants. Antwerpen: Institute of Tropical Medicine.
Verheijden MW, Ton A, James AL, Wood M, Musk AW (2002) Respiratory morbidity and lung function in two Aboriginal communities in Western Australia. Respirology 7: 247–253.
Wasserman GA, Liu X, Parvez F, Ahsan H, Levy D, Factor-Litvak P, Kline J, van Geen A, Slavkovich V, LoIacono NJ, Cheng Z, Zheng Y, Graziano JH (2006) Water manganese exposure and children's intellectual function in Araihazar, Bangladesh. Environ Health Perspect 114: 124–129.
Water and Rivers Commission (2002) Investigation of soil and groundwater acidity, Stirling: Report to the Minister for the Environment and Heritage. Perth: Water and Rivers Commission, Department of Environmental Protection.
Western Australia Legislative Committee (2007) Inquiry into the cause and extent of lead pollution in the Esperance area. Report No. 8 in the 37th Parliament. Education and Health Standing Committee. Western Australia: Government Printer, State Law Publisher.
White I, Heath L, Melville M (1999) Ecological impacts of flood mitigation and drainage in coastal lowlands. Aust J Emerg Manag 14:9–15.
White I, Melville M, Macdonald B, Quirk R, Hawken R, Tunks M, Buckley D, Beattie R, Heath L, Williams J (2006) From conflict to industry – regulated best practice guidelines: A case study of estuarine flood plain management of the Tweed River, eastern Australia. In C Hoanh, T Tuong, J Gowing, B Hardy (eds) Environment and Livelihoods in Tropical Coastal Zones. Wallingford: CAB International.
White I, Melville M, Wilson B, Sammut J (1997) Reducing acid discharge from coastal wetlands in eastern Australia. Wetl Ecol Manag 5:55–72.
White N (2003) Melioidosis. Lancet 361:1715–1722.
World Health Organisation (2004) Manganese in drinking water – background document for development of WHO guidelines for drinking-water quality. WHO/SDE/WSH/03.04/104. Geneva: World Health Organisation.

Medical Geology in the Middle East

Humam Misconi and Maryam Navi

Abstract The Middle East region has reported some endemic diseases which are more prominent than in other parts of the world and in some cases have been seen only in this region. This chapter provides information and addresses various cases from all over the Middle East.

The Middle East is well known for its arid and semi-arid environment with frequent and severe dust- and sand storms. This has affected human health in the southern provinces of Iran like the southwestern Khuzestan Province and the southeastern Sistan and Baluchistan Provinces. Health effects of dust storms in Khuzestan Province include asthma in some cities, especially for people with chronic respiratory and cardiovascular diseases. In spite of the fact that dust and sand storms endanger the lives of over 3 million of Khuzestan Province inhabitants, no detailed studies exist on the nature, type, and health effects of wind-blown dust and sand. Also, two main regions in Iraq including Baghdad and Al-Basra are stricken frequently by dust and sand storms, but little attention has been paid to study the health impact of dust and sand storms in Iraq. This section shows health impacts of dust storms in four case studies.

Numerous examples from all over the Middle East are illustrated on element toxicities such as arsenic and fluorine and deficiencies, especially iodine deficiency. This chapter emphasizes health problems resulting from arsenic toxicities in the region and shows cases of the long-existing iodine deficiency disorders in the Middle East.

Several areas from the Middle East face high natural fluoride levels in drinking water. The levels of natural fluoride in drinking water of up to 3 mg l^{-1} were found in the Negev Desert region in Israel. In some parts of Saudi Arabia like Hail region a strong association is seen between fluoride levels in well water used for drinking and the severity of dental fluorosis. Mecca was also reported to be an area with endemic fluorosis. There are also some cities in Iran where high levels of fluoride in drinking water account for incidence of fluorosis.

The relationship between radon exposure and health effects in the Middle East is also described. Many epidemiological studies have shown a relationship between radon exposure and lung cancer. Case–control studies have also been carried out in Ramsar in northern Iran indicating inverse relationship between residential radon exposure and lung cancer.

Geophagia associated with iron and zinc deficiency, short stature, delayed sexual maturity, hepatosplenomegaly, and delayed bone age is reported from Middle East countries. This includes Shiraz in Iran, parts of Saudi Arabia, and Arab children from Gaza Strip.

Keywords Iran · Iraq · Jordan · Israel · Saudi Arabia · Yemen · Dust · Arsenic · Fluorine · Iodine · Geophagia

Minerogenic Aerosols

Minerogenic aerosols are a heterogeneous mixture of air-suspended particles that can be solid, liquid, or both, organic and inorganic, and of varying size and shape. Sources include sand and dust storms, combustion of fossil fuels, industrial aerosols from

H. Misconi (✉)
Consultant, Baghdad, Iraq
e-mail: humammisconi@yahoo.com

O. Selinus et al. (eds.), *Medical Geology*, International Year of Planet Earth,
DOI 10.1007/978-90-481-3430-4_6,

construction, mining and mineral processing activities, both natural and man-made fires (e.g., oil well fires of Al Kuwait of 1991 and the fire of Al Mishraq State Sulfur Mine in northern Iraq in 2003), incineration of waste, and the detonation of explosive ammunition and devices, especially depleted uranium ammunition (DU dust).

The Middle East is well known for its arid and semi-arid environment with frequent and severe dust and sand storms. Arabia is recognized as one of five world regions where dust storm generation is especially intense (Idso, 1976) and characterized by a wide distribution of aeolian sediments. This applies also to southwestern and southern Iran. Until recently, this has been related mostly to the wind action in response to the fluctuation in temperature and precipitation and the absence of vegetation in two-thirds of this region (Ghahreman, 2003). However, with the intensive utilization of natural resources, urbanization, industrialization, armed conflicts, and other anthropogenic activities, land's surface disturbance became inevitable and the frequency, intensity, complexity, and duration of dust and sand storms have increased substantially, imposing heavy damages to the economy, society, and public health.

Dust and sand storms in the Middle East occur when the sub-tropical jet stream pushes up from south of the Arabian Peninsula and a polar front jet stream pushes down from the European continent (Fig. 1) and prevail mostly in the spring and summer when the northwesterly winds dominate.

The high winds and soil moisture deficiency depicts all needed criteria to the creation of dust storms. Nowadays the anthropogenic disruption of land surface increases the spreading of deserts in the Middle East. Knowing the source and distribution process is a key factor in drawing the distribution patterns and anticipating the affected areas where life is endangered. Dust storm brings microbes, insects, organic and inorganic particles, and also organic chemicals such as dyes and pesticides which result in a wide range of respiratory and infectious diseases.

Health effects of wind-blown sand and dust as well as minerogenic aerosols on human pulmonary and cardiovascular functions vary with a number of factors. These include particle size, composition, levels and duration of exposure, and the health status of the exposed population. Particles of a diameter of less than 10 μm may reach the upper part of the airways and lung while fine particles can penetrate more deeply and may reach the alveolar region. Health effects are also a function of timescale, which range from long-term exposure (months to years) influencing the incidence of chronic disease and susceptibility and short-term exposure (days) causing acute health events. On both scales, health effects may range in severity from subclinical to deadly. The proportion of the population affected decreases as the severity of the effect increases. Specific health effects of chronic exposure include increase in lower respiratory symptoms, chronic obstructive pulmonary disease (COPD), reductions in lung function, and reduction in life expectancy (primarily related to cardiorespiratory mortality) (Weese and Abraham, 2009).

The amount of dust and the kinds of particles involved influence how serious the lung injury will be.

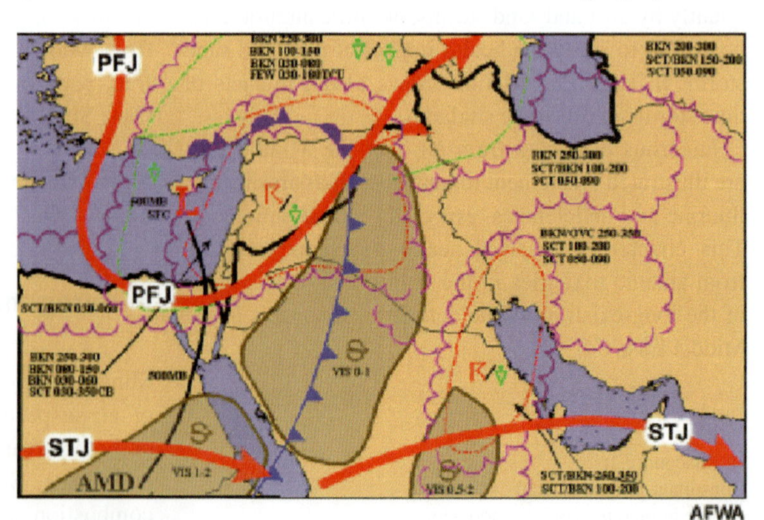

Fig. 1 Weather map for the Middle East showing conditions for a prefrontal occurrence of dust storms (Source: Taghavi and Asadi, 2008)

Dust is organic and inorganic (Sullivan and Krieger, 2001). Organic dusts consist of particles of biological, animal, and microbial origin and contain bacteria and fungi. Allergic alveolitis is caused by organic dust (Sullivan and Krieger, 2001). Histoplasmosis, parrot disease (psittacosis), and Q Fever are diseases that people can get if they breathe in organics that are infected with certain microorganisms. Dusts can also occur from organic chemicals (e.g., dyes, pesticides). Inorganic dusts can occur from grinding metals or minerals such as rock or soil. Examples of inorganic dusts are silica, asbestos, and coal (CCOHS, 2005). The influence of inorganic particles when inhaled varies with the size and nature of the particles (Collis and Greenwood, 1977). The changes which occur in the lungs vary with the different types of dust. For example, the injury caused by exposure to silica is marked by islands of scar tissue surrounded by normal lung tissue. Some particles dissolve in the bloodstream. The blood then carries the substance around the body where it may affect the brain, kidneys, and other organs. The table below summarizes some of the most common lung diseases caused by dust (Table 1).

Little information is available on the mineralogical composition of sand and dust particles in the Middle East. Al-Ali (2000) indicated that over central and southern Iraq sand and dust particles comprised 8–45% carbonates, 18–63% quartz, 4–27% feldspar, 1–16% mica, and 1–8% gypsum in addition to 4–28% multi-mineral rock fragments. Texturally, most of the fallout consists of very fine sand (125–62 μm) and coarse silt (62–31 μm). Alternatively, dust fallout over the Dead Sea consists of soluble salts, carbonates (6.7–47.9%), apatite (1–5%) (derived from the phosphate mining activity 45 km away), and clay minerals (Singer et al., 2003).

Table 1 Some types of pneumoconiosis according to dust and lung reaction

Inorganic dust	Type of disease	Lung reaction
Asbestos	Asbestosis	Fibrosis
Silica (quartz)	Silicosis	Fibrosis
Coal	Coal pneumoconiosis	Fibrosis
Beryllium	Beryllium disease	Fibrosis
Tungsten carbide	Hard metal disease	Fibrosis
Iron	Siderosis	No fibrosis
Tin	Stenosis	No fibrosis
Barium	Baritosis	No fibrosis
Organic dust		
Moldy hay, straw, and grain	Farmer's lung	Fibrosis
Droppings and feathers	Bird Fancier's lung	Fibrosis
Moldy sugarcane	Bagassosis	Fibrosis
Compose dust	Mushroom worker's lung	No fibrosis
Dust or mist	Humidifier fever	No fibrosis
Dust of heat-treated sludge	Sewage sludge disease	No fibrosis
Mold dust	Cheese washers' lung	No fibrosis
Dust of dander, hair particles, and dried urine of rats	Animal handlers' lung	No fibrosis

Source: Data from CCOHS, 2005

Health Effects of Dust Storm in Khuzestan Province – Iran

Dust storms blow out of Iraq and Saudi Arabia over western, southwestern, and central provinces of Iran causing severe health effects, especially in Khuzestan Province, southwestern Iran with Ahvaz as its center. Khuzestan Province lies between 47° 40′ E and 50° 33′ E longitude and 29° 57′ N and 33° 00′ N latitude and covers an area of 64,055 km^2.

In the period 2006–2008, the frequency, intensity, and duration of dust and sand storms have increased substantially from a record annual average of 2–3 to up to 50 in 2007 with 31 occurring in fall and winter of 2007 lasting 46 h reflecting a changing seasonal pattern. During some of these dust and sand storms, high density of up to 17 times the standard limit of suspended particles or aerosols caused low visibility and endangered inhabitants' health. In 2008, a record of 57 dusty days occurred until July 5 with 27 dust and sand storms lasting over 60 h. The density of particulate matter ranged between 450 and 3,786 μg m^{-3} on April 19, 2008 (Figs. 2, 3, 4, and 5).

Several factors stirred up these dust and sand storms including the presence of the vast desert region of western and southern Iraq and the Arabian Peninsula, decrease in precipitation, lack of vegetation in part of Khuzestan Province, drying up part of Hoor-Al-Azim Wetland, and decreasing discharge of Tigris and Euphrates related to global climate change (Zarasvandi and Mokhtari, 2008).

Health effects of dust in Khuzestan Province include increasing asthma in some cities especially for people

Fig. 2 A massive dust plume blowing out of south-central Saudi Arabia in a counter-clockwise direction over central and southern Iraq, southwestern and southern Iran, and the Persian Gulf on April 18, 2008. The dust narrowly misses the United Arab Emirates (UAE) and leaves the eastern part of the Persian Gulf clear. The dust appears particularly thick over Qatar (NASA, 2008)

Fig. 3 A street in Khoramshahr, April 30, 2008 (*Source*: Fars News Agency)

with chronic respiratory and cardiovascular diseases. Allergic diseases are more frequent in children and adolescence compared to other age groups. On June 16 and 17, 2008, 205 people went to hospital and 2 people perished due to respiratory poisoning of dust (Fazlollahi, 2008).

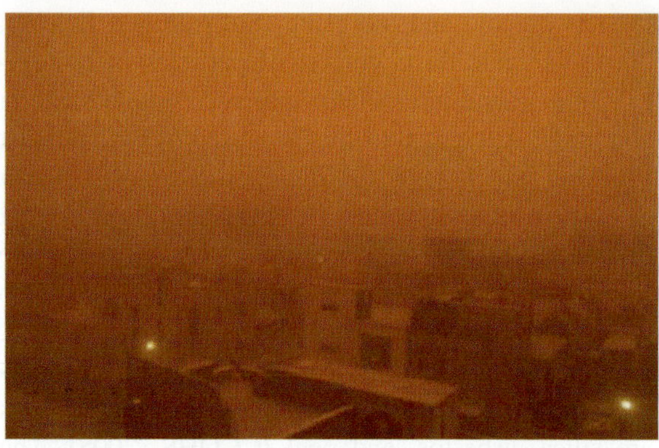

Fig. 4 Abadan, April 30, 2008

Fig. 5 Ahvaz, April 19, 2008 (*Source*: Islamic Republic News Agency)

In spite of the fact that dust and sand storms endanger the lives of over 3 millions of Khuzestan Province inhabitants, no detailed studies exist on the nature, type, and health effects of wind-blown dust and sand. Zarasvandi and Mokhtari (2008, unpublished data) have done sampling of dust fallout from Ahvaz and along the Iraq–Iran border and results showed chemical, mineral, and microbial contaminants in many samples. The contents of uranium, thorium, arsenic, lead, zinc, nickel, and cobalt exceeded the natural background level. The dust consisted predominantly of clay and quartz silt. The dust storm sediments analyzed consisted of a large amount of clay and silt (quartz silt). Clay soil is lighter and can move farther. Dusts originating from Khuzestan are of this type. Although the levels of chemical, mineral, and microbial contaminants are not very high, the contaminants could enter into the food chain and cause health problems; therefore more studies on contaminants in food and water resources in the region are required.

Health Effects of Dust Storms in Sistan and Baluchistan Provinces

Sistan and Baluchistan Provinces are located in eastern and southeastern Iran, between latitudes 25° 4′ N and 31° 29′ N and longitudes 55° 58′ E and 63° 20′ E (Fig. 6) and covers an area of 181,785 km^2. Sistan region in the northern part of the province lies within a dry temperate zone with lowlands. It is an arid region with very low annual precipitation (61 mm), low air humidity, and frequent droughts and dry winds. Sistan Basin once used to accommodate the 2,000 km^2 wetland ecosystem known as the Hamoun Wetlands which consists of several lakes (Dhondia and Diermanse, 2006). But with unprecedented population growth in the region throughout the 20th century, coupled with a relatively sudden and dramatic increase of irrigation of the Helmand River, in addition to human mismanagement of the river, the Hamoun Wetlands have almost completely dried up within the last 5 years. The light, silty lake floor is now vulnerable to the intense heat and strong winds often experienced in this region that, after the 1999 drought (Miri et al., 2007), started to generate some very impressive dust storms that appear to be increasing in both frequency and severity (Fig. 6).

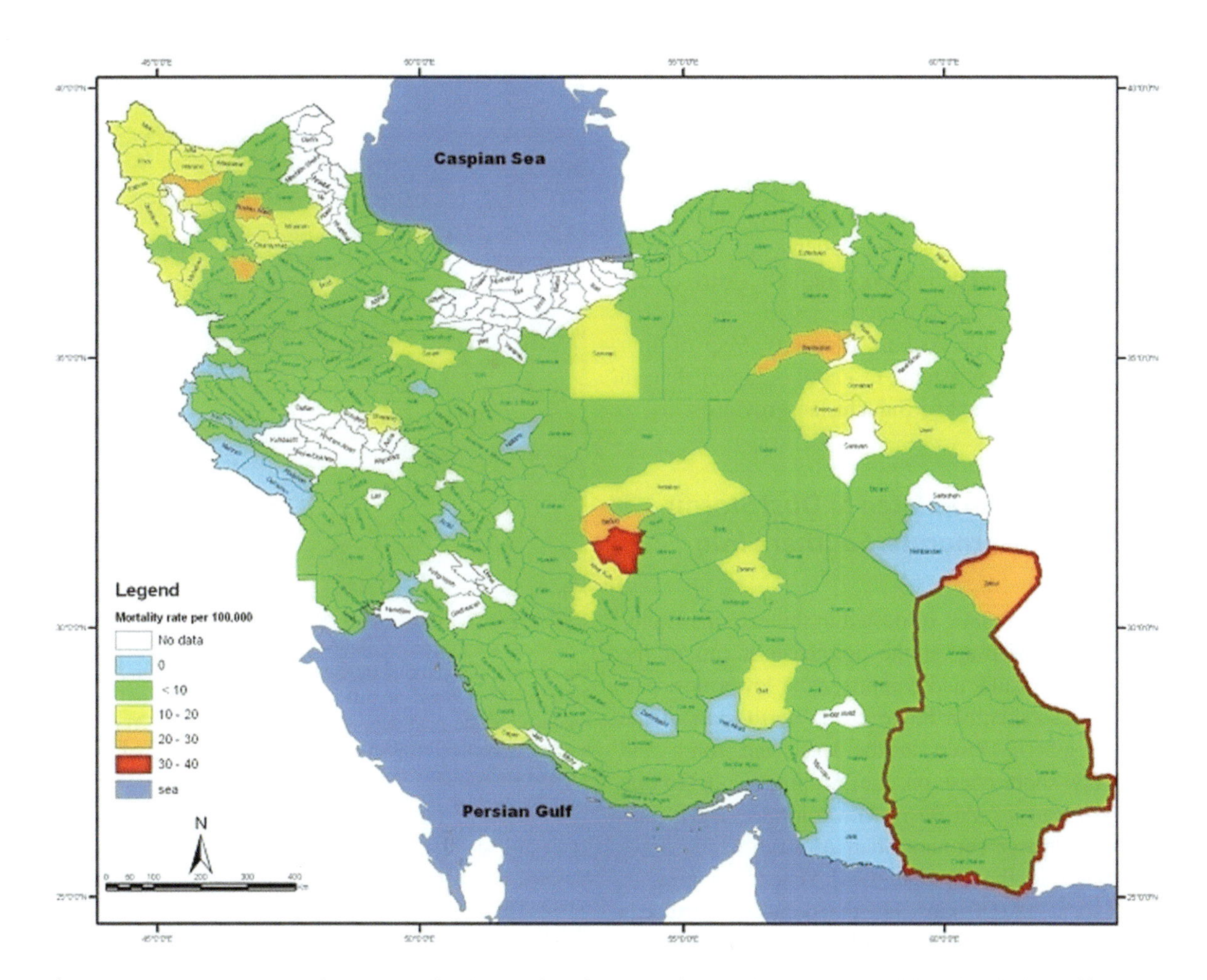

Fig. 6 Asthma mortality map of Iran. Mortality data obtained from the Ministry of Health. The *dark boundary* indicates Sistan and Baluchistan Province and the *dark strip* shows Zabol District (prepared by Navi, 2009)

Tens of thousands of people have been suffering through months of devastating sand storms in Sistan Basin, especially the cities Zabol and Zahak and surrounding villages. A severe sandy dust storm occurred in Zahak and its 80 villages on June 30, 2008 and resulted in closed schools and businesses. With the storm lasting about 5 days, more than 3,000 people suffering from allergy and respiratory diseases went to hospitals or health centers. Miri et al. (2007) showed that 63% of the people of Zabol suffer from respiratory diseases with the majority coming from villages rather than the city. Miri et al. (2007) indicated that 132,000 people have been considered as patients suffering from respiratory diseases related to the dust storms. The health damages to the population were estimated at over US $66.7 million in the period 1999–2004. The information obtained from hospitals indicated that most of the patients who visited hospitals suffered from chronic obstructive pulmonary disease (COPD) and asthmatic diseases with the peak of incidence during the summer season (June, July, and August) when the severest dust storms occur. It is estimated that 90% of the population living in the region suffered from respiratory problems in June, July, August, and September. Medical costs for patients for the period of study exceeded US $166.7 million (Miri et al., 2007).

According to the Asthma Mortality Map of Iran (Fig. 6), the rate of asthma in Zabol is higher than in other cities in Sistan and Baluchistan Provinces.

Health Problem Related to Dust Storms Al-Basra Governorate, Southern Iraq

Al-Basra Governorate is located in the extreme southeastern part of Iraq, at the northeastern rim of the Arabian Peninsula, and comprises the lowermost part of the Mesopotamian Plane in the north, part of the Iraqi Southern Desert in the west and south and relatively short coast on the Arabian Gulf. In the northern part of Al-Basra Governorate, Tigris and Euphrates merge forming Shatt Al Arab which flows southward to the Arabian Gulf. The total area of Al-Basra is 19,070 km^2 and the population exceeds 3 million, including 1.7 million in Al-Basra City (Fig. 7a–c).

Fig. 7 **a–c**. The city of Baghdad ground to a halt on August 8, 2005, under a cloud of suffocating dust that lingered over the region until the next day. According to The New York Times and the Website TerraDaily, reduced visibility slowed traffic to a crawl among those determined to brave the storm while many commuters stayed home. A number of Iraqis were quoted as saying that this dust storm is the worst they have seen in years. The storm also reportedly overwhelmed Al Yarmuk Teaching Hospital, which treated more than a thousand people with respiratory distress. City officials shut down Baghdad's main airport, and Iraq's constitution talks were postponed (*Source*: NASA MODIS Aqua, August 7, 2005)

Fig. 7 (continued)

Dust and sand storms occur frequently presenting an acute environmental problem with impact on human health, economy, and safety. Al-Najem (1975) identified two main regions stricken by dust and sand storms: one is the central part of Iraq with Baghdad as its center and another in the southern part of Iraq with its center west of Al-Basra. The major source of dust is the Southern Desert (the northern extension of the Great Arabian Desert) with the Northern Desert playing a secondary role. Estimates of the regional rate of dust fallout sedimentation ranged between 1.5 mm per year in central and southern Iraq and 0.88 mm per year in the northwestern part of the Arabian Gulf. Khalaf and Al-Hashash (1983) estimated dust fallout over Al Kuwait as 1,002.7 tons/km^{-2} in July 1979 compared to 9.8 tons/km^{-2} in November 1979.

In spite of the increasing frequency, duration, and intensity, especially since 1991, little attention has been paid to study the health impact of dust and sand storms in Iraq. In Al-Basra Governorate Al-Kabi (1999) studied the geographical pattern of some chronic, non-communicable diseases while Al-Huwaider (2001) studied the effect of industrial pollution on geographic distribution of allergy and asthma diseases. Respiratory system allergies gained the first rank among other categories of allergies and asthma in Al-Basra Governorate while bronchial asthma gained the third rank. Geographically, Al-Zubair District in the southwest comprising mostly desert plains recorded the highest incidence (3.95% of its total population) while Al Qurna District located in the north within the Mesopotamian Plain recorded the lowest incidence (0.34% of its total population) with the other five districts ranging between (Al-Huwaider, 2001). An exception is Um Qasr Sub-District situated in the extreme southeast on the head of the Arabian Gulf. There, Al Marsoumi (2009) attributed the unexpectedly high rate of incidence to the abundance of dust particles less than 5 μm in size which represent a serious health risk. Figure 8 illustrates the great similarity between the monthly distribution of dust fallout over Al-Basra Governorate and the number of infections by bronchial asthma.

Al-Huwaider (2001) attributed this variation to the geographical, geological, and climatological factors while Al Marsoumi (2009) considered the textural composition of dust as another factor.

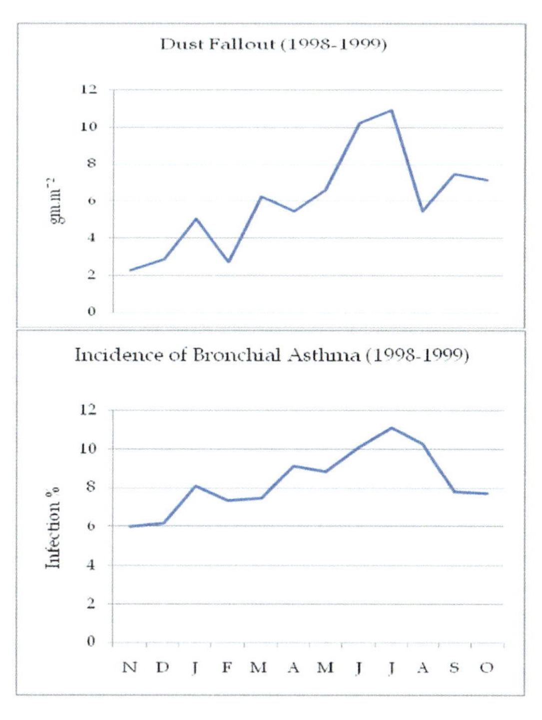

Fig. 8 Comparison between the monthly average weight at dust fallout and the number of infections during the period 1998–1999 (*Source*: Al-Ali, 2000; Al-Huwaider, 2001)

Health Implications Associated with Particular Matter Exposure for US and Australian Military in Southwest Asia

Weese and Abraham (2009) reviewed and documented the potential health implications associated with exposure to air pollution, including dust, on the US military deployed in Iraq and Afghanistan as follows:

"Sanders et al. (2005) conducted a survey of 15,000 military personnel deployed to Afghanistan and Iraq and estimated that 69.1% reported experiencing respiratory illnesses, of which 17% required medical care. The frequency of respiratory conditions doubled from a pre-combat period to a period of combat operations in this group.

Newly incident and previously unrecognized or undiagnosed asthma is a substantial source of reported respiratory symptoms among military personnel (Gunderson et al., 2005; Nish and Schwietz, 1992). In this study, asthmatics and non-asthmatics reported statistically significantly increased respiratory symptoms (wheezing, cough, sputum production, chest pain/tightness, and allergy symptoms) during deployment relative to symptom prevalence prior to deployment. In an abstract presented at the 2008 American Thoracic Society (ATS) Meeting, Szema et al. (2008a) reported results of an investigation of medical records of US soldiers discharged from active duty. They found that soldiers deployed to the Persian Gulf for 1 year or longer had a higher incidence of asthma (6.6%), relative to soldiers stationed in the USA (4.3%). In an abstract presented at the 2008 annual meeting of the American Academy of Allergy, Asthma and Immunotherapy, Szema et al. (2008b) also reported an association between deployment in Iraq and newly diagnosed allergic rhinitis. Results showed that 9.9% of soldiers deployed to the Persian Gulf for a year or more had allergic rhinitis vs. 5.1% of homeland-stationed personnel.

In an abstract presented at the 2008 American Thoracic Society, King et al. (2008) presented a case series of 47 soldiers with recent deployment in Iraq who presented with unexplained dyspnea. All of the cases had sand storm, smoke, and dust exposure, and many, but not all, of the soldiers were exposed to a sulfur fire in Mishraq, which occurred in 2003. Al Mishraq Native Sulfur Mine is located in northern Iraq, 40 km south of Al Mosul. On June 24, 2003, a fire ignited in a 400,000 tons native sulfur dump there that lasted approximately for weeks. The smoke plume contained various contaminants including particulate matter and varying concentrations of sulfur dioxide (SO_2) and hydrogen sulfide (H_2S). Health effects extended to the population of Al Mosul (1.17 million) to the north and Al-Qayara, 25 km south. This event is probably the largest non-volcanic SO_2 emission incident recorded to date.

Twenty-four of these soldiers had pathologic findings consistent with bronchiolitis, an uncommon inflammatory and fibrotic lesion of the terminal bronchioles of the lungs.

Eighteen cases of acute eosinophilic pneumonia (AEP) occurred from March 2003 to March 2004 among 183,000 military personnel deployed in or near Iraq. There were two deaths. AEP is a rare disease of unknown etiology characterized by respiratory failure, radiographic infiltrates, and eosinophilic infiltration of the lung. Extensive evaluation failed to demonstrate an infectious etiology or association with known causes. No geographic clustering was evident."

Kelsall et al. (2004) studied the respiratory health status of Australian veterans of the 1991 Gulf War and the effects of exposure to oil fire smoke and dust storms using a cross-sectional study comparing 1,456 Australian Gulf War veterans with a randomly sampled military comparison group ($n = 1588$). Australian Gulf War veterans had a higher than expected prevalence of respiratory symptoms and respiratory conditions suggesting asthma and bronchitis first diagnosed since the Gulf War, but did not have poorer lung function or more ventilatory abnormalities than the comparison group. Veterans exposed to dust storms had a slightly better peak expiratory flow rate than veterans who did not report exposure.

Kelsall et al. (2004) concluded that increased self-reporting of respiratory symptoms, asthma, and bronchitis by veterans was not reflected in poorer lung function. The findings do not suggest major long-term sequelae of exposure to oil fire smoke or dust storms.

Arsenic

Arsenic is a ubiquitous element in the environment, and can be mobilized due to various natural processes and human activities. Arsenic is relatively mobile at typical pH values of natural waters (pH 6.5–8.5) and under both oxidizing and reducing conditions. However, arsenic exhibits maximum mobility under high pH and oxidizing conditions and under strongly reducing conditions.

Although most environmental arsenic problems result from mobilization under natural conditions, anthropogenic impacts have been locally significant due to such activities as mining, fossil fuel combustion, and use of arsenic chemicals.

Human exposure to arsenic may be through a number of pathways, including air, food, water, and soil. Of these, drinking water represents the greatest threat to human health, with the highest aqueous arsenic concentrations tend to be found in groundwaters because of the natural water–rock interaction processes and the high solid/solution ratios found in aquifers (Smedley and Kinniburgh, 2005).

Chronic exposure to arsenic may cause skin disorders (melanosis and keratosis), chronic cardiomyopathy (especially preclinical microcirculatory defects and arterial insufficiency), interstitial inflammation with fibrosis, hypertension, and diabetes mellitus, probable lower IQ, and a variety of cancers involving the skin, lungs, bladder, kidney, and liver. Acute arsenic toxicity may result in a toxic myocarditis (Smedley and Kinniburgh, 2005; Fowles et al., 2005). Arsenic may cause also brain neuronal apoptosis which lead to Alzheimer's disease (Gharibadzeh and Hoseini, 2008).

Arsenic is associated with the metallic mineralizations in the Eastern Iran Zone (Khorasan Razavi and Khorasan South Provinces), Central Iranian Micro-Continent (Lut Block) (Yazd and Kerman Provinces), Alborz Mountains Zone (Khorasan North, Golestan, and Mazandaran Provinces), Sanandaj–Sirjan Zone (Azarbayjan East, Azarbayjan West, and Kurdistan Provinces), and the Arabian Shield. Arsenic is also associated with sulfide minerals in coal and bitumen deposits in the Central Iranian Micro-Continent (Lut Block) and Alborz Mountains Zone and with thermal springs in Sanandaj–Sirjan Zone, Central Iranian Micro-Continent and Zagros Fold Belt (Kermanshah, Ilam, Lorestan, Khuzestan, Charmahal-e-Bakhtiyari, Kohgiluyeh, Bushehr, Fars, and Hormozgan Provinces).

In a statistical modeling of global geogenic arsenic contamination in groundwater, Amini et al. (2008a) considered the Middle East region of prevailing high pH/oxidizing conditions where arsenic is soluble in its oxidized state. The statistical model indicated that substantial parts of the Middle East region have 0.25–0.50 probability of $>10\ \mu g\ l^{-1}$ with higher probability of 0.50–0.75 in Al Hijaz (the Arabian Shield), the central part of the Arabian Plateau (Hayil Arch), and a strip in north central Iraq corresponding to an evaporate terrain. The highest probability corresponds to a strip within the central part of the Mesopotamian Plain comprising Baghdad and Babil Governorates as well as to the Precambrian igneous complexes of the Arabian Shield and probably to Hadramawt Group (Paleocene–Eocene) in the northern and eastern parts of Yemen, respectively. However, this model is questionable at least within the Mesopotamian Plain; neither the geology of the terrain which consists of very thick sequences of recent sediments with no possible sources of geogenic arsenic nor available analyses of underground water for arsenic support that (Fig. 9).

Arsenic Distribution and Mortality Rates of Related Diseases in Iran

The Arsenic Distribution and Mortality Rates of Related Diseases (National Geoscience Database of Iran, 2007) (Fig. 10) shows the relationship between arsenic and some diseases in Iran. The map illustrates the geochemical distribution of arsenic in various geological units and in hot springs, mineralizations, arsenic containing coal and bitumen deposits, and the geographical distribution of arsenic mortality rates obtained from the Ministry of Health and Medical Education in the period 2001–2003.

The map shows that the mortality rate is higher in Isfahan, Semnan, Kerman, Khorasan Razavi, and Azerbaijan (particularly Azerbaijan West) Provinces than in other parts of Iran. Maximum recorded mortality rates of skin and liver cancers are, respectively, 5 and 25 per 100,000 while the maximum mortality rate of all cancers is 125 per 100,000. There is no significant change in the mortality rates of skin and liver cancers in the period 2001–2003. Also, mortality rate of cardiovascular diseases increased in the period 2001–2003 to over 400 per 100,000.

The geochemical distribution of arsenic demonstrates significant correlation with mortality rate distribution of all cancers in East Azerbaijan, Khorasan

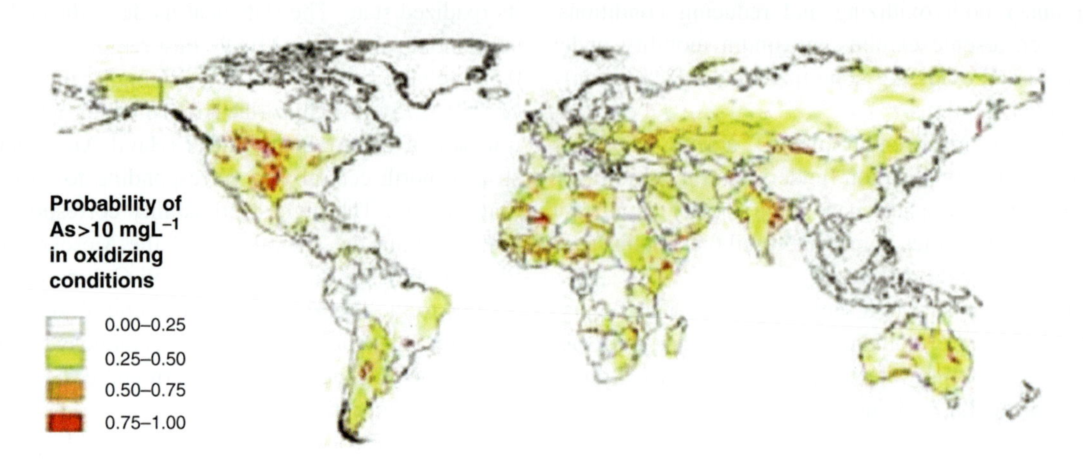

Fig. 9 Global probability of geogenic arsenic contamination in groundwater for high-pH/oxidizing conditions where arsenic is soluble in its oxidized state (*Source*: Amini et al., 2008a)

Razavi, Kerman, southern parts of Yazd, and particularly Azerbaijan West, Semnan, and Isfahan.

In addition to arsenic mineralizations, significant amounts of arsenic may also be associated with gold, silver, lead–zinc, and cobalt ores. Mining of these may result in mobilization and/or transportation of arsenic into drinking water.

Geothermal spring waters mainly in Azerbaijan West, Kerman, and Fars Provinces contain relatively high concentrations of arsenic, hence the positive

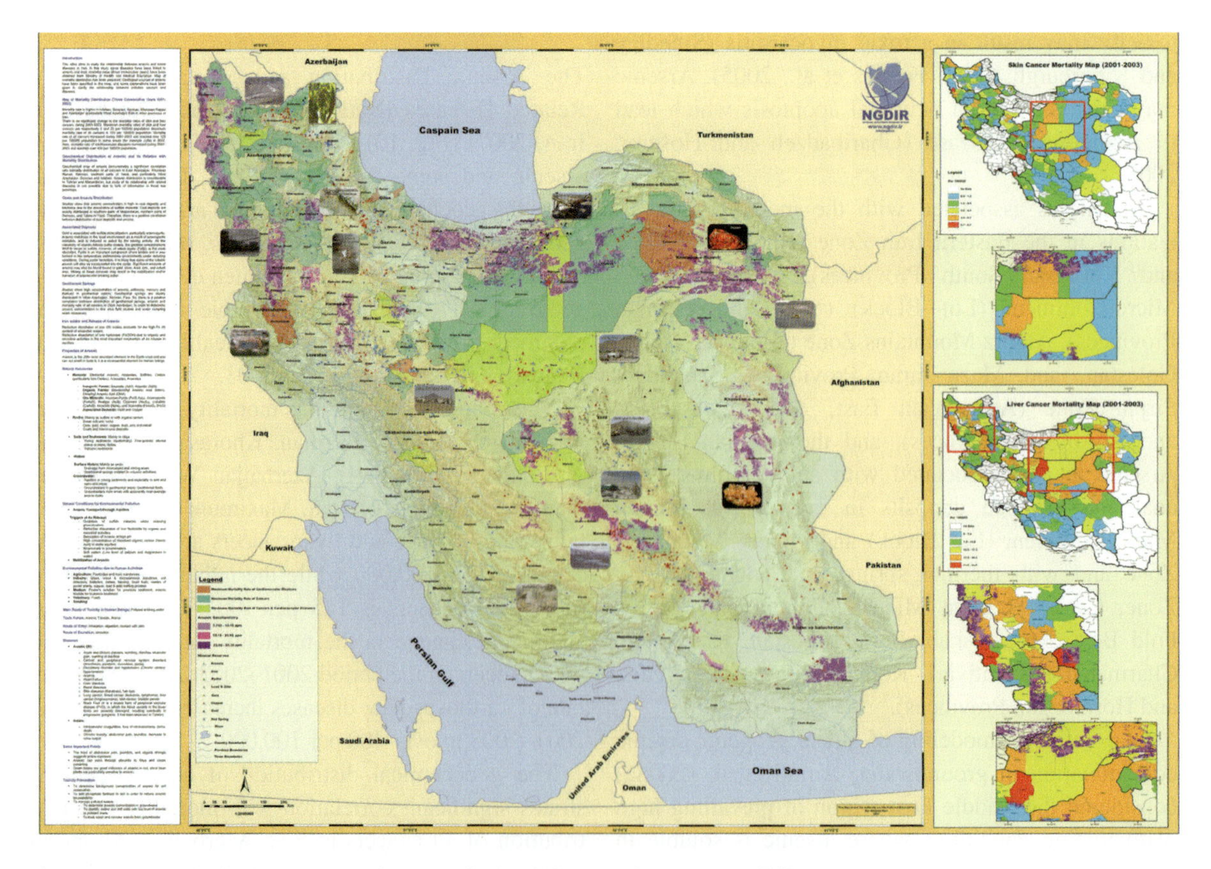

Fig. 10 Arsenic distribution and mortality rates of related diseases (*Source*: Navi, 2007)

correlation between the distribution of geothermal springs, arsenic, and mortality rate of all cancers in Azerbaijan West.

Arsenic Occurrence in Drinking Water of Iran – The Case of Kurdistan and Azerbaijan Provinces

Mosaferi et al. (2008) indicated that several parts of Iran suffer from natural arsenic contamination of groundwater, one of these is Bijar District in Kurdistan Province (Fig. 9). Bijar District is 580 km^2 and situated 1,750 m, above sea level. The total annual consumption of water for municipal, industrial, and agricultural purposes is about 30 million cubic meter, most of it from groundwater resources through 42 springs, 9 subterranean channels, 17 deep wells, and 616 hand-dug wells (Fig. 11).

Field studies indicate elevated levels of arsenic in drinking water ranging (for 18 villages) between 10.7 and 1,480 $\mu g\ l^{-1}$ (Table 2) with an average value of 290 $\mu g\ l^{-1}$ compared to a national upper limit of 50 $\mu g\ l^{-1}$.

The source of arsenic is definitely geogenic due to the dissolution of arsenic and arsenic containing minerals. Although mining and the application of pesticides as anthropogenic sources also might contribute, the study did not identify industrial wastewater or pesticide application as sources of pollution.

Mosaferi et al. (2008) indicated that the first case of chronic arsenic poisoning in Iran was recognized in Kurdistan Province in 1986. This particular case was a woman with intense skin lesions who lived in a village in Bijar District and she had lost her legs as a result of gangrene due to the consumption of water containing high levels of arsenic. Some limited monitoring of drinking water resources had shown a concentration of arsenic >1 mg l^{-1} in some villages in Iran.

In a later, comprehensive study, Mosaferi et al. (2008) investigated skin lesions among 752 participants in eight villages in Bijar District with emphasis on total lifetime intake of arsenic (TLIA) from drinking water. Arsenic concentrations ranged between 0.01 ($\pm$0.006) and 0.46 ($\pm$0.02) mg l^{-1} with a mean of 0.138 ($\pm$0.154) mg l^{-1}. The study concluded that there is significant linear relationship between TLIA and hyperkeratosis and hyperpigmentation (Table 3). The study concluded also that for a more accurate estimation of exposure it is necessary to investigate the history of water sources with regard to arsenic concentration for the whole period of exposure, exposure

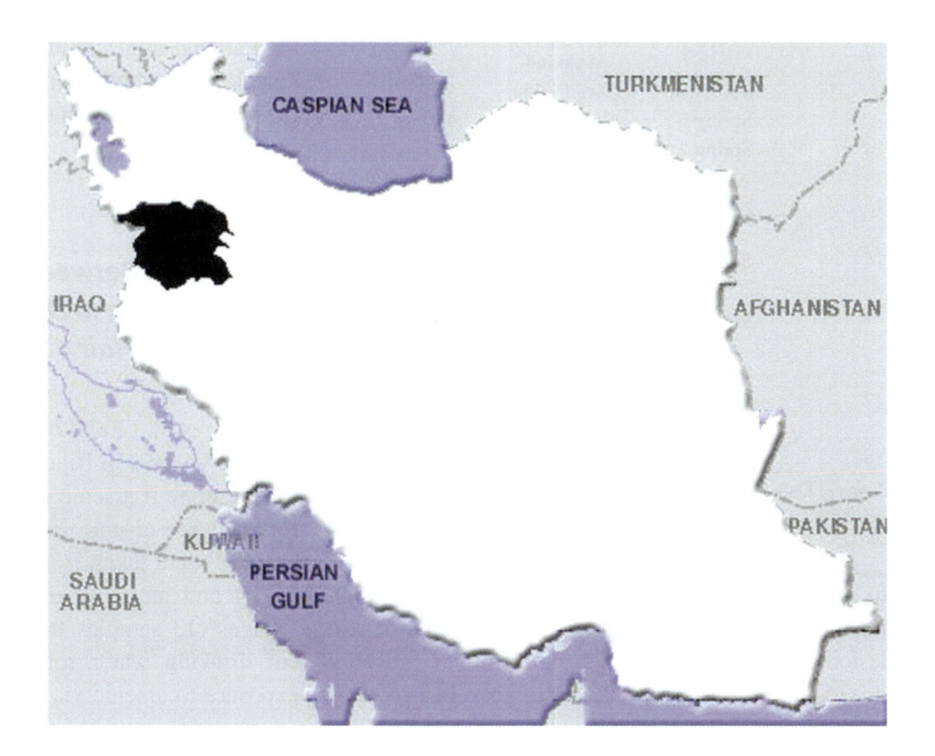

Fig. 11 Location map of Kurdistan Province (*Source*: Mosaferi et al., 2008)

Table 2 Arsenic concentration of some water sources in rural areas of Kurdistan Province. Sampling conducted in 2001 and arsenic determined using neutron activation

No. of village	Type of water source	Type of use	As concentration ($\mu g\ l^{-1}$)
1	Spring	Piped for drinking	32
2	Spring	Piped for drinking	10.7
3	Well	Piped for drinking	21
4	Spring	Piped for drinking	187
5	Well	Piped for drinking	205
	Underground channel	Non-drinking	206
6	Spring	Piped for drinking	11.6
	Spring	Non-drinking	11
7	Spring	Piped for drinking	15.9
8	Spring	Non-drinking	28
9	Spring	Non-drinking	15
10	Spring	Non-drinking	28
11	Spring	Non-drinking	45
12	Spring	Piped for drinking	90
	Spring	Piped for drinking	30
13	Spring	Non-drinking	1,040
	Spring	Piped for drinking	210
	Spring	Piped for drinking	801
	Spring	Non-drinking	440
	Spring	Non-drinking	470
14	Spring	Usually used for drinking	43
15	Spring	Piped for drinking	23.9
16	Spring	Piped for drinking	1,480
	Spring	Usually used for drinking	20.8
	Spring	Non-drinking	118
17	Spring	Piped for drinking	422
18	Spring	Non-drinking	408

Source: Data from Mosaferi et al., 2008

to arsenic in drinking water for long periods which increases mortality from cancer. The prevalence of skin lesions varied among villages and families probably due to genetic and nutritional differences. In an overall conclusion, the study found a clear dose–response relationship between the prevalence of skin lesions and TLIA as a reliable indicator of exposure. In similar epidemiological studies where the sources of drinking water have been changed, TLIA may prove to be a better indicator of arsenic exposure than the current level of arsenic in drinking water. The study recommended further investigation of other health problems of arsenic in drinking water, e.g., internal cancers, in polluted areas of Kurdistan Province.

Table 3 Logistic regression analysis between hyperkeratosis, hyperpigmentation, intake of arsenic, and age

	Variable	OR[a] (e^b)	95% CI for e^b
Hyperkeratosis	*TLIA* (g)	1.14	1.039–1.249
	Age (year)	1.024	1.009–1.04
Hyperpigmentation	*TLIA* (g)	1.254	1.112–1.416
	Age (year)	1.029	1.006–1.052

[a]For hyperpigmentation adjusted for age and EC. [b]For hyperkeratosis adjusted for age, EC, TDS, hardness, alkalinity, Cl^-, $HC0_3^-$, Na+, Al^{3+}, and Fe^{3+}
Source: Data from Mosaferi et al., 2008

Mosaferi et al. (2008) studied the arsenic content in drinking water in Hashtrood District in Azerbaijan East Province to the north of Kurdistan Province and related health effects. The study covered all urban and rural centers in Hashtrood District. Analysis of drinking waters indicated that arsenic was present in the water supply of 50 villages, with 9 villages higher than the national upper limit of 50 $\mu g\ l^{-1}$. The source of arsenic is geogenic and its health impact on 11,087 persons (21.96% of the total population) have been investigated.

Correlation Between Arsenic Concentration in Drinking Water and Human Hair

Mosaferi et al. (2008) conducted a follow-up research in three villages in Bijar District, Kurdistan Province, to determine if arsenic in hair can be used as a proxy for total life arsenic intake via drinking water. Hair samples were taken from 39 female participants (age 11–74 and mean 32 ± 17 years) residing in three villages (13 samples from each village) with different drinking water sources and different levels of exposure to arsenic via drinking water for a duration of 2–74 years. One village, Najafabad, has had no arsenic in drinking water and, accordingly, considered unexposed. The second, Gheshlaghnoruz, has been exposed to low-level (85 μg l^{-1} annual average) arsenic while the third village had high levels (455 μg l^{-1} annual average) of arsenic in drinking water but the exposure was eliminated 4 years ago. Hair samples were analyzed for arsenic using neutron activation analysis.

Mosaferi et al. (2008) indicated that there is no statistical significance between age and residence duration of the three villages. Alternatively, Fig. 10 shows that there is a consistency in arsenic concentrations in hair and total life intake. A strong positive correlation also exists between hair arsenic and total arsenic content via drinking water and between current arsenic content of drinking water and arsenic concentration in hair samples. For each milligram increase in hair arsenic there has been a 2.69 increase in total arsenic intake via drinking water irrespective of age (Figs. 12, 13, and 14).

The study concluded that (1) the concentration of arsenic in hair clearly increased in people consuming drinking water with high arsenic concentration; (2) there is close relationship between arsenic concentration in hair and total lifetime intake (via ingestion of contaminated water); and (3) significant correlation exists between arsenic concentration of drinking water and in hair. While arsenic concentration in hair can be used as proxy for chronic arsenic ingestion via

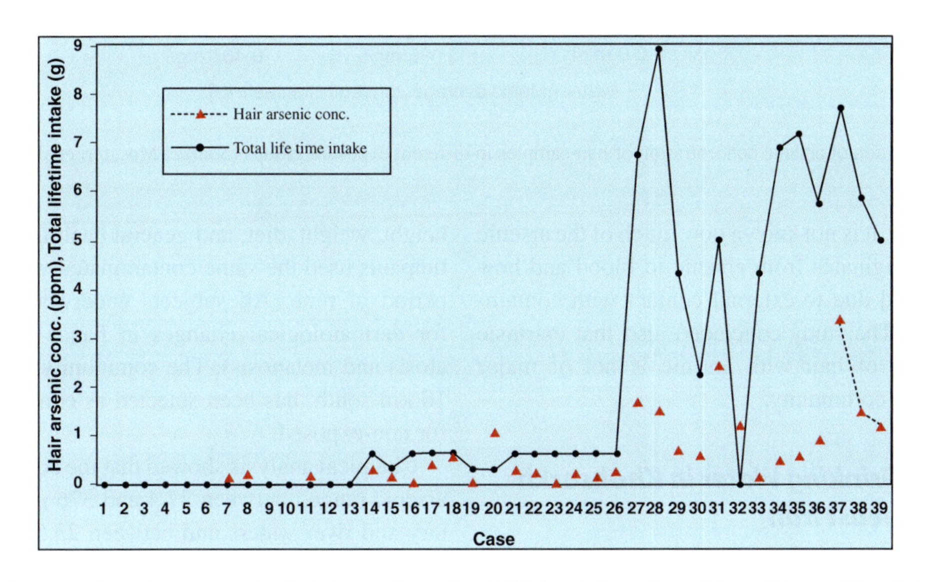

Fig. 12 Consistency of arsenicconcentration in hair samples and total life intake in each participant (*Source*: Mosaferi et al., 2008)

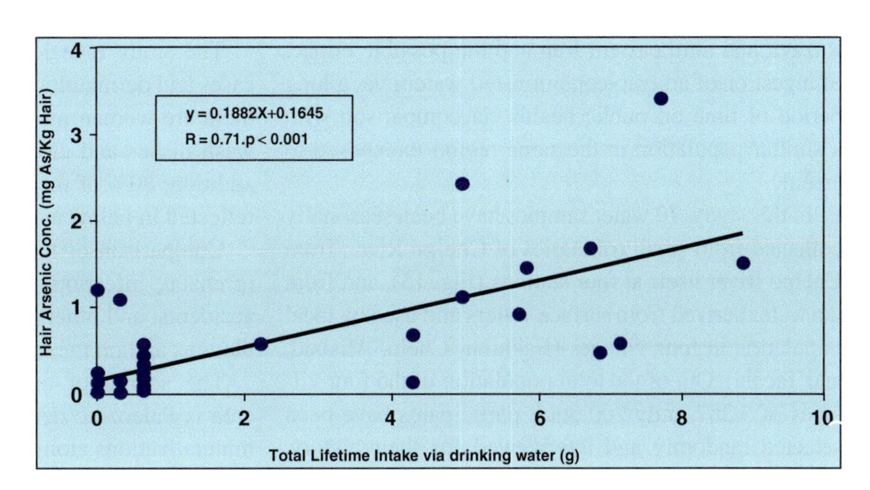

Fig. 13 Arsenic concentrationin hair vs. total arsenic intake via drinking water. (*Source*: Mosaferi et al., 2008)

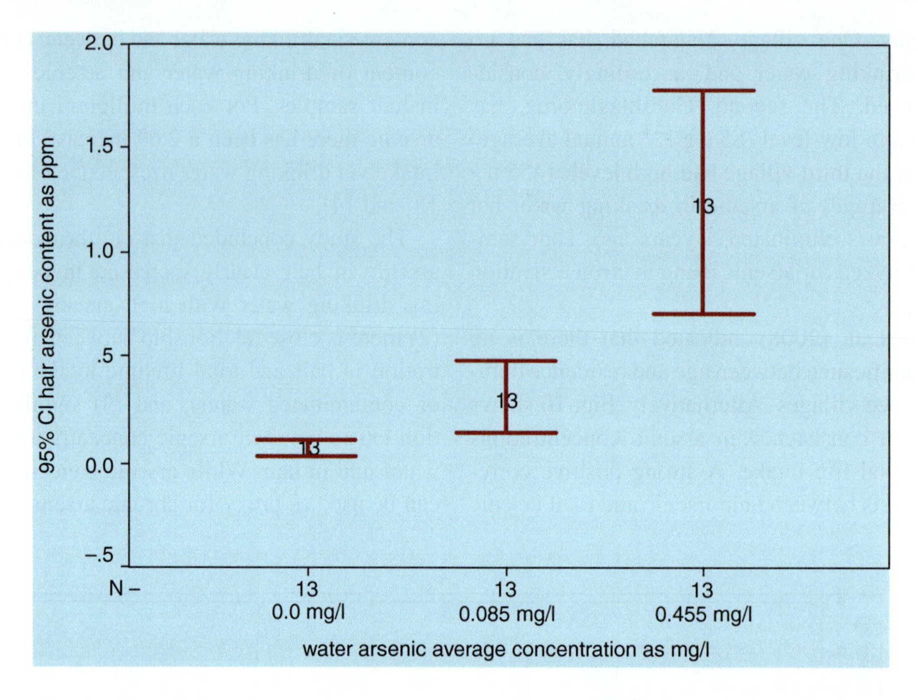

Fig. 14 Distribution of arsenic concentration of hair samples in different exposure groups (*Source*: Mosaferi et al., 2008)

drinking water, it is not known how much of the arsenic in hair that originates from arsenic in blood and how much is bound due to external contact with contaminated water. The study concluded also that extrinsic contamination of hair with arsenic is not of major concern in the community.

Arsenic in Drinking Water in Khohsorkh Area, Northeast Iran

Ghassemzadeh et al. (2008) studied the distribution of arsenic in surface waters of Khohsorkh area, Khorasan Razavi, and northeastern Iran and the possible impact of ingestion of arsenic-contaminated water over a long period of time on public health via comparison with a similar population in the same region unexposed to arsenic.

In this study, 79 water samples have been seasonally collected from small tributaries of Chelpo River, from Chelpo River itself at four stations (Fig. 15), and from tap water derived from surface waters and used by local population in four villages (Bakhtiar, Chelp, Aliabad, and Tacala). Out of the total population of the four villages of 3,287, only 200 adult participants have been selected randomly and interviewed for their history, height, weight, diet, and general health. All of the participants used the same contaminated water for a long period of time. All subjects underwent examination for dermatological changes in hands and feet (keratosis and melanosis). The community of Akbarabad, 10 km south, has been selected as control population for non-exposed.

Chemical analysis showed that the concentrations of arsenic ranged between 37.4 and 376 $\mu g\ l^{-1}$ in tributary and river waters and between 25.5 and 104.0 $\mu g\ l^{-1}$ in tap water with an overall mean value of 97.1 $\mu g\ l^{-1}$ compared to a mean of 7.8 $\mu g\ l^{-1}$ in the drinking water of unexposed population.

The study revealed that out of the 200 cases, 20 cases had dermatological problems (keratosis), 65% of them are women more than 40 years old who usually wash dishes and clothes with contaminated water. In addition, 80% of the cases suffered from malnutrition reflected in below average height and weight.

Comparison of total deaths in six different groups (prenatal, infectious diseases, cardiovascular, cancer, accidents, and others) in exposed and unexposed populations and on the national level is given in Table 3.

The source of arsenic in this naturally polluted area is Paleozoic deposits cut by faults with metallic mineralizations along this fault zone.

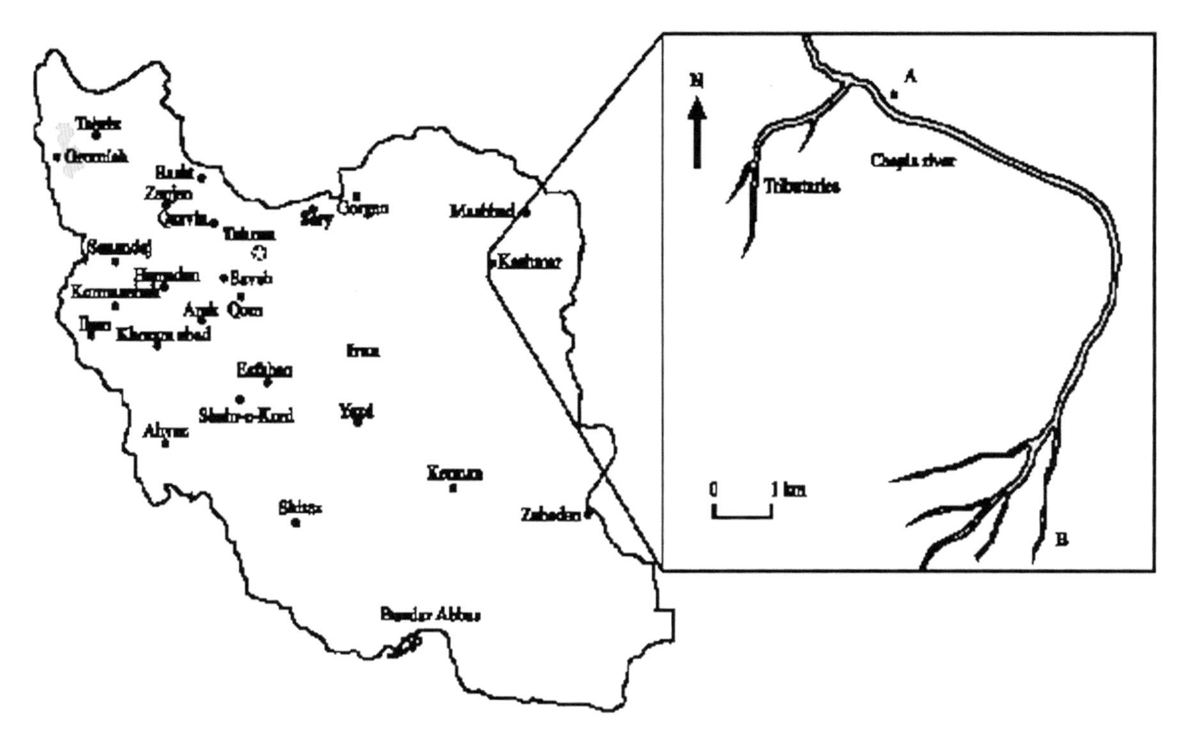

Fig. 15 Location map of Khohsorkh area, Khorasan Razavi (*Source*: Ghassemzadeh et al., 2008)

As indicated in Table 4, a 10% prenatal death rate in the exposed population is much higher than that in the unexposed population (3.8%). Also, 95% of the newly born babies are 2.5 kg or less in weight. There is a substantially higher death rate of cancer in the exposed population (26.0%) compared to the unexposed population (8.3%) and the national average for rural communities in Iran (11.7%). However, out of this 26% mortality rate, none is caused by skin lesions, while digestive track cancer and pulmonary cancers account for 36.7 and 16.7% of cancer mortalities, respectively. Also 52% of the study participants, mostly men in the age range of 30–65 years, suffered from other pulmonary effects, specifically cough and shortness of breath. Considering that Kohsorkh is located in an arid region where dust particles might be contaminated with arsenic from nearby ore bodies, pulmonary cancer and effects can be shown, especially for farmers and shepherds working in open air. Malnutrition, widespread among the study participants, and the presence of antimony in association with arsenic might increase the toxicity of arsenic and its health impact.

Table 4 Percentage of deaths by cause in the exposed and unexposed communities and average for rural communities in Iran

Cause of death	Exposed population (%)	Unexposed population (%)	Average for rural communities (%)
Prenatal	10.0	3.8	4.2
Infectious diseases	2.5	5.8	1.6
Cardiovascular	12.5	19.6	33.8
Cancer	26.0	8.3	11.7
Accident	9.5	15.2	12.4
Others	39.5	47.0	35.9

Source: Data from Ghassemzadeh et al., 2008

Iodine

Iodine has long been known as an essential element for humans, and mammals in general, where it is concentrated in the thyroid gland. It is a component of the thyroid hormone thyroxine. Deprivation of iodine results in a series of iodine deficiency disorders (IDD), the most common of which is endemic goiter, a condition where the thyroid gland becomes enlarged in an attempt to be more efficient. Iodine deficiency during fetal development and in the first year of life can result in endemic cretinism, a disease which causes stunted growth and general development along with brain damage. However, the more insidious problem is that iodine deficiency impairs brain development in children even when there is no obvious physical effect.

Indeed it has been suggested that iodine deficiency is the most common preventable cause of mental retardation (Combs, 2005; Nordberg and Cherian, 2005).

Endemic goiter and cretinism along with related IDD have long been recognized as serious health problems and consequently much work has been carried out on the etiology and geographical distribution of these diseases. It became generally agreed that the primary cause of IDD is a lack of iodine in the diet. Iodine was the first element recognized as being essential to humans and endemic goiter was the first disease to be related to environmental geochemistry (Fuge, 2005). Iodine remains the classic success story in medical geology as far as human health is concerned (Davies et al., 2005)

Endemic goitrous areas where IDD are concentrated tend to be geographically defined. Thus many of the most severe occurrences of endemic goiter and cretinism have been found to occur in high mountain ranges, rain shadow areas, and central continental regions. This distribution of IDD results from the unique geochemistry of iodine. Most iodine is concentrated in seawater and derived from volatilization from the oceans with subsequent transport onto land. Since most of the iodine in soils is derived from the atmosphere and ultimately the marine environment, the proximity of an area to the sea therefore is likely to exert a strong influence on the iodine content of soils in that area. This also results in considerable geographic variation of soil iodine content.

IDDs in the Middle East affect those parts remote from marine influence, including continental inlands, mountainous regions, and their rain shadow areas such as the Alburz Mountains in northern Iran, Taurus–Zagros Mountains in northern–northeastern Iraq and northwestern Iran, Lebanon, and Anti-Lebanon Mountains in western Syria and Lebanon and Al Hijaz and Asir Mountains in western Saudi Arabia and its southern extensions in Yemen. In all of these parts, IDD can be explained according to the classic explanations of low iodine supply and hence low iodine availability. However, the prevalence of IDD cannot be simply explained in this context in at least one case where a near coastal area suffers from IDD. This is likely governed by the geochemistry of iodine and its bioavailability. Iodine retention capacity of soils is related to composition with organic matter and iron and aluminum oxides are the most important retentive components. So, sandy soils deficient in these components and prevailing in coastal areas are deficient in iodine. Another example of environmental control on IDD is the apparent strong association of some goitrous regions with limestone bedrock widely exposed in northwestern Iran, northern and northeastern Iraq, and Lebanon Mountains. IDD occurs in some of these limestone terrains despite relatively high iodine in soils. This would imply that iodine is not bioavailable (Fuge, 2005).

In a review of IDD in the Middle East, Azizi (2001) indicated that before 1987, IDD had not been recognized as a serious health problem in the Middle East. In the mid-1980s, IDDs in the Middle East were regarded as strictly limited to certain geographical areas and thus not a public health problem of national interest.

In Asia, where the largest global concentration of population affected by IDD is located, salt fortification and iodized oil distribution, training, and education have produced improvements in some countries (ACC/SCN, 1992). Following maps show the effects of implementing programs to eliminate iodine deficiency, including the purchase of salt iodization equipment (Fig. 16).

The International Council for Control of the Iodine Deficiency Disorders (ICCIDD) has issued detailed national and regional reports on the status of the IDD in the Middle East.

Although Iran was announced IDD-free since 2001 with a total goiter rate (TGR) dropped from 68.0% in 1989 to 54.6% in 1996 and 9.8% in 2001, some investigations of IDD are still underway. Monajemzadeh and Moghadan (2008) studied the prevalence of goiter among 11–16-year-old children in Ahwaz City, Khuzestan Province, southwestern Iran. The studied population comprised of 1,950 (1,050 males and 900 females) children, 146 children with goiter. Prevalence of goiter was 4.4 and 11.1% for male and female students, respectively. The study observed a significant difference in height and weight between students of

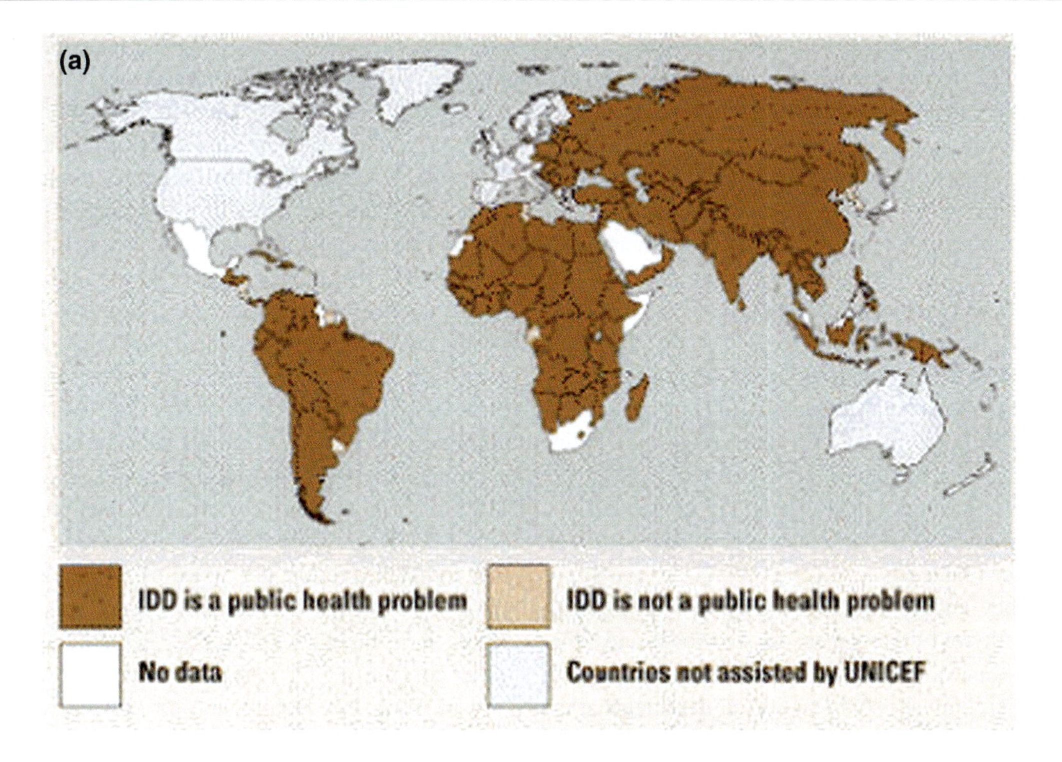

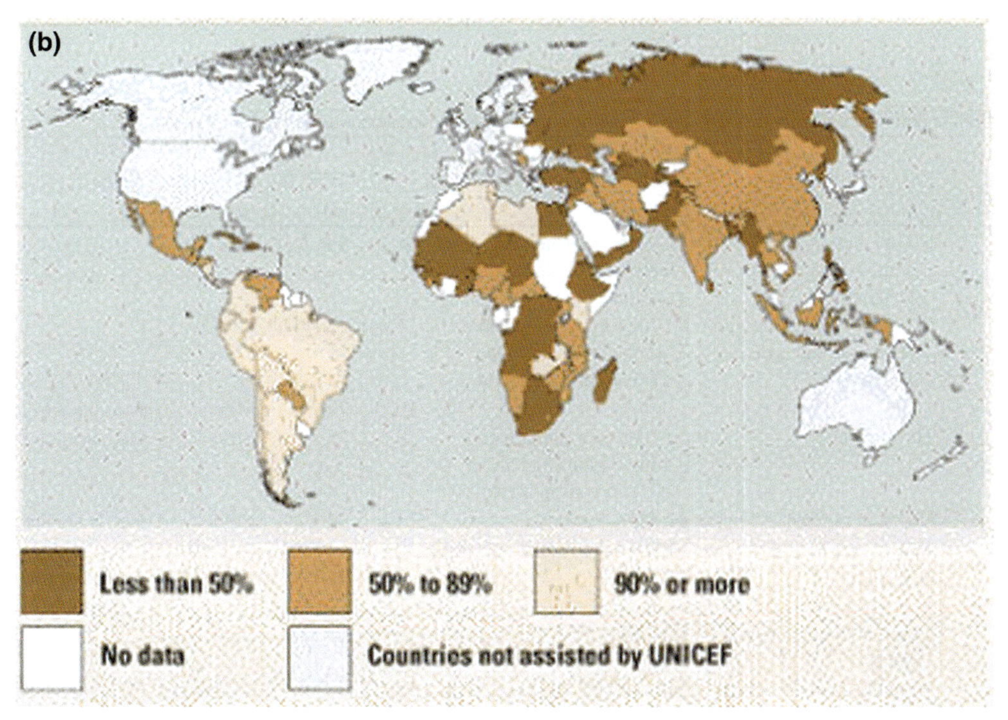

Fig. 16 **a**. Global Prevalence of IDD (circa 1990); **b**. Percentage of households consuming iodized salt (1992–1996) (*Source*: IDD data from UNICEF 'Report on Progress Towards Universal Salt Iodization', 1994. Salt iodization data compiled by UNICEF, 1997). (http://www.unicef.org/sowc98/approach4.htm)

grade 1 goiter (83 cases) and those with grades 2 (55 cases) and 3 (8 cases) goiter. The study concluded that Ahwaz City is not endemic with IDD due to its geographic location, low altitude, and nutritional habits of its people. Najafi et al. (2008) conducted a pilot neonatal thyroid screening in the mild IDD endemic region of eastern Iran. In the period May 2006–February 2007 a total of 59,436 neonates have been screened and recall rate and incidence of hypothyroidism recorded. The study indicated that the recall rate of 3.6 is high compared to other countries in the region while the frequency of neonatal blood thyroid stimulate hormone (TSH) was above 3% indicating mild iodine deficiency. The incidence of hypothyroidism (2 per 1,000) was relatively high. The study concluded that hypothyroidism is of the transient type and is of high prevalence (Table 5).

Caughey and Follis (1965) observed high prevalence of thyroid enlargement in Al Mosul and other cities in Ninawa Governorate, northern Iraq, and conducted a preliminary study on 500 adult patients at the Republican Hospital in Al Mosul. Goiter was found in 273 (55%) of them, 85 men and 188 women. Age and sex were clearly significant, since the prevalence was high in females aged 10–30 years. In men and women over 50 years, the prevalence fell and thyroid glands were notably smaller. The majority of the patients were Arabs or Kurds, the remainder being Turkman, Assyrian, or Armenian. Race did not appear to be important. The study investigated also the prevalence of goiter at a girls school and a boys school, both intermediate with an age range of 11–18 (Table 6). The results confirm the experience of Marine and Kimbell (1917) (in Caughey and Follis, 1965) that adolescent girls form the most sensitive group in any population. Similar studies were conducted in Tell Afar, 120 km northwest of Al Mosul, on 60 teenage girls with an age range of 10–16 and in Dohuk and Akra, 55 and 110 km north and northeast of Al Mosul, respectively (Table 7). Data on urinary iodine excretion indicated that iodine excretion per gram of creatinine is low (16.8–24.0 $\mu g\ g^{-1}$ creatinine) except in Tell Afar where it is higher (33.2 $\mu g\ g^{-1}$ creatinine). Iodine water content ($\mu g\ l^{-1}$) ranged between 3 in Al Mosul to 4 in Akra, 7 in Dohuk and, for comparison, 3.5 in Baghdad while in Tell Afar the iodine content is 17.5 since its source is natural springs. The study explained the situation in Al Mosul as of very high prevalence

Table 5 Iodine deficiency disorder (IDD) status in some countries of the Eastern Mediterranean Region

Country (1)	Endemic region (2)	Prevalence (before intervention) (3)	MUI $\mu g\ dl^{-1}$ (4)	TGR % (5)	IDD status (6)	Notes (7)
Iran	Highly endemic area southwest of Tehran (Shahryar)	17–100% prevalence among schoolchildren in various regions	16.7	9.8	Free	
Iraq	Northern mountain region (Ninawa) and around Baghdad	60–85% in hyperendemic northern area. About 30% around Baghdad	2.4	22–44	Severe	
Syria			< 10	70	Mild	Data need to be updated
Lebanon	Throughout	40–75% in mountain and hilly regions and about 12% in coastal areas	9.4	25.7	Mild	Data need to be updated
Jordan	Northern region	37.7 (18–76%)	15.4	32.1	Mild	Needs regular monitoring
Palestine	NA	14.9 in schoolchildren				
Oman	Only mild	TGR 10%	9.5	10	Mild	No national IDD control program
UAE	Desert and mountainous areas	1.5–20%	10	1.5–20	Mild	
Saudi Arabia	Probably southern region	NA	18	4–30	Moderate	Needs IDD control program, especially in southern province
Yemen	Mountainous region	TGR 32%, 60–100% in some locations	17.3	16.8	Mild	Needs regular monitoring

Source: Compiled from Azizi (2001) (columns 2 and 3) and Azizi and Mehran (2004)

Table 6 Goiter in studied populations in Ninawa Governorate, northern Iraq

Location	No. examined	No. with goiter	Classification Group 0 (normal)	Group 1 (small)	Group 2 (medium)	Group 3 (large)
Al Mosul School (Girls)	436	429 (98%)	7	374	54	1
Al Mosul School (Boys))	471	303 (64%)	168	298	5	0
Tell Afar	60	43 (72%)	14	43	0	0
Akra	68	53 (79%)	15	44	8	1
Dohuk	100	95 (95%)	5	77	18	0

Source: Data from Caughey and Follis, 1965

Table 7 Prevalence of goiter in Baghdad and Al-Basra

Location	Sex	Age group (years)	Total no. examined	Positive goiter (%) Grade 1	Grades 2 and 3	All grades
Baghdad (urban)	M	6–12	2,102	16.8	4.5	21.3
	F		1,910	22.0	5.6	27.6
	M	12–16	1,619	22.7	2.8	25.5
	F		1,459	33.3	7.9	41.2
	M	17–20	1,227	19.7	1.7	21.4
	F		1,208	38.4	17.7	46.1
	M	Over 21	550	16.5	4.2	20.7
	F		264	27.3	10.6	37.9
Baghdad (rural)			412	27.4	2.9	30.0
			548	25.7	4.0	29.7
			503	3.6	Nil	3.6
Al-Basra	M	6–12	1,179	2.2	Nil	2.2
	F		1,044	3.8	0.8	4.6
	M	12–16	723	4.2	0.1	4.3
	F		1,025	8.4	2.4	10.8
	M	17–20	568	2.6	0.3	2.9
	F		242	10.7	3.2	13.9

Source: Data from Demarchi et al., 1969

of goiter and recommended an immediate response through a program of goiter prophylaxis.

Demarchi et al. (1969) conducted a comprehensive investigation of goiter and its etiology in Baghdad and Al-Basra, the largest cities in Iraq and the surrounding rural areas. The study covered the examination for thyroid enlargement of 11,852 persons in the Baghdad and 4,789 persons in the Al-Basra, representing about 1% of the population of the areas surveyed (Table 6). The majority of the subjects were students of primary and secondary schools. The study concluded that Baghdad is endemic in goiter while Al-Basra is not and explained, on basis of iodine radionuclide uptake by the thyroid gland, that there is iodine avidity of the thyroid gland as an attempt to make up either for a deficient supply of iodine or for a block in iodine metabolism by goitrogenic factors. In addition the study indicated that deficiency in vitamin A might be another factor.

In the most recent study on the sedimentary and environmental geochemistry of iodine in, Iraq, Al-Jumaily (2001) indicated that iodine content in the exposed rock units in Al Mosul and the adjacent areas ranges between 3.1 and 9.0 $\mu g\ g^{-1}$ while its average content in soil and drainage sediments is 10.7 and 14.0–16.2 $\mu g\ g^{-1}$, respectively. In water, average iodine content varied between 0.39 $\mu g\ g^{-1}$ in rain water (the content decreased with successive rain fall), 0.63 $\mu g\ g^{-1}$ in mineral springs, 0.5 $\mu g\ g^{-1}$ in shallow wells, and 0.44 $\mu g\ g^{-1}$ in Tigris River. The study identified different sources of iodine in soil including washed iodine from atmospheric dust particulate (wet precipitation), liberated iodine from hydrocarbons associated with mineral springs activity,

selected, while Marabah area was selected as a relatively low altitude (500 m asl). The study population comprised 940 students underwent clinical examination for thyroid gland. The study revealed an overall prevalence of goiter of 24% among all schoolchildren. This prevalence was significantly higher among schoolchildren of high-altitude areas than among their counterparts in low-altitude areas. Schoolchildren in the high altitudes were 2.5 times more likely to develop goiter as compared to their counterparts in low altitudes. The study referred that to geological processes etiological effects. Heavy rains and the thawing of snow sweep soil away which is replaced by new soil derived from iodine-low crystalline rocks. Also, mountainous areas may not have access to seafood or food with high iodine content, making its people susceptible to the development of goiter.

Before 1991, the health authorities of Yemen did not perceive IDD to be a public health problem. However, since then, much progress has been made. Zein et al. (2000) studied the epidemiology of IDD in two ecological zones: Zone I comprising the mountains and highlands with an estimated TGR of 32% and zone II comprising the submountainous and lowlands (the coastal plains and the eastern plains) with a maximum TGR of 10%. The study populations comprised of 1,170 children (age 6–12, 811 males and 373 females) from zone I and 1,800 children (6–12 years, 1,192 males and 608 females) from zone II. The study involved examination by goiter palpation and determination of urinary iron excretion (UIE). Table 8 shows a summary of the results of the study. The prevalence rates of goiter in zone I being greater than those in zone II are not surprising as zone I is mountainous and a previous survey has documented very high goiter rates in this zone. However, the magnitude of the goiter prevalence rate in zone II is more than expected. An earlier survey in parts of this zone reported a very low goiter prevalence rate of 0.15%. The TGR of 16.8% for the whole country found in this survey is half of the TGR reported for Yemen previously. The study concluded, based on TGR alone, that zone I is an area where IDD still constitutes a severe public health problem. However, another indicator, a UIE level above 10 mg dl^{-1} in all ages and sexes and 70% of children excreting >100 mg l^{-1} iodine (thus exceeding the recommended value of 50% as the core indicator for monitoring progress toward IDD elimination), provides unequivocal evidence of the rapid progress made in Yemen toward elimination of IDD in a period just over 30 months following the introduction of universal salt iodization. Al-Hureibi et al. (2004) studied also the epidemiology, pathology, and management of goiter in Yemen. The study population comprised 667 patients admitted to the hospital, including 617 females, with an age range of 13–90 years and mean age of 35.2 years complaining of thyroid swelling in the period January 1997–December 2001. Most patients came from the highlands with an average elevation of 2,000–2,600 m, above sea level. The study indicated that the TGR in Yemen dropped to half of that reported before launching the salt ionization program. Although the study concluded that endemic goiter is related to iodine deficiency, it did not exclude other causes like dyshormonogenesis.

Fluoride

Fluorine is an essential element in the human diet. Deficiency of fluorine has long been linked to the incidence of dental caries. However, optimal doses of fluorine in humans appear to fall within a very narrow range. The detrimental effects of ingestion of excessive doses are well documented also. Chronic ingestion of high doses has been linked to the development of dental fluorosis, and in extreme cases, skeletal fluorosis. Concentrations of fluoride ions in natural waters span more than four orders of magnitude, although values typically lie in the 0.1–10 mg l^{-1} range. Where concentrations are high, drinking water can constitute the dominant source of fluorine in the human

Table 8 Total goiter by zone and gender

	Zone I			Zone II		
Gender	No. examined	No. with goiter	%	No. examined	No. with goiter	%
Male	811	266	32.8	1,192	97	8.1
Female	373	102	27.3	608	36	5.9
Total	1,184	368	31.1	1,800	133	7.4

Source: Zein et al., 2000

diet. Concentrations in drinking water of around 1 mg l^{-1} are often thought to be optimal; however, chronic use of drinking water with concentrations above about 1.5 mg l^{-1} is considered to be detrimental to health. The World Health Organization (WHO) (1993) guideline value for fluoride in drinking water is 1.5 mg l^{-1}.

High fluoride concentrations are most often associated with groundwaters as these accumulate fluoride as a result of water-rock interaction. However, in most groundwaters are below the upper concentrations considered detrimental to health.

Despite the uncertain health effects of fluoride in drinking water at low concentrations (0.7 mg l^{-1} or less), the chronic effects of exposure to excessive fluoride in drinking water are well established. The most common symptom is dental fluorosis (mottled enamel). With higher exposure to fluoride, skeletal fluorosis can result. This manifests in the early stages as osteosclerosis, involving hardening and calcifying of bones and causes pain, stiffness, and irregular bone growth. At its worst, the condition results in severe bone deformation and debilitation (Table 9).

Fluorosis has been described as an endemic disease of tropical climates (Fawell et al., 2006). One of the reasons for that in the tropics is temperature (Ekstrand et al., in Aminabadi et al., 2007). People in tropical regions drink more water, hence more fluoride intake (Fejerskov et al., in Aminabadi et al., 2007). The prevalence of dental fluorosis has been reported in children living at an altitude of 2,800 m (with 2.5 mg F/L in the drinking water) is higher than children living at an altitude of 1750 m (with the same level of F in the drinking water) (Rwenyonyi in Poureslami et al., 2008). Severe dental fluorosis is endemic in parts of the world where extremely high concentrations of fluoride occur naturally in drinking water.

Table 9 Health effects of fluoride concentrations in drinking water

Fluoride concentration range (mg l^{-1})	Chronic health effects
Nil	Limited growth and fertility
0.0–0.5	Dental caries
0.5–1.5	Promotes dental health, prevents tooth decay
1.5–4.0	Dental fluorosis (mottled teeth)
4–10	Dental fluorosis, skeletal fluorosis
>10	Crippling fluorosis

Source: Data from Dissanayake, 1991

In the Middle East region, fluorite deposits occur within Sanandaj–Sirjan Zone (metamorphic rocks with intrusive igneous bodies) (Kurdistan and Markazi Provinces), Central Iran Domain (crystalline basement) (Semnan and Yazd Provinces), and within the Precambrian Arabian Shield (crystalline basement) (Al Hijaz). Disseminated fluorite is probably associated with the volcaniclastic sediments of Zagros fold Belt in northern and northeastern Iraq and western Iran and the clastic and evaporate sequences of the Mesopotamian Basin.

Amini et al. (2008b) constructed a statistical model of global geogenic fluoride contamination in groundwater aiming at providing a global probability map indicating the risk of fluoride contamination in groundwater (Fig. 18). The model indicates that Middle East region has a 0.8–1 probability of fluoride concentrations exceeding 1.5 mg l^{-1}. However, this probability model is based on measured points for 92 samples in Iran (F^- range 1.49–41.71 mg l^{-1}, median 4.7 mg l^{-1} with 98% of the concentrations less than 1.5 mg l^{-1}) and 12 samples in Palestine (F^- range 0.02–0.56 mg l^{-1}, median 0.13 mg l^{-1}) with all of the concentrations less than 1.5 mg l^{-1}. However, this model is questionable at least within the Mesopotamian Plain; neither the geology of the terrain which consists of very thick sequences of recent sediments with no possible sources of geogenic fluoride nor available analyses of groundwater for fluorine support that.

Fluoride is found in both surface waters and groundwater. In surface freshwater, however, fluoride concentrations are usually low (0.01–0.3 mg l^{-1}). In groundwater, the natural concentration of fluoride depends on the geological, chemical, and physical characteristics of the aquifer, the porosity and acidity of the soil and rocks, the temperature, the action of other chemical elements, and the depth of wells. Because of the large number of variables, the fluoride concentrations in groundwater can range from well under 1 mg l^{-1} to more than 35 mg l^{-1} (UNICEF, 2000). Figure 18 indicates countries with endemic fluorosis due to excess fluoride in drinking water. The map coincides well with the map of predicted probability of fluoride concentration in the groundwater except for the Arabian Peninsula.

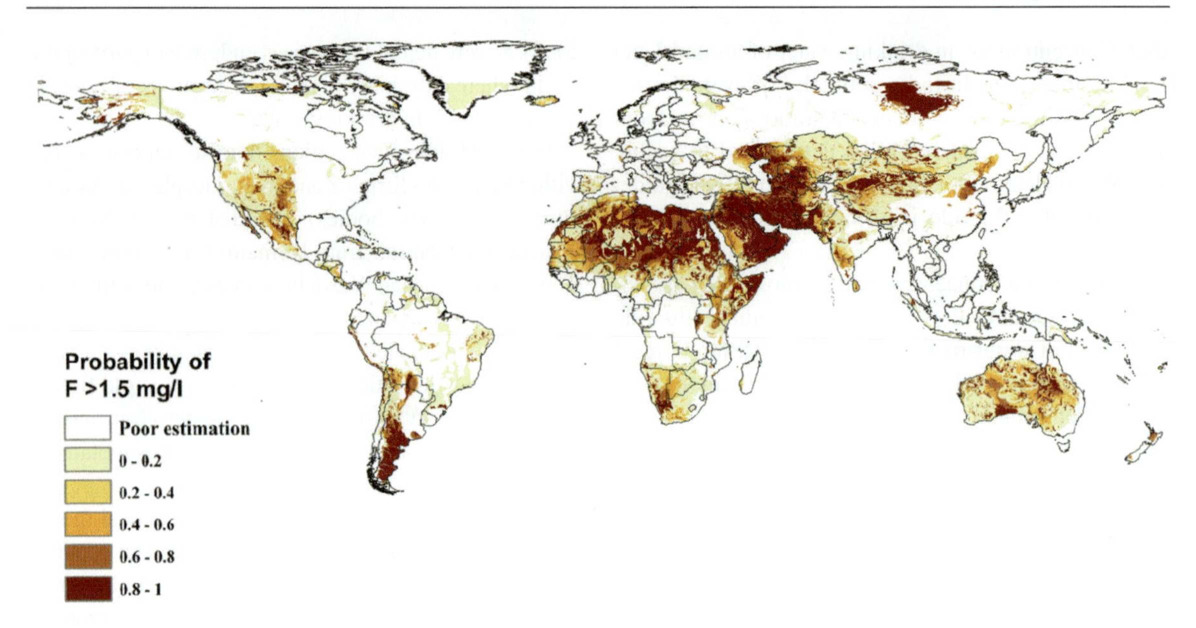

Fig. 18 Predicted probability of fluoride concentration in the groundwater exceeding the WHO guideline for drinking water of 1.5 mg l^{-1} (*Source*: Amini et al., 2008b)

Fluorosis Case Studies in the Middle East

There are some regions in Iran such as Maku, Koohbanan, Dayer in Bushehr, Bandarlenge, and Larestan, some parts of Isfahan province, Karaj, and Poldasht that have high levels of fluoride in drinking water and fluorosis. The map below shows locations of the areas with incidence of fluorosis (Fig. 19).

A study in four areas of Isfahan province has shown incidence of fluorosis among 7–12-year-old schoolchildren who using drinking water containing fluoride. The levels of drinking water fluoride in four areas of Najafabad, Joozdan, Rahmatabad, and Filor were 0.23, 0.6, 1.35, and 0.78 ppm, respectively. There was a direct relationship between the drinking water containing fluoride and incidence of fluorosis (Khademi and Taleb, 2000).

High-fluoride groundwater occurs in Maku area, in the north of West Azerbaijan Province, northwest of Iran. Groundwater is the main source of drinking water for the residents. Inhabitants of the area who obtain their drinking water supplies from basaltic springs and wells are suffering from dental fluorosis. The population of the study area is at a high risk due to excessive fluoride intake (Asghari Moghaddam and Fijani, 2008). Water fluoride levels were measured in different rural areas in 1983 ranging from 0 to 9 ppm. In 33.8% of the villages, water fluoride concentration was higher than standard limit (Gholamhoseini in Ramezani et al., 2004).

A study in Koohbanan (Kuh-e Banan) which is located in Kerman Province in SE Iran has determined high fluoride content in foods. The fluoride content varied from a very low level of 0.02 mg kg^{-1} to a relatively high level of 8.85 mg kg^{-1}, and in water sources ranging from 2.36 to 3.10 mg l^{-1}. The small city of Koohbanan (Kuh-e Banan, population ca. 20,000) is situated in a cold mountainous region 2,000 m above sea level. Most mountains and hills around the city have coal mines. In this region, vegetables and fruits are grown by the inhabitants, and drinking water and agricultural water are provided from separate sources. In a survey of secondary school students in Koohbanan, the prevalence of dental fluorosis was found to be 93%. It appears that in Koohbanan at 2,000 m above sea level and 2.38 mg F l^{-1} in the drinking water, altitude is an influencing factor on the prevalence and severity of dental fluorosis (Poureslami et al., 2008).

Observations in some regions such as Dayer city in Bushehr province in the south of Iran and the Persian Gulf areas have shown that fluorosis is probably endemic in the region. Dayer city is considered a

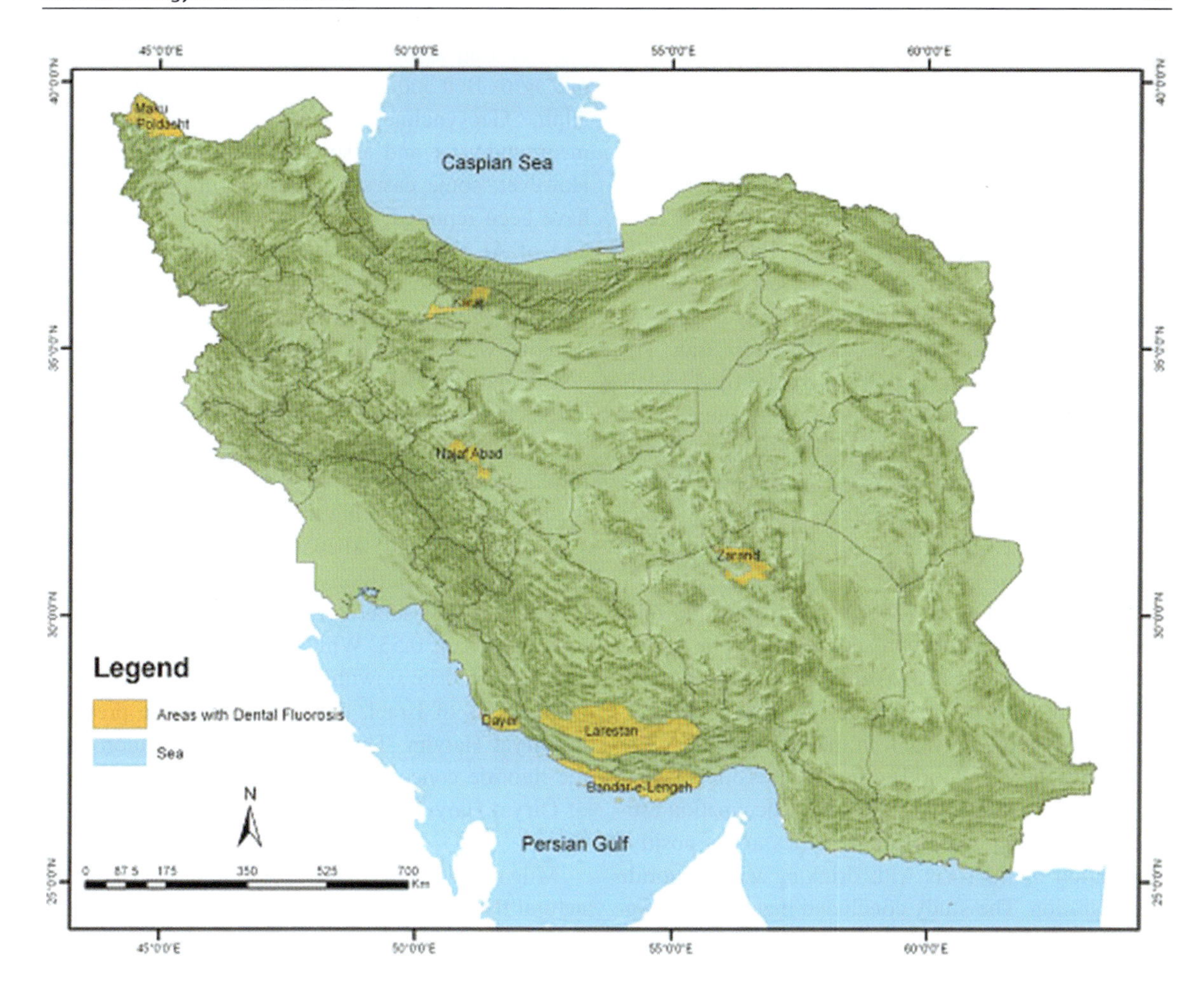

Fig. 19 Map of some areas with incidence of fluorosis in Iran (*Source*: National Geoscience Database of Iran)

hot region. The water fluoride concentration was determined from six supplies including three city supplies and three village supplies. The most of Dayer is supplied from springs and semi-deep wells. The mean fluoride concentration was 2.23 ± 0.25, ranging from 2 to 2.5 ppm in city areas and 2.63 ± 0.21, ranging from 2.4 to 2.8 ppm in rural areas. The fluoride concentration was higher than the standard limit (0.6–1.2 ppm) in all of the drinking water supplies, and in rural areas 0.40 ppm (17.9%) higher compared to urban areas (Ramezani et al., 2004).

Reports also show fluorosis prevalence up to 67% in some regions of Iran such as Bandarlenge and Larestan regions (Aminabadi et al., 2007). Drinking water fluoride concentration was 1.27 ± 0.68 (Eftekhari in Ramezani et al., 2004) and fluorosis rate was 76% (Eftekhari and Mazloum, 1999) in Larestan and its suburb in the year 1996–1997 (Eftekhari in Ramezani et al., 2004). The average water fluoride level in Bandarlenge was 0.15–1.35 ppm in the year 1999 (Airemlou in Ramezani et al., 2004).

Five hundred and ten samples of 84 popular foods and drinks were collected from three areas of Iran where water fluoride concentrations were 0.32, 0.58, and 4.05 mg l^{-1}. The concentration of fluoride in water influences positively fluoride concentration in foods cooked in water, but the increase in food was less than the increase in fluoride concentration in water (Zohouri and Rugg-Gunn, 1999). In a report on fluoride in vegetables in Fars Province of Iran, more fluoride was found in stem parts than in roots and that the use of pesticides and herbicides increased fluoride levels (Poureslami et al., 2008). Some areas of Iran such as Karaj and Poldasht have shown fluorosis prevalence 69% in Karaj (Esmailpourin Aminabadi et al., 2007) and 100% prevalence in Poldasht with 3 ppm fluoride (Aminabadi et al., 2006). In all of the above-mentioned, a direct relation between water fluoride and fluorosis prevalence has been observed.

Dobaradaran et al. (2008, 2009) examined the relationship between dental caries in children, fluoride content, and some inorganic constituents of groundwater used for drinking in 14 villages in Dashtestan District of Bushehr Province, southwestern Iran. The study population comprised 2,340 children 6–11 years old. Fluoride concentration in groundwater varied between 0.99 and 2.50 mg l^{-1}. The study concluded that there is direct but weak linear correlation between increasing fluoride concentration in drinking water and dental caries in permanent and deciduous teeth (Fig. 20).

Meyer-Lueckel et al. (2006) investigated the relationship of caries and fluorosis prevalence in adolescents from an area with high- and low-fluoride levels in the drinking water in Iran and indicated that fluorosis is significantly more prevalent in the high-fluoride compared with the low-fluoride areas. Regression analysis adjusted for age and status of father's occupation estimated a negative association of caries and a positive association of fluorosis with drinking water fluoride concentration. The study concluded that caries prevalence in the examined areas in Iran is quite low. Water fluoride concentration is negatively related to caries experience and strongly positively related to fluorosis severity and prevalence under the conditions examined.

In spite of the fact that there are potential fluoride-rich drinking waters in the northern and northeastern parts of Iraq associated with the igneous complexes of Zagros Belt and the volcaniclastic sediments of the Alpine Geosyncline, no systematic study of fluoride in groundwater and associated fluorosis is available. However, some cases of fluorosis in elderly people have been reported in Bashiqa, a town 20 km northeast of Al Mosul, Ninawa Governorate. These cases relate, most probably, to relatively high fluoride contents in hand-dug groundwater wells used as source of drinking water until late 1960s.

Hamdan (2003) investigated dental fluorosis among 1,878 children of 12 years age in urban and rural areas in Jordan and concluded that prevalence of fluorosis in rural areas is higher than in urban areas and is related to the sources of drinking water.

Rosenzweig and Abewitz (1963) studied dental caries and endemic fluorosis in 262 young students in Qiryat Haiyim, a suburb of Haifa and determined a fluorosis index of 0.53. While the study concluded that caries prevalence is within the limits observed in the major cities of Israel, fluorosis is far more prevalent in Qiryat Haiyim. The study related the fluorosis to the fluoride concentration in the groundwater supply of Qiryat Haiyim varying between 0.35 and 0.95 mg l^{-1}.

Milgalter et al. (1974) (Fawell et al., 2006) indicated natural fluoride levels in drinking water of up to 3 mg l^{-1} in Yotvata in the Negev Desert region in Israel, probably related to Yotvata Salt Flat. Yet, there is no reference in the study to the health effects of such high concentrations of fluoride.

Mann et al. (1990) investigated fluorosis and dental caries in 152 children 6–8-year-old in a 5 mg l^{-1} fluoride drinking water in Israel and indicated that 41%

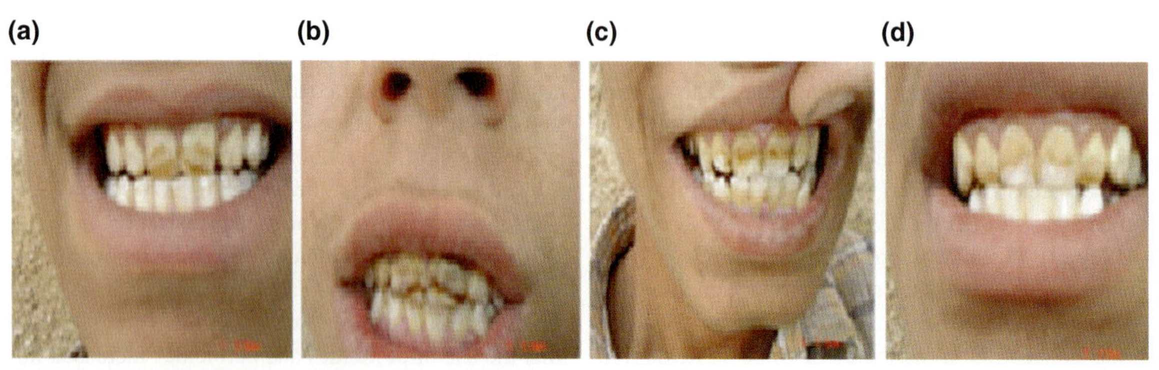

Fig. 20. Examples of dental fluorosis in three villages of Dashtestan District with 1.87–2.07 mg l^{-1} fluoride in drinking water. **a**: age: 8, sex: male, village: Dalaki; **b**: age: 9, sex: female, village: Dalaki; **c**: age: 10, sex: male, village: Tang-e Eram; **d**: age: 11, sex: female, village: Bushkan (*Source*: Dobaradaran et al., 2008)

of the investigated cases had mild fluorosis in both primary and permanent dentition, 20% with moderate fluorosis and 36% with severe fluorosis, mostly in permanent dentition.

Al Khateeb et al. (1991) examined 1,400 children of ages 6–15 years from schools in Jeddah, Rabagh, and Mecca, all in Mecca Province of Saudi Arabia with <0.3, 0.77, and 2.47 mg l^{-1} fluoride in drinking water, respectively. However, the study did not refer to fluorosis in Mecca in connection with the relatively high fluoride level.

In Ha'il Province, northern Saudi Arabia, Akpata et al., 1997 investigated the relationship between fluoride levels in drinking water from wells, severity of dental fluorosis, and dental caries in 2,355 rural children 12–15 years old. The study revealed that over 90% of the examined subjects showed dental fluorosis and that a strong association exists between fluoride level (0.5–2.8 mg l^{-1}) in well water used for drinking and the severity of dental fluorosis.

Almas et al. (1999) examined fluorosis in 400 subjects in four age groups ($12 \geq 65$ years in age) in Al Qaseem Province in central Saudi Arabia and indicated that fluorosis is more prevalent in rural areas, especially among the age group 35–44 years and related this to the fluoride level of 2–3 mg l^{-1} in the water supply. Later, Al-Dosari et al. (2004) indicated that there is no linear correlation between drinking water fluoride level and caries in Al Qaseem and the neighboring Al Riyadh Provinces.

In 2006, King Abdulaziz City of Science and Technology initiated the "Fluoride Mapping of Saudi Arabia" with the objectives of establishing a fluoride distribution map and correlate it with prevalence of dental caries and fluorosis. Al-Dosari et al. (2007) indicated that 5,134 water samples, including 4,419 drinking water samples, have been collected from all of the 13 provinces of Saudi Arabia. Preliminary results revealed that Ha'il Province exhibited the highest level of 1.27 mg l^{-1}.

In Al Yemen, people in rural areas use deep wells as source of water drinking and household uses. However, some of these underground waters contain 2.5–32 mg l^{-1} of fluoride with the highest levels recorded in Sana'a, Ibb, Dhamar, Ta'iz, Al Dhalei, and Raimah Governorates. In Sanhan District, southeast of Sana'a, there has been an abnormal increase in the incidence of rickets among children in addition to osteomalacia in adults and decay and corrosion of teeth (Figs. 21 and 22). In Al Maqula village (population 3,100), the fluoride level varies between 5 and 6 mg l^{-1} while in She'an (population 2,850) and Al Daram (population 950) villages the level is 5.3 and 1.5–1.9, respectively.

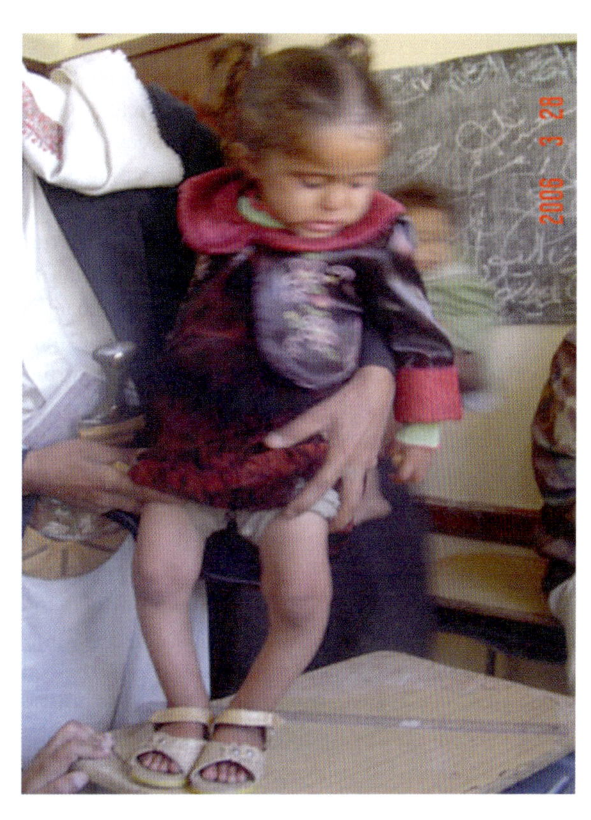

Fig. 21 Rickets in a 3-year-old child in Sanhan (*Source*: Ghaleb, 2007)

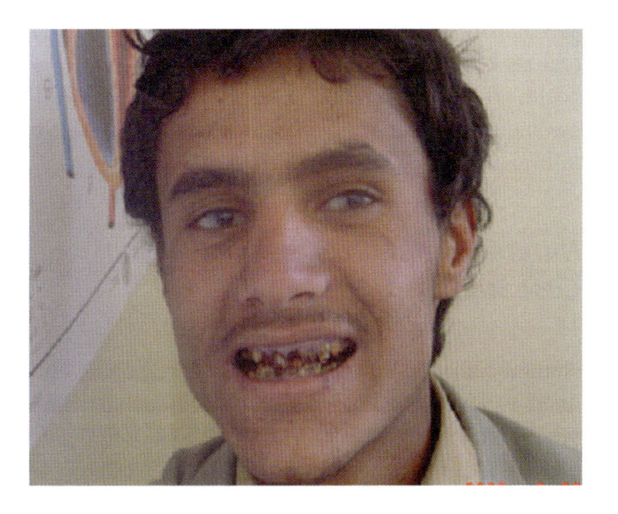

Fig. 22 Dental fluorosis in an adult in Sanhan (*Source*: Ghaleb, 2007)

The groundwater is there associated with aquifers within Al Yemen Volcanic Group

Natural Radiation

Sources of natural radiation include radon, terrestrial gamma rays, and cosmic rays. Geological sources constitute the major proportion of sources of natural radiation with radon isotopes contributing about 60%. Radon is a natural radioactive gas produced by the radioactive decay of radium which in turn is derived from the radioactive decay of uranium. Radon decays to form alpha and beta radioactive particles and gamma radiation. Geology is the most important factor controlling the sources and distribution of radon (Appleton, 2005). Relatively high levels of radon are associated with particular types of bedrock (some granites, phosphate rocks, and shales rich in organic matter associated with unconsolidated deposits). Everyone is exposed in daily life to radon. Radon and its decay products are well established as lung carcinogens (UN, 2006). The health effects due to radon exposure have been well studied in mining populations and provide firm evidence of the link between radon exposure and lung cancer (Makofske and Edelstein, 1988). In addition to uranium miners and other hard rock miners, workers in the oil–gas industry are exposed to high level of radon gas by inhalation. The number of workers subject to technologically enhanced natural radiation exposure is significantly higher in the Middle East and Central Asia as compared to the number of workers of all other regions combined (Steinhausler, 2003). But studies of health effects due to radon exposure in workers of oil–gas industry in the Middle East are limited by lack of information.

Natural Radiation in the Middle East: Case Studies

Natural radiation in the Middle East region is confined to igneous complexes and the associated fault systems in the northwestern, northern, and central parts of Iran and in the Arabian Shield in the western part of Saudi Arabia (Al Hijaz), Yemen, and the extreme southern parts of Jordan and Israel on the Gulf of Aqaba. Natural radiation is associated with the relatively active Dead Sea Transform Fault (the northern extension of the Great Rift Valley) extending from the Gulf of Al Aqaba through Wadi Araba, the Dead Sea, Jordan River Valley westward to the Golan Heights in Syria and Lebanon, and Anti-Lebanon Mountains. Natural radiation is associated with the widespread phosphate deposits in western (Azerbaijan West, Chahar Mahal, and Bakhtiari and Kohkilyueh and Buyer Ahmad Provinces), central (Semnan Province), and southern (Kerman Province) of Iran, in the Western and Southern Deserts of Iraq, and its southern extension in Sarhan–Al Turaif Basin of northern Saudi Arabia, in Jordan, and in Negev Desert, Israel. However, almost all the investigations carried out until now concentrated on radon as the major source of natural radiation and the associated health effects.

Iran

In some parts of Iran there is an association between radon exposure, lithology, and geological structures, the most significant of which is Ramsar, a northern coastal city of Iran (Fig. 23). In Ramsar there are several high-level natural radiation areas (HLNRAs) and one of the highest known background radiation levels in the world occurs here. The high background radiation is primarily due to the presence of very high amounts of ^{226}Ra and its decay products brought to the earth's surface by hot springs. Groundwater is heated by subsurface geological activity and then passes through relatively young, uraniferous igneous rocks dissolving radium. When reaching the surface, dissolved $CaCO_3$ is precipitated as travertine with radium substituting for calcium in the minerals (Karam et al., 2002; Ghiassi-Nejad et al., 2002). A secondary source of high radiation levels is travertine deposits with high thorium concentration. Since soils are derived from the weathering of local bedrock, the radioactivity of these soils and food grown on them is high as well. At least nine hot springs with various levels of radioactivity occur around Ramsar and used as health spas. Locals use the residue of these hot springs as building material for their houses. The indoor and outdoor gamma radiation-absorbed dose rates vary between 50 and 900 μm Gy h^{-1} with the annual dose to monitored individuals ranges up to 132 m Gy. Compared

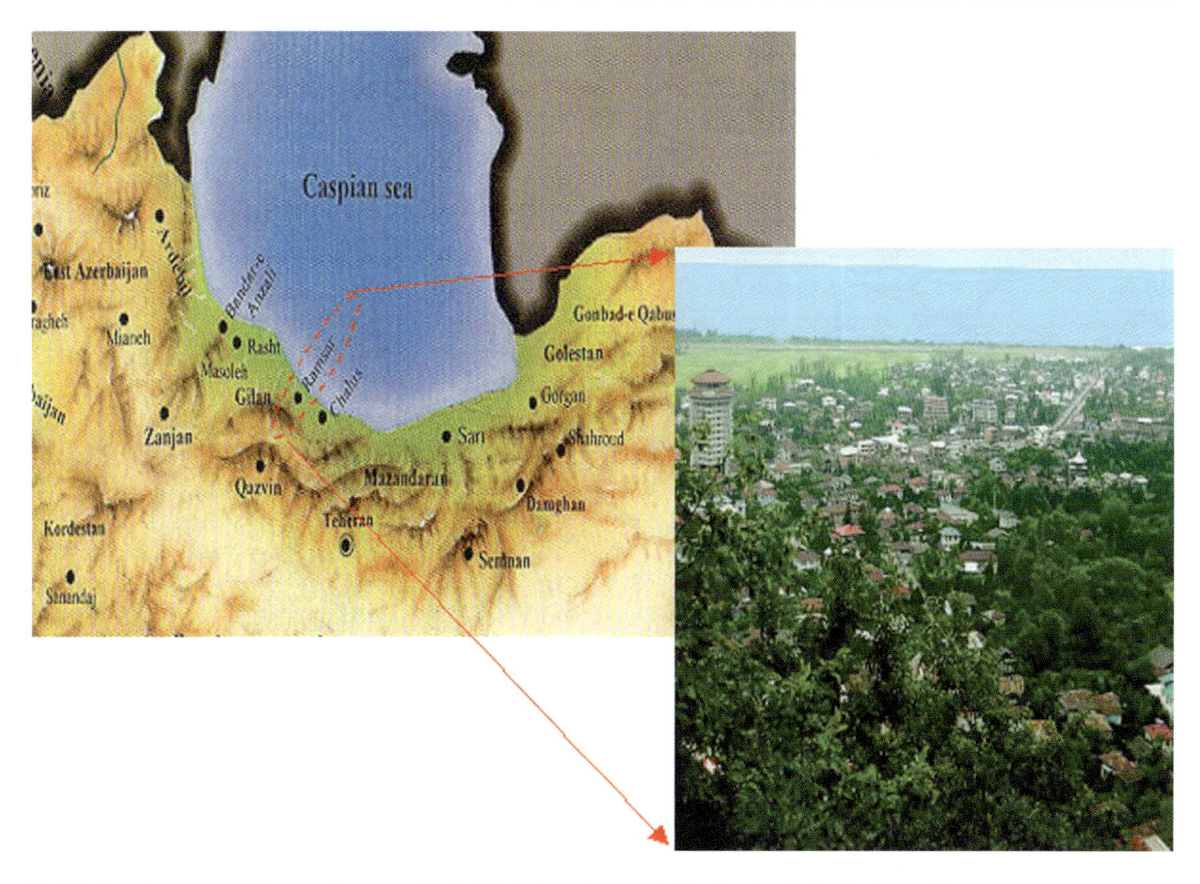

Fig. 23 Location map of Ramsar, the area with the highest environmental levels of radiation in the world

to the recommended dose limit for workers in Iran of 20 mSv year^{-1} (Karam et al., 2002), some residents of Ramsar are exposed to levels of natural radiation as high as 55–200 times higher than the average global dose rate. Furthermore, radon levels in some regions of Ramsar are up to 3,700 Bq m^{-3} (Mortazavi et al., 2005a).

Out of the total population of Ramsar city of about 69,000, 1,000 people are living in HLNRAs, namely Chaparsar, Ramak, and Talesh Mahalleh (Fig. 24) (Karam, 2002). The population of Talesh Mahalleh is about 500 (118 families) and around 10% of this population are called "critical group." This group receives the highest total effective dose and possesses reasonably homogeneous exposure in the region (Sohrabi, 1993 in Ghiassi-Nejad et al., 2003). In Talesh Mahalleh the maximum and minimum vegetable-to-soil concentration ratio (CR) was measured in leafy and root vegetables with average values of 1.6×10^{-2} and 4.0×10^{-3}, respectively. The mean effective dose resulting from ^{226}Ra due to consumption of edible vegetables by adults in the critical group in this region was estimated to be 72.3 μSv year^{-1}. This value is about 12 times greater than the average of effective dose resulting from this radionuclide due to combined intake of all foods and drinking water in normal background areas. The concentrations of natural radionuclides in food and drinking water of the residents are higher than the world average and are correlated with the high concentration of these radionuclides in soil and water (Samavat et al., 2006).

The preliminary studies of Ghiassi-Nejad et al. (2002) show no significant differences between residents in high background radiation areas (HBRAs) compared to those in normal background radiation areas (NBRAs) in the areas of life span, cancer incidence, or background levels of chromosomal abnormalities. Further, when administered an in vitro challenge dose of 1.5 Gy of gamma rays, donor lymphocytes showed significantly *reduced* sensitivity to

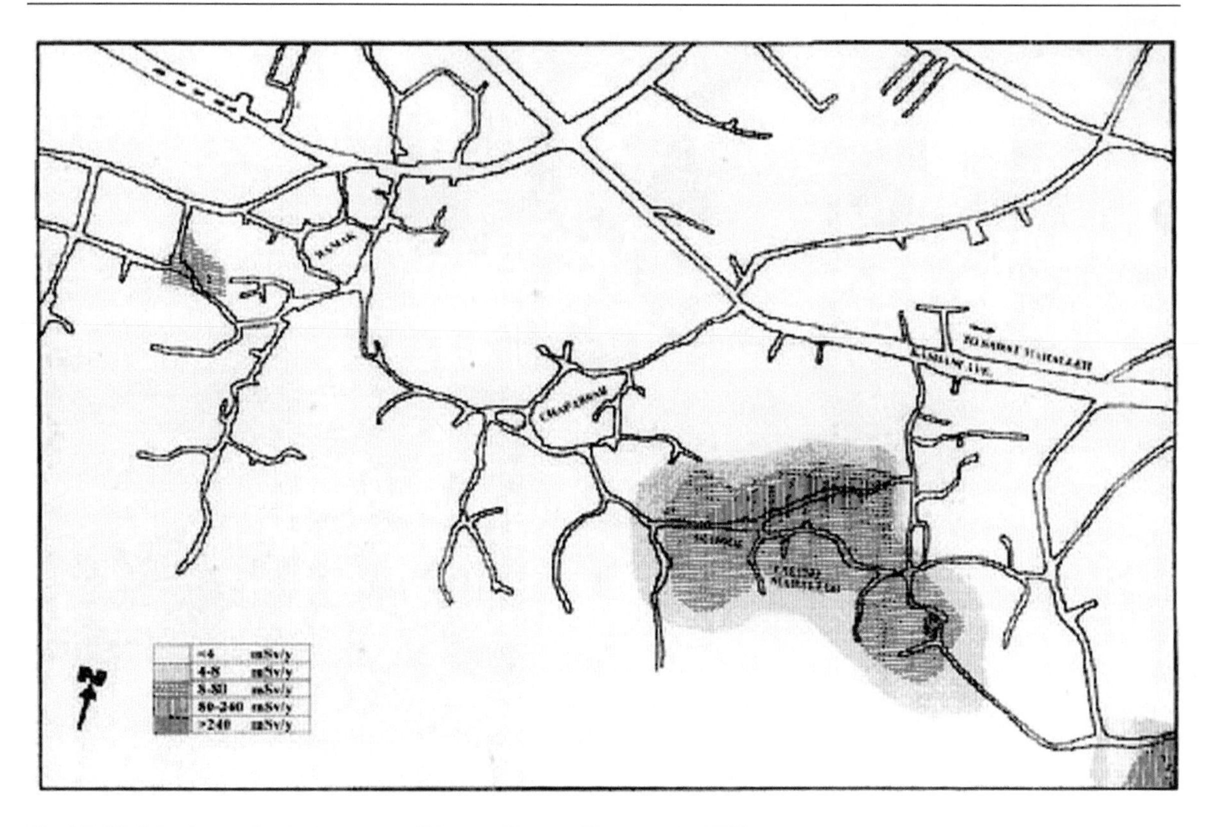

Fig. 24 High background radiations areas of Ramsar (*Source*: Karam et al., 2002)

radiation as evidenced by their experiencing *fewer* induced chromosome aberrations among residents of HBRAs compared to those in NBRAs. Specifically, HBRA inhabitants had 44% fewer induced chromosomal abnormalities compared to lymphocytes of NBRA residents following this exposure.

Several case–control studies and recently published data suggest that there is *no* detectable chromosomal damage from the high levels of natural background radiation found in Ramsar and other HBRAs, contrary to the predictions of linear, no-threshold, or supralinear models of radiation dose–response (Ghiassinejad et al., 2005). In fact, these investigations indicated inverse relationship between residential radon exposure and lung cancer. Mortazavi et al. (2002) showed that the highest lung cancer mortality rate was in Galesh Mahalleh, where the radon levels were normal. Interestingly, the lowest lung cancer mortality rate was recorded in Ramak, where the highest concentrations of radon in the dwellings were found. The study also indicated that crude and adjusted lung cancer rates in the district with the highest recorded level of radon concentration is lower than those of the other districts and concluded that lung cancer may show a negative correlation with natural radon concentration. Karam et al. (2002) explained that although there is no solid epidemiological information yet, most local physicians in Ramsar believe that there is no increase in the incidence rates of cancer or leukemia in their area, and the life span of the high-level natural radiation areas residents also appears no different than that of the residents of a nearby normal background radiation area. Accordingly, overall data showed a significant radioadaptive response in the residents of high-level natural radiation areas (Mortazavi et al., 2005b).

Several other investigations of natural radiation in Iran exist, but none have dealt with health effects. Beitollahi et al. (2007) investigated five hot springs called Abegarm-e-Mahallat located in Markazi Province, central Iran. Water of these springs is of mixed origin (magmatic geothermal), sulfated with a mean temperature of 46 ± 1°C. While these hot springs are used as spas, the precipitated travertine is used as building material and decorative stone. Abegarm-e-Mahallat is of high natural radiation background due to the presence of ^{226}Ra and its decay products in the deposited travertine while radiation levels above that of normal background (~100 nGy h^{-1}) are mainly limited to the Quaternary travertine formations in the vicinity of the hot springs. Alirezazadeh (2005) studied radon concentrations in Tehran's public water supply system and concluded that although groundwater consisting 30% of Tehran's water supply contains relatively high concentrations of radon (range 27.7–74.3, mean 40.4 Bql^{-1}), the concentration of radon in tap water is much less (range 2.7–6.0, mean 3.7 Bq l^{-1}) due to mixing with surface waters. Fathabadi et al. (2006) studied the exposure of miners to radon and its decay products in 12 non-uranium underground mines and indicated that the radon exposure dose is relatively high (10–31 mSv $year^{-1}$) in three mines only. Talaeepour et al. (2006) determined the average concentration of radon in Tehran's subway system as 14.19 Bq m^{-1}, which is much lower than the threshold dose of 148 Bq m^{-1} and concluded that variation in radon concentration is related to lithology not to depth. Mowlavi and Binesh (2006) measured radon concentrations in eight springs, deep wells, and river water resource of Mashhad Province, northeastern Iran, and indicated that some samples have radon concentration of more than 10 Bq l^{-1}. Hadad et al. (2007) monitored indoor radon in Lahijan, Ardabil, Sar–Ein and Namin cities in northern Iran and indicated that Ardabil and Lahijan have the highest radon concentrations in Iran following Ramsar with 163 and 240 Bq m^{-3} corresponding to annual effective doses of 3.43 and 5.00 mSv y^{-1}, respectively, with maximum and minimum concentrations of 2.386 (during winter) and 55 Bq m^{-3} (during spring). The study also dealt with the relation between radon concentration and building material and room ventilation and concluded that the four cities can be categorized as Average Natural Radiation Zones. Gilmore and Vasal (2009) conducted a reconnaissance study on radon concentrations in dwellings in Hamadan, western Iran built on permeable alluvial fan deposits. Indoor radon level varied between 4 and 364 Bq m^{-3} with a mean value of 107.87 which is 2.5 times the global average. The study concluded that the use of hammered clay floors, the widespread use of subsurface channels locally called "kariz" to distribute water, the textural properties of surficial sediments, and the human behavior affect radon concentrations in old buildings.

Syria

In Syria, Othman et al. (1996) conducted a nationwide investigation of radon levels in Syrian houses in the period 1991–1993, which indicated that the mean radon concentration is 45 Bq m^{-1} with the highest values recorded in Damascus Governorate in the south. The study concluded that old houses built with mud with no tiling exhibited higher radon concentrations. However, there is no indication of health impact associated with radon exposure.

Jordan

In Jordan, Matiullah and Khatobeh (1997) studied indoor radon levels and natural radioactivity in soils in some cities and indicted that indoor levels of radon vary from 27 to 88 Bq m^{-3} and elates directly to radon contents in soils. Abumurad et al. (1997) determined radon concentrations at different depths in soils of Irbid Governorate in the north. The study concluded that radon levels vary considerably according to type of soil and underlying bedrock from 4,000 Bq m^{-3} in limestone to 400,000 Bq m^{-3} in chalky marl and increase exponentially with depth. Al-Tamimni and Abumurad (2001) investigated radon levels in soil air along traceable fault planes and indicated that radon levels are higher by a factor of 3–10 away from faults and recommended a detailed study on the implications of the fact that several cities in Jordan lie on intensive fault zones. Abumurad and Al-Tamimim (2005) investigated natural radioactivity related to radon in Soum region, northwest of Jordan, and indicated that average indoor radon concentration is 144 Bq m^{-3} which is less than the action level recommended by ICRP but about three times the national Jordanian average. Namrouqa (2009) reported that the radiation level of Disi aquifer south of Jordan is 20 times more than what is considered safe for drinking, mostly related to radon. This aquifer is being developed to be the major source of water for Jordan, one of the driest 10 countries in the world, in the coming 100 years,

Israel

Haquin et al. (2006) conducted a radon survey of Israel in 1998 and in 2003 to determine the radon prone areas in Israel. The survey calculated that the national average radon concentration is 47 Bq m^{-3}, identified six areas as radon prone, and concluded that there is a strong correlation between geological conditions and indoor radon concentration. Shirav-Schwarzz (2004) commented on the natural terrestrial dose rate map (Fig. 25) and indicated that the average dose rate for the entire territory of Israel is 0.28 mGy $year^{-1}$, with some 20% of the territory having ranges between 0.47 and 5.1 mGy $year^{-1}$. Most of the high natural background radiation terrains covering about 10% of the territory coincide with the exposure of Mount Scopus Group, a Senonian–Palaeocene rock unit characterized by phosphorite layers with appreciable amounts (90–100 ppm) of uranium. Elevated radiation levels of up to 3.0 mGy $year^{-1}$ in other areas are associated with high concentrations of thorium and potassium in soils and outcrops of young basalts in central and northern Israel. Although in most cases the outcrops of uranium-rich phosphorites are located in desert areas, some heavily populated cities are located in these regions – Maale Adumim, Arad, and the eastern part of Jerusalem. High radon levels have been recorded in these urban communities. However, a GIS analysis of health-related data for the period 1984–1999 published by the Israeli Cancer Registry did not reveal, so far, any correlation between the geographic distribution of malignant diseases and high natural background radiation terrains or high radon levels. In Jordan, some of the most populated areas are built on rocks equivalent to the Mount Scopus Group (i.e., Amman, capital of Jordan), where traditionally a large portion of the population is living on ground floor dwellings. All over the region, workers in the phosphate mining industry should be monitored, as well as people living in temporary ground level dwellings.

Saudi Arabia

In Saudi Arabia, Vincent (2008) explained that although detailed epidemiological studies are still to be carried out there is no doubt that lung cancer cases are well above expected norms in western Saudi Arabia and these are probably linked to Rn-222. Data for Saudi Arabia are few and far between. Abu-Jarad and Al-Jarallah (1986) conducted the first study on indoor radon in 400 houses all over Saudi Arabia and indicated that radon concentrations varied from 5 to 36 Bq m^{-3} with a mean of 16 Bq m^{-3}. Al-Jarallah (2001) measured radon exhalation for 50 selected samples of construction material used in Saudi Arabia and found that granite is the main source of radiation and a source of indoor radon. Al-Ghorabie (2004) reported indoor gamma radiation dose in At-Taif city in Al Hijaz as of ranging between 90 and 221 nGy h^{-1} for the time interval September 2002–September 2003 and that the measured dose varied with the type of house and season. The highest average value 192 $\pm$ 7 nGy h^{-1} was measured inside apartments made of cement and brick (157–221 nGy h^{-1}) and the lowest average value 92 $\pm$ 6 nGy h^{-1} in mud houses (58–117 nGy h^{-1}). The average indoor gamma radiation dose rate received by the population of At-Taif city is 138 nGy h^{-1} and its corresponding annual dose is 1,211 μGy y^{-1}. Al-Jarallah and Fazal-ur-Rehman (2006) studied indoor radon concentrations in 136 dwellings of Al-Jauf region, northern Saudi Arabia, and indicated that the average, minimum, maximum radon concentrations and standard deviation were 35, 7, 168, and 30 Bq m^{-3}, respectively. Recent preliminary work by Elliot and Hashem (in Vincent, 2008) indicated that Rn-222 is potentially a serious hazard. They have investigated Jabal Said mineralized granite consisting of an aplite–pegmatite radioactive body approximately 35 km northeast of Mahd adh Dhahab (N 23° 50′, E 40° 57′). Values of radon in soil gas ranged from 55 to 35, 182 pCi L^{-1} with the extremely high values associated with alluvium derived directly from the aplite–pegmatite body. The level of Ra-222 in the poorly ventilated village houses along the floors of the widyan draining this body clearly needs to be investigated as a matter of some urgency. Hisham (2006) indicated that the Saudi Arabian Geological Survey started geological radon mapping of rocks and soil in 2000. The geological and structural data together with radiometric data from airborne survey and geochemical analyses of 280 samples were used to describe the potential existence of radon gas. Soil permeability data were also included to show the gas mobility in soil. All the previous data were used to set up the first draft of the base potential map of the Arabian shield (crystalline map). The final map shows areas of interest,

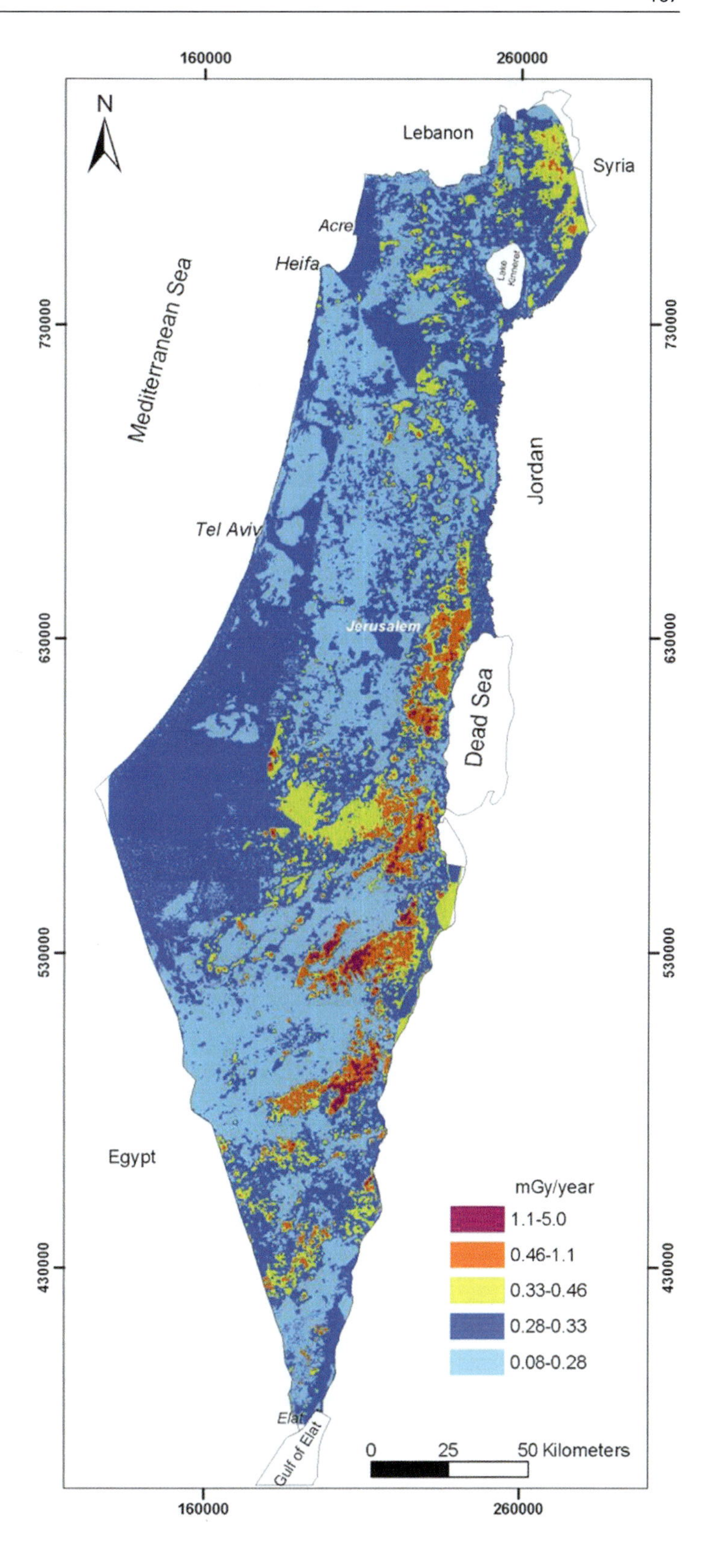

Fig. 25 Terrestrial dose rate map of Israel and the Palestinian territories (*Source*: Shirav- Schwarzz, 2004)

including Ha'il in the north part of the KSA, which consists mainly of uranium-bearing granitic rocks. The maximum radon concentration of the collected data was 105 Bq l^{-1} and the minimum was 20 Bq l^{-1} at 50 cm depth from the ground surface. The final map shows good correlation with radon concentration in soil. Further work has to be carried out to complete the remaining zones, prior to the final production of the radon risk map within the culmination of the project.

Yemen

Khayrat et al. (2003) conducted an indoor radon survey in 241 dwellings distributed in Dhamar, Taiz, and Hodeidah in Yemen and indicated that radon concentrations varied from 3 to 270 Bq m^{-3} with an average of 42 Bq m^{-3} and that radon concentration increases with altitude since the highest concentrations occur in Dhamar, a mountain city, and the lowest in Hodeidad, a seaport.

Geophagia

Geophagia can be defined as the deliberate ingestion of soil (Abrahams, 2005). Many people around the world practice geophagia. Geophagia may be stimulated by dietary deficiencies (Coreil et al., 2000), a cultural phenomenon passed from generation to generation, the consumption of soil for physiological, psychiatric, and psychological causes (Abrahams, 2005) the way to satisfy the appetite (Coreil et al., 2000), or as a way to detoxify foods containing poisoning substances (Johns, 1991 in Coreil et al., 2000). It can also be considered as a behavioral symptom of structural brain disorders (Nejat and Kazmi, 2005). Intake of soil is usually associated with risk of human diseases. Examples are pathogens such as tetanus and worm infections (Norman and Sparks, 1992). The clinical manifestations of prolonged geophagia are iron deficiency, microcytic anemia, hepatosplenomegaly, short stature, delayed puberty (Quattromani et al., 2008), and mental disorders (NRC, 2007). Radiographic manifestations of geophagia are delayed skeletal maturation, radiopaque material in the GI tract (lead) (Quattromani et al., 2008). In addition geophagia has been clearly associated with both iron and zinc deficiency; debate persists as to whether pica is the cause or result of the deficiency (Korman., 1990). The following table reveals potential benefits and negative consequences of geophagia (Table 10).

Geophagia associated with iron and zinc deficiency, short stature, delayed sexual maturity, hepatosplenomegaly, and delayed bone age is reported from Middle East countries (Korman, 1990). This includes Turkey's villages (Çavdar et al., 1980 in Abrahams, 2005), villages around Shiraz in Iran (WHO, 1973) (Prasad, 1989), among adults in Saudi Arabia (Hawass et al., 1987), and Arab children from Gaza (Korman, 1990).

Table 10 Overview of potential benefits and negative consequences of geophagia

Mechanism	Potential benefits	Potential negative consequences
Adds elements	Adds useful micronutrients, e.g., iron, zinc, or calcium	Adds poisonous elements, e.g., lead, mercury Adds helpful minerals in excess, e.g., potassium, zinc
Binds substances	Binds with plant toxins Binds with pathogens: bacteria, viruses, fungi, protozoa Binds with dietary iron (causing nutritional immunity)	Binds with dietary iron (causing iron deficiency) Reduces the effectiveness of pharmaceuticals)
Creates a barrier	Protective coating of intestine Slows gastric motility	Damages intestinal mucosa Gastric/intestinal obstruction, constipation
Quells gastrointestinal upset	Reduces upper and lower gastrointestinal upset	Inhibits useful mechanisms for detoxification (e.g., diarrhea)
Increases pH	Soothes heartburn	Makes iron less available More hospitable environment for enterotoxins and/or helminthes
Introduces organisms	Live micro organisms	Geohelminthis

Source: Data from Young, 2008

In Iran, Geophagia appears to be linked to zinc deficiency. Zinc response growth failure and sexual infantilism have been studied in detail in southern Iran and observed in both sexes (WHO, 1973). Prasad (1989) indicated that the habit of geophagia in the villages around Shiraz was fairly common. Excess amount of phosphate in the clay may have prevented absorption of both dietary iron and zinc (Sigel, 2004). Analysis of clay of the type eaten by some villagers was obtained from an area in southern Iran revealed a high calcium level, 25.4% as CaO, or 18.6% calcium. Since the clay was supplemented at a 20% level, this resulted in 3.7% additional calcium being added to the basal diet. Preliminary experiments feeding untreated clay showed that this high level of calcium resulted in growth inhibition with a high incidence of urolithiasis (Smith and Halsted, 1970).

Halsted and Prasad (1961) studied a syndrome occurring in males characterized by severe iron deficiency anemia, hypogonadism, dwarfism, hepatosplenomegaly, and geophagia, observed in villagers in Iran suffering from malnutrition, with 11 such patients, studied in detail (Figs. 26 and 27). Despite hepatosplenomegaly the liver function tests were uniformly normal except for alkaline phosphatase which was consistently elevated. The anemia was not associated with blood loss, hookworm infestation, or intestinal malabsorption and responded promptly to oral iron therapy. Although the diet contained adequate amounts of iron it is believed that the predominantly wheat diet with a high phosphate content interfered with its absorption because of formation of insoluble iron complexes. Correction of the anemia resulted in marked decrease in the size of liver and spleen. Prolonged follow-up of patients receiving a well-balanced diet

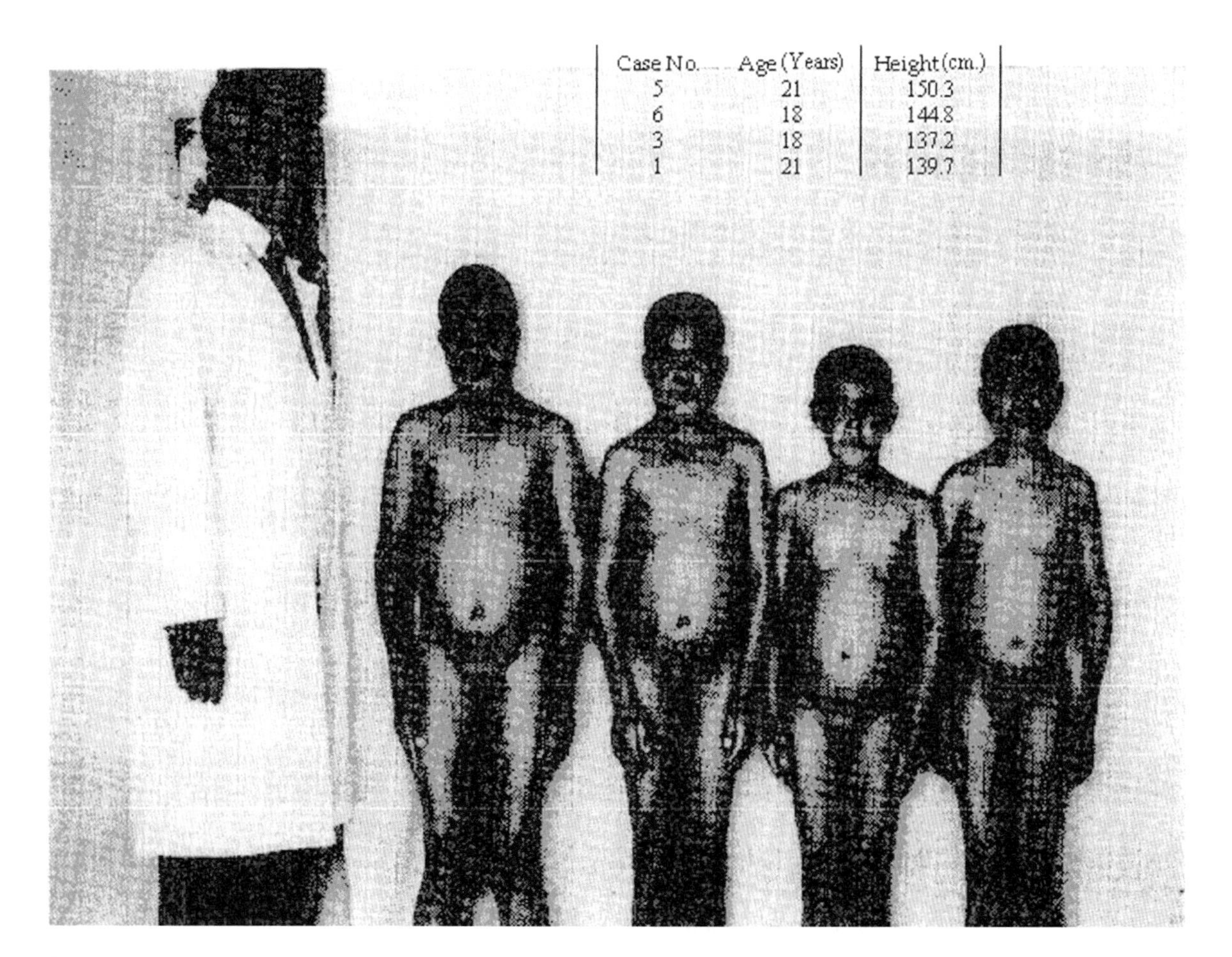

Case No. Age (Years) Height (cm.)
5 21 150.3
6 18 144.8
3 18 137.2
1 21 139.7

Fig. 26 Some of the cases studied. Staff physician at left is 183 cm in height (*Source*: Halsted and Prasad, 1961)

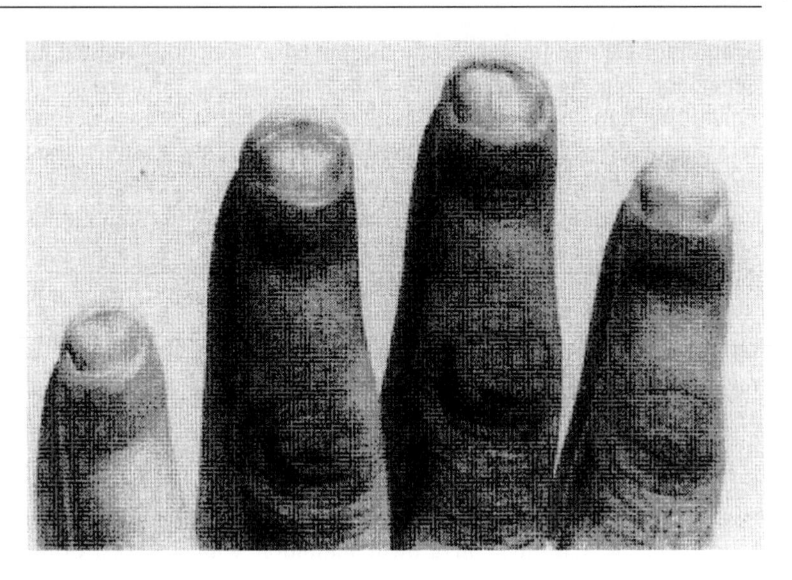

Fig. 27 Characteristic spoon nails (koilonychia), present in a majority of the patients (*Source*: Halsted and Prasad, 1961)

indicates that the endocrine abnormalities of growth and sexual development are reversible. Although the study indicated that the relationship of geophagia (which existed in nearly all patients) to this syndrome is not clear, and has been discussed, the fact that 9 of the 11 patients continued to eat clay in large amounts for 10–13 years cannot be ignored. The study considered the possibility of zinc deficiency as an explanation of hypogonadism, dwarfism, and changes in alkaline phosphatase.

Prasad et al. (1963) reported syndrome of hypochromic anemia due to iron deficiency, hepatosplenomegaly, dwarfism, and lack of development of primary and secondary sexual characteristics in relation to prolonged history of geophagia in villages surrounding Shiraz, central Iran. They explained that the entire range of clinical features of this syndrome was revered following adequate nutrition and iron therapy for anemia. Prasad (1992) explained that the major constituent of diet of individuals in these areas is cereal proteins high in phosphates and phytates, which adversely affected the availability of iron and zinc. Once dwarfs who were hospitalized in Saadi Hospital received adequate nutrition, they did not revert to geophagia after discharge from the hospital (Prasad, 1991). Clinical observation in Shiraz figured out a patient who was 21-year-old male but he looked like a 10-year-old boy. In addition to severe growth retardation and anemia he had hypogonadism, hepatosplenomegaly, rough and dry skin, mental lethargy, and geophagia. The patient ate only bread made from wheat flour and consumed one pound of clay daily (Prasad, 1989).

Among some subgroups, the practice of pica is greatest during pregnancy (Sayers et al., 1974 in Hooper et al., 1993). In recent years prevalence of pica was reported among pregnant women in Iran. Mortezavi and Mohammadi (2002, 2003) indicated that the prevalence of pica among 560 pregnant women who referred to health centers in Zahedan, southeastern Iran, was 15.5%, of which 25.3% ate soil. This rate was 22.3% in pregnant women in Kermanshah, western Iran (Rezavand, Rawshani, 2006). In Tehran, the prevalence of pica among the studied population was 17.5% and soil was the most common non-nutritive substance (Rostami, 2003). Childbearing women ate soil probably for perceived purposes of stress medication (Edwards et al., 1994), enhanced digestion, preventing frequent vomiting in early pregnancy, and as a mineral nutrient supplement, though the causes of pica prevalence in pregnant women in Iran are an important gap which has to be investigated.

Hawass et al. (1987) recorded the first cases of geophagy undertaken by adults in Saudi Arabia. The study dealt with three female cases, two with a history of eating mud and one with a history of eating pebbles. Systematic examination revealed hepatomegaly, weak lower limbs with hypotonia and hyperreflexia, gallstones, and intra intestinal stones.

Korman (1990) described three Arab children from Gaza with long-standing geophagia and iron deficiency in whom celiac disease was ultimately diagnosed. The dramatic resolution of the pica in these cases after institution of a gluten-free diet is consistent with a mechanism of pica having been induced by iron deficiency secondary to the celiac disease.

References

Abrahams PW (2005) Geophagia and the involuntary ingestion of soil. In O Selinus, BJ Alloway , JA Centeno, RB Finkelman, R Fuge, U Lindh, P Pauline Smedley (eds) (2005) Essentials of Medical Geology. Elsevier Academic Press, 812 p.

Abu-Eshy SA, Abolfutouh MA, Al Naggar YM (2001) Endemic goiter in school children in high and low altitude areas of Asir Region, Saudi Arabia. Saudi Medical Journal 22(2): 146–149.

Abu-Jarad F, Al-Jarallah MI (1986) Radon in Saudi houses. Radiat Protect Dosim 14:243–249.

Abumurad KM, Al-Tamimim H (2005) Natural radioactivity due to radon in Soum region, Jordan. Radiat Meas 39(1):77–80.

ACC/SCN (Administrative Committee on Coordination/Subcommittee on Nutrition (1992) 2nd Report on the World Nutrition Situation – Volume I: Global and regional results, October 1992.

Al-Ali JT (2000) Study of Textural, Mineralogical Composition and Amount of Aeolian Deposits in Al Basra, Unpublished M.Sc Thesis, Al Basra University, 95 p. (in Arabic).

Al-Dosari AM, Wyne AH, Akpata ES, Khan NB (2004) Caries prevalence and its relation to water fluoride level for school children in Central Province of Saudi Arabia. Int Dent J 54(6):424–428 (Abstract).

Al-Dosari AM, Akpata ES, Al-Shalan TA, Khan N (2007) Fluoride map of different cities, towns and villages in Saudi Arabia. Saudi Dent J 19 (SI) (Abstract).

Al-Ghorabie FHH (2004) Measurements of indoor gamma radiation dose in At-Taif city, Saudi Arabia using CaSO4:Dy (TLD-900). Radiat Protect Dosim 113(2):178–184.

Al-Hureibi KA, Abdulmughni YA, Al-Hureibi MA (2004) The epidemiology, pathology and management of goiter in Yemen. Ann Saudi Med 24(2):119–123.

Al-Huwaider AJM (2001) The Effect of Industrial Pollution of the Geographic Distribution of Allergy and Asthma Diseases in Al Basra Governorate, Unpublished Ph.D Thesis, Al Basra University, 221 p. (in Arabic).

Alirezazadeh N (2005) Radon concentrations in public water supplies in Tehran and evaluation of radiation dose. Iran J Radiat Res 3(2):79–83.

Al-Jarallah M. (2001) Radon exhalation from granites used in Saudi Arabia. J Environ Radioactiv 53:91–98 (Abstract).

Al-Jarallah MI, Fazal-ur-Rehman (2006) Indoor radon concentration measurement in the dwellings of Al-Jauf region of Saudi Arabia. Radiat Protect Dosim 12(3):293–296 (Abstract).

Al-Jumaily HAAH (2001) Sedimentary Geochemistry of Iodine and Its Environmental Significance in Al Mosul and Adjacent Areas – Northern Iraq, Unpublished Ph.D Thesis, Al Mosul University, 171p. (in Arabic).

Al-Kabi ASA (1999) The Geographical Pattern of Some Chronic Diseases (non-communicable) in Al Basra Governorate, Unpublished Ph.D Thesis, Al Basra University, 185 p. (in Arabic).

Al-Khateeb TL, Al-Marsafi AI, O'Mullane DM (1991) Caries prevalence and treatment amongst children in an Arabian community, Community Dent Oral Epidemiol 19(5):227–280 (Abstract).

Al-Marsoumi AMH (2009) Dust fallout and bronchial asthma diseases in Al Basra Governorate. Al Basra J Sci 27(C)(1):96–106.

Almas Kh, Shakir ZF, Afzal M (1999) Prevalence and severity of dental fluorosis in Al-Qaseem province – kingdom of Saudi Arabia. Odonto- Stomatologies Tropicate 85:44–47.

Al-Najem F (1975) Dust storms in Iraq, Bulletin of the College of Science, University of Baghdad, 16(2):437–451.

Al-Nuaim AR, Al-Mazrou Y, Kamel M, Al-Attas O, Al-Daghari N, Silimani R (1997) Iodine deficiency in Saudi Arabia. Ann Saudi Med 17(3):293–297.

Al-Sarayrah L Izzat M, Takrouri H, Nader M, Abu-Zaid A (2005) Prevalence of iodine deficiency among goiterous school children. J Jordan Royal Med Serv 12(2): 42–44.

Al-Tamimi MH, Abumurad KM (2001) Radon anomalies along faults in north of Jordan. Radiat Meas 34(1):397–400 (Abstract).

Aminabadi N, Taghizdeh Gangi A, Balayi E, Sadighi M (2007) Prevalence of fluorosis in 5–12 year-old children in the North-Western villages of Makoo. J Dent Res, Dent Clin, Dent Prospects, 1(1):33–41.

Amini M, Abbaspour KC, Berg M, Winkel L, Hug SJ, Hoehn E, Yang H, Annette Johnson C (2008a) Statistical modeling of global geogenic arsenic contamination in groundwater. Environ Sci Technol 42:3669–3375.

Amini M, Muller K, Abbaspour KC, Rosenberg T, Afyuni M, Moller KC, Sarr M Annettejohnson C (2008b) Statistical modeling of global geogenic fluoride contamination in groundwaters. Environ Sci Technol 42:3662–3668.

Appleton JD (2005) Radon in air and water. In O Selinus, BJ Alloway, JA Centeno, RB Finkelman, R Fuge, Lindh, U. and Pauline SmedleyP (Eds.) (2005) Essentials of Medical Geology (812 p). Elsevier Academic Press.

Asghari Moghaddam A, Fijani E (2008) Distribution of fluoride in groundwater of Maku area, northwest of Iran. J Environ Geol 56:281–287.

Azizi F (2001) IDD in the Middle East, IDD Newsletter. International Council for Control of Iodine Deficiency Disorders 17(3).

Azizi F, Mehran L (2004) Experiences in the prevention, control and elimination of iodine deficiency disorders: A regional perspective. East Mediterr Health J 10(6):761–770.

Barzilai D, Harris P (1965) The problem of endemic goiter in Wadi Ara and the Jordan valley – Northern Israel. Israeli J Med Sci 1:62–70.

Beitollahi M, Ghiassi-Nejad M, Esmaeli A, Dunker R (2007) Radiological studies in the hot spring region of Mahallat, Central Iran. Radiat Protect Dosim 123:505–508.

Benbassat C, Tsvetov G, Schindel B, Hod M, Blonder Y, Sela BA (2004) Assessment of iodine intake in the Israel coastal area. IMAJ, Israel Medical Association 6(2):75–77.

Brand N, Gedalia I, Jungreis E, Levitus Z, Maayan M (1961) Endemic Goiter in northern Galilee. Israeli Medical J 20:206–214.
Canadian Centre for Occupational Health and Safety (CCOHS) (2005) What are the Effects of Dust on the Lungs?. http://www.ccohs.ca/oshanswers/chemicals/lungs_dust.html. Retrieved on 18 April, 2002
Caughey JE, Follis RH (1965) Endemic goiter and iodine malnutrition in Iraq. Lancet I:1032–1034.
Collis EL, Greenwood M (1977) The Health of the Industrial Worker. Ayer Publishing, 450 p.
Combs GF (2005) Geological impacts of nutrition. In O Selinus, BJ Alloway, JA Centeno, RB Finkelman, R Fuge, U Lindh, P Pauline Smedley (eds) Essentials of Medical Geology (812 p.). Elsevier Academic Press.
Coreil J, Bryant CA, Henderson JN, Forthofer MS, Quinn GP (2000) Social and Behavioral Foundations of Public Health. Published by SAGE, 360 p.
Demarchi MB, Al-Hindawi A, Abdulnabi M, El-Din Tj H (1969) Prevalence and etiology of goiter in Iraq. Am J Clin Nutr 22:1660–1666.
Dhondia J, Diermanse F (2006) Integrated Water Resources Management for the Sistan Closed Inland Delta, Iran. Annex F – FFS Manual. Delft Hydraulics. http://www.wldelft.nl/cons/area/rbm/wrp1/pdf/annex_f.pdf. Retrieved on 19 April 2009.
Dissanayake, CB (1991) The fluoride problem in the groundwater of Sri Lanka—Environmental management and health. Int J Environ Stud 38:137–156.
Dobaradaran S, Mahvi AH, Dehdashti S, Vakil Abadia DR (2008) Drinking water fluoride and child dental caries in Dashteshan, Iran. Fluoride 41(3):220–226.
Dobaradaran S, Mahvi AH, Dehdashti S, Dobaradaran S, Shoara R (2009) Correlation of fluoride with some inorganic constituents in groundwater of Dashtestan, Iran. Fluoride 42(1):50–53.
Edwards CH, Johnson AA, Knight EM, Oyemade UJ, Cole OJ, Westney OE, Jones S, Laryea H, Westney LS (1994) Pica in an urban environment. J Nutr 124:954S–962S (Abstract).
Eftekhari M, Mazloum Z (1999) Fluorosis prevalence study and its relation to drinking water among the 7–11 year old students in Larestan town and its suburb, Shahid Baheshti University of Medical Sciences Dental Faculty Research Magazine, 17:75–79.
Fathabadi N, Ghiassi-Nejad M, Haddadi B, Moradi M (2006) Miners' exposure to radon and its decay products in some Iranian non-uranium underground mines. Radiat Protect Dosim 118(1):111–116 (Abstract).
Fawell JK, Bailey K, Chilton J, World Health Organization, Dahi E (2006) Fluoride in Drinking-Water. World Health Organization, 134 pages.
Fazlollahi MR (2008) Member of Asthma and Allergic Diseases Committee of Iran, Interview with Mehr News, 10 July 2008. (In Persian).
Fuge R (2005) Soils and iodine deficiency. In O Selinus, BJ Alloway, JA Centeno, RB Finkelman, R Fuge, U Lindh, P Pauline Smedley (eds) Essentials of Medical Geology (812 p.). Elsevier Academic Press
Ghahreman A (2003) Flora of Iran (Vol.1–23). University of Tehran Press.
Ghaleb T (2007) Yemen: Fluoride contamination is putting bones at risk. Yemen Observer, June 05, 2007. http://www.yobserver.com/sports-health-and-ifestyle/10012356.html
Gharibadzeh S, Hoseini SS (2008) Arsenic exposure may be a risk factor for Alzheimer's disease. J Neuropsychiatry Clin Neurosci 20:501.
Ghassemzadeh F, Yousefzadeh H, Arbab-Zavar MH (2008) Arsenic phytoremediation by phragmites Australis: Green technology. Int J Environ Stud 65(4):587–594.
Ghiassi-Nejad M, Mortazavi SMJ, Cameron JR, Nirooman-Rad A, Karam PA (2002) Very high background radiation areas of Ramsar, Iran: Preliminary biological studies. Health Phys 82(1):87–93.
Ghiassi-Nejad M, Beitollahi MM, Asefi M, Reza-Nejad F (2003) Exposure to ^{226}Ra from consumption of vegetables in the high level natural radiation area of Ramsar-Iran. J Environ Radioact 66:215–225.
Ghiassi-Nejad M, Beitollahia MM, Fallahiana N, Saghirzadehb M (2005) New findings in the very high natural radiation area of Ramsar, Iran. Int Congr Series 1276:13–16.
Gilmore GK, Vasal NJ (2009) A reconnaissance study of radon concentration in Hamadan city, Iran. Geophys Res Abstr 11 (Abstract).
Gunderson EK, Garland CF, Gorham ED (2005) Health surveillance for asthma in the US Navy: Experience of 9,185,484 person-years. Ann Epidemiol 15(4):310–315.
Hadad K, Doulatdar R, Mehdizadeh S (2007) Indoor radon monitoring in Northern Iran using passive and active measurements. J Environ Radioact 95(1):39–52 (Abstract).
Halsted JA, Prasad AS (1961) Syndrome of iron deficiency anemia, hepatosplenomegasly, hypogonadism, dwarfism and geophagia. Trans Am Clin Climatol Assoc 72: 130–148.
Hamdan MA (2003) The prevalence and severity of dental fluorosis among 12-year old school children in Jordan. Int J Pediatr Dentistry 13(2):85–92.
Haquin G, Reimer T, Shamai Y, Margaliot M, Shirav-Schwartz M, Kenett R (2006) Radon Survey of Israel, Proceedings of the Conference of Nuclear Societies in Israel, Dead Sea, February 2006.
Hawass NE, Alnozha MM, Kolawole T (1987) Adult geophagia – report of three cases with review of the literature. Trop Geograph Med 39(2):191–195.
Hisham H (2006) General Radon Potential Map of the Kingdom of Saudi Arabia, Radon Investigations in the Czech Republic XI and the 8th International Workshop on the Geological Aspects of Radon Risk Mapping, Prague, Czech Republic (Abstract).
Hooper SR, Hynd GW, Mattison RE (1993) Child Psychopathology: Diagnostic Criteria and Clinical Assessment. Lawrence Erlbaum Associates, 504 pages.
Idso SB (1976) Dust storms. Sci Am 235(4):108–11, 113–14.
Karam PA, Mortazavi SMJ, Ghiassi-Nejad M, Ikushima T, Cameron JR, Niroomand-Rad A (2002) ICRP evolutionary recommendations and the reluctance of the members of the public to carry out remedial work against radon in some high-level natural radiation areas. Int Congr Series 1236:35–37.
Kelsall HL, Sim MR, Forbes AB, McKenzie DP, Glass DC, Ikin JK, Ittak P, Abramson MJ (2004) Respiratory health status of Australian veterans of the 1991 Gulf War and the effects

of exposure to oil fire smoke and dust storms, Thorax 59: 897–903.

Khademi H, Taleb M (2000) Dental caries and fluorosis in different levels of drinking water fluoride. J Res Med Sci (JRMS) Fall 5(3):213–215.

Khalaf F, Al-Hashash M (1983) Aeolian sedimentation in the northwestern part of the Arabian Gulf. J Arid Environ 6(4):319–332.

Khayrat AH, Al Jarallah MI, Fazal-ur-Rahman X, Abu-Jarad F (2003) Indoor radon survey in dwellings of some regions in Yemen. Radiat Meas 36:449–451 (Abstract).

King MS, Miller R, Johnson J, Ninan M, Lambright E, Shorr AF, Sheller MJ (2008) Bronchiolitis in Soldiers with Inhalational Exposures in the Iraq War. Presented at the American Thoracic Society, May, 2008, Toronto, Canada.

Korman SH (1990) Pica as a presenting symptom in childhood celiac disease. Am J Clin Nutr 51:139–141.

Makofske WJ, Edelstein MR (1988) Radon and the Environment. William Andrew Inc., 465 pages.

Mann J, Mahmoud W, Ernest M, Sgan-Cohen H, Shoshan N, Gedaia I (1990) Fluorosis and dental caries in 6–8 year-old children in a 5 ppm fluoride area. Community Dent Oral Epidemiol 18(2):77–79 (Abstract).

Matiullah AN, Khatobeh AJAH (1997) Indoor radon levels and natural radioactivity in Jordanian soils. Radiat Protect Dosim 71:231–233 (Abstract).

Meyer-Lueckel H, Paris S, Shirkhani B, Hopfenmuller W, Kielbassa AM (2006) Caries and fluorosis in 6- and 9-year-old children residing in three communities in Iran. Community Dent Oral Epidemiol. 34(1):63–70.

Miri A, Ahmadi H, Ghanbari A, Moghaddamnia A (2007) Dust torms impact on air pollution and public health under hot and dry climate. Int J Energy Environ 2(1):101–105.

Monajemzadeh SM, Moghadan AZ (2008) Prevalence of goiter among children aged 11–16 years in Ahwaz, Iran. Med Princ Prac 17(4):331–333,

Mortazavi SMJ, Ghiassi-Nejadb M, Rezaieanc M (2005a) Cancer risk due to exposure to high levels of natural radon in the inhabitants of Ramsar, Iran. Int Congr Series 1276: 436–437.

Mortazavi SMJ, Shabestani-Monfaredb A, Ghiassi-Nejadc M, Mozdarani H (2005b) Radioadaptive responses induced in lymphocytes of the inhabitants in Ramsar, Iran. Int Congr Series 1276:201–203.

Mortazavi SMJ, Ghiassi-Nejad M, Ikushima T (2002) Do the findings on the health effects of prolonged exposure to very high levels of natural radiation contradict current ultra-conservative radiation protection regulations?. Int Congr Series 1236(1):9–21.

Mortezavi Z, Mohammadi M (2003) The prevalence of pica in pregnant women referring to health centers in Zahedan. Med J Reproduct Infert 4(16) (In persian).

Mosaferi M, Yunesian M, Dastgiri S, Mesdaghinia A, Esmailnasab N, (2008) Prevalence of skin lesions and exposure to arsenic in drinking water in Iran. Sci Total Environ 390(1):69–76.

Mowlavi AA, Binesh A (2006) Radon concentration measurement in the some water sources of Mashhad region in Iran. In S Kim, T Suk Suh (2007)World Congress on Medical Physics and Biomedical Engineering 2006: "imaging the Future Medicine".

Najafi M, Khodaee GH, Bahari M, Sabahi Farsi MM. Kiani F (2008) Neonatal thyroid screening in a mild iodine deficiency endemic area in Iran. Indian J Med Sci 62(3):113–116.

Namrouqa H. (2009) No hidden agenda behind water study – researcher. The Jordan Times, 2 March 2009.

National Aeronautics and Space Agency (NASA) (2005) Iraq dust storms, AQUA, August 9, 2005. http://www.nasa.gov/vision/earth/lookingatearth/iraq_dust_storm.html. Retrieved on 22 September 2008.

National Aeronautics and Space Agency (NASA) (2008) Dust over the Persian Gulf, http://earthobservatory.nasa.gov/NaturalHazards/view.php?id=19813&oldid=14792. Retrieved on 22 Sep 2008

National Research Council (NRC) (U. S.) (2007) Earth Materials and Health: Research Priorities for Earth Science and Public Health. Published by National Academies Press, 176 p.

Navi M (2007) Map of arsenic distribution and mortality rates of related diseases. National geoscience database of Iran (NGDIR), geological survey of Iran, Ministry of Industries and Mines. Official Report pressed by NGDIR.

Navi M (2009) The health effect of dust storms in Iran, National geoscience database of Iran (NGDIR), Geological survey of Iran, Ministry of Industries and Mines. Official Report pressed by NGDIR.

Nejat F, Kazmi SS (2005) Pica in neurosurgery: Report of an infant with pica and posterior fossa dermoid tumor. J Ped Neonatal 2(3):CR60–61.

Nish WA, Schwietz LA (1992) Underdiagnosis of asthma in young adults presenting for USAF basic training. Ann Allergy 69(3):239–242.

Nordbeg M, Cherian MG (2005) Biological responses of elements. In O Selinus, BJ Alloway, JA Centeno, RB Finkelman, R Fuge, U Lindh, P Pauline Smedley (eds) Essentials of Medical Geology (812 p.). Elsevier Academic Press.

Norman AG, Sparks DL (1992) Advances in Agronomy. Academic Press, 332 pages.

Othman I, Hushari M, Raja G, Al Sawaf A (1996) Radon in Syrian houses. J Radiol Protect 16:45–50 (Abstract).

Poureslami HR, Khazaeli P, Nooric GR (2008) Fluoride in food and water consumed in Koohbanan (Kuh-e Banan) Iran 2008 International Society for Fluoride Research, Research Report, Fluoride, 41(3):216–219.

Prasad AS (1989) Zinc deficiency in humans. Citation classic. Current Contents/ Clinical Medicine 17(38):14.I8 September 1989.

Prasad AS (1991) A diet of zinc or clay. Citation classic. Current Contents/Life Science 34(28):11.

Prasad A S (1992) Deficiencies of iron and zinc. Current Contents/ Clinical Medicine 44:10.

Quattromani F, Handal G, Lampe R (2008) Pediatric Imaging: Rapid-Fire Questions and Answers. Published by Thieme, 436 pages.

Ramezani GH, Valaei N, Eikani H (2004) Prevalence of DMFT and fluorosis in the students of Dayer City (Iran). J Indian Soc Pedo Prev Dent June (2004) 22(2):49–53.

Rezavand N, Rawshani D (2006) Pica and its related iron deficiency anaemia in pregnant women in Kermanshah. Med J Behbood 10(3):226–234 (In persian).

Rosenzweig KA, Abewitz I (1963) Prevalence of endemic fluorosis in Israel at medium fluoride concentration. Pblic Health Repos 78(1):77–80.
Rostami A (2003) The prevalence of pica in pregnant women referring to Rasul-Akram and Akbarabadi hospitals in Tehran. Iran University of Medical Science, graduation thesis (In persian).
Sack J, Feldman I, Kaiserma I (1998) Congenital hypothroidism screening in the West Bank: At test case for screening in developing regions. Hormone Research.
Samavat H, Seaward MRD, Aghamiri SMR, Reza-Nejad F (2006) Radionuclide concentrations in the diet of residents in a high level natural radiation area in Iran. Radiat Environ Biophys 45:301–306.
Sanders JW, Putnam SD, Frankart C, Frenck RW, Monteville MR, Riddle MS, Rockabrand DM, Sharp TW, Tribble DR (2005) Impact of illness and non-combat injury during Operations Iraqi Freedom and Enduring Freedom (Afghanistan). Am J Trop Med Hygiene 73(4):713–719.
Shirav-Schwarzz M (2004) Some Remarks on High Natural Background Radiation Terrains, Geology and Public Health in Israel and Adjacent Countries, Presentation Dedicated to Dr. Uzi Vulcan (1938–1999).
Sigel H (2004) Metal Ions and Their Complexes in Medication: Metal Ions and Their Complexes in Medication (Vol. 41, 519 p). Published by CRC Press.
Singer A, Ganor E, Dultz S, Fischer W (2003) Dust deposition over the dead sea. J Arid Environ 53(1):41–59 (Abstract).
Smedley P, Kinniburgh DG (2005) Arsenic in groundwater and the environment. In O Selinus, BJ Alloway, JA Centeno, RB Finkelman, R Fuge, U Lindh, P Pauline Smedley (eds) Essentials of Medical Geology (812 p.). Elsevier Academic Press.
Smith JC, Halsted JA (1970) Clay Ingestion (Geophagia) as a source of zinc for rRats. J Nutr 100:973–980.
Steinhausler F (2003) Radiological impact on man and the environment from the oil and gas industry : Risk assessment for the critical group. In MK Zaidi, I Mustafaev(eds) Radiation Safety Problems in the Caspian Region: Proceedings of the NATO Advanced Research Workshop on Radiation Safety Problems in the Caspian Region, Baku, Azerbaijan, 11–14 September 2003, Springer, 249 p.
Sullivan JB, Krieger GR (2001) Clinical Environmental Health and Toxic Exposures. Lippincott Williams and Wilkins, 1323 pages.
Szema AM, Peters MC, Weissinger KM, Gagliano CA, Chen JJ (2008a) Increased Rates of Asthma Among U.S. Military Personnel after Deployment to the Persian Gulf. Presented at the American Thoracic Society, May, 2008, Toronto, Canada.
Szema AM, Peters MC, Weissinger KM, Gagliano CA, Chen JJ (2008b) Increased allergic rhinitis rates among U.S. Military Personnel after deployment to the Persian Gulf. J Allergy Clin Immunol 121(2):S230.
Taghavi F, Asadi A (2008) The Persian Gulf 12th April 2007 dust storm: Observation and model analysis. EUMETSAT Meteorological Satellite Conference, Darmstadt, Germany 8–12 Sep. 2008
Talaeepour M, Moattar F, Atabi F, Borhan Azad S, Talaeepour AR (2006) Investigation on radon concentration in the Tehran subway stations in regard with environmental effects. J Appl Sci 6(7):1617–1620.
(UN) United Nations, General Assembly. United Nations Scientific Committee on the Effects of Atomic Radiation: Report of The Fifty-Fourth Session (29 May-2 June 2006) Published by United Nations Publications, 2006, 18 pages.
UNICEF (United Nations Children Fund) (2000) UNICEF's Position on Water Fluoridation available, http://www.nofluoride.com/Unicef_fluor.htm
Vincent, P. (2008) Saudi Arabia: An Environmental Overview, Routledge, 332 p.
Weese CB, Abraham JH (2009) Potential health implications associated with particulate matter exposure in deployed settings in Southwest Asia. Inhal Toxicol 21(4):291–296.
WHO Expert Committee (1973) Trace Elements in Human Nutrition, World Health Organization. Technical Report Series No. 532.
Young SL (2008) A vile habit? In J MacClancy, J Henry, H Macbeth (eds) Consuming the Inedible: Neglected Dimensions of Food Choice. Berghahn Books, 208 pages.
Zarasvandi A, Mokhtari B (2008) A Scientific Overview of the 50th Dusty Days in Khuzestan, Unpublished Data. http://www.tabnak.ir/pages/?cid=13646. Retrieved on 10 July 2008 (In Persian).
Zein AZ, Al-Haithamy S, Obadi Q, Noureddin S (2000) The epidemiology of iodine deficiency disorders (IDD) in Yemen. Public Health Nutr 3:245–252.
Zohouri FV, Rugg-Gunn AJ (1999) Fluoride concentration in foods from Iran. Int J Food Sci Nutr 50(4):265–274 (Abstract).

Some Aspects of the Medical Geology of the Indian Subcontinent and Neighbouring Regions

C.B. Dissanayake, C.R.M. Rao, and Rohana Chandrajith

Abstract The tropical terrains of the Indian subcontinent and the neighbouring regions provide some of the best examples in medical geology where the impact of the geosphere on human health is markedly seen. The life styles of millions of people in these regions are such that there is a very close and intimate association between the rocks, minerals, soils, water and the human population. The case studies described in this chapter, as illustrated from India, Pakistan, Sri Lanka, Bangladesh and Nepal, notably on fluorides, iodine deficiency disorders (IDD), arsenic pollution, selenium-based health issues among others, demonstrate this close geochemical interaction. The problem of contamination of groundwater by arsenic in Bangladesh and West Bengal, termed as one of the world's largest environmental disasters, as discussed in this chapter typifies the dangers that could threaten countries and extremely large populations, when geochemistry of the environment and the associated medical geology are not properly understood and neglected. The negative impact of such gross neglect will be seen for many years to come and in Bangladesh and West Bengal for many future generations.

Keywords India · Sri Lanka · Pakistan · Bangladesh · Nepal · Iodine · Fluorine · Arsenic · Selenium · Podoconiosis · Radiation

General Introduction

The Indian subcontinent and the neighbouring regions provide some of the best examples of the impact of geology on the health and well-being of human beings and animals. Unlike the developed countries of the temperate climates, the millions of people of the regions in and around the Indian subcontinent live in intimate association with the general geological environment. The impact of the rocks, minerals, soils, water and the vegetation on the health of these people is far more profound than that observed in most other parts of the world. This is clearly due to the fact that the vast majority of the population obtain their drinking water supplies directly from the ground and surface waters and locally grown crops. The geochemical pathways of the essential trace nutrients that enter the food chain are clear and being intimately linked to the immediate geological environment, the impact of the mineral excesses, deficiencies and imbalances on humans are very well marked. The clear demarcation of "geochemical provinces" is also closely linked to the incidence of several regionally distributed diseases.

The economic status of the general population, many of them living in poverty, contributes to the direct impact of the environmental chemistry on health. Proper sanitation, pipe-borne clean water, etc. are seriously lacking and this clearly enhances the spread of many diseases. Contrary to the popular belief that environmental poisoning due to toxic metals and trace elements is an endemic problem of highly industrialized countries, it is the population of developing countries like India, Pakistan, Sri Lanka and Bangladesh among many others that are more vulnerable to such threats. This is mainly due to

C.B. Dissanayake (✉)
Institute of Fundamental Studies (IFS), Kandy, Sri Lanka
e-mail: cbdissa@hotmail.com

O. Selinus et al. (eds.), *Medical Geology*, International Year of Planet Earth,
DOI 10.1007/978-90-481-3430-4_7, © Springer Science+Business Media B.V. 2010

- a very high population and poor hygienic conditions in the crowded cities
- a high percentage in population of vulnerable groups such as children and pregnant women
- general poverty, leading to poor nutrition and health status
- consumption of food grown in contaminated areas, since food imported from non-contaminated areas is costly
- poor implementation of pollution control laws
- low level of awareness about health hazards due to environmental toxins
- limited availability of health care

It should also be emphasized here that even though the World Health Organization (WHO) has given guidelines for limits of essential and toxic elements in water, food, etc. many of these guidelines do not strictly apply to the people living in areas mentioned above in view of a variety of factors including poor resistance against disease and lack of other essential elements necessary to counteract the ill effects of some of the disease-causing trace elements. A good case in point is that an average American would be able to withstand a higher degree of fluoride in drinking water without any ill effects while a poverty stricken adult from the Indian subcontinent may suffer visible ill effects even at a considerably lower fluoride dosage, presumably due to lack of other essential elements, notably calcium. The danger limits for disease-causing elements for such countries should therefore be much lower. Thus, in view of the vulnerability of the population, geomedical studies in general pertaining to adverse health effects of low-dose, high-duration exposure to toxic elements in particular are essential.

The tropical environment, in which the lands under discussion lie, is characterized by seasonal heavy rainfall (exceeds over 5,000 mm/yr in some cases) with long periods of drought and high ambient temperatures. The intensity of chemical weathering is therefore extreme and some soils are weathered to such an extent that potassium, an essential plant nutrient, usually found in appreciable concentrations, is present only in trace amounts (Fyfe et al., 1983). Coincidentally, the world's most underdeveloped countries also lie in such tropical environments, where poor agricultural productivity caused by impoverishment of soil in essential nutrients has, over the years, had a marked impact on the economy and now on the health of the population of these countries. The very thick laterite formations and extensive leaching out of essential nutrients have, over several decades, caused severe mineral imbalances. The rock-soil-water-plant links are thus extraordinarily marked in the lands of the Indian subcontinent, making tropical medical geology truly a fascinating subject.

Arsenic in Groundwater

The presence of arsenic in groundwater notably in South and Southeast Asia has become a major health issue. The prevalence of skin disorders, cancer and cardiovascular diseases among others, is caused mainly by the consumption of arsenic-laden drinking water and also food grown in arsenic-rich soils. Hyperpigmentation, melanosis, plantar and palmer keratoses, carcinoma of the hand (Fig. 1), broken nails, cracked soles, gastroenteritis and oedema of legs are some of the common clinical symptoms related with arsenic poisoning.

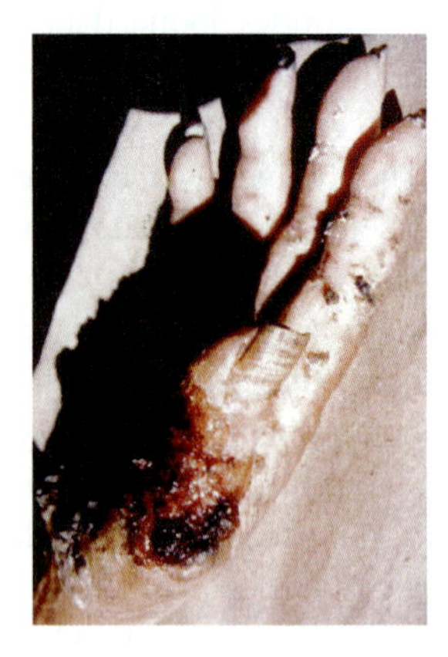

Fig. 1 Bowen's carcinoma of the hand due to arsenic poisoning (photo courtesy of Arsenic Foundation Inc.)

Smedley (2005) in a publication of the World Bank (2005) noted that even though arsenic in natural waters globally is usually low (<10 μg/L), under certain specific geochemical and hydrogeological conditions, the arsenic concentration can reach much higher levels, bearing in mind that most occurrences of high-arsenic groundwater are of natural origin.

It has been noted that sediments of Quaternary age constituents, the large alluvial deltaic plains and inland basins as seen in South Asia are most vulnerable to arsenic enrichment. The Bengal Basin of Bangladesh and West Bengal are typical examples. These regions are also extremely densely populated and any contamination of ground and drinking water resources will undoubtedly have an enormous impact on the health of these large populations. As shown by Smedley (2005), the increasing incidence of arsenic-related health problems in the South Asian regions coincided with the change to using groundwater from tube wells about 30 years ago. Slow groundwater movement and stagnant conditions within the sedimentary basins enhance the arsenic enrichment, and variability of the actual arsenic content within a small distance is very large. This creates a necessity for the testing of virtually each and every well in a region. Prediction of arsenic-rich regions is therefore difficult though some older aquifers in Bangladesh, West Bengal and Nepal appear to have much lower arsenic concentrations. Table 1 shows the summary of the distribution, nature and scale of documented arsenic problems (>50 μg/L) in aquifers in South and East Asia.

Arsenic-Contaminated Aquifers in India

The arsenic problem in Bangladesh and the West Bengal state of India is a major groundwater pollution problem and has been classified as one of the world's greatest environmental disasters. The arsenic-induced public health hazard was first noticed around 1978 in West Bengal, when frequent cases of patients with symptoms of dermatological disorders attracted the attention of medical specialists in Murshidabad district. Later, a link between arsenical dermatosis and arsenic-contaminated groundwater, the main source of drinking water, in the region was established (Chakraborty and Saha, 1987). Research carried out by various state, national and international geological organizations in India, over the last two decades in West Bengal, have generated a wealth of geoscientific data which can be effectively utilized by medical experts for conducting further epidemiological studies on arsenic-related health problems (Fig. 2).

Table 1 Summary of the distribution, nature and scale of documented arsenic problems (>50 μg/L) in aquifers in South and East Asia (World Bank, 2005)

Location	Arial extent (km^2)	Population at risk[a]	Arsenic range (μg/L)
Alluvial/deltaic/ lacustrine plains			
Bangladesh	150,000	35,000,000	<1.0–2,300
China (Inner Mongolia, Xinjiang, Shanxi)	68,000	5,600,000	40–4,400
India (West Bengal)	23,000	5,000,000	<10–3,200
Nepal	30,000	550,000	<10–200
Taiwan	6,000	(?) 10,000[b]	10–1,800
Vietnam	1,000	10,000,000[c]	1–3,100
Myanmar	3,000(?)	3,400,000	na
Cambodia	<1,000(?)	320,000[d]	na
Pakistan	na	na	na

Source: World Bank Regional Operational Responses to Arsenic Workshop in Nepal, 26–27 April 2004
na – Not available
[a]Estimated to be drinking water with arsenic >50 μg/L (from Smedley, 2003 and data sources therein). [b]Before mitigation. [c]United Nations Children's Fund (UNICEF) estimate. [d]Maximum

Arsenic is naturally occurring and is not derived from any anthropogenic source. Its primary source has been ascribed mainly to the rock formations of the Himalayan Mountain system and partly to the rock formations of Shillong Plateau (Acharyya et al., 1999a). The recently reported occurrence of arsenic in the Quaternary formations and groundwater in parts of Nepal and in groundwaters of a few sporadic locations in the Himalayan foothill zone of India strongly supports this contention. However, according to some workers, at least a part of the arsenic was derived from rock formations of Bihar Plateau region (Chakrabarti, 2004). The recent detection of high arsenic values in the groundwaters of Bhojpur district (Bihar), located on Ganga flood plain deposits of Quaternary age, indicates that from the primary sources arsenic was transported either in solution or in suspension along with detrital sediment particles, by the rivers originating in the Himalayan mountains, Shillong Plateau and Bihar Plateau and flowing into the Ganga–Brahmaputra deltaic region.

The area in which transported arsenic was finally deposited is referred to as the arsenic sink. Among the geological formations occurring within the sink, arsenic is localized within a particular formation. This formation comprises semi-consolidated and unoxidized sequences of silt, clay and mostly fine to medium grained sands. These sediments were laid down as flood plains, meander belts and deltaic deposits related to the present day river systems. Arsenic mainly occurs in an adsorbed state in hydrous ferric oxides (HFO) associated with these sediments,

to the Dongargarh rift zone. This geological control makes the demarcation of the sink and identification of target areas for epidemiological studies, comparatively easy (Acharyya et al., 2000).

The toxicity of arsenic depends upon its chemical form, route of entry and dose/duration of exposure. The nutrition level of the person exposed to arsenic toxicity plays a key role on the magnitude of effect. Although high values of arsenic are reported in some regions of India, no cases of arsenic-related diseases were reported. In the Gurwandi area, for instance, where arsenopyrite is present in the rhyolitic country rock in which arsenic values are as high as 250–680 μg/L in groundwater, no arsenic-related symptoms were recorded.

Remedial Measures

Short-term measures:

(i) Use of conventional surface water sources with treatment for arsenic.
(ii) Construction of rain water-harvesting ponds.
(iii) Replacement of arsenic-affected water sources.
(iv) Attachment of arsenic removal filters in the existing hand pumps.

Long-term measures:

(i) Deep tube well tapping of arsenic-free water.
(ii) Deep tube wells with community arsenic removal plants.
(iii) Surface water and conventional water treatment plants.

Arsenic in Bangladesh

The problem of arsenic poisoning in Bangladesh where an estimated population of 35 million is at risk is considered to be the largest mass poisoning in history. The British Geological Survey in a very detailed report (BGS, 2001) has discussed all aspects of the problem extensively.

Geologically, the Bengal Basin consists of a 15 km thick sequence of Cretaceous to Recent sediments. This covers an area of about 10,000 km^2 of lowland flood plain and delta making it one of the largest in the world. The sediment load carried by the rivers Ganga, Brahmaputra and Meghna (GMB) river systems surpasses that of any other river system in the world. The extensive aquifer system underlying these sediments is therefore a result of the deposition of Pleistocene to Recent fluvial and estuarine sediments.

The work of the British Geological Survey (BGS, 2001) has estimated that most elements including arsenic are found in the fine-grained sediments, notably clayey sediments. This therefore supports the hypothesis that iron oxides are a primary source of arsenic in the groundwater of Bangladesh. It was observed that the process of adsorption and co-precipitation of arsenic on iron oxides yielded the net arsenic-rich groundwater. Hence areas rich in these iron oxides had markedly high arsenic contents in the water.

In view of the fact that a large number of tube wells in Bangladesh are contaminated, an estimated 95% of the population of Bangladesh is highly susceptible to arsenic poisoning indicating the magnitude of the problem. A total of 4.7 million tube wells in Bangladesh have been tested for arsenic and 1.4 million of those were found to contain arsenic above the Bangladesh Government drinking water limit of 50 μg/L. About 20 million people in Bangladesh use these tube wells containing arsenic (UNICEF, 2008). This was also shown by the British Geological Survey (BGS, 2001) from among 2,022 samples of water analysed as given below:

(i) 51% were above 10 μg/L (WHO guideline value)
(ii) 35% were above 50 μg/L
(iii) 25% were above 100 μg/L
(iv) 8.4% were above 300 μg/L
(v) 0.1% were above 1,000 μg/L

Only 20% of the wells were considered to be "arsenic free".

Smedley and Kinniburgh (2002) had shown that arsenic tends to be released slowly from recently buried sediments in rivers, lakes and ocean. Arsenic is adsorbed on the surface of the iron oxide/hydroxide phase making them arsenic rich. Microbial reduction of FeOOH brings about a change from arsenate to arsenite. Even though the actual mean concentration of arsenic in the sediments may not be extremely high, the vast amount of sediments involved in this process makes even low levels of arsenic quite important.

Apart from the drinking water, the rice grains, which form the staple diet of the Bangladesh population, also show markedly high arsenic contents particularly in areas where the paddy soil is contaminated. Some rice samples had arsenic contents as high as 1.7 μg/g, and among the crops, Arum (*Colocasia antiquorum*) and boro rice had high arsenic contents (Huq et al., 2006).

As in the case of West Bengal in India, arsenic manifests itself in a variety of diseases in Bangladesh and from among these, patients with melanosis (93%), leucomelanosis (39.1%), keratosis (68.3%), hyperkeratosis (37.6%), dorsum, non-petting oedema, gangrene and skin cancers have been identified (Karim, 2000).

Arsenic in Tube Wells of Bangladesh

The fact that about 130 million people comprising of 97% of the population relay on 8.6 million tube wells in Bangladesh indicates the importance of groundwater as the major source of potable water. In a major study of the spatial variability of arsenic in 6,000 tube wells in a 25 km^2 area in Bangladesh, van Geen et al. (2003), noted that arsenic ranges from <5 to 900 μg/L in groundwater obtained from these tube wells. They also observed that the arsenic increases from 25% with depth between 8 and 10 m to 75% between 15 and 30 m, with a decline to less than 10% at 90 m. This indicates a geologically related arsenic concentration over a period of time. van Geen et al. (2003) were of the view that the oxic low-As aquifers accessible by drilling were of Pleistocene age while the shallower reducing aquifers which had higher and variable As concentrations had been deposited since the last Glacial Maximum, approximately 20,000 years ago.

Arsenic in Groundwater of Nepal

Arsenic in groundwater is now seen as a major health issue in Nepal with many shallow tube wells containing arsenic at concentrations higher that the safe limit of WHO. As shown by the population census data of 2001, the total population of Nepal was 23.4 million of which about 45% live in 20 Terai districts where 8% of tube wells were found to be contaminated (Panthi et al., 2006).

As in the case of Bangladesh and West Bengal, arsenic is considered to be clearly associated with the oxidation–reduction processes of iron oxide and pyrite. Bhattacharya et al. (2003) had shown that in a strongly reducing environment (Eh: −110 to −200 mV) of groundwater in Nepal, the reductive desorption theory is perhaps more likely. The arsenic-rich iron oxides are known to enter the aqueous phase easily.

When arsenic contamination in groundwater of the Bengal Delta Plain in the neighbouring Indian state of West Bengal and Bangladesh was detected, the Department of Water Supply and Sewage (DWSS) with assistance from WHO, Nepal, conducted several studies (Panthi et al., 2006). Table 2 summarizes the arsenic concentrations recorded in Nepal.

A very approximate estimate of the tube wells of Nepal gives a figure of 400,000 of which 25,058 had been tested for arsenic in 2003. Of this 1916 tube wells (about 8%) had arsenic concentrations more than 50 μg/L while 23% had more than 10 μg/L. As shown in Table 2, Nawalaparasi is the most affected district

Table 2 Arsenic concentrations in Nepal (after NASC, 2003)

District	Total no. of tests	Samples with arsenic <10 μg/L	10–50 μg/L	>50 μg/L	Max. conc. detected
Jhapa	571	493	77	1	79
Morang	341	264	72	5	70
Sunsari	675	566	105	4	75
Saptari	772	669	94	9	98
Siraha	289	191	70	28	90
Dhanusha	331	267	55	9	140
Mahottari	202	177	21	4	80
Sarlahi	532	402	114	16	98
Rautahat	2,485	740	1,520	225	324
Bara	2,124	1,783	291	50	254
Parsa	2,207	1,895	253	59	456
Chitwan	219	219	0	0	8
Nawalparasi	3,833	1,385	1,340	1,108	571
Rupandehi	2,725	2,191	410	124	2,620
Kapilvastu	4,099	3,471	466	162	589
Dang	667	639	25	3	81
Banke	1,835	1,316	486	33	270
Bardiya	652	472	160	20	181
Kailali	299	149	106	44	213
Kanchanpur	200	167	21	12	221
Total sample tested	25,058	17,456	5,686	1,966	2,620
% of total tested	100	69	23	8	

in Nepal with Rautahat, Kailali and Siraha also being seriously affected in which 69% wells had greater than 50 μg/L of arsenic.

Maharjan et al. (2005) conducted a community-based dose–response study on arsenic contamination in three communities in Terai in lowland Nepal. The arsenic concentration of all the tube wells in use (n = 146) and the prevalence of arsenic-induced skin manifestations among 1,343 (approx 80%) of the subjects, indicated the existence of a higher contamination in Terai. The authors observed a rate of 6.4% of arsenosis in the population older than 15 years, nearly the same as that found in Bangladesh. Further, they observed that males had a prevalence twice as much as females.

The Arsenic Problem in Pakistan

As in the case of West Bengal, Nepal and Bangladesh, Pakistan also faces a health hazard due to arsenic concentration of groundwater. In major cities, such as Lahore, thousands of people are susceptible to diseases caused by excessive quantities of arsenic in their drinking water supplies.

In the Muzaffargah district of southwestern Punjab, central Pakistan, for example, concentration of arsenic exceeded the WHO permissible guideline value of 10 μg/L, in 58% of 49 samples of groundwater (Nickson et al., 2000). A maximum arsenic level of 906 μg/L had also been detected. The authors attributed this high level of arsenic in the semi-arid region to canal irrigation that caused water logging and evaporative concentration of salts. It was observed that in rural areas, the levels remain below 25 μg/L, in view of fact that arsenic is in the shallow oxic groundwater where sediments sorb the arsenic.

Arsenic in the Indus Plain of Pakistan

The Indus Plain of Pakistan, mainly Punjab and Sindh Province, comprising of Quaternary sediments of alluvial and deltaic origin has shown significant arsenic concentrations in the groundwater. As in the case of Bangladesh and West Bengal, these sediments are also derived from the Himalayan sources. Climatically, however, this region is more arid with greater unconfined and greater aerobic aquifer conditions (Mahmood et al., 1998). The different redox conditions of the Indus plain as compared to that of Bangladesh make arsenic mobilization in the groundwater more difficult.

Table 3 Range of arsenic concentrations found in groundwater samples from northern Punjab (Iqbal, 2001)

District	No. of samples	<10–11 (%)	11–20 (%)	21–50 (%)	>50 (%)
Gujrat	38	33 (87)	2 (5)	1 (3)	2 (5)
Helum	37	32 (86)	1 (3)	3 (8)	1 (3)
Chakwal	72	63 (88)	9 (12)	0 (0)	0 (0)
Sargodha	59	49 (83)	5 (8)	2 (3)	3 (5)
Attock	74	68 (92)	6 (8)	0 (0)	0 (0)
Rawalpindi	84	81 (96)	3 (4)	0 (D)	0 (0)
Total	364	326 (90)	26 (7)	4 (1)	6 (2)

Unlike in Bangladesh and West Bengal, surveys for arsenic in groundwater in Pakistan are limited. The World Bank Report (2005) on arsenic mentions a testing program conducted by the Provincial Government of Punjab with UNICEF in 2000. From a total of 364 samples analysed, 90% had arsenic concentrations less than 10 μg/L with 2% having concentrations above 50 μg/L (Table 3) (Iqbal, 2001). The levels of arsenic concentrations of groundwater in the Indus Plain appear to be relatively small though further detailed surveys are clearly required.

Fluoride in Groundwater and Dental Health

Dental Fluorosis in Sri Lanka

In studies on the distribution of diseases linked to geology and geochemistry, the concept of geochemical province becomes very useful. Very broadly, one could define a geochemical province as a geological terrain consisting of a characteristic chemical composition in soil, rock, water and stream sediments and which could be delineated from other such provinces. Even though it was originally used in mineral exploration, due to the imbalances of chemical elements in the delineated geochemical province, the different populations living on them could show symptoms of different diseases and hence its use in medical geology.

In Sri Lanka, four major geochemical provinces can be demarcated, based on the hydrogeochemical maps of the country (Dissanayake and Weerasooriya, 1986).

The climate plays the most significant role in Sri Lanka with the climatic zones, namely dry, intermediate and wet zones showing clear geochemical differences and different weathering rates, clay mineralogy and leaching of essential elements. This intense leaching, so characteristic of humid tropical terrains, is clearly seen in Sri Lanka. The dry zone in particular is geochemically very different from the wet zone and the impact is clearly seen in the distribution of some diseases, namely dental fluorosis and iodine deficiency disorders (IDD).

The dry zone of Sri Lanka is markedly rich in fluoride-rich groundwater (Fig. 3). Over 2 million people, the majority of them children, are susceptible to dental fluorosis caused by the consumption of water with high levels of fluoride, as obtained from the groundwater. It should be emphasized here that even at rather low levels of fluoride in the groundwater (e.g. 0.4–0.6 mg/L), dental fluorosis is still common (Fig. 4). This indicates possibly a higher consumption of water, bearing in mind the hot dry climate, and a deficiency of other essential elements (e.g. calcium, which may retard the onset of dental fluorosis). The WHO recommended levels are not appropriate in this case and a maximum level of 0.6 mg/L has been recommended for fluoride in drinking water in Sri Lanka

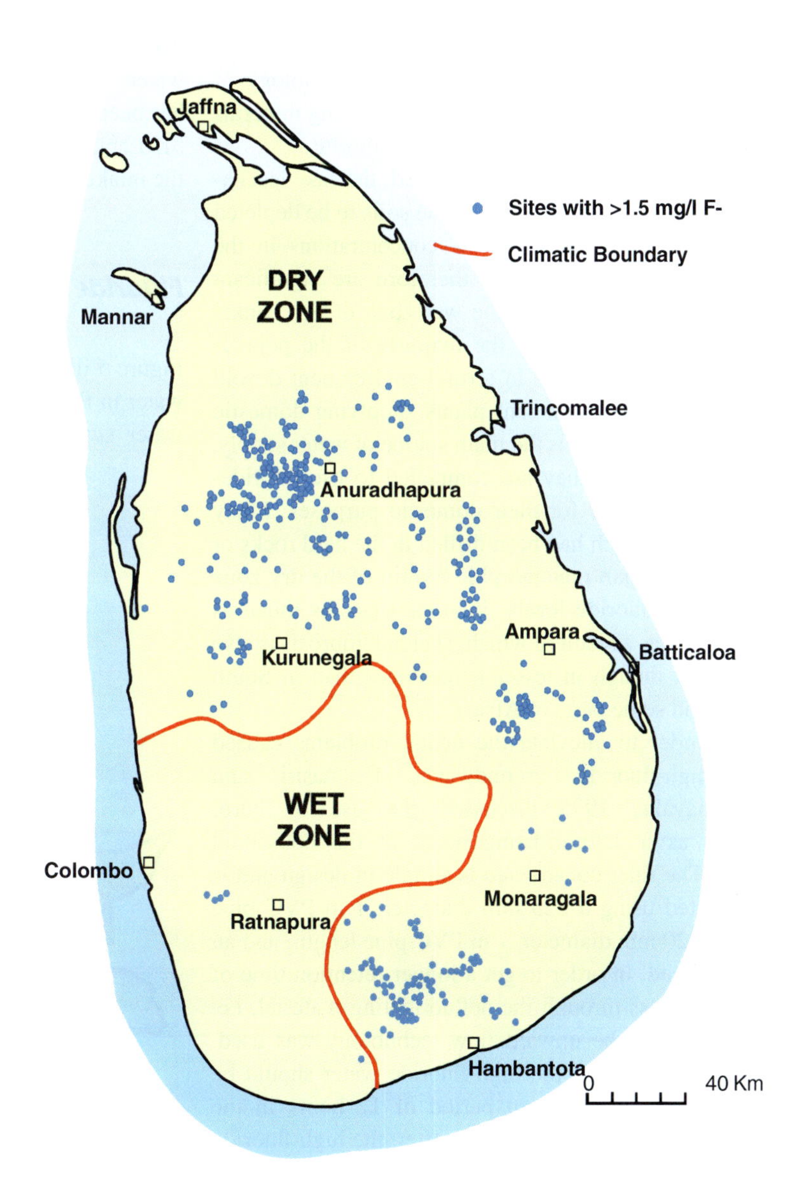

Fig. 3 Distribution of fluoride in groundwater of Sri Lanka (after Dissanayake and Weerasooriya, 1986)

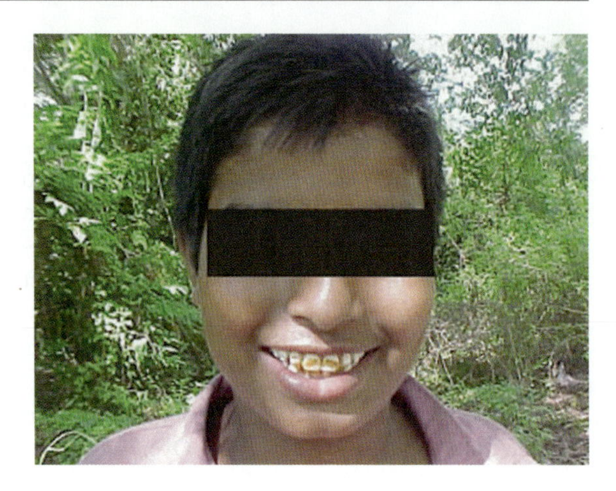

Fig. 4 A young boy affected by dental fluorosis which is common in dry zone of Sri Lanka (photo by Rohana Chandrajith)

(Warnakulasuriya et al., 1992). Precambrian metamorphic rocks constitute about 90% of the lithology of Sri Lanka, and these contain fluoride-bearing minerals such as hornblende, mica, topaz, and fluorite.

In the wet zone, on the other hand, the easy leachability of the fluoride ion causes the soils to be depleted of the ion with resulting low concentrations in the groundwater. Dental caries therefore are significant in the children living in the wet zone of Sri Lanka. In view of the fact that the majority of the population in Sri Lanka lives in a rural environment devoid of central water treatment plants supplying domestic water, groundwater is the main source of water supply. In the dry zone, they are compelled to use fluoride-rich groundwater for their domestic purposes. Many deep wells which had been drilled in the hard rocks of the Precambrian metamorphic terrain of the dry zone have high fluoride levels (in some cases as much as 10 mg/L), as a result of which, skeletal fluorosis is also prevalent though in lesser percentages than in South India and some parts of Africa.

In order to alleviate the dental problems caused by high-fluoride groundwater, Padmasiri and Dissanayake, 1995 discussed the use of burnt bricks as a defluoridating agent at the household level. The filter constructed is simple in design and is fabricated using a 225 mm diameter, 1 m PVC pipe length; 20 mm diameter, 1 m PVC pipe length; and an elbow bend. In order to get a longer retention time of water to pass through the defluoridating material, i.e. burnt bricks, the upward flow technology was used. In using this filter, the high-fluoride water should be retained for a minimum period of 12 hours in the defluoridator at the start. Thereafter, the high-fluoride water is fed into the column and the defluoridated water comes from the outlet. This simple technique has been quite successfully used in the remote areas of Sri Lanka affected by dental fluorosis resulting from the intake of high-fluoride water.

Fluoride in Drinking Water in Pakistan

Figure 5 illustrates the fluoride distribution in groundwater in Pakistan (Khan et al., 2002). Analysis of 987 water supplies showed that they are predominantly

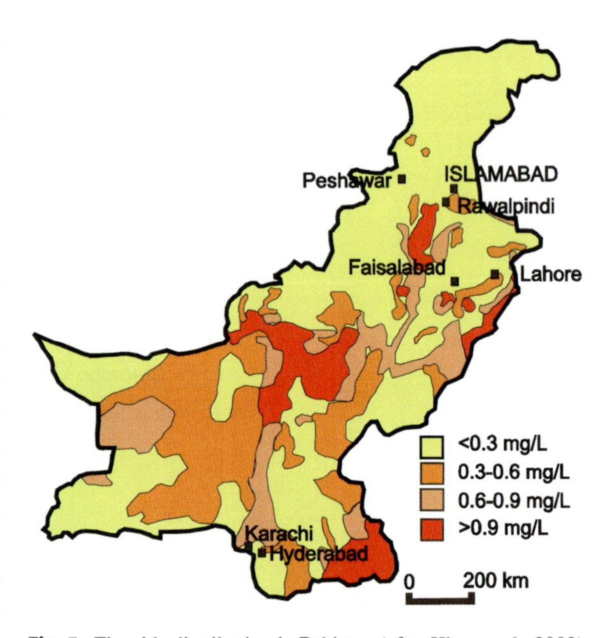

Fig. 5 Fluoride distribution in Pakistan (after Khan et al., 2002)

low in fluoride content, 84% containing less than 0.7 mg/L fluoride. The majority (64%) of the drinking water sources of Pakistan had fluoride levels less than 0.3 mg/L, while 20% ranged from 0.3 to 0.7 mg/L. A range of 0.7–1.00 mg/L fluoride was observed in 6% of the samples and another 6% had 1.00–2.00 mg/L. Fluoride in the range of 2.00–3.00 mg/L was found in 1.5% of the samples while 2% had fluoride levels higher than 3.00 mg/L.

Determining the optimal levels of fluoride in drinking water in hot dry climates of the tropical belt has some inherent difficulties. Yet, the community as a whole requires a knowledge of the most appropriate conditions, in view of the fact that dental diseases are a major health concern.

The areas in the Indian subcontinent and neighbouring regions, as discussed in this chapter, have extremely large populations with nearly all of them having a very low rate of pipe-borne water supply, generally 20–30%. Such large populations depending on groundwater and surface water mainly for their drinking water supplies are therefore highly vulnerable to any mineral excesses, deficiencies and imbalances in water, fluoride being one of the best examples.

The case of Pakistan was studied by Khan et al. (2004) and this study clearly proved the need for consideration of a variety of factors needed for the establishment of optimal drinking water standards. Khan et al. (2004) carried out clinical dental examinations of 1,020 school children aged 12 years, in 19 cities of Pakistan (Table 4). Correlations between concentrations of water fluoride, caries and fluorosis levels were studied by analysing the data of fluoride.

Based on the observation of the trend lines for fluorosis and dental caries (Fig. 6), it was shown that at 0.35 mg/L fluoride in water, the maximum reduction in dental caries was observed while the level of fluorosis was 10%. Only mild degrees of fluorosis were obtained at 0.35 mg/L level fluoride. Khan et al. (2002) also recommended a value of 0.39 mg/L as the optimal fluoride in drinking water for Pakistan.

This recommendation should be viewed against some background facts, namely 80% of the population has no access to a piped water supply and that

Table 4 Level of dental caries (as DMFT: D = Decayed, M = Missing due to caries, F = Filled, T = Teeth) and fluorosis (Dean's index) for 12-year-olds, according to the varying concentrations of fluoride in drinking water of 19 cities in Pakistan (Khan et al., 2004)

City	Fluoride level (mg/L)	DMFT	Dean's index (%)				Total fluorosis (%)
			Very mild	Mild	Moderate	Severe	
Mirpur Khas	0.03	2.20	0	0	0	0	0
Islamabad	0.07	1.86	7	3	0	0	10
Jehlum	0.10	0.64	0	0	0	0	0
Sialkot	0.14	1.05	3	3	0	0	6
Karachi	0.15	1.08	10	10	0	0	20
Lahore	0.15	1.07	3	10	0	0	13
Peshawar	0.15	1.02	3	0	0	0	3
Hyderabad	0.16	0.97	3	0	0	0	3
Faisalabad	0.20	0.78	12	3	0	0	15
Rahimyar Khan	0.22	0.76	3	3	0	0	6
Gujranwala	0.29	0.51	0	0	0	0	0
Sukkur	0.30	0.46	3	2	0	0	5
Sahiwal	0.33	0.42	7	3	0	0	10
Khanpur	0.40	0.50	7	3	3	0	13
Sammundari	0.53	0.60	12	15	12	3	42
Hasilpur	0.60	0.60	10	12	10	3	35
Khairpur	0.73	0.60	10	20	10	18	58
Quetta	0.91	0.60	7	30	18	10	65
Mianwali	1.40	0.90	3	15	45	30	93
Pakistan (total)		0.90	6	7	5	3	21

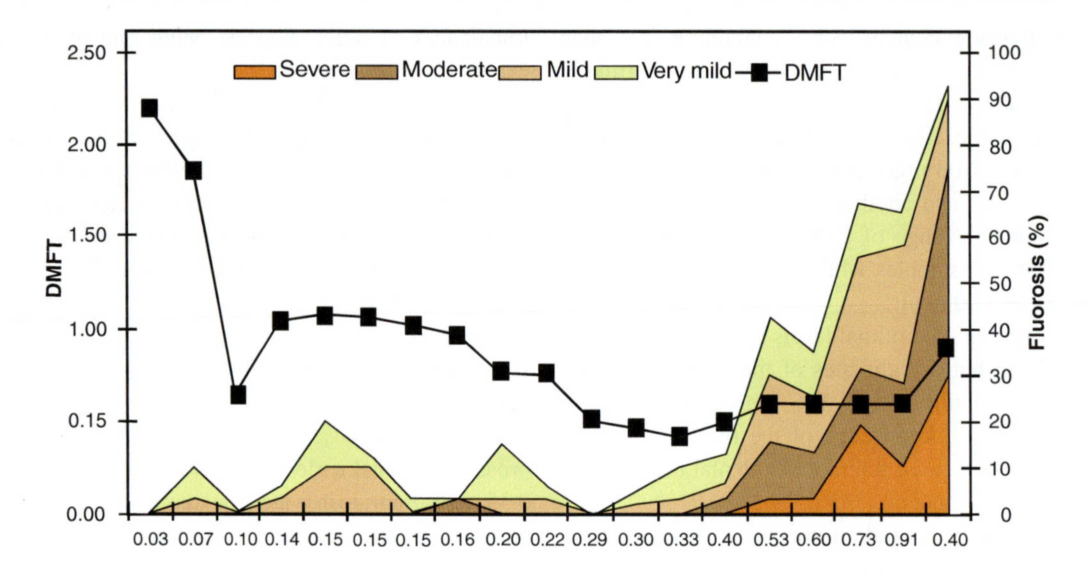

Fig. 6 The relationship of varying degree of dental fluorosis and dental caries among 12-year-olds to the fluoride concentration in drinking water of 19 cities in Pakistan (DMFT refers to permanent teeth: D = decayed, M = missing due to caries, F = filled, T = teeth; The DMFT index is most commonly used indices in epidemiologic surveys of dental caries) (after Khan et al., 2004)

the provision of piped water to the majority of the population cannot be a possibility in the near future. The case of Pakistan illustrates further the importance of each country having its own recommended standard for fluoride in water, some examples of which are shown in Table 5.

The distribution of fluoride in groundwater under varying geological conditions is an interesting aspect of medical geology (Edmunds and Smedley, 2004). For example, the Thar desert located in the southwestern corner of Pakistan in the Sindh Province covers an area of 22,000 km^2. Except for the occurrence of a few granites and other igneous rocks, the desert is mostly a sand dune terrain. Water is available only at depths around 52–93 m and this is brackish and rich in fluoride (Rafique et al., 2008). The high fluoride levels in the water are attributed to the granites which have fluoride-bearing minerals and which leach out fluoride into the groundwater upon weathering in which fluoride contents vary from 0.09 to 11.60 mg/L with a mean of 3.64 mg/L and median of 3.44 mg/L (Rafique et al., 2008).

Table 5 Recommended standard for fluoride in water from different countries

Country	Recommended F level (mg/L)	Reference
Austria	0.7–1.00	Nell and Sperr (1994)
Botswana	<0.5	Zietsman (1991)
Sudan	<0.5	Ibrahim et al. (1995)
Senegal	0.6	Brouwer et al. (1988)
Sri Lanka	0.6	Warnakulasuriya et al. (1992)
Chile	0.5–0.6	Locker (1999)
Western India	0.34	Chandra et al. (1980)
WHO value	0.50–1.50	WHO (1994)

It was observed that based on WHO values for fluoride in drinking water, 88.4% (n = 108) out of a total of 122 groundwater samples were above the value of 1.5 mg/L. With fluoride levels reaching levels as high as 11.63 mg/L (Mau area) and 8.23 mg/L (Pabe jo Tar area) and 9.05 mg/L (Mithi Town) even skeletal fluorosis is prevalent.

Rafique et al. (2008) also comment on the extreme and arid climatic conditions prevalent and the high water consumption. Thus, as shown in the work of Khan et al. (2004), the recommended value for fluoride needs to be lowered drastically, bearing in mind the other factors such as inputs of other essential elements and dietary habits.

Fluoride in India

Fluorosis is a countrywide problem in India and it has reached alarming proportions in 17 states (Fig. 1) (Susheela, 1999); these are

- States having 50–100% districts affected – Andhra Pradesh, Tamil Nadu, Uttar Pradesh, Gujarat and Rajasthan.
- States having 30–50% districts affected – Bihar, Haryana, Karnataka, Maharashtra, Madhya Pradesh, Punjab, Haryana, Orissa and West Bengal.
- States having <30% districts affected – Jammu and Kashmir, Delhi and Kerala.

Fluorosis was first reported from India by Shortt et al. (1937) and since then there have been a number of reports of fluorosis from different parts of the country (Reddy 1985; Reddy et al., 2004). Even today fluorosis persists as a disease without proper treatment and cure. Endemic fluorosis known as genu valgum is common in parts of Andhra Pradesh and Tamil Nadu.

Fluorspar (fluorite), apatite (fluorapatite) and phosphorite are generally the main minerals contributing to the presence of fluoride in soil and water. Several thermal springs are known to contain very high fluoride levels ranging from 10 to 17 mg/L. Secondary dispersion of fluoride is known in soil, river and groundwaters, rain and atmosphere. The secondary dispersions correlate favourably with primary occurrences.

Fluorspar deposits in India are of three types: (i) those associated with the Precambrian granites – gneisses, pegmatites, quartz veins, mostly in prominent shear zones as in Tamil Nadu, Andhra Pradesh, Madhya Pradesh, Gujarat, Rajasthan and Bihar; (ii) those associated with the carbonatites as in Gujarat (Early Tertiary) in the Narmada tectonic lineaments; and (iii) sedimentary digenetic types as in Ramanwara Rewa (Proterozoic–Early Cambrian) in Madhya Pradesh (Kariyanna, 1987).

The major apatite deposits in India are associated with Singhbhum shear zones in Bihar and Eastern Ghats, orogenic belt, in Vishakhapatnam, Andhra Pradesh. The second type is associated with carbonatites as in Sattur, Tamil Nadu, in prominent shear zones and the third type is with the lamprophyre intrusives (late Cretaceous – Early Tertiary) in the lower Gondwana coal fields in Bihar and along the Damodar fault/graben in West Bengal.

The phosphorite occurrences are in Uttar Pradesh, Himachal Pradesh, Jammu and Kashmir, Rajasthan, in the northwest, and Tamil Nadu, in the south, besides a few minor occurrences in other states. Low-grade phosphorite and guano of recent origin occur in the Laccadive–Amindivi group of Islands. The fluorine content of the phosphorite varies from 0.21 to 1.65% (Kariyanna, 1987).

Fluoride-Related Diseases in India

Fluorosis is of four types, viz.: (a) dental fluorosis, (b) skeletal fluorosis, (c) osteoporosis or genu valgum and (d) dental caries.

(a) Dental fluorosis: The teeth lose their shiny appearance and chalky white patches develop. This is known as "mottled enamel" and is an early sign of dental fluorosis. White patches later become yellow and sometimes brown or black. In several cases loss of enamel is accompanied by pitting which gives the tooth corroded appearance. Mottling is best seen on incisors of the upper jaw.

(b) Skeletal fluorosis: In older people, the disease affects the bones and ligaments. Early symptoms are tingling sensation in legs and feet followed by pain and stiffness of the back, later, joints of both limbs and limitation of neck movements. Early detection of the disease may escape till radiological help is sought. Radiological changes are quite characteristic and include formation of new bones (exostotis) and calcification of tendons and ligaments membranes. Skeletal fluorosis has been reported to be a public health problem of considerable magnitude in several parts of Andhra Pradesh, Haryana, Karnataka, Kerala, Punjab, Rajasthan and Tamil Nadu (Susheela, 1999).

(c) Genu valgum: Recently scientists working at the National Institute of Nutrition, Hyderabad, found a new form of fluorosis characterized by genu valgum (knock knee) in South India, particularly in Andhra Pradesh and Tamil Nadu. The prevalence of genu valgum was as high as 3–4% especially in Nalgonda and Prakasam districts of Andhra Pradesh and Pollachi area, Coimbatore district of Tamil Nadu. Circumstantial evidence strongly suggests that building of dams in and around soil rich in fluoride contents may have a role to play in the etiopathogenesis of endemic genu valgum in fluorosis affected communities (Reddy, 1985).

(d) Dental caries: Fluoride levels below 0.5 mg/L are usually associated with high prevalence of dental caries.

Treatment of Fluorosis: Possible Measures

The study of fluorosis has assumed great importance in a developing country like India. The element fluorine has been used as the therapeutic agent in dental caries, atherosclerosis and osteoporosis in human beings, but because of its profound affinity for skeletal tissues "fluorosis" results if taken in excess amounts (Reddy, 1985). An attempt has been made by Rao et al. (1974) to treat human beings suffering from endemic fluorosis, using serpentine which is a silicate of magnesium and capable of taking up colossal amounts of fluoride over a wide range of pH in aqueous solutions. Oral as well as intravenous administration of $Mg(OH)_2$ has also been suggested (Reddy et al., 1974). According to Rao and Murthy (1974), the mechanism by which serpentine and $Mg(OH)_2$ operate is not clear but suggestions are offered for the role of $Mg(OH)_2$ in complexing the free fluoride ions. Thus the possible utility of other defluoridating agents such as aluminium hydroxide and ferric hydroxide in the treatment of fluorosis is also indicated (Meenakshi and Maheswari, 2006).

Gupta and Gupta (2004) indicate that fluorosis could be reversed, at least in children, by a therapeutic regimen (calcium, vitamin C and vitamin D) which is cheap and easily available. Treatments take a lot of time and also have a limited application for wide scale use. Taking precautionary early measures is the only effective way for prevention of fluorosis. For fluorosis, change or modification of the water supply is the only long-term practical solution. Digging deep bore wells has been shown to provide lower, but not appropriate fluoride contents. Defluoridation of water can also be performed by adding activated alumina or by the Nalgonda technique (alkalinization and chlorination followed by adding aluminium salts). However, these methods, though effective, require strong motivation and cooperation of local population and administration (Singh and Maheswari, 2001).

Meenakshi et al. (2004) have proposed nano filtration as a cost- effective deflouridation measure, which can be applied to field conditions in third world countries. Nano filtration (also called low-pressure reverse osmosis) is a membrane-based process, which can selectively remove small molecules and ions including fluoride and can sterilize water. It is a low-cost technique, is effective and requires a small amount of energy, which can be supplied by a photo-electric cell.

Fluoride affects virtually all the components of bone metabolism. The pathogenesis of skeletal fluorosis is still not clear. The role of micronutrients and macronutrients still needs to be well defined. Skeletal fluorosis is symptomatic only in later stages of the disease. Presently, no effective treatment is available and provision of fluoride-free water seems to be the only solution. Several deflouridation techniques are available but has not attained wide acceptance for want of public cooperation and bureaucratic apathy. Creating awareness among the public about the safe drinking water and toxicity of fluoride is therefore, very much required for fluorosis prevention.

Crippling deformities characteristic of vitamin D deficiency were also associated with the villagers consuming water containing high level of fluoride (Hari Kumar et al., 2004). The unique features indicate that children suffered from these deformities, while the adults suffered from stiffness of the spine. Environmental changes in recent years suggest that very high levels of fluorine in water could lead to secondary vitamin D deficiency. It is possible that deficiency of several nutrients may modify and aggravate the toxic effects of chronic fluoride intoxication, while playing a part in the causation of these deformities. Supplementation of the entire population of the affected village with calcium tablets (400 mg/day for the children aged between 1 and 9 years; 600 mg/day from 10 to 15 years, 500 mg/day from 16 to 18 years and 1,000 mg/day in case of pregnant and lactating women) may be used as a remedy. Vitamin D may also be given as an empirical treatment (Gupta and Godbole, 2004).

Medical Geology of Iodine Deficiency Disorders

Iodine is a vitally important trace element present in soil, water, and vegetables/cereals, whose deficiency is usually a harbinger of the disease known as endemic goitre. Endemic goitre normally does not occur when the adult iodine intake ranges above 0.075 mg/day level. Ideally, the daily requirement of iodine is 0.14 mg for an adult man and 0.10 mg for an adult woman. Growing children and pregnant or lactating woman may need more. Iodine is essential for the synthesis of the thyroid hormones, thyroxin and triiodothyronine.

Endemic goitre (Fig. 7) is a health hazard to both human and animals, commonly due to lack of dietary iodine and environmental deficiency. It has a crippling effect upon the affected communities and leads to high incidence of endemic cretinism in the severely affected areas. Endemic cretinism represents developmental abnormalities including decreased intellectual potential.

It is generally observed that except for marine rocks, which are rich in iodine, most rocks of continental origin are conspicuously deficient in iodine and have concentrations ranging from 0.2 to 0.8 mg/kg and in basic rocks, it ranges from 0.5 to 0.8 mg/L and acid volcanic rocks have an average of 0.4 mg/kg. Metamorphic rocks (schist) and certain sedimentary rocks (clay) contain slightly higher concentrations of iodine (1–2 mg/kg). However, sedimentary rocks of marine origin contain extremely high concentrations (100–1,000 mg/kg) of iodine (Vinogradov, 1959). Sands, sandy glacial deposits being more porous and permeable and limestone have low contents of iodine (0.2–0.3 mg/kg).

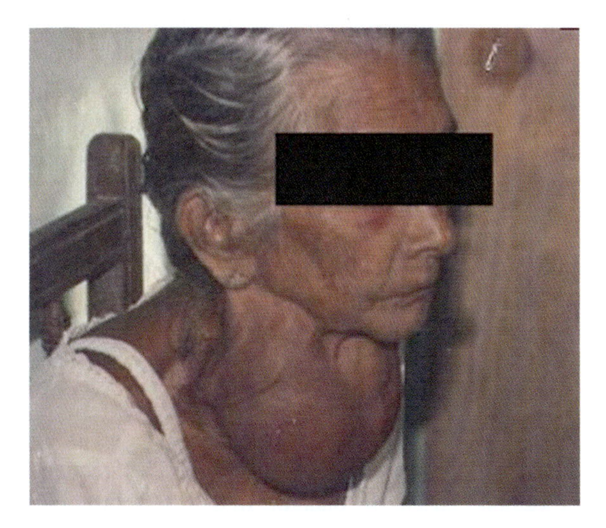

Fig. 7 Typical case of endemic goitre (photo by Dr. Rohana Chandrajith)

Water is an important medium for the iodine supply to humans and animals with hardness of water playing a vital role in the disease proliferation. Goitre is commonly associated with limestone terrains and in areas yielding high calcium. Rains, water logging, and biological activities within the soils also contribute iodine. The distribution of iodine in the soil profile increases with enhanced organic matter and humus accumulation. The soils of maritime regions have higher iodine contents, the principal source being atmospheric iodine coming from the evaporation of seawater.

Iodine Deficiency Disorders in Sri Lanka

As shown by Dissanayake and Chandrajith (2007), the geochemistry of iodine and its bioavailability influence the prevalence of iodine deficiency disorders (IDDs) and geological and mineralogical factors play a key role in this process. Among the factors noted were

(a) intense rainfall
(b) high temperature and their marked diurnal variations
(c) very high rates of weathering and leaching of rocks and soils
(d) clay–organic interactions
(e) greater fractionation of elements in the physical environment

Poverty and poor dietary habits were considered to be the main non-geological factors that influence IDD.

As illustrated in Fig. 8, the endemic goitre belt in Sri Lanka is more influenced by the climate which influences the heavy leaching of soils. Lateritic types of soils are abundant in the wet zone of Sri Lanka. The groundwater fluctuations are quite marked and any iodine present in the already leached soils is removed by the groundwater.

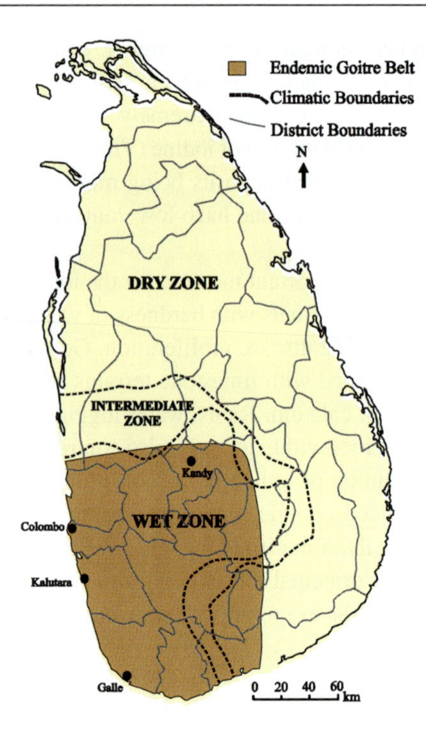

Fig. 8 Endemic goitre region of Sri Lanka

Iodine Deficiency Disorders in Pakistan

Pakistan is another country in the South Asian region where IDDs are prevalent. It has been shown that about 20 million people live in iodine-deficient areas of Pakistan and of these 9 million have shown symptoms pertaining to IDD including mental retardation (Khan, 2008). In the early 1990s, an "Iodine Deficiency Project" was jointly carried out by the Government of Pakistan, UNICEF and the Population Services International (PSI). The study showed that in 1993, an estimated 70% of the population was at risk with iodine deficiency.

More recently, Elahi et al. (2004) studied the status of iodine-deficiency in Lahore, Pakistan, by estimation of neonatal cord serum thyrotropin (CST). They analysed the blood of 1,295 neonates randomly selected from 25,000 live births during a 5-year period from 1998 to 2002. The results showed that in areas where there was sufficient iodine, there was an incidence of less than 3% with CST values >10 mlU/L where a frequency of 3–19.9% indicated mild IDD. On the average, 13.5% of the neonates had a CST level of >10 mlU/L indicating mild IDD.

Iodine Deficiency Problems in India

Geological studies reveal that goitre endemicity is more pronounced in the lesser Himalayas where metasedimentary rocks, dolomite, limestone, granites, and crystalline rocks are exposed, have low iodine contents or are completely devoid of it (Fig. 1). The rocks of the Siwalik belt mainly comprising of sandstone and the Tarai region of alluvial plain are covered with sandy soils. It is observed that clay sediments have higher concentrations of iodine (100–1,000 mg/kg) as compared to sandy soil (0.4–5 mg/kg) (Prasad and Srivatsava, 2004). Deficiency in iodine in this area is directly related to the chemical composition of rocks and soil. Distribution of goitre shows a close correlation with the areas and extent of Quaternary glaciations where soils have not yet been sufficiently saturated with post-glacial airborne oceanic iodine.

It was reported that goitre incidence is high in terrains of magnesium limestone sequences and in areas where soil and water are rich in calcium contents (Prasad and Srivatsava, 2004 and the references therein). Geomorphologically, steep gradients in the Himalayan terrain, ranging in elevation from 100 to 1,000 m; between Siwalik belt and alluvial plain elevation ranging from 200 to 1,000 m; promote quick run off of the surface water, with the result that water finds little time to transfer its iodine to the soil and in turn to the groundwater.

Migration of the rivers like Gandak, Ghaghra, etc. has led to the development of extensive, wide flood plains in the alluvial plain, covered with thick piles of sandy sediments and subjected to frequent and severe flooding due to natural geographical locations, eroding top clayey soil of the flood plain that contains requisite trace elements including iodine.

Geoenvironmental Factors Responsible for Goitre Endemicity in India

Flood Menace

The goitre endemic area experiences moderate to high rainfall (119 cm average of 50 years). The frequency of rainfall based on data of Indian Meteorological Department for the period 1901 and 1950 indicates 100 cm in 8 years, 120–130 cm in 7 years, 130–140 cm in 5 years, 160 cm in 3 years and 160–170 cm in 1 year.

Torrential rain of short spells in the Himalayan region and longer spells in the Tarai region and alluvial plains saturate the soil. This oversaturation of the ground results in increased run off and flash floods. Intense glaciations in the higher Himalaya and frequent floods together with sandy soil in the area are responsible for iodine-deficient soils and shallow aquifer water (Prasad and Srivatsava, 2004).

Socioeconomic Conditions

Sandy soil deficient in iodine, irrigated through iodine-deficient dug well water, contributes towards an iodine-deficient food chain. Poverty and faulty diet are also the reasons for the disease. Inhabitants dependent upon locally produced iodine-deficient food are more prone to the disease as compared to the populace who consume food/vegetable transported from goitre free areas.

Endemic goitre is therefore a serious health hazard to both human and animal resources. Its incidence transgresses all social barriers and observations are consistent with the hypothesis that environmental iodine deficiency is the primary cause for its endemicity. During the course of studies between 1981 and 1984 and again in 1988, it was observed that there has been a marked decline in the prevalence rate of goitre from 65% in 1965 to 15% in 1985 and now it is even less than 10% (Prasad and Srivatsava, 2004). This is attributed to (a) besides supply of iodized salt, supply of drinking water from a deeper aquifer (iodine content ranging from 2–10 μg/L) through hand pumps and tube wells and (b) control of flood menace and soil conservation measures like construction of check dams/bunds has reduced soil erosion and run off of surface water which in turn has facilitated conservation of top soil of the flood plain including atmospheric iodine.

Remedial Measures

In order to make the program more effective and successful, long- and short-term measures are suggested:

(i) Drinking water and irrigation may be supplied from deeper aquifers.
(ii) Control of flood-menace and soil erosion by constructing barriers/check dams may be taken up more rigorously for preserving atmospheric iodine to the soil.
(iii) Use of iodized fertilizer in the goitrogenic areas.

These measures will help in replenishing iodine in the system itself. Endemics in most part of the area have been controlled and eliminated by better living standards and high intake of better quality food. Apart from the above measures, supply of iodized salt in the diet to optimum intake of 0.075 mg level per day may be continued.

Selenium Geochemistry

Selenium Geochemistry in Sri Lanka

There had been an extreme scarcity of research on selenium in rocks, water and soils in Sri Lanka. Fordyce et al. (1998) for the first time studied the selenium geochemistry of these geomaterials. In recent years it has been suggested that selenium deficiency may be an important factor in the onset of goitre and other iodine deficiency disorders (IDD). The pioneering work of Fordyce et al. (2000) in Sri Lanka focused on the selenium status of the human population (demonstrated by hair samples from women) from 15 villages. The villages were characterized by low (<10%), moderate (10–25%) and high (>25%) goitre incidence (NIDD, MIDD and HIDD, respectively). Their results from the analysis of 75 soil samples showed that concentrations of total Se and iodine were highest in the HIDD villages. Even though the actual amount of iodine is high, factors such as clay content may have reduced the bioavailability of the iodine and selenium. Fordyce et al. (2000) also showed that a significant proportion of the Sri Lankan female population may be selenium deficient (24, 24 and 40% in the NIDD, MIDD and HIDD villages, respectively), as summarized in the data in Table 6.

Selenium Geochemistry in India

Selenium ranging from 0.85 to 4.52 mg/kg is reported in soils from parts of Hoshiarpur, Nawanshahr and Jalandhar districts of Punjab. The reported seleniferous

Table 6 Summary of iodine and selenium determinations in all sample types in each goitre incidence village group (Abbreviations: NIDD, no/low goitre incidence; MIDD, moderate goitre incidence; HIDD, high goitre incidence; nd, no data) (Fordyce et al., 2000)

Group	Sample type	Min. Se	Max. Se	Mean Se	No.	Min. I	Max. I	Mean I	No.
NIDD	Soil (ng/g)	113	663	226	25	130	10,000	2,260	25
	Rice (ng/g)	6.8	150	42	25	45	58	51	5
	Water(μg/L)	0.06	0.24	0.11	5	53	84	66.5	5
	Hair (ng/g)	104	765	294	25	nd	nd	nd	
MIDD	Soil (ng/g)	310	5,238	875	24	130	6,600	2,008	25
	Rice (ng/g)	0.1	776	55	25	<38	<38	<38	5
	Water(μg/L)	0.06	0.09	0.07	5	3	23.5	5.5	5
	Hair (ng/g)	118	2,652	389	25	nd	nd	nd	
HIDD	Soil (ng/g)	276	3,947	1,124	25	1,000	9,600	3,914	25
	Rice (ng/g)	0.1	127	25	25	<38	<38	<38	5
	Water (μg/L)	0.06	0.09	0.07	5	3.3	20.2	7.02	5
	Hair (ng/g)	111	984	302	25	nd	nd	nd	

area lies in the interfluves area of Satluj and Beas rivers in the proximity of Siwalik Foot Hills. Plants grown on seleniferous soil show high selenium contents. The livestock of the area has been found to be affected by consumption of excess selenium through water, plant, cereals, vegetables and fodders. Diseases like cracks in the hoofs and their gradual detachment, necrosis of tails and premature abortions, etc. are observed (Dhilon and Dhilon, 1991).

The Quaternary sediments in which selenium toxicity has been reported are mostly derived from Siwaliks as well as Lesser- and Higher-Himalayas rock formations. The geology, geomorphology, geochemical environment and tectonic control of the area show an association with the selenium enrichment in the soil (Dhilon and Dhilon, 1991; Prasad and Kar, 2004).

Podoconiosis (Non-filarial Elephantiasis)

It was in 1988 that Ernest Price observed clinical features of swelling and deformity of the legs, very much akin to what otherwise is known is elephantiasis, but associated with enlargement of the draining lymph nodes. He could clearly show that this particular type of clinical features were of non-filarial type and named the cases as "Podoconiosis" or non-filarial elephantiasis (Price, 1988, 1990). He further confirmed that the podoconiosis cases came from many regions in Africa with very similar climatic conditions and soil type (Price and Bailey, 1984). Superimposing the prevalent data for podoconiosis on a geological map of Africa reveals a correlation between alkali basalts and podoconiosis.

In Bhiwapur area of Nagpur district in India, one very frequently comes across people with clinical features of swelling and deformity of legs. The problem is more frequent in the ladies working as farm labourers. A cursory look in the records of Bhiwapur Public Health Clinic (PHC) catering to the health needs of the residents of 72 nearby villages shows that during the year 2002–2003, 864 patients with clinical features of so-called elephantiasis reported to the PHC. On medical examination of these patients, only 388 cases were confirmed as cases of "filaria" with micro-filarial germs detected in their blood samples. For the remaining 476 patients, filaria could not be established and they have been treated as "unexplained cases" as per the medical terminology. However, of late it has been established in various parts of the world that these clinical features are not linked to filaria but are caused by geogenic factors the cases being "podoconiosis". The geogenic factors for the disease are presence of submicron size particles in clayey soil containing aluminosilicate and silica. Such particles enter into the blood through abrasions in the feet and affect the cells of the immune system. The pathogenicity of such particles depends on its enrichment in Ti and Fe content, near surface composition, surface micro-topography, surface charge and pH. Geologically the Bhiwapur area comprises of a thick succession of basic and acid volcanic and metapelites. The basic volcanic and associated tuffs have a very high Ti content besides high Fe. The area of active cultivation is over a flat land developed over the basic volcanic-tuff terrain. The geological set-up of Bhiwapur area thus matches very well with the established zones and case histories affected with "podoconiosis". It is time that the unexplained cases of elephantiasis from Bhiwapur area are clinically tested for "podoconiosis" (Sinha and Saha, 2004).

At present, it is suspected that there could be a large population of victims of podoconiosis in Bhiwapur and surrounding areas. Hence extensive geochemical studies need to be carried out on the soil profile developed in the flat terrain as referred to above, to confirm a positive area for podoconiosis. It is also necessary to carry out in-depth studies as to why the women engaged in paddy and chilly transplantation working in this area suffer more from the "swelled feet" syndrome. It is also a fact that almost all the women carry out their farm-related jobs barefooted and hence more prone to abrasion and cuts which in turn make them more susceptible to podoconiosis.

Natural Radiation

The United Nations Scientific Committee on Effects of Atomic Radiation Report (UNSCEAR, 2000) states that while the natural background radiation is the largest contributor to human exposure, the worldwide average annual effective dose per capita is 2.4 mSv. Radioactivity in natural materials is due to uranium (^{238}U), thorium (^{232}Th) and potassium (^{40}K) present naturally in the geological materials. Besides the nature of the geological materials, external sources and life style also contribute to the naturally available radiation in the environment (Kathren, 1991). The anomalous concentration of radioactive minerals in rocks at certain places may give rise to radiation enhancement, but they are also natural and the people living in such environments are adapted to it over centuries. Dissanayake (2006) in his essay on "Of stones and health–medical geology in Sri Lanka" commented on the seemingly lack of health issues among people living in high background radiation areas (HBRAs) such as Ramsar (Iran), Kerala (India) and Guarapari (Brazil). Cameron (2001) stated that disease can be attributed to excesses of essential trace elements such as potassium, calcium and phosphorus in the diet and disease (e.g. radiation sickness) can be attributed to a massive dose of ionizing radiation absorbed over a short period of time. It is important, however, to distinguish between an acute dose of 3–5 Gy (1 Gray = 100 rads), the LD_{50} for nuclear radiation and a chronic exposure of 260 mGy/yr as reported for Ramsar (Iran), which would lead to an accumulated dose of 3–5 Gy in a mere 12–20 years. The people in Ramsar survive to old age and exhibit none of the symptoms of radiation disease, although these residents are exposed to extraordinary levels (up to 260 mGy/y) of natural radiation (Mortazavi et al., 2003). Ramsar residents showed fewer chromosome aberrations compared to residents in a nearby control area. This implies the existence of some sort of radioadaptive mechanism which repairs the damage done to the body by the ionizing radiation over a period of time (Cameron, 2001). Every one of us, no matter where we live, is subjected to "natural" or "background" radiation and this account for 82% of the total exposure that an average human being receives during a life time (Hall, 2001). Facius (2005) was of the view that the "radiation paradox" (that people living in certain areas with high levels of background radiation do not seem to suffer adverse effects from exposure to radiation) resides entirely in the minds of those who are convinced that any small amount of chronic exposure to ionizing radiation constitutes a health risk, a notion devoid of any empirical corroboration.

In natural surroundings, certain radioactive substances occur in abundances far greater than their normal level of concentration, for example, some vast stretches of coastal India where the beach sands are enriched in placer deposits of "black sands" which carry heavy minerals such as ilmenite, garnet and monazite. Monazite is a highly insoluble rare mineral and the radionuclides in the monazite are from the ^{232}Th series, but there is also some uranium and its progeny ^{226}Ra. The external radiation levels and the black sands range up to 5 mrad/hr (50 mGy/hr), almost 400 times the normal background (Paul et al., 1998).

The monazite deposits of Kerala, in the southwest, and Bhimilipatnam in east India are larger than those in Brazil. The radiation dose from external radiation is on average similar to the doses reported in Brazil, 500–600 m rad/yr (5–6 mGy/hr), but individual doses up to 3,260 m rad/yr (32.6 mGy/yr) have been reported (Mohanty et al., 2004a, b). In India it will be worthwhile to carry out radon surveys in areas over granite/gneiss terrains, or in areas traversed by major lineament and groundwater movement. In some parts of Hyderabad, the granites are enriched in uranium and these areas need to be specifically studied by geochemists and town planners.

There are numerous species of uranium-bearing minerals occurring in nature, viz. pitchblende, uraninite, samarskite and thorianite. These are known to occur in the Singhbhum copper belt, Jharkhand, Udaipur, Bilwara, Alwara, Durgapur and Jhunjhunu district in Rajasthan, Jhansi areas in Uttar Pradesh, Kalsapur in Chikamagalur district and Karnataka. The known monazite occurrence includes Gaya district in Bihar, coastal tracts of Cuttack, Ganjam, Bhimilipatnam in Andhra Pradesh and beach sands of Kerala, Tamil Nadu, Quilon and Kanya Kumari coasts. Mining of radioactive minerals like uranium, monazite creates health hazards due to ionized radiation hazards and causes acute burns, dermatitis, etc. Chronic exposure causes malignancies, congenital defects and also lung cancer due to inhalation of radioactive dust.

The Kanya Kumari coast has very high radiation of more than 1,200 mrem as in Quilon Coast. Thorium, a radioactive mineral producing alpha radiation is found in very high quantities. Due to the high incidences of cancer in the district, an International Cancer Centre was established at Neyyoor Hospital and a recent survey showed that more than 2,800 cases of cancers were reported from Kanya Kumari Coast at Regional Cancer Research Centre, Tiruvanthapuram and about 2,000 cases in Nagercoil Hospital. More intensive studies are in progress on the Kanya Kumari and Quilon coastal population on the relationship of natural radiation and occurrence of cancer (Saxena, 2004).

Collaborative Medical Geology Projects

Several collaborative projects with reputed organizations like Apollo Hospitals, Hyderabad; Nizam Institute of Medical Sciences (NIMS), Hyderabad; National Institute of Nutrition (NIN), Hyderabad; Sanjay Gandhi Postgraduate Institute of Medical Sciences, Lucknow; Andhra Pradesh Panchayathi Raj Department, Hyderabad at present are involved in several medical geology research projects such as (a) kidney diseases in some parts of the rural areas in Andhra Pradesh, (b) toxic elements in fluoride-contaminated waters, (c) reasons for several kidney stone problems in some parts of the rural areas in Lucknow and Gorakhpur districts in the state of Uttar Pradesh and (d) study to collect information on trace elements/essential minerals as well as toxic metals in soil, food and fodder as well as human blood samples..

It has been observed that, when groundwater sources are used for drinking purposes and which are contaminated with high amounts of silica, strontium and uranium, these areas have been acutely affected by various kidney problems. In some villages near Nalgonda and Prakasam districts of Andhra Pradesh, where fluoride levels in drinking waters are near or marginally higher than the permissible limit, the people have serious fluorosis problems in comparison to the people using higher fluoride levels in drinking water. To find out the causative factors scientists and doctors in Hyderabad are carrying out detailed investigations and it was observed that in areas where people have serious fluorosis problems the drinking water also contain appreciable concentrations of nitrate, total dissolved solids (TDS) and toxic elements like Cr, Ni, Pb, Cd, rare earth elements, etc. Their presence may have resulted in the rapid deterioration of kidney functioning thereby causing serious health problems even at lower levels of fluoride contents. The National Institute of Nutrition (NIN), India, is undertaking a pilot project with the collaboration of the Department of Science and Technology to study trace elements/essential minerals as well as toxic metals in soil, food, fodder as well as human blood samples at village level. This information will be used for predicting disease as well as help in prevention.

It has been observed that the number of patients seeking health care for chronic kidney diseases (CKD) in the north central regions of Sri Lanka has shown a marked upwards trend. This is a recent phenomenon and appears to be having an uncertain aetiology and common among the farming community. It was noteworthy that not only the prevalence appeared to be increasing during the 8–10 years, but the disease also appears to have a heavy geographical bias towards the north central dry zone region of the country. The Department of Geology and the Department of Pharmacology of the University of Peradeniya, Sri Lanka, jointly are carrying out research studies pertaining to the occurrence of CKD with uncertain aetiology in the affected districts. Thus population-based studies were carried out in different geographical areas in Sri Lanka to understand the disease profile and to identify the possible risk factors. Based on the studies carried out so far, there is an indication of an involvement of geoenvironmental factors and therefore a multidisciplinary study has been initiated to find the possible causative factors. Population screening and analysing geoenvironmental samples in affected and non-affected regions are currently being carried out. With systematic studies of populations for CKD, in-depth histological, biochemical investigations and site investigations for geoenvironmental analysis, medical scientists and geochemists would be in a better position to contribute towards the origin of the CKD.

Use of Minerals and Metals in Ayurvedic Medicine

The *Ayurvedic* (*Ayu* = life, *Veda* = science, in Sanskrit) system of medicine as practiced in Southeast Asia, mainly in India and Sri Lanka has been in existence since the Vedic period or even earlier. This is the period in the second and first millennia of BCE (Before Common Era) continuing up to the 6th century BCE as based on literary evidence. The *Ayurvedic* system of medicine endeavours to create a balance between the physical, mental and spiritual functions of the human today. The ancient Indians had a deep understanding of both the human body as well as the physical environment and one could therefore say that the Science of Medical Geology was, in a way, well known to these ancient Indians (Rao, 1993).

As far back as the 7th Century BC, minerals, including gems had been used in the treatment of diseases and prior to this, herbal medicines were used amidst prayers and chanting of Mantras (Prakash, 1997). The gems and other minerals found in ancient India has been well documented (Mitra and Sayakhare, 1993). While the beneficial effects of minerals and metals were known to the ancient practitioners of *Ayurvedic* medicine, they were also well aware of their toxic effects when used in excessive concentrations.

The preparations of the minerals and metals to be introduced to the medicines were therefore, carefully carried out using elaborate ancient chemical procedures as noted in the texts. The methodology, preparations and extractions of metals from minerals is described in the ancient test "Rasa Sastra" (Prakash, 1997). From among the metals, mercury, zinc, copper, tin, gold, silver, lead, iron, arsenic and antimony had been used in the medicines.

Prakash (1997) describes the process of manufacturing metals into the digestible "bhasma" as (a) *Satrapatana* (metal extraction); (b) *bhasmikarana* (conversion to non-toxic chemical form); (c) other processes such as conversion to metallic chemicals using sublimation, distillation, etc. The processing of minerals including gems includes (a) selection and control of mineral, (b) *Sodhana* or purification, (c) *Marana* or conversion to non-toxic powder, (d) *Mardan* or preparation of intermediate mixture or paste, (e) *Jarana* or alchemical reactions to high temperature.

Interestingly, substances such as organic juice mixtures, lime, water, urine, oil and herbal extracts were used in the mineral and metal purification stages. Biotite in particular had been extensively used for the extraction of iron, with cow urine being used to quench the biotite. Smelting had been done in a crucible at temperatures 1,450–1,500°C. Likewise, copper had been extracted from chalcopyrite. As described by Jha (1990), the process consists of roasting the sulphides mineral to remove sulphur and conversion of the copper sulphides into copper oxide by the use of lime juice. The production of the gold *bhasma* is also an interesting procedure using organic bile juices to form organometallic compounds (Parish and Cottrill, 1987). The selection of the suitable minerals for the extraction of metals was done by skilled workers using the natural mineral properties of colour, streak, lustre, refraction, etc. Cow dung figured prominently as a fuel for the furnaces used for heating the materials.

Medically, the benefits of the indigenous system of medicine, *Ayurveda*, are well known to millions of people in the Indian subcontinent. Among the diseases treated by *Ayurveda* are acne, allergies, asthma, anxiety, arthritis, chronic fatigue syndrome, colds, colitis, depression, diabetes, flu, gastric ailments, heart diseases, hypertension, problems of immunity, inflammation, insomnia, nervous disorders, obesity, skin problems, and ulcers.

Copper is thought of as being beneficial to the spleen, liver and lymphatic system and is used to treat obesity, oedema and anaemia. Gold is used for the treatment of heart ailments, epilepsy and hysteria and for improving stamina (Lad, 1984). The metal contents in *Ayurvedic* medicines have been studied by Saper et al. (2004) and Dargan et al. (2008) and they warn of the toxicity caused by the excess of metals in the medicines. Shukhla and Jain (2006) discuss these issues and they are of the opinion that even though the actual metal contents may appear to be high, the non-toxic chemical form of the metal in the *Ayurvedic* medicine needs to be emphasized.

Future of Medical Geology in the Indian Subcontinent

Though at present, animal and human health problems related to the anomalous association of toxic elements in water, soils and sediments have been reported in a limited way, their exact inter-relationship needs to be systematically established. The awareness of certain diseases of animal/human/plant life related to toxicity by As, F, Hg, etc. requires a coordinated approach between geological, medical and agronomic fraternities. Geochemical mapping programs on a national scale (as envisaged in recent years) as an aid to mineral exploration, soil fertility assessment, human and animal health, establishing valid environmental baseline data and understanding the chemistry of our environment are expected to guide in the detection of anomalies for various elements. Though the main objective is to identify new targets for mineral search (near surface and subsurface ore bodies) this database should be utilized to clearly delineate anomalies for As, F, Hg, etc. which may prove hazardous for human, animal, or plant life. The modern tools of scientific analysis will help in the detection of various elements at very low levels thus facilitating easy and accurate delineation of geochemical anomalies. Filtering of the anomaly maps to generate toxic element distribution patterns will go a long way to guide in this important aspect of medical geology.

Toxic element maps will help in delineating regional geological and human factors, which can then be related to prevalent health problems as well as anticipated diseases in different areas by integrating the information collected by geologists and health experts. A coordinated effort by geoscientists and medical scientists can lead to the identification of the causative factors for health hazards and their geological locales (soil, surface and groundwater, alluvium, older sediments, etc.). This can then help in mitigating the health hazards by scientific use of preventive measures, biogenetic and chemical treatments and any other modern techniques. This alone will reduce the likely dangers being caused or which may be caused to the health of human beings, plants and animals.

In view of the unique geological and climatic features of the Indian subcontinent with the associated health problems medical geology will prove to be an invaluable asset in all future health programs.

References

Acharyya SK, Chakravarti P, Lahiri S, Raymahashay BC, Guha S, and Bhowmik A (1999a) Arsenic poisoning in the Ganga delta. Nature 401: 545.

Acharyya SK, Chakravarti P, Lahiri S, Mukherjee (1999b) Ganga-Bhagirathi delta development and arsenic toxicity in groundwater. International Conference on Man and Environment Silver Jubilee Celebration, Center for Study of Man and Environment, Nov 99, Calcutta, 178–192.

Acharyya SK, Lahiri S, Raymahashay BC, Bhowmik A (2000) Arsenic toxicity of groundwater in parts of the Bengal basin in India and Bangladesh: The role of Quaternary stratigraphy and Holocene sea-level fluctuation. Environ Geol 39(10):1127–1137.

Acharyaa SK, Ashyiya ID, Pandey Y, Lahiri S, Khangan VW, Sarkar SK (2001) Arsenic Contamination in Groundwater in Parts of Ambagarh Chowki – Korse Kohri belt (Dongargarh – Kotri Rift zone) Chattisgarh. GSI Sp Pub. No 65 (I): 8–18.

Bhattacharya P, Tandukar N, Neku A, Valero AA, Mukherjee AB, Jacks G(2003) Geogenic arsenic in groundwaters from Terai Alluvial Plain of Nepal. J Phys IV France 107: 173–176.

British Geological Survey (2001) Arsenic Contamination of Groundwater in Bangladesh. BGS/DFID Technical Report WC/00/19, Keyworth, UK.

Brouwer ID, DeBruin A, Backer-Dirks O, Hautvast JGAJ (1988) Unsuitability of World Health Organisation guidelines for fluoride concentrations in drinking water in Senegal. Lancet 1:223–225.

Cameron J (2001) Is radiation an essential trace energy? Forum on Phys Soc 30(4):1–4.

Chakrabarti P (2004) Arsenic and mercury pollution- A geomedical perspective in the Indian context. Proceedings of the Workshop on Medical Geology IGCP-454 held at Nagpur, India, 3–4 Feb.2004, GSI Sp. Pub. No 83, pp. 20–28.

Chakraborty AK, Saha KC (1987) Arsenical dermatosis from tubewell water in West Bengal. Indian J Med Res 85: 326–334.

Chandra S, Sharma R, Thergaonkar VP, Chaturvedi SK (1980) Determination of optimal fluoride concentration in drinking water in an area in India with dental fluorosis. Community Dent Oral Epidemiol 8:92–96.

Dargan PI, Gawarammana IB, Archer JRH, House IM, Shaw D, Wood DM (2008) Heavy metal poisoning from Ayurvedic traditional medicines, an emerging problem? Int J Environ Health 2(3/4):463–474.

Dhilon KS, Dhilon SK (1991) Selenium toxicity in soils, plants and animals in some parts of Punjab, India. Int J Environ Studies 37:15–24.

Dissanayake C (2006) Of stones and health: Medical geology in Sri Lanka. Science 309:883–885.

Dissanayake CB, Chandrajith R (2007) Medical Geology in Sri Lanka. Environ Geochem Health 29(2):155–162.

Dissanayake CB, Weerasooriya SVR (1986) The Hydrogeochemical Atlas of Sri Lanka (103 pp). Colombo: Natural Resources Energy and Science Authority of Sri Lanka.

Edmunds WM, Smedley PL (2004) Fluoride in natural waters. In O Selinus (ed) Essentials of Medical Geology (pp. 301–329). Amsterdam: Elsevier.

Elahi S, Syed Z, Nagra S (2004) Status of iodine deficiency disorders as estimated by neonatal cord serum thyrotropin in Lahore, Pakistan. Nutrient Research 24:1005–1010.

Facius R (2005) The paradox of radiation's effects. Science 310 (5752):1279.

Fordyce F, Johnson CC, Navaratne URB, Appleton JD, Dissanayake CB (1998) Studies of Selenium Geochemistry and Distribution in Relation to Iodine Deficiency Disorders in Sri Lanka. Tech Report WC/98/28, Overseas Geology Series BGS-UK.

Fordyce F, Johnson CC, Navaratne URB, Appleton JD, Dissanayake CB (2000) Selenium and iodine in soil rice and drinking water in relation to endemic goiter in Sri Lanka. Sci Total Environ 263:127–141.

Fyfe WS, Kronberg BI, Leonardos OH, Olorunfemi N (1983) Global tectonics and agriculture a geochemical perspective. Agric Ecosys Environ 9:383–399.

Gupta S, Godbole MM (2004) Endemic Fluorosis and Its Medical Implications. Proceedings of the Workshop on Medical Geology IGCP-454 held at Nagpur, India, 3–4 Feb. 2004, GSI Sp. Pub. No 83, pp. 41–45.

Gupta AB, Gupta SK (2004) Recent Advances in Fluorosis and Defluoridation with Special Referenc to Rajasthan. Proceedings of the Workshop on Medical Geology IGCP-454 held at Nagpur, India, 3–4 Feb. 2004, GSI Sp. Pub. No 83, pp. 38–40.

Hall JE (2001) Radiation in life, World Nuclear Association, Introduction to Nuclear Energy, Education papers, 1–9 pp.

Hari Kumar R, Khandare AL, Siva Kumar B (2004) Vitamin–D Deficiency in Fluorosis Affected Village in Nawada District of Bihar State. Proceedings of the Workshop on Medical Geology IGCP-454 held at Nagpur, India, 3–4 Feb.2004, GSI Sp Pub No. 83, pp. 177–181.

Huq SMI, Joardar JC, Parvin S, Correll R, Ravi Naidu R (2006) Arsenic Contamination in food-chain: Transfer of arsenic into food materials through groundwater irrigation. J Health Popul Nutr 24(3):305–316.

Ibrahim Y, Affan A, Bjorvatn K (1995) Prevalence of dental fluorosis in Sudanese children from two villages with 0.25 and 2.56 ppm fluoride in the drinking water. Int J Paediatr Dent 5:223–229.

Iqbal SZ (2001) Arsenic contamination in Pakistan. Presentation at a UN-ESCAP meeting, Geology and Health: Solving the Arsenic Crisis in the Asia-Pacific Region; UNESCAP, Bangkok, May 2001

Jha CB (1990) Study of Satvapatana with Special Reference to Abharaka and Makshika. Ph.D. Thesis, B.H University, Varanasi, India.

Karim MM (2000) Arsenic in groundwater and health problems in Bangladesh. Water Research 34(1):304–310.

Kariyanna H (1987) Geological and geochemical environment and causes of fluorosis: Possible treatment – A brief review, Paper presented at National Symposium on the Role of Earth Science in Environment held at IIT, Mumbai, pp. 113–122.

Kariyanna H, Sitaram GS (2007) Endemic diseases of South India-A medical geology perspective. Appl Geochem 9(1):142–149.

Kathren R (1991) Radioactivity in the Environment. USA: Taylor & Francis.

Khan AA, Whelton H, O'Mullane D (2002) A map of natural fluoride in drinking water in Pakistan. Int Dent J:291–297.

Khan AA, Whelton H, O'Mullane D. (2004) Determining the optimal concentration of fluoride in drinking water in Pakistan. Community Dent Oral Epidemiol 32:166–172.

Khan M (2008) Pakistan's Iodine Deficiency Program; http://socialmarketing.wetpaint.com (assessed on 8-03-2009).

Korte NE (1991) Naturally occurring groundwater in the midwestern United States. Env Geol Water Sci 18:137–141.

Lad V (1984) *Ayurveda* – The Science of Healing (175 pp). India: Lotus Press.

Locker D (1999) Benefits and Risks of Water Fluoridation. Toronto: Ontario Ministry of Health.

Maharjan M, Watanabe C, Ahmad A, Ohtsuka R (2005) Arsenic contamination in drinking water and skin manifestations in lowland Nepal: the first community-based survey. Am J Trop Med Hyg 73(2):477–479.

Mahmood SN, Naeem S, Siddiqui I, Khan FA (1998) Metal contamination in groundwater of Korangi Indutrial Area, Karachi. J Chem Soc Pakistan 20:125–131.

Meenakhi, Maheswari, RC, Jain SK, Gupta A, (2004) Use of membrane technique for potable water production. Desalination 170(2):105–112.

Meenakshi, Maheswari RC (2006) Fluoride in drinking water and its removal. J Hazardous Materials B137:456–463.

Mitra S, Sayakhare K (1993) Metals and Minerals in Ancient India. Souvenir Volume of First National Workshop on 'Rasa Sastra'.

Mohanty AK, Sengupta D, Das SK, Saha SK, Vans KV (2004a) Natural radioactivity and radiation exposure in the high background area at Chhatrapur beach placer deposit of Orissa India. J Environ Radioact 75:15–33.

Mohanty AK, Sengupta D, Das SK, Vijayan V, Saha SK (2004b) Natural radioactivity in the newly discovered high background radiation area on the eastern coast of Orissa India. Radiat Meas 38:156–165.

Mortazavi SMJ, Ghiassi-Nejad M, Ikushima T, Assaie R, Heidary A, Varzegar R, Zakeri F, Asghari K, Esmaili A (2003) Are the inhabitants of high background radiation areas of Ramsar more radioresistant? Iran J Radiol June: 37–43.

NASC (2003) Nepal's Interim Arsenic Policy Preparation Report, Report prepared for Nepal Arsenic Steering Committee. Nepal: DWSS.

Nell A, Sperr W (1994) Fluoridgehaltuntersuchung des Trinkwassers in Osterreich 1993 (Analysis of the fluoride content of drinking water in Austria 1993). Wien Klin, Wochenschr 106:608–614 (abstract).

Nickson RT, McArthur JM, Ravenscroft P, Burgess WG, Ahmed KM (2000) Mechanism of arsenic release to groundwater,

Bangladesh and West Bengal. J Appl Geochem 15(4): 403–413.

Padmasiri JP, Dissanayake CB (1995) A simple defluoridator for removing excess fluorides from fluoride-rich drinking water. Int J Environ Health Res 5:153–160.

Panthi SR, Sharma S, Mishra AK (2006) Recent status of arsenic contamination in groundwater of Nepal-a review. Katmandu University J Sci, Engineering Technol II(1): 1–11.

Parish RV, Cottrill SM (1987) Medicinal gold compounds. Gold Bulletin 20:1–12.

Paul AC, Pillai PMB, Haridasan PP, Radhakrishnan S, Krishnamony S (1998) Population exposure to airborne thorium at the high natural radiation areas in India. J Environ Radioact 40(3):251–259.

Prakash B (1997) Use of metals in Ayurvedic Medicine. Indian Journal of History of Science 32 (1): 1–27

Prasad M, Srivatsava VC (2004) Goitre Endemicity in Deoria and Gonda Districts of Uttar Pradesh and its Geological Linkage, Proceedings of the Workshop on Medical Geology IGCP-454 held at Nagpur, India, 3–4 Feb. 2004, GSI Sp Pub No. 83, pp. 139–143.

Prasad S, Kar SK (2004) Geological control of selenium concentration in the soil of Punjab plain, India, Proceedings of the Workshop on Medical Geology IGCP-454 held at Nagpur, India, 3–4 Feb.2004, GSI Sp Pub No. 83: pp. 306–313.

Price EW (1988) Non-filarial elephantiasis-confirmed as geochemical disease, and renamed podoconiosis. Ethiop Med J 26:151–153.

Price EW (1990) Podoconiosis Non-Filarial Elephantiasis (144 pp). Oxford: Oxford University Press.

Price EW, Bailey D (1984) Environmental factors in the aetiology of endemic elephantiasis of the lower legs in tropical Africa. Tropical Geogr Med 36:1–5.

Rafique T, Naseem S, Bhanger MI, Usmani TH (2008) Fluoride ion contamination in the groundwater of Mithi sub-district, the Thar desert, Pakistan. Environ Geol 56(2): 317–326.

Rao SR, Murty KJR, Murty TVSD, Reddy SS (1974) Cervical Spondylosis in Fluorosis. Proceedings of the Symposium on Fluorosis (Hyderabad) Paper 45, 441–448.

Rao KV, Murty BVSR (1974) On the Possible Role of Defluoridating Agents in Treatment of Fluorosis. Proceedings of the Symposium on Fluorosis (Hyderabad) Paper 49, 471–475.

Rao TS (1993) 'Atharva Veda': An Ulterior Modern Science. Congress on Traditional Sciences and Technologies of India. 8th Nov. Bombay.

Reddy DR (1985) Some observation on fluoride toxicity. Nimhans J 3:79–86.

Reddy DR, Rao CRM, Nayak RK, Prasad BCM (2004) Estimation of Trace Elements in Drinking Water Supplies of Podili, Darsi and Kanigiri Areas of Andhra Pradesh. Proceedings of the Workshop on Medical Geology IGCP-454 held at Nagpur, India, 3–4 Feb 2004, GSI Sp Pub. No. 83, pp. 384–390.

Reddy DR, Satyanarayana K, Khader SA, Narasinga Rao G, Pentaiah P (1974) Trial of Intravenous Magnesium Hydroxide in Fluorosis. Proceedings of the Symposium on fluorosis (Hyderabad) Paper 46, pp. 449–453.

Saper RB, Kales SN, Paquin J (2004) Heavy metal content of Ayurvedic herbal medicine products. J Am Med Assoc 292(23):2868–2873.

Saxena VP (2004) Radioactive Minerals, Radiations and Health Hazards. Proceedings of the Workshop on Medical Geology IGCP-454 held at Nagpur, India, 3–4 Feb 2004, GSI Sp Pub. No. 83, pp. 46–58.

Shortt HE, Mcrobert GR, Barnard TW, Nayar ASM (1937) Endemic fluorosis in Madras presidency. Ind J Med Res 25:553.

Shukhla K, Jain V (2006) Heavy Metals in Ayurvedic Formulations – Safety Issue. The Pharmacological Magazine (Vol. 1, pp. 1–4). Raipur, India: Institute of Pharmacy, Pt. Ravishankar Shukla University.

Singh R, Maheswari RC (2001) Defluoridation of drinking water- a review. Ind J Environ Protec 21(11):983–991.

Sinha M, Saha AK (2004) A possible linkage of non-filarial elephantiasis "Podoconiosis" with geochemical factors, Bhiwapur area, Nagpur district, Maharashtra, Proceedings of the Workshop on Medical Geology IGCP-454 held at Nagpur, India, 3–4 Feb.2004, GSI Sp Pub. No. 83, pp. 391–396.

Smedley, PL (2003) Arsenic in groundwater – South and East Asia. In AH Welch, KG Stollenwerk (eds) Arsenic in Ground Water: Geochemistry and Occurrence (pp. 179–209). Boston, MA: Kluwer Academic Publishers.

Smedley PL (2005) Arsenic occurrence in groundwater in South and East Asia. In K Kemper, K Minatullah (eds) Towards a More Effective Operational Response (pp. 20–98). Washington, DC: World Bank.

Smedley PL, Kinniburgh DG (2002) A review of the source, behaviour and distribution of arsenic in natural waters. Appl Geochem 17:517–568.

Susheela AK (1999) Fluorosis management program in India. Curr Sci 77(10):1250–1256.

UNICEF (2008) Arsenic Mitigation in Bangladesh, UNICEF, 4 pp.

UNSCEAR (2000) Source and Effects of Ionizing Radiation. New York: United Nations Scientific Committee on the effect of atomic radiation, United Nations.

van Geen A, Zheng Y, Versteeg R, Stute M, Horneman A, Dhar R, Steckler M, Gelman A, Small C, Ahsan H, Graziano JH, Hussain I, Ahmed KM (2003) Spatial variability of arsenic in 6000 tube wells in a 25 km^2 area of Bangladesh. Water Resources Res 39:1140.

Vinogradov AP (1959) The Geochemistry of Rare and Dispersed Chemical Elements in Soil (2nd ed.). (Eng. Translation). London: Chapman Hall.

Warnakulasuriya KA, Balasuriya S, Perera PA, Peiris LC (1992) Determining optimal concentrations of fluoride in drinking water for hot, dry climates – a case study in Sri Lanka. Community Dent Oral Epidemiol 20:364–367.

WHO (1994) Expert Committee on Oral Health Status and Fluoride Use. Fluorides and Oral Health. Technical Report Series, 846 pp.

World Bank (2005) Towards a more effective operational response arsenic contamination of groundwater in South and East Asian Countries. Report No. 31303 Vol. II:219 pp.

Zietsman S (1991) Spatial variation of fluorosis and fluoride content of water in an endemic area in Bophuthatswana. J Dent Assoc S Afr 46:11–15.

Medical Geology in Africa

T.C. Davies

Abstract A large body of evidence points to significant health effects resulting from our interactions with the physical environment and we continue to recognise connections between geological materials and processes and human and animal disease. In Africa, these relationships have been observed for many years, but only recently have any real attempts been made to formalise their study. Africa is a continent with a diverse geography, characterised by a range of altitudes, a peculiar hydrological network created in part by the formation of the Great Rift Valley on the eastern flank and arid lands typified by the Sahara and the Mega Kalahari. Volcanic activity accompanying rifting and formation of most of the highlands and mountains has released various trace elements, mostly above background levels, into the environment. A unique distribution pattern of these elements has developed in more recent geological times, following pronounced separation due to extreme tropical conditions of weathering, leaching and eluviation. It is therefore possible to delineate large areas of the continent containing element deficiencies or toxicities, which are closely related to the local geology and/or geographical location. In a region where rural communities are still largely dependent on water and food sources that are locally derived, the above setting provides an attractive opportunity for studying the influence of geochemical factors on the distribution of diseases in man and animals. This pursuit constitutes a large part of the study of the rapidly emerging science of "medical geology".

According to this definition, the influence of the indoor environment, for example in factories and offices, thus falls outside the scope of medical geology and comes within the area of occupational medicine. To attain completeness in the present review, however, industrially derived exposure to known toxic elements (originating from mining or ore processing, such as are contained in mineral dust, for instance) is also briefly considered. Of the elements for which there are proven or suspected direct causal relationships with man's health, significant data exist for fluorine, iodine, silicon, arsenic and certain trace metals (both micronutrients and potentially harmful elements). The reader is presented with a summary of available information including current hypotheses and an illustration of why improved understanding of these relationships, based on interdisciplinary studies, would lead to better diagnoses and therapy.

T.C. Davies (✉)
Department of Mining and Environmental Geology, University of Venda, Thohoyandou 0950, Limpopo Province, Republic of South Africa
e-mail: daviestheo@hotmail.com

Keywords Africa · Geochemical factors · Diseases · Diagnoses · Therapy

Introduction

Medical geology is a rapidly emerging discipline that examines links between geological materials and processes and the incidence and spatial/temporal distribution of human (and other animal) diseases. In Africa health problems arising from our interactions with the natural environment have been observed for many years. Indeed, many of the leading causes of death in Africa, as in other developing regions of the world, could be linked to factors of the environment (Table 1). Pervasive dust from geogenic sources causes respiratory infections that include asthma; water pollution

O. Selinus et al. (eds.), *Medical Geology*, International Year of Planet Earth,
DOI 10.1007/978-90-481-3430-4_8, © Springer Science+Business Media B.V. 2010

Table 1 Top ten causes of death in low-income countries in 2004

Causes	Deaths in million	Percentage of deaths
Lower respiratory infections	2.94	11.2
Coronary heart disease	2.47	9.4
Diarrhoeal diseases	1.18	6.9
HIV/AIDS	1.51	5.7
Stroke and other cerebrovascular diseases	1.48	5.6
Chronic obstructive pulmonary disease	0.94	3.6
Tuberculosis	0.91	3.5
Neonatal infections	0.90	3.4
Malaria	0.86	3.3
Prematurity and low birth weight	0.84	3.2

Source: WHO (2008)

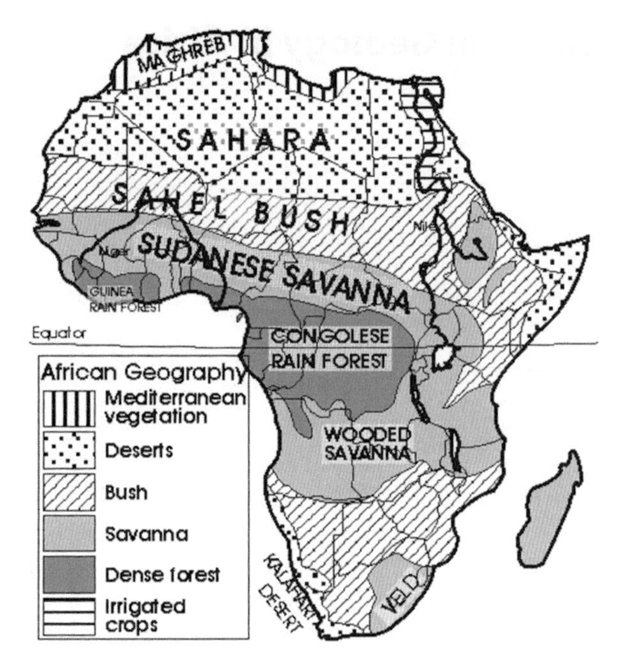

Fig. 1 Africa vegetation overview. (*Source*: Xenakis (2004))

causes diarrhoeal diseases and other water-related illnesses; water hardness has been implicated in some forms of cardiovascular disorders (e.g. Leary, 1985; Derry et al., 1990). The crust of Africa has undergone widespread episodes and styles of ancient mineralisation dating as far back as the Archaean. Substantial quantities of elements, both nutritional and toxic, that were involved in ore forming processes, have been released into the environment, either by later geological processes (e.g. weathering, leaching and eluviation) or by mining and ore-processing operations, with sometimes severe public health consequences.

Again, volcanic activities (both ancient and modern) in tectonically active regions of the continent continue to impact the lives of geographically marginal populations, e.g. the Lake Nyos disaster of 1986 (Lanigan, 1989), the Mount Nyiragongo eruption of 2003 (Allard et al., 2003). Geologically related health problems on the continent are also more keenly felt due to the added stress of factors such as poverty and malnutrition.

Additionally, while most of the population of the developed countries have diets that include food sources from geographically diverse regions, in African countries most of the population live close to the land, relying on locally produced food and water. Large tracts of cultivated land in Africa are arid, semi-arid or lack essential trace elements for healthy plant growth (Fig. 1). In these settings, therefore, the probability of detecting relationships between the geological environment and certain diseases is greatly increased.

Although there is now a burgeoning interest in this study, research on medical geology in Africa is still considered to be in its infancy. A significant body of knowledge exists for the halogens (fluorine and iodine), arsenic, silicon and a small number of trace metals (Table 2), but here too, data interpretation and identification of meaningful correlations between geochemistry and epidemiology are hampered by the use of non-multidisciplinary approaches and lack of high-quality assurance data for pertinent elements in environmental samples (soils, plants, natural waters, etc). Nor are there enough trained personnel for these types of analyses.

To draw tangible correlations, we need input from a battery of professionals including physicians, nutritionists, geochemists, botanists, biochemists, atmospheric chemists and others. We need to improve our medical record-keeping, sources from which reliable epidemiological data about incidence, prevalence and trends in disease occurrence can be extracted. We need a more enhanced analytical capacity that would enable the contents of nutritional and potentially harmful elements (PHE) to be measured at the very low concentrations needed for making tangible correlations between geology and environmental health conditions (Davies, 1996).

Table 2 Some examples of proven or suspected correlations between geochemical factor and environmental disease/public health condition in Africa

Geochemical factor	Possible health effects	Distribution	Mitigation	Reference	Remarks
Fluoride deficiency (less than 0.5 mg l^{-1})	Dental caries and dental cavities may occur; weak bones and teeth	A continent-wide problem	Increased fluoride intake can be achieved by controlled direct additions to drinking water or to specific foods	Shike (2006); WHO (2004)	Adding fluorides to soils that produce food crops is ineffective because fluorides in the soil are inactivated and do not usually find their way into the food chain
Fluoride toxicity (greater than 1.5 mg l^{-1})	Mottling of teeth and dental fluorosis may occur; association with skeletal fluorosis at higher concentrations	Populations living within the African rift zones and associated volcanic centres (Ethiopia, Kenya, Tanzania, etc); areas of crystalline bedrock (e.g. Ghana, parts of Tanzania and Malawi), and some sedimentary basins with high-fluoride groundwaters (e.g. Senegal and parts of North Africa)	Defluoridation, for which there are a number of simple and readily available techniques, e.g. the Nalgonda technique	Davies (2008); Edmunds (2008)	–
Iodine deficiency	Iodine deficiency disorders (IDD), namely goitre and its sequelae	These disorders are found in almost every country right across the continent	Mainly fortification of table salt by addition of potassium iodate and use of oil	WHO (2007)	A noteworthy achievement of WHO is the recent reduction of IDD in African countries that have implemented the universal salt iodisation programme
High arsenic concentrations in domestic water supply	Skin disorders and malignancy; internal cancers	Populations living in the vicinity of metal sulphide mines (e.g. South Africa, Zimbabwe) and those consuming arsenic-rich groundwaters from mineralised Proterozoic basement rocks (e.g. Burkina Faso)	Methods for removal of arsenic from contaminated water by use of activated alumina, iron oxide adsorption, membrane and ion exchange; chelation therapy	James et al. (2005); Smedley et al. (2007); WHO (2004)	The new maximum acceptable concentration of arsenic in drinking water is 10 $\mu g\, l^{-1}$ (WHO 2004)

Table 2 (continued)

Geochemical factor	Possible health effects	Distribution	Mitigation	Reference	Remarks
Selenium deficiency in soils and food crops	IDD and high diffusion rate of HIV-1 in southern Africa	IDD incidence is continent-wide; HIV-1 incidence highest in southern Africa	Selenium supplementation; research on-going on the most effective means of supplementation	Taylor et al. (1997); Utiger (1998)	–
Other micronutrient deficiencies, e.g. iron, zinc, selenium, phosphorous, potassium, calcium and manganese	Possible examples are Mseleni joint disease (MJD) and geophagy	The only documented cases of MJD are found in Maputaland in South Africa; geophagy is practiced continent-wide	MJD is treated with non-steroid-containing drugs and anti-inflammatory medication; sometimes a total hip replacement is needed	Ceruti et al. (2003); Abrahams (2005)	–
Mineral dust	(i) Asbestosis	Populations in and around asbestos mining and processing plants, mainly in southern Africa, e.g. Swaziland and Zimbabwe (chrysotile); South Africa (amphibole fibre)	–	Mcculloch (2003)	Most asbestos mines have now closed, but the latency period for asbestosis can be up to 40 years
	(ii) Silicosis; can predispose to tuberculosis	Found among miners and mining communities in countries where mining and processing of ore deposits associated with free crystalline silica take place, e.g. gold and gemstones (Ghana and South Africa); diatomite (Algeria, Ethiopia, Kenya and South Africa)	–	Churchyard et al. (2004)	The occupational exposure limit (OEL) for silica dust of 0.1 mg m^{-3} seems not to be protective against silicosis, as a large burden of silicosis among older black South African gold miners is likely to worsen
	(iii) Coal worker's pneumoconiosis (CWP)	Coal-producing countries of Africa, of which South Africa is by far the largest	–	Naidoo et al. (2004)	–
High environmental levels of cerium	Endomyocardial fibrosis	The only documented cases are found in Uganda	–	Masembe et al. (1999)	–

Table 2 (continued)

Geochemical factor	Possible health effects	Distribution	Mitigation	Reference	Remarks
Fine reddish-brown volcanic soils high in elements such as beryllium and zirconium	Podoconiosis (non-filarial elephantiasis)	Endemic in Africa, especially among populations living in areas of heavy exposure to volcanic clay minerals, such as around the African Rift Valley	The main preventive measure is wearing protective shoes from childhood	Gomes and Silva (2007)	–
High environmental levels of iron taken in as inorganic iron in the diet	Siderosis (iron overload); the deposition of iron occurs mainly in the liver and reticuloendothelial system	Most notable example found among the South African Bantu living in areas of iron-rich soils in Johannesburg	–	Joffe (1964); Wapnick et al. (1971)	The iron-rich soils are thought to have been formed by the metamorphism of ultrabasic rocks by the Johannesburg Dome, a granitic intrusion lying between Johannesburg and Pretoria
Exposure to radon gas and its decay products	Lung cancer in uranium miners and associated communities	Uranium deposits occur in several countries all around the continent, but South Africa, Niger, Namibia and Gabon have the largest resources	Adherence to the code of practice and safety guide for radiation protection and radioactive waste management in mining and mineral processing, including (i) hermetic sealing of galleries; (ii) elimination of infiltrating water; (iii) powerful ventilation and (iv) normal dust prevention techniques	Abdelouas (2006)	–

However it may be, it is evident from the scenario described that geochemists working in Africa are in a unique position to contribute to a better understanding of the role of minerals in diseases. With continued improvement in analytical protocol, characterisation of naturally occurring trace elements and toxic organic compounds in groundwater will help to explain patterns of diseases such as various cancers, cardiovascular diseases and osteoporosis. In partnership with public health professionals, geoscientists working on environmental health problems in Africa will begin to understand the role of earth materials and systems in the spread of infectious diseases such as West Nile virus fever.

This chapter sets out to summarise and synthesise the geomedical information so far available for Africa, as well as emphasise the potential effectiveness of more holistic, multi-disciplinary, multi-element studies in advancing this emergent field.

Historical Development of Medical Geology in Africa

Realisation of the links between the geological environment and disease grew rapidly in Africa since the 1960s, at about the same time that the principles of geochemical exploration began to be incorporated in mineral exploration programmes across the continent. Originally referred to as "geomedicine" in the North, but now referred to as "medical geology" by international consensus (Skinner and Berger, 2003), the foundations of this science were firmly laid down by the Norwegian Academy of Sciences and Letters (NASL). Under the chairmanship of Prof. Jul Låg, NASL convened a symposium in Oslo in May 1978 to provide a reciprocal orientation for and contact between researchers engaged in this field. The symposium rightly concluded that scientists in other countries will, at a later stage, be interested in joining an organised research effort.

Organised efforts aimed at clarifying relationships between geology and health in Africa have triggered two exciting workshops in the last two decades. The first real attempt to coordinate research in this field took place in Nairobi in 1999, bringing together over 60 interdisciplinary scientists from the region, who reached the conclusion that considerable knowledge gaps still exist in the identification and quantification of the mineral and chemical forms of certain nutritional and toxic trace elements entering water and the food chain (Davies and Schlüter, 2002). It was noted by the Nairobi Workshop that natural trace element abundances can vary by up to three orders of magnitude through a combination of geological processes carried to extremes in tropical Africa (weathering, leaching, eluviation), as well as through other environmental factors. Despite the obvious need for detailed knowledge, in whatever country or locality within Africa, up to 80 percent of the continent's land surface is still unknown in terms of relevant geochemical data (Davies, 2008). Perhaps, the first stage in bridging these knowledge gaps is the production of geochemical maps not only of those chemical elements of known biological significance but also of PHEs and those with potential economic value.

Another outcome of this first workshop was the constitution of the "East and Southern Africa Association of Medical Geology" (ESAAMEG) as a branch of the "International Working Group on Medical Geology" (IWGMG), which was itself established in Athens in June 1997. This has now been reconstituted to the "International Medical Geology Association" (IMGA, www.medicalgeology.org).

The second African workshop, "GEOMED 2001", was held in Lusaka, Zambia, in 2001 (Ceruti et al., 2001). A main part of that meeting was a short course and seminar, titled "Metals, Health and the Environment", held under the auspices of the International Geological Correlation Programme (IGCP), Project 454: Medical Geology. An extensive syllabus of about 300 pages was produced, covering many aspects of the subject. Following the short course and seminar was a 1-day workshop on "The Role of Geomedicine in the 21st Century", at which many researchers presented ongoing work in several African countries (Ceruti et al., 2001). A massive stride has been made since, in explaining a number of revelations coming out of these workshops, though admittedly more needs to be done.

The phenomenal growth in interest in medical geology has continued since these two workshops (e g. Finkelman et al., 2007; Selinus et al., 2008), with many geology departments around the continent now concluding plans for the inclusion of the subject in their postgraduate curricula. This is a welcome sign on both sides of what has often been an unbridged chasm between geology and health in Africa.

Current Status of Research on Medical Geology in Africa

This section attempts to summarise the present state of knowledge based on work done and studies that are ongoing in Africa on medical geology. The need to address the types of data needed for drawing up meaningful correlations between patterns of trace element distribution and environmental diseases in humid tropical settings is strongly emphasised throughout.

Nutrient Deficiencies and Excesses

Fluorine

It is now well established that fluoride mineralisation takes place mostly in geologically unstable areas and has been associated with rifting and rise of fluorine from the lower crust or the upper mantle. Accordingly, most of the cases of environmental health conditions involving dietary fluorine imbalance are reported for populations living within the African rift zones and associated volcanic centres (Davies, 2008). Excessive fluorine (mainly in the form of fluoride) is present in parts of the hydrological system and the biosphere, including soils and food crops.

In the East African Rift Valley, for instance, fluoride in groundwater, lakes and rivers is derived mainly from mixing with fluids from hot springs and volcanic gases, which can contain concentrations of several tens to hundreds of milligrams per litre. The highest natural fluoride concentration ever found in water, 2,800 mg l^{-1} (Bakshi, 1974), was in Lake Nakuru in the Kenya Rift Valley.

However, the occurrence of high-fluoride groundwaters (and health problems) has also been reported in areas of crystalline bedrock (especially granite) and sedimentary basins (Edmunds, 2008). Examples of such granitic areas can be found in Ghana, parts of Tanzania and Malawi. Sedimentary basins with high-fluoride groundwaters occur in Senegal and parts of North Africa (Edmunds, 2008).

Fluoride in minor amounts (around 1.3 ppm in drinking water) reduces dental decay and enhances the proper development of the bone (Table 2). A similar level of fluoride intake may also be beneficial to animals. When the amount of fluoride consumed is either too low or much too high, undesirable physiological consequences appear, such as dental caries, mottled staining of the teeth and malformed bone structure in man and animals (Fendall and Grounds, 1965).

Health conditions arising from nutritional fluoride imbalance have been reported from several other African countries, including Ethiopia, Tanzania, Nigeria and Zimbabwe. Though most reported cases of fluorosis are among residents of high-fluoride areas, there exist a significant number of individuals around the Rift Valley zones and other volcanic centres with fluorosis from exposure to relatively low levels of fluoride (Manji et al., 1986). An explanation for this would almost certainly invoke interplay of metabolic and nutritional factors. Essentially, the metabolic argument for the epidemiology of fluorosis is based on the chemistry of the fluoride ion, which largely determines its bioavailability in mammals. But this question, and that of whether fluorosis is the cause or the result of metabolic dysfunction, is still open-ended and could form the basis of research that would greatly benefit from a complete geochemical baseline map of fluorine for Africa.

Iodine

Iodine was the first trace element to be recognised as essential for humans. It is required by adult mammals for proper functioning of the thyroid gland (Laurberg, 2009). Trace amounts of the order of 0.0004 weight percent must be established and maintained throughout the life of the animal as a result of ingesting small quantities of the element from drinking water and various feed inputs.

Volcanic emanations and sea salt are among the most important associations of iodine, but additional sources are formation waters and fluid inclusions. Traverses from the Moroccan coast, for instance, show that strong soil-iodine enrichment is limited to about 100–200 km from the coast (Johnson, 2003). However, some would say that the link between iodine in soils and distance from the coast is not that strong (see e.g. Fuge, 2005).

Africa is characterised by huge differences in elevation, the continent having undergone a complex tectonic episode. The mountainous areas thus constitute a barrier to the transport of iodine to central continental areas. The leaching of iodine originally supplied by volcanic activity, from acid mineralised soils, is also an important determinant of the element's geographical distribution. Thus, the most notorious goitre areas in Africa are the lee sides of high-altitude areas such as the Kerio district in the Kenya Rift Valley, which records the highest goitre prevalence in the region (72 percent of children have goitre; Hanegraaf and McGill, 1974) and where the soil has been depleted by leaching by heavy iodine-poor rain.

Many of the iodine-deficient regions of the world have already been identified and the widespread occurrence of goitre and related conditions firmly established. These are collectively referred to as iodine deficiency disorders (IDD) and have been recognised as serious problems in several areas in Africa. Consequently, much work has been carried out on the aetiology and geographical distribution of these diseases within the region.

Although Africa ranks third among regions of the world most affected by IDD, after the Western Pacific and South-East Asia, the toll of IDD in Africa has been enormous. The estimated population at risk is 220 million, of whom 95 million are goitrous (WHO/UNICEF/ICCIDD, 1993). But, thanks to a combination of efforts over the last two decades, the spectre of IDD in Africa is now diminishing. Many recent articles have already looked at the advances made in the elimination of IDD in Africa (e.g. Kavishe, 1996; Davies, 1994a). As advocated by the 1990 "World Summit for Children", the 1992 "International Conference on the Assistance to the African Child" (ICAAC) and the 1992 "Global Plan of Action of the International Conference on Nutrition" (ICN), the establishment of universal salt iodisation programmes has been by far the leading approach, although in some countries targeted iodised oil and iodisation of water have been used as short-term measures. These efforts have resulted in what is generally considered to be a success story. But, there is another side of the coin!

Expanding the Spectrum of IDD

There is now ample evidence to suggest that iodine supplementation programmes in Africa, however systematic and far-reaching, are unlikely to produce a foolproof solution, as they have done in other parts of the world. Among possible militating factors to salt iodisation programmes are the potential role of "goitrogens", substances present in the African diet (Davies, 1994a) and the development of "thyrotoxicosis", from application of excessive levels of iodine supplementation (Bourdoux et al., 1996).

The Possible Role of Goitrogens

The physiological and biochemical relationships between iodine and goitrogens in relation to thyroid function are still unclear. However, the goitrogenic effect of vegetables such as Brassica (e.g. kale or "sukumawiki" (in Kenya), cabbage, radish, broccoli) has now been well established (Michajlovskij, 1964) and could be due to a depression of the thyroid hormone formation, which causes a decrease in the circulating thyroid hormone. This leads to increased thyroid stimulating hormone (TSH) secretion and thyroid enlargement as a compensating mechanism.

Other goitrogens that may directly or indirectly account for some of the goitre cases in Africa include the univalent complex ions, thiocyanate (SCN^-) and fluoroborate, (BF_4^-), whose ionic sizes are similar to that of iodide; these may competitively inhibit the active transport of iodide both in the thyroid and in the extrathyroidal iodine-concentrating tissues (Davies, 1994a).

Thiocyanate is present in cassava (*Manihot utilissima*), which is consumed in large quantities by several African tribes. The goitrogenic activity of thiocyanate found in cassava has been demonstrated on several occasions, such as on the Idjwi Island, Lake Kivu, Republic of Congo, where cassava is a common food staple (Thilly et al., 1977). It decreases the penetration of iodide into the thyroid (Delange et al., 1973). It is therefore possible that if consumers of cassava or kale are living in marginally iodine-deficient environments, or have congenital biosynthetic defects in their thyroids, these goitrogens may play a role in the development of goitre.

Similarly, the fluoroborate ion is thought to be present in natural waters in parts of the African Rift Valley (Davies, 1994b). This univalent ion has a similar size to that of iodide and its goitrogenic properties have been reported by Langer and Greer (1977). If the importance of the BF_4^- species in quantitative terms can be established, it will have far-reaching implications in the discussion of goitre endemia in the Rift Valley areas of Africa.

Thyrotoxicosis

The term "thyrotoxicosis" refers to the hypermetabolic clinical syndrome resulting from serum elevations in thyroid hormone levels, specifically free thyroxine (T4), triiodothyroxine (T3) or both. Hyperthyroidism is a type of thyrotoxicosis in which accelerated thyroid hormone biosynthesis and secretion by the thyroid gland produce thyrotoxicosis. However, hyperthyroidism and thyrotoxicosis are not synonymous. In Africa, most cases of thyrotoxicosis can be ascribed to hyperthyroidism from the application of too-high levels of iodine supplementation.

In Kivu, Zaire (now the Democratic Republic of Congo, DRC), clinical providers noted thyrotoxicosis in 191 individuals with goitre after introduction of iodised salt (as much as 148 mg kg^{-1} iodine) in an area with previously severe iodine deficiency (Bourdoux et al., 1996). In these individuals, the urinary iodine level increased from a median of 0.13 to 1.89 mcmol l^{-1}, but only a few of these exhibited any minor clinical signs. Sometimes, the individuals needed antithyroid drug treatment. Iodine-induced thyrotoxicosis had also occurred in Zimbabwe in a moderate to severe iodine-deficient area after introduction of iodised salt (30–90 mg iodine kg^{-1}) (Bourdoux et al., 1996). It is often difficult to clinically recognise the sometimes life-threatening thyrotoxicosis.

No attentive consideration of local nutritional conditions, eating habits and medical assistance is applied when deciding the increase in the salt iodisation level. Iodine prophylaxis programmes in Africa sometimes apply a salt iodisation level 5–10 times greater than that applied in Europe, i.e. 20–40 mg kg^{-1} (EuSalt, 2009). The effects of high levels of iodine supplementation have never been tested and are likely unsafe for some segments of the treated population. International agencies may also need to implement monitoring procedures as well as reconsider the salt iodisation level recommended for African countries.

Selenium as a Co-factor in IDD

Selenium, as selenocysteine, is an integral component of two important enzymes – glutathione peroxidase and type I 5′-iodothyronine deiodinase – that are present in many tissues, including the thyroid gland where they play an indispensable role in thyroid hormone synthesis. Glutathione peroxidase catalyses the breakdown of hydrogen peroxide, thereby protecting against oxidative damage. Iodothyronine deiodinase catalyses the deiodination of thyroxine to triiodothyronine, a more potent thyroid hormone. Iodine is not an integral component of any enzyme, but it is an integral component of thyroxine and triiodothyronine. Concurrent selenium and iodine deficiencies may result in a modified thyroid hormone metabolism in animals (Arthur et al., 1987; Arthur et al., 1991; Vanderpas et al., 1990). Selenium and iodine are thus linked biochemically because both are involved in thyroid hormone production.

What then are the clinical consequences of selenium deficiency? Though suspected correlations abound (hypothyroid cretinism, suppressed growth factors in juveniles, increased diffusion rate of HIV-1), conclusive evidence for linking specific disorders with selenium deficiency is lacking for Africa. Although this essential trace element is a highly probable co-factor in myxoedematous cretinism (Goyens et al., 1987; Thilly et al., 1993), the administration of selenium to iodine-deficient and selenium-deficient schoolchildren and myxoedematous cretins in the DRC resulted in a decompensation of the thyroid hormone synthesis that was particularly apparent in the cretin subjects (Thilly et al., 1991; Vanderpas et al., 1993).

Selenium and the Diffusion Rate of HIV-1 in Southern Africa

There are many reasons why dietary selenium may be influencing the diffusion of HIV-1 in southern Africa. Selenium in vitro blocks HIV-1 replication (Hori et al., 1997). It is also clear that selenium plays key roles in the operation of the immune system, as mentioned previously (Section “Selenium as a Co-factor in IDD”). Taylor and his co-workers (Taylor, 1997; Taylor et al., 1997) have demonstrated that HIV-1 encodes for the selenium-containing protein, glutathione peroxidase. This means that the virus cannot be replicated without competing with its host for this element. However, there appears to be a dietary selenium threshold. Above it, the human immune system is effective enough to avoid HIV-1 infection. Below it, the immune system is so weakened by selenium deficiency that on exposure, infection is possible, if not probable.

The Role of Nutrient Deficiency in Endemic Osteoarthritis (Mseleni Joint Disease)

Unusually high incidences of dwarfism and endemic osteoarthritis, termed Mseleni joint disease (MJD), afflict up to 25 percent of the local population of adults living on the sandy coastal plain of Maputaland in South Africa (Ceruti et al., 2003). Maputaland, with an area of about 50 km by 100 km, is located on the northeast corner of South Africa.

MJD is a rare disease which begins with stiffness and pain in the joints and progresses to varying degrees of disability, with some of the afflicted requiring aid in walking and others without the ability to walk at all. The disease is typified by a combination of multiple epiphyseal dysplasia, polyarticular osteoarthritis and protrusion acetabuli, with earlier stages of MJD displaying osteoarthritis and advanced cases typified by severe joint surface irregularity. Medical studies have, until recently, been unable to determine conclusively the aetiology of MJD or the dwarfism, but work at the University of Cape Town (Ceruti et al., 1999) has indicated the possible role of environmental factors, in particular extremely variable concentrations of available nutrient elements, as causative factors. This suggests the possibility of the existence of isolated pockets of land where deficiencies (e.g. available soil calcium, potassium, phosphorus, zinc and copper) may be acute. Plant growth trials in the Mseleni area, using a subtractive element technique, confirmed these deficiencies in maize (*Zea mays*) for phosphorus, potassium, calcium, magnesium, copper and zinc (Ceruti et al., 2003).

Environmental Exposure to Cerium and Endomyocardial Fibrosis

In the late 1980s, medical workers from India noted a strong link between endomyocardial fibrosis (EMF), a fatal coronary heart condition in children throughout the tropics, and enhanced environmental levels of cerium, related to the presence of monazite sands in Kerala province (Smith, 2000).

In a collaborative study, Ugandan scientists from Mulago Hospital, the Institute for Child Health, Makerere University, and colleagues from the British Geological Survey have investigated whether similar environmental exposure could account for the occurrence of endemic EMF in Uganda (Masembe et al., 1999). Case–control studies performed at Mulago Hospital, Kampala, indicate a high incidence of EMF among patients from Mukono and Luwero districts over the past 40 years. Model calculations based on the data collected in this work (Smith et al., 1998) indicate that the most important exposure route is through the ingestion of soil-bound cerium as a result of either the inadvertent oral ingestion of soil or the habitual eating of soil (geophagy).

Is Nutrient Deficiency a Causal Factor for Geophagy?

Geophagy – the involuntary or sometimes the deliberate eating of earth – is extremely common among traditional African societies and has been recognised since ancient times. Among the many groups active in geophagy research in Africa are the School of Physical and Mineral Sciences, University of Limpopo (South Africa), researching under a sub-component of IGCP project 545: "Clays and Clay Minerals of Africa: Geneses and Palaeoenvironmental Considerations, Mineralogy, Geochemistry and Applications" (Ekosse and de Jager, 2008) and the Department of Geology and Mining, University of Jos, Nigeria (Davies et al., 2008a, 2008b). Building on earlier work by the British Geological Survey in Uganda (Smith, 2000), the work of these groups is directed specifically at (a) increasing our understanding of the risks and benefits associated with soil ingestion (deliberate and/or inadvertent); (b) determining the bioavailability of a particular contaminative source, so that this is taken into account during site-specific risk assessment; and (c) determining how the practice can be carried out in a safe way.

The reasons why soils are being deliberately consumed can be difficult to establish, but known causative explanations include the use of soil during famine, as a food detoxifier, as a pharmaceutical and for neuropsychiatric and psychological (comforting) reasons (Abrahams, 2005). It remains a matter of conjecture whether ingestion of soil actually satisfies a nutritional deficiency. Nevertheless, soils do have the potential to supply mineral nutrients, especially iron, where the ingestion of soil can account for a major proportion of the recommended daily intake.

Typical quantities of soil eaten by practising geophagics in Kenya in the order of 20 g per day have been recorded (Smith, 2000). Although the eating of such large quantities of soil increases exposure to essential trace nutrients, it also significantly increases exposure to potentially toxic trace elements and biological pathogens. The former is particularly likely in mineralised areas, associated with mineral extraction, or polluted urban environments, where levels of potentially toxic trace elements are high.

Geogenic Dust

Worries about African population exposed to particulate matter (PM) in ambient air, especially in cities, continue to grow. While the impact of dust from industrial processes on human health is well recognised and widely regulated, less attention has been paid to dust from geogenic sources and to less visually obvious finer particulates. This is probably partly because research is in many respects at an early stage and because of practical difficulties of observing, monitoring and sampling diffuse aerosols in the complexly moving atmospheric column (B. Marker, personal communication).

It is thought that the greater proportion of particulate load in the African urban atmosphere is from geogenic sources (e.g. Davies, 2008), and these may cause more toxicity problems in humans than dust components from other sources. Known sources of atmospheric dust in Africa include the following: (i) Harmattan dust sometimes tinged with radioactive particles (such as dust blown from Niger Republic across northern Nigeria); (ii) mines and quarries; (iii) treatment plants; (iv) solid waste disposal sites; (v) unpaved roads; (vi) disturbed land; (vii) construction sites (viii) sawdust and (ix) particulates from all types of combustion.

Exposed soils are eroded by the wind and wind-borne particles, wherever these are exposed for any length of time in dry conditions, giving rise to aerosols containing dust ranging from very fine particulates to silt grade material. Once in the air-stream, dust can be carried for very long distances before it is deposited on land or in water. For instance, it is now well established that dust from North Africa reaches the Americas and northern Europe depending on weather conditions (B. Marker, personal communication).

Dust also arises from point sources such as erupting volcanoes, fires and events such as atmospheric nuclear tests. Deposits of material that have not been stabilised, for instance, tips and lagoons containing mineral wastes, in dry conditions, also give rise to dust.

Most roads in rural environments in Africa are made of "murram" (untarred or non-tarmac); hence, dust admixed with toxic metal particulates (e.g. lead and arsenic) as well as other pathogens are freely inhaled, causing asthma and other respiratory disorders (Table 1). We also know that long-term exposure to airborne particulate matter for years or decades is associated with elevated total, cardiovascular and infant mortality (e.g. Delfino et al., 2005). Dust storms from the Sahara also come with toxic inorganic particulates and other pathogens, which could stress the respiratory system when inhaled (Policard and Collet, 1952; Prospero, 1999). Therefore, the control of dust and reduction of risks is of importance to society.

Asbestos

Asbestos is the general industrial term encompassing six different natural fibrous silicates. Amosite [grunerite, $(Fe,Mg)_7Si_8O_{22}(OH)_2$], crocidolite [riebeckite, $Na_2(Fe,Mg)_3Fe_2Si_8O_{22}(OH)_2$], tremolite [$Ca_2Mg_5Si_8O_{22}OH_2$], anthophyllite [$(Mg,Fe)_7Si8O_{22}(OH)_2$] and actinolite [$Ca_2(Fe,Mg)_5Si_8O_{22}(OH)_2$], all belonging to the amphibole group, while chrysotile [$(Mg_6(Si_4O_{10})(OH)_8$] is a serpentine.

These minerals were exploited largely in the past century in industrial applications because of their versatile and unique properties. Applications include their use in insulation, pipe lagging, roofing and brake linings. Nowadays, asbestos is associated with its potency to cause asbestosis, a debilitating and often fatal lung disease, and malignancies such as lung cancer and pleural mesothelioma, which may appear several decades after exposure. It is noteworthy that a strong synergistic effect between cigarette smoke and asbestos has been proven for lung cancer but not for mesothelioma (Kane, 1996).

Asbestos refinement and use is being restricted progressively or banned by several countries, including the entire European Union, which banned asbestos use in 2005. Conversely, despite all available evidence on the occupational hazards related to asbestos exposure, asbestos is still being produced and used in African countries as well as a number of other developing countries (Vogel, 2005).

The mining of amphibole asbestos is almost completely confined to South Africa. Crocidolite and amosite, locally abundant and mined asbestos minerals, occur in metamorphosed Precambrian sedimentary strata (banded ironstones) in the Transvaal Supergroup. About 75 percent of the asbestos mineral production is used in the manufacture of asbestos-containing cement pipe.

Silica

Silicosis, the most ancient occupational disease, is not caused by all types of silica, but only by a subset of its crystalline polymorphs, namely quartz, tridymite and cristobalite. The high-pressure forms, coesite and stishovite, are less toxic and may even be inert. Lung cancer (IARC, 1997) and some autoimmune diseases have also been associated with exposure to silica. In spite of a massive body of work since the early 1950s on the pathogenicity of silica-related diseases, there is no consensus on the mechanism of action of crystalline silica particles at the molecular level.

Silicon is a component of diatomite (diatomaceous earth), a deposit mined and processed in Algeria, Ethiopia, Kenya and South Africa. It is used in filters, as a filler and sometimes as a mild abrasive. The dust from the processing and handling of diatomite is very dangerous to humans and can cause serious and fatal illness, not only in those with heavy occupational exposure but also in people who live near the industrial site or even simply in the same household with an exposed worker. During calcining in particular, highly reactive, free crystalline silica (tridymite, cristobalite) is produced, resulting in silicosis. The illnesses include nodular pulmonary fibrosis (silicosis), secondary heart disease, lung cancer and bronchitis. The general term to describe lung disease due to inhalation of inorganic dusts is "pneumoconiosis", defined as the accumulation of dust in the lungs and the tissue reactions to its presence.

Other Mineral Dusts

Other mineral dusts that may cause adverse pulmonary responses but have not been officially reported as human carcinogens include coal (coal workers' pneumoconiosis (CWP), reviewed recently in Huang et al., 2006), kaolinite, talc (talcosis) and iron oxides (siderosis). The aetiology of diseases that follow the inhalation of such minerals is often not well defined, because epidemiological studies are complicated by co-exposure to other minerals such as quartz and asbestos (Ross et al., 1993).

Volcanoes produce nano- to micron-sized particles, including crystalline silica, that may have toxic potential. The particles of volcanic ash are fragmented upon eruption, producing fresh, reactive surfaces that are not oxidised – a rarity in the natural environment.

Potentially Harmful Elements (PHEs) from Mining Operations

Arsenic

Large tracts of the African Precambrian (especially the greenstone belts) contain metal sulphide mineralisation and have released considerable quantities of arsenic into aquifers and the surficial environment not only through the natural processes of weathering, erosion and transportation, but also through mining and processing of metal sulphide ores. For instance, arsenic in groundwater from mineralised Proterozoic basement rocks of Burkina Faso has directly affected the quality of drinking water (Smedley et al., 2007). Gold mining (e.g. in Ghana and South Africa), metal sulphide mining (e.g. in South Africa and Zimbabwe) and coal mining (e.g. in South Africa) have also released considerable amounts of arsenic into waterways mainly through acid mine drainage (AMD).

Arsenic, a metalloid, is not an abundant element in the Earth's crust, only approximately 2 mg kg^{-1}. It has a strong association with sulphide-bearing minerals. Oxidative weathering, such as is common in humid tropical environments in Africa, results in formation of arsenite [As(III)] and arsenate [As(IV)]. pH and Eh are the dominant controls on mobility, speciation and solubility. Sorption by oxides and hydroxides of iron, aluminium and manganese limits arsenic concentration and controls its mobility and bioaccessibility. Recent studies suggest that some organic forms are highly toxic. However, in spite of decades of research on arsenic chemistry, detailed mechanisms of the element's toxicity are still not yet well understood.

Human exposure occurs primarily through ingestion of water and food, although airborne particulates containing arsenic may be locally important. Environmental arsenic exposure is a causal factor in human carcinogenesis and numerous non-cancer health disorders. Chronic exposure symptoms most commonly include skin disorders such as hyperkeratosis, hyperpigmentation, peripheral arteriosclerosis ("black foot disease") and skin malignancies; other ailments like internal cancers may also occur. All of these disorders are known in populations consuming water with 100–1,000 $\mu g\ l^{-1}$ arsenic (Reeder et al., 2006). The present WHO guideline value for arsenic is 10 $\mu g\ l^{-1}$ (WHO, 2004) although many countries continue to use the pre-1993 value of 50 $\mu g\ l^{-1}$ (WHO, 1993) as their national standard. In reality, however, the level at which toxicity symptoms appear may depend on other factors as well, such as the presence of the element selenium, co-exposure to which is held to reduce toxicity of arsenic.

Arsenic in Mine Waters

A major concern confronting land management agencies in Africa is the generation of arsenic-rich acid drainage from both abandoned and active metalliferous mines and from naturally occurring sulphide-bearing geological units.

Extreme examples with drainage acidities (below pH 1.0) are relatively rare, but AMD with a pH of 0.52 was encountered at Iron Duke mine near Mazowe, Zimbabwe, in February 1994 during an investigation of the environmental geochemistry of mine waters in the Harare, Shamva and Midlands greenstone belts of Zimbabwe. Arsenic values of up to 72 mg l^{-1}, recorded at Iron Duke, constitute the highest dissolved arsenic concentration published to date for mine waters worldwide (Williams and Smith, 2000). This site provides a valuable opportunity for studies of acutely acid mine waters and may assist in validating and refining models for such systems.

Mine waste remediation, extraction techniques, including buffering methods to control AMD (e.g. Breward and Williams, 1994), and modelling have been undertaken by many and are amply described in the literature.

Mercury Exposure in Small-Scale Gold Panning

Mercury is released into the environment during artisanal gold mining in a variety of ways. When it is used to amalgamate gold, some escape directly into water bodies as elemental mercury droplets or as coatings of mercury adsorbed onto sediment grains. The mercury that forms the amalgam with gold is emitted into the atmosphere when the amalgam is heated – if a fume hood or retort is used. Naturally occurring mercury in soils and sediments that are eroded by sluicing and dredging becomes remobilised and bioavailable in receiving waters (Telmer and Veiga, 2008). Where a combination of cyanide and mercury is used, the formation of water-soluble cyano-mercuric complexes enhances transport and bioavailability. However, the fate of mercury in any of these processes as well as the interactions of cyanide and mercury are still poorly understood (Telmer et al., 2006).

Increasing health problems associated with mercury contamination, as a result of its use as an amalgam agent by small-scale gold miners, have been reported in many African countries during the past two decades (Ghana (Tschakert and Singha, 2007); Kenya (Ogola et al., 2002); South Africa (Eisler, 2003); Zimbabwe (Bose-O'Reilly et al., 2008); etc). Approximately 2 tons of mercury is used for each ton of gold recovered. Less efficient operations may use even more. With global gold production by small-scale miners now at a rate of thousands of tons per annum, the design and implementation of appropriate methods for monitoring mercury contamination in areas of alluvial gold mining are urgently required. The growth of independent small-scale gold mining provides employment and generates wealth, but economic gains are frequently offset by health degradation and the destruction of vital industries such as fishing.

In Africa, problems due to mercury and arsenic contamination associated with gold mining are a result of lack of environmental concern (or lack of enforcement of regulations) and, in some cases, inappropriate mining and operating practices. Geochemical surveys prior to operation and close monitoring during mining coupled with good planning and working procedures can minimise contamination, while remediation schemes can also be devised and monitored by the application of sound geochemical science (Breward and Williams, 1994).

Radiation and Radon Gas

Radon is a colourless, radioactive gas formed naturally by the radioactive decay of uranium that occurs in all rocks and soils. There is a direct link between the levels of radon generated at the surface and the underlying rocks and soils. Radon generated in shales and granite can reach the surface via faults and fractures.

Variations in radon levels are evident between different parts of the African continent and beyond. Country average for Egypt is given as 9 Bq m^{-3} (UNSCEAR, 2000) but can be much higher in the other uranium-producing countries.

South Africa has Africa's largest identified uranium resources (currently estimated at over 241,000 metric tons; Mudd and Diesendorf, 2008), followed by Niger, Namibia and Gabon. Uranium has also been found in Algeria, Botswana, the Central African Republic, Chad, the DRC, Egypt, Guinea, Lesotho, Madagascar, Malawi, Mali, Mauritania, Morocco, Nigeria, Somalia, Tanzania, Togo and Zambia. However, under current economic conditions, it is unlikely that any of these deposits will be fully exploited in the immediate future.

Uranium occurs in a variety of ores, often in association with other minerals such as gold, phosphate and copper, and may be mined by open-cast, underground or in situ leach methods, depending on the circumstances. South Africa produces uranium concentrates as a by-product of gold mining and copper mining and possesses uranium conversion and enrichment facilities. Uranium is chiefly used as a fuel in nuclear reactors for the production of electricity.

The processes involved in radon gas generation and its movement to the surface and into buildings are complex. These include interactions between rocks, soils and groundwaters. It is important to determine the natural levels of radon produced from rocks and soils as these will provide essential data indicating the potential for high levels of radon entering buildings. Transport of radon-containing soil air into a home can be through cracks in solid floors and walls below construction level; through gaps in suspended concrete and timber floors and around service pipes; and through crawl spaces, cavities in walls, construction joints and small cracks or pores in hollow-block walls. The design, construction and ventilation of the home affect indoor radon levels.

The principal basis for present concern about radon and its decay products focuses on exposures resulting from industrial processes – primarily the mining and milling of uranium – that increase the accessibility of radon to the outdoor atmosphere or to indoor environments, leading to the incidence of lung cancer. A large body of epidemiological data have accumulated over several decades relating to studies of the incidence of lung cancer in miners, and risk estimates have been derived from this data (National Academy of Sciences (NAS) 1998), but most of these studies have involved subjects outside Africa.

An ongoing activity is the investigation of health impacts of uranium exploration in the Poli region of northern Cameroon (F. Toteu, personal communication). These deposits, hosted by Late Proterozoic granitoid, are being explored by a private company. No information is yet available on the possible impact that exploration activities may be having on the health of nearby residents, but reports abound of miscarriages in pregnancy and growth defects in children that may be due to uranium exposure.

Similarly, the environmental and health impacts from exploration and mining activities in two uranium prospects in Zambia are currently being assessed (O. Sikazwe, personal communication). One occurrence of uranium mineralisation is in southern Zambia and is located within Karoo sandstones, but here, exploration work is not yet at an advanced stage. The second uranium mineralisation is in north-western Zambia, where uranium is associated with copper; and here, mining is ongoing.

Kaposi's Sarcoma and Podoconiosis

The clinical picture described by Kaposi in 1872 (cited by England, 1961) is now quite familiar. Sarcoma and podoconiosis remained a curiosity of dermatology for many years in Europe and America, until about half a century ago, when it became evident that it was a common tumour of the skin throughout Africa, especially in the moist tropical regions, and comparatively rare in the drier sandy areas. Ninety percent of cases occur in males and the lesions are most commonly found in the extremities, especially on the feet and legs.

Many theories of the aetiology of endemic Kaposi's sarcoma (KS) have been advanced, among which is a geographically determined environmental factor. This nodular condition is believed to arise in the lymphatic endothelium and is associated with chronic lymphoedema. As such, KS bears a superficial resemblance to podoconiosis (non-filarial elephantiasis). The geographical prevalence of both conditions in highland areas of moderate rainfall in proximity to volcanoes suggests a pathogenetic relationship to exposure to volcanic soils. The geographical endemicity of KS in the Congo–Nile watershed (western branch of the East African Rift System) and Nigeria–Cameroon border (Benue Trough and Cameroon Volcanic Line) is particularly striking (Ziegler, 1993).

Podoconiosis too is largely endemic in Africa, especially among populations living in areas of heavy exposure to volcanic clay minerals (Price, 1990). These iron-oxide-rich kaolinitic soils overlie regions of alkaline basalt associated with volcanism of the major African intercontinental rift systems. Podoconiosis is characterised microscopically by neo-volcanic submicron mineral particles (mainly aluminium, silicon, titanium and iron) stored in the phagolysosomes of

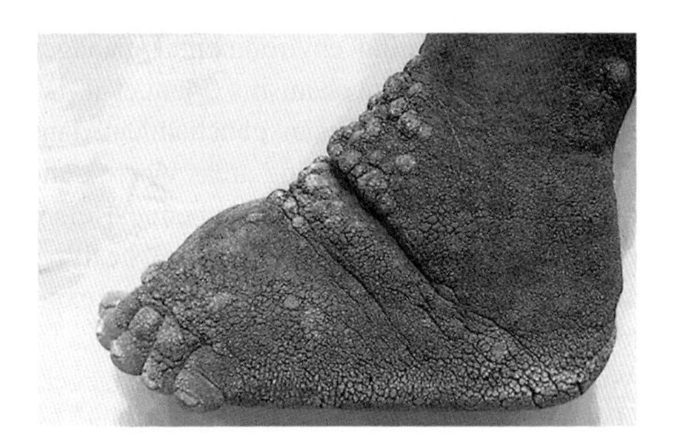

Fig. 2 A view of the lower left leg and foot of a patient with podoconiosis. There is thick diffuse hyperkeratotic swelling confined to the lowest part of the leg, ankle and foot. The calf above the knee was almost normal. (*Source*: Itakura, 1995)

macrophages within lymphoid tissues of the lower limbs. This pathogenesis is consistent with the theory that ultra-fine clay minerals from the soil are absorbed through the feet (Gomes and Silva, 2007). The resulting chronic lymphatic irritation, inflammation and collagenesis cause obstruction and lymphoedema of the affected limb (Fig. 2). The geographical proximity of endemic KS to areas containing volcanic clay minerals, its lympho-endothelial origin, predilection for the feet and legs and its prevalence among rural dwelling peasants and cultivators suggest a common aetiology, namely chronic dermal exposure to volcanic clay minerals.

Trace Elements in Soils and Plants – Implications for Wildlife Nutrition

The soils of Africa are complex. A unique combination of pedogenetic and hydrological factors brings about mineral imbalances that are of consequence to crop, livestock and wildlife development. The soil represents a likely candidate as an environmental reservoir for many reasons, such as its potential role in the transmission of "prion" diseases (Mathiason et al., 2006; Schramm et al., 2006).

In the area of the Kenya Rift Valley around the city of Nakuru (east of Lake Victoria), mineral problems in soils have led to copper deficiency in wheat (Pinkerton, 1967) and copper and cobalt deficiencies in cattle (Hudson, 1944; Howard, 1970). At Lake Nakuru National Park in Kenya, the elevated pH of soils has been linked with the high molybdenum content of several plant species (Maskall and Thornton, 1992). Molybdenum-induced copper deficiency has been reported in other wild animals including Grant's gazelle (*Gazelle granti*) from the Kenya Rift Valley; in this case the plant molybdenum content ranged from 0.5 to 5.6 μg^{-1} (Hedger et al., 1964). Cases of "molybdenosis" (molybdenum toxicity), a disease in ruminants (stiffness of legs, loss of hair) feeding on forage containing more than 10–20 mg kg^{-1} molybdenum, have been reported from Kenya (McDowell, 1992).

Low concentrations of copper and cobalt have also been reported in soils developed on volcanic ash and lake sediments in the Kenya Rift Valley (e.g. Chamberlin, 1959; Nyandat and Ochieng, 1976; Maskall and Thornton, 1991). Ruminants require cobalt in order to synthesise vitamin B12. Cobalt deficiency arises on soils of diverse geological origin and climatic conditions, including coarse volcanic soils around the town of Nakuru in the Kenya Rift Valley, where the disease is referred to by the local name of "nakuruitis". The most appropriate scientific designation for the condition is enzootic marasmus (McDowell, 1992). In this region, sheep and cattle were unable to thrive unless they were periodically moved to "healthy" areas. Ruminants remaining in the deficiency areas stopped eating, became weak and emaciated and usually died. Because diseased animals could be cured by a change to healthy ground, it was rightly assumed that this was a nutritional deficiency disease. Cobalt deficiency in ruminants can be cured or prevented by treatment of the soils or pastures with cobalt-containing fertilisers or by direct oral administration of cobalt to the animal (Underwood, 1981).

Working in an area of north-eastern Zimbabwe, Ndiweni and Finch (1995) have provided evidence to suggest that supplementation of cattle with vitamin E and selenium may optimise resistance to "mastitis" (an inflammation of the udder in cattle, with implications for reproductive health), by enhancing the functions of resident macrophage and lymphocyte populations.

Perhaps, worth mentioning too is the disproportionate growth of body parts observed in juvenile members of a population around the "coltan" (columbite/tantalite) mines in the Republic of Rwanda (W. Pohl, personal communication). A connection with excess of some toxic element from coltan mining getting into the food chain via the soil is suspected, but conclusive evidence is still lacking.

Africa Geochemical Database

For medical geology to succeed in Africa, an obvious first step would be the preparation of a high-quality geochemical database for the entire region, preferably in conformity with guidelines laid down in the "Global Geochemical Baselines Initiative" (Darnley et al., 1995).

At present, complete geochemical databases do not exist for any of the countries of Africa. Some form of conventional geochemical coverage (i.e. samples collected and analysed by various methods in a laboratory) exists for a percentage of the surface area of some countries such as Uganda and South Africa. However, there is considerable variation between data sets because sample media, methods of sample collection and preparation, extraction and analytical techniques and levels of quality control were generally decided according to perceived mineral exploration targets of each country and a desire to minimise costs. As a result of this diversity, numerical values can only be compared within areas where consistent methods and standards have been applied. Variation between data sets with respect to the assemblage of elements also makes it difficult to compare such data sets or use them as a basis for studying processes operating in the region's surficial environment. In addition, few elements of environmental importance have been determined in past survey work. To find ways to overcome these problems, or at least prevent their perpetuation into the future, is one major reason for advocating a comprehensive database (Plant et al., 2000).

Data obtained from a geochemical database for Africa will have a range of environmental applications: they will provide quantitative information on natural levels of the chemical elements over different geological environments in the region, which may be used to assess the effects of past, present and future human influence; they will have an important contribution to make in land use planning issues; the data will also have applications in mineral exploration, resource evaluation, geological mapping and research into the fundamental processes of crustal evolution and ore formation. In addition, such a compilation will facilitate the modelling of environmental systems linked to issues of formulating environmental management and legislative control mechanisms (Plant et al., 2000). Most importantly, an African geochemical database will be invaluable in identifying areas where deficiencies or potentially toxic levels of chemical elements may present risk to human and animal health or to livestock or crops. It can be used for tracing exposure routes for toxicants in soils and groundwater.

Work on the construction of a formal geochemical database for Nigeria started off in 2008, led by the Nigeria Geological Survey Agency in collaboration with the British Geological Survey (C. Johnson, personal communication). The main source of financial support for this project is the World Bank. Reconnaissance surveys in 2008 were followed by a strategic workshop in Kaduna, northern Nigeria, in January 2009. Preparation for detailed drainage and soil surveys is in progress.

Future Perspectives

Rapid development of medical geology followed concretisation and formalisation of the study by the International Medical Geology Association (IMGA) meeting in Uppsala in 2000, when it was agreed by international consensus that the former name of "geomedicine" be replaced by "medical geology", a term that was deemed more likely to gain acceptance by the medical profession. Several short courses have since been organised worldwide and a variety of awareness campaigns mounted to promote the study.

The countries of Africa are also experiencing an unprecedented growth in this field. Because of the continent's unique geo-environmental situation, it is considered likely that the discovery of remarkable correlations between Earth materials and processes and human health will continue into the next decades, as research activities in this emerging area of study proliferate. Modern geo-analytical techniques offer promise of developing innovative solutions to prevent or minimise exposure to potentially deleterious natural materials and geological processes.

Today, an important question lies in how we can map geochemical information and correlate this with environmental epidemiology and disease registry data. Present medical registry records are still inadequate in most African countries, bedevilling efforts in providing evidence required for substantiating suspected correlations. But as efforts in medical record-keeping improve across the continent, research in medical geology would in turn also gather pace, spurning the emergence of further exciting revelations in this field. One of the most challenging problems for the future will be the understanding and characterising of the environmental and natural factors that contribute to emerging and re-emerging infectious diseases.

In the above-stated context, it is thought that Africa will benefit immensely from a complete geochemical map of the entire region for all non-gaseous elements and other chemical parameters in pertinent environmental media, based on agreed guidelines. For now, although some geochemical mapping is being undertaken by geological surveys of some African countries, few are using the internationally agreed standard methods as set forth in the "Global Geochemical Baselines" initiative (Darnley et al., 1995). Countries such as Sierra Leone, Tanzania and Uganda simply lack the analytical infrastructure and funding for such a high-precision and systematic exercise. As such, earlier geochemical surveys show many possible intrinsic differences in data sets for any given element, which hinder or do not justify the assembly of most existing data into broader regional compilations. However, as analytical capacity continues to improve and collaborative links strengthened with high-performance regional laboratories, these gaps will soon begin to disappear.

Monitoring of the levels and compositions of dust emissions is now widely undertaken in relation to industry, for instance within factories, processing plants and mining and quarrying operations (Fanning et al., 2009). However, even in these relatively constrained settings, it has, until recently, been challenging to then discriminate between particles from different sources within mixtures, emissions or deposited dust. This remains an important area for future research.

What do we know about CWP in coal mining regions of Africa (especially South Africa)? Naidoo et al. (2004) have examined radiographic outcomes among South African coal miners, but much more information is needed to understand the bigger picture, such as whether or not some physico-chemical parameters play a role in the observed differences in the prevalence of CWP among the various coal mines in Africa.

Is there a tangible correlation between dust inhalation and the incidence of meningitis in Africa (A. Konare, personal communication)? Satellites are now being used to monitor the movement of huge dust clouds moving across oceans that carry pathogenic microbes, kill coral and cause asthma. However, an important additional step in elucidating the role of natural dust as a contributor of disease would be to facilitate cooperation among medical professionals, geologists, climatologists and microbiologists. This would assist in the characterisation of the properties of geogenic particulates, their dispersal and the toxicological pathways of infectious agents they may transport.

In spite of the large-scale monitoring by remote sensing in tracking the dispersal of dust clouds, it is not generally possible, at present, to investigate the three-dimensional structure and variations within dust clouds to any extent. Here is scope for more work at both the local and regional levels to improve research techniques. Also, few studies have been published on the health effects caused by long-term inhalation of dusts generated in desert areas.

During small-scale gold mining activities in Africa, miners often use a combination of cyanide with mercury to amalgamate the gold. Cyanide dissolves not only gold but also mercury, forming cyano-mercury complexes. These complexes are easily mobilised by rain and often, due to poor containment practices, quickly reach stream waters. It is expected that water-soluble mercury cyanide is either more bioavailable or easier to be biomethylated than elemental mercury (Telmer and Veiga, 2008). This possibility deserves more attention, but indirect evidence collected by the "Global Mercury Project Team" in Zimbabwe, Indonesia and Brazil (Telmer and Veiga, 2008) suggests this is the case.

Conclusion

Africa represents a natural field laboratory for studies in medical geology, since the majority of its people live close to the land, depending on food and drinking water sources in their immediate vicinity. Again, extreme conditions of weathering, leaching and eluviation in a humid, tropical setting have led to a marked redistribution of elements, both nutritional and toxic, in the surficial environment, carving out distinct zones of element excesses and deficiencies. These factors enhance the credibility of correlations drawn between certain environmental diseases and the biochemical, biophysical or geochemical interactions within the bedrock–groundwater–soil–food crop–human/animal tissue continuum.

Drawing up such correlations, however, relies on the availability of very high-quality analytical data for element concentrations and their distribution in environmental samples. The content of certain critical elements (sometimes down to microgram per litre level) needs to be accurately known. But the need for quality assurance still bedevils analytical work in many African geochemical laboratories. There are too few technicians with sufficient expertise for proper installation, maintenance and repair of today's analytical instruments.

One important consequence of the rather poor geochemical database is the fact that many environmental diseases that result from excess or deficiency of certain key elements are poorly diagnosed. The production of a high-precision geochemical map of Africa would be a cost-effective method of indirectly investigating the chemical composition of crops; and rural communities in Africa will offer a particularly valuable opportunity for examining the relationship between geochemistry and health. Such baseline data would aid in the understanding of the hydrological, chemical and biological processes that determine the behaviour of chemical elements in this part of the Earth's surface, in relation to how they may affect the life of man and animals. Significant correlations could then be conveyed to the medical profession, to heighten awareness of the huge potential significance that factors of the geo-environment can have on disease causation and thereby serve to broaden the diagnostic spectrum.

References

Abdelouas A (2006) Uranium tailings: Geochemistry, mineralogy and environmental impact. ELEMENTS 2(6):335–341, Mineralogical Society of America.

Abrahams PW (2005) Geophagy and the involuntary ingestion of soil. In O Selinus, B Alloway, JA Centeno, RB Finkelman, R Fuge, U Lindh, P Smedley (eds) Essentials of Medical Geology (Chapter 17, pp. 435–458). Amsterdam: Elsevier Academic Press.

Allard P, Baxter P, Halbwachs M, Kasareka M, Komorowski JC, Joron JL (2003) The most destructive effusive eruption in modern history: Nyiragongo 2003. Geophys Res Abstr 5:11970–11970.

Arthur JR, Nicol F, Boyne R, Allen KGJ, Hayes JD, Beckett GJ (1987) Old and new roles for selenium. In DD Hemphill (ed) Trace Substances in Environmental Health (Vol. XXI, pp. 487–498). Columbia, MO: University of Missouri Press.

Arthur JR, Nicol F, Beckett G (1991) The role of selenium in thyroid hormone metabolism and effects of selenium deficiency on thyroid hormone and iodine metabolism. Biol Trace Element Res 33:37–43.

Bakshi AK (1974) Dental conditions and dental health. In LC Vogel, AS Müller, RS Odingo, Z Onyango, A de Geus (eds) Health and Disease in Kenya (pp. 519–522). Nairobi, Kenya: East African Literature Bureau.

Bose-O'Reilly S, Lettmeier B, Gothe RM, Beinhoff C, Siebert U, Drasch G (2008) Mercury as a serious health hazard for children in gold mining areas. Environ Res 107(1):89–97.

Bourdoux PP, Ermans AM, Mukalay wa Mukalay A, Filetti S, Vigneri R (1996) Iodine-induced thyrotoxicosis in Kivu, Zaire (now Democratic Republic of the Congo). Lancet 347 (9000):552–553.

Breward N, Williams M (1994) Arsenic and mercury pollution in gold mining. Mining Environ Manag:25–27.

Ceruti PO, Davies TC, Selinus O (2001) GEOMED 2001. Medical geology: The African perspective. EPISODES 24(4):268–270.

Ceruti PO, Fey M, Pooley J (2003) Soil nutrient deficiencies (Mseleni Joint Disease) and dwarfism in Maputaland, South Africa. In HCW Skinner, AR Berger (eds) Geology and Health: Closing the Gap (Chapter 24, pp. 151–154). New York: Oxford University Press.

Ceruti P, Pooley J, Fey M (1999) Do Environmental Factors Play a Role in the Aetiology of Mseleni Joint Disease? Proc XV Meeting Int Epidem Assoc II: 355. Florence, Italy, 31 August–04 September.

Chamberlin GT (1959) Trace elements in some East African soils and plants: 1. Cobalt, beryllium, lead, nickel and zinc. East African Agric J 25:12–125.

Churchyard GJ, Ehrlich R, teWaterNaude JM, Pemba L, Dekker K, Vermeijs M, White N, Myers J (2004) Silicosis prevalence and exposure-response relations in South African gold miners. Occup Environ Med 61:811–816.

Darnley AG, Björklung A, Bølviken B et al. (1995) A Global Geochemical Database for Environmental and Resource Management: Recommendations for International Geochemical Mapping. Earth Science Rep 19. Paris: UNESCO Publishing.

Davies TC (1994a) Combating iodine deficiency disorders in Kenya: The need for a multidisciplinary approach. Int J Environ Health Res 4:236–243.

Davies TC (1994b) The construction of regional maps showing the distribution of iodine in natural waters of Kenya in relation to the epidemiology of iodine deficiency disorders. Project proposal document; UNICEF Kenya Country Office, 12 p. Kenya: University of Nairobi.

Davies TC (1996) Geomedicine in Kenya. J Africa Earth Sci 23(4):577–591.

Davies TC (2008) Environmental health impacts of East African Rift volcanism. Environ Geochem Health 30(4):325–338.

Davies TC. Schlüter T (2002) Current status of research in geomedicine in East and Southern Africa. Environ Geochem Health 24:99–102.

Davies TC, Solomon AO, Lar P, Abrahams PW (2008a) A socio-economic study of geophagy in the Jos Plateau, Nigeria. Abstract Vol. 38. First International Conference on Human and Enzootic Geophagia in Southern Africa, Bloemfontein, South Africa, 22–24 October, 2008; 64 p.

Davies TC, Solomon AO, Lar P, Abrahams PW (2008b) Mineralogy and geochemistry of geophagic materials consumed in the Jos Plateau, Nigeria. Abstract Vol. 27. First International Conference on Human and Enzootic Geophagia in Southern Africa, Bloemfontein, South Africa, 22–24 October, 2008; 64 p.

Delange F, van der Velden M, Ermans AM (1973) Evidence of an antithyroid action of cassava in man and animals. Chronic cassava toxicity. Int Res Centre Monograph IDRC-OIO C. London.

Delfino R, Sioutas C, Malik S (2005) Potential role of ultrafine particles in associations between airborne particle mass and cardiovascular health. Environ Health Perspect 113: 934–946.

Derry CW, Bourne DE, Sayed AR (1990) The relationship between the hardness of treated water and cardiovascular disease mortality in South African urban areas. South African Med J 77(10):522–524.

Edmunds WM (2008) Groundwater in Africa: Palaeowater, climate and modern recharge. In S Adelana, A MacDonald, T Alemayehu, C Tindimugaya (eds) Applied Groundwater Studies in Africa (pp. 305–322). London: Taylor and Francis.

Eisler R (2003) Health risks of gold miners: A synoptic review. Environ Geochem Health 25(3):325–345.

Ekosse G-I, de Jager L (2008) Introduction. Abstract Vol. 1–8, First International Conference on Human and Enzootic Geophagia in Southern Africa, Bloemfontein, South Africa, 22–24 October, 2008; 64 p.

England NWJ (1961) Histological determination of three cases of Kaposi's sarcoma in West Africans: One showing metastasis in an inguinal node. Trans Roy Soc Trop Med Hygiene 55:301.

European Salt Producers' Association (EuSalt) (2009) EuSalt Position Paper on Fortification. Brussels, 5 p. Also at: www.eusalt.com.

Fanning EW, Froines JR, Utell MJ, Lippmann M, Oberdörster G, Frampton M, Godleski J, Larson TV (2009) Particulate matter (PM) research centres (1999–2005) and the role of interdisciplinary centre-based research. Environ Health Perspect 117(2):167–174.

Fendall NRE, Grounds TG (1965) The incidence and epidemiology of disease in Kenya, Part I: Some diseases of social significance. J Trop Med Hygiene 68:77–84.

Finkelman RB, Selinus O, Centeno JA (eds) (2007) Medical geology in developing countries, Part I. Special Issue, Environ Geochem Health 29(2): 81–167.

Fuge R (2005) Soils and iodine deficiency. In O Selinus, B Alloway, JA Centeno, RB Finkelman, R Fuge, U Lindh, P Smedley (eds) Essentials of Medical Geology (pp. 417–433). Amsterdam: Elsevier Academic Press.

Gomes CSF, Silva JBP (2007) Minerals and clay minerals in medical geology. Appl Clay Sci 36(1–3):4–21.

Goyens P, Golstein J, Nsombola B, Vis H, Dumont JE (1987) Selenium deficiency as a possible co-factor in the pathogenesis of myxoedematous endemic cretinism. Acta Endocrinol 114:497–502.

Hanegraaf TAC, McGill PE (1974) Thyroid diseases: Population based studies of endemic goiter. In LC Vogel, AS Müller, RS Odingo, A de Geus(eds) Health and Disease in Kenya (pp. 395–403). Nairobi, Kenya: East Africa Literature Bureau.

Hedger RS, Howard DA, Burdin M.L (1964) The occurrence in goats and sheep of a disease closely similar to swayback. Vet Record 76:493–497.

Hori K, Hatfield D, Mandarelli F, Lee BJ, Clouse KA (1997) Selenium supplementation suppresses tumour necrosis factor alpha induced human immunodeficiency virus type 1 replication in vitro. AIDS Res. Hum Retroviruses 13(15): 1325–1332.

Howard DA (1970) The effects of copper and cobalt treatment on the weight gains and blood constituents of cattle in Kenya. Vet Record 87:771–774.

Hudson JR (1944) Notes on animal diseases, 13: Deficiency diseases. East African Agric J 10:51–55.

Huang X, Gordon T, Rom WN, Finkelman RB (2006) Interaction of iron and calcium minerals in coals and their roles in coal dust-induced health and environmental problems. In N Sahai, MAA Schoonen (eds) Medical Mineralogy and Geochemistry, Reviews in Mineralogy & Geochemistry (Vol. 64, pp. 153–178). Virginia: The Mineralogical Society of America.

IARC (International Agency for Research on Cancer) (1997) Silica. IARC Monograph 68 on the Evaluation of the Carcinogenic Risk of Chemicals to Humans. Geneva: WHO.

Itakura H (1995) Idiopathic elephantiasis. In W Doerr, G Seifert (eds) Tropical Pathology (2nd ed., pp. 1328–1331). Berlin, Heidelberg, New York: Springer.

James W, Berger T, Elston D (2005) Andrews' Diseases of the Skin: Clinical Dermatology, (10th ed.). Oxford, UK: WB Saunders.

Joffe N (1964) Siderosis in the Bantu. Brit J Radiol 37:200–209.

Johnson CC (2003) The Geochemistry of Iodine and Its Application to Environmental Strategies for Reducing the Risks from Iodine Deficiency Disorders (IDD). Report CR/03/057 N. British Geological Survey, Keyworth, Nottingham, UK, 54 p.

Kane AB (1996) Mechanisms of mineral fibre carcinogenesis. In AB Kane, P Boffetta, R Saracci, JD Wilburn (eds) Mechanisms of Fibre Carcinogenesis (Vol. 140, pp. 11–35). Int Agency Res Cancer Sci Publ.

Kavishe FP (1996) Can Africa Meet the Goal of Eliminating Iodine-Deficiency Disorders by the Year 2000? www.unu.edu/unupress/foo/8F173e/8F173E0n.htm Downloaded 01 June, 2009.

Langer P, Greer MA (1977) Antithyroid Substances and Naturally Occurring Goitrogens. S Karger, AG Basel, 178 p.

Lanigan C (1989) Lake Nyos disaster. Brit Med J 299(6692):183.

Laurberg P (2009) Thyroid function: Thyroid hormones, iodine and the brain – an important concern. Nature Revs Endocrin 5(9):475–476.

Medical Geology in Russia and the NIS

Iosif F. Volfson, Evgeny G. Farrakhov, Anatoly P. Pronin, Ospan B. Beiseyev, Almas O. Beiseyev, Maxim A. Bogdasarov, Alla V. Oderova, Igor G. Pechenkin, Alexey E. Khitrov, Oxana L. Pikhur, Julia V. Plotkina, Olga V. Frank-Kamenetskaya, Elena V. Rosseeva, Olga A. Denisova, Georgy E. Chernogoryuk, Natalia Baranovskaya, Leonid P. Rikhvanov, Igor M. Petrov, Armen K. Saghatelyan, Lilit V. Sahakyan, Olga V. Menchinskaya, Tamara D. Zangiyeva, Murat Z. Kajtukov, Zukhra H. Uzdenova, and Anastassia L. Dorozhko

Abstract Systematic and regular epidemiological studies on endemic diseases and the natural environment in Russia and NIS are scarce and sporadic. However, there have been some studies of the links between health of the population and the geological background. Information on fluorine, iodine, arsenic, selenium, and other elements' behavior in natural environment and their effect on human health is presented in this chapter and is the first attempt to synthesize the interdisciplinary knowledge on some geological factors which affect human health in Russia and NIS. Also anthropogenic factors are mentioned, however, of geological origin.

Currently, the most important areas of the study in the field of Medical Geology in Russia – NIS are

- Geological and geochemical aspects of medical geology in terms of modeling and mapping of the spreading of endemic diseases, toxic elements, such as uranium, fluoride, radon, arsenic, in subsurface, geosphere, and the atmosphere, and its effect on human health.
- Urban and mining medical geology.
- Crystal chemistry and crystal genesis of biogenic minerals of different origin.
- The therapeutic usage of the minerals in terms of biological functions of the elements metals in medicine and industry, and economic minerals in medicine.

This chapter, written by leading experts of Russia and NIS, will be of interest to a wide audience of geologists, geochemists, physicians, as well as historians.

Keywords Geochemical anomalies · Human health · Russia and NIS · Arsenic · Fluorine · Iodine · Selenium · Radon · Mining

The International Medical Geology Association was established in 2005. The Regional Division of Russia-NIS was established in March 2006 at the annual meeting of the Russian Geological Society Medical Geology Division (MGD ROSGEO). The founding of the IMGA (International Medical Geology Association) Regional Division of Russia-NIS (IMGA RD Russia-NIS) was organized through cooperation between the MGD ROSGEO and the IMGA council. Since 2005, MGD ROSGEO has maintained an organization to promote contact between the IMGA and the NIS medical geology communities.

IMGA RD RUSSIA – NIS is a voluntary association of scientists – geologists, geochemists, mineralogists, geophysicists, physicians, chemists, biologists, microbiologists, medical scientists, and others who study the effects on human health of geological processes, materials (minerals, ores, volcanic emissions, atmospheric dust, water, and soils), and other natural effects. The purpose of IMGA RD RUSSIA – NIS is to bring together specialists, scientists, and practitioners working in different areas of knowledge, to unite their efforts, and to encourage collaboration on this rapidly developing field of the science. The main purpose of IMGA RD RUSSIA – NIS activities is to support the studies in the field of medical geology in the Russian Federation and Newly Independent States. Scientists from NIS are occupied in many fields of

I.F. Volfson et al. (✉)
Russian Geological Society, h.10, st. 2-d Roshchenskaya, Moscow 115191, Russia
e-mail: iosif_volfson@mail.ru

O. Selinus et al. (eds.), *Medical Geology*, International Year of Planet Earth,
DOI 10.1007/978-90-481-3430-4_9, © Springer Science+Business Media B.V. 2010

medical geology. Some results of these studies have been introduced to the users on the IMGA web site and to the IMGA Medical Geology Newsletter.

Currently, the most important areas of the study in the field of Medical Geology in Russia – NIS are

- Geological and geochemical aspects of medical geology in term of mechanisms of element concentrations as a base for the modeling and mapping of the spreading of endemic diseases including the study of toxic elements, such as uranium, fluoride, radon, arsenic, in subsurface geospheres, and the atmosphere, and the effect on human health.
- Urban- and mining medical geology.
- Crystal chemistry and crystal genesis of biogenic minerals of different origin.
- The therapeutic usage of the minerals in terms of biological functions in medicine and industry, and economic minerals in medicine.

History

Iosif F. Volfson, Evgeny G. Farrakhov, Ospan B. Beiseyev, Almas O. Beiseyev, and Maxim A. Bogdasarov

The first applications of medical geology in Russia were on the use of medical properties of different minerals and rocks. In ancient times minerals and rocks were used as a medical treatment among the population of the highlands, while in lower areas mainly plants were used for healing purposes. Different branches of medicine had their specific medicines and traditions for treatment created during ancient times, reflecting geographical and geological properties of the different areas (for example, Tibetan, Eastern, West-European) (Pelymskij, 2008).

Russia and later the Commonwealth of Newly Independent States adopted the experience of people of the Near East, the Middle East, the Far East, and Western Europe; and a vast knowledge was accumulated during long periods of time. The research of prominent scientists of the Middle Ages such as Avicenna, Biruni, and others was characterized by detailed investigations of minerals, including studies of their physical, chemical, and medical properties. Avicenna in his famous book "Canon of Aesculapian Science" (11th century) recommended the use of more than 30 minerals for medical treatment and described the recipes of many medical preparations from these minerals. Biruni published his medieval study of semiprecious stones for medical use in his book "Congregation of the Lore for Cognition of the Valuables. The Mineralogy", including medical information for about 450 minerals and natural chemical compounds (Yushkin, 2003, 2004a, 2004b; Beiseev, 2004).

The history of pharmacopeia in Russia began with the introduction of Christianity, in the epoch of Prince Vladimir (10th century). The first literature on medical properties of minerals can be found in "Izbornik" by Svyatoslav (1073), in "Vetrogrady" (15th–17th century), in "Torgovaja kniga" ("Trade book", 16th–17th century), and in "Azbukovnik" (17th–19th century). The pharmacy came to Russia much later, at the beginning of the 17th century, and was widely spread by the time of Peter the Great. There were many drugstores in the 16th century, where it was possible to buy such cures, as salpeter, potash, rosin, sulfur, borax, mercury, verdigris, wax, as well as jewels – , sapphire and ruby, which were used for preparation of medicines (Pelymskij, 2008).

Some examples of domestic mineral recipes were

Emerald: "Who often look at the emerald, then human eye is strengthened, protected and healthy. And besides emerald brings a cheer to whom who carries it".

Turquoise: "After it is grinded, take it, and it helps from bite of snake".

Quartz: "Crystal-stone to grind and stirred with fresh honey, that mother's milk multiplies" (Yushkin, 2004a, 2004b).

These recipes correspond almost to the recipes of medieval lapidaries (British Bishop Renskij 11th century, Arabic scientists Tabib and Shirazi – 12th century) (Walters, 1996).

At the beginning of the 17th century these pharmaceutical recipes were prescribed to the Russian army.

The Russian scientist V.M. Severgin played an important role in the development of the pharmaceutical chemistry and mineral use for medical purposes. In 1798 he translated the book of "Natural history" by Plinij the Senior, which contained information about minerals, including their medical properties. He then for the first time established the term "esculapian mineralogy" (medical geology) in everyday life.

The traditions of "esculapian mineralogy" in NIS are still alive and getting stronger every year. For example, scientists and doctors from the states of Central Asia and Kazakhstan currently make significant contributions with the use of natural minerals and their mixtures that possess valuable properties for human health. These minerals are used in cosmetics, in preventive medicine and treatment of different diseases, such as skin lesions, and many diseases of internal organs, traumas, and fractures, and to treat damages from ecological catastrophes. At present almost 500 known minerals are used in medicine, more than 350 are well defined in the territory of NIS and particularly in Kazakhstan. Among these are ore minerals and economic minerals such as clays, shungites, flints, zeolites, salts, and others (Beiseev, 2004; Volfson, 2004; Volfson et al., 2010; Florinsky, 2010).

The successful use of bentonite clays in medicine in medieval times was shown by Avicenna in his book "Canon of Esculapian Science" (11th century). He recommended its use as an efficient cure for the treatment of the diseases of the gastrointestinal tract, mouth, drinking water detoxication, and filtering. At the present time our contemporaries successfully used bentonite clay as preventive measures and treatment of caries and various manifestations of dermatitis, and in making complex medical preparations (Kariev, 2007).

For centuries, mumie (mumie, or shilajit, a natural dark mineral substance found in rock crevices at altitudes of 2,000–5,000 m high in the mountainous regions of Asia) has been known to possess an array of uses related to health and longevity. Specifically, mumie shows benefits to both the human metabolic and immune systems and has had big success among the population. Resources of this mineral, consisting of organic material, formed by bacteria living on rocks and excrements of animals are significant in South Kazakhstan, Central Asia, in the Altai Mountains. Research in mumie has revealed a large variety of free amino acids (from 0.03 to 3.55%) dominated by amino-acetic acid and glutamic acid (6.07–99.75% and 0.02–19.96%), respectively, and also a large amount of all amino acids, as well as elements such as S, C, H, Al, Mo, P, Hg, Cu, S, and Mn from which the antimicrobial and other important properties of mumie are dependent (Frolova and Kiseleva, 1996; Savinykh, 2003; Volfson et al., 2010). The ability of mumie to activate phagocyte activity of human leukocytes and to promote healing of wounds and fractured bones has been revealed (Schepetkin et al., 2002; Ghosal, 2006). There is a contemporary theory on the endogenic genesis of mumie as a result of degassing of Earth along abyssal structures (Savinykh, 2006).

Another popular and efficient geological material is the Kola shungites (shungites are pre-Cambrian carbon-rich, silicate rocks known from deposits in the north-western part of lake Onega, near the city of Petrozavodsk in Karelia, north of St. Petersburg, Russia). Shungites are extremely rich in carbon, with carbon accounting for 98% in vein shungites. The use of this material for medical treatments dates back to Peter the Great. Resting after a long march in one of the Kola counties, Tsar Peter, after advice from the local population, took a bath and drank water drained from outcrops of the shungite rocks. He felt relief and the impression was so strong that he located the first Russian spa resort there. The active use of the Kola shungites as antibacterial treatment for water purification in the Russian army also became possible after an edict of Tsar Peter. One of the reasons for the success of Tsar Peter in the battle of Poltava in 1709 against Swedish troops was the possibility to purify drinking water by means of natural shungite (Rysyev, 2001). This has continued to this day. Scientists of Saint-Petersburg University in collaboration with experts of the City enterprise Vodokanal (water supplier) have confirmed antibacterial properties of the Kola shungites. They proved antagonism of bacteria in shungite rocks to test cultures of pathogenic microorganisms (Charykova et al., 2006). Serious investigations of shungites mineralogical and geochemical features were carried out by Russian scientists in collaboration with foreign scientists (Buseck et al., 1997).

During the following years of the Russian empire, as well as in modern Russia and NIS, valuable experience of water- and mud cure has been developed, based on the experience from the late Middle Ages. Examples include the water- and mud cure resorts created in Moscow by the German doctor Christian Loder in 19th century, the resorts of the Russian Caucasian mineral waters Essentuki and Pyatigorsk, the resort Borzhomi in Georgia, the resort Archman in Turkmenia, the resort Arzni in Armenia, and many others. It should be noted that there are now 273 such sites of mineral waters in the NIS. They are used for treating patients suffering from cardiovascular diseases, for whom carbonate waters are most helpful; such waters are currently found at 57 sites. There are 38 locations of sulfide waters which are used for patients with diseases of joints and spine, as well as for patients with diseases of the peripheral nervous system (Tsarfis, 1985, 1991; Volfson et al., 2010).

Endemic Disease Areas in Russia and NIS

Iosif F. Volfson, Evgeny G. Farrakhov, Anatoly P. Pronin, Alla V. Oderova, Igor G. Pechenkin, and Alexey E. Khitrov

It is important to note that systematic and regular epidemiological studies on endemic diseases and the natural environment in Russia and NIS are scarce and sporadic. The reason could be the low density of the population in much of the regions and accordingly quite complicated conditions for epidemiological research. On the other hand, intensive industrialization has resulted in increasing density of the population in large industrial centers, where people are subject to effects of high concentrations of anthropogenic contaminants. This also requires greater efforts in determining the origin of anomalous concentrations of elements in the natural environment as well as their speciation. However, there have been some studies on the links between the health of the population and the geological background. Valuable information on fluorine, iodine, arsenic, selenium, and other elements in the natural environment and their effect on human health has been obtained (Ermakov, 1992; Skal'ny, 1999; Agadzhanjan et al., 2001; Golovin et al., 2004; Saet et al., 1990). Some results of these studies will be introduced in this chapter.

Society develops and exists in contact and interaction with naturally occurring elements and minerals; and there is an obvious relationship between the health of the people and the different geological backgrounds that they live in (Avtsyn, 1972; Aggett and Rose, 1987; Abrahams, 2002; Hough, 2007; Peeters, 1987; Persinger, 1987; Yushkin, 2004a, 2004b). A. P. Vinogradov (Vinogradov, 1938, 1939; Vinogradov, 1964) revealed a relationship between the concentrations of certain elements in the environment and health problems of humans and animals. He introduced a biogeochemical hypothesis on the formation of the disease discussing biogeochemical provinces and biogeochemical endemicity. As demonstrated by the publications issued by the Biogeochemical Laboratory headed long ago by Vinogradov and his follower V. V. Kovalsky later in 1970s, relatively enhanced background levels (2–3 times above the clarke values), may cause serious health problems. The concept of geochemical endemicity was proposed in 1998 (Golovin et al., 2004). Geochemical endemicity depends on the parameters and features of regional backgrounds having enhanced or deficient contents of elements relative their clarke values.

The Kashin–Beck (Urov) disease is a disease mainly caused by geochemistry. This serious disease (deforming osteochondroarthritis) is located in Eastern Siberia (Transbaikalia), China, and Korea. Some researchers believe that it is caused by toxicity of phosphorous and manganese that suppresses calcium activity in humans and results in a critically low Ca/P ratio in bone tissues. However, most researchers are in favor of the hypothesis of selenium deficiency. The areas within these territories promote formation of increased concentrations of these elements in soils (Saet et al., 1990; Kravchenko, 1998; Golovin et al., 2004).

Another example is manganese which can be found in elevated contents in sedimentary geological environments of the East-European platform. Manganese is actively absorbed by plants and is capable of entering into organisms through the food chain. Manganese is one of the essential elements, both for animals, and humans. It plays an important role in the process of formation of tissues and bones, in the process of growth, and in lipid and carbohydrates metabolism and it is necessary for the reproductive function. The main path of intake for manganese is the respiratory and gastrointestinal tract. The degree of absorption of manganese entering by inhalation is unknown, but in the gastrointestinal tract not more than 5% is taken up (Revich, 2001; Aggett and Rose, 1987). A special role of manganese is noted as a pathogenic factor on the origin of goiter (Agadzhanjan et al., 2001). Manganese poisoning is caused mainly by the chronic effect of its oxides, which cause manganese toxicosis. The chronic results of this poisoning are effects on the central nervous system. Bronchial asthma, pneumonia, allergy of the upper respiratory tract, eczema, and dermatitis are connected with overexposure to manganese (Revich, 2001).

Manganese was found in partially burnt wood of aspen and birch at excavations of the medieval layers in the ancient settlement of Moscow. This firewood was broadly used by the inhabitants for heating and cooking as well as in technological processes. Probably the concentration of manganese in firewood exceeded the upper limited concentrations and could lead to serious health problems of the population. The content of manganese in bones of Moscow residents of the medieval period was as high as 10 mg/100 g. The common people had 30–70 mg/100 g in their bones (Aleksandrovskaia and Aleksandrovskij, 2003).

Figure 1 shows the areas of endemic diseases caused by environmental factors in Russia and NIS. By using modern methods of treatment of geological, geochemical, and geophysical information we have also revealed abyssal fractures which control halos of helium and fluorine as well as strontium, bromine, and lithium in water-bearing formations of sedimentary rocks of the East European platform (Pronin, 1997; Pronin et al., 1997; Klimas and Mališauskas, 2008; Farrakhov et al., 2008).

There have also been found three interrelated ore-forming systems playing a significant role in epigenetic ore formation within the borders of the Central Asian metal-bearing and oil-bearing sedimentary basins (Grushevoy and Pechenkin, 2003; Pechenkin et al., 2005): catagenetic

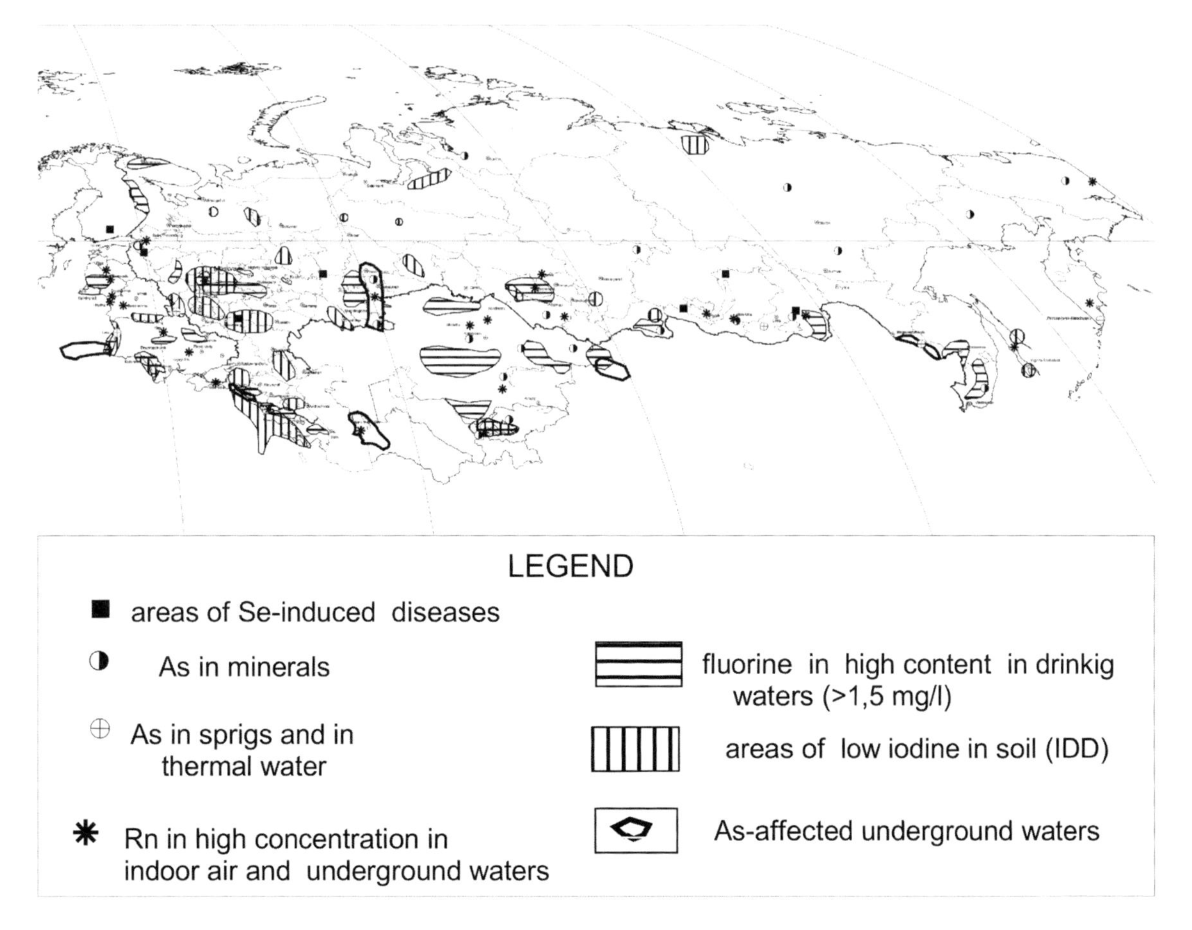

Fig. 1 Areas of endemic diseases in Russia and NIS

(stadial), exfiltration, and infiltration. Each system forms specific geochemistry and hence epidemiological settings (Grushevoy and Pechenkin, 2003; Pechenkin and Pechenkin, 2005; Pechenkin et al., 2005; Volfson et al., 2006). On the basis of this research a relationship between the distribution of U, Se, As, F, Br, B, Rn, and potential endemic diseases has been established.

Fluorine

Oxana L. Pikhur, Julia V. Plotkina, Olga V. Frank-Kamenetskaya, and Elena V. Rosseeva

Introduction

Fluorosis can be found all over Russia and NIS and adjacent territories of the Baltic States particularly in Lithuania (Narbutaite et al., 2007). The most serious fluorosis is found in Ukraine, Azerbaijan, Moldova, and in Kazakhstan (Admakin, 1999) (Fig. 1). Fluorine is also identified in groundwaters in the Central Region of Russia.

In Ukraine fluorosis is widespread in the Transcarpathian region, in the Poltava territory.

Investigation of 25,595 rural people in the Republic of Kazakhstan has been carried out and has shown various diseases of hard tooth tissues, among which fluorosis dominated. Interesting regional distributions of fluorosis have been found. The absence of modern technologies in the treatment of fluorosis, the poor technical equipment, and the low level of prophylactic work are serious. It is worth mentioning that in areas with hot climate, acute fluorosis can be observed with contents of fluorine in drinking water of 0.5–0.7 mg/l (Esembaeva, 2006).

More than 90% of the population of Russia do not get fluorine in necessary amounts (Revich, 2001; Kuz'mina, 1999; Kuz'mina and Smirnova, 2001). Fluorine is particularly low in surface drinking water supplies in the Kol'skij and Karel'skij region (the Kola Peninsula), in the Archangelsk region, in the Leningrad region, in the Republic of Komi, and in North Kaukasus. Fluorine deficiency in water is a matter of concern for dental diseases such as tooth decay and caries (experienced by 60% of the population). On the other hand excess of fluorine in groundwater has caused fluorosis among children in the Upper Volga region and in the Volga region, particularly in Nizhny Novgorod, in the capital of Republic of Chuvashia Cheboksary, Republic of Mordovia Saransk, and elsewhere. In the region east of Moscow fluorine haloes with a concentration of more than 0.16 mg/l are revealed in melting water from snow (Fig. 2). Concentrations of fluorine (over 19 mg/l) and bromine (125 mg/l) (more than 10 times the maximum limited concentration of 1.5 mg/l) are found in groundwater within the Kaluga volcanic structure and within a number of similar geological structures in the East-European platform which are covered by Paleozoic sedimentary deposits (Pronin et al., 1997).

Fluoride Content in Drinking Water and Dental Morbidity in Different Regions of Russia

Drinking water is one of the main sources of fluoride for humans (Smith, 1988; Freatherstone, 1993; WHO Technical Report Series, 1994; Edmunds and Smedley, 2005). Fluoride concentration in drinking water in Russia varies from 0.01 to 2.90 ppm. Epidemiological dental studies of the Russian populations allow us to draw conclusions about dental morbidity and the necessity of dental treatment of patients of certain age groups (6, 12, 15, 35–44, 65 years and older) in different Russian regions (Pachomov, 1982; Admakin, 1999; Kuz'mina, 1999; Kolesnik, 1997; Sherbo, 2002). Results of dental investigations of populations from 56 cities can be found in Table 1.

Chemical Composition of Teeth Enamel and Hair of Citizens in Centers of North West Russia

The chemical composition of teeth enamel and hair of persons of different age groups living in centers of North-West region of Russia (Saint Petersburg, Apatites, Monchegorsk) has been studied using ICP-MS.

The deterioration of the environment in industrial cities steadily increases the abundance of human

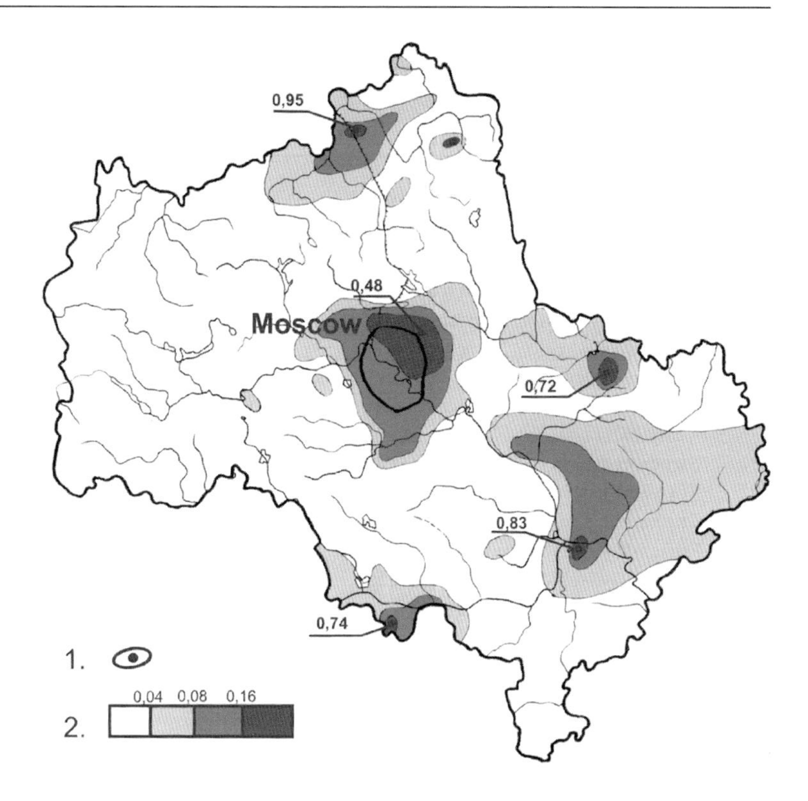

Fig. 2 Map of fluorine distribution in melting water from snow 1993 in Moscow and surroundings. Sampling site with higher concentration of fluorine in snow water. Concentration of F (mg/l)

Table 1 Fluoride content in drinking water and dental diseases of 12-year-old residents of Russian regions

No.	Region (*n* – number of persons)	Fluoride in drinking water (ppm)	Frequency of teeth caries	Intensity of teeth caries (index CFE)[a]	Frequency of teeth fluorosis
1	Arkhangelskaya region, Arkhangelsk city (49)	0.15	76	2.04	0
2	Republic Adygeya, Maikop city (47)	0.07	79	3.24	2
3	Astrakhanskaya region, Astrakhan' city (100)	0.16	74	3.19	2
4	Republic Bashkortostan, Ufa city (50)	0.28	84	2.60	4
5	Bryansky region, Bryansk city (102)	0.18	66	1.96	4
6	Republic Buryatia, Ulan-Ude city (49)	0.48	76	2.00	12
7	Volgogradskaya region, Volgograd city (50)	0.22	70	2.42	2
8	Voronezhskaya region, Voronezh city (101)	0.25	90	3.87	5
9	Dagestan Republic, Buinaksk city (101)	0.20	84	3.49	8
10	Ingushetia Republic, Nazran' city (50)	0.17	94	3.58	8
11	Irkutskaya region, Irkutsk city (50)	0.21	82	2.38	0
12	Kabardino-Balkaria Republic, Nal'chik city (77)	0.41	70	3.12	1
13	Kalmykia Republic, Elista city (50)	0.32	66	1.82	0
14	Kaluzhskaya region, Kaluga city (83)	0.22	68	1.79	1
15	Republic Karelia, Petrozavodsk city (50)	0.08	70	1.94	6
16	Republic Karelia, Nadvoitsy village (50)	0.90	74	2.56	44
17	Kemerovskaya region, Kemerovo city (54)	0.30–0.50	86	3.26	0
18	Krasnodarsky land, Krasnodar city (50)	0.42–0.54	90	3.34	4
19	Krasnoyarsky land, Krasnoyarsk city (50)	0.13	74	3.50	0
20	Kurskaya region, Kursk city (50)	0.41	66	1.72	8
21	Lipetskaya region, Lipetsk city (50)	0.20	82	3.32	0

Table 1 (continued)

No.	Region (*n* – number of persons)	Fluoride in drinking water (ppm)	Frequency of teeth caries	Intensity of teeth caries (index CFE)[a]	Frequency of teeth fluorosis
22	Moscow (1161)	0.16–0.22	82	3.59	3
23	Moscow region, Mozhaisk city (50)	0.41–0.61	68	3.10	10
24	Moscow region, Podolsk city (30)	1.2	60	1.07	32
25	Moscow region, Odintsovo city (30)	1.80	37	0.50	74
26	Moscow region, Zheleznodorozhny city (30)	1.0	53	1.30	6
27	Murmanskaya region, Murmansk city (50)	0.10	76	2.38	4
28	Murmanskaya region, Monchegorsk city (50)	0.20	84	2.90	8
29	Murmanskaya region, Apatity city (50)	0.65	82	2.38	10
30	Nizhegorodskaya region, Nizhny Novgorod city (201)	0.10–0.50	96	4.43	0
31	Nizhegorodskaya region, Arzamas city (200)	0.86–1.2	37	1.50	5
32	Novosibirskaya region, Novosibirsk city (50)	0.17	90	3.34	0
33	Omskaya region, Omsk city (50)	0.25	86	3.10	0
34	Penzenskaya region, Penza city (103)	0.41	69	2.32	0
35	Penzenskaya region, Serdobsk city (50)	2.85–2.90	46	0.92	78
36	Permskaya region, Perm' city (50)	0.50	72	2.20	0
37	Primorsky land, Vladivostok city (50)	0.11	91	3.48	0
38	Rostovskaya region, Rostov-na-Donu city (50)	0.28	78	1.66	4
39	Samarskaya region, Samara city (52)	0.52–1.3	88	3.67	6
40	Saint-Petersburg City (120)	0.04	77	1.94	0
41	Sakhalinskaya region, Yuzhno Sakhalinsk city (110)	0.1–0.2	84	3.36	0
42	Sverdlovskaya region, Ekaterinburg city (87)	0.70	85	3.42	0
43	Smolenskaya region, Smolensk city (50)	0.28–0.35	76	2.06	2
44	Tatarstan Republic, Kazan' city (101)	0.16	82	3.11	0
45	Tverskaya region, Tver' city (140)	0.79–2.00	84	2.36	72
46	Tomskaya region, Tomsk city (100)	0.08–0.29	74	1.97	0
47	Tul'skaya region Tula, city (75)	0.26–0.28	49	3.64	3
48	Tuva Republic, Kyzyl city (50)	0.22	74	2.52	6
49	Tyumen region, Tumen' city (105)	0.15	76	2.90	0
50	Udmurtia Republic, Izhevsk city (50)	0.12	86	3.22	4
51	Khabarovsky land, Khabarovsk city (49)	0.15	89	4.00	2
52	Khanty-Mansiisky region, Khanty-Mansiisk city (100)	0.19	87	3.71	0
53	Khanty-Mansiisky region, Nizhnevartovsk city (52)	0.10	90	4.54	0
54	Chelyabinskaya region, Chelyabinsk city (50)	0.15	58	1.56	0
55	Chitinskaya region, Chita city (50)	0.30	70	1.50	0
56	Yaroslavskaya region, Yaroslavl' city (50)	0.14–0.17	80	2.37	2

[a]Index CFE: sum of caries (component «C»), filling (component «F»), extracted (component «E») teeth of examined person. Index CFE in group = $\Sigma CFE_{personal}/n$, where *n* – number of persons. (Kolesnik, 1997; Kuz'mina, 1999; Nikolishin, 1989; Pikhur, 2003)

dental diseases. Teeth caries is one of the most important dentistry problems because of its wide distribution (Zyryanov, 1981; Breus, 1981). During the past 10–15 years the frequency of dental diseases associated with non-caries lesions has dramatically increased. Thus the early diagnostics of these diseases and their prevention is now a very important issue. Major and trace elements in the dental enamel of healthy and sick patients have been extensively investigated by many authors (Breus, 1981; Elliott, 1994; Dowker et al., 1999). However, papers devoted to complex investigations of the chemical composition of dental hard tissues of persons of a single age group and living permanently in regions with different environmental conditions are scarce. It is therefore not possible in Russia to analyze the resistance of dental hard tissues to disease depending on environmental conditions.

It is, however, well known that environmental changes influence the chemical composition of hair in some respect, and therefore hair, in contrast to teeth, is more available for research and is thus more widely used as an indicator of changing environmental conditions. For instance, humans living in areas chronically exposed to volcanic emissions show high concentrations of essential and nonessential trace metals in scalp hair. It is suggested that this type of exposure may be as harmful as living close to industrial facilities (Amaral et al., 2008). The relations between the hair composition and diseases have been studied in Russia more than the composition of human dental hard tissues.

One investigation has, however, been a detailed investigation of the chemical composition of dental enamel and hair of persons of different age groups permanently living in centers of North-West region of Russia (Saint Petersburg, Apatites, Monchegorsk) and the comparative analysis of environmental influence on chemical composition of human dental hard tissue and hair. In order to exclude the effect of age changes on the composition of the hard tissues samples of hard tissues are used, obtained from persons of different age groups: I – 10–15 years, II – 16–30, III – 31–45, IV – 45–65 years. Teeth were sampled by medical evidence (chronic parodontitis, orthodontic causes). The enamel samples were prepared as a powder. Samples of hair were represented with a length of about 5–7 cm. ICP-MS has been used to determine the chemical composition (Larina, 2006). In order to determine the differences between chemical compositions of teeth and hair from different regions cluster analysis was used (Soshnikova et al., 2007). The concentration of major elements (Ca, P, Na, Mg) of the studied enamel samples (Table 2) is typical for the human teeth hard tissue (Frank-Kamenetskaya et al., 2004; Dowker, 1999; Elliott, 1997). The result of cluster analysis of the investigated samples has not allowed the classification of samples into well-defined groups.

The significant distinctions of chemical composition of teeth enamel of inhabitants from different cities are apparent. The variations in concentration of Fe, V, Mo, Ta, Ag (in St. Petersburg), Ni, Cr, Ba, Mo (in Apatity) are probably connected to working conditions in industries.

Cluster analysis (Fig. 3, Tamashevich and Makhnach, 2004) confirms the differences in trace elements concentration of teeth enamel of Kola Peninsula and St. Petersburg inhabitants. Teeth of inhabitants of Monchegorsk of age groups II and IV are closest to the composition of enamel. The grouping of samples in cluster at high D values reflects essential influence of endogenous factors on trace element composition of human teeth enamel.

The results of the investigations on chemical composition of hair of inhabitants from different regions (Table 3) demonstrate that calcium is the major element of hair ash residue. Calcium contents vary from 0.02 to 0.13 wt.% with a maximum in the hair of St.Petersburg inhabitants. Furthermore, the calcium content of human hair decreases with increasing patient age.

The result of the cluster analysis shows (Figs. 3 and 4) significant differences in trace element composition of hair samples from inhabitants of different age groups living in North-West centers of Russia. At $D = 26\%$ three clusters corresponding to hair samples of inhabitants from St. Petersburg, Apatity, and Monchegorsk are observed. The results demonstrate the smaller variability of the chemical composition of the hair of the inhabitants of cities of the Kola Peninsula compared to St. Petersburg. It is important to note that inside of the second and the third clusters, corresponding to Apatity and Monchegorsk, successive grouping of samples occurs by the way of reduction of resident's age. At the following stage of clusterization (at $D = 46\%$) the grouping of the second and the third clusters in one cluster is observed, which reflects the similarity of a chemical composition of the hair of the inhabitants from the Kola Peninsula towns in contrast to hair samples from St. Petersburg.

The results of the investigation of teeth enamel and hair of inhabitants from St. Petersburg, Apatity, and Monchegorsk demonstrate the difference of their trace element composition. Differences in chemical composition of the investigated tissues of the inhabitants from Kola Peninsula (Apatity and Monchegorsk) are less in comparison to St. Petersburg inhabitants. Furthermore the chemical composition of hair and especially dental enamel samples of inhabitants from St. Petersburg is very heterogeneous. This can probably be explained by influences of many factors (social, physical, chemical, biological, and other).

Table 2 Chemical composition of enamel of inhabitants from different industrial centers (mas.%)

Element	St. Petersburg		Monchegorsk		Apatity	
Age group	16–30	45–65	16–30	45–65	16–30	45–65
Ca	35.66	34.66	35.38	34.02	32.02	31.80
P	17.32	16.71	17.41	17.02	16.36	15.62
Na	0.63	0.73	0.50	0.63	0.58	0.52
Mg	0.24	0.37	0.18	0.32	0.22	0.27
Zn	3,000.00	1,800.00	4,300.00	4,600.00	3,200.00	1,800.00
Sr	828.00	457.00	1,300.00	742.00	1,200.00	533.00
K	3,071.00	2,656.00	1,245.00	1,992.00	1,743.00	4,067.00
Fe	9,092.63	839.32	1,468.81	307.75	0.00	0.00
Ti	131.89	3,177.35	3,357.20	3,417.15	3,417.15	2,817.65
Cu	189.00	78.50	35.90	6.20	30.80	63.40
Ni	130.00	10.00	37.80	26.90	1,700.00	60.90
Al	0.00	0.00	0.00	0.00	0.00	1,005.53
W	615.00	93.20	90.20	79.90	107.00	122.00
Co	530.00	10.70	22.00	21.30	16.50	25.90
Cr	147.00	20.20	45.30	<10.00	750.00	<10.00
Mn	30.98	15.49	38.72	46.47	54.21	38.72
Pb	19.20	51.70	<10.00	28.10	39.00	36.60
Zr	17.40	38.90	18.90	<10.00	<10.00	<10.00
Ba	52.90	25.90	19.50	30.50	<10.00	135.00
Sn	33.20	6.20	10.70	11.00	20.00	6.50
Mo	17.70	2.40	<1.00	<1.00	14.00	<1.00
Ga	<0.50	<0.50	<0.50	<0.50	<0.50	2.10
Be	<0.30	<0.30	0.47	0.47	<0.30	3.20
Rb	3.20	<0.30	1.40	3.40	0.99	15.50
Y	<0.10	0.15	<0.10	<0.10	<0.10	<0.10
Nb	2.00	1.10	<0.10	<0.10	6.70	<0.10
Ag	3.900	0.000	0.000	0.110	0.000	0.000
Cd	<0.10	<0.10	<0.10	<0.10	8.50	<0.10
In	0.09	0.36	0.03	<0.01	0.06	0.23
Sb	0.24	3.80	<0.10	<0.10	<0.10	<0.10
I	0.63	3.60	1.50	0.18	<0.10	0.53
Cs	0.07	0.22	0.08	0.07	<0.05	<0.05
La	6.90	2.90	<0.10	<0.10	<0.10	<0.10
Ce	15.40	5.40	<0.10	<0.10	<0.10	<0.10
Pr	1.20	0.63	<0.10	<0.10	<0.10	<0.10
Nd	2.30	1.90	<0.10	<0.10	<0.10	<0.10
Sm	<0.10	0.28	<0.10	<0.10	<0.10	<0.10
Eu	<0.01	0.03	<0.01	<0.01	<0.01	0.03
Gd	0.05	0.21	<0.01	<0.01	<0.01	<0.01
Tb	<0.01	0.02	<0.01	<0.01	<0.01	<0.01
Dy	<0.01	0.03	<0.01	<0.01	<0.01	<0.01
Ho	0.02	0.03	<0.01	<0.01	<0.01	<0.01
Hf	0.08	0.19	0.07	<0.05	<0.05	<0.05
Ta	7.50	1.20	1.30	1.10	1.60	2.60
Th	<0.01	0.40	<0.01	<0.01	<0.01	<0.01

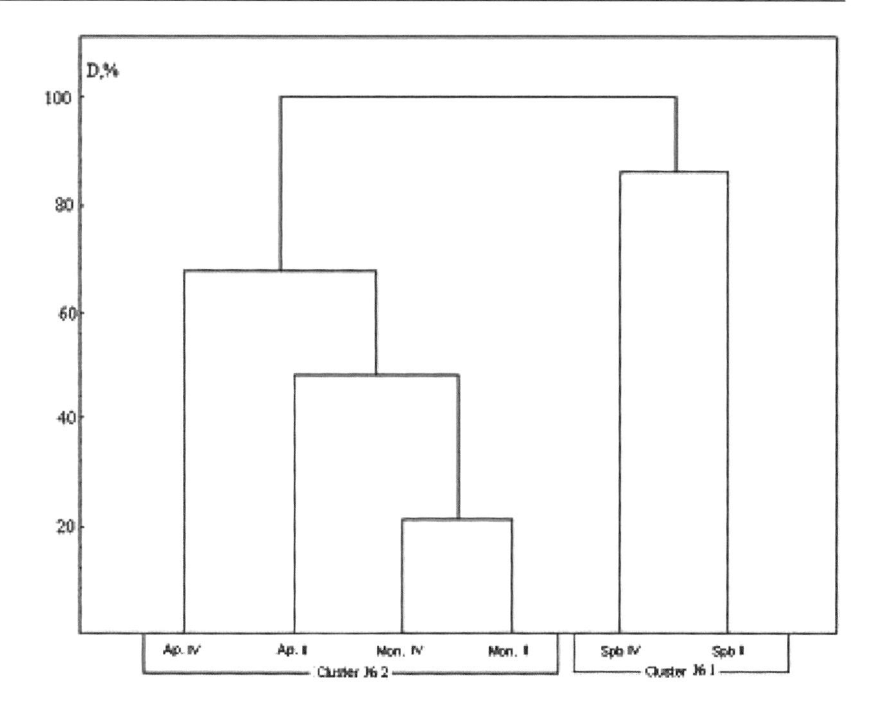

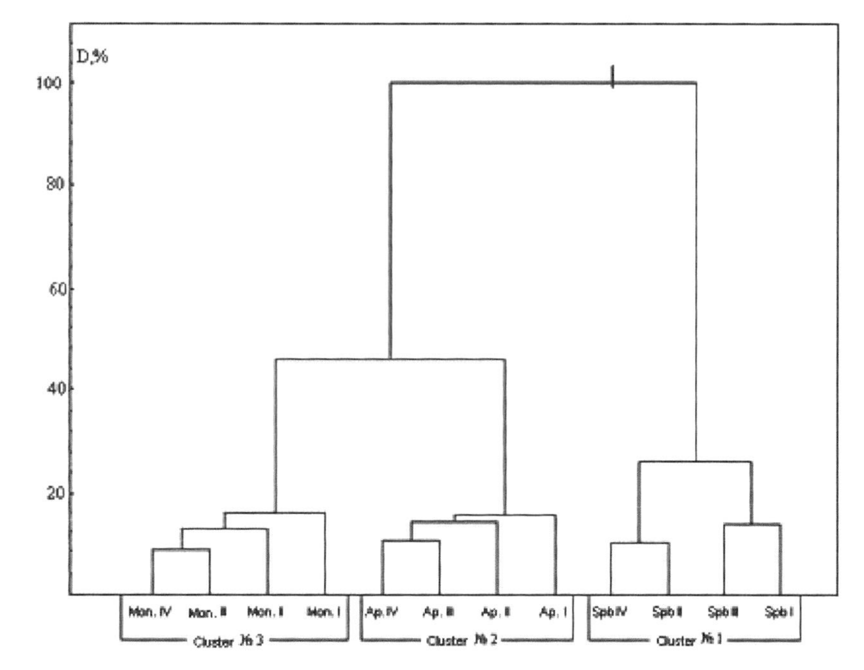

Fig. 3 Results of cluster analysis of trace element content in samples of teeth enamel (**a**) and hair (**b**) of residents of different industrial centers. Ap – Apatity, Mon – Monchegorsk, SPb – St. Petersburg

Table 3 Element composition of hair of inhabitants from different regions (mas.%$*10^4$)

Element	St. Petersburg				Monchegorsk				Apatity			
Age group	10–15	16–30	31–45	45–65	10–15	16–30	31–45	45–65	10–15	16–30	31–45	45–65
Ca	1,254	1,210	1,230	1,213	268	216	252	220	369	310	330	290
Na	285	367	320	298	210	275	244	235	324	453	344	335
Zn	210	161	177	165	176	155	165	150	172	149	168	148
K	69	167	143	105	169	184	182	175	243	291	282	275
Mg	126.0	119.0	120.5	120.0	32.1	25.8	24.5	24.0	45.8	40.3	38.5	38
Cu	15.1	7.8	12.2	7.2	11.3	10.0	11.0	10.2	9.7	8.7	8.0	8.2
Fe	23.1	19.1	21.3	19.8	13.4	8.6	11.6	9.2	21.3	17.5	20.1	18.1
Al	15.3	14.3	15.0	14.8	9.4	7.3	9.2	8.8	12.6	10.7	12.0	11.8
Si	24.1	24.1	24.3	24.0	7.5	7.8	8.2	8.0	13.1	13.3	14.3	14.0
Sr	6.42	6.41	5.24	6.17	0.60	0.29	0.56	0.50	3.01	2.31	2.80	2.60
B	1.10	1.05	1.34	1.20	2.27	3.00	2.70	2.60	1.28	1.52	1.70	1.64
Ba	3.90	3.80	3.80	3.98	0.93	1.36	1.32	1.20	1.23	1.53	1.72	1.62
Pb	0.48	0.36	1.20	0.45	1.90	1.80	3.64	2.31	0.93	0.84	2.18	1.52
Sb	0.40	0.30	0.32	0.30	0.56	0.31	0.52	0.45	0.57	0.34	0.52	0.47
Se	0.31	0.22	0.30	0.23	0.42	0.40	0.50	0.40	0.60	0.52	0.68	0.63
Ni	0.75	0.22	0.61	0.33	1.05	0.60	0.90	0.80	0.86	0.52	0.62	0.56
Mn	0.80	0.78	1.01	0.78	0.51	0.62	0.74	0.60	0.60	0.75	0.95	0.80
Mo	0.05	0.02	0.01	0.01	<0.1	<0.1	<0.1	<0.1	0.1	0.1	0.1	0.1
Co	0.03	0.01	0.02	0.02	<0.1	<0.1	<0.1	<0.1	<0.1	<0.1	<0.1	<0.1
As	0.38	0.34	0.32	0.46	0.24	0.30	0.38	0.36	0.44	0.42	0.48	0.46
Sn	0.5	0.4	0.4	0.4	0.8	0.7	0.8	0.8	0.8	0.7	0.8	0.8
Ag	0.4	0.5	0.4	0.4	0.7	0.8	0.7	0.7	0.7	0.8	0.7	0.7
Cd	0.12	0.11	0.12	0.11	0.16	0.17	0.16	0.16	0.13	0.15	0.15	0.18
Cr	0.010	0.016	0.010	0.013	0.012	0.016	0.014	0.015	0.014	0.015	0.014	0.018

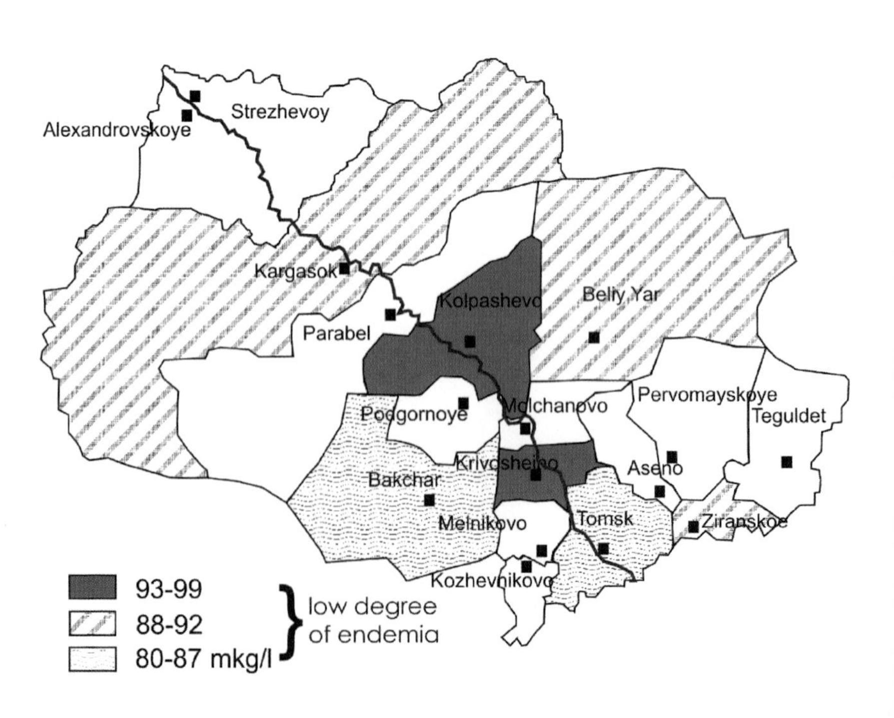

Fig. 4 Distribution of ioduria in children in Tomsk region

Iodine: Trace Elements and Diseases of the Thyroid Gland in the Population in Tomsk Region

Olga A. Denisova, Natalia Baranovskaya, Leonid P. Rikhvanov, and Georgy Chernogoryuk

The problem of iodine∗deficiency disorders (IDD) is significant in Russia. Goiter, of mild-and-moderate degree, is recorded in most regions of the Russian Federation (Dedov and Sviridenko, 2001). West Siberia has moderate iodine deficiency. Areas of severe IDD are to be found in the Smolensk region, beyond and near the polar circle, near the Urals and south of Krasnoyarsk region where the estimated incidence of goiter in children of prepubertal age is 38.6%, compared to 93% in teenagers. Iodine-deficiency studies in Nizhny Novgorod have revealed goiter in 6.38–18% of all children, with levels of iodine ranging from 92.7 to 18.8 mcg/l. The data have confirmed a strong correlation between iodine contents in water and the severity of goiter and prevalence of hypothyreosis. In almost all districts goiter frequency in children varied from 30.1 to 76.8% which fell into the high degree of endemia (Kravetz et al., 2000).

The epidemiological data and level of ioduria does not correspond to the distribution of iodine in soils, plants, and water in the region (Volkotrub et al., 2000). This implies a possible effect of other factors on goiter development. An investigation has been carried out (Rihvanov, 2005; Denisova et al., 2008) on the ecological and geochemical factors controlling the thyroid pathology in Tomsk region to reveal the imbalance of trace element levels in the thyroid gland in different forms of thyroid pathology. The research material was a database of the Tomsk regional endocrinological register from 2001 till 2005. It contains data for about 18,500 patients with various diseases of the thyroid gland. The level of morbidity of diffuse non-toxic goiter, Graves's disease, and nodular goiter in various districts of Tomsk region has been studied. The incidence of thyroid pathology over this period has been analyzed in different parts of the region and compared with the regional mean findings.

Geochemical data consisted of 2,100 measurements of 24 elements determined by neutron activation analysis (Na, Ca, Sc, Cr, Fe, Co, Zn, Br, Rb, Ag, Sb, Au, La, Th, U, Hf, Se, Hg, Ce, Sm, Eu, Yb, and Lu) in soils. Over a 5-year period of observation the incidence of nodular goiter, diffuse non-toxic goiter, Hashimoto's disease, and acquired hypothyreosis has increased by 43–52%. A high incidence rate of all these diseases is noted in the Tomsk district. The Aleksandrovsky, Bakcharsky, Teguldetsky, and Kozhevnikovsky districts are areas with lower incidence rates of all studied thyroid diseases. The area has been divided in three zones due to ecological, climatic, and geographical characteristics (Fig. 5). The first group is around the city of Tomsk which is exposed to anthropogenic factors (fuel and oil industries and nuclear fuel industries). The number of people with diseases is highest here.

The second group includes the districts to the north of 58°. It is affected by cold climate and the pathology level is medium in these districts. The third group is mainly an agrarian district with low disease rate. The content of iodine in soils does not correlate with the thyroid disease level. Strong positive associations were revealed between the chrome content in soil and nodular goiter and between the chrome content in soil and Graves's disease. The chrome content in the soils of this area exceeds the median contents in world soils with 1.5–2 times (Fig. 6).

The southern part of Tomsk region was studied in detail and this is a densely populated part of the region. The main industrial activities are concentrated to this region. The highest incidences (except for nodular goiter) were found in the northeastern part of the district in the wind direction from the industrial activities, the Tomsk-Seversk Industrial Agglomeration. The incidence of thyroid diseases is low in the southwestern area which is mainly agricultural. There is no significant correlation between the level of iodine in drinking water and diseases here. The iodine content in the southeastern part is high but the incidence of nodular goiter in this part is also high. Statistically significant associations have been found between a diffuse non-toxic goiter and the elements Cr, Br, La, U, Fe of the southern part of Tomsk region (Fig. 7).

Elements were analyzed by neutron activation in 97 patients with different thyroid nodules. Among them eight patients had undergone surgery for cancer. For the rest of the patients, 41 had surgery for nodular colloid goiter, 30 for benign adenomas, and 18 for Hashimoto's disease. The age of patients ranged from 20 to 75 years. For a control group 11 thyroid glands

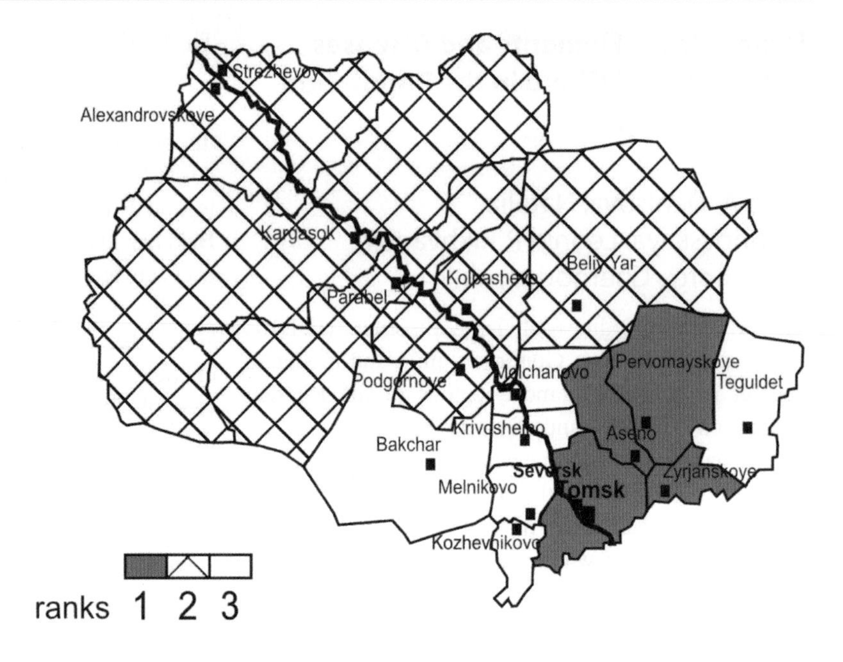

Fig. 5 The zoning of Tomsk region by the sum of indexes of thyroid pathology

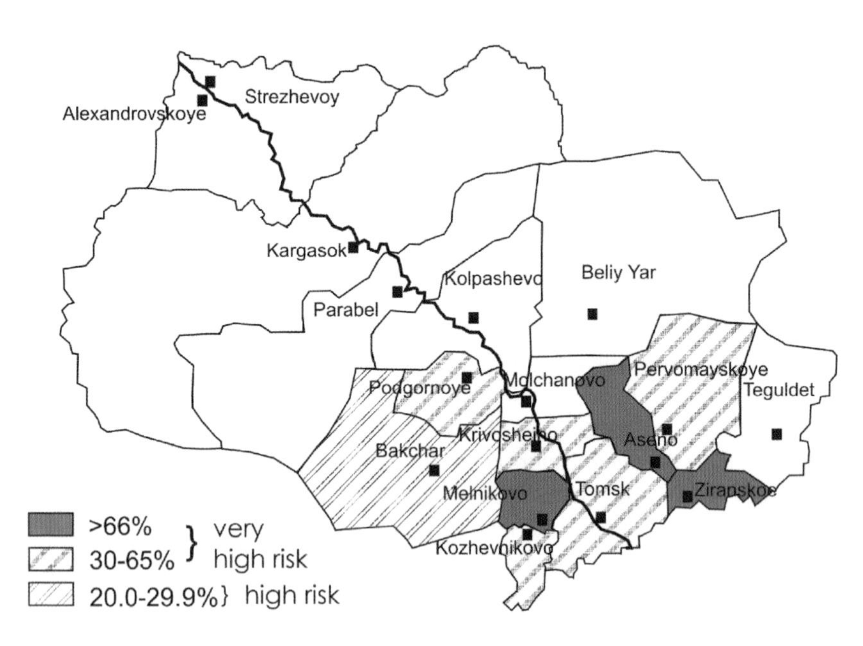

Fig. 6 The correlation between chromium concentrations in soils and thyroid pathology in Tomsk region

were taken during autopsies of people. The criteria for including the patients were a resident of the Tomsk region for not less than 20 years. All groups of thyroid diseases (Table 4) differ significantly from the control in that they have increased accumulation of sodium and iron and decreased levels of calcium and selenium. In thyroid carcinoma and Hashimoto's disease a much greater increase of chromium was observed. Probably both deficiency and excess of this element play a role.

The studies confirmed literature data about the protective role of selenium in the development of thyroid pathology (Veldanova, 2000). Deficiency of this trace element was observed in all forms of thyroid pathology, particularly in carcinoma. In patients with carcinoma its concentration was significantly lower compared not only with the control but also with the group of patients with nodular colloid goiter. The increase of its deficiency versus the excess of many

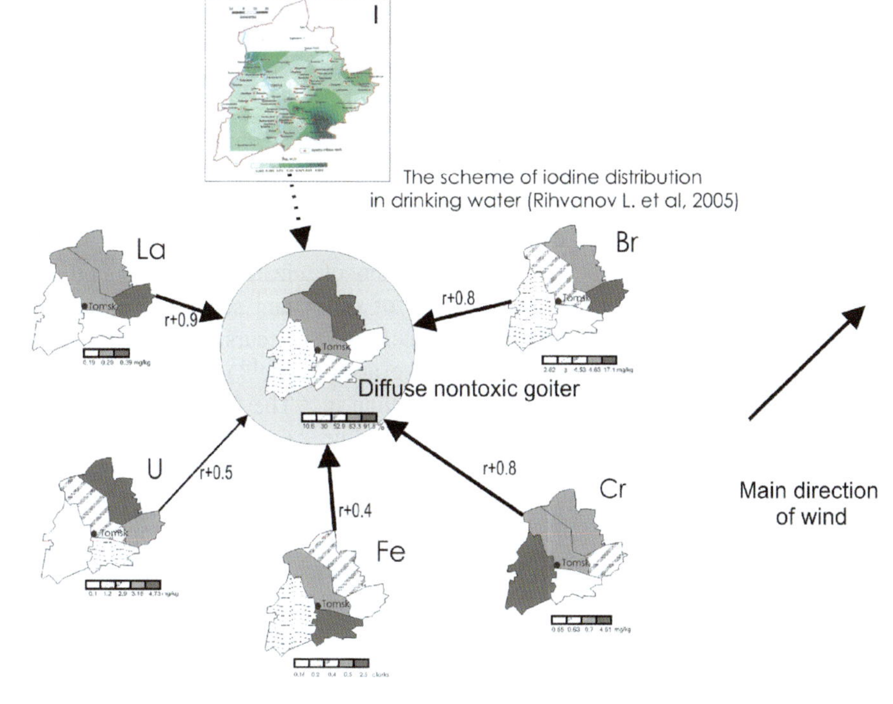

Fig. 7 The correlation between element concentrations and thyroid pathology in the southern part of the Tomsk region

Table 4 Chemical elements in the thyroid gland with different forms of pathology and control, mg/kg (mean)

Element	Forms of the thyroid pathology: Nodular colloid goiter	Benign adenoma	Hashimoto's disease	Thyroid carcinoma	Control
Na	3,113*	3,305*	2,733*	2,600	1,326
Ca	2,904*	1,850*	2,011*	2,785*	5,383
Sc	<0.002	0.004*	0.006	0.006	0.009
Cr	0.63	0.46	0.76*	1.03	0.34
Fe	301*	410*	256*	301*	103
Co	0.07	0.1036	0.087	0.064	0.095
Zn	91	96	85	97.4	88
Br	5.71*	14*	5.34*	6.6*	2.9
Rb	2.02	3.3	3.87*	1.0	1.15
Sb	0.06	0.0414	0.1269*	0.054	0.04
Au	0.01	0.009	0.012	0.01	0.01
La	0.25	0.21	0.24	0.31	0.26
Sm	0.06*	0.05	0.04	0.08*	0.04
Th	0.02	<0.005	0.022	0.018	<0.005
U	<0.1	0.20	0.21	<0.1	<0.1
Hf	0.02	0.022	0.013	0.014	0.012
Se	1.69*	1.63*	1.62*	1.49*	2.8
Hg	0.35*	0.3*	0.22	0.71*	0.16

Note: * – significantly different from the control ($p < 0.05$)

other trace elements is observed at the age of risk for the development of thyroid pathology at ages of 41–55.

In all patients (except patients with carcinoma) the levels of selenium in the thyroid tissue correlate positively with the levels of Zn and Ca. The level of these elements in the thyroid tissue correlates negatively with the iron level in nearly all groups. Besides, in some groups negative correlations between selenium concentration and those of bromine, cobalt, chrome, mercury, antimony, and thorium were found. There may be two explanations of these correlations. Either selenium acts as antagonist for the above-mentioned trace elements or these trace elements substitute for selenium. It is possible that via this mechanism of selenium substitution a known negative role of certain elements in regard to the thyroid gland is found. Selenium deficiency is accompanied with the accumulation of many nonessential elements. Bromine in particular is the most powerful competitor of iodine in the enzymes (Kudo et al., 2006). The unfavorable effect of mercury and cobalt on the thyroid gland was also revealed (Ellingsen et al., 2000; Veldanova, 2000). The positive associations of selenium with these trace elements can be explained by formation of complexes with them as well as their common sources and chains of technogenic migration (Glazovsky, 1982). By substituting for and combining with selenium they aggravate its deficiency which worsens the thyroid metabolism and hormonal synthesis, and results in the decrease of cell resistance to oxidant stress.

Thus, iodine deficiency depends on both anthropogenic and natural geochemical factors in the Tomsk region. This information may give an opportunity to predict thyroid pathology in the population and to plan for preventive measures. Goiter multielement imbalances necessitate the correction of the bodýs microelement profile not only with iodine. It is necessary to plan for nutrition recommendations for selenium and calcium. Correction of chromium deficiency would, however, be difficult because of an excessive level of this element in the thyroid gland. In Tomsk there are institutes collaborating and carrying out research in medical geology, for example, in the hospital and departments of Siberian State Medical University and the Geoecology and Geochemistry Department of Tomsk Polytechnic University. There are plans for studying the influence of geochemistry on morbidity, monitoring of iodine deficiency in the region, microelement composition of the environment, and creating regional databases on the microelement composition of human organs and tissues.

Selenium and Strontium

Iosif F. Volfson

Selenium is a trace element with very low crustal abundance (Fordyce, 2005; Taylor and McLennan, 1985; Frankenberger and Benson, 1994). Phosphatic rocks, organic-rich black shales, coals, and sulfide mineralization, as well as other sedimentary rocks are important seleniferous sources (Fleming, 1980). Selenium occurs in soils in the form of selenites (Se^{4+}), selenates (Se^{6+}), selenides (Se^{2-}), and elemental selenium (Se^{0}). The form of Se is dependent on pH and redox conditions of soils and type of underlying rocks (Kabata-Pendias, 1998). The main natural source of selenium is from food (Fordyce, 2005; Skal'ny, 1999; Skal'ny and Kudrin, 2000). Selenium is accumulated in the kidneys, liver, marrow, myocardium, pancreas, lungs, skin, and hair. The normal adult daily demand is 20–100 μg of selenium. At parental feeding, the human organism must uptake no less than 30 μg selenium daily, because it is essential for the metabolism (WHO, 1987, 1996). Deficiency of selenium occurs if the daily intake of this element is 5 μg or less. The threshold of selenium toxicity is 5 mg/day. Current recommendations of the World Health Organization (WHO, 1987, 1996) concerning the daily Se uptake are 30 μg for females and 40 μg for males. In Russia, the value is established at 63 μg (Ermakov, 2004). Selenium is an important element that has multiple defensive functions in the human organism, strengthens the immune defense, and increases the life time of humans.

Both the Keshan disease (named after the Keshan County afflicted by this disease), which is manifested as degeneration of the heart muscle and chronic cardiomyopathy, and the Kashin–Beck disease (or Urov disease in Russia), which is manifested as dystrophy of muscles, endemic osteoarthropathy, oncologic disease, and iodine deficiency (goiter) are conditioned by low contents of selenium in the food chain (Chen et al., 1980; Fordyce, 2005). In addition, the deficit of selenium in animals leads to the "white muscle disease," which is characterized by the dystrophy of muscles, necrosis of liver, and deficit of protein (James et al., 1989).

An excess Se concentration has been detected in some regions of Yakutia, Tuva, and the southern Urals. Selenium toxicosis is recorded among animal and human population and is conditioned by the surplus uptake of this element together with plants (Astragalus, Stanlea, Happlopappus, and others), which serve as Se accumulators. Such selenium toxicosis is manifested as the "alkaline disease." The main manifestations of selenium excess are unstable emotional conditions; garlic scent from mouth and skin (due to the formation of dimethyl selenide); nausea and retching; abnormalities of the function of liver; erythema of skin; rhinitis, bronchoalveolitis, and edema of lungs (due to breathing the selenium vapor); hair loss; and nail fragility (Hough, 2007; Selinus and Frank, 2000; Ermakov, 2004; Fordyce, 2005).

Calcium, phosphorus, and iodine are a group of elements, the efficiency of which in the body depends on selenium. Disorder of phosphorus and leaching of calcium from the bones under acute deficiency of selenium in the food chain provokes the development of ostheoarthropatic disorders (Urov or Kashin–Beck disease) (Chen et al., 1980).

In Russia, selenium-deficiency belts are found in the northwestern and eastern regions, such as Yakutia and Transbaikalia. In Russia and NIS much interest is focused on selenium as an essential element and its effect on humans. Regions deficient in selenium are Transbaikalia, Karelia (the Kola Peninsula), the Leningrad region, and the Republic of Udmurtia. The Moscow area and other areas of the Central part of Russia are characterized as a belt of selenium deficiency. The deficiency of selenium in humans is caused by low concentrations of selenium in water, soil, and food (Hough, 2007).

The endemic Urov disease (Kashin–Beck disease) was for the first time described in Transbaikalia and later identified in the China and Kazakhstan highlands (Vinogradov, 1938; Vinogradov, 1939; Kovalsky, 1979). There the deficiency of iodine and selenium was revealed in soil and thus in the food chain. The Gazimurozavodsky and Uletovsky areas in the Transbaikalia region are considered to be endemic in selenium and related cardiovascular diseases, retardation of growth, and osteopathy (osteoarthritis deformas endemica) (Ermakov, 1992, 2002, 2004).

The significance of strontium for the spreading of an endemic disease such as the Kashin–Beck (or Urov) disease is a subject of discussion between Russian and foreign scientists (Allander, 1994; Vinogradov, 1939, 1964; Chan et al., 1998; Cabrera et al., 1999; Fordyce, 2005; Kovalsky, 1974, 1977). As was shown above, recent studies of the essentiality of selenium reveal a strong dependence between the four essential elements Se, I, P, and Ca in the human body and the spreading of some endemic and somatic (cardiovascular, oncological, endocrine, bone, etc.) diseases (Fordyce, 2005). Because of high contents of strontium in groundwater and surface waters, the Transbaikalia region is more susceptible to the Urov endemic (Kashin–Beck) disease and osteoporosis (increased fragility of bone system) in comparison to other areas of the world (Avtsyn, 1972; Vinogradov, 1939, 1964; Kovalsky, 1974, 1977). Since strontium, an analogue of calcium, replaces Ca in bone tissues, endemic osteoarthropathy (bone ossification) is quite possible (Avtsyn, 1972; Avtsyn et al., 1991; Kovalsky, 1974, 1977; Kravchenko, 1998; Golovin et al., 2004). The Sr/Ca ratio in the endemic area at the Urov basin river is more than 36 times higher than in chernozem soils. In addition to the Kashin–Beck disease by excessive strontium in soils and waters (up to $5.5 \times 10^{-3}\%$) and low calcium ($<3 \times 10^{-4}\%$), cobalt and iodine in the food chain in clayish and sandy soils of the endemic areas are represented by goiter, vitamin D-resistant form of rickets as well as chondrodystrophy in children and bone fragility (ossification) (Vinogradov, 1939; Kovalsky, 1974).

Arsenic

Igor M. Petrov and Iosif F. Volfson

Introduction

Arsenic (As) is an element detected in low concentrations in virtually every part of the environment. Small quantities of arsenic (about 0.1 mg) are always found in the human body. Arsenic is concentrated in the lungs, liver, skin, nails, and small intestine (Versieck, 1985; Finkelman et al., 1999; Mandal and Suzuki, 2002; Smedley and Kinniburgh, 2005). The estimated safe dose of arsenic is 12–25 μg/day (Nielsen, 2000). Arsenic is efficient as a hormone material and activator. Arsenic seems to indirectly stimulate the function of the hypophysis. In small doses (less than 0.1 mg), As_2O_3 is efficient as a remedy (Paul and Paul, 2002). Using small quantities of this element, one can reduce basal metabolism and oxygen consumption and enhance the assimilation of fat and formation of new red blood corpuscles in the bone marrow. In accordance to W. Paul and H. Paul (Paul and Paul, 2002) an Arabic physician (850–923 AD) recommended As_2O_3 as a healing remedy in cases of blood deficiency, neurosis, and skin diseases. As-rich mineral waters are also efficiently used in cases of chlorosis, rachitis, neuralgia, nervous asthma, body weight reduction due to the basedow-type disease, physical debility, and nervous conditions of inanition (Paul and Paul, 2002). Arsenic can concentrate in the human body due to the shortage of selenium and, thus, promote the deficiency of selenium. Antagonists of As are S, P, Se, vitamins C and E, and amino acids. Arsenic blocks the intake of Zn, Se, vitamins A, C (ascorbic acid), and E, and amino acids (Skal'ny, 1999; Agadjanyan et al., 2001). It is known that a single dose of 0.1 g of As_2O_3 can have a deadly effect, but the sensitivity of human organism for As_2O_3 fluctuates very much. For instance, mountain climbers in the Tyrol used to eat bread with lard or butter containing traces of metallic arsenic to promote appetite, produce rosy complexion, and prevent altitude sickness (Paul and Paul, 2002; Bentley and Chasteen, 2002). Most cases of human toxicity from arsenic have been associated with exposure to inorganic arsenic. Inorganic trivalent arsenate (As^{3+}) is 2–10 times more toxic than pentavalent arsenate (As^{5+}).

High contents of arsenic are found in water, soil, and air in regions of sulfide deposits. There are some areas in Russia and NIS which are known to be high in arsenic: the Central and South Urals, the Far East, the Fergana Valley as well as areas of young orogeny and volcanic activity – the Caucasus, the Kamchatka Peninsula, the East-European platform, the West-Siberian platform, the Scythian-Turan plates, and other (Fig. 8). But there are few epidemiological studies in Russia and NIS on arsenic. It has been possible to forecast unfavorable regions in Russia and NIS. A significant health concern is the enormous deteriorated areas of metallurgy industries in Russia, Kazakhstan, Tadzhikistan, and Armenia, providing enormous amounts of arsenic oxides, formed by roasting of gold-bearing ores and released into the atmosphere, soil, and groundwater.

Arsenic Emissions of Russian Metallurgy and Its Environmental Impact

Arsenic danger for human health is clearly underestimated in Russia especially considering the harmful industrial emissions of the element. This section will pay attention to the gaseous metallurgical emissions containing arsenic mainly in the form of trioxide (As_2O_3), which originates from roasting and smelting of copper, zinc, lead, gold, and other concentrates of nonferrous and precious metals. Regular monitoring of arsenic pollution is not conducted in Russia; data on the arsenic content in the atmosphere in large industrial cities are not available (permissible level of arsenic trioxide content in air is 0.3 mg/m^3). Arsenic precipitated from air is accumulated in soil and snow and is transferred to surface and groundwater and is finally found in crops and food, grown near metallurgical industries (in private gardens and farms and private homestead lands). The content of arsenic in ores ranges from 0.01% in copper–nickel ores to 1.5–2% in gold–arsenic ores (Table 5).

Copper and copper–zinc ores are characterized by intermediate arsenic contents (0.05–0.3%), and similar contents are characteristic also for copper and zinc concentrates obtained from the ores. However, because of the large volumes of concentrate processing, the amount of arsenic from the Ural copper smelters is considerable (900–1,000 tonnes per year from our estimates). A large part of this arsenic goes directly into the atmosphere. It is estimated that arsenic from each Ural smelter emitted into the atmosphere ranged from 100 to 300 tonnes per year until (Petrov et al., 2007). The maximum arsenic emissions are from JSC Svyatogor, Krasnouralsk, Sverdlovsk regions. In recent years, however, some smelter operators have realized that it is necessary to reduce the atmospheric emissions. For instance the reconstruction at SUMZ (Revda, Sverdlovsk region) resulted in considerably lower arsenic atmospheric emissions – from 310 tonnes in 1993 to 14 tonnes in 2004.

It should be noted, however, that most of the Ural smelters have operated for many years (more than 80–90 years) and have caused catastrophic arsenic pollution of the natural environment in these regions. For instance, Karabashmed (former Karabash copper smelter in the Soviet era) operating since 1907, its arsenic atmospheric emission totaled about 96,000

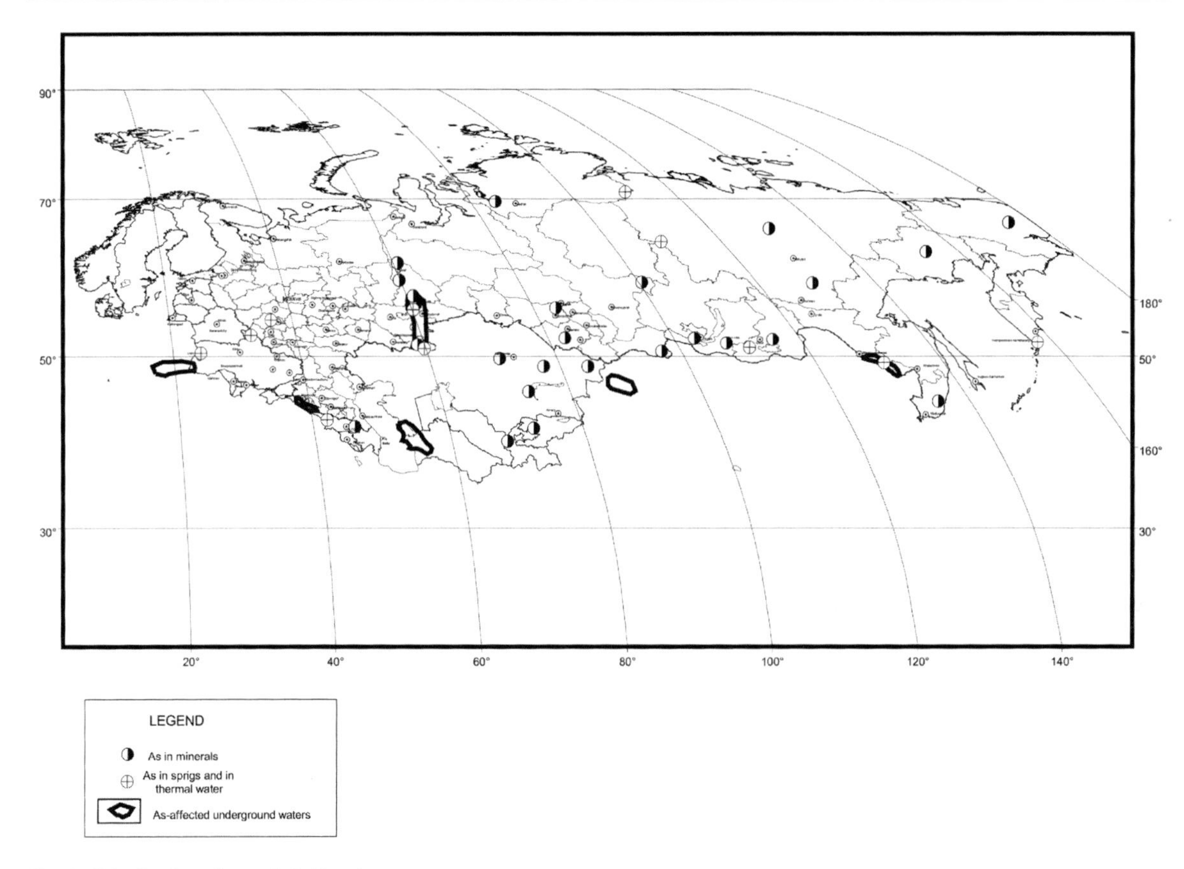

Fig. 8 Distribution of arsenic in Russia

Table 5 Arsenic content in some ore deposits in Russia

Deposit	Ore type	Degree of development	Operator	Arsenic content, %
Nezhdaninskoe	Au	Mined	Nezhdaninskoe Zoloto (Polyus)	1.72
Maiskoe	Au	Being prepared		0.64–0.97
Dzhusinskoe	Cu	Standby		0.25
Uzel'ginskoe	Cu-Zn	Mined	Uchaly GOK	0.21–0.39
Kotsel'vaara-Kammikivi	Cu-Ni	Mined	Pechenganickel (Norilsk Nickel)	0.18
Priorskoe	Cu-Zn	Standby		0.18
Podol'skoe	Cu	Standby		0.11–0.23
Molodezhnoe	Cu-Zn	Mined	Uchaly GOK	0.1–0.3
Uchaly	Cu-Zn	Mined	Uchaly GOK	0.1–0.2
Gai	Cu-Zn	Mined	Gai GOK	0.1–0.15
Pokrovskoe	Au	Mined	Peter Hambro Mining	0.08
Burgachanskoe	Sn	Standby		0.08
Zapolyarnoe	Cu-Ni	Mined	Pechenganickel (Norilsk Nickel)	0.066
Krasnogorskoe	Pb-Zn	Standby		0.06
Novo-Shemurskoe	Cu-Zn	Being prepared	UMMC	0.035–0.05
Vostok	Cu-Ni	Mined	Pechenganickel (Norilsk Nickel)	0.01
Semiletka	Cu-Ni	Mined	Pechenganickel (Norilsk Nickel)	0.01

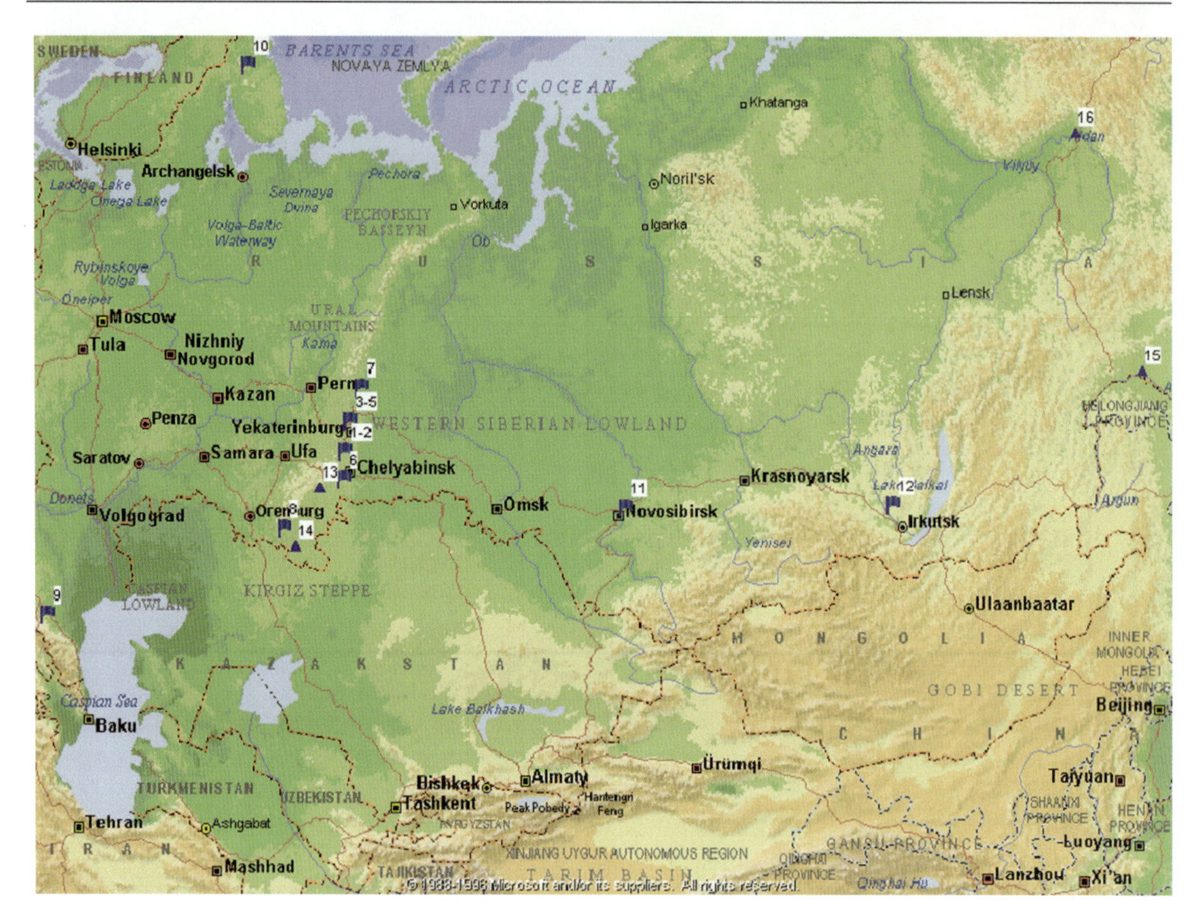

Fig. 9 The territories of Russia, polluted with arsenic from mining-metallurgical activities. Areas of technological pollution of arsenic: 1 – Karabash, 2 – Verkhny Ufalei, 3 – Kirovgrad , 4 – Revda, 5 – Rezh, 6 – Plast, 7 – Krasnouralsk, 8 – Mednogorsk, 9 – Vladikavkaz, 10 – Monchegorsk, 11 – Novosibirsk, 12 – Svirsk. Natural objects with arsenic in high content: 13 – Chelyabinsk territory, Uzelginskoye cooper and zinc deposit, 14 – Orenburg territory, Dzhusinskoye cooper and zinc deposit, 15 – Amur territory, Pokrovskoye gold deposit, 16 – Republic of Sakha – Yakutia, Nezhdaninskoye gold deposit

tonnes! Statistics of arsenic atmospheric emission of the industry is seen in Figs. 9 and 10. The highest volume of emission are from 1970s to 1980s. By now the levels have decreased significantly but nevertheless remains considerable (Petrov et al., 2007).

The ecological situation in Karabash city (Chelyabinsk region), adjacent to the smelter is simply catastrophic. For instance, in 2001 the mortality rate in the city was 20.7 per 1,000 persons (compared to 15.4 in Russia as a whole); in 2005, the figure slightly decreased to 18.8 (15.2 in Russia). The frequency of neoplasms of the adult population of the city is 1.18 times higher compared to the baseline, endocrine system diseases – 1.56 times, hemopathies – 2.85 times, and congenital abnormalities – 1.4 times. Cases of birth of infants without kidneys have occurred. In 2004, neonatal mortality in Karabash city was 20.2 per 1,000 infants (Revich, 2001).

Mednogorsk copper smelter (Orenburg region) produces annual atmospheric emissions of around 50 tonnes arsenic, accompanied by other pollutants. Investigations in 2,000 revealed 1,555 cancer patients per 10,000 persons (and 226 patients died of cancer). Neonatal mortality in the city is 1.5–2 times higher than in Orenburg region as a whole. Note that health hazard in cities, where copper smelters are located, are caused by a spectrum of toxic components in addition to arsenic, for instance, lead, sulfur compounds, etc. (Belan, 2005a, 2005b, 2005c, 2007). Arsenic emission levels at Russian industries manufacturing zinc, lead, and tin metals are lower compared to copper smelters and ranges 0.05–0.6 tonnes per annum (Belan, 2005a, 2005b, 2005c; Gaev et al., 2006; Petrov, 2008).

The "leader" in arsenic atmospheric emission is the Novosibirsk tin combine (NOK), processing tin concentrates from Far East, containing 1–6% As. The

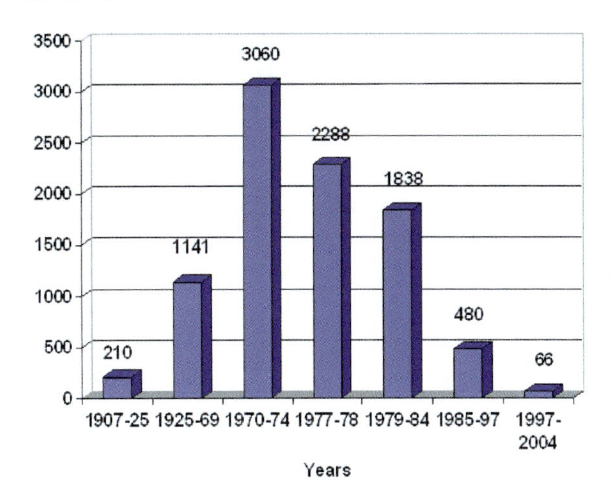

Fig. 10 Average annual arsenic emission for the period of the Karabash copper smelter operation. *Source*: «InfoMine» on the basis of data from web-site of Karabash city (http://www.karabash.ru/)

combine is located in Novosibirsk city with a population above 1 million inhabitants. Outside the city the combinate stores arsenic-containing cakes, originated from the tin concentrate processing. By now, about 6,000 m^3 has been stored, and the dump does not meet modern requirements. The ground at the NOK dump, now used for buildings, contained on the surface 589 mg/kg As (the permissible level is 10 mg/kg) (Revich, 2001).

Arsenic containing zinc concentrates, with up to 0.1% As, are processed by the plant Electrozinc (Vladikavkaz, North Ossetia-Alania). At the present time the arsenic atmospheric emission of the plant is below 0.1 tonnes per annum. In recent years, instead of processing lead concentrates (also containing As), the plant is involved in battery scrap processing, also environmentally dangerous. Electrozinc has been operating for about 100 years but has paid attention to environmental protection only in recent years. In Vladikavkaz city, in addition to atmospheric emission, environmental hazards are also caused by solid wastes produced by Electrozinc (and Pobedit plant), covering around 40 Ha in the city. These wastes contain more than 8,000 tonnes arsenic. In dry season, the waste dumps actively pollute the atmosphere and soils with arsenic over large areas. The ecological situation in Vladikavkaz is also considered to be close to catastrophic. In the city, where around a half of North Ossetia's population lives, birth of infants with congenital abnormalities is 65–69 per 1,000 infants. Neonatal mortality in the city is now above 20 (per 1,000). (Menchinskaya and Zangiyeva, 2007).

Because of larger volumes of nickel ore processing, compared to the Ural smelters, other nickel combinates in Russia – Norilsk, Pechneganickel, Severonickel – produce large arsenic atmospheric emissions. However, reliable data on the combinates' emission are not available, except data on arsenic content in dust atmospheric emission of Severonickel – 7–11 tonnes per year. Among other places with increased arsenic content in the natural environment are roasting plants (Plast city, Chelyabinsk region) processing dusts from nonferrous metallurgy activities (containing up to 70% As) and As-containing wastes of gold mining over the whole of Russia. From 1950s to 1990s the industrial atmospheric emission was up to 100 tonnes per year but in recent years, after introducing wet cleaning of exhaust gases, the emission decreased to 7–10 tonnes per year.

Investigations have shown arsenic in hair and nails of the whole population of Novotroitsk settlement, where a roasting plant was located. Arsenic is also found in mince and berries, collected nearby. Concerning cancer, Plast city is among the leaders in the Chelyabinsk region. Arsenic atmospheric emissions are also generated from ferrous metallurgy. Arsenic is contained in coke (5–7 ppm) and iron ore sinter (up to 20 ppm) is used in iron production. In the process, large amounts of arsenic are transferred to gaseous phase in the form of As compounds, but monitoring of arsenic atmospheric emission of ferrous metallurgy industries is not conducted. Besides arsenic atmospheric emissions, health hazards are produced by technogenic arsenic, accumulated in the form of compounds in tailing ponds, metallurgical wastes, ash dumps, etc. For instance the content of arsenic in effluents of some concentrators is as high as 400 mg/l. Concentration tailings containing pyrite with arsenic are actively oxidized in the tailing ponds that can transfer arsenic to groundwater. The largest tailing ponds containing pyrite-bearing tailings are found at JSC Gai GOK (Orenburg region), JSC Ychaly GOK, and JSC Bashkir copper–sulfur combine (JSC BCSC, Bashkortostan) (Belan, 2000, 2005a, 2005b, 2005c, 2007; Petrov, 2008). Large sources of arsenic pollution can be found in wastes of the Angarsk metallurgical plant, Irkutsk region. The plant manufactured the so-called "white arsenic" in 1934–1949. The arsenic pollution sources are the plant ruins and wastes of arsenopyrite cinders (around 130 kt). Areas of dangerous soil pollution stretches along the water storage basin shore (about 9 km long and 3–3.5 km wide). In this area, As content in the soils is 300 times above the permissible level. This arsenic penetrates into the groundwater: drilling reveal arsenic at a depth of 20 m close to a permeable aquifer.

In conclusion arsenic and health in Russia is crucially important. Besides a need for permanent monitoring of arsenic in atmosphere, water, and soil, one key problem is the introduction of effective technologies for gas cleaning, as well as processing of As-containing materials, and the building of reliable storage facilities for As-containing wastes. Of special importance is the need for a research program on the impacts of arsenic and its compounds on human health.

Medical Geology in Connection to Mining

Armen K. Saghatelyan, Lilit V. Sahakyan, Olga V. Menchinskaya, Tamara D. Zangiyeva, M.Z. Kajtukov, and Z.H. Uzdenova

Medical Geology in Armenia and Mining

The assessment of the impact of the mining industry on the environment and on trophic chains and humans was performed at the Kajaran copper–molybdenum deposit located in south Armenia. Monitoring of heavy metal pollution of the soils of Armenia's regions (Grigoryan and Saghatelyan, 1990) has shown that Mo, Cu, Pb, and a number of other metals in Armenia have high natural background values often exceeding the established maximum acceptable concentrations (MAC) for these elements. When assessing anthropogenic pollution, it is also necessary to consider the element concentrations in the geological environment. Thus assessments should include the elements in both the natural and the anthropogenic environment.

An example of such an approach is the situation around the Kajaran deposit. The deposit is situated in the boundaries of the ore region identified by V. Kovalsky (Kovalsky, 1974) as a biogeochemical province enriched by Mo and Cu. The concept of "biogeochemical province" was suggested by Vinogradov as a biogeochemical region based on the homogeneity of life and geochemistry. According to Vinogradov "Deficiency or the excess of an element in the environment vs. the accepted Clark value, induces changes in the flora and fauna of the given region or biogeochemical province" (Vinogradov, 1938). To obtain a generalized characteristic of the geochemical situation of Armenia's ore regions, Saghatelyan studied the element composition of alluvial formations because the composition of such a matrix reflects typomorphic components and their contents in the biogeochemical province (Saghatelyan, 2004). For certain ore regions, ranked series of natural associations of chemical elements have been calculated through standardizing element contents by their background contents. For the Kajaran ore region, the following range of typomorphic elements was established: Mo–Cu–Sn–Pb–Zn–Mn–Be. As seen from the series, it is dominated by the two basic ore elements of the Kajaran copper–molybdenum deposit. The deposit is huge with a reserve of over 1 billion tons of ore, containing about 8% of the global reserves of Mo. The deposit was mined for copper as far back as the 1850–1860s. In 1951 the Zangezur mining and dressing plant was constructed and has been operating with daily output and treatment up to 24,000 tons of ore containing 0.055% of Mo and 0.21% of Cu. Besides base elements, Re, Se, Te, Au, Ag, Pb, Zn, Bi, Cd, Sb, As, Co, Ni, Hg are also found in the ore.

The mining and dressing plant is a dominant enterprise at the city of Kajaran. Studying the city's soil pollution concerning heavy metals (Saghatelyan et al., 2008b) the background-standardized mean has been reported: $Mo_{(29.8)}$–$Cu_{(5.1)}$–$Pb_{(2.2)}$–$Co_{(1.7)}$–$Zn_{(1.4)}$–$Mn_{(1.2)}$.

In brackets, the ratio of excess vs. the background is given. This clearly demonstrates the anthropogenic impact on the natural background which is important to know when establishing values of maximum pollution. To identify the risks connected with anthropogenic pollution the same has been done for drinking water and the food chain. The risk factors are shown in Fig. 11.

During a period of time, mining waste was deposited in three tailing dams located close to the

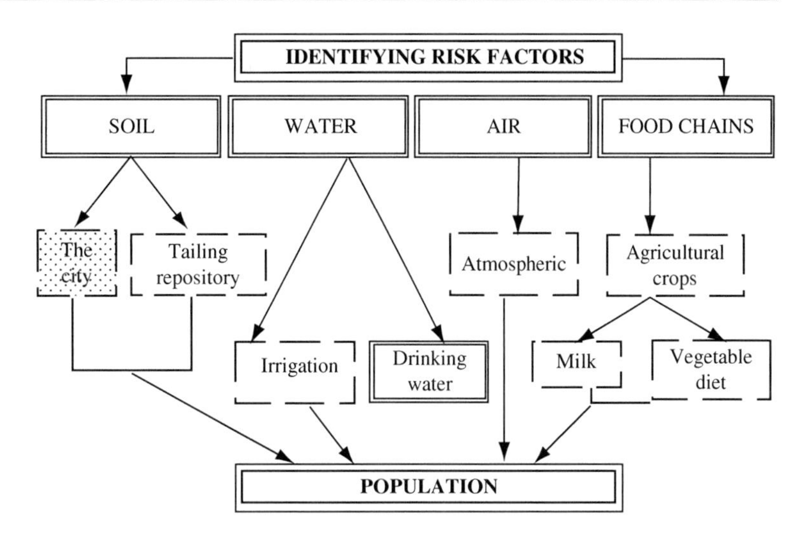

Fig. 11 Identification of risk factors to the population (*dotted line* indicates significant risk factors)

city. In the 1980s the tailing dams were re-cultivated and covered with a 30-cm soil layer. After the former USSR disintegration and the following social and economic crisis the tailing dams were abandoned and gradually local people started using them for agricultural purposes as pastures and for home gardening. Studies of the former tailing dams indicate that in some places the recultivation layer has been eroded and the material is actively weathered and is transported as dust into the city. Thus the tailings are pollution sources adding to the quarry, the mining, and the dressing plant. Ecological and geochemical assessments of the contents of heavy metals in the soils of the tailings indicate high contents of Mo, Cu, V, Hg, As. The analysis of agricultural crops growing in these areas (Saghatelyan et al., 2007a, b) indicate MAC-exceeding contents of Cu, Mo, Pb, Ni, Cr, Hg in the basic dietary products such as potatoes, beans, beet, herbs. A substantial part of the tailing repositories are used as pastures (Fig. 12) and the analysis of the fodder plants shows excessive contents of many elements vs. MAC. In particular, in the most widespread grass-clover (*Trifolium pretense L.*) Mo, Ni, Cr, Hg showed 131.0, 6.3, 20.2, 28.8 time excess vs. MAC. In order to monitor further migration of heavy metals in the food chain also the quality of milk of the cows grazing on the studied site have been studied. The results show excessive Pb, Hg, As.

This herd of cows grazing on the Voghchi tailing repository site indicates that many families living in the city of Kajaran regularly consume agricultural crops high in toxic elements. Studies of drinking water supplied to the city indicate no excess of heavy metals above the accepted standards.

The major air pollutant in the city is from dust. The dust load in the city shows a season-dependent variation: the mean deposition in winter is 101.7 kg/sq km, in summer – 418 kg/sq.km reaching 800 kg/sq.km in eastern and western parts of the city. The main elements are $Mo_{(7.5)}$–Cu, $Zn_{(4.1)}$–$Ni_{(3.3)}$–$Pb_{(2.6)}$–$Mn_{(1.1)}$. In brackets, the ratio of the excess vs. the background is given. Also Hg, As, Cd were found, posing an additional risk factor for public health (Saghatelyan et al., 2008a).

The assessment of risk factors indicates that the citizens are exposed to a heavy load of environmental pollution with heavy metals that can enter the body through respiratory tracts and the food chain. To establish the impact level on humans, children were selected as the most sensitive risk group in the population (Saghatelyan et al., 2007a, b). Hair samples of 12 children of preschool age were collected. The results of hair analyses compared to biologically allowable levels (Skal'ny and Kudrin, 2000) are shown in Fig. 13.

Based on the biological role of microelements and their involvement practically in all biological processes running in the human organism, to express all pathologic processes induced by microelement deficiency, excess, or disbalance, a notion called "microelementosis" was introduced (Avtsyn et al., 1991). BAL determines lower and upper limits of standard contents of

Fig. 12 Herd of cows grazing on the Voghchi tailing repository site

elements in the organism. The research has provided evidence of significant risk factors to the public health as a result of the operation of the Kajaran mining and dressing plant. The considerable excess of heavy metal contents in the hair of children vs. biologically allowable levels is an alarming fact supporting the urgency of public health protection actions.

Health Effects of Tungsten–Molybdenum Mining

Russia is one of the few countries in the world that is self-supported with metals but this has also caused widespread effects on the environment and human health by mining and processing. In this section the effects on the environment and on human health in Tyrnyauz (Republic Kabardino-Balkaria) are discussed using the technology of *ECOSCAN*. The ECOSCAN technology was developed in IMGRE for such projects in the early 1990s (Menchinskaya and Zangieva, 2004). It is based on the exposure of connection between environmental geochemical parameters and epidemiological markers of disturbance in population's health. Ecological assessment of environment can be supplemented by modeling and calculating the ecological hazards on given habitats, as well as recommendations for rehabilitation of the disturbed natural habitats and preventive medical measures for recovering the population's health. The *ECOSCAN* technology uses geochemical and epidemiological data to assess the health impacts. Data for fertile women and children of 0–14 years were used in this work.

Tyrnyauz is a tungsten–molybdenum deposit including elements such as Cu, Bi, As, Ag, Sn, Zn, Se, Te, Re, Ge, Ga, Cd, and In. Before 1994 Tyrnyauz was second in order in Russia concerning air pollution but is now shut down. The industrial activities are located on a slope 300 m above the city. Detailed geochemical samples of soil (A and B layers), sediments, surface water, and plants have been carried out (a total of 829 samples). As a result of data treatment of the soil sampling it was detected that 30% of the examined territory was heavily contaminated, many times more than limited permissible concentrations (LPC) for several of the elements. Forty percent of the soil was slightly contaminated, and only 30% of the territory is characterized by background dispersion of the elements, not exceeding LPC.

The analysis found three genetically different groups of elements: technogenic, natural, and urbanogenic. The first group includes ore elements and related elements such as W, Mo, Cu, Bi, As, Sn, Zn, Ag, and Ge (high concentration and general dispersion). The maximum is in the areas of the tailing depots or in the central part of the city (W 1,500 ppm, Mo > 300 ppm, Cu < 8,000 ppm, As < 6,000 ppm, Bi > 80 ppm, Sn > 100 ppm, Ge 15 ppm). Increased Ag values (0.15–10.0 ppm) are typical for areas of tailing deposits and for the part of the city intended for building. Unfortunately, it is only possible to estimate how many times the concentrations exceed normal values for As, Cu, and Sn. For Cu, As, and Sn, LPC is exceeded 240, 3,000, and 22 times, respectively.

The second group of elements is Cr, Co, Ni, V, and B. These reflect the natural geochemical background in the area. Increased concentrations are found in the valleys and increase when moving away from the city. Pb is typical for the third group. Its concentrations are very low in the ore and also in the tailings. However, increased concentrations of Pb, exceeding LPC 2–3 times, are typical for the housing areas with a maximum of 5,000 ppm.

Hydrochemical results from Bucksun river after the Ore Mining and Processing Enterprise (OM&PE) had stopped did not detect any high values of the main pollutants. The composition of water was analyzed for 73 elements at the analytical centre of IMGRE (Institute of Mineralogy, Geochemistry and Crystal Chemistry

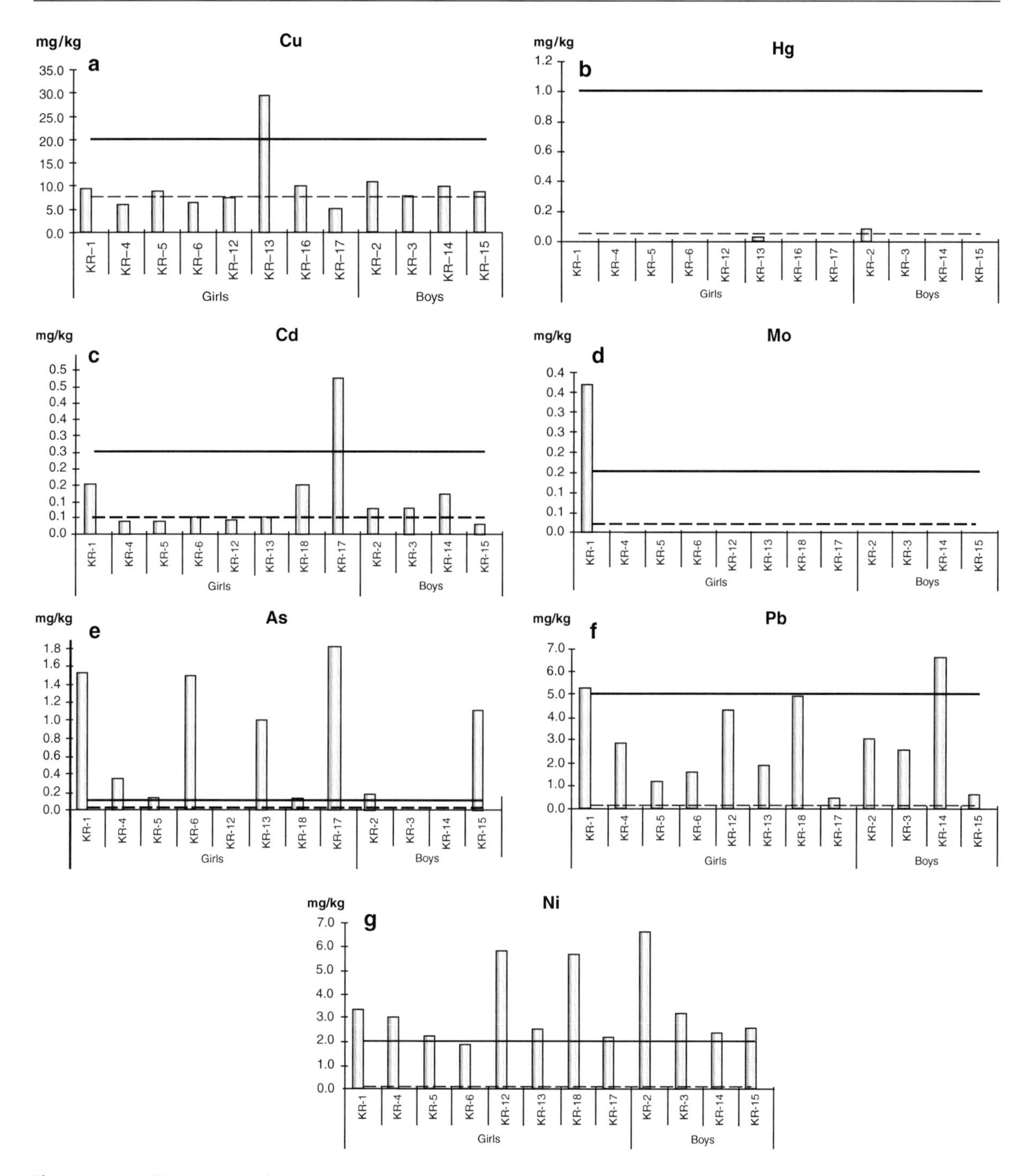

Fig. 13 (a–g). The contents of heavy metals in the hair of children living in the city of Kajaran. Biologically allowable levels: Lower limit of BAL (__); Upper limit of BAL (- -)

of Rare Elements). Analysis of drinking water was also done. All the elements are related to the second—third danger class (classification of elements, based on hazard degree for human, accepted in Russia) except for those elements which have no danger class (Ga, Cs, La, etc.). Concentrations exceed LPC only for Li and B. Maximum levels for Li (by 6 times) and B (by 2.6 times) were found in drinking water. Also high levels of Mo were found in drinking water. All samples also have high iron levels.

For biogeochemical sampling, trees (poplars) were chosen as they are spread widely in the region. Twenty-one leaf samples were taken. The distance between the samples was 500 m inside the city and 1,000 m outside. All the samples have high concentrations of Mo (20–30 ppm) and W (10 ppm), on the right bank of Bucksun opposite the OPF (Ore-Processing Factory). The highest concentrations (1,500 ppm) are for Zn and B. High Zn is typical for urban areas. High B (<1,500 ppm) is typical for the all the valley and is explained by turmalinization and high B in granites located in the valley. There are also high levels of Pb and Cu.

Health Aspects

The most sensitive persons to xenobiotics (a chemical compound, drug, pesticide, or carcinogen that is foreign to a living organism) are fertile women and children. An indication of the bad environmental conditions is pathological changes in people's health. The emphasis is on detection of the ecology-depended pathology. Examples of indexes are perinatal and infant mortality, miscarriages, inherent defects, childish malignant growths.

The analysis of the health indexes of Tyrnyauz was done by medical data from the Elbrus district in the republic Kabardino-Balcaria which includes Tyrnyauz. The indexes of 1997 and 1999 were used (Bolshakov et al., 1999). The examined group included fertile women and 1-year-old children. The perinatal mortality rate is divided into two groups: the perinatal death and death on the 1st week of life (28th week of the pregnancy – 7th day of life). In 2 years the perinatal mortality increased five times (Table 6). During the same time the neonatal mortality rate increased three times. The causes of mortality are inherent defects and anomalies of developing. Examination of the fetuses detected inherent defects such as unencephalopathy (2), enteric athrasy (1), and hydrocephaly (1). All the fetuses were aborted. There was also a tendency of increase of the extra-genital pathology and complex pregnancy (Kulakov et al., 1996).

Table 6 Data on perinatal mortality

Year of birth	Year 1997	Year 1999
Number of children born	348	283
Prematurely born	6.1%	8.1%
Perinatal mortality	3.5%	20.1%
Early neonatal mortality	3.5%	11.5%

Ecological Hazards Affecting People's Health

Ecological hazards were calculated in the examined areas according to the method of the Environment Protection Agency (EPA) in the United States. This method is based on counting the hazard index for people's health when non-cancerogenic matters affect it. The main pathways of xenobiotics are ingestion and inhalation. An assessment of the consequence of the dose was made for the most toxic elements entering the organism. Low sensitivity of the spectral analysis did, however, not allow for the full picture of the soil arsenic. Only areas with >100 ppm were detected (more than 50 times of limit permissible concentration [LPC]) covering a third of the area.

During the assessment of the carcinogenic effects, it is assumed that xenobiotics cause hazards only when exceeding safe effect limits. Such doses are marked by EPA as reference doses (RfD) or reference concentrations (RfC). Non carcinogenig hazards were calculated for Mo. RfD was taken from the site of IRIS (Integrated Risk Information System – EPA's Office of Research and Development, National Center for Environmental Assessment) – http://epa.gov/iris/

Hazard Quotient for the non-cancerogenic matters is calculated as a ratio: $HQ = C/RfC$ or $HQ = D/RfD$. HQ for Mo = *6.4*.

In the State Epidemiological Control criteria of the health hazard value it is recommended to calculate the hazard twice: by the RfC/RfD and by Russian hygienic standards, and in case of big differences to use stricter standards. Using this Russian system to calculate HQ for Mo resulted in $HQ = 2.6$. The same calculation was made also for B. Using the IRIS standard gave $HQ = 14.2$; using the Russian system gave $HQ = 51.2$. In case of B the Russian standard is stricter. When chemical matters enter the organism simultaneously in the same way Hazard Index – $HI = \sum HI_i$ should be calculated. Using stricter standards this gave $HI = 57.6$.

Conclusions

Geochemical research of the Tyrnyauz tungsten–molybdenum industrial complex shows significant technogenic pollution connected with dumps and drains by the factory during a long time (since 1940). Main pollutants are As, W, Mo, Zn, Pb, Cu with maximum levels of 6,000, 1,500, 300, 3,000, 5,000, and 8,000 ppm, respectively. Arsenic pollution with soil concentrations over 100 ppm covers third of the territory. W pollution covers all the territory and the average concentration is higher than 40 times its Clark value. In the Bucksun River Li exceeds LPC, in drinking water – Li and B exceeds LPC, but this is connected with the natural geochemical background.

Assessment by "dose–effect" has some problems caused by the very restricted experimental data base. Absence of Russian and international standards of W does not allow for estimating the hazard. Absence of data related to air As concentrations does not allow for calculating carcinogenic hazards, but the presence of As in high concentrations in soil and bottom sediments allows us to link it with serious inherent defects detected during the perinatal period. Nevertheless, calculated hazard assessment even for non-main hazard factors detected exceeds the safety limit. The calculated hazard index for non carcinogenic chemicals is high and implicates that the hazard should not be ignored. In the authors view, the most rational courses for improvement of the health status in the region are following measures: elimination of the harmful trace elements and their toxic metabolites, deletion of the detected xenobiotic toxic affects, changes in people's allergic, immunologic, and somatic status, and deletion of extra-sensitive reactions.

Natural Radon Distribution in Soils, Influence on Living Organisms, and Radon Safety Measures in Russia

Anastassia L. Dorozhko

Introduction

Many people within Russia and NIS were moved in the 19th–20th century to prison camps and institutions of repression which also included harsh mining areas. The hard forced labor of the miners and disastrous ecological conditions in underground mining were used as alternatives for punishment for different crimes, including antiauthority actions. One example is the rebels of the Decembrists in Saint-Petersburg in 1825 who protested against the Tsarś autocracy and were sentenced to these terrible conditions. Most of the rebellions were exiled in remote regions of Transbaikalia, to silver mines where the prisoners were subject to unfavorable environment. The most serious health danger of the Transbaikalia's miners and prisoners was radon. In addition to this, also dust from the lead-, silver-, cadmium-, and zinc mines.

There are many high radon areas within Russia and NIS. For example, the Alpine and Himalayan tectonic belt, Mongolia and Okhotsk tectonic belt, Big Caucasian ridge, Kopet-Dag ridge, Kazakh and Turkmen Caspian region, the Thuran Plate region as well as tectonic blocks of the East European Platform (Pechenkin and Pechenkin, 1996) (Figs. 14, 15, 16, 17, and 18). Radon is also typical for the regions with uranium deposits. At the copper- and molybdenum Kadzharan mine (Armenia) and in Akchatau mining county (deposit of tungsten Central Kazakhstan) high radon can, for example, be found in local schools and daycare centers (Revich, 2001; Saghatelyan et al., 2008).

The general population is exposed to radiation from natural radioactive nuclides. The most important are potassium-40 and nuclides of the uranium and thorium series. Based on estimates of the average external gamma-radiation in Finland and Sweden, the annual effective dose per person related to various sources of radon is as follows: 0.2 mSv (millisievert) from potassium in the body, 0.3 mSv cosmic radiation, and 0.5 mSv from nuclides of the uranium and thorium series. Domestic radon accounts for an annual effective dose of 2 mSv from inhalation estimated at 100 Bq/m^3.

Radon has effective half-life $T^1\!/_2 = 3.825$ days, with α-radiation. There are three forms of radon: actinon (^{219}Rn), thoron (^{220}Rn), and radon (^{222}Rn), but the use of the term refers specifically to radon-222 (^{222}Rn). All of them are members of the natural radioactive series. Radon contributes up to two-thirds of the dose from natural sources. It is believed that radon accounts for 40% dose of the irradiation received by population from natural sources of radiation. It is important to note

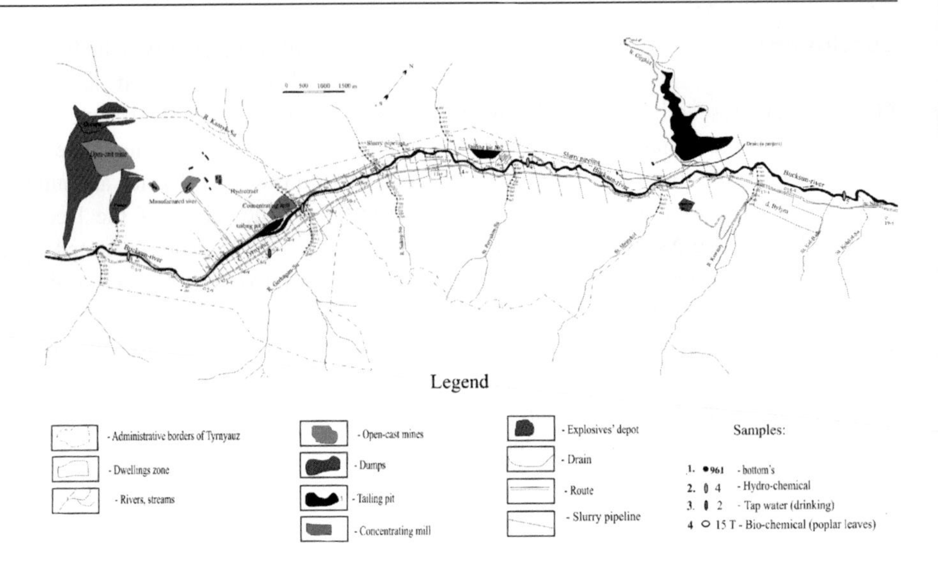

Fig. 14 Tyrnyauz ore-processing enterprise

that this gas represents the second most important risk factor for lung cancer after smoking. The annual mortality rate due to radon irradiation from lung cancer is 13,000 persons in the United States, around 3,000 in England, and more than 20,000 in India. In Russia, this rate can be even higher (Bogdasarov, 2008). Negative and positive (in small dose) impacts of natural radon on humans, animals health, and plants have been studied all over the world (Komov and Frolov, 2004; Appleton, 2005; Låg, 1987, 1990).

According to radiation safety norms accepted in Russia, average annual radon volumetric activity (VA) should be no more 200 Bk/m^3 (Volfson and Bakhur, 2007). Otherwise, buildings require anti-radon protection, which is aimed at underpinning and installation of additional ventilation systems in basements. This measure is absolutely necessary, because the influence of domestic radon on human health has been confidently confirmed. In human organism, which intakes radon in the form of aerosols and accumulates it in bronchi and alveolus, a detrimental effect is induced on lungs tissue.

Despite the many negative aspects, radon is broadly used in medical practice (Tsarfis, 1991; Finkelman, 2008). Radioactive waters are divided into four groups based on radon activity (kBq/l): very weak – 0.75, weak – 1.5, medium – 7.5, and high – above 7.5. First of all, the radon-containing water of natural sources is used for radon baths, which have a healing effect. Radon decay produces new, short-lived isotopes. Of special interest for medicinal purposes are the short-life products of radon decay. Together with minerals and salts, these decay products induce intrinsic transmutation in the organism of a sick person. The so-called active pellicle formed on the human body during radon bath acts for around 3 h, rendering beneficial effect on human organism. Radon baths can be prepared from water saturated with radon. The radon baths can render a dualistic influence on the human organism, e.g., reduction of the uric acid concentration in blood and hematogenic organs, metabolism disorders, diseases of peripheral nervous system, gynecological and skin diseases, hypertensions and overactivity of thyroid gland, and problems of locomotor apparatus and joints, on the one hand, and promotion of anti-inflammatory and analgesic action, prophylaxis and treatment of cardiovascular diseases on the other hand (Tsarfis, 1991).

High radon concentrations are detected in nitrogen- and oxygen-rich mineral and carbonic acid waters. Sources of natural mineral water containing radon, broadly used as SPAs, are widespread all over the world. The most famous among them are Choltubo (Georgia), Jalal Abad (Kyrgyzstan), Baranovichi (Belarus), Baden-Baden (Germany), Karlovy Vary and Ioahimov (Czech Republic), and Belokurikha (Russia) (Volfson et al., 2010).

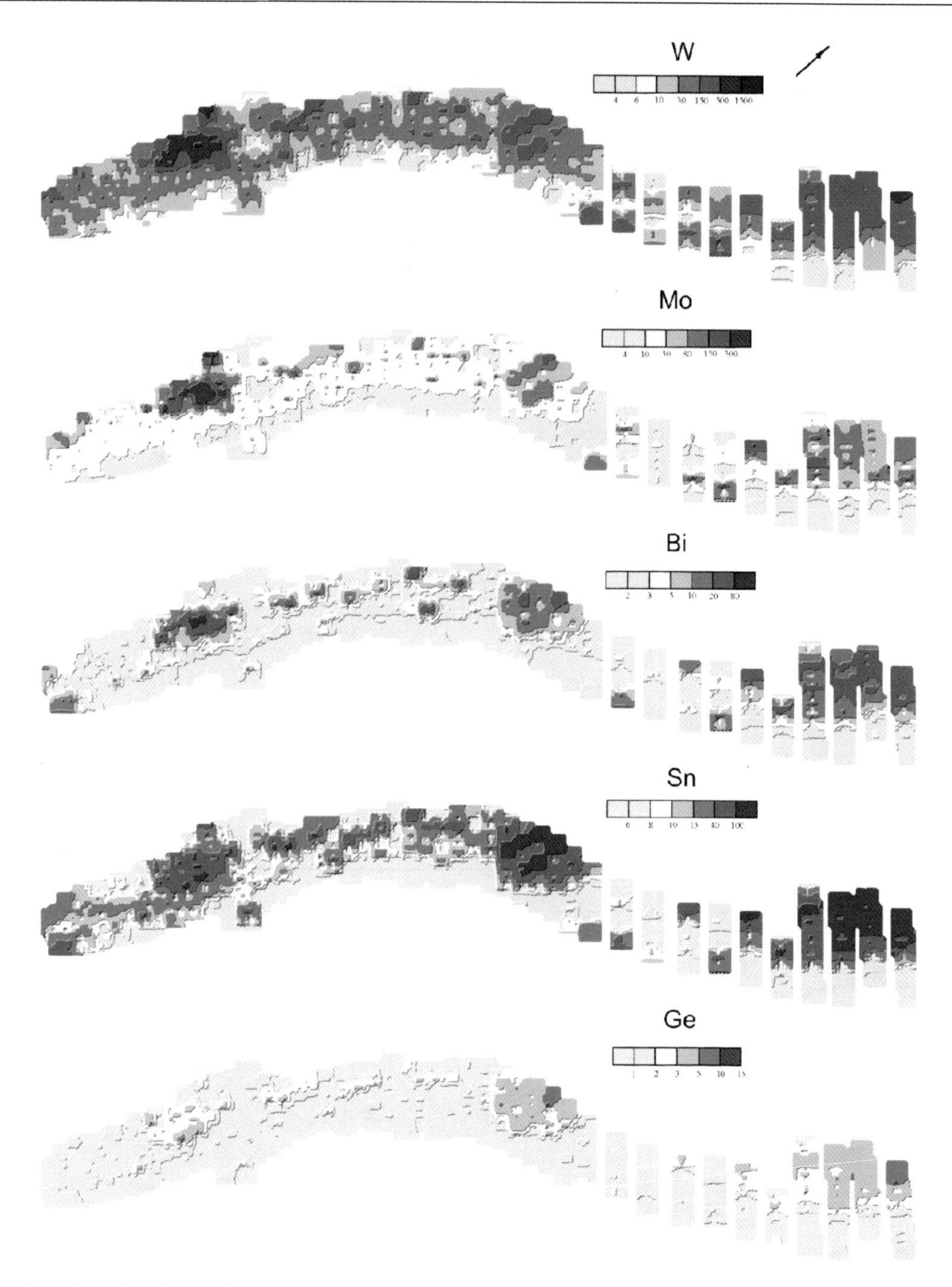

Fig. 15 Dispersion of W, Mo, Bi, Sn, and Ge in soil of the Tyrnyauz area, ppm

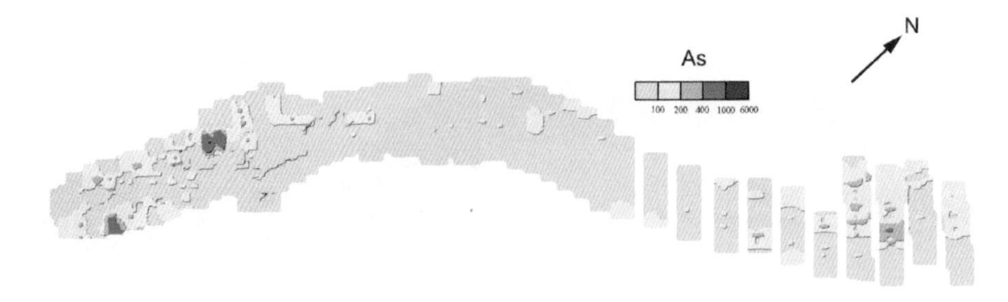

Fig. 16 Dispersion of as in soils of the Tyrnyauz area, ppm

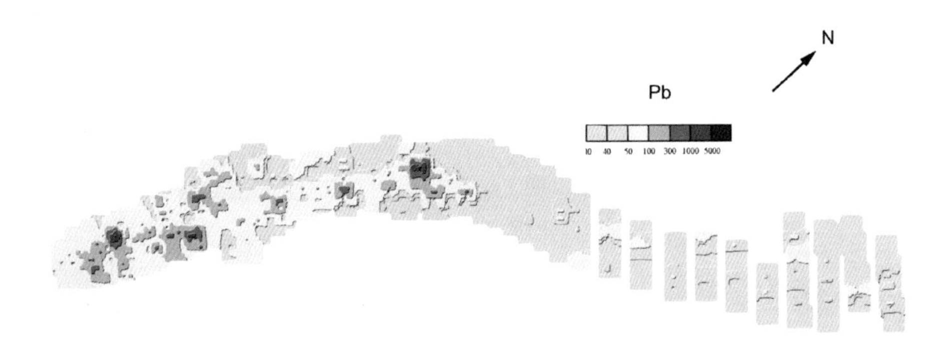

Fig. 17 Dispersion of Pb in soils of the Tyrnyauz area, ppm

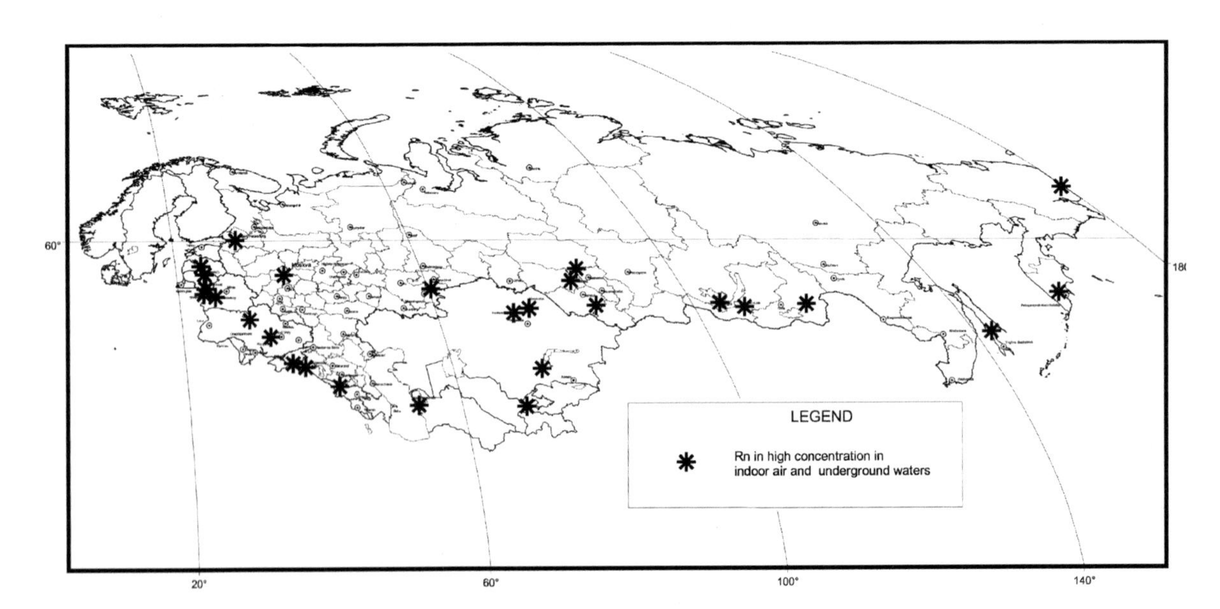

Fig. 18 Areas of high radon in indoor air and groundwater

Natural Radon Distribution and Radon Safety Measures in Russia

The effects of natural radon on the population caught the attention of Russian scientists for the first time in the beginning of the 1980s. The WHO pays increased attention to the radon problem (WHO, 1996; Komov and Frolov, 2004). Until recently there has been no medical multifactorial statistical research in Russia on the correlation between radon concentration in houses and cancer. The most significant research has recently been carried out at the Institute of Industrial Ecology in Yekaterinburg. The research has been carried out in houses with different radon concentrations in three cities including Pervouralsk, Karpinsk, and Lermontov (the Lermontov city has a uranium mine). Two groups of people were chosen in each city (112–237 people in each group): a group of healthy people and a group of people with lung cancer. The health of each participant was studied for 25 factors. Among them were radon volumetric activity (VA) and thoron (thorium emanation) equilibrium volumetric activity (EVA), which were constantly measured in dwellings for more than a month. Factors as smoking, age, gender, and chronic illnesses were highly ranked in all three regions. In Lermontov which is situated in an area of high concentration of uranium in bedrock, working exposure to radon was another important factor. If the sum of all studied factors contributing to cancer is 100%, the share of natural radioactive sources causing lung cancer will be 0.5% in Pervouralsk, 1% in Karpinsk, and 10% in Lermontov. Despite the fact that these cities are located in "dangerous radon zones," the radon concentration level in dwellings in these areas cannot be a cause of lung cancer. Nevertheless, preventive measures of radon danger are necessary, especially in such megalopolises as Moscow. According to radiation safety norms in Russia an average annual radon EVA should be no more than 100 Bk/m^3 when designing new dwellings and 200 Bk/m^3 in old ones otherwise installation of additional ventilation systems in basements are required. The only criteria to measure radon safety on building land used in Russia is radon flux density (RFD) at the surface, which should be measured before a foundation pit is dug. This measure may not be correct because RFD at the bottom of a foundation pit can differ from that at the surface. According to the sanitary code, radon building protection should be carried out if the average of RFD is more than 80 mBk/m^2 on the surface, and for industrial buildings if it is more than 250 mBk/m^2.

RFD is a very variable factor and can vary within small homogeneous areas for no apparent reason. RFD space distribution is logarithmically normal and its fluctuation in clayey soils can be as high as 270 mBk/m^2. In addition, it has been proven that RFD measured in soil depends on the season. The most constant radon flux is observed during summer time. The most significant RFD fluctuations take place in autumn and spring due to frequent rains which temporarily block radon flux. RFD flaring (more than 500 mBk/m^2) can be observed at this time, probably due to radon flux redistribution caused by uneven soil moisture. Besides that, greater RFD value (up to 1,000 mBk/m^2) can be registered during air pressure falls. Therefore, the RFD value is determined by uranium concentration in soil and underlying rocks, and radon diffusion as well as convection processes associated with soil air flows caused by weather changes. Thus, single RFD measuring on the surface is not enough for a reliable estimation of radon safety.

Research has been carried out on RFD data during 8 years in Moscow (Dorozhko and Makarov, 2007; Miklyaev, 2008; Ziangirov et al., 1999; Miklyaev and Ziangirov, 2004). Four hundred sample points in these areas vary from 6 to 1,583 mBk/m^2, with a normal interval of 10–40 mBk/m^2. Geodynamic active zones are characterized by fissures and porosity of rock and such zones can serve as pipes not only for radon but also for radium. In the Moscow region Jurassic clay is the most radioactive geological agent and after that in descending order are Quaternary clay, glacifluvial clay and loam, covering loam, carboniferous limestone and cretaceous, and Quaternary sands. In analyzing the RFD distribution within Moscow, four areas are identified which differ in distribution and intensity of anomalies (Fig. 19). In this figure the correlation between RFD values and Quaternary rocks with different lithologic characteristics in Moscow territory can be seen.

The most intensive area of RFD intensity I is mainly associated with covering loams in the northern part which are Quaternary. The least intensive area II of RFD intensity is associated with prevalent glacifluvial sands. Area III is characterized by average RFD intensity and in general is associated with Moscow river

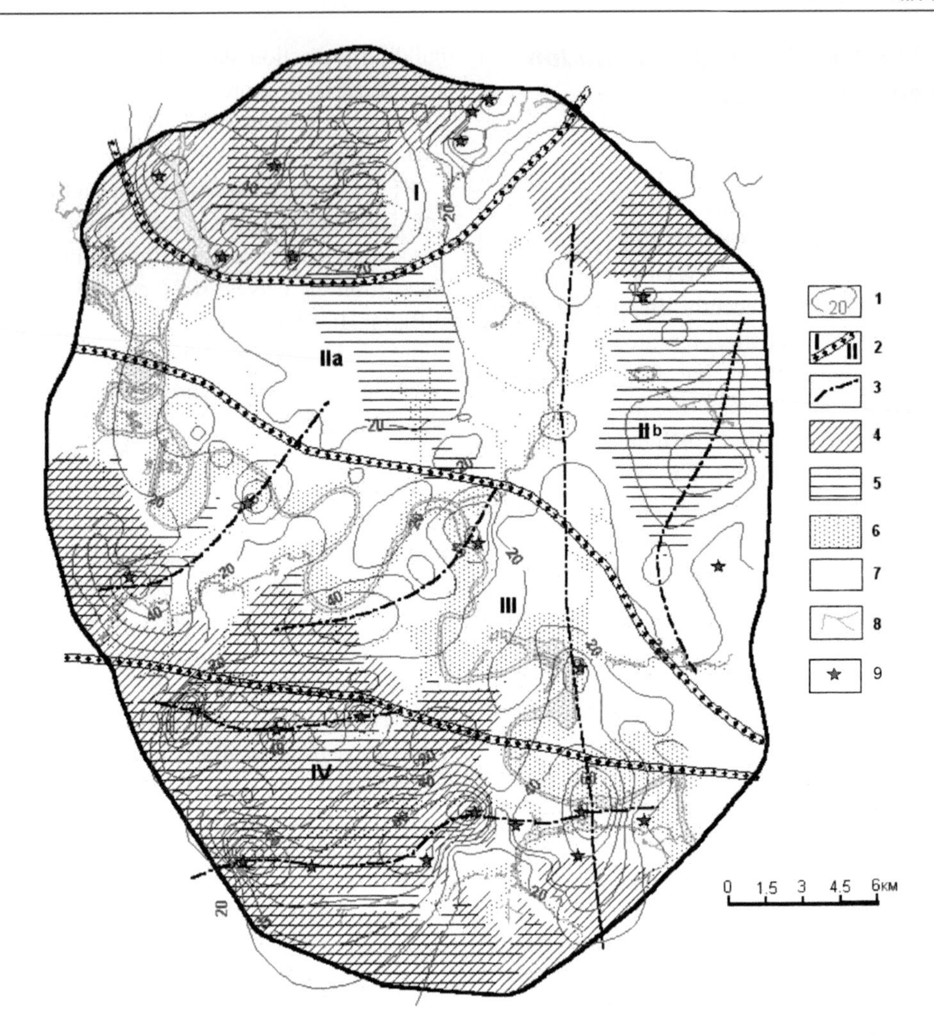

Fig. 19 Correlation between RFD values and quaternary rocks with different lithologic characteristics in Moscow territory: 1 – isolines of RFD with section 10 mBk/m^2 s; 2 – area boundaries; 3 – axes of anomalies; 4–7 – pleistocene–holocene rocks: 4 – covering loam, 5 – moraine clay, 6 – alluvial sands, 7 – fluvioglacial sands; 8 – rivers; 9 – points with high RFD values

valley. Intensive area IV with a prevailing maximum value of RFD intensity (up to 115 mBk/m^2) is associated with loams, except the eastern part where the anomaly (max = 73 mBk/m^{2}s) is found in the area of alluvium (Miklyaev and Ziangirov, 2004). There is a clear correlation between RFD intensity and radioactive composition of near-surface soil, but no correlation with pre-Quaternary rocks. However, this does not explain the presence of several anomalies with high RFD values on the general background of high or low RFD values caused by lithology. Therefore a study was carried out on the correlation between the RFD field and structural, geomorphological and geodynamic characteristics. The geomorphological map of Moscow is used for this. Different crustal blocks are identified on this map as well as structural lineaments (Fig. 20). Area I is associated with the Rublev-Yauzskaya (RYa) zone, and the majority of the RFD maximum values are found in the area of boundaries between regional geoblocks, as well as other geodynamic active zones. Maximum RFD values that are found on a certain distance from such zones can be explained as follows:

1. Exposure of underground geodynamic active zones, as well as other heterogeneities on the surface can be widespread, and this dispersion depends on depth and morphology of these heterogeneities. This explains the fact that radon maximum values are found some distance from the zones;

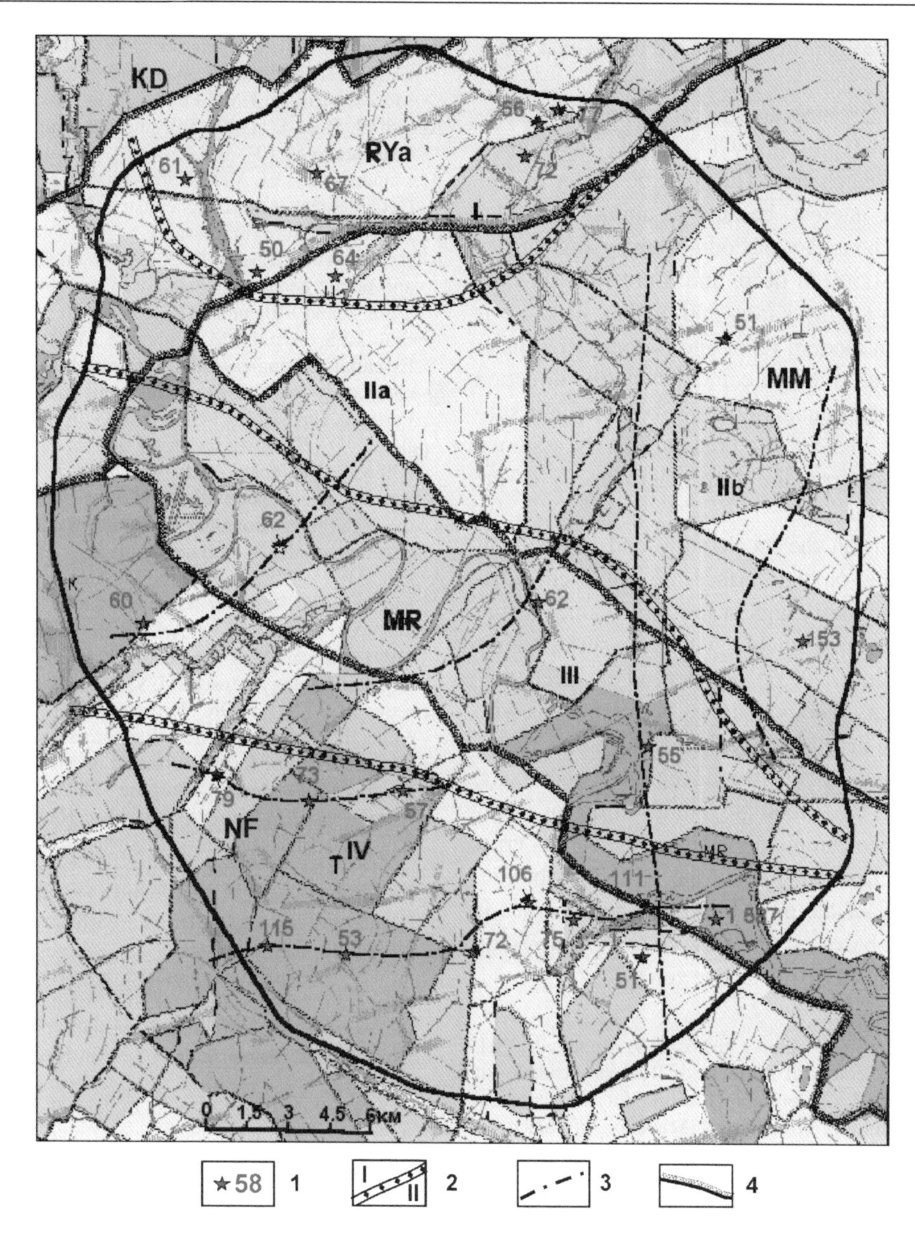

Fig. 20 Structural geomorphological map of Moscow with RFD anomalies: KD – Klinsko-Dmitrovsky geoblock, MM – Moskovsko-Mescherky geoblock, NF – Narofominsky geoblock, RYa – Rublev-Yauzskaya regional zone, MR – Moskvoretskaya regional zone. C – central geoblock, T – Teplostansky geoblock. 1 – points with RFD values more than 50 mBk/m^2 s, 2 – area boundaries, 3 – axes of anomalies, 4 – regional geoblock boundaries

2. Radon (or radium), as an element soluble in water, can be transferred by groundwater some distance from its original source (zones of extensive jointing).

Area II can be associated with Moscow-Meschersky (MM) regional geoblock. One of the longest axes of RFD anomalies, which passes through the areas IV, III, and II, coincides with the boundaries of the low block considered to be an earth crust stretching zone. Area III is characterized by local RFD anomalies belonging to the Moscow river valley. This area can be associated with the Moscow regional structural zone, which is characterized by the lowest absolute heights and continued tendency to depression of the Moscow river valley structure. The majority of the

RFD maximum values in this area are found near block boundaries.

The most interesting area IV almost entirely coincides with the Teplostansky (T) block, the highest block in the research area. The north boundary of area IV is of most interest, because it almost coincides with the north boundary of Podmoskovny ancient rift, which is 800–1,500 m deep, and a pre-Jurassic flexure (Dorozhko and Makarov, 2007). This structure is not observed in modern relief and is considered inactive. The anomalies do not coincide with the direction of neither blocks boundaries, however, they are subparallel to the ancient rift boundary. It is assumed that these axes mirror the deep heterogeneity of the sedimentary cover.

As a conclusion the localization and intensity of radon anomalies on the surface are not evenly distributed. The primarily high or low RFD values within each areas are conformable with near-surface soils containing radium and do not depend on deep pre-Quaternary rocks. The boundaries between areas with different RFD distribution coincide to some extent with the boundaries between regional geoblocks. Together with the lithologic composition of near-surface soils, neotectonic divisions of the earth's crust also play an important role in the RFD distribution. Future forecast of the most important radon zones is possible if it is based on the knowledge of neotectonic structures and geodynamic active zones.

Acknowledgments The authors thank D.V. Rundkvist (V.I. Vernadsky State Geological Museum of Russian Academy of Sciences, Russia), I.V. Florinsky (Institute for Mathematical problems of Biology of Russian Academy of Sciences, Russia), L.N. Belan (Bashkir State University, Russia), W. Paul (Miner's Christian Mission, Celifodina, Spokane, USA) for fruitful discussions and materials, H.C.W. Skinner (Departments of Geology and Geophysics and Orthopaedics and Rehabilitation, Yale University, New Haven, USA), M.I. Savinykh (Sibdalmumie, Novokuznetsk, Russia) for a useful suggestion, as well as N.A. Serper and I.E. Lyubimova (Fedorovsky All-Russian Research Institute for Mineral Resources, Moscow, Russia) for librarian assistance.

References

Abrahams PW (2002) Soils: their implications to human health. Sci Total Environ 291:1–32.

Admakin OI (1999) Dental Morbidity of Residents on Different Climatic-Geographical Regions of Russia. Moscow. 256 p (in Russian).

Agadzhanjan NA, Veldanova MV, Skal'nyj AV (2001) Ecological portrait of man and role of microelements. Moscow, Russia: KMK Publisher (in Russian).

Aggett PJ, Rose S (1987) Soil and congenital malformations. Experientia 43:104–108.

Aleksandrovskaia EI, Aleksandrovskij AL (2003) Historical and Geographical Anthropochemistry. Moscow, Russia: RAS IG–NGA-Nature (in Russian).

Allander E (1994) Kashin-Beck disease: an analysis of research and public health activities based on a bibliography 1849–1992. Scand J Rheumatol 23(S99):1–36.

Amaral AFS, Arruda M, Cabral S, Rodrigues AS (2008) Essential and non-essential trace metals in scalp hair of men chronically exposed to volcanogenic metals in the Azores, Portugal. Environ Int 34:1104–1108.

Appleton JD (2005) Radon in air and water. In O Selinus, B Alloway, JA Centeno, RB Finkelman, R Fuge, U Lindh, P Smedley (eds) Essentials of Medical Geology: Impacts of the Natural Environment on Public Health (pp. 227–262). Amsterdam, the Netherlands: Elsevier.

Avtsyn AP (1972) Introduction in Geographical Pathology. Moscow, USSR: Medicina (in Russian).

Avtsyn AP, Zhavoronkov AA, Rish MA, Strochkova LS (1991) Human Microelementoses: Etiology, Classification, Organopathology. Moscow, USSR: Medicina (in Russian).

Beiseev AO (2004) Ore Formations of Natural Medical Minerals, their Properties, Resources in Kazakhstan and Prospects of use for Manufacture of Medical Products and Products of Medical Purpose Ph.D. Thesis. Almaty, Kazakhstan: Satpaev Kazakh National Technique University (in Russian with English abstract).

Belan LN (2007) Geoenvironmental Essentials of Natural and Technological Ecosystems of Mining Areas of Bashkortostan. Ph.D. Thesis. Bashkir State University, Ufa (in Russian).

Belan LN (2005a) Environmental and Hydro Geochemical Condition of the Territory of Mining Enterprises (by the example of JSC BCSC) Vestnik of Voronezh State University, 2, Voronezh, 173–177 (in Russian with English abstract).

Belan LN (2005b) Medical and Biological Habits of Mining Areas. Vestnik of Orenburg State University, 5, Orenburg, 112–118 (in Russian with English abstract).

Belan LN (2005c) Environmental and geochemical condition of mining areas of the Bashkir trans-Ural. Vestnik of Orenburg State University, 6, Orenburg, 113–117 (in Russian with English abstract).

Belan LN (2000) Noninfectious Pathology and Environmental and Geochemical Habits of Mining Areas. Proceedings of USC RAS, Ufa, 34–35 (in Russian).

Bentley R, Chasteen TG (2002) Arsenic curiosa and humanity. Chem Educat 7:51–60.

Bogdasarov AA (2008) Radon Curiosa. Advantages and Disadvantages of Invisible Sly. JSC "Brestskaya Tipographya", Brest (Republic of Belarus). 64 p (in Russian).

Bolshakov AM, Krutjko VN, Pucillo EV (1999) Assessment and Control of Environment Effect Hazards on the Peoples' Health. Moscow, Editorial URSS, p. 256.

Breus VE (1981) Ca, P and F concentration in human teeth enamel. Stomatology 6:52–54.

Buseck PR, Galdobina LP, Kovalevski VV, Rozhkova NN, Valley JW, Zaidenberg AZ (1997) Shungites–the C-rich rocks of Karelia, Russia. Can Mineral 35:1363–1378.

Cabrera WE, Schrooten I, Debroe ME, D'Haese PC (1999) Strontium and bone. J Bone Miner Res 14:661–668.

Chan S, Gerson B, Subramaniam S (1998) The role of copper, molybdenum, selenium, and zinc in nutrition and health. Clin Lab Med 18:673–685.

Charykova MV, Bornyakova II, Polekhovskii YS, Charykov NA, Kustova EV, Arapov OV (2006) Chemical composition of extracts from shungite and "shungite water". Russ J Appl Chem 79:29–33.

Chen X, Yang G, Chen J, Chen X, Wen Z, Ge K (1980) Studies on the relations of selenium and Keshan disease. Biol Trace Elem Res 2:91–107.

Criteria of Assessment of the Population's Health Hazards for Predominated Environment Elements-Pollutants: Methodical Recommendations. M.: Sisyn Scientific Research Institute of Human Ecology and Environmental Hygiene RAMS (2001) Sechenov Moscow Medicine Academy, Centre of the State Epidemiology Control in Moscow.

Dedov I, Sviridenko I (2001) Strategy of eliminating iodine-deficiency diseases in the Russian federation. Probl Endocrinol 48:3–12 (in Russian with English abstract).

Denisova O, Chernogorjuk G, Rikhvanov L, Baranovskaja N (2008) Ecogeochemical factors and diseases of a thyroid gland of the population in Tomsk region In Proceedings of the 33rd International Geological Congress, Oslo, Norway, Aug. 6–14, 2008. X-CD Technologies (CD-ROM).

Dorozhko A, Makarov V (2007) The distribution of natural radon in the Moscow region. Medical Geology Newsletter, 10.

Dowker SEP, Anderson P, Elliott JC, Gao XJ (1999) Crystal chemistry and dissolution of calcium phosphate in dental enamel. Mineral Mag 63(6):791–800. 7.

Edmunds M, Smedley P (2005) Fluoride in natural waters. In O Selinus, B Alloway, JA Centeno, RB Finkelman, R Fuge, U Lindh, P Smedley (eds) Essentials of Medical Geology: Impacts of the Natural Environment on Public Health (pp. 301–329). Amsterdam, Netherlands: Elsevier.

Ellingsen DG, Efskind J, Haung E et al (2000) Effects of low mercury vapour exposure on the thyroid function in chloralkali workers J Appl Toxicol 20:483–489.

Elliott JC (1997) Structure, Crystal Chemistry and Density of Enamel Apatites. Dental En. Wily, Chichester (Ciba Foundation Symposium 205), 54–72.

Elliott JC (1994) Structure and Chemistry of the Apatites and Other Calcium Orthophosphates. Amsterdam, Netherlands: Elsevier, 398 p.

Ermakov VV (1992) Biogeochemical regioning problems and the biogeochemical selenium provinces in the former USSR. Biol Trace Elem Res 33(3):171–185.

Ermakov VV (2004). Biogeochemistry of selenium and its role in prevention of endemic diseases of man. Vestnik Otdelenia nauk o Zemle RAN, 22. http://www.scgis.ru/russian/cp1251/h_dgggms/1-2004/scpub-4.pdf (in Russian, with English abstract).

Ermakov VV (2002) Urov Kashin-Beck disease: Ecological and Environmental Aspects/Mengen- und Spurenelemente. 21 (pp. S. 899–909), Leipzig: Schubert-Verlag.

Esembaeva S (2006) The Prevalence of Non Caries Damages of Hard Teeth Tissues at Population of Rural Areas of Republic of Kazakhstan. Vestnik of Youzhno-Kazakhstanskoi Medicinskoi Academii, 2 (28) (in Russian with Kazakh and English abstract).

Farrakhov E, Miletenko N, Pechenkin I, Pronin A, Volfson I, Beiseyev O, Bogdasarov M, Komov I (2008) Sedimentary basins: Medical and geological aspects of the studies In Proceedings of the 33rd International Geological Congress, Oslo, Norway, Aug. 6–14, 2008. X-CD Technologies (CD-ROM).

Finkelman RB (2008) Health benefits of geological materials and geologic processes. In International Medical Geology Association, *Short course on medical geology, the 33rd International geological congress, Oslo, Norway, Aug. 10, 2008* (CD-ROM).

Finkelman RB, Belkin HE, Zheng B (1999) Health impacts of domestic coal use in China. Proc Natl Acad Sci 96:3427–3431.

Fleming GA (1980) Essential micronutrients II: Iodine and selenium. In BE Davis (ed) Applied Soil Trace Elements (pp 199–234). New York: John Wiley & Sons.

Florinsky IV (ed) (2010) Man and the Geosphere. Hauppauge, NY: Nova Science Publishers (in press).

Fluorides and oral health (1994) WHO Technical Report Series, No 846, Geneva. 37 p.

Fordyce F (2005) Selenium deficiency and toxicity in the environment. In O Selinus, B Alloway, JA Centeno, R. Finkelman, R Fuge, U Lindh, P Smedley (eds) Essentials of Medical Geology: Impacts of the Natural Environment on Public Health (pp. 373–415). Amsterdam, the Netherlands: Elsevier.

Frankenberger WT, Benson S (eds) (1994) Selenium in the Environment. New York, NY: Marcel Dekker.

Frank-Kamenetskaya OV, Golubtsov VV, Pikhur OL, Zorina ML, Plotkina YV (2004) Non-stoichiometric apatite of human teeth hard tissues. Age changes. Proc All-Russ Mineral Soc 5:120–130, 6.

Freatherstone JD (1993) Prevention and reversal of dental caries: Role of low level fluoride. Community Dent Oral Epidemiol: 280.

Frolova LN, Kiseleva TL (1996) Structure of chemical compounds, methods of analysis and process control chemical composition of mumijo and methods for determining its authenticity and quality (a review) Pharm Chem J 30(8). Translated from Khimiko-Farmatsevticheskii Zhurnal, 30(8):49–53, August, 1996. Original article submitted July 25, 1995.

Gaev AY, Albakasov DA, Gatskov VG, Belan LN, Blinov SM, Mikhailov YV, Alferova NS (2006) Geoenvironmental Tasks of Mining Areas (by the example of Orenburg territory, the South Urals). Vestnik of Orenburg State University, 6, 77–84 (in Russian).

Ghosal S (2006) Shilajit in Perspective. Oxford, UK: Alpha Science International.

Glazovsky N (1982) Technogenic substance streams in biosphere (4th ed.). In The Mining Operations and Geochemistry of Natural Ecosystems. Moscow: Science (in Russian).

Golovin AA, Krinochkin LA, Pevzner VS (2004) Geochemical specialization of bedrock and soil as indicator of regional geochemical endemicity. Geologija 48:22–28.

Grigoryan SV, Saghatelyan AK (1990) Monitoring of Heavy Metal Pollution of the Soils of Armenia's Industrial Regions.

Abstract Book of the Conference "Ecological Problems of Living Nature Protection", Moscow, 116–117 (in Russian).

Grushevoy GV, Pechenkin IG (2003) Metallogeny of Uranium-Bearing Sedimentary Basins of the Central Asia. Moscow, Russia: Fedorovsky All-Russian Research Institute for Mineral Resources (in Russian, with English abstract).

Hough RL (2007) Soil and human health: An epidemiological review. Eur J Soil Sci 58:1200–1212.

James LF, Panter KE, Mayland HF, Miller MR, Baker DC (1989) Selenium poisoning in livestock: a review and progress. In LW Jacobs (ed) Selenium in Agriculture and the Environment (pp. 123–131). Madison, WI: Soil Science Society of America.

Kabata-Pendias A (1998) Geochemistry of selenium. J Environ Pathol Toxicol Oncol17(3–4):137–177

Kariev AR (2007) Medical aspects of Tadzhikistan' bentonite application. In V Gavrilenko, E Panova (eds) Proceedings of the 2nd International Symposium "Bio-inert Interaction: Life and Stone", Saint Petersburg, Russia, June 26–29, 2007 (pp. 124–126). Saint Petersburg, Russia: Saint Petersburg University Press (in Russian).

Klimas A, Mališauskas A (2008) Boron, fluoride, strontium and lithium anomalies in fresh groundwater of Lithuania. Geologija 50:114–124

Kolesnik AG (1997) Fluoride Monitoring on Dentistry. Moscow. 119 p (in Russian).

Komov IL, Frolov OS (2004) Methods and Facilities for the Assessment of the Radon-Hazard Potential. Kiev, Ukraine: Logos.

Kovalsky VV (1979) Geochemical ecology and problems of health. Phil Trans Roy Soc Lond Series B Biol Sci 288:185–191.

Kovalsky VV (1977) Geochemische Ökologie, Biogeochemie. Berlin: Deutscher.

Kovalsky VV (1974) Geochemical Ecology. Moscow: Nauka (in Russian).

Kravchenko SM (1998) Environmental impact on the tolerance of hydroxylapatite of human bone: setting up a problem. In IV Melnikov (ed) Mineralogical Studies in Handling Environmental Problems: Proceedings of the Conference on Ecomineralogy, Moscow, Russia, Jan. 29–30, 1996 (pp. 97–105). Moscow, Russia: Institute of Geology of Ore Deposits, Petrography, Mineralogy and Geochemistry, Russian Academy of Sciences (in Russian).

Kravetz E, Gratsianova N, Oleynik O et al (2000) State of children and teenagers health with the thyroid pathology. Russian Pediatric J 80:14–16.

Kudo Y, Yamauchi K, Fukazawa H et al (2006) In vitro and in vivo analysis of the thyroid system-disrupting activities of brominated phenolic and phenol compounds in Xenopus laevis. Toxicol Sci 92:87–95.

Kulakov VI, Karetnikova NA, Stigar AM (1996) Searching the ways of introwomb correction of the inherent defects. Russian Bulletin of Perinatalogy and Pediatry No 3 (in Russian).

Kuz'mina EM, Smirnova TA (2001) Fluorides on Clinical Dentistry. Moscow. 32 p (in Russian).

Kuz'mina EM (1999) Dental Morbidity of Russia Residents. Moscow. 228 p (in Russian).

Låg J (ed) (1990) Geomedicine. Boca Raton, FL: CRC Press.

Låg J (1987) Soil properties of special interest in connection with health problems. Experientia 43:63–67.

Larina OV (2006) Treatment and Recovery of Hair. Moscow, Etherna. 288 p.

Mandal BK, Suzuki KT (2002) Arsenic round the world: A review. Talanta 58:201–235

Menchinskaya O, Zangiyeva T (2007) Ecological and medical habitat quality assessment. International Medical Geology Association Medical Geology Newsletter, Newsletter NO. 11, ISSN 1651-525, September 2007, pp. 16–18.

Menchinskaya OV, Zangieva TD (2004) Technology ECOSCAN in Ecological-Geochemical Assessment of Metallurgic Centers Territories. Applying Geochemistry, No 5, M.: IMGRE, p. 178–188 (in Russian)

Miklyaev P (2008) Causes of space-time fluctuations of soil radon flux density. Proceedings of the 33rd International Geological Congress, Oslo, Norway, 6–14 August, 2008 (#10-94140). http://www.x-cd.com (CD-ROM).

Miklyaev PS, Ziangirov RS (2004) Regularity of radon volatilization from soils into the atmosphere on the territory of Moscow. J Geoecology 3:244–250

Narbutaite J, Vehkalahti MM, Milčiuviene S (2007) Dental fluorosis and dental caries among 12-yr-old children from high- and low-fluoride areas in Lithuania. Eur J Oral Sci 115:137–142.

Nielsen FH (2000) Possibly essential trace elements. In JD Bogden, LM Klevay (eds) Clinical Nutrition of the Essential Trace Elements and Minerals: The Guide for Health Professionals (pp. 11–36). Totowa, NJ: Humana Press.

Nikolishin AK (1989) Teeth Fluorosis. Moscow. 45 p (in Russian).

Pachomov GN (1982) First Prevention on Dentistry. Moscow. 240 p.

Patching SG, Gardiner PHE (1999) Recent developments in selenium metabolism and chemical speciation: A review. J Trace Elem Med Biol 13:193–214.

Paul W, Paul H (2002) Mineralogical and Bio-geo-chemical Research Works 1949–1959 (A contribution to mining engineering and health research). Oakesdale, WA: Arts & Crafts Book Bindery & Publishing.

Pechenkin IG, Volfson IF, Sysoev AN, Pechenkin VG, Grushevoy GV (2005) Metallogeny of the uranium-bearing sedimentary basins. In J Mao, FP Bierlein (eds) Mineral Deposit Research: Meeting the Global Challenge (pp. 307–310). Berlin, Germany: Springer.

Pechenkin IG, Pechenkin VG (1996) Metallogeny of the turan plate sedimentary cover. Lithol Min Res 31:325–332.

Pechenkin VG, Pechenkin IG (2005) Exfiltrative mineralization in the Bukan-tau ore district (Central Kyzyl Kum region, Uzbekistan)./Lithol Miner Resour 40(5):462–471.

Peeters E-G (1987) The possible influence of the components of the soil and the lithosphere on the development and growth of neoplasms. Experientia 43:74–81.

Pelymskij (2008) Minerals on Medicine History, Represented in the Exposition of Lomonosov MGU Earth Sciences Museum. Proceedings of Readings from Lomonosov. Museology Section, 34–37. Lomonosov MGU Publisher.

Persinger MA (1987) Geopsychology and geopsychopathology: Mental processes and disorders associated with geochemical and geophysical factors. Experientia 43:92–104.

Petrov I (2008) Sulphuric acid research in Russia/CIS, Infomine, Report, 2008.

Petrov I, Burstein M, Volfson I (2007) Exposure in the East: curing Russia's. A Report from Infomine on Health-Related Problems and Arsenic Pollution by Russian Smelters. Mining Environmental Management, January 2007: 8–9

Pikhur OL (2003) Influence of Chemical Composition of Drinking Water at Condition Hard Teeth Tissues. St. Petersburg. 21 p.

Pronin AP, Bashorin VN, Zvonilkin BD (1997) Geological features and fluid activity of the Kaluga ring structure. Trans Russ Acad Sci–Earth Sci Sec 356:960–964.

Pronin AP (1997) Using a Helium Method in the Geoenvironmental Research and Protection of the Earth's Bowels. Moscow, Russia: Geoinformmark (in Russian).

Revich BA (2001) Environmental Pollution and Human Health. Introduction to Environmental Epidemiology. Moscow, Russia: MNEPU (in Russian).

Rihvanov L, Yazikov Ye, Suhih Yu et al (2005) Ecogeochemical Features of Natural Environments of Tomsk District and Diseases of the Population. Italics, Tomsk (in Russian).

Rysyev OA (2001) Shungite–The Stone of Health. Saint Petersburg, Russia: Tessa (in Russian).

Saet YE, Revich BA, Janin EP (1990) Environmental Geochemistry. Moscow, Russia: NEDRA (in Russian).

Saghatelyan AK, Gevorgyan V Sh, Arevshatyan SG, Sahakyan LV (2008a) Ecological and Geochemical Assessment of Environmental State of the City of Kajaran. Yerevan, Edition of the Center for Ecological-Noosphere Studies NAS RA) (in Russian and Armenian).

Saghatelyan AK, Arevshatyan SH, Sahakyan LV (2008b) Mercury in urban ecosystem as a risk factor. Mater of International Geological Congress (Abstract CD-ROM), Oslo, August 6–14, 1204902

Saghatelyan AK, Arevshatyan SH, Sahakyan LV (2007a) Heavy metals in system "soil-farm produce-organism" within the area of environmental impact of ore-mining production. Geophysical Research Abstracts CD-ROM of EGU General Assembly 2007, Vol. 9, Vienna, April 15–20, 2007, EGU2007-A-00765

Saghatelyan AK, Gevorkyan VSh, Arevshatyan SH, Sahakyan LV (2007b) Assessing environmental impact of mining production and revealing risk groups in children. Mater of International Youth Scientific Conference "Mountain Areas-Ecological Problems of Cities", 29–30 May, 2007, Yerevan, Edition of the Center for Ecological-Noosphere Studies NAS RA, 63–77

Saghatelyan AK (2004) The Peculiarities of Heavy Metals Distribution on Armenia's Territory. Yerevan, Publisher: Center for Ecological-Noosphere Studies NAS RA) (in Russian).

Savinykh MI (2006) Classification of shilajit ore deposits. Izvestiya Vuzov, Geologia i Razvedka. 5:39–41 (in Russian, with English abstract).

Savinykh MI (2003) Regularities of Displacement, Formation Condition and Classification of Shilajit Ore Deposits. PHd thesis (in Russian).

Schepetkin I, Khlebnikov A, Kwon BS (2002) Medical drugs from humus matter: focus on mumie. Drug Dev Res 57:140–159.

Selinus O, Frank A (2000) Medical geology. In L Möller (ed) Environmental Medicine (pp. 164–183). Stockholm: ArbSkNämnden.

Severgin V (ed) (1819) Plinij the senior (Plinij Kai the Second) natural history of fossilized bodies. Proceedings of Russian Academy of Sciences, St. Petersburg, p. 364.

Sherbo AP (2002) Environment and Health: Approaches at Risk Mark. St.Petersburg. 246 p.

Skal'ny AV, Kudrin AV (2000) Radiation, Microelements, Anti-Oxidants and immunity. Moscow: Lir Make (in Russian).

Skal'ny AV (1999) Human Microelementoses (Diagnosis and Therapy). Moscow, Russia: KMK Press (in Russian).

Smedley P, Kinniburgh DG (2005) Arsenic in groundwater and the environment: In O Selinus, B Alloway, JA Centeno, RB Finkelman, R Fuge, U Lindh, P Smedley (eds) Essentials of Medical Geology: Impacts of the Natural Environment on Public Health (pp. 263–299). Amsterdam, Netherlands: Elsevier.

Smith GE (1988) Fluoride and fluoridation. Soc Sci Med 26(4):451–462.

Soshnikova LA, Narbutaite, J, Vehkalahti, MM, Milčiuviene S (2007) Dental fluorosis and dental caries among 12-yr-old children from high- and low-fluoride areas in Lithuania. Eur J Oral Sci 115:137–142.

Tamashevich VN, Makhnach LA (2004) Multivariate Statistic Analysis. Minsk, BGEU. 162 p.

Taylor SR, McLennan SM (1985) The Continental Crust: Its Composition and Evolution. Oxford, England: Hackwell.

Tsallagova LV (1999) Reproductive Health of Women Working in the Industrial Sphere. Principles of Rehabilitation. Med. Sc. Doctor Dissertation. M., p. 315 (in Russian) .

Tsarfis PG (ed) (1991) Resorts, 2 Vols. Moscow, USSR: Profizdat (in Russian).

Tsarfis PG (1985) Nature and Health: Treatment and Rehabilitation by Natural Factors. Moscow, USSR: Mir Publishers.

Veldanova M (2000) Effect of some goitrogenic factors of environments on endemic goiter. Trace Elem Med 1:17–25 (in Russian).

Versieck J (1985) Trace elements in human body fluids and tissues. Critical Rev Clin Lab Sci 22: 97–184.

Vinogradov AP (1938) Biogeochemical provinces and endemies. Comptes Rendus de l'Académie des Sciences de l'URSS 18:283–286 (in Russian).

Vinogradov AP (1939) Geochemical research in the field of Urov endemy invasion. Comptes Rendus de l'Académie des Sciences de l'URSS 23:67–71 (in Russian).

Vinogradov AP (1964) Provinces biogéochimiques et leur rôle dans l'évolution organique. In U Colombo, GD Hobson (eds) Advances in Organic Geochemistry: Proceedings of the International Meeting, Milan, Italy, 1962 (pp. 317–338). New York: Macmillan.

Volfson IF, Paul W, Pechenkin IG (2010) Geochemical anomalies: Sickness and health. In IV Florinsky (ed) Man and the Geosphere. Hauppauge, NY: Nova Science Publishers (in press).

Volfson IF, Bakhur AE (2007) Medical radiogeology. ANRI 48:25–33 (in Russian).

Volfson IF, Pechenkin IG, Kremkova EV (2006) Medical geology of today: Targets and the ways of their realization. In EG Panova, VV Gavrilenko (eds) Bio-inert Interactions: Life and Stone (pp. 107–123). Saint Petersburg, Russia: Saint Petersburg University Press (in Russian).

Volfson IF (2004) Medical and biological aspects of the study and application of biologically active flints of the Russian platform. In V Gavrilenko, E Panova (eds) Proceedings of the 2nd International Symposium "Bio-inert Interaction: Life and Stone", Saint Petersburg, Russia, June 23–25, 2004 (pp. 157–160). Saint Petersburg, Russia: Saint Petersburg University Press (in Russian, with English abstract).

Volkotrub L, Karavaev N, Zinchenko N et al (2000) Hygienic aspects of iodine-deficiency diseases. Hygiene and Sanitation 102:28–31

Walters RJL (1996) The Power of Gemstones. London, UK: Carlton Books.

World Health Organization (1996) Trace Elements in Human Nutrition and Health. Geneva, Switzerland: World Health Organization.

World Health Organization (1987) Environmental Health Criterion 58-Selenium. Geneva, Switzerland: World Health Organization.

Yushkin NP (2004a) World of minerals and human health. Vestnik Otdelenia nauk o Zemle RAN, 22, http://www.scgis.ru/russian/cp1251/h_dgggms/1-2004/scpub-1.pdf (in Russian).

Yushkin NP (2004b) Mineral Factors of the Human Health: Concepts of Medical Mineralogy. Zapiski VMO (Proceedings of the Russian Mineralogical Society) PtCXXXIII, No 4 (in Russian with English abstract).

Yushkin NP (2003) Human health effects of "table" minerals. In N Kamata (ed) Proceedings, 1st International Symposium of the Kanazawa University 21st century COE program, Kanazawa, Japan, March 17–18, 2003, (Vol. 1, pp. 261–265). Kanazawa, Japan: Kanazawa University.

Zyryanov BN (1981) Extreme North Medical-Geographic Features Influence on Teeth Tissues State and Teeth Caries Morbidity Among Aboriginal Population and Immigrants. Ph.D. Thesis. Moscow. 20 p (in Russian).

Medical Geology in Europe

Olle Selinus, Mark Cave, Anne Kousa, Eiliv Steinnes, Jaques Varet, and Eduardo Ferreira da Silva

Abstract Medical geology or earth and health has a long history in Europe. Also the newer development of medical geology has a firm base in Europe. Health problems associated with geologic material and geologic processes occur quite frequently in Europe. The problems associated with the geologic environment in Europe are generally chronic, caused by long-term, low-level exposures. Examples are exposure to trace elements such as fluorine, arsenic, radon, mineral dust and naturally occurring organic compounds in drinking water. The chapter provides the history of medical geology in Europe, many examples of these environmental health problems and a look into the future.

Keywords Europe · Portugal · France · UK · Sweden · Finland · Norway · Arsenic · Asbestos · Dusts · BEN · Iodine · Fluorine · Radon · Cardiovascular diseases

History in Europe

Medical geology or earth and health has a long history in Europe. The Greek philosopher Hippocrates (400 B.C.) is considered to be the founder of medical geology. He recognized that environmental factors affected the distribution of disease. He noted in his treatise, *On Airs, Waters, and Places*, that under certain circumstances, water "comes from soil which produces thermal waters, such as those having iron, copper, silver, gold, sulphur, alum, bitumen, or nitre", and such water is "bad for every purpose". Vitruvius, a Roman architect in the last century B.C. recognized potential health dangers related to mining, noting that the water and pollution near mines posed negative health threats. Later in the first century A.D., the Greek physician Galen reaffirmed the potential danger of mining activities when he noted that acid mists are often associated with the extraction of Cu from the ground (Bowman et al., 2003).

The use of heavy metals in everyday ancient society introduced the negative affects of toxicity-related problems. During the Roman Empire it has been estimated that the annual production of Pb approached 80,000 tonnes. During the Roman Empire Pb usage exceeded 550 g per person per year but the main uses were for plumbing, architecture and shipbuilding. Lead salts were used to preserve fruits and vegetables and Pb was also added to wine to stop further fermentation and to add colour or bouquet. Large amounts of Pb usage in the daily life of Roman aristocracy had a number of negative health implications, including: epidemics of plumbism and saturnine gout, high incidence of sterility and stillbirths as well as mental incompetence. Physiological profiles of Roman Emperors dating between 50 B.C. and 250 B.C. suggest that the majority of individuals suffered from Pb poisoning. In turn, it is generally believed that a contributing factor to the fall of the Roman Empire, in 476 A.D., may have been the result of the excessive use of Pb.

Besides Pb other heavy metals such as Hg, Cu and As were also widely used in Roman and pre-Roman times. For instance, Hg was used during the Roman Empire to ease the pain of teething infants as well as in the recovery of Au and Ag, a method also widely used in Egypt in the twelfth century and in Central and South America in the sixteenth century. Mercury was also used to treat syphilis during the sixteenth

O. Selinus (✉)
Geological Survey of Sweden, Uppsala SE-751 28, Sweden

O. Selinus et al. (eds.), *Medical Geology*, International Year of Planet Earth,
DOI 10.1007/978-90-481-3430-4_10, © Springer Science+Business Media B.V. 2010

century and in the felting process in the 1800s. Greeks, Romans, Arabs and Peruvians used As for therapeutic purposes, since small doses were thought to improve the complexion; however, it has also long been used as a poison.

Formal recognition of the medical geology subdiscipline appears to reside with Zeiss who first introduced the term "geomedicine" in 1931 and at the time considered it synonymous with "geographic medicine" which was defined as a "a branch of medicine where geographical and cartographical methods are used to present medical research results". In the 1970s Låg in Norway redefined the term geomedicine as "the science dealing with the influence of ordinary environmental factors on the geographic distribution of health problems in man and animals" (Låg, 1990). Early in the 1970's initiative was taken in several countries to organize activities in medical geology. In USA, Canada and Great Britain investigations on the relation between geochemistry and health were carried out. Geochemistry has for a long time also a strong position in the former Soviet Union, and basic knowledge in this science is applied in medical connections.

In 1997 an International Union of Geological Sciences working group on medical geology was established. Since then there has been a rapid development of this field. Discussions focused on the organization and the feasibility of preparing a new textbook on medical geology. In September 2000, about 50 people participated in a meeting and workshop in Uppsala, Sweden. A 2-day seminar was held on medical geology "*The Geochemical Environment and Human Health*" resulting in a proceedings volume (Skinner and Berger, 2003). In 2006 the International Medical Geology Association, IMGA was established (Selinus et al., 2008).

Medical Geology Issues in Europe

Arsenic

Arsenic and arsenic-containing compounds are human carcinogens. Exposure to arsenic may occur through several anthropogenic processes, including mining residues, pesticides, pharmaceuticals, glass and microelectronics, but the most prevalent sources of exposure today are natural sources. Exposure to arsenic occurs via the oral route (ingestion), inhalation and dermal contact. Drinking water contaminated by naturally occurring arsenic remains a major public health problem (Centeno et al., 2002). The source of arsenic is geological, the element being present in many rock-forming minerals. There is growing concern about the toxicity of arsenic and the health effects caused by exposure to low levels in the geochemical environment. The danger to human health due to arsenic poisoning has been recognized by World Health Organization (WHO) and the provisional guideline value for arsenic in drinking water has been lowered from 50 to 10 $\mu g\ L^{-1}$ (WHO, 1993). Acute and chronic arsenic exposure via drinking water has been reported in many countries (Centeno et al., 2007). Among the countries that have well-documented case studies of arsenic poisoning are several countries in Europe, for example, UK, Hungary, Romania, Sweden, Slovakia. The common symptoms of chronic arsenic poisoning are depigmentation, keratosis and hyperkeratosis (Smedley and Kinniburgh, 2005; Centeno et al., 2002). Figure 1 shows an overview of arsenic in stream water in Europe from the FOREGS atlas (Salminen et al., 2005).

Asbestos

Asbestos and natural asbestiform compounds are among the most significant fibrous minerals with implications for health. The major forms of asbestos extracted commercially were chrysotile (white asbestos) and the asbestiform amphiboles crocidolite (blue asbestos), amosite (brown or grey asbestos), anthophyllite asbestos and tremolite. Elevated exposures to asbestos are known to be associated with asbestosis. It is generally thought that the asbestiform amphiboles are more carcinogenic than chrysotile asbestos, with lower level exposures of the former associated with increased risk of mesothelioma and lung cancer.

Asbestiform compounds may be present in soils or rocks from which they can be dispersed to adjacent human populations. In general, the risks posed by exposure to dusts containing naturally occurring asbestos have not been as thoroughly investigated or described as those arising in occupational contexts. The relationship is best established in certain regions in Turkey, where inhabitants are exposed to erionite,

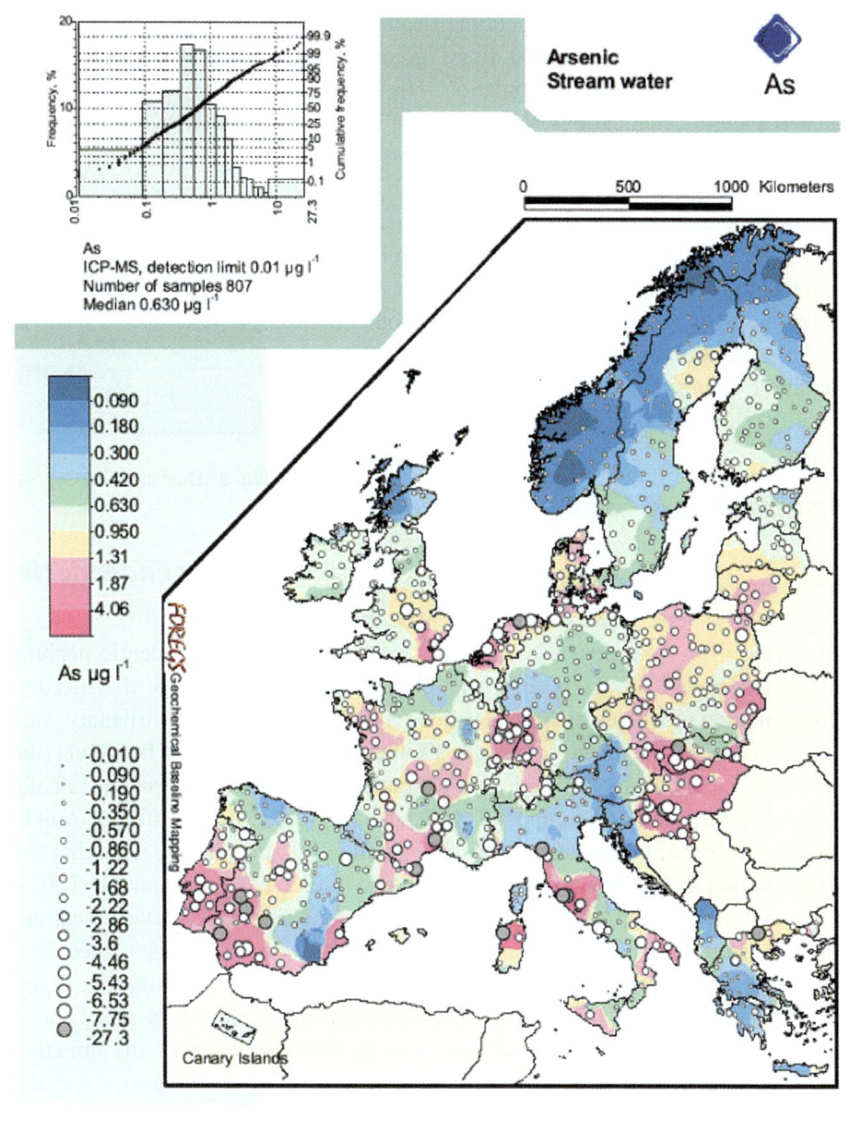

Fig. 1 Arsenic in stream water Europe (figure from the FOREGS atlas, Salminen et al. 2005). These maps of Europe in black and white can be seen in colour at the following website: http: //www.gsf.fi/publ/foreg satlas/ map_compare.html

a form of fibrous zeolite, present in the volcanic tuffs used as building stone. Exposure to erionite fibres, which circulate through the houses and streets of affected villages, has been linked to markedly elevated rates of mesothelioma. In one village, the incidence rates are many times higher than expected compared with the general population. In some zeolite-rich villages, malignant mesotheliomas are responsible for more than 50% of the total deaths (Dogan et al., 2008).

A number of fibrous minerals, including anthophyllite and tremolite, have also been identified in the soils of Mediterranean and central Europe and Turkey. Inhaled materials from these sources have been linked to pleural plaques in neighbouring agricultural communities. A common exposure pathway is through the domestic production and use of products made from local soils. One such product is the whitewash made from "luto soil" in the Metsovo region of Greece. Analysis of samples of the whitewash material revealed high levels of asbestiform minerals, particularly, tremolite. Radiographic screening has detected pleural calcifications in almost half of inhabitants in some northwestern Greece villages with no known occupational exposure to asbestos fibres. In the early

1980s, the incidence of mesothelioma deaths in the Metsovo region was about 300 times greater than expected in a non-exposed population. High prevalence of pleural calcifications have also been reported for inhabitants of several rural regions of northeastern Corsica, Turkey and Cyprus where whitewash containing tremolite asbestos has been used. Another fibrous mineral from the soil, sepiolite, has also been linked to pleural calcification in Bulgarian populations. An association has also been reported between mesothelioma and volcanic dust containing fluoro-edenite, a fibrous amphibole chemically similar to tremolite, in a region of Sicily (Dogan et al., 2008).

Dust Particles of Geological Origin

These are a widely dispersed component of the Earth's atmosphere, often forming extensive plumes that derive from volcanoes, dust storms, long-range transport episodes of desert dust and displacement through natural processes such as landslides and earthquakes. These phenomena occur on all the major continents, including mobilization of Saharan dust to Europe. The elements and compounds which are transported by dust phenomena and which may potentially impact on human health are diverse and include trace metals and metalloids, radioactive elements, fluoride, silicates, natural asbestiform compounds and alkali salts. Natural dust events may act to disperse pathogens and bioallergens, and the health implications of such phenomena – often occurring over considerable distances – have only recently been acknowledged. Geogenic dusts also have indirect socio-economic impacts on health, with ongoing effects on water quality, food production and infrastructure, such as transport networks (Derbyshire, 2005).

Dominant sources of natural dusts around the world are largely in or adjacent to the major drylands of the northern hemisphere. World dust emissions from dryland terrains may reach as much as 5 billion tons/year. The greatest of these includes a broad swathe of land across North Africa, The Middle East, north-west India and central and eastern Asia – from the western Sahara to the Yellow Sea. Africa is regarded as the greatest single source of minerogenic dust on Earth. This affects Europe significantly, especially southern Europe but also most other parts of the continent are affected (**Fig. 2**).

Fig. 2 Dust cloud in Cyprus 2008. Photo O Selinus

Balkan Endemic Nephropathy (BEN)

Balkan endemic nephropathy(BEN) is a serious kidney disease that occurs in clusters of rural villages located in tributary valleys of the Danube River in the Balkan Peninsula (the former Yugoslavia, Romania and Bulgaria). It is believed that at least 25,000 persons are suffering from BEN or are suspected of having the disease and that the total number of people at risk may exceed 100,000. BEN was first described as a distinct medical entity in 1956, but it may have existed undescribed for many centuries. Work by the U.S. Geological Survey (USGS) and others in the early 1990s noted the close correlation between the location of the affected villages and the occurrence of coal deposits, specifically lignites deposited in the Pliocene Epoch 5.3–1.6 million years ago. Further research showed that the well water in affected villages has measurable amounts of organic compounds such as polycyclic aromatic hydrocarbons (PAHs) and aromatic amines that may be toxic. The nearby Pliocene lignite deposits are unusual in that these coals release large amounts of organic substances when leached with water or other polar solvents. These coal deposits, therefore, could be a source of toxic organic compounds such as those found in the well water from the affected villages. Consumption of well water contaminated with toxic organic compounds derived from the coal may be implicated in the onset of BEN. The study of BEN demonstrates the value of close working relations between the geoscience and medicine communities. Medical geology research is needed to study whether these cancers are caused by toxic organic compounds leached from coal. Coal deposits are present worldwide and several direct links between coal combustion and disease in humans have been documented. Studies of BEN indicate that contamination resulting directly from coal deposits could affect human health in susceptible population groups. It has been found that this is the case also in Greece, Turkey and Portugal in Europe. (Feder et al., 1991; Orem et al., 1999; Orem et al., 2007; Tatu et al., 1998).

Iodine

Prior to about 1950 iodine deficiency disease (IDD) had affected virtually every country in the world. However, since the problem is known to be essentially due to the lack of iodine in the diet, it has been possible to introduce schemes for the mass treatment of affected populations such as the addition of iodine to the diet through the use of iodized salt and bread, injections of iodized oil or the addition of iodine to irrigation waters resulting in the alleviation of the symptoms in many countries. It was noted in the 1970s that IDD had been effectively eradicated in the developed world and endemic goitre was described as a disease of the poor being largely confined to third world countries. However, it is important to point out that in the 1990s iodine deficiency has been reported to occur in several affluent countries in western Europe, probably as a result of dietary changes, suggesting that such changes could cause the re-introduction of IDD in many countries.

Many of the areas affected by iodine deficiency are remote from marine influence. Many of the areas highlighted are the mountainous regions and their rain shadow areas such as the European Alps region. In all of these situations IDD can be explained according to the classic explanations of low iodine supply and hence low iodine availability. However, many endemias are not explicable in these simplistic terms and several countries and regions which are close to the coast have been known to suffer from IDD problems. For example large regions of the United Kingdom are known to have histories of IDD, despite the strong maritime influence on the country. In some endemias the involvement of the sulphur-containing goitrogens has been invoked as a reason. Some of these may derive from geological sources and be incorporated into drinking water or food, while others occur naturally in vegetables such as those of the *Brassica* genus.

There are many examples of limestone-associated endemias occurring in the UK and Ireland with limestone regions of north Yorkshire being historically renowned for its severe goitre and cretinism problems. Similarly County Tipperary in Ireland, underlain by Carboniferous Limestone, was historically one of the major areas of IDD in that country. One of the most well-documented areas of IDD in the UK is the Derbyshire region of northern England, an area that is between 150 and 180 km from the west coast, in the direction of the prevailing wind. Here endemic goitre was rife, being known as "Derbyshire Neck", with the endemia being confined to areas underlain by limestone bedrock. Analyses of soil in the former goitrous region have shown that iodine concentrations range up to 26 mg kg^{-1}, with a mean value of 8.2 mg kg^{-1}. Similarly, an area of north Oxfordshire, England, where endemic goitre was prevalent and where iodine deficiency problems in school children were recorded as recently as the 1950s, is underlain by limestone and soil iodine ranges between 5 and 10 mg kg^{-1}. IDD occurs in these limestone areas despite relatively high iodine in soils, which would imply that iodine is not bioavailable. Soils over the limestones would generally be well drained and circumneutral to alkaline in nature. In these conditions any soluble soil iodine is likely to be present as the iodate anion. It has been shown that iodate uptake through plant roots is more limited than the uptake of iodide so that in neutral to alkaline soils root uptake of iodine may be low.

The distribution of IDD reflects the geochemistry of iodine and with large areas of iodine deficiency occurring in central continental regions, mountainous and rain shadow areas, this distribution fits in with the classical explanation of IDD being governed by the external supply of iodine from the marine environment via the atmosphere. However, the geochemistry of iodine is more complex than this simplistic approach, and from a closer scrutiny of the distribution of IDD it is apparent that, in many cases, iodine deficiency problems are related to the bioavailability of iodine in soils and are not related directly to the external supply of iodine.

Figure 3 shows an overview of Iodine in soil in Europe from the FOREGS atlas (Salminen et al., 2005)

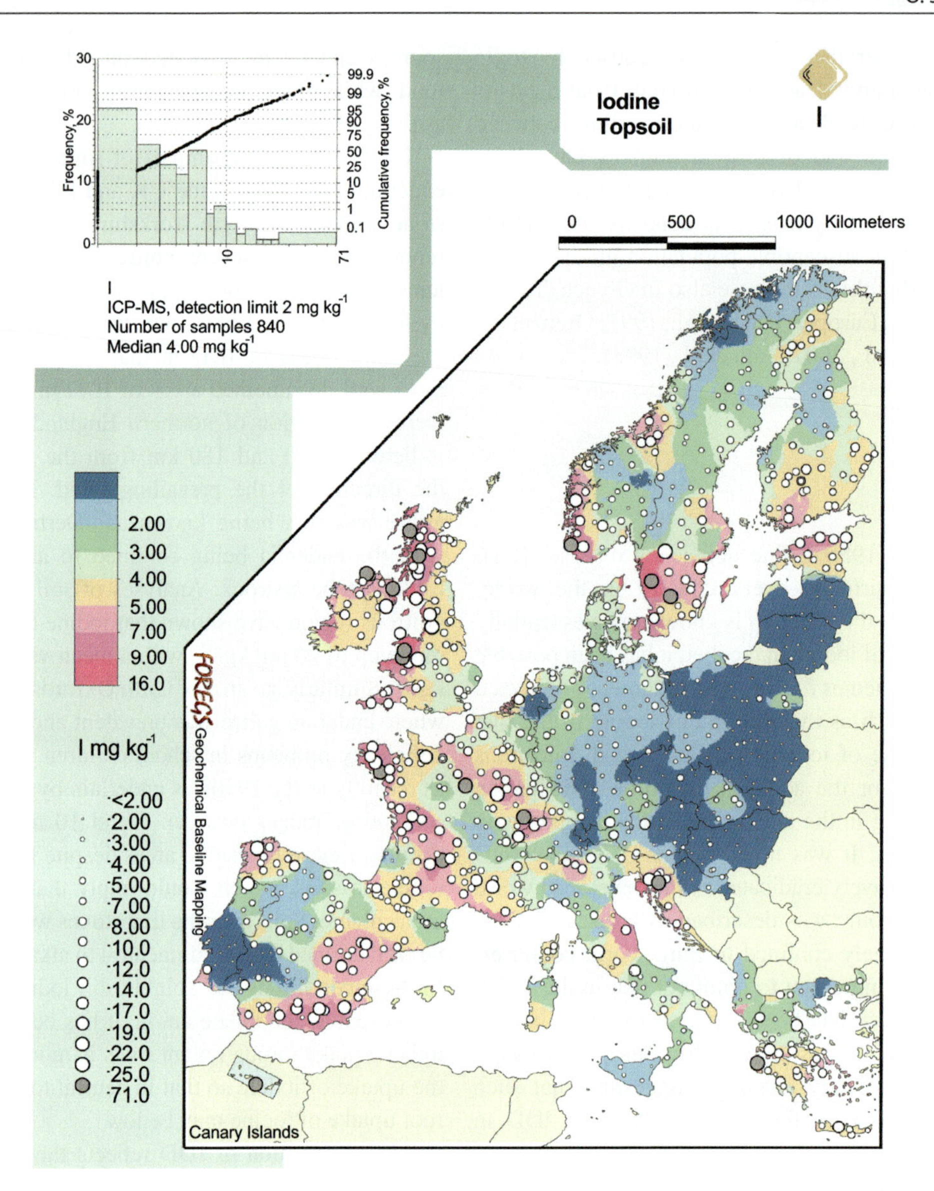

Fig. 3 Iodine in soil in Europe (figure from the FOREGS atlas, Salminen et al. 2005)

Fluorine

Fluorine is an essential element in the human diet. Deficiency in fluorine has long been linked to the incidence of dental caries and the use of fluoride toothpastes and mouthwashes has been widely advocated in mitigating dental health problems. Fluoridation of water supplies to augment naturally low fluoride concentrations is also undertaken in some countries. However, despite the essentiality of fluorine in humans, optimal doses appear to fall within a narrow range. Chronic ingestion of high doses has been linked to the development of dental fluorosis, and in extreme cases, skeletal fluorosis. High doses have also been linked to cancer, although the association is not well established. Drinking water is particularly sensitive in this respect because large variations in fluoride concentration exist in water supplies in different areas. Concentrations in natural waters span more than four orders of magnitude, although values typically lie in the 0.1–10 mg L^{-1} range. Where concentrations are high, drinking water can constitute the dominant source of fluorine in the human diet. Concentrations in drinking water of around 1 mg L^{-1} are often taken to be optimal. However, chronic use of drinking water with concentrations above about 1.5 mg L^{-1} is considered to be detrimental to health. The WHO (1993) guideline value for fluoride in drinking water is 1.5 mg L^{-1}.

A common situation in many aquifers is to encounter increasing fluoride concentrations in the groundwaters down the flow gradient. This situation arises initially as a result of continuous dissolution of fluoride from minerals in the carbonate aquifers, up to the limit of fluorite solubility, followed by ion-exchange reactions involving removal of calcium. The phenomenon is particularly well illustrated in the Jurassic Lincolnshire Limestone aquifer of eastern England.

In many aquifers, the evolution of groundwater down the flow gradient has taken place over centuries or millennia, with water being abstracted having had significant opportunity for equilibration with host aquifer minerals. The accumulation of fluoride in water is ultimately limited by mineral solubility. In groundwaters where calcium is abundant, fluoride concentrations are limited by saturation with the mineral fluorite. In cases where calcium concentrations are low, or where calcium is removed by ion exchange, fluoride may build up to excessive and dangerous concentrations. Given these key controls on fluoride occurrence and distribution, it is possible to anticipate broadly where areas of regionally high-fluoride concentrations are likely to exist. Such an understanding of the fluoride occurrence is important for the management of the fluoride-related epidemiological problems. Water-supply programmes in potentially high-fluoride areas should have geological and hydrological guidance, including chemical analyses and geological maps.

Cardiovascular Diseases (CVD)

In Europe, water hardness is often expressed in terms of degrees of hardness. One French degree is equivalent to 10 mg L^{-1} as $CaCO_3$, one German degree to 17.8 mg L^{-1} as $CaCO_3$ and one English of Clark degree to 14.3 mg L^{-1} as $CaCO_3$. One German degree of hardness (dH) is equal to 1 mg of calcium oxide (CaO) or 0.72 mg of magnesium oxide (MgO) per 100 mL of water. Despite the wide usage of the term, the property of hardness is difficult to define exactly. Water hardness is not caused by a single substance but by a variety of dissolved polyvalent metallic ions, predominantly calcium and magnesium, although other ions, for example, aluminium, barium, iron, manganese, strontium and zinc, also contribute. The source of the metallic ions is typically sedimentary rocks, the most common being limestone ($CaCO_3$) and dolomite ($CaMg(CO_3)_2$).

One of the most comprehensive studies of the geographic variations in cardiovascular mortality was the British Regional Heart Study. The first phase of this study (Pocock et al., 1980) applied multiple regression analysis to the geographical variations in CVD for men and women aged 35–74 in 253 urban areas in England, Wales and Scotland for the period 1969–1973. The investigation showed that the effect of water hardness was non-linear, being much greater in the range from very soft to medium-hard water than from medium to very hard water. The adjusted SMR decreased steadily in moving from a hardness of 10 to 170 mg L^{-1} but changed little between 170 and 290 mg L^{-1} or greater. After adjustment, CVD in areas with very soft water, around 25 mg L^{-1}, was estimated to be 10–15% higher than in areas with medium-hard water, around 170 mg L^{-1}, while any further increase in hardness beyond 170 mg L^{-1} did not additionally lower CVD mortality. Hence, it appeared that the maximum effect on CVD was principally between the very soft and medium-hard waters. Importantly, adjusting for climatic and socio-economic differences considerably reduced the apparent magnitude of the effect of water hardness (Pocock et al., 1980).

A problem with correlation studies such as the British Regional Heart Study is the failure of much of the research to consider the causal mechanism that links independent variables to the disease outcome. Also, many of the calibrated models presented in the literature are socially blind in including only variables pertaining to the physical environment, often a large number of water quality elements.

A more detailed overview on cardiovascular diseases in Sweden and Finland can be seen later in this chapter.

Radon

Radon is a natural radioactive gas produced by the radioactive decay of radium, which in turn is derived from the radioactive decay of uranium. Uranium is found in small quantities in all soils and rocks, although the amount varies from place to place. Radon decays to form radioactive particles that can enter the body by inhalation. Inhalation of the short-lived decay products of radon has been linked to an increase in the risk of developing cancers of the respiratory tract, especially of the lungs. Breathing radon in the indoor air of homes contributes to about 20,000 lung cancer deaths each year in the United States and 2,000–3,000 in the UK. In Sweden 400 people die every year of radon. Only smoking causes more lung cancer deaths.

Geology is the most important factor controlling the source and distribution of radon. Relatively high levels of radon emissions are associated with particular types of bedrock and unconsolidated deposits, for example some, but not all, granites, phosphatic rocks and shales rich in organic materials. The release of radon from rocks and soils is controlled largely by the types of minerals in which uranium and radium occur. Once radon gas is released from minerals, its migration to the surface is controlled by the transmission characteristics of the bedrock and soil; the nature of the carrier fluids, including carbon dioxide gas and groundwater; meteorological factors such as barometric pressure, wind, relative humidity and rainfall; as well as soil permeability, drainage and moisture content.

Examples from Specific European Regions

Health Issues at the British Geological Survey in the UK

The geology of Great Britain is varied and complex and gives rise to the wide variety of landscapes found across the islands. For its size, the geology of the UK is probably one of the most complex in the world. In the twenty-first century, the interaction of humans with geology in the UK is perhaps lower than in the past. Drinking water comes from treated public supplies, not from local boreholes, and much of the food eaten is supplied by large retail outlets and is not grown locally. With concerns about genetic modification of foodstuffs, chemical pollution and contamination of food and the desire for the ultimate in freshness there is, however, a small but significant cohort of the population that grow their own fruit and vegetables in back gardens and allotments. In the UK, allotments are small parcels of land rented to individuals usually for the purpose of growing food crops. There is no set standard size but the most common plot is c. 250 m^2 and there are probably of the order of 300,000 plots in cultivation in the UK. For this section of the population, the geology underlying their soils may well have a significant effect on their health and well-being. The other large cohort of the population which come into contact with the soil environment and are perhaps more vulnerable to health problems are children. Play activities in gardens and parks can bring children in contact with soil and its subsequent oral ingestion due to hand-to-mouth activity.

Interestingly, it is the concern of government with the contamination of land from anthropogenic activity such as industrial, mining and waste disposal activities that has brought the natural geochemistry to the fore. In 1987 the Interdepartmental Committee on the Redevelopment of Contaminated Land set out trigger values for a number of contaminants in soils including potentially toxic metals. These indicated concentrations of acceptable and unacceptable risk. More recently (Department for the Environment Food and Rural Affairs and the Environment Agency, 2002) new guidelines known as CLEA (Contaminated Land Exposure Assessment) superseded the ICRCL values by more comprehensive soil guideline values (SGV) based on toxicology and human exposure to soil contamination under different land use scenarios. The SGV are indicators for "intervention" either in the form of further detailed risk assessment and/or remediation. Legislation relating to the identification of contaminated land and the suitability of land for redevelopment has led local authorities to apply the principles specified in CLEA and in certain areas of the UK where the concentrations of some potentially toxic metals in soils, of which arsenic was most common, were causing some concerns. Over a similar timescale there was increasing interest in the natural geochemistry of the UK driven, in the first instance, by the need to identify mineral resources. Stream sediment geochemical surveys of the UK (Johnson et al., 2005) were showing clear patterns of potentially harmful elements (PHE) at elevated concentrations. Subsequent soil mapping (Johnson et al., 2005) aimed more at environment and health investigations showed that the patterns for arsenic could be linked to the geology of the underlying parent material and were geogenic rather than man-made phenomena. A recent publication (Appleton et al., 2008) has combined the stream sediment and soil data to produce a national scale UK estimate of soil arsenic concentrations. This study showed that the largest areas with the highest concentrations (>30 mg kg^{-1}) occur in the English Lake District, western Wales, the northeast midlands going into the north midlands and southwest England.

The British Geological Survey has carried out a number of studies (Cave et al., 2003; Palumbo-Roe et al., 2005; Wragg, 2005; Cave et al., 2007; Wragg et al., 2007) investigating the human bioaccessibility of arsenic in soils with elevated arsenic from the NE midlands and in SW England. The investigations have been based on an in vitro physiologically based extraction test (PBET) which mimics the physical and chemical environment of the human stomach and the upper intestine. The test measures the bioaccessibility of the arsenic in the soil (the amount of arsenic liberated into solution from the soil during the PBET test). The NW midland soils developed over iron stone parent materials were found to have <10% of their total arsenic content in bioaccessible form. In SW England where the arsenic was associated with natural mineralization, total concentrations generally higher than in NW midlands, especially where mining activities were carried out, had higher bioaccessible arsenic (c.10–20%). Geochemical testing including sequential extractions shows that the bioaccessible arsenic is associated with carbonates and fine-grained Fe/Mn oxy-hydroxides whereas the unavailable fraction is often bound to highly crystalline Fe oxides (goethite, hematite). The studies of the ironstone soils have shown that the bioaccessible fraction of arsenic in these soils is unlikely to be a health risk and that in the mineralized areas in the southwest especially where the highest total arsenic concentrations occur there could be potential health risks. It has been shown that humans living in a high arsenic area in SW England do have elevated As in urine (Kavanagh et al., 1997).

To illustrate the potential problems with arsenic in the UK, the BGS recently completed a geochemical soil survey of the river Tamar catchment in the south west of the UK (Rawlins et al., 2003). Figure 4 shows the location and the geology of the catchment along with the areas where mining activity has taken place. Figure 5 shows the concentration of arsenic in soil interpolated from chemical analysis of 450 samples taken over the catchment. Clearly, large proportion has soil concentrations exceeding the current UK soil SGV of 20 mg kg^{-1}. The very high concentrations of arsenic found in the south of the area coincide with the location of the Devon Great Consols arsenic mine which finished operation in 1930. Work is currently under way, using the archive of soils collected from this area, to measure and map the bioacessibility of the arsenic in the soil.

Medical Geology in Portugal

Mainland Portugal is located in the western edge of Europe between latitudes 37 and 42°N. The latitude, together with the orography, and the proximity to the Atlantic Ocean are considered to be the key factors conditioning the climate. The northern part of the country is mountainous, with 90% of the land above 400 m, while rolling plains prevail in the south, where 60% of the land is below 400 m. The geology is considerably diverse and complex. Briefly, the country can be divided into two large units: the Hesperian Massif and the Epi-Hercynian cover. The Hesperian Massif, occupying more than half the country, is of Precambrian and Palaeozoic ages. Granitoids and a flysch-type series of schists and greywackes are the dominant lithologies. The Epi-Hercynian cover includes the western and southern Meso-cenozoic margins and the basins of Tejo and Sado rivers. Limestones, marls, shales, sandstones and conglomerates are very common.

Soil Geochemical Atlas of Portugal

The relevance of geochemical mapping as a method of providing multi-element databases documenting the present surface environment is now generally recognized. The continental area of Portugal is now entirely covered by a soil geochemical survey (1 site/135 km^2), taking as the sampling media topsoils (upper mineral horizons, A) and organic horizons (humus, O). The main purpose was to obtain baseline levels of several

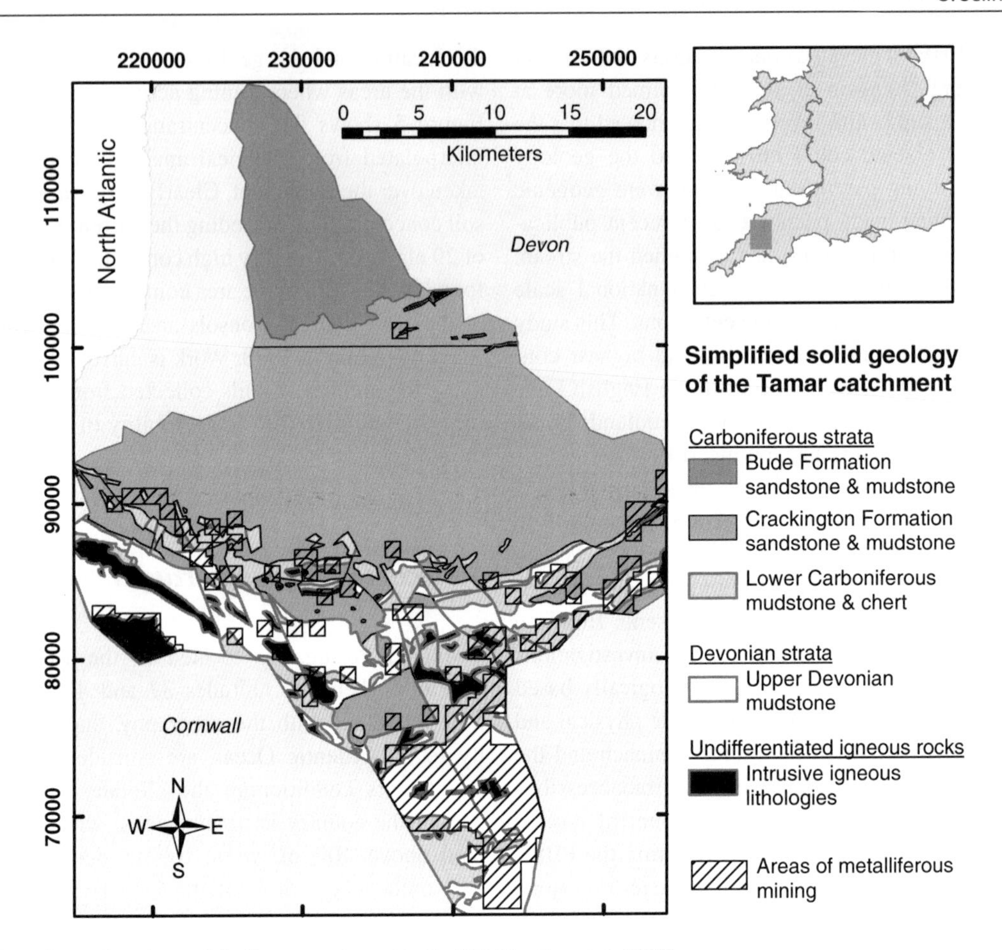

Fig. 4 Location and geology of the Tamar catchment in the UK (Rawlins et al. 2003)

elements in Portuguese soils, leading to a complete coverage of continental Portugal with a low-density geochemical survey. The sampling and analytical programme was performed according to international recommendations (IGCP Project 259 "International Geochemical Mapping"). The compilation of all data (nearly 45,000 individual data) in an organized way led to the production of the first Soil Geochemical Atlas of Portugal. In this Atlas it is possible to find for each chemical element a set of information statistics, maps of spatial distribution, among other information of geochemical and environmental interest. Among other findings the soil survey identified (1) strong regional differences within the geochemical baseline, (2) relationships between the geochemical patterns and lithology, soil type and mineral occurrences, (3) areas with low concentration of many elements or where concentrations of potentially harmful elements are high and (4) the influence of anthropogenic factors. These data are also of interest for geochemical exploration purposes in the sense that they may constitute a general framework for normalizing existing soil and stream sediment surveys which are sparse all over the country (Inácio et al., 2008).

Some identified health problems in Portugal related to geogenic and anthropogenic sources.

Arsenic

It is well known that the risk associated with As exposition is a fact and can affect populations in particular urban regions or with significant drop in water levels in certain geological formations. The highest values of As are related with some mineralizations normally associated with the contact of granites and metasediments

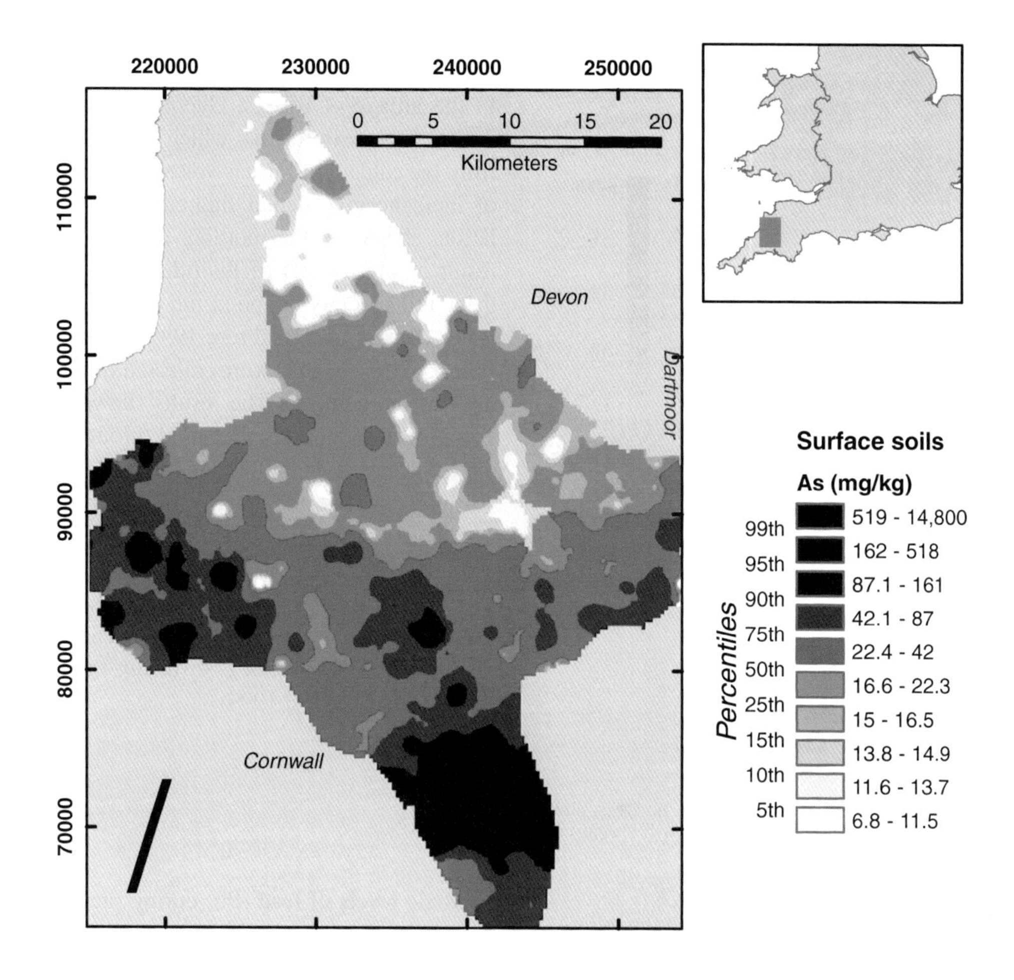

Fig. 5 Soil arsenic geochemical map of the Tamar catchment (Rawlins et al., 2003)

(Reis et al., 2005). The Soil Geochemical Atlas of Portugal revealed that the highest values are located in the Central Iberian Zone and are related to a granitic basement, associated with the contact between granites and metasediments. Drinking water contaminated by naturally occurring arsenic minerals was identified in several areas of the Central-North region of Portugal (Fig. 6).

Arsenic was reported in high levels in soils through atmosphere deposition and As containing water around some industrial areas due to mining and smelting activity. Human exposure risk to environmental arsenic was evaluated by combining drinking water and atmospheric deposition, assessed tap water analysis and a moss biomonitoring survey. The Portuguese survey was developed during 2002 as part of the activities of the UNECE ICP Vegetation Program and reported, for the first time, the concentration of arsenic at a national scale (Figueira et al., 2007).

Asbestos

Data on the incidence of occupational asbestos-related diseases in Portugal are only available for the period 1985–1993. During this time 71 cases of asbestos-related diseases were recorded. In 1992, six deaths from asbestos-related diseases were registered. In 2003 the Social Affairs Ministry reported that there were 161 cases of asbestos-related diseases as well as many more cases of pulmonary complications due to the inhalation of asbestos dusts. Since the implementation

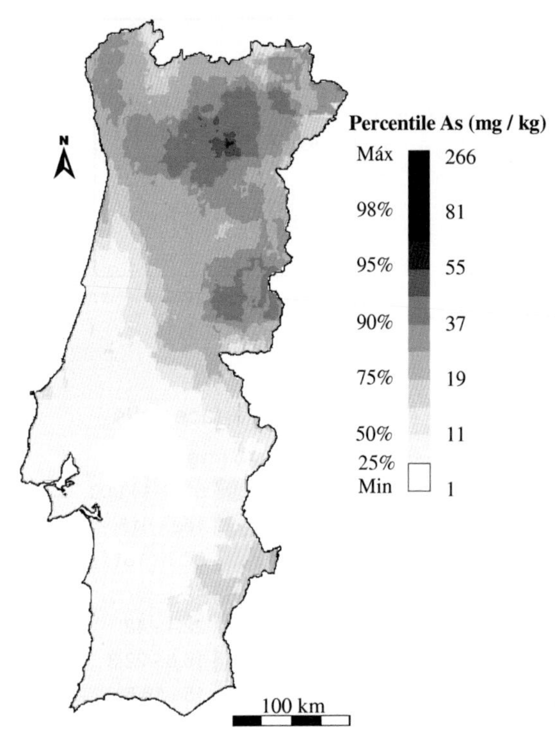

Fig. 6 Spatial distribution map of As in the A horizons plotted by kriging using a variographic model (Inácio et al., 2008)

of the EU Asbestos Directive, industry has been replacing asbestos with safer alternatives such as PVA and cellulose (Kasan-Allen, 2006).

Fluoride

Fluorosis is a widespread disease related to ingestion of high levels of fluorine through water and food. Although sometimes of anthropogenic origin, high levels of fluorine are generally related to natural sources. The problem of fluorosis related to volcanic activity, which releases magmatic fluorine generally as hydrogen fluorine through volcanic degassing, was first recognized in Japan where this pathology was called "Aso volcano disease" (Kawahara, 1971) due to the fact that fluorosis was widespread in the population living at the foot of this volcano. The major pathway of magmatic fluorine to humans is in the form of fluoride ion (F^-), through consumption of contaminated vegetables and drinking water. Contamination occur either through direct uptake of gaseous HF or through rainwater's and volcanic ashes. The direct effects of volcanogenic fluoride on human populations are less well recorded. In the village of Ribeira Quente (São Miguel Island in the Azores, Portugal), which is located only 5 km from the caldera of Furnas volcano there are a number of incidences of dental fluorosis in the local population, shown by the mottling of teeth, which is the result of high levels of fluoride in the groundwaters that until recently supplied the village with drinking water (Baxter and Baubron, 1999). To investigate the source of this, lichens were collected from the area and used as biomonitors. The results shown that gaseous volcanogenic fluoride is released through the soil into the environment in the Furnas Caldera. These gaseous volcanogenic fluoride are the most likely source of the high levels of fluoride in the groundwaters. The main conduits controlling the release of the gases are faults and ring structures associated with the formation of craters. This study has shown that lichens may be used to detect fluoride which is diffusing through the soil (Notcutt and Davies, 1999).

Lead

Subtoxic levels of lead (Pb) during pregnancy can be responsible for intrauterine delays in foetal development and thus increased risk of morbidity and mortality of newborns. Identified most vulnerable individuals should not live in conditions that expose them to newly evidenced hazardous Pb levels and those potentially exposed need to be identified through blood Pb screening, preferably targeted instead of universal (Fátima Reis et al., 2008). The project "Foetal exposure to lead" as determined by human and environmental biomarkers, investigation of influence on human reproductive outcomes and autonomic nervous system in rats (FEXHEBIO) is a multidisciplinary project, involving several national institutions, aiming (i) to assess human exposure to environmental lead through the determination of lead relative bioavailability and bioassessibility; (ii) to determine human exposure to lead through blood lead levels (BLL) in umbilical cord; only blood will be used as human sample media; (iii) to test the following hypothesis: (1) umbilical cord BLL of concern are positively associated with higher environmental levels of lead, determined in mosses collected in residence areas of corresponding women, and (2) there is a positive association between human exposure to lead and exposure to subtoxic levels of lead during pregnancy; (iv) to investigate the putative effect of lead exposure during the prenatal period in the activity of the autonomic nervous system in rats.

Radon

By far, radon is known to be the main contributor to exposure from natural background radiation. Radon is a radioactive gas that occurs in rocks, soils, air and water in variable concentrations and is considered an environmental health hazard. Because this radioactive gas is considered to be the second leading cause of lung cancer (first is smoking), most European countries have adopted a number of regulations to identify radon-prone areas. From 1987 to 1992, the Radiological Protection and Nuclear Safety Department (DPRSN) of the Nuclear and Technological Institute (ITN) carried out in Portugal a national wide radon survey. In the framework of that study, indoor radon measurements were first carried out in dwellings located in technologically enhanced naturally radioactive zones and it was later enlarged to the entire country (Faísca et al., 1992). The measurements were performed on a statistical basis of one detector for 2,000 inhabitants and the detectors distribution were performed in order to cover, whenever possible, rural and urban areas, in a total of 4,200 measurements. Indoor radon concentration levels are affected not only by the underlying geological composition, but also by soil permeability and porosity, the occurrence of rock cracks, pressure differences between the gas in the soil and at the surface and other kinds of parameters related to the construction type of the buildings, ventilation rates and living habits of the inhabitants (Pereira et al., 2003). A new approach is being developed in order to obtain indoor radon risk maps through the use of geostatistic simulation techniques. This is being done by using radon data already available for Portugal (4,800 results) and integrating data from new sampling campaigns, aiming a good cover of the Portuguese territory. On the basis of geology, in situ gamma-ray measurements (rocks, soils and water) and emanometry techniques a methodology for the elaboration of small-scale radon risk mapping (1:5,000 scale) was proposed (Pereira and Neves, 2005).

Radon concentrations in groundwater are highly variable and are mainly related with geological factors. The radon concentrations in groundwater from the Hesperian massif of Central Portugal were evaluated by Pereira et al. (2007). A survey carried out between Viseu (north limit) and Nisa (south limit) showed that the radon concentrations are disperse over a wide range of variation, between a minimum of 19 and a maximum of 8,830 $Bq.L^{-1}$. In general, the variability in radon concentrations is high, with the highest variance detected in the Beiras granite sample set which also displays the highest absolute values, often above 1,000 $Bq.L^{-1}$(16%). This value is the limit recommended by the European Union (recommendation 2001/928/Euratom) for water used for human consumption. In general, within each lithology, the highest values relate to waters that percolate through U-enriched faults that crosscut not only the igneous rocks but also the metasedimentary ones.

Uranium

In granites of the Centre of Portugal, uranium mineralizations are known since the early twentieth century and several deposits were extracted for radium and uranium production. The mining and milling works of radioactive ores originated a large amount of residues accumulated in spoil heaps and tailings piles. These solid residues contain high concentrations of radionuclides belonging mainly to the uranium decay series (Carvalho et al., 2005a; Carvalho and Oliveira, 2007). In Portugal the recent closure of uranium mines (2001) has raised concerns regarding the possible chemical and radiological (from ^{238}U daughters alpha-emitting radionuclides) effects on the health of the populations living around the mining areas. During the last few years, the radioactive contamination of this region has been investigated in a systematic manner (Carvalho et al., 2005a). In particular, a detailed investigation on the exposure of human populations to uranium mining and milling wastes was performed in the framework of a multidisciplinary project *MinUrar* – Uranium mines and their wastes: health effects in a Portuguese population (Falcão et al., 2005; Carvalho et al., 2007).

Carvalho et al. (2007) studied the enhancement of radioactivity in selected freshwater ecosystems (Vouga, Dão, Távora and Mondego rivers) and conclude that, in spite of the large number of uranium mines exploited in this region, there is no widespread and lasting accumulation of radionuclides in the watersheds of the studied rivers. The overall results of fish analyses show an enhancement of radionuclide concentrations in fish from Mondego river, collected in the area receiving the discharges from Ribeira da Pantanha. Depending on the radionuclide, concentrations in fish generally are 2–10 times above naturally occurring concentrations in fish from the other rivers of the region. However, according to currently accepted radiological safety standards, these concentrations do not pose a significant radiological risk to human consumers and to aquatic fauna.

The contamination of agricultural soils in the rural area of Cunha Baixa (Mangualde, Central Portugal) due to irrigation with U-contaminated water and the consequent phyto-accumulation in food crops associated with dietary exposure to U has also been investigated (Neves and Abreu, 2009).

Volcanic Soils and Related Health Problems

Volcanic regions are important scenarios for the study of heavy metal contents in soils and the effects of those on living organisms because (a) they are densely inhabited in some areas of the Earth; (b) soils effectively retain chemicals acting as a reservoir affecting agriculture; (c) the volcanic islands are more closed geochemical systems than continental land; (d) little has been said concerning natural pollution related especially with volcanic activity. Studies carried out in S. Miguel Island (Azores) showed that

(i) inhabitants of the São Miguel Island (Azores) revealed that humans living chronically exposed to volcanic emissions show high concentrations of essential and non-essential trace metals in scalp hair (Amaral et al., 2008).
(ii) humans present high incidence of chronic bronchitis and some cancer types, which may be related to volcanic emissions (Amaral and Rodrigues, 2007). Baxter et al. (1999) demonstrated a probable health hazard associated with ground gas emissions in Furnas (São Miguel). According to the Direcção-Geral de Saúde the cancer mortality in the Azores (1999–2201) was higher than in Madeira Island and Portugal mainland. Amaral et al. (2006) suggest that the much higher rates of lip, oral cavity and pharynx cancer, and breast cancer in Furnas may be partially explained by the chronic exposure to environmental factors resulting from volcanic activity, such as hazardous gases, trace elements and radon. Since the cancer chemopreventive role of Se has been well documented, the authors also speculate to a possible association between the low concentrations of Se, the high concentrations of Cd, which is carcinogen, and Pb, which is a probable carcinogen, and some of the cancers found previously (Viegas-Crespo et al., 2000; Pavão et al., 2006).
(iii) the high Mn/Mg ratio in animals and humans (blood) can also be the cause of ataxia. Some researchers are trying to establish the association of abnormal Mn/Mg ratio to the Machado–Joseph neurodegenerative disease, with a high incidence in Azores (Purdey, 2004).
(iv) cattle population in the Azores Island revealed deficiencies of micronutrients. There were significant differences in the levels of glutathione peroxidase (enzyme indicator of the levels of Se in the body) in populations of cattle with and without problems of enzootic hematuria and acute foetus by the common (Pinto et al., 2000). Also the studies demonstrated that there is a high mortality rate and a quite high incidence rate of cancer in the cattle (Carvalho et al., 2005b, 2006).

The results show that an association between the incidence of cancer and the concentrations of essential and toxic elements in volcanic soils (and consequently in the agricultural product deriving from those soils) can exist. The possibility of other health problems cannot be neglected, namely respiratory diseases as a result of the aerial transport of soil particles enriched in toxic elements. Although some of these associations are speculative, they should be investigated and so, according to toxicologists and epidemiologists, it is necessary to have a detailed knowledge of the physico-chemical characteristics of soils and the global distribution of the element concentrations.

Beneficial Effects of Geological Materials

In Portugal balneology and hot springs therapy as a part of preventative medicine is widely recognized and encouraged. Medical prescriptions are given for the treatment of a wide range of health problems utilizing mineral waters. In Portugal different chemical classes of hot springs have been recognized (simple, chloride, hydrogen carbonate, sulphate, carbon dioxide, iron, sulphur, acid and radioactive). Each class of hot spring is reputed to be effective in treating a wide range of health problems such as burns, skin diseases (psoriasis), respiratory diseases (asthma, rhinitis), hypertension, diabetes, gout, muscle aches, arthritis.

In addition to hot springs, geological materials such as mud and sand are actually investigated regarding their application in pelotherapy, mud therapy or fangotherapy. Beach sand baths in Porto Santo Island, Portugal, are popular therapeutic practice. Gomes and Silva (2006) believe that essential elements in the carbonate and clay-rich sands are incorporated into the bodies of the people covered by the sand by percutaneous mechanisms. Porto Santo, rather appreciated by very many tourists, has been considered as a natural health resort due to the healing assets of their natural resources, utilized in thalassotherapy centres and spas.

Presently, there is much scientific information on the assets or properties of clay minerals that justify their incorporation in pharmaceutical formulations and cosmetics as well. Gomes and Silva (2007) design and develop clay-based products, in the form of creams, ointments or lotions with balsamic or curative properties, which could be used for topical applications in balneotherapy, dermopharmacy and dermocosmetics. Also, there is some investigation regarding the processing or maturation that is required for the improvement of volcanic mud or fango properties.

Slovakia

In Slovakia, as well as in many other countries, Medical Geology issues are being addressed by geological projects financed from the state budget. One such project being conducted by the Geological Survey of the Slovak Republic is the evaluation of potential influence of geochemical environment on the health state of population in the region of the Spišsko-Gemerské Rudohorie Mts. (Rapant and Krcmova, 2008). The aim of the project is to verify methodical principles of evaluation of the influence of either excess or deficiency of chemical elements within geological environment on the health of the population in one of the most contaminated regions of the Slovak Republic. Methodological interconnections of geochemical and medical data and their further mutual assessment have been developed, aiming at the final analysis of environmental and medical risks. A methodology has also been developed for evaluating and systematically mapping of environmental health risks from harmful elements in Slovak groundwaters. Applying national geochemical databases Rapant and Krcmova (2008) have dealt with the feasibility of calculating and visualizing health risk from arsenic groundwater contamination. Potential health risks have been assessed based on existing geochemical data in accordance with present methodological procedures established for human health risk assessment. Screening analysis has been used to estimate the contribution to total chronic risk from groundwater contamination by potentially toxic elements including As, Ba, Cd, Cu, Hg, Pb, Sb, Se and Zn. These results point out a significant contribution of arsenic to total risk in about 10% of the Slovak territory. The areas characterized with high health risk levels are mainly geogenically contaminated. High carcinogenic risk was determined in 34 of 79 districts and in 528 of 2,924 municipalities.

The environmental risk, defined as the possibility of impairment of biotic or abiotic components of the environment, derived from geological sources has also been estimated for the whole of Europe. The assessment was based on data obtained from the Geochemical Atlas of Europe (De Vos and Tarvainen, 2006; Salminen et al., 2005). The assessment method was based on the calculation of two-step environmental risk index values (IER) for individually analysed water, soil and sediment samples. Environmental risk quotients are first calculated for each analysed chemical element or compound that exceeds the limit for risk values, and their sum is then calculated. The risk at each site was defined and classified as negligible, low, medium, high, very high or extremely high. The results are presented as risk maps covering the 26 countries of Europe. The calculated maps delineate regions where highly elevated concentrations pose a risk to the environment, or where the action limits for soil, sediment or water contamination should be revised using the high local baseline (Rapant et al., 2008).

Sweden

In Sweden, research started in the 1960s on coronary heart disease and water hardness (Nerbrand et al., 1992). Later, a large study was carried out on diabetes type 1 in children, which resulted in evidence for a correlation between high contents of Zn in drinking water and this type of diabetes (Haglund et al., 1996). In the following some examples on medical geology research in Sweden are presented. They are just a few examples undertaken as a close collaboration between geochemists and medical scientists, epidemiologists, toxicologists, veterinarians, etc. Activities have also been carried out with the Royal Swedish Academy of Sciences which resulted in a special publication of their journal *AMBIO* (AMBIO, 2007).

In the 1980s, when thousands of moose died in Sweden, close collaboration between veterinarians and geochemists showed that this disease was the result of liming of acidified areas (Selinus and Frank, 1999; Selinus et al., 1996). The liming mobilizes molybdenum in bedrock and soils, causing a disturbed Cu/Mo ratio which is critical for the health of ruminants. In connection with a new type of moose disease of unknown etiology in southwestern Sweden, re-examinations of tissue metal concentrations were performed in 1988 and 1992. Decreased hepatic Cu concentrations in 1988 and 1992 (30 and 50%, respectively) were observed in comparison with the corresponding value in 1982. Decrease of other essential and toxic metal concentrations was noted in the liver such as Cr (80%), Fe, Zn, Pb, Cd. In the kidneys, decreases of Ca, Mg, Mn indicate severe metabolic disturbances (The Cd concentration in the kidneys decreased by about 30%.) (Fig. 7).

Fig. 7 Moose

Decreasing metal concentrations in organ tissues of the moose, after such a short period of 5–6 years indicate changes in the concentration of metals via plants in the upper soil surface layer. Thus, the moose appears to be a sensitive monitor of environmental changes. In ruminants, the utilization and availability of Cu in feed is greatly influenced by interactions between Cu, Mo and S. Increased Mo concentrations with respect to Cu in the feed causes secondary copper deficiency in ruminants with clinical signs in close agreement with the signs reported to relate to the unknown moose disease. When livers from yearlings, collected in 1982 and 1992, were analysed for Mo, increased hepatic Mo concentrations were found in northern and southern districts of Älvsborg County (24 and 21%, respectively) during the 10 years between sample collections (Frank et al., 1994).

When pH in soils decreases, Mo becomes less available to plants and its uptake diminishes. Nonetheless, the increased Mo uptake by plants in the strongly acidified region is difficult to explain. A small pH increase in the environment causes a marked increase of Mo uptake by plants (Mills and Davis, 1987). Thus, the results evidently indicate elevated pH in the soils. The intensive liming of lakes, wetlands, fields and pastures in the western part of Sweden during the later 1980s, and to some extent the liming of forest areas during recent years, is suggested as a possible explanation of increased hepatic Mo concentrations in this exceptionally acidified region. This side effect of liming and its harmful consequences for domestic and wild ruminants appears to have been overlooked.

Cardiovascular Studies in Sweden

In Sweden, three case–control studies have been conducted during the last decades. First, the relation between death from acute myocardial infarction (AMI) and the level of magnesium and calcium in drinking water was examined among men, using mortality registers (Rubenowitz et al., 1996). A few years later a study with a similar design was made comprising women (Rubenowitz et al., 1999). The studies were conducted in a part of southern Sweden, in a relatively small geographic area where there was a great difference between and within the municipalities regarding magnesium and calcium content in drinking water. The advantage with this limited study area was that the possible risk of such confounding factors as climate, geographical, cultural and socio-economic differences was minimized.

A prospective interview study was conducted in the same area, where men and women who suffered from AMI during the years 1994–1996 were compared with population controls (Rubenowitz et al., 2000). The results showed that magnesium in drinking water protected against death from AMI, but the total incidence was not affected. In particular, the number of deaths outside hospitals was lower in the quartile with high magnesium levels. This supports the hypothesis that magnesium prevents sudden death from AMI rather than all CHD deaths. The mechanisms that could explain these findings are discussed below.

The data collected in the Swedish case–control studies can be used to estimate the impact of water magnesium on the incidence of myocardial infarction in the study population. If everyone in the male study base were to drink water from the highest quartile ($\geq$9.8 mg L^{-1}), the decrease in mortality from AMI would be about 19%. This means that the age-specific incidence of death from myocardial infarction in the study area would change from about 350/100,000 year^{-1} to 285/100,000 year^{-1}. The decrease of the incidence per mg L^{-1}magnesium can be calculated to be approximately 10/100,000 year^{-1}, which is an even larger decrease than estimated. For women, the corresponding decrease in mortality from AMI would be about 25% if everyone were to drink water from the highest quartile ($\geq$9.9 mg L^{-1}). The age-specific mortality among women in the study area would change from about 92/100,000 year^{-1} to 69/100,000 year^{-1}.

Multiple Sclerosis

Multiple sclerosis (MS) is a chronic neurological illness that affects nerve cells in the central nervous system (CNS) belonging to a group of illnesses called autoimmune diseases where the immune system attacks the body's own tissue. There is a general belief that the epidemiological patterns of MS vary with geography. Present research in Sweden has focused on possible links between MS and geochemistry.

In a recent study the Swedish MS-register was used, which includes almost all MS-patients in Sweden, spatially distributed census data, geochemical data from soil (till), stream water and groundwater. All data were spatially re-allocated to post code areas, the smallest spatial administrative unit in Sweden. Analysis indicates that a geographical pattern could be found with higher prevalence of MS in parts of the county. No north–south or east–west gradient of the prevalence was found. However, visual interpretation of prevalence measures is strongly biased towards large post code areas, masking the variation of prevalence measures of small areas. Multivariate and univariate analyses (zinc) were performed on geochemical data and prevalence of MS. But no statistically significant correlations were found. One important conclusion drawn is that administrative divisions (i.e. parishes, post code areas, etc.) are less appropriate for this kind of study. Divisions with respect to natural (geographical) borders such as catchment areas would be more useful for when combining epidemiological data with geophysical data. The use of average values over districts is problematic. A high density of sampling in an area does not necessarily mean that the calculated mean value is representative for the whole area.

One study has also started, continuing the described one, looking into MS and acid sulphate soils in Sweden and Finland.

Morbus Gaucher's Disease

Gaucher's disease (GD) is a lipidosis; a disorder of the lipid metabolism depending on a deficiency of an enzyme, beta-glucosidase. The first symptom of GD is an enlargement of the spleen, which often begins as early as the first year of life and proceeds rapidly. The heaviest spleen reported in a patient with GD had a weight of 12 kg; that is, 100 times the weight of a normal spleen.

Researchers have made a survey of the occurrence of GD in all Sweden and found 55 cases. The incidence of GD is 13 times higher in northern Sweden than in southern. Gaucher's disease is inherited with an autosomal, recessive gene. It is assumed that the GD gene does not occur in any other family in the same area. Furthermore, the assumption is made that the GD gene has been created by a mutation in a germinal cell of the man in generation 1. The son of this man, in generation II, is the first carrier of the GD gene, the first CD heterozygote. It is important to keep it in mind that if a gene of recessive type is created in an individual living in a certain geographical area, it will take at least four generations or about 100 years until the first individual with the manifest disease is horn in that family. At that time it might be the case that most members of the family have moved to other districts of the country. If the incidence of a disease is high in a certain geographical area, there is, as a matter of fact, a possibility that the responsible gene was created by a mutation 100 years or several hundred years ago in quite another part of the country. Of importance for the discussion is the fact that in such districts where the rate of marriages between near relatives is high, the incidence of recessive diseases existing in the district is also high. In many parts of northern Sweden the rate of such cousin marriages has been unusually high during several hundred years.

In order to discover the places where the GD ancestors lived five generations ago, researchers went to the parish records. Of the 176 forefathers in this "generation of the grandparents' grandparents", 96 lived in the little village of Överkalix at the Kalix river. The interest was then directed to Överkalix as the district where the GD mutation could possibly have taken place, possibly in several different families.

A study of 30 Swedish families with the Norrbottnian type of Gaucher's disease provides evidence for two clusters of the gene, one close to Överkalix in the northern part of the county of Norrbotten and the other in the vicinity of Arvidsjaur in the southern part of the same county. The gene for Gaucher's disease in Överkalix appeared later than that in Arvidsjaur. A founder moving from Arvidsjaur to Överkalix during the seventeenth century or two different mutations are the most likely explanations of this finding.

In collaboration with the Geological Survey of Sweden a possible cause of the disease in these areas was found. In the beginning of the seventeenth century an intensive mining was carried out in Nasafjäll in the high mountain areas in northern Sweden. As long as these heavy metals lie biologically inactive in ore bodies, they are harmless to man, but when they are mined and refined a genetical risk can occur especially in the harsh environmental conditions such as in Nasafjäll where there also were furnaces and toxic fumes. The hypothesis is that such heavy metals might be the mutagen agent responsible for the mutation causing GD in northern Sweden (Hillborg, 1978).

Radon

Radon-222 in dwellings is the dominant source of exposure to ionizing radiation in most countries. Nationwide measurement programmes suggest that the average radon concentration in Sweden is about 2.7 pCi L^{-1} (100 Bq m^{-3}), a level that appears higher than those in many other countries. Current standards in Sweden correspond to about 3.8 pCi L^{-1} (140 Bq m^{-3}) for new houses and 10.8 pCi L^{-1} (400 Bq m^{-3}) for existing houses, whereas in the United States the recommended level at which action should be taken is 4 pCi L^{-1} (148 Bq m^{-3}). In Sweden it is estimated that about 400 people die every year because of radon.

Arsenic

Around the world arsenic in drinking water has become a major issue concerning serious health problems. Arsenic is an element found naturally in rocks and soil, also in Sweden. As a rule, however, yields are very low. SGU has examined arsenic levels in water from the individual, rock drilled wells throughout the country. The risk of elevated levels is, not surprisingly, in areas where the bedrock is arsenic rich. The concentrations of arsenic in the Swedish bedrock and soil are generally low. Elevated and high arsenic levels are, however, found in quite many with sulphide rocks and also older sedimentary rocks. In these areas the groundwater may also have elevated arsenic levels.

Several local studies have been carried out in recent years on arsenic in groundwater. The geological survey is now in a joint project with the Swedish Radiation Protection Authority (SSI) analysing arsenic levels in well water from 750 wells from all over Sweden. Arsenic concentrations are generally low, below or near detection limits, but elevated levels can occur almost everywhere. In areas where bedrock or soil has high levels of arsenic, it is more common with elevated arsenic levels in well water. However, it is not always the case that an arsenic-rich bedrock means elevated arsenic levels in water. The highest concentrations are found in northern Sweden up to 800 μg L^{-1} (the guideline value is 10) (Selinus et al., 2005).

Geomedical Investigations in Norway

In Norway problems related to the geographical distribution of trace elements in nature have been known for a long time in human as well as veterinary medicine. As an example the connection between iodine deficiency and goiter, most prevalent in areas situated far from the ocean, was known already a 100 years ago in man as well as in animals (Løken, 1912). A number of investigations have been performed over the years on deficiency problems in animal husbandry related to low abundance of essential trace elements in pasture soils. Essential trace elements such as cobalt, copper, selenium and molybdenum have been subject to considerable problems in areas where the natural abundances are sub-optimal, particularly in sheep. Many of these findings are described in the textbook edited by J. Låg (1990) and in proceedings of numerous symposia arranged by the Norwegian Academy of Science and Letters, e.g. Låg (1992) and Steinnes (2004).

Ever since the elimination of goiter by adding iodide to table salt, however, no well-documented geomedical problems have been shown in the human population of Norway. A number of investigations have been carried out in order to elucidate possible connections between natural factors and disease in humans, and in some cases statistically significant correlations have been found. Still these findings remain as hypotheses until the present time.

In the 1970s some interesting occurrences of extremely high concentrations of toxic metals such as Pb and Cu were observed in natural soils in several places in Norway (Låg et al., 1974). They were observed spot-wise where ore minerals had been exposed to weathering and the released metals had been supplied to the surface by groundwater movement. This purely natural phenomenon is found to be more common in some areas than others, but until present no associated problems in human or animal health have been disclosed.

Problems in Veterinary Medicine

Problems related to natural deficiency or excess of given essential elements have been well documented in Norway, most frequently with Co, Cu and Mo and Se. In many cases these problems appear more frequently in some geographical areas than elsewhere in the country.

Cobalt deficiency in sheep was recognized already 60 years ago (Ender, 1946). Ruminants are particularly sensitive to inadequate supply of this metal, which they need for the synthesis of vitamin B_{12}. In spite of the great attention to this problem among Norwegian veterinary scientists (Ulvund and Øverås, 1980; Ulvund, 1995; Sivertsen and Plassen, 2004) and the countermeasures introduced, the problem is still evident in some geographical areas, particularly in the southwest.

Because of metabolic interaction the elements Cu and Mo are interdependent in ruminant animals (Frøslie, 1990). Although Mo is not particularly toxic per se, an excess supply of molybdenum may lead to copper deficiency in sheep and cattle. On the contrary Cu poisoning may occur in sheep even at normal copper levels in the feed if the intake of molybdenum is low. Problems in Norway related to the Cu/Mo balance in sheep have been discussed in several papers (Frøslie, 1977; Frøslie and Norheim, 1977; Søli, 1980; Frøslie et al., 1983). The main problem here seems to be associated with low levels of Mo in natural pastures (Frøslie et al., 1983). Since Mo uptake in plants increases with increasing pH whereas the Cu uptake is less influenced by pH, the largely acidic soils in many natural pasture areas in Norway may accentuate the problems observed in sheep.

Problems related to Se and vitamin E deficiency are also well known in Norway (Frøslie et al., 1980). There are appreciable regional differences in Se status within the country, most clearly demonstrated in sheep (Sivertsen and Plassen, 2004). Winter-fed lactating cows, however, were found to have an adequate plasma level of vitamin E and a marginal to adequate level of Se (Sivertsen et al., 2005).

One particular development followed closely in Norway with respect to possible trace element imbalances is the rapid development of organic farming (Steinnes, 2004). Organic farmers are dependent on approaches different from those used in conventional agriculture in order to compensate for deficiencies of essential nutrients in the soil and may therefore be more vulnerable to geomedical problems in their livestock. Govasmark et al. (2005), in a study of the animal nutritional status of Cu, Mo and Co at 27 Norwegian dairy and sheep farms, found that the herbage concentrations of these micronutrients were not sufficient to meet the dietary needs of ruminants and concluded that extra supplements were required.

From the early 1990s a high frequency of bone fractures in moose (*Alces alces*) was reported from the counties of Aust-Agder and Vest-Agder in southern Norway (Ytrehus et al., 1999). The fractures were seen most often in young, growing animals and occurred eight times more frequently in these counties than in reference populations in other counties farther north. Since the southernmost part of the country suffers from problems related to transboundary air pollution from other parts of Europe, such as extensive acidification of soils and fresh waters and contamination with heavy metals, it was speculated that the problems in moose might be connected with effects of air pollution. It was found, however, that the high frequency of bone fractures in southern Norway was not associated with trace element imbalances, but may have resulted from inadequate nutrition following general overcrowding and high pressure on feed resources.

Possible Geomedical Associations in Human Medicine

At the Geological Survey of Norway geomedical research has been carried out since 1971. This activity has included multi-element countrywide geochemical mapping and correlating spatial distributions of geochemical and other types of natural data with the epidemiology of endemic human diseases. An example of this work is a comparison of the composition of drinking water from main water works all over Norway with the geographical distributions of various types of cancer and other diseases (Flaten and Bølviken, 1991). Interesting regional distributions of elements were disclosed, and in some cases statistically significant correlations were evident between chemical and corresponding epidemiological data on a geographical basis. A novel method for spatially moving correlation was developed (Bølviken et al., 1997a). By the application of this method several geomedical associations were revealed, among them (Bølviken et al., 2003) significant correlations for rates of *multiple sclerosis* versus Rn in indoor air (positive) and atmospheric deposition of marine salts (inverse). By extending their studies to literature data from China (Bølviken et al., 1997b) these scientists observed strong associations for rates of *nasopharyngeal carcinoma* versus the soil contents of Th and U (positive) and Mg (inverse).

Geochemical Studies in Norway of Potential Worldwide Significance to Human Health

Professor J. Låg was probably one of the very first scientists to realize the possible importance of the oceans as a significant source of elements to soils in coastal areas, and he was the first to demonstrate this experimentally. Based on an extensive collection of forest soils from different parts of southern and central Norway (Låg, 1968), he was able to show that exchangeable concentrations of Na and Mg in the surface soil depended strongly on the distance from the ocean and the corresponding chemical composition of precipitation. The same soil material was studied with respect to the halogens Cl, Br and I (Låg and Steinnes, 1976) and Se (Låg and Steinnes, 1974; 1978). A distinct and regular decrease with increasing distance from the ocean was also evident for these elements, and this finding was later confirmed in a transect study of natural surface soils in northernmost Norway (Steinnes and Frontasyeva, 2002) and in a separate study of Norwegian agricultural soils (Wu and Låg, 1988). Interestingly the I/Cl ratio in the soil was more than 1,000 times that in ocean water. The main reason for this is apparently the enrichment of iodine at the ocean surface caused by biogenic emission of CH_3I (Yoshida and Muramatsu, 1995). Considering the current distribution of iodine deficiency disorders worldwide (Fordyce, 2005), it is striking that the problem areas are mostly confined to areas located at significant distances from the nearest ocean. Possibly the marine supply of iodine from ocean to land is much more important to human health than what has been generally assumed so far.

Extension of the trace element studies of the above-mentioned forest soils to selenium, a geographical distribution similar to that of iodine was evident indicating that the ocean might be a significant source even for Se (Låg and Steinnes, 1974; 1978). This seemed very unlikely, considering the fact that the Se concentration in seawater is only around 0.1 $\mu g\ L^{-1}$. Later, however (Cooke and Bruland, 1987), it has been shown that biogenic emissions of volatile compounds such as dimethylselenide lead to enhanced Se concentrations in the marine atmosphere, which may explain the above findings in Norwegian soils. Again this biogeochemical cycling could be of geomedical importance in some areas of the world with naturally low selenium in the bedrock.

Medical Geology in Finland

Geographical variation of occurrence of certain non-communicable diseases has been reported in Finland. Goiter, dental caries, multiple sclerosis, cardiovascular diseases, diabetes and Parkinson's disease among others are examples of those diseases (Adlercreutz, 1928; Pärkö, 1990; Sumelahti et al., 2000; Kannisto, 1947: Karvonen et al., 2002; Rytkönen et al., 2001; Havulinna et al., 2008). The etiology of diseases is usually multifactorial resulting from several biologic, behavioural, genetic and also environmental risk factors. For example, certain geological risk factors, which are fairly stable, may have a meaningful role in the etiology and regional variation of certain non-communicable diseases. Several studies of the associations of the conceivable risk factors of geological environment on human health have been carried out in Finland during the past decades.

Iodine and Fluoride

Adlercreutz (1928) was probably among the first who presented the hypothesis about the geographical association of iodine in drinking water with the occurrence of goiter in Finland (Adlercreutz, 1928). In the 1950s goiter was common in Finland, especially in eastern part of the country where the intake of iodine was low (Lamberg, 1986). At present, deficiency of iodine is controlled by iodization of table salt.

Fluoride is a naturally occurring element that is essential to human health in small doses but is harmful in excess (Plant et al., 1996). The highest average fluorine contents in Finland are found in rapakivi granites in southeastern and southwestern parts of the country. Fluoride anomalies in groundwater reflect the high fluorine content of bedrock in rapakivi area (Lahermo and Backman, 2000). Pärkö (1990) reported that caries prevalence was lower in rapakivi area where fluoride content in drinking water was higher than in control area. Naturally higher fluoride concentrations in drinking water have consistently associated lower caries rates. In contrast, the exposure to natural fluoride in well water was associated with the increased risk of hip fracture among women (Kurttio et al., 1999). The association of fluoride in drinking water with coronary heart disease (CHD) occurrence has also studied in Finland. Luoma et al. (1983) reported that fluoride intake was inversely associated with risk of myocardial infarction (MI) in men. They suggested that fluoride concentrations at the level of 1 mg L^{-1} in household water were beneficial. Finnish studies also suggest that the geographical pattern of coronary heart disease (CHD) in Finland is consistent with the concentration of fluoride content in well water (Kousa and Nikkarinen, 1997; Kaipio et al., 2004). The authors suggest that one mechanism could be that fluoride prevents dental infections, which in turn reduces mortality from CHD (Kaipio et al., 2004). However, the widely used toothpaste with fluoride may influence to results.

Arsenic

Arsenic is considered one of the most toxic naturally occurring elements. Some evidence for an increased risk of bladder cancer associated with arsenic exposure from drilled wells 2–9 years before the diagnosis has been reported in Finland (Kurttio et al., 1999). The association between arsenic exposure and kidney cancer risk was not found. Locally high arsenic concentrations in deep groundwater, soils and bedrock may pose a risk to public health and the environment in the Pirkanmaa region in southern part of the country. About 23 % of the 965 wells exceeded the limit value of 10 $\mu g\ L^{-1}$ (Backman et al., 2006). The arsenic content of hair correlated well with the past and chronic arsenic exposure. About 10 $\mu g\ L^{-1}$ increase in the arsenic concentration of the drinking water or 10–20 $\mu g\ day^{-1}$ increase in arsenic exposure corresponded to a 0.1 mg kg^{-1} increase in hair arsenic (Kurttio et al., 1998). Muscle cramps, mainly in the legs, were associated with elevated arsenic exposure in drinking water (Kurttio et al., 1998).

Selenium

Selenium concentration in soil and groundwater is exceptionally low for geochemical reasons in Finland (Koljonen, 1975; Alfthan et al., 1995; Lahermo et al., 1998; Tarvainen et al., 2001). Mortality from cancer (Salonen et al., 1985) and cardiovascular diseases has been associated with a low serum selenium concentration in Finland (Salonen et al., 1982). Eastern Finland is a low selenium area (Tarvainen et al., 2001) with high copper content in drinking water (Punsar et al., 1975). The data of Finnish study indicate synergistic effect of high serum copper, low serum selenium content and low-density lipoprotein (LDL) cholesterol in atherogenesis (Salonen et al., 1991). Efficient intake of selenium is ensured by supplementing sodium selenate to all agricultural fertilizers used in Finland since 1984 (Ministry of Agriculture and Forestry, 1984).

Asbestos

Occurrences of asbestos minerals and their environmental impact have been studied in eastern Finland (Nikkarinen et al., 2001). Asbestos mineral here is anthophyllite and this was mined from 1918 to 1975. Kokki et al. (2001) found an increased risk of lung cancer around the former asbestos mine using a method called SMASH (small area statistics on health) developed by the Finnish Cancer Registry and the National Institute for Health and Welfare (THL).

Natural Radiation

The risk of lung cancer due to indoor radon in Finland has been studied. Results of the studies did not indicate increased risk of lung cancer from domestic radon exposure (Auvinen et al., 1996; Piispanen, 2000). However, collaborative analysis of individual data from 13 case–control studies (including Finnish study) suggested that residential radon accounts for about 9% of deaths from lung cancer and about 2% of all deaths from cancer in Europe (Darby et al., 2005). The map of the indoor radon in Finland is available in the web site of The Radiation and Nuclear Safety Authority (STUK) (http://www.stuk.fi/sateilytietoa/sateily_ymparistossa/radon/kartat/fi_FI/radon_koko_suomi_3_2008/). The association between well water radioactivity and risk of cancers of the urinary organs has also been studied in Finland (Kurttio et al., 2006). The authors concluded that even though ingested radionuclides from wells drilled in bedrock are a source of radiation exposure, they are not associated with a substantially increased risk of bladder or kidney cancers in concentrations occurring in drilled wells. The map animations of the spatial and temporal variations of the incidence of a number of common cancer forms in Finland (Pukkala and Patama, 2008) and in the Nordic countries (Pukkala et al., 2007) are available from the web page of Finnish Cancer Registry (www.cancerregistry.fi).

Calcium, Magnesium and Water Hardness

The geographical variation of the occurrence of coronary heart disease (CHD) is well known over 60 years in Finland (Kannisto, 1947; Karvonen et al., 2002) and it has been studied also from the perspective of medical geology in Finland (Fig. 8). Punsar and Karvonen (1979) studied mortality from CHD and water quality in two rural areas in western and eastern Finland. They suggest that CHD may be associated with low concentrations of magnesium and chromium in the drinking water, but definite relationship between water quality and sudden death was not found. Kaipio et al. (2004) found that in rural Finland CHD mortality declined with increasing magnesium concentration in drinking water, but slightly increased with calcium content. Piispanen (1993) found that in the western part of Finland with the low cardiovascular disease (CVD) mortality, the hardness and magnesium concentration in well water were markedly higher than in the eastern part of the country.

During the last 10 years, the Geological Survey of Finland (GTK) and the National Institute for Health and Welfare (THL) (former National Public Health Institute KTL) have carried out studies of the spatial variation of the incidences of acute myocardial infarction (AMI) and childhood type 1 diabetes (T1DM) in relation to the geochemistry of local groundwater in rural Finland. Results of recent studies suggested that water hardness, especially regionally high concentration of magnesium was associated with decreased AMI incidence in Finland (Kousa et al., 2004a, b, 2008) (Table 1). The high concentration of calcium in relation to magnesium concentration was associated with increased incidence of AMI (Kousa et al., 2006) (Fig. 9, Table 1). The geological circumstances have a great impact in well water quality in rural areas. The magnesium concentration in water, especially in Finland, is lower than in most other countries due to

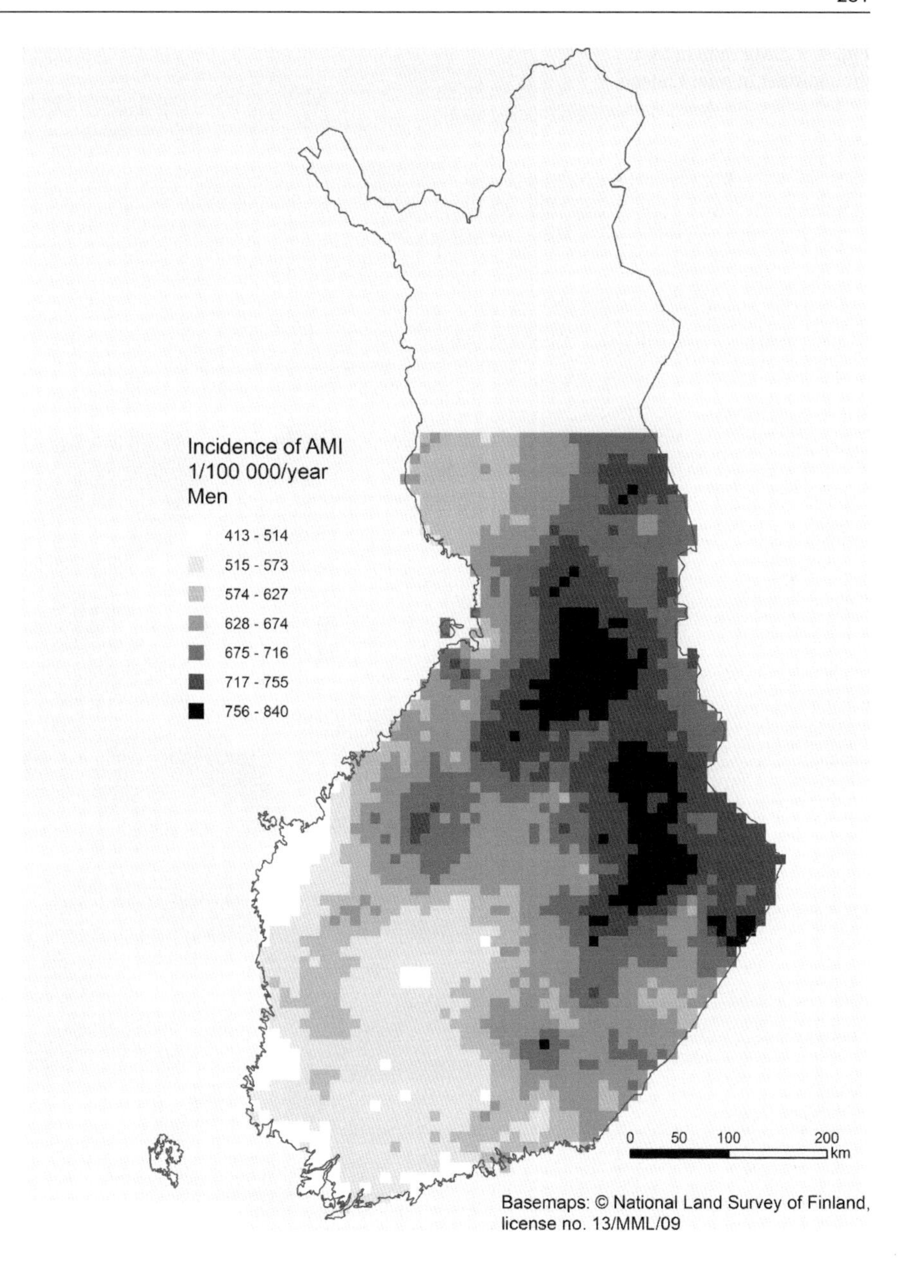

Fig. 8 The incidence of acute myocardial infarction in men aged 35–74 years in rural Finland (*Source*: THL. GTK)

Table 1 The estimated regression coefficients of Mg, Ca and Ca/Mg ratio on the incidence of the first acute myocardial infarction among Finnish men in 1983, 1988 and 1993 (pooled data)

Element	Posterior mean (%)	95% HDR[a]
Mg mg/l[b]	– 4.9	– 8.8, – 0.9
Ca mg/l[b]	0.9	– 0.1, 2.1
Ca/Mg[c]	3.1	0.5, 5.7

[a] Highest density regions
[b] Full spatial model
[c] A separate spatial model

geological circumstances (Lahermo et al., 1990). Although Finnish groundwater is in general soft, all studies including those with a different study design conducted in Finland have suggested the association of soft ground (drinking) water, low in Mg, with an increased risk of CHD.

Moltchanova et al. (2004) studied the association between spatial variation of type 1 diabetes and its putative environmental risk factors such as zinc and nitrates. Results of the ecological study did not show

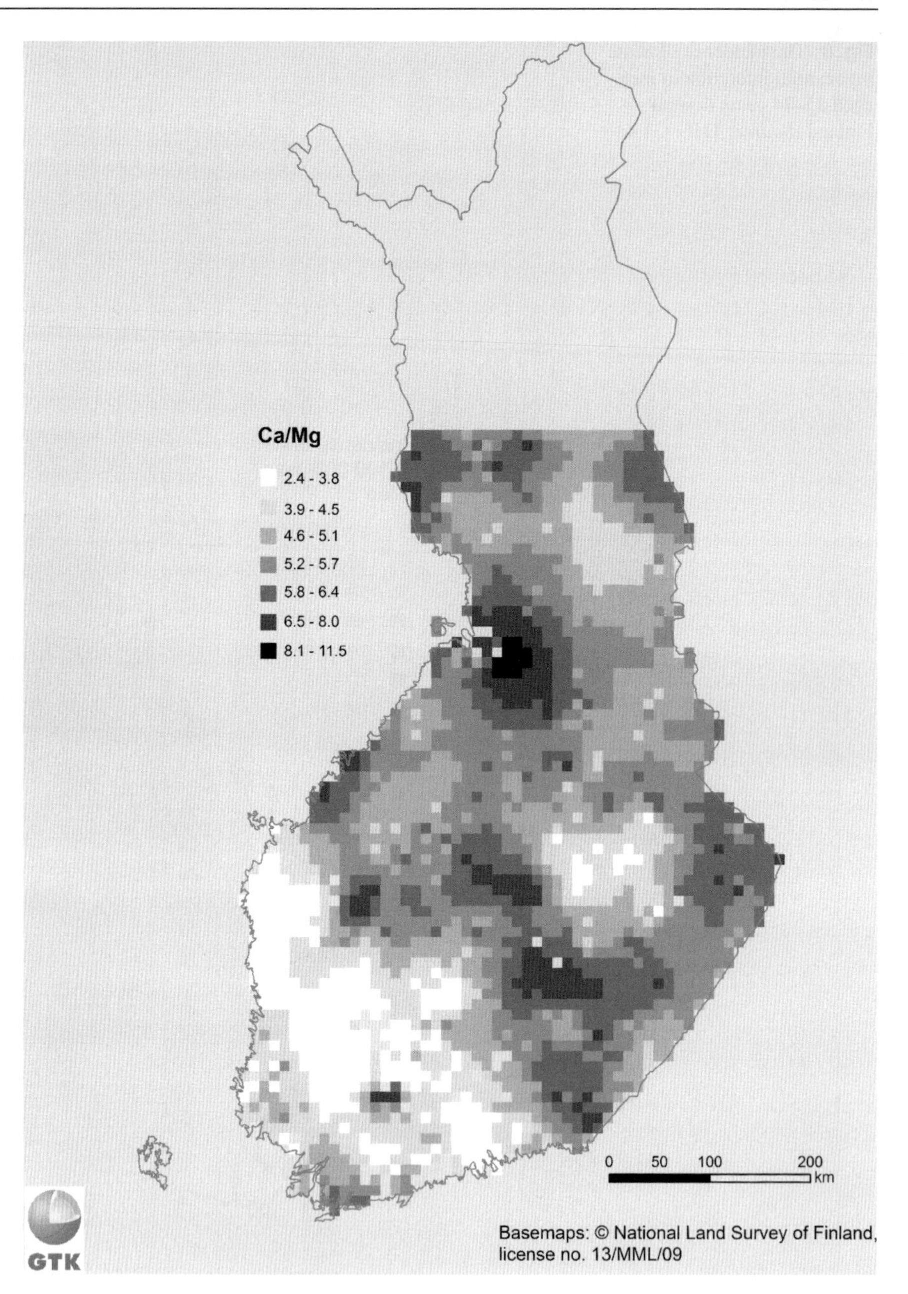

Fig. 9 Ca/Mg ratio in local groundwater in rural Finland

clear spatial association between zinc or nitrates and the incidence of childhood type 1 diabetes at the population level in Finland (Moltchanova et al., 2004). The geographical association between type 2 diabetes in young adults and magnesium, calcium and water hardness is the topic of ongoing collaboration study of GTK and THL.

Bayesian spatial statistics and Geographical Information Systems (GIS) were applied to estimate the regional patterns of the incidence of AMI and diabetes and geochemical elements in groundwater. A special feature of these studies is that 10 × 10 km grid cells were used when defining the study areas instead of administrative areas. The geographical association between the incidence of AMI and diabetes with geochemical elements of groundwater was for the first time studied using flexible geographical scale without administrative boundaries at a whole country level (excluding Lapland and Åland). In Finland, every resident has a unique personal identification number (ID). ID numbers assigned at birth enabled the location of individuals according to map coordinates of the place of residence at the time of diagnosis. Development of GIS enables a researcher to link data from several databases.

Ecological and Health Risk Assessment – Multidisciplinary Network

Environmental risk assessment methods are increasingly applied to investigate small- and large-scale environmental problems and their impact on human health. GTK, the Finnish Environment Institute (SYKE), THL (former KTL) and the University of Eastern Finland (former the University of Kuopio) have founded the Environmental Risk Assessment Centre (ERAC) to conduct scientific research and to develop new projects (http://www.eracnet.fi). The ERAC is based on multidisciplinary networking and co-operation, ranging from geochemistry, geology, biogeochemistry and ecology to environmental sciences, toxicology, epidemiology, risk analysis and the political sciences. One objective of ERAC, in co-ordination with The Finnish Cancer Registry, is to investigate relationships between cancer incidence and metal concentrations in soil. Another multidisciplinary project, FINMERAC (Integrated Risk Assessment of Metals in Finland), aims to improve methodology of health risk assessment for metals. The general model of metal risks developed by FINMERAC is described in Fig. 10. Two metal industry areas and one mining target area are selected examples of environmental pollution, thus providing different challenges for risk assessment and management (Nikkarinen et al., 2008).

ERAC network offers also environmental risk assessment education (ERAC edu). Environmental risk assessment education is implemented via research network (GTK, THL and the University of Eastern Finland) and with help of both national and international experts. Primary target group is researchers, experts, municipal officials, industrial experts and postgraduate students located in eastern Finland. Applicable environmental risk assessment postgraduate and supplementary education is offered to target groups (http://risk.eracnet.fi/edu/index.php/ERAC_Edu).

Geology and Health: The French Case

Due to a diversified geology and a long history, the French case is probably emblematic for a view of geology and health issues in western Europe. The present contribution is based on a recent survey published by the French geological survey (BRGM) in the spring issue of the biannual journal *Geosciences*, dominantly published in French with extended abstracts and figure captions in English. This allows us to consider the case seen through geological and hydrogeological approaches, but also considering anthropogenic effects (pollution) affecting bedrock, soil and groundwater, as well as beneficial aspects of geology and mineralogy on human health.

Located in the western part of Europe, France represents in itself a good part of geology and health questions as the geology is quite diversified, with anciently eroded metamorphic basement, well-developed Mesozoic sedimentary basins, alpine orogenic zones as well as quaternary volcanic districts. If Corsica and overseas areas and territories are also considered, this diversity is increased with active volcanoes (in French West Indies and Reunion Islands) and vast tropical forest areas of French Guyana. In addition to natural effects on health from geology, one should also consider the case of pollution induced by former human activities, which have modified the soil composition and induced other types of health effects. Groundwater is also worth examination in itself, as it displays all categories from health caring thermal springs to toxic waters of either natural or anthropogenic origin.

In a recent issue of *Geosciences* (N°5, March 2007), a wide coverage of the subject was presented in a 110 pages edition covering geology and health issue in general with a focus on French cases. This presentation is an updated summary of this issue and offers to English readers a renewed and coherent entry in the subject of a west-European perspective for medical geology.

Hard Rock Geology and Related Health Issues

The oldest geological formations of Proterozoic and Palaeozoic ages are generally the richest in health-bearing minerals and metals. A quick look at the numerous maps of the Geochemical Atlas of Europe (GTK – FOREGs, 2005 and 2006) show how most

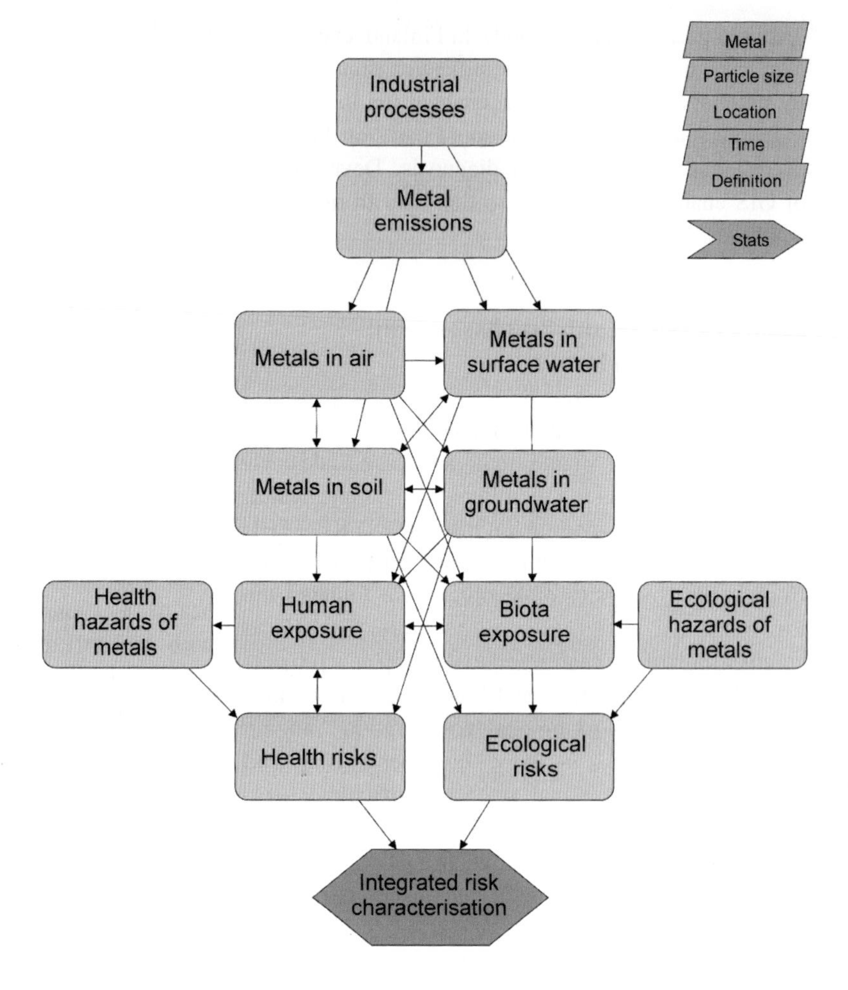

Fig. 10 The general model of metal risks (Nikkarinen et al., 2008)

metals possibly toxic for health (Al, As, Cd, Sn, Sb, Hg, Ti, Pb and U) are exposed at relatively high concentrations in the French basement areas of Britain and Massif Central (Fig. 11).

More specifically, health effects from geological sources can be expected or even documented in the following regions:

- lead in the SE border of Massif Central (Cevennes metallogenic province);
- arsenic in eastern Massif Central and southern Brittany;
- uranium in a few natural or formerly exploited sites in Brittany and Massif Central.

Besides these metals, radon and asbestos need specific considerations:

- Radon is emitted from granitic environments (more specifically leuco-granites), and areas where radon risk is identified by French health authorities (Research Institute for Radioactive Safety (Perrin, 2007) correspond with the granitic basement regions (Figs.12, 13 and 14).
- Asbestos occurs naturally in metamorphic environments which are found in Britanny, Massif Central, the Alps and in Corsica (Fig. 15). Theses sites are studied by the French Institute of Sanitary Survey (Billon-Galland et al., 2007).

The sedimentary basins do not generally generate any health risks for the population, except for the Permo–Triassic early cover where uranium- as well as fluorine- and lithium (southern rim of Paris basin)-rich deposits may occur.

The volcanic districts of Massif Central do not display geochemical anomalies of concern for health. Of course, this is not the case for active volcanic districts in the overseas islands of Réunion and Guadeloupe. The Piton de Fournaise is frequently emitting basaltic flows, and when winds blow, sulphurous gases reach

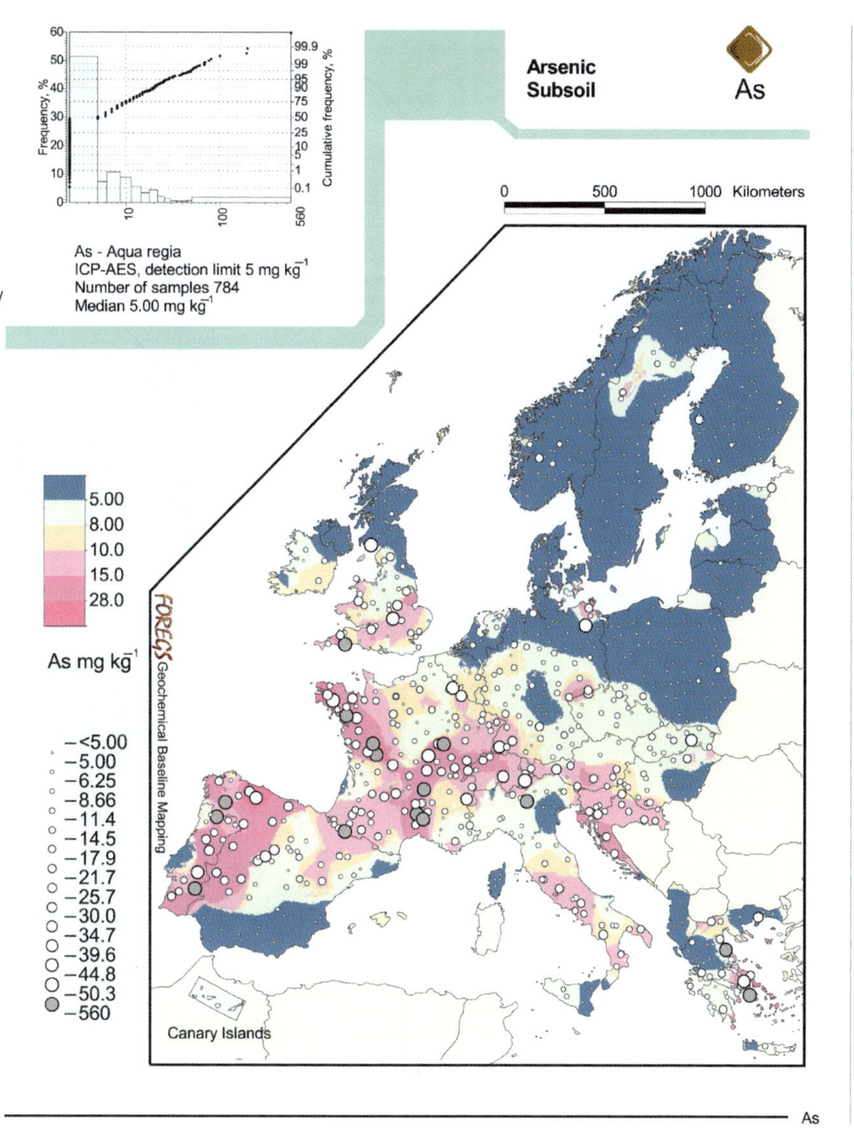

Fig. 11 a–e maps showing the high values in As, Pb, Ta, U and Zr found in French basement regions from the geochemical survey of Europe (FOREGS, Salminen et al., 2004, 2005). These maps of Europe in black and white can be seen in colour at the following website: http://www.gsf.fi/publ/foregsatlas/map_compare.html

neighbouring villages, as observed during the summer 2006 eruption. The less accessible Soufriere of Guadeloupe also emits chlorine gases, imposing restriction of access for visitors due to health risks: for safety reasons the top of the volcano is not accessible to tourists.

Polluted Soils and Health Effects from Former Human Activities

France, as in most west-European industrial areas, is rich in former industrial sites. On such sites where industrial processes concerning heavy metals or toxic components have been used for tens or hundreds of years, including historical periods for which the health and safety regulations were not as strict as now, one can expect effects of these activities on the underlying and surrounding natural environment. This is the reason why a detailed and exhaustive inventory of the former industrial sites has been undertaken at France national on regional and local levels. This allowed producing a database now accessible to the public and very regularly used for planning purposes, Fig. 15.

Polluted areas need to be known and eventually surveyed, as pollutants in soils and waters are clearly subject to alteration through a variety of mechanisms,

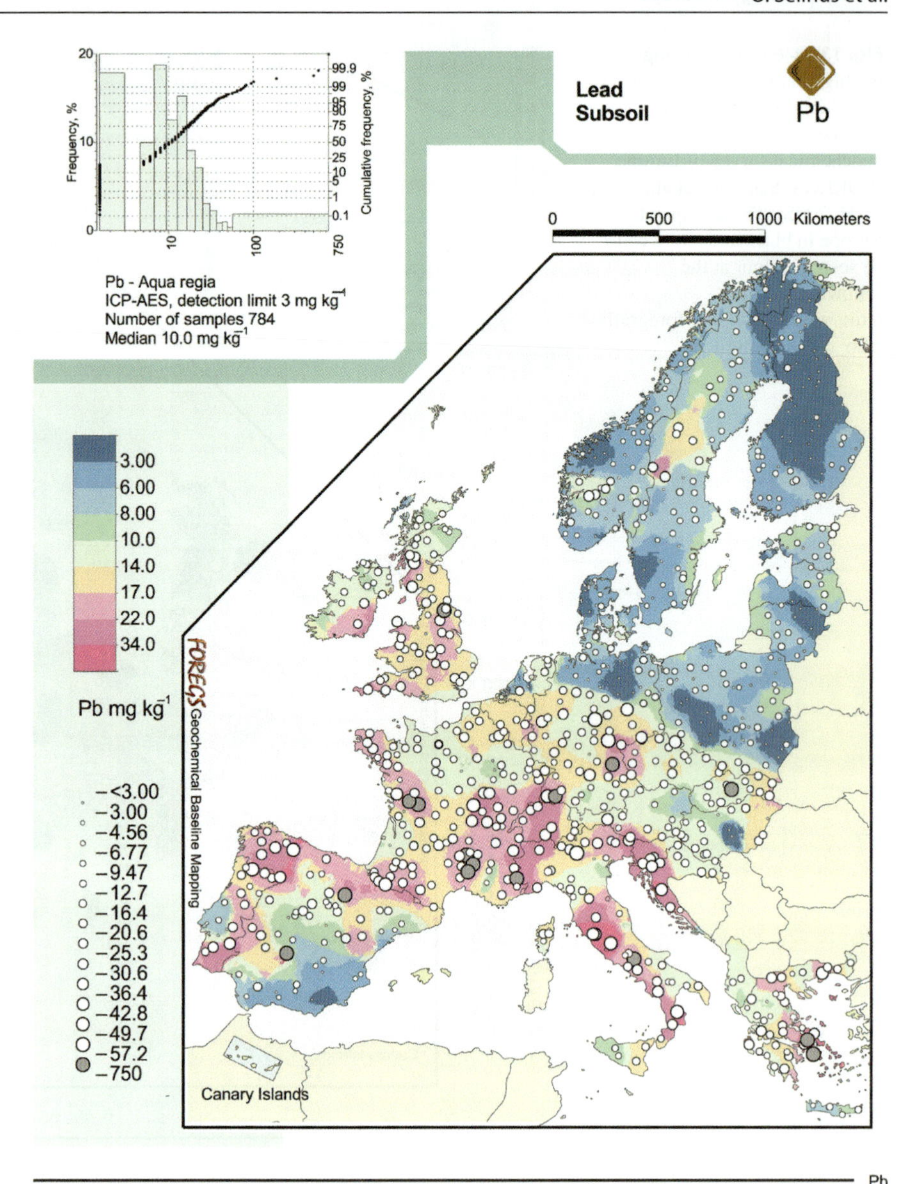

Fig. 11 (continued)

among them transport by water, sorption, degradation and redox reactions. Using mathematical models accounting for all these mechanisms is a complex undertaking. First, the biogeochemical conditions at the site must be determined so as to ascertain whether biodegradation can occur. For example, chlorinated solvents need reducing conditions, whereas hydrocarbon and PAHs require oxygen and nitrates. Once identified, the source of pollution must be removed and the plume must be monitored in order to determine the plume movement and to calculate the mass balance. More or less complex predictive models are then used to estimate the future extension of the plume (Atteia and Saada, 2007).

Another issue is the storage of industrial and domestic wastes. In France as in many other European countries, storage remains one of the main methods for waste management. An in-depth study was carried out in 2002 under the guidance of concerned public agencies, aimed at assessing the impact of municipal solid waste disposal sites on health. A quantitative health risk assessment method was developed and applied, showing that – from existing data – the present situation was not alarming, but that conditions of

Fig. 11 (continued)

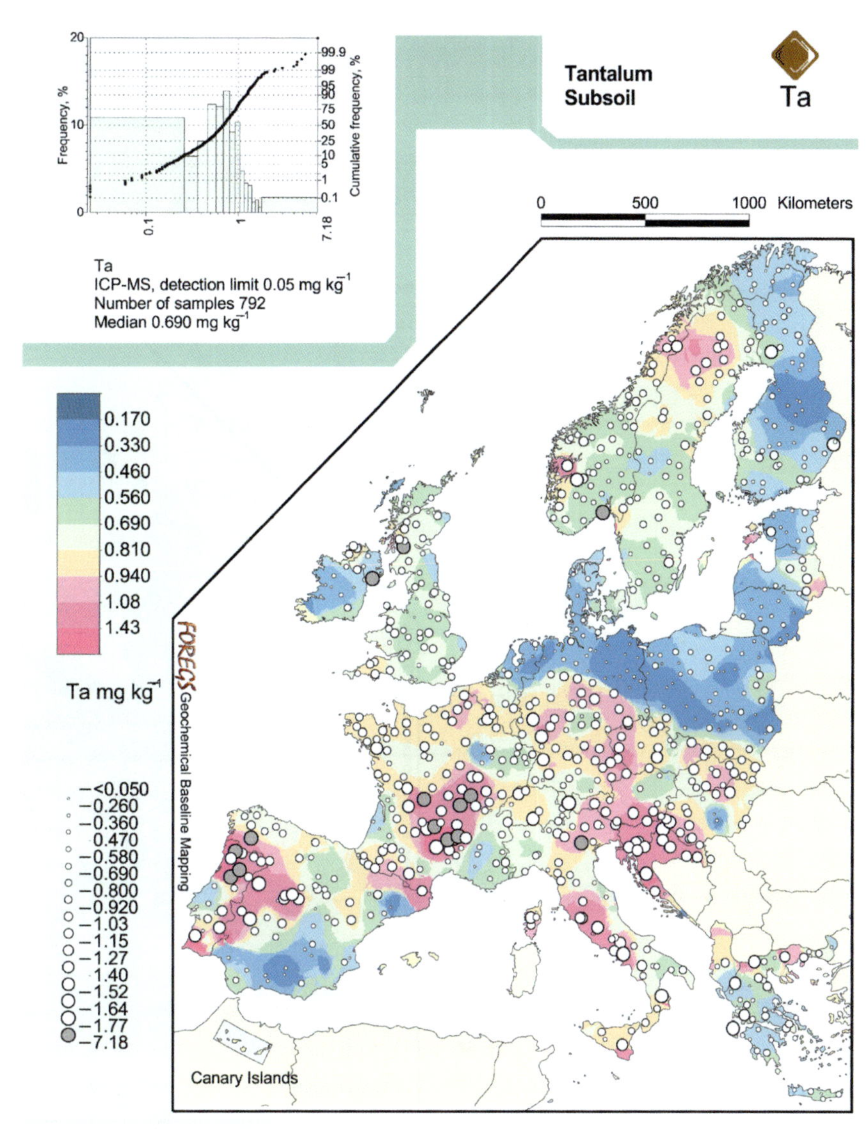

Ta

monitoring must be improved due to information gaps, especially concerning leachates (Dor and Guyonnet, 2007).

The Case of Groundwater

The case of groundwater is the most important in terms of health effects, as water is a large part of human needs, either for drinking or as preserving or cooking agent. Therefore, norms of quality have been defined by health authorities in France and in the European Union. These norms concern the following elements: Al, As, B, Ba, Cd, Cr, Cu, F, Fe, Hg, Mn, Ni, Pb, Sb, Se and NH^{4+}.

As shown by Blum et al. (2007), some toxic or undesirable elements for human health may occur in natural groundwater in France with values exceeding the standards for quality set by French as well as European regulations. This applies in particular for arsenic, antimony, selenium, nickel and fluorine. Therefore, a research programme is developed in order to identify natural high concentrations of these elements and to understand the phenomena implying the dissolution of these elements in groundwater. The objective of these studies is to develop management

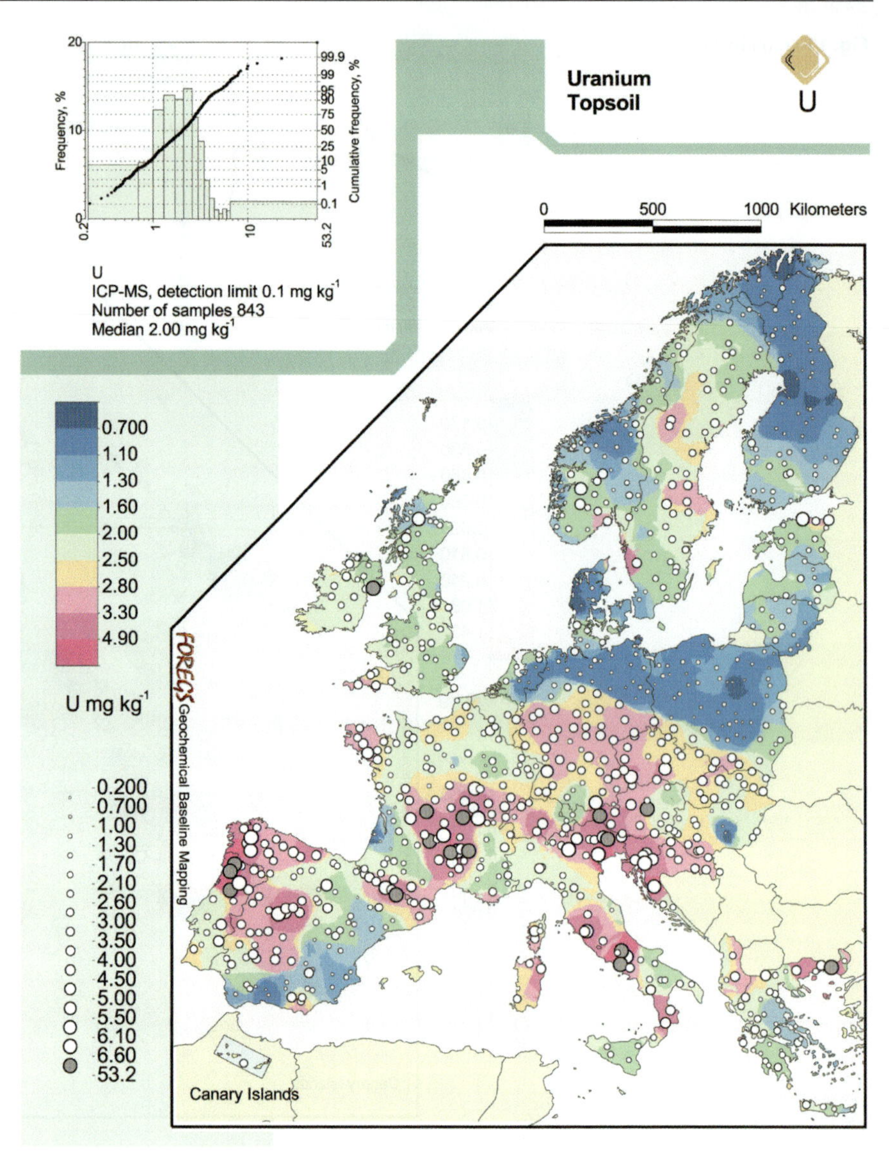

Fig. 11 (continued)

tools for concerned agencies to comply with the European water framework directive, which defines the "proper chemical state" of groundwater, and with the National Health Environment Plan, in particular for locating new catchments for producing drinking water.

In particular, BRGM coordinated the European BRIDGE project (2005–2006), which allowed elaborating the criteria necessary to define the good chemical state of groundwater. At national level, the zones at risk from background geochemistry high in trace elements in ground and surface water were precisely mapped in the Rhone-Mediterranean and Corsica regions.

Of course, natural geochemistry is not the only harmful subject in terms of health effects of groundwater. Due to intensive human impact, whether due to industry, agriculture or even transport, polluted groundwater is certainly the most worrying for sanitary policies. The question of lead (Glorennec et al., 2007) can be considered as presently mastered due to the introduction of unleaded fuels in Europe (Fig. 16), but the mercury issue is not yet solved in French Guyana,

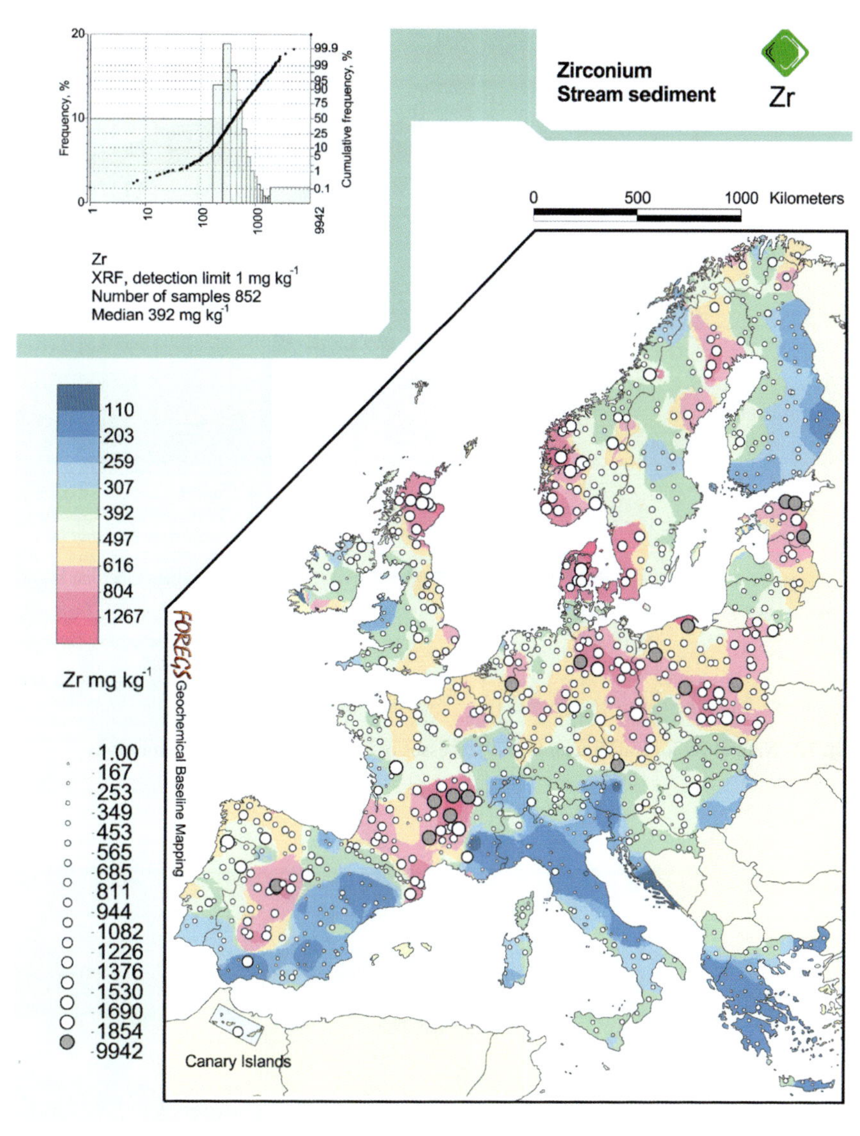

Fig. 11 (continued)

due to gold illegal extraction (so-called orpaillage), with health effects on Indian tribes consuming carnivore fishes. But in continental France, it is certainly the effects of intensive agriculture on groundwater which is the most harmful. The case of nitrates is particularly severe as the tendencies for nitrate contents in groundwater are not yet oriented towards a decay but mostly still increasing in the most concerned areas such as Brittany (pork feeding) and Beauce plain south of Paris (agriculture).

In such conditions, the protection of groundwater in general, and the protection of capture areas in particular (Figs. 17, 18) is a critical issue for hydrogeology.

Various Uses of Geological Products Beneficial to Human Health

When considering the case of geology and health issue, one has the tendency to only focus on the effects deteriorating the health, neglecting all the positive effects of geology on human health. Certainly the most positive and unevidenced effect of geology on human health is the role of geology in mountain building. The healthiest conditions are frequently found in altitude, and living in mountain regions, or benefiting from mountains for holidays is certainly one of the most beneficial effects of geology on health. Similarly, the natural

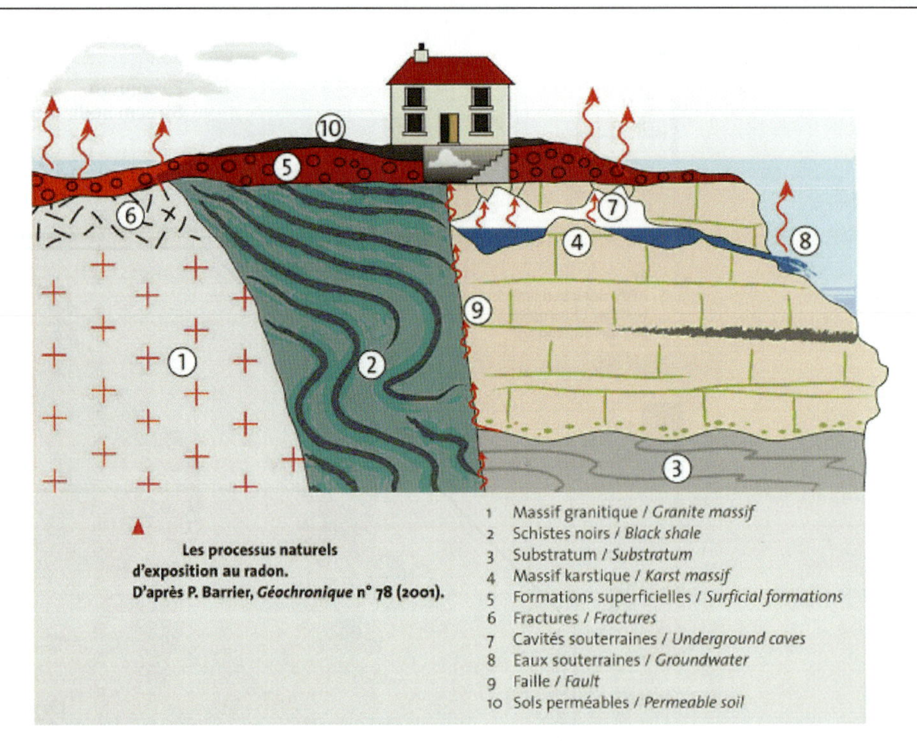

Fig. 12 Natural geological structures and processes of exposure to radon gas emissions

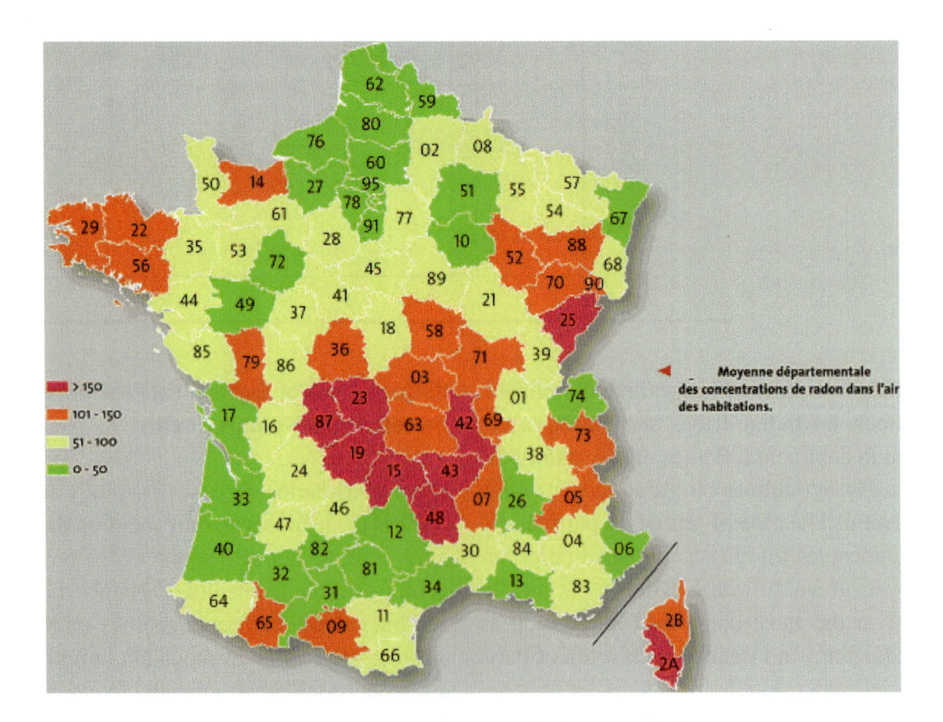

Fig. 13 Average concentration of radon in the air of housing in French departments (IRSN, 2008)

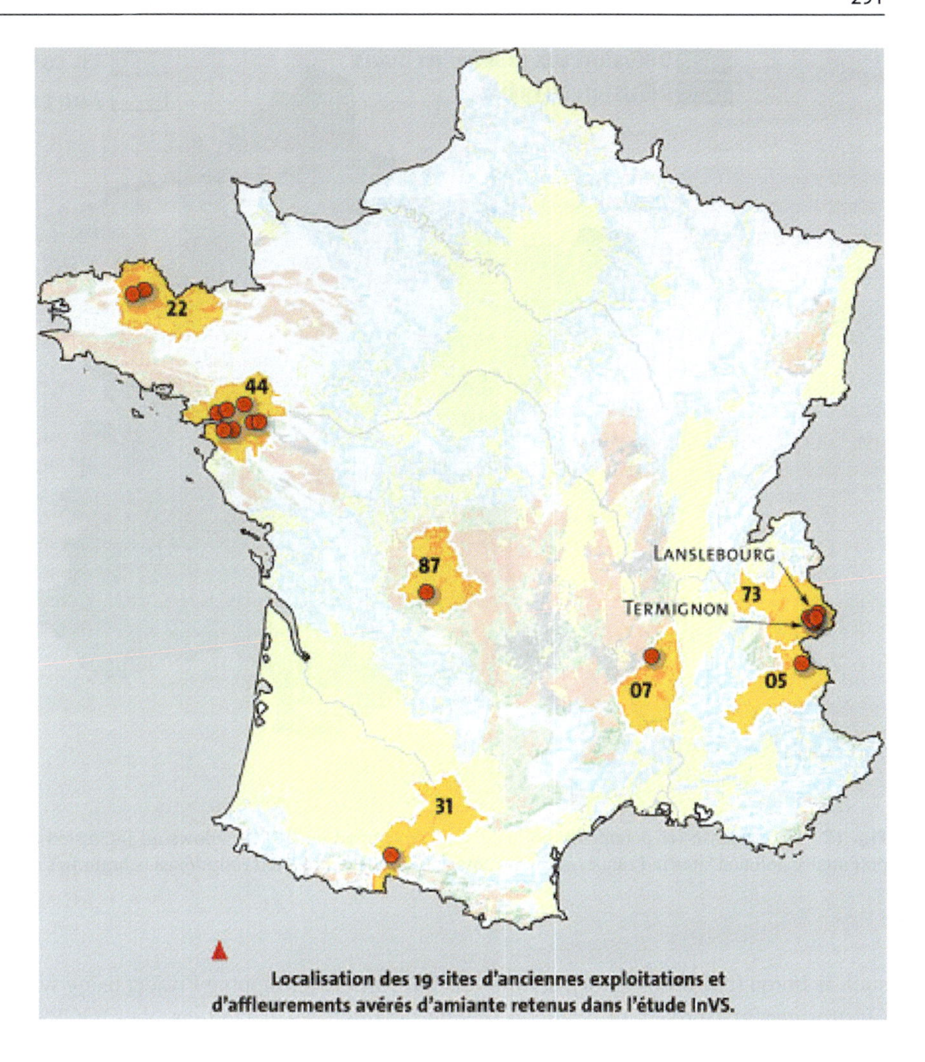

Fig. 14 Location of former asbestos mines in France (BRGM, 2005)

filter offered by soils and porous rocks in the percolation of surface polluted waters in the aquifers is also a very important contribution of geology to health. Considering the large proportion of groundwater in the drinking water production in France, and the very positive effect of groundwater sources on prices of distributed waters, this is also certainly another major – although underestimated by the economy – contribution of geology to human health.

Besides these general cases, there are more specifically identified contributions of geology to health. These are the thermal springs and waters, used in thermal stations recognized as such by health and sanitary authorities (and refunded as such by the French public social security system), or simply consumed as commercially labelled mineral waters. As shown by Roques (2007) there are nearly 100 thermal stations in metropolitan France (Fig. 19), with documented effects on rheumatism, respiratory diseases, dermatology, neurology, stomach or urinal diseases, gynaecology and development troubles, etc. More than 100,000 jobs are offered for these health cares.

Health and geology would not be complete without a word on the direct role of solid minerals on health, which are used in pharmacy. As shown by des Ylouses, many mineral elements enter medical products, whether as active principles or recipients. This is the case for alkaline elements such as lithium (maniacal depressions), sodium (physiologic serum) and potassium (cardiovascular), but also earth-alkaline as magnesium (nervous system), calcium (bones and teeths) or barium (radiology), transition elements as titanium (UV filter), chromium (metabolism of glucoses), manganese (antiseptic solutions), iron (blood), cobalt (vitamin B2), copper (antiseptic, cellular defence), zinc (enzymes metabolism) and elements

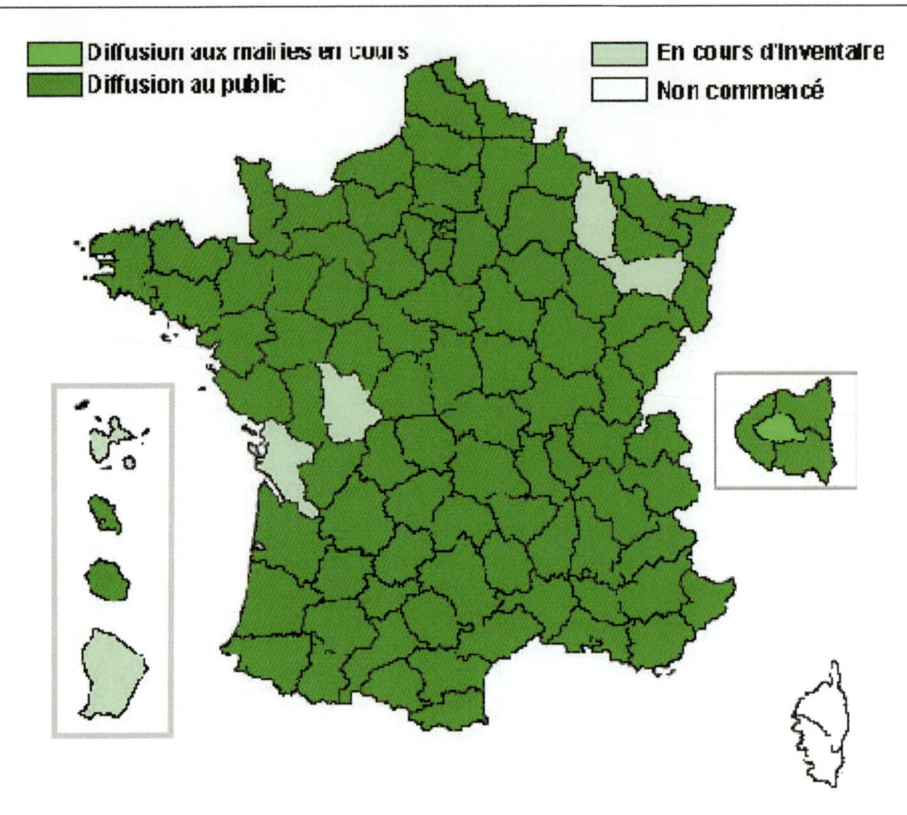

Fig. 15 The database for potentially polluted soils developed by BRGM allows all interested parties to have a free access to all potentially polluted sites in France, as the inventory is almost completed (http://basias.brgm.fr/)

such as boron (buffering agent), aluminium and carbon (gastro-intestinal protection), silicon (dental creams), sulphur (protein structures, bones, articulations, etc.), fluor (bones and teeths), chloride (blood equilibrium), iodine (thyroid metabolism, antiseptic). The need for very pure elements and minerals used for medical and pharmaceutical applications is a challenge for geologists and mineralogists.

European Databases

In the 1970s and 1980s regional or national geochemical mapping projects were carried out in many countries. However, due to lack of standards the results were not satisfactory, thus standardization for geochemical mapping was particularly needed. The International Geochemical Mapping Project was accepted as part of the International Geological Correlation Program (IGCP259) in 1988. Plans for a Global Geochemical Mapping Project using wide-spaced sampling was then accepted as the IGCP360 project entitled "Global Geochemical Baselines" in 1995. The major resulting aspects of the project were to determine the natural global distribution of various chemical elements and compounds and the present state of pollution in the surface environment, delineate areas and geochemical provinces especially enriched in elements of economical importance and provide data on the regional distribution of compounds that are linked to human and animal health. As one further step the "Geochemical Atlas of Europe" was carried out by the Geological Surveys of the European Union (EuroGeoSurveys).

The European survey covered 26 countries and provided information in different sample media of the near-surface environment (topsoil, subsoil, humus, stream sediment, stream water and floodplain sediment). This was the first multi-national project, performed with a harmonized sampling, sample preparation and analytical methodology, producing

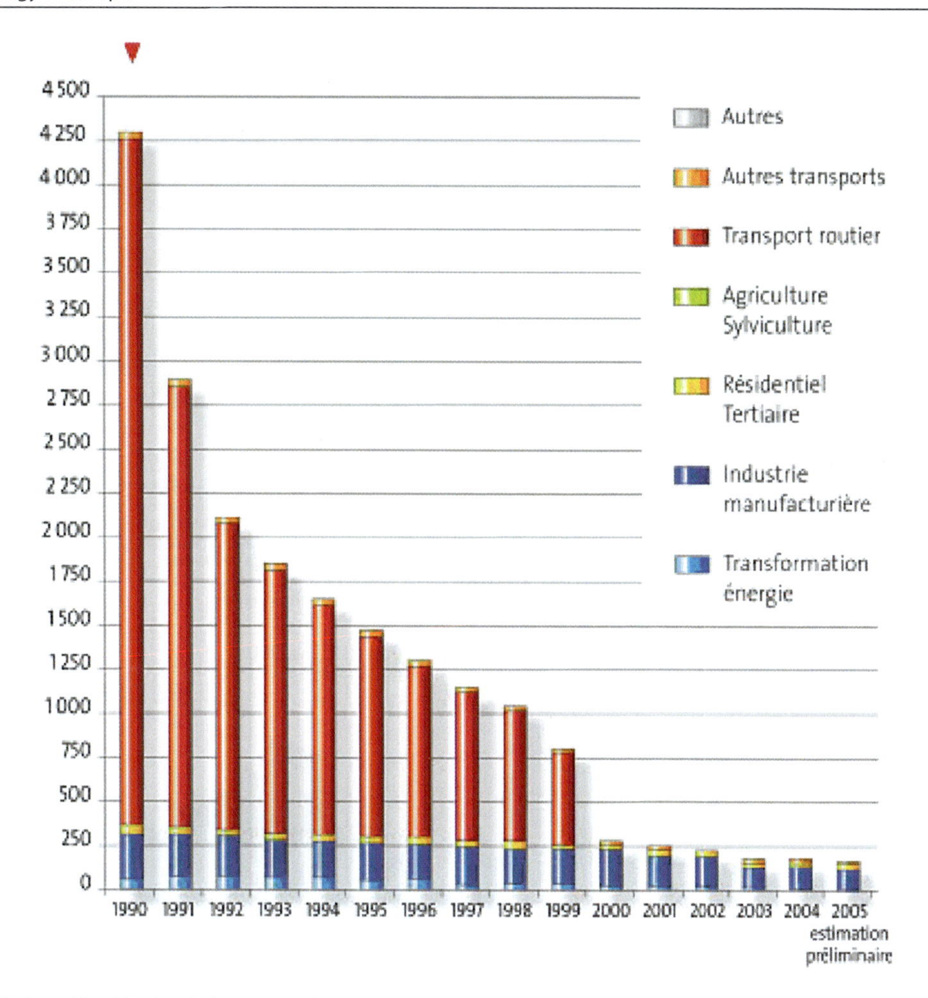

Fig. 16 Evolution of lead in the air in metropolitan France, 1990–2005. Observe the decrease of the road transport source (*Source*: OTEPA 2006)

high-quality compatible data sets across national borders. Over 60 determinants were determined, 400 maps plotted and interpreted, most for total and aqua regia extractable concentrations. The first phase of the project was completed, and the results published in a two-volume set, which is also freely available for viewing and downloading. Also the whole database can be downloaded free of charge (Salminen et al., 2005).

Another example of database is the Baltic soil survey covering total concentrations of major and selected trace elements in arable soils from 10 countries around the Baltic Sea. Agricultural soils were collected from 10 European countries over a 1,800,000 km^2 area surrounding the Baltic Sea. Large differences between element concentrations and variations can be observed for most elements when the different countries are compared. The database can be downloaded for free (Reimann et al., 2003).

In addition, the big Kola Ecogeochemistry Project should be mentioned. This is an Environmental Investigation in Arctic Europe. The Kola Ecogeochemistry Project is concerned with regional mapping of heavy metals and radioactivity pollution of terrestrial and aquatic ecosystems of an area of 188,000 km^2 in the Barents Region, sulphur and trace element deposition, the impact of major industrial activity in the western Kola Peninsula, and the degradation of a particularly vulnerable arctic environment (Reimann et al., 1998).

Fig. 17 Block diagram showing the major issues of groundwater protection (BRGM, 2008)

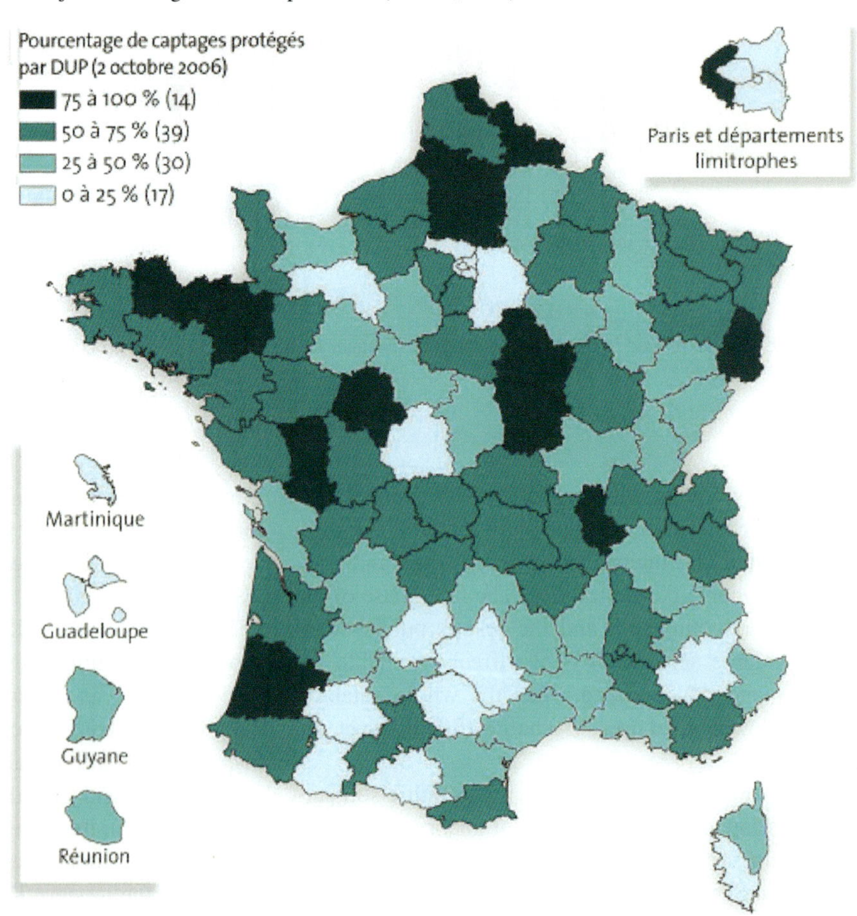

Fig. 18 Status of measures taken for protection of the quality of water sources protection by departments in France (*Source*: French Ministry of health)

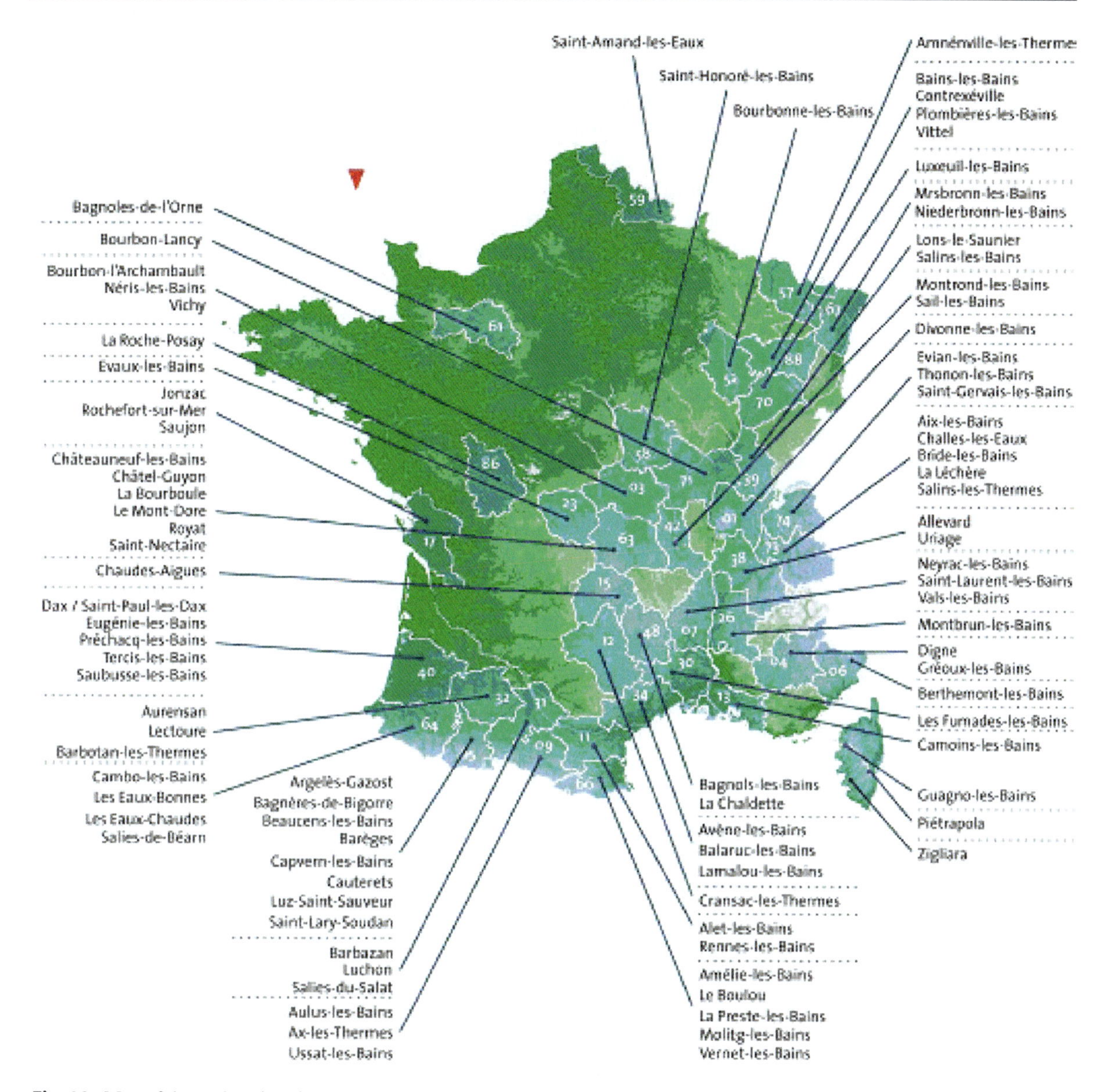

Fig. 19 Map of thermal stations in metropolitan France (BRGM, 2008)

Publications and Outreach in Europe

Information and news on medical geology has been published in many journals and books.

Some few selected examples from Europe can be mentioned.

On of the first important books in the subject was published in Norway in 1990 "Geomedicine" edited by J. Låg (1990). This book covers many aspects of geology and health.

A book on Medical Geology has been published by Elsevier (Academic Press) in 2005. O. Selinus is chief editor and there are six associate editors (four from Europe): Brian Alloway, Jose Centeno, Bob Finkelman, Ron Fuge, Ulf Lindh and Pauline Smedley (Selinus et al., 2005). There are almost 60 distinguished authors from all around the world. About 50% are geoscientists and about 50% are medics, veterinarians and other public health scientists. In November 2005 *Essentials of Medical Geology* was recognized as a "Highly Commended" title in the Public Health category by the British Medical Association, a very prestigious acknowledgement. The book was also cited as one of the best published books in Public Health in 2005. They bestow awards upon publications "which are deemed to best fulfil the criteria of clinical accuracy and currency and which maintain a high standard of design and production". *Essentials of Medical Geology* also won a second prestigious reward in January 2006. The book has now thus been recognized in both communities for which it was intended.

Environmental Medicine, a book published in 2000 at the Karolinska Institute, Sweden, included a chapter on medical geology. This book is used for courses for medical students at the Karolinska Institute and elsewhere. The BGS magazine, *Earthwise*, has published a thematic issue on Geology and Health (Earthwise, 2001). A book, "Geology to Health", edited by A.R. Berger and C. Skinner published by Oxford University Press, covered all the presentations given at the meeting in Uppsala, Sweden in 2000 (Skinner and Berger, 2000). A special volume of the Journal of *BRGM* (Geological Survey of France) on Medical Geology was published in 2007 (Geosciences et Sante, 2007). Two special issues in *Journal of Environmental Geochemistry and Health* on medical geology in developing countries have been published (*Medical Geology in Developing Countries* 2007, 2008). These include several European authors.

The Future of Medical Geology in Europe

The International Medical Geology Association (IMGA) was launched in January 2006 with at present one European chair and two European councillors (UK and Turkey). The association is the umbrella for regional divisions around the world. Among the divisions are those for Europe, Russia and NIS and southern Mediterranean. It is always risky to anticipate what the future holds. Nevertheless, we are confident that the future for medical geology still looks promising, notwithstanding the already rapid growth of the sub-discipline. The book, *Essentials of Medical Geology*, has received an overwhelmingly positive response. The reviews have been uniformly positive and the first printing has nearly sold out in less than a year. We anticipate that the book will stimulate the teaching of medical geology in colleges and universities. The International Medical Geology Association should provide a stable platform for the exchange of ideas and dissemination of information. The raft of other medical geology activities enumerated above should maintain enthusiasm and momentum for the next few years. After that the medical geologist will have to demonstrate that what we have to offer will indeed benefit society by helping to improve the quality of life for people around the world.

References

Alfthan G, Wang D, Aro A, Soveri J (1995) The geochemistry of selenium in groundwaters in Finland. The Sci Total Environ 162:93–103.

Adlercreutz E (1928) Orientierende Untersuchung über die Verbreitung des Kropfes in Finnland und über deren Zusammenhang mit dem Jodvorkommen im Wasser. Acta Med Scand 69:1–45.

Amaral AFS, Arruda M, Cabral S, Rodrigues AS, (2008) Essential and non-essential trace metals in scalp hair of men chronically exposed to volcanogenic metals in the Azores, Portugal. Environ Internat 34:1104–1108.

Amaral AFS, Rodrigues AS (2007) Chronic exposure to volcanic environments and chronic bronchitis incidence in the Azores, Portugal. Environmen Res 103:419–423.

Amaral A, Rodrigues V, Oliveira J, Pinto C, Carneiro V, Sanbento R, Cunha R, Rodrigues A (2006) Chronic exposure to volcanic environments and cancer incidence in the Azores, Portugal. Sci Total Environ 367:123–128.

AMBIO (2007) Special issue on medical geology. 36(1).

Appleton JD, Rawlins B, Thornton I (2008) National-scale estimation of potentially harmful element ambient background concentrations in topsoil using parent material classified soil:stream-sediment relationships. Appl Geochem 23: 2596–2611.

Atteia O, Saada A, (2007) Etude des risques sur les sites pollués: le rôle des processus naturels. Géosciences 5:52 à 57.

Auvinen A, Mäkeläinen I, Hakama M, Castrén O, Pukkala E, Reisbacka H et al. (1996) Indoor radon exposure and risk of lung cancer: a nested case-control study in Finland. J Natl Cancer Inst 88:966–972.

Backman B, Luoma S, Ruskeeniemi T, Karttunen V, Talikka M, Kaija J (2006) Natural Occurrence of Arsenic in the Pirkanmaa region in Finland. Espoo: Geological Survey of Finland, Miscellaneous Publications. Geological Survey of Finland

Baxter PJ, Baubron JC (1999) Health hazards and disaster potential of ground gas emissions at Furnas Volcano, São Miguel, the Azores. J Volcanol Geotherm Res 92(1–2):219.

Baxter PJ, Baubron JC, Coutinho R (1999) Health hazards and disaster potential of ground gas emissions at Furnas volcano, São Miguel, Azores. J Volcanol Geotherm Res 92: 95–106.

Blum L, Chery L, Legrand H (2007) L'eau souterraine est-elle toujours potable à l'état naturel? Géosciences 5:58 à 67.

Billon-Galland MA, Daniau C, Martinon L, Pascal M, (2007) L'amiante dans l'environnement en France: de l'exposition au risque. Géosciences 5:30 à 39.

Bølviken B, Nilsen R, Ukkelberg Å (1997a) A new method for spatially moving correlation analysis in geomedicine. Environ Geochem Health 19:143–153.

Bølviken B, Flaten TP, Zheng C, (1997b) Relations between nasopharyngeal carcinoma and magnesium and other alkaline earth elements in soils in China. Med Hypotheses 48:21–25.

Bølviken B, Celius EG, Nilsen R, Strand T (2003) Radon: A possible risk factor in multiple sclerosis. Neuroepidemiology 22:87–94.

Bowman C, Bobrowski PT, Selinus O (2003) Medical Geology: New relevance in the Earth Sciences. Episodes 26(4): 270–278.

Carvalho FP, Madruga MJ, Reis MC, Alves JG, Oliveira JM, Gouveia J, Silva L, (2005a) Radioactive survey in former uranium mining areas in Portugal. In International Workshop on Environmental Contamination from Uranium Production Facilities and Remediation Measures, 11e13 February 2004, Lisbon. International Atomic Energy Agency, Vienna, pp. 29–40.

Carvalho T, Vala H, Pinto C, Pinho M, Peleteiro MC (2005b) Immunohistochemical Studies of Epithelial Cell Proliferation and p53 Mutation in Bovine Ocular Squamous Cell Carcinoma. Vet Pathol 42:66–73.

Carvalho T, Pinto C, Peleteiro MC (2006) Urinary bladder lesions in bovine enzootic haematuria. J Comp Pathol 134:336–346T. doi:10.1016/j.jcpa.2006.01.001

Carvalho FP, Oliveira JM (2007). Alpha emitters from uranium mining in the environment. J Radioanal Nucl Chem 274:167–174.

Carvalho FP, Oliveira JM, Lopes I, Batista A, (2007) Radionuclides from past uranium mining in rivers of Portugal. J Environ Radioact 98:298–314.

Cave M, Taylor H, Wragg J (2007) Estimation of the bioaccessible arsenic fraction in soils using near infrared spectroscopy. J Environ Sci Health, Part A 42:1293–1301.

Cave MR, Wragg J, Palumbo B, Klinck, BA (2003) Measurement of the Bioaccessibility of Arsenic in UK soils. Environment Agency P5-062/TRI.

Centeno JA, Mullick FG, Martinez L, Gibb H, Longfellow D, Thompson C (2002) Chronic arsenic toxicity: An introduction and overview. Histopathology 41(2):324–326.

Centeno JA, Tseng CH, van der Voet GB, Finkelman RB (2007) Global Impacts of Geogenic Arsenic – A Medical Geology Research Case. Ambio 36(1):78–81.

Cooke, TD, Bruland KW (1987) Aquatic chemistry of selenium: Evidence of biomethylation. Environ Sci Technol 21: 1214–1219.

Darby S, Hill D, Auvinen A, Barros-Dios JM, Baysson H, Bochicchio F, Deo H, Falk R, Forastiere F, Hakama M, Heid I, Kreienbrock L, Kreutzer M, Lagarde F, Mäkeläinen I, Muirhead C, Obereigner W, Pershagen G, Ruano-Ravina A, Ruosteenoja E, Schaffrath-Rosario A, Tirmarche M, Tomasek L, Whitley E, Wichmann H-E, Doll R (2005) Radon in homes and lung cancer risk: Collaborative analysis of individual data from 13 European case-control studies. Br Med J 330:223–226.

Department for the Environment Food and Rural Affairs and the Environment Agency (2002) Assessment of risks to human health from land contamination: An overview of the soil guideline values and related research, CLR7.

Derbyshire E (2005) Natural aerosolic mineral dusts and human health: Potential effects. In O Selinus, B Alloway, JA Centeno, RB Finkelman, R Fuge, U Lindh, P Smedley (eds) Essentials of Medical Geology (820 p). Elsevier: Amsterdam.

De Vos, W, Tarvainen, T, (Chief eds) (2006) Geochemical Atlas of Europe. Part 2: Interpretation of Geochemical maps, additional tables, figures, maps and related publications. 690 p.

Dogan AU, Doga M, Hoskins JA (2008) Erionite series minerals: Mineralogical and carcinogenic properties. En Geochem Health 30(4):367–381.

Dor F, Guyonnet D (2007) L'évaluation des risques sanitaires: Application au stockage des déchets ménagers. Géosciences 5:100 à 105.

Earthwise (2001) Geology and health. British Geological Survey. Issue 17.

Medical Geology in Developing Countries (2007) Two special issues of Environmental Geochemistry and HealthVol 29, No 2 and Vol 30 No 4 2008.

Ender F (1946) The significance of cobalt deficiency as a cause of disease in cattle and sheep elucidated by therapeutic experiments. Norsk Vet Tidskr 58:313 (In Norwegian, English summary).

Faísca MC, Teixeira MMGR, Bettencourt AO, (1992) Indoor radon concentrations in Portugal – a national survey. Radiation Protection Dosimetry 45:465–467.

Falcão JM, Carvalho FP, Leite MM, Alarcão M, Cordeiro E, Ribeiro J (2005). MINURAR project d'uranium mines and their residues: Health effects in a Portuguese population. Scientific Report I (in Portuguese), Report Published by INSA, INETI, ITN. Available from: <www.itn.pt>.and <www.insa.pt>.

Fátima Reis M, Segurado S, Brantes A, Simões HT, Melim JM, GeraldesV Miguel JP (2008) Ethics issues experienced in HBM within Portuguese health surveillance and research projects. Environmen Health 7(Suppl 1):S5 doi:10.1186/1476-069X-7-S1-S5.

Feder GL, Radovanovic Z, Finkelman RB (1991) Relationship between weathered coal deposits and the etiology of Balkan endemic nephropathy. Kidney International 40(Suppl. 34):s-9–s-11.

Figueira R, Sérgio C, Leal Lopes J, Sousa AJ (2007) Detection of exposition risk to arsenic in Portugal assessed by air deposition in biomonitors and water contamination. Int J Hyg Environ-Health 210:393–397.

Flaten TP, Bølviken T, (1991) Geographical associations between drinking water chemistry and morbidity of cancer and some other diseases in Norway. Sci Total Environ 102:75–100.

Fordyce F (2005) Selenium deficiency and toxicity in the environment. In O Selinus, B Alloway, JA Centeno, RB Finkelman, R Fuge, U Lindh, P Smedley (eds) Essentials of Medical Geology–Impacts of the Natural Environment on Public Health (pp. 373–415). Elsevier – Academic Press, London.

Frank A, Galgan V, Petersson L (1994) Secondary copper deficiency, chromium deficiency and trace element imbalance in the moose (Alces alces L.): Effect of anthropogenic activity. Ambio 23(4–5).

Frøslie A (1977) Copper status in sheep in Norway. Norsk Vet Tidskr 89:71–79 (In Norwegian, English summary).

Frøslie A (1990) Problems on deficiency and excess of minerals in animal nutrition. In Låg J(ed) Geomedicine (pp. 37–59). Boca Raton: CRC Press.

Frøslie A, Norheim G (1977) Concentrations of molybdenum and zinc in liver in relation to copper accumulation in normal and copper poisoned sheep. Acta Vet Scand 17:307–315.

Frøslie A, Karlsen JT, Rygge J (1980) Selenium in animal nutrition in Norway. Acta Agric Scand 30:17–25.

Frøslie A, Norheim G, Søli NE (1983) Levels of copper, molybdenum, zinc and sulphur in concentrates and mineral feeding stuffs in relation to chronic copper poisoning in sheep in Norway. Acta Agric Scand 33:261–267.

Geosciences et sante (2007). Special issue of Geosciences No 5, March 2007. BRGM. 118 pp.

Glorennec P, Guyonnet D, Laperche V (2007) Plomb et santé. Géosciences n5:40 à 45.

Gomes CSF, Silva JBP (2006) Minerals and Human Health: Benefits and Risks/ Os Minerais e a Saúde Humana: Benefícios e Riscos/ (published a bilingual Portuguese and English). In C Gomes, J Silva (eds) Litografia da Maia. Maia. 300 p.

Gomes CSF Silva JBP (2007) Minerals and clay minerals in medical geology. Appl Clay Sci 36:4–21.

Gomes CSF Silva JBP (2006) Published a bilingual (Portuguese and English) book entitled "Os Minerais e a Saúde Humana: Benefícios e Riscos/Minerals and Human Health: Benefits and Risks that presents a fair balance of both positive and negative effects of minerals on human health. Gomes CSF Silva JBP (eds) 316 p.

Govasmark E, Steen A, Bakken AK, Strøm T, Hansen S, Bernhoft A (2005) Copper, molybdenum and cobalt in herbage and ruminants from organic farms in Norway. Acta Agr Scand Section A:21–30.

Haglund B, Ryckenberg K, Selinus O, Dahlqvist G (1996) Evidence of a relationship between childhood-onset type 1 diabetes and low groundwater concentration of Zinc. Diab Care 19(8) August 1996.

Havulinna AS, Tienari PJ, Marttila RJ, Martikainen KK, Eriksson JG, Taskinen O, Moltchanova E, Karvonen M (2008) Geographical variation of medicated parkinsonism in Finland during 1995 to 2000. Mov Disord 23(7): 1024–1031.

Hillborg PO (1978) Gauchers disease in Sweden and its hypothetical dependance on metal anomalies in the soil. Geomedical Aspects in Present and Future Research. Symposium at the Norwegian Academy of Science and Letters, pp. 173–181.

Inácio M, Pereira V, Pinto M, (2008) The soil geochemical atlas of Portugal: Overview and applications. J Geochem Explor 98:22–33.

Johnson CC, Breward N, Ander EL, Allt L (2005) G-BASE: Baseline geochemical mapping of Great Britain and Northern Ireland. Geochem ExplorEnviron Anal 5:1–13.

Kaipio J, Näyhä S, Valtonen V (2004) Fluoride in the drinking water and the geographical variation of coronary heart disease in Finland. Eur J Cardiovasc Prev Rehabil 11(1):56–62.

Kannisto V (1947) The causes of death as demographical factors in Finland. [In Finnish, English summary]. Helsinki. Kansantaloudellisia tutkimuksia-Economic studies XV.

Karvonen M, Moltchanova E, Viik-Kajander M, Moltchanov V, Rytkönen M, Kousa A, Tuomilehto J (2002) Regional inequality in the risk of acute myocardial infarction in Finland: A Case Study of 35- to 74-Year-Old Men. Heart Drug 2:51–60.

Kasan-Allen L (2006) Asbestos – The Human cost of corporate greed. GUE/NGL European United Left/Nordic green Left. European Parliamentary Group, Brussels: 43p.

Kavanagh P, Farago M, Thornton I, Elliott P, Goessler W Irgolic KJ (1997) Urinary arsenic concentrations in a high arsenic area of south west England. Occup Environ Med 54:840.

Kawahara S (1971) Odontological observations of Mt. Aso-volcano disease. Fluoride 4:172–175.

Kokki E, Ranta J, Penttinen A, Pukkala E, Pekkanen J (2001) Small area estimation of incidence of cancer around a known source of exposure with fine resolution data. Occup Environ Med 58(5):315–320.

Koljonen T (1975) The behavior of selenium in Finnish soils. Annales Agriculturae Fenniae 14: 240–247.

Kousa A, Nikkarinen M (1997) Geochemical environment in areas of low and high coronary heart disease mortality. Geological Survey of Finland. Special Paper 23:137–148.

Kousa A, Moltchanova E, Viik-Kajander M, Rytkönen M, Tuomilehto J, Tarvainen T, Karvonen M (2004a) Geochemistry of ground water and the incidence of acute myocardial infarction in Finland. J Epidemiol Community Health 58(2):136–139.

Kousa A, Moltchanova E, Taskinen O, Nikkarinen M, Tuomilehto J, Karvonen M, (2004b) Geographical variation of Acute Myocardial Infarction (AMI) and geochemistry of local groundwater: Application of medical geology. GFF, Vol 126, part 1.

Kousa A, Havulinna AS, Moltchanova E, Taskinen O, Nikkarinen M, Eriksson J, Karvonen M (2006) Calcium:magnesium ratio in local groundwater and incidence of acute myocardial infarction among males in rural Finland. Environ Health Perspect 114(5):730–734.

Kousa A, Havulinna AS, Moltchanova E, Taskinen O, Nikkarinen M, Salomaa V, Karvonen M (2008) Magnesium in well water and the spatial variation of AMI incidence in rural Finland. Appl Geochem 23:632–640.

Kurttio P, Komulainen H, Hakala E, Kahelin H, Pekkanen J (1998) Urinary excretion of arsenic species after exposure to arsenic present in drinking water. Arch Environ Contam Toxicol 34(3):297–305.

Kurttio P, Pukkala E, Kahelin H, Auvinen A, Pekkanen J (1999) Arsenic concentrations in well water and risk of bladder and kidney cancer in Finland. Environ Health Perspect 107(9):705–710.

Kurttio P, Salonen L, Ilus T, Pekkanen J, Pukkala E, Auvinen A (2006) Well water radioactivity and risk of cancers of the urinary organs. Environ Res 102(3):333–338.

Kurttio P, Salonen L, Ilus T, Pekkanen J, Pukkala E, Auvinen A (2006) Well water radioactivity and risk of cancers of the urinary organs. Environ Res 102(3):333–338.

Lahermo P, Alfthan G, Wang D (1998) Selenium and arsenic in the environment in Finland. J Environ Pathol Toxicol Oncol 17(3–4):205–216.

Lahermo P, Ilmasti M, Juntunen R, Taka M (1990) The geochemical atlas of Finland, Part 1. The hydrogeochemical mapping of Finnish groundwater. Geological Survey of Finland, Espoo.

Lahermo P, Backman B (2000) The occurrence and geochemistry of fluorides with special reference to natural waters in Finland. Geological Survey of Finland. Report of Investigation 149. 40p.

Lamberg BA (1986) Endemic goitre in Finland and changes during 30 years of iodine prophylaxis. Endocrinol Exp 20(1):35–47.

Luoma H, Aromaa A, Helminen S, Murtomaa H, Kiviluoto L, Punsar S, Knekt P (1983) Risk of myocardial infarction in Finnish men in relation to fluoride, magnesium and calcium concentration in drinking water. Acta Med Scand 213: 171–176.

Løken, A (1912) Goiter. Norsk Vet Tidskr 24:178 (In Norwegian).

Låg J (1968) Relationships between the chemical composition of the precipitation and the contents of exchangeable ions in the humus layer of natural soils. Acta Agric Scand 18:148–152.

Låg J (ed) (1990) Geomedicine (278 p.). Boca Raton, USA: CRC Press.

Låg J (ed) (1992) Chemical climatology and geomedical problems (226 p.). Oslo: The Norwegian Academy of Science and Letters.

Låg J, Steinnes E (1974) Soil selenium in relation to precipitation. Ambio 3:237–238.

Låg J, Steinnes E, (1976) Regional distribution of halogens in humus layers of Norwegian forest soils. Geoderma 16: 317–325.

Låg J, Steinnes E (1978) Regional distribution of selenium and arsenic in humus layers of Norwegian forest soils. Geoderma 20: 3–14.

Låg J, Hvatum OØ, Bølviken B (1974) Some naturally heavy-metal poisoned areas of interest in prospecting, soil science, and geomedicine. Norges Geol Unders 304:73–96.

Mills CF, Davis GK (1987) Molybdenum. In W Mertz (ed) Trace Elements in Human and Animal Nutrition (Vol. 1, pp. 429–457). New York: Academic Press, Inc.

Ministry of Agriculture and Forestry (1984) Proposal for amounts of selenium to be added into fertilizers. Working Group Report. No. 7, Helsinki (in Finnish).

Moltchanova E, Rytkönen M, Kousa A, Viik-Kajander M, Karvonen M (2004) Zinc and nitrate in the ground water and the incidence of type 1 diabetes in Finland. Diabet Med 21:256–261.

Nerbrand CH, Svärdsudd K, Ek J, Tibblin G (1992) Cardiovascular mortality and morbidity in seven counties in Sweden in relation to water hardness and geological settings. Europ heart j 13:721–727.

Neves O, Abreu MM (2009) Are uranium-contaminated soil and irrigation water a risk for human vegetables consumers? A study case with Solanum tuberosum L., Phaseolus vulgaris L. and Lactuca sativa L. Ecotoxicology DOI 10.1007/s10646-009-0376-4.

Nikkarinen M, Aatos S, Teräsvuori E (2001) Asbestin esiintyminen ja sen vaikutus ympäristöön Tuusniemellä, Outokummussa, Kaavilla ja Heinävedellä. In Finnish. Summary: Occurrences of asbestos minerals and their environmental impact in the area of municipalities of Tuusniemi, Outokumpu, Kaavi and Heinävesi, eastern Finland. Report of Investigation 152. Espoo:Geological Survey of Finland. 41 p.

Nikkarinen M, Kollanus V, Ahtoniemi P, Kauppila T, Holma A, Räisänen ML, Makkonen S, Tuomisto JT (eds) (2008) Metallien yhdennetty kohdekohtainen riskinarviointi. In Finnish. Abstract: Integrated site-specific risk assessment of metals. University of Kuopio, Department of Environmental Science: Seminar Publications 3/2008. 401 p.

Notcutt G, Davies F (1999) Biomonitoring of volcanogenic fluoride, Furnas Caldera, São Miguel, Azores. J Volcanol Geotherm Res 92:209–214.

Orem WH, Feder GL, Finkelman RB (1999) A possible link between Balkan endemic nephropathy and the leaching of toxic organic compounds from Pliocene lignite by groundwater: preliminary investigation. Int Jour Coal Geol 40(2–3):237–252.

Orem W, Tatu C, Pavlovic N, Bunnell J, Lerch H, Paunescu V, Ordodi V, Flores D, Corum M, Bates A (2007, February) Health effects of toxic organic substances from coal: Toward "panendemic" nephropathy. Ambio 36(1):98–102.

Palumbo-Roe B, Cave MR, Klinck BA, Wragg J, Taylor H, O'Donnell K, Shaw RA (2005) Bioaccessibility of arsenic in soils developed over Jurassic ironstones in eastern England. Environ Geochem Health 27:121–130.

Pavão ML, Figueiredo T, Santos V, Lopes PA, Ferin R, Santos MC, Nève J, Viegas-Crespo AM (2006) Whole blood glutathione peroxidase and erythrocyte superoxide dismutase activities, serum trace elements (Se, Cu, Zn) and cardiovascular risk factors in subjects from the city of Ponta Delgada (Island of San Miguel, The Azores′ Archipelago, Portugal). Biomarkers 11(5):460–471. (doi: 10.1080/13547500600625828).

Pereira AJSC, e Neves, LJPF, Gonçalves CVM (2007) Radon in groundwater from the Hesperian massif (Central Portugal). XXXV IAH Congress, International Association of Hydrologists, Lisbon, Abstract book, Ribeiro L, Chambel A, Condesso de Melo MT (eds) p.190.

Pereira AJSC, Neves LJPF, Godinho MM, Dias JMM (2003) Natural radioactivity in Portugal: influencing geological factors and implications for land use planning. Radioprotecção 2(2–3):109–120.

Pereira AJSC, e Neves LJPF (2005) – Radon risk maps: the Portuguese experience. Annual Meeting of the Geological Society, Salt Lake City, Abstracts with Programs, vol. 37 (7), p. 354.

Perrin J (2007) La géologie du radon. Géosciences 5:22 à 23.

Piispanen R (1993) Water hardness and cardiovascular mortality in Finland. Environ Geochem Health 15:201–208.

Piispanen R (2000) Radon and lung cancer in Finland: Are there signs of radiation hormesis? Environ Geochem Health 22(2):113–130.

Pinto CA, Lima R, Louza AC, Almeida V, Melo M, Vaz Y, Neto Fonseca I, Lauren DR, Smith BL (2000) Bracken fern-induced bovine enzootic haematuria in São Miguel Island, Azores Bracken-fern: Toxicity, Biology and Control: Proceedings of IV International Bracken 99. Conference. pp. 136–140.

Purdey M (2004). The Pathogenesis of Machado Joseph Disease: A High Manganese/Low Magnesium Initiated CAG Expansion Mutation in Susceptible Genotypes? J Am Coll Nutr 23(6):715S–729S.

Plant JA, Baldock W, Smith B (1996) The role of geochemistry in environmental and epidemiological studies in developing countries. In JD Appleton et al. (ed) Environmental Geochemistry and Health with special reference to developing countries. A review. Geological Society Special Publication 113: 7–22 London.

Pocock SJ, Shaper AG, Cook DG, Packham RF, Lacey RF, Powell P, Russell PF (1980) British regional heart study: Geographic variations in cardiovascular mortality, and the role of water quality. Br Med J 24;280(6226): 1243–1249.

Pukkala E, Patama T (2008) Small-area based map animations of cancer incidence in the Finland, 1957–2006. Finnish Cancer Registry http://astra.cancer.fi/cancermaps/suomi19572006/en/.

Pukkala E, Patama T, Engholm G, Ólafsdóttir GH, Bray F, Talbäck M, Pasanen K (2007) Small-area based map animations of cancer incidence in the Nordic countries, 1971–2003. Nordic Cancer Union http://astra.cancer.fi/cancermaps/Nordic.

Punsar S, Erämetsä O, Karvonen MJ, Ryhänen A, Hilska P, Vornamo H (1975) Coronary heart disease and drinking water. A search in two Finnish male cohorts for epidemiologic evidence of a water factor. J Chron Dis 28:259–287.

Punsar S, Karvonen MJ (1979) Drinking water quality and sudden death: Observations from West and East Finland. Cardiology 64(1):24–34.

Pärkö A (1990) Longitudinal study of dental caries prevalence and incidence in the rapakivi (high fluoride) and olivine diabase (low fluoride) areas of Laitila, Finland. Proc Finn Dent Soc 86(2):103–106.

Rapant S, Krcmova K, (2008) Environmental and health risk estimation for potentially toxic elements in groundwater in Slovakia. European Geologist 25. June 2008. pp. 13–16.

Rapant S, Salminen R, Tarvainen T, Krcmovál K, Cvecˇkovál V (2008) Application of a risk assessment method to Europewide geochemical baseline data. Geochem Explor Environ Anal 8(3–4):291–299.

Rawlins BG, O'Donnell K, Ingham M (2003) Geochemical survey of the Tamar catchment (south-west England). British Geological Survey, CR/03/027.

Reimann C, Äärăs M, Chekushin V et al. (1998) Environmental Geochemical Atlas of the Central Barents Region. Special Publication, NGU-GTK-CKE. Geological Survey of Norway, Trondheim.

Reimann C, Siewers U, Tarvainen T et al. (2003) Agricultural Soils in Northern Europe: A Geochemical Atlas. Geologisches Jahrbuch, Sonderhefte, Reihe D, Heft SD 5. Schweizerbart'sche Verlagsbuchhandlung, Stuttgart.

Roques CF (2007) Le thermalisme, la médecine que la Terre nous a donnée. Géosciences 5:74 à 79.

Reis AP, Sousa AJ, Ferreira da Silva E, Cardoso Fonseca E, (2005) Application of geostatistical methods to arsenic data from soil samples of the Cova dos Mouros mine (Vila Verde-Portugal). Environ Geochem Health 27:259–270.

Rubenowitz E, Axelsson G, Rylander R (1996) Magnesium in drinking water and death from acute myocardial infarction. Am J Epidemiol 143:456–462.

Rubenowitz E, Axelsson G, Rylander R (1999) Magnesium and calcium in drinking water and death from acute myocardial infarction in women. Epidemiology 10:31–36.

Rubenowitz E, Molin I, Axelsson G, Rylander R (2000) Magnesium in drinking water in relation to morbidity and mortality from acute myocardial infarction. Epidemiology 11:416–421.

Rytkönen M, Ranta J, Tuomilehto J, Karvonen M (2001) Bayesian analysis of geographical variation in the incidence of Type I diabetes in Finland. Diabetologia 44(Suppl 3):B37–B44.

Salminen R (Chief ed), Batista MJ, Bidovec M, Demetriades A, De Vivo B, De Vos W, Duris M,Gilucis A, Gregorauskiene V, Halamic J, Heitzmann P, Lima A, Jordan G, Klaver G, Klein P, Lis J, Locutura J, Marsina K, Mazreku A, O'Connor PJ, Olsson SÅ, Ottesen R-T, Petersell V, Plant JA, Reeder S, Salpeteur I, Sandström H, Siewers U, Steenfelt A, Tarvainen T (2005) Geochemical Atlas of Europe. Part 1 – Background Information, Methodology and Maps. Geological Survey of Finland, Espoo.

Salonen JT, Alfthan G, Huttunen JK, Pikkarainen J, Puska P (1982) Association between cardiovascular death and myocardial infarction and serum selenium in a matched-pair longitudinal study. Lancet 24;2(8291):175–179.

Salonen JT, Salonen R, Lappetelainen R et al. (1985) Risk of cancer in relation to serum concentrations of selenium and vitamins A and E: Matched case-control analysis of prospective data. Br Med J 290:417–490.

Salonen JT, Salonen R, Seppänen K, Kantola M, Suntioinen S, Korpela H (1991) Interactions of serum copper, selenium, and low density lipoprotein cholesterol in atherogenesis. Br Med J 302:756–760.

Selinus O, Frank A, (1999) Medical geology. In L Möller (ed): Environmental medicine (pp. 164–183). Stockholm: Joint Industrial Safety Council.

Selinus OC, Frank A, Galgan V (1996) Biogeochemistry and metal biology – an integrated Swedish approach for metal related health effects. In D Appleton, R Fuge, J McCall (ed) Environmental Geochemistry and Health in Developing Countries Special Publication (Vol. 113, pp. 81–89). London: Geological Society. , Chapman and Hall.

Selinus O, Alloway B, Centeno JA, Finkelman RB, Fuge R, Lindh U, Smedley P (eds) (2005) Essentials of Medical Geology (820 p). Amsterdam: Elsevier.

Selinus O, Alloway B, Centeno JA, Finkelman (2008) The medical geology revolution – the evolution of an IUGS initiative. Episodes 30(4).

Sivertsen T, Plassen C (2004) Hepatic cobalt and copper levels in lambs in Norway. Acta Vet Scand 45:69–77.

Sivertsen T, Øvernes G, Østerås O, Nymoen U, Lunder N (2005) Plasma vitamin E and blood selenium concentrations in Norwegian dairy cows: Reginal differences and relations to feeding and health. Acta vet Scand 46: 177–191.

Skinner C, Berger A (eds) (2003) Geology and Public Health – Closing the Gap. Oxford: Oxford Press.

Smedley P, Kinniburgh DG (2005) Arsenic in groundwater and the environment. In O Selinus, B Alloway, JA Centeno, RB Finkelman, R Fuge, U Lindh, P Smedley (eds), Essentials of Medical Geology (820 p). Amsterdam: Elsevier.

Steinnes E (ed) (2004) Geomedical aspects of organic farming (138 p.). Oslo: The Norwegian Academy of Science and Letters.

Steinnes E, Frontasyeva MV (2002) Marine gradients of halogens in soil studied by epithermal neutron activation analysis. J Radioanal Nucl Chem 253:173–177.

Søli NE (1980) Chronic copper poisoning in sheep. Nord Vet Med 32:75–89.

Sumelahti M-L, Tienari PJ, Wikström J, Palo J, Hakama M (2000) Regional and temporal variation in the incidence of multiple sclerosis in Finland 1979–1993. Neuroepidemiology 19:67–75.

Tarvainen T, Lahermo P, Hatakka T, Huikuri P, Ilmasti M, Juntunen R, Karhu J, Kortelainen N, Nikkarinen M, Väisänen U (2001) Chemical composition of well water in Finland – main results of the "One thousand wells" project. In S Autio (ed) Geological Survey of Finland, Current Reseasrch 1999–2000. Special Paper 31: 57–76.

Tatu CA, Orem WH, Finkelman RB, Feder GL, (1998) The etiology of Balkan Endemic Nephropathy: Still more questions than answers. Environ Health Pers 106(11):689–700.

Ulvund MJ (1995) Cobalt/vitamin B-12 deficiency in sheep. Norsk Veterinærtidskr. 107, 489–501 (In Norwegian, English summary).

Ulvund MJ Øverås J (1980) Chronic hepatitis in lambs in Norway, a condition resembling ovine white liver disease in New Zealand. N Z Vet J 28:19.

Varet J (2007) Tribune – Volcanisme et santé. Géosciences 5:110 à 111.

Viegas-Crespo AM, Pavfio ML, Paulo O, Santos V, Santos MC, Nève J (2000) Trace element status (Se, Cu, Zn) and serum lipid profile in Portuguese subjects of San Miguel Island from Azores'archipelago. J Trace Elements Med Biol 14:1–5.

WHO (1993) Guidelines of drinking-water quality. Volume 1: Recommendations. – 2nd ed. Geneva: World Health Organization.

Wragg J (2005) A study of the relationship between arsenic bioaccessibility and its solid phase distribution in Wellingborough soils. Nottingham.

Wragg J, Cave M, Nathanail P (2007) A Study of the relationship between arsenic bioaccessibility and its solid-phase distribution in soils from Wellingborough, UK. J Environ Sci Health, Part A, Vol. 42, 1303–1315.

Wu X, Låg J (1988) Selenium in Norwegian farmland soils. Acta argic Scand 38:271–276.

Yoshida S, Muramatsu Y (1995) Determination of organic, inorganic, and particulate iodine in the coastal atmosphere of Japan. J Radioanal Nucl Chem – Articles 196, 295–302.

Ytrehus B, Skagemo H, Stuve G, Siversen T, Handeland K, Vikøren T (1999) Osteoporesis, bone mineralization, and status of selected trace elements in two populations of moose calves in Norway. J Wildlife Diseases 35, 204–211.

Medical Geology in China: Then and Now

Zheng Baoshan, Wang Binbin, and Robert B. Finkelman

Abstract The impact of the natural environment on human health has been a subject of study in China for at least 5,000 years. China's varied geology and geography and its large population living off the land have resulted in the presence of virtually every known environmental health problem and some of the most serious medical geology problems. Fluorosis in China has been caused by drinking fluoride-rich waters, burning coal briquettes using fluorine-rich clay binders, and eating high-fluorine salt. Iodine deficiency disorders such as goiter and cretinism were common in central China but the introduction of iodized salt has greatly reduced these health problems. Arsenic poisoning has been widespread in China, mainly caused by drinking arsenic-rich groundwater but burning arsenic-rich coals and metal mining have also contributed to this problem. Selenium deficiency that is believed to be a causative factor in Keshan disease and Kashin-Beck disease was widespread in a zone from northeast China to Tibet. Human selenosis, thallium poisoning, and various respiratory problems caused by exposure to minerals have also been reported from China.

Keywords China · Keshan disease · Kashin-Beck disease · Selenium · Fluorine · Arsenic · Dusts · Thallium · Respiratory problems · Coal · IDD

Z. Baoshan (✉)
The State Key Laboratory of Environmental Geochemistry
Institute of Geochemistry, Chinese Academy of Sciences,
Guiyang 55002, P. R. China
e-mail: zhengbaoshan@vip.skleg.cn

Introduction

China's Natural Environment, Lifestyle, and Their Relation to Geochemical Diseases

China is a vast country embracing a wide variation of topography, climatic conditions, rocks and soil types, animals, and plants. China's land area is about 9.6 million km^2, with distances of 5,500 km from north to south and 5,200 km from east to west. In eastern China, alluvial plains formed by the Heilongjiang River, Liaohe River, Haihe River, Yellow River, Huaihe River, Yangtze River, and Pearl River are distributed from north to south with average altitude of 200 m. The alluvial plains are home to two-thirds of China's 1.3 billion people and are the traditional Chinese agricultural areas.

In western China, there are more prominent mountains, plateaus, and basins. The Junggar Basin, Turpan Basin, and Tarim Basin are located in the Tianshan Mountains in the northwest. The Turpan Basin has the most arid, hot climate in China, and the Aiding Lake in the basin has the lowest elevation (155 m below sea level). In the middle of the Tarim Basin is China's largest and the world's third largest desert, the Taklimakan Desert.

In western China there is the largest and highest plateau in China, Qinghai-Tibet Plateau, known as the "Roof of the World" with an average altitude of more than 4,000 m. To the southeast of the Qinghai-Tibet Plateau, there is the Yunnan-Guizhou Plateau with average elevations from 1,000 to 2,000 m and the Sichuan Basin surrounded by mountains. The Sichuan Basin, with altitudes up to 500 m and the best climatic

O. Selinus et al. (eds.), *Medical Geology*, International Year of Planet Earth,
DOI 10.1007/978-90-481-3430-4_11, © Springer Science+Business Media B.V. 2010

conditions for agricultural development, is the most densely populated area in western China. Northeast of the Qinghai-Tibet Plateau is the Loess Plateau and Inner Mongolia plateau, with altitudes between 500 and 1,500 m. The surface of the Loess Plateau is a loose accumulation of loess with a thickness of a few hundred meters and has the most serious soil erosion problems in China. There are low hills, mountains, and the ramps of transitions from plateaus to plains between China's eastern plains and western mountains.

Separated by the Qinghai-Tibet Plateau, the warm and humid air from the Pacific Ocean forms the area with the Chinese mainland's largest annual rainfall in western Sichuan Province and the area with the world's largest annual rainfall (more than 10,000 mm annual) in the western slope of the Qinghai-Tibet Plateau. For this reason, the world's major rivers, including the Yangtze River, the Yellow River, Mekong River, Salween River, Brahmaputra River, and Ganges River, originate in the Qinghai-Tibet Plateau.

The geology of China is impressively diverse. The southeast coastal region is a large area of granite bedrock and the eastern plains are covered by river alluvium with a thickness of more than 1,000 m. From compression by the India plate, geothermal areas formed in Yunnan Province and Tibet in southwest China. From compression by the Pacific plate, geothermal areas formed in Taiwan Province, Fujian Province, and Guangdong Province. A large number of geothermal hot springs and some active volcanoes occur in these areas. There are also some hot springs distributed along some major faults in mainland China.

Due to a variety of different geological processes, some elements harmful to human health are enriched in some rocks, soils, surface, and groundwater and some elements necessary for human survival are depleted in some rocks, soils, surface and groundwater. These elements, both harmful and useful, could be taken up by plants and animals and ultimately by humans. From the late Neolithic period, 8,000 years ago to the late 1970s, the period of China's reform and opening up, the majority of Chinese people had always lived a settled farming life, drinking the local water and eating crops grown on the land. Poor nutrition combined with exposure to harmful elements or to the deficiency of essential elements has led to a wide range of environmental health problems affecting hundreds of millions of people, perhaps the most serious examples of medical geology problems in the world.

Early Medical Geology Research in China

Foreign doctors and missionaries were the first to conduct studies of medical geology in China using modern scientific ideas and methods. Kashin-Beck disease (KBD) is an endemic osteoarticular disorder. Following the Japanese occupation of Northeast China in 1931, the Japanese government sent medical workers to study KBD in northeast China. Keshan disease (KD) is a cardiovascular disease of unknown etiology and long common in northeast China. In 1936, the Japanese government sent medical researchers to investigate the disease, conduct a detailed description of the disease, and named the disease "Keshan disease" (Sokoloff, 1989; Chinese Science Press, 1990).

In 1930, Anderson, an American doctor working in China reported 398 cases of dental fluorosis in the Xiaotangshan hot spring area in Beijing. In 1946, Oliver Lyth, a doctor working in a church-run hospital reported endemic fluorosis in the Shimenkan area in Weining County, Guizhou Province, which is China's first report of skeletal fluorosis. He also was the first to report fluorosis from coal combustion, although at that time he thought that the disease was induced by drinking water (Lyth, 1946).

In 1926, Bernard E. Read and George K. How, working in the Department of pharmacology, Peking Union Medical College, measured the concentrations of iodine, arsenic, calcium, iron, and sulfur in seaweed used in Chinese medicines. The results showed that iodine concentrations ranged from 0.21 to 1.23% that proved reasonable for treatment of iodine deficiency disorders. During the period of late 1920s, William H. Adolph and Shen-Chao Ch'en, working in the Department of Chemistry, Yenching University, conducted a systemic determination of iodine in drinking water and foodstuff in iodine deficiency diseases (IDD) and control areas in North China. These were China's first studies of IDD (Adolph and Ch'en, 1930; Read and How, 1927).

Development of Medical Geology in China

In a modern sense the object of medical geology is the study of the impacts of geological materials and geological processes on animal and human health. In this sense, the earliest discussion of medical geology in China began in the China University of Science and Technology in the early 1960s. Li Changsheng, still a student at that time, began to study the relationship between different chemical compositions of crustal rocks and the health of residents. The study was limited to theoretical discussion and speculation. In 1961, on the basis of extensive investigations, Yu Weihan, professor of Harbin Medical University, pointed out that KD was a biogeochemical disease due to the lack of certain minerals, amino acids, or vitamins in the diet.

In 1959, 1964, and 1970, the incidence of KD in China was exceptionally high, with ratios all higher than 40/100,000 and with an annual mortality of more than 2,000. To address this problem a large number of research institutions and researchers focused their attention on KD. With the expansion of the research, studies on KBD, endemic fluorosis, and endemic goiter were also carried out. In these institutions there were scholars engaged in epidemiology, pathology, toxicology, nutrition, and scholars engaged in the geology, geochemistry, and geography. In China, the research was referred to as endemic studies by scholars from medical profession, as medical geology by geologists, as medical geography or chemical geography by geographers, and as environmental geochemistry or geochemistry and health and disease by geochemists. Now all of these studies can be attributed to the scope of medical geology.

Endemic Diseases in China

During the past 60 years, in order to control endemic diseases, the Chinese government has made great efforts and the incidence of endemic diseases and mortality rates have been significantly reduced. With economic development, living standards and nutritional status of residents in endemic areas have greatly improved and the trace element-related diseases such as KD and KBD have been almost eradicated. Now there are cases of KBD in only some areas in Tibet. Iodine deficiency diseases have been well controlled by the policy of nationwide use of iodized salt. In drinking water fluorosis areas, government-funded construction of public drinking water facilities has provided low-fluoride drinking water for local residents. In areas of coal combustion fluorosis, government-funded stoves have discharged harmful gases outside the houses and greatly reduced the dangers of fluorosis. Although the situation has greatly improved, problems caused by long-term endemic hazards remain and are still very serious for these environmental health issues that have been recognized for thousands of years.

Understanding the Relationship Between Disease and the Natural Environment by the Ancient Chinese

The ancient Chinese people noted the relationship between some diseases and climate, soil, water, and diet and also found that certain diseases were prevalent in particular areas. Modern Chinese medicine focuses on the effects of geological materials and processes on human and animal health, including two harmful diseases – endemic fluorosis and iodine deficiency that are widely distributed and have long been known and described by ancient Chinese people who studied the natural characteristics, distribution, causes, treatments, and prevention of these diseases.

Records of Iodine Deficiency Disorders and Endemic Fluorosis in Ancient Chinese Books

Jiaguwen is a kind of inscription carved on bones or tortoise shells used by shaman to predict good or bad fortune during the Shang and Zhou dynasties 3,000 years ago. Since 1899, a large number of these oracle bones have been unearthed in the suburbs of Anyang City, once the capital of the Shang Dynasty (16th century BC–11th century BC), in Henan Province, China. More than a hundred thousand oracle bones have been unearthed, in which about 4,500 different words were discovered and 1,700 of them have been interpreted, in which "Ying" was included.

"Ying" is a person's neck bump, the expression of endemic goiter due to the lack of iodine in the diet. The Chinese word "Ying" consists of two parts: one part indicates a disease and the other part is the description of the image of the bump of a patient's neck.

There are four very important books on Chinese traditional medicine, and "Huang Di Nei Jing" (475 BCE–AD 220) is one devoted to the cause and prevention of diseases. "Huang Di Nei Jing" consists of two parts: "Su Wen" and "Lingshu." In many parts "Su Wen" involves the relationship between diseases and water quality, soil, and climate [Huang Di Nei Jing].

The book was likely to have been started during the Warring States period (475–221 BCE). The original author is not known. Over the next 1,000 years, this book had gone through various amendments and additions by scholars. But its main part may have been formed in the Warring States-Eastern Han Dynasty period (475 BCE–AD 220), since Zhang Zhongjing, a person living in the late Han Dynasty, first quoted this book but the version he quoted has been lost. The widely used version was finalized in the years 1057–1067.

"Lu Shi Chun Qiu" (239 BCE) also known as "Lu Lan" is the first book with a clear description of the relationship between drinking water and health in China. This book was collectively completed under the organization of Lu Buwei, the prime minister of the Qin State during the late Warring States Period (circa 221 BCE). This book was finished in the year 239 BCE, the eve of the unification of China by the Qin State. The following words are written in the chapter of "Jin Shu": "More people of bald, and iodine deficiency lives in the place with light (mineral deficient) water; more people of swollen feet and crippled legs live in the place with heavy (mineral surplus) water; more healthy and beautiful people live in the place with sweet water; more people with dermatosis live in the place with spicy water; and more people with rickets and humpbacks live in the place with bitter water" (Lu Bu Wei, circa 221 BCE).

In ancient China, the so-called light and heavy water refers to the difference in the amount of minerals in the water. Generally the water in the upper reaches of the river and in mountainous areas is called light water, and downstream river water and water in plains or coastal areas is called heavy water. Sweet water has a slightly sweet taste and is uncontaminated, generally with good water quality and low salinity. Spicy water might have been seriously polluted by organic matter and sewage. Bitter water must come from saline areas with a high degree of mineralization.

The above passage in "Lu Shi Chun Qiu" can be interpreted as follows: in areas with low-salinity drinking water and early human hair loss, thyroid disease is common; in areas with high-salinity water more people suffered joint swelling and lameness; in areas with good water quality people are beautiful and noble; in areas with severely polluted water people are prone to skin abscesses; and people drinking brackish water tend to have rickets and "humpbacks." If we accept that the symptoms of swollen feet, crippled legs, rickets, and "humpbacks" recorded in the book "Lu Shi Chun Qiu" to be a description of skeletal fluorosis, this book is the earliest Chinese medical book recording this disease.

"Huai Nan Zi" (180–123 BCE) is a book of collective works later than "Huang Di Nei Jing." Liu An (180–123 BCE) was the King of the Huainan State in the Han Dynasty. He is said to have invited thousands of scholars to jointly prepare a book entitled "Hong Lie," now only 21 papers from it called "inside the book" have been handed down to form the book "Huai Nan Zi" (Liu An, 180–123 BCE). In the book, there are a lot of descriptions of the relationship between people's appearance, health status, and geographical conditions, drinking water, food, climate, and season. In one of the papers called "falling-shaped articles," it is recorded "...More 'Ying' in dangerous areas, more die in hot areas and more long lives in cold areas..." Translated into modern Chinese, those words can be interpreted as living in the mountain areas, people easily suffer from thyroid disease, people living in hot climate regions will die at young ages, while in a cold climate region, people live longer.

Zhang Hua (AD 232–300), a minister in the Western Jin Dynasty, was known as a knowledgeable person. In his book "Bo Wu Zhi," he wrote that people living in the mountains easily suffered from thyroid disease because they drank spring water that did not flow. The disease was prevalent in mountain areas in west Hubei Province. People living along the Yangtze River did not have that disease due to thin soil layers with low content of salt (Zhang Hua, AD 232–300).

Lu Wenchao from Qing Dynasty did not agree with Zhang Hua and explained that only people living in the south had that disease because there was black-colored water in the south and not because the water did not flow. He pointed out that the disease was due to the qualities of water and soils but not static spring and well water.

Based on modern scientific knowledge, both arguments are partially correct, but contain errors based on speculation.

What Zhang Hua saw, that thyroid disease is common in the mountains areas or thin soil areas, especially the "halogen-free" soil area, is completely correct. Iodine is very easy to move with air and water, therefore is very low in mountains areas with serious soil erosion by leaching, where the plants also have low contents of iodine that leads to the prevalence of the disease.

It is remarkable that, 1,800 years ago, the ancient Chinese considered the relationship between iodine deficiency diseases and soils. Only in recent times has it been confirmed that the direct cause of iodine deficiency disorders is lack of iodine in soils.

Ji Kang (AD 223–262) was a writer, philosopher, and musician who lived in the Three Kingdoms period (AD 220–280). He noted that the teeth of a lot of people who lived in Shanxi Province were yellow and recorded this discovery in his book, "Yang Sheng Lun" (The Health Theory). It is now considered the earliest record of endemic fluorosis in ancient China. According to the statistics in 2000, 7.4% of the total population (2.43 million) of the people in Shanxi Province are still suffering from fluorosis because of high-fluorine groundwater, one of largest endemic fluorosis areas in China (Ji Kang, AD 223–262).

The book, "Zhu Bing Yuan Hou Lun" (Source of the Disease), was published by Chao Yuanfang, the former imperial doctor of the Sui Dynasty in 610. This book summarized the previous ancient Chinese medical descriptions of various diseases and treatment experiences including "Ying." It was written in the book that the disease was common in the people who lived many years in mountains areas, spring areas, or areas with a black soils rich in humus materials (Chao Yuanfang, 610).

Prevention and Treatment of Endemic Diseases in Ancient China

In contrast to fluorosis, endemic goiter can be treated and prevented. The ancient Chinese may have found ways to treat goiter at a very early age. The earliest written record is the medical book "Qian Jin Yao Fang," written by the famous physician Sun Simiao (581–682). The book discusses hundreds of kinds of diseases and collected more than 10,000 prescriptions for the treatment of various diseases. To treat thyroid disease, drugs including deer and sheep thyroid, seaweed, sea clams, and oysters were listed in the book (Ge Hong, 284–364). Wang Tao, generally lived in the same era of Sun Simiao and recorded in his medical book "Wai Tai Mi Yao" similar drugs. Those treatments clearly formed prior to the completion of those books (Sun Simiao, 581–682).

Zhang Zihe (about 1156–1228) in the Jin Dynasty wrote a book named "Ru Men Shi Qin," featuring diagnosis and treatment of a variety of diseases. In the part discussing "Ying," he pointed out that it can be cured by drinking the water dipping from any two of the three kinds of seafood, kelp, algae, and seaweed. In fact, this method not only can treat patients who have been suffering from endemic goiter but also adds iodine for the people living in the same iodine deficiency environment and can help prevent iodine deficiency diseases. Since seafood are all natural foods and not chemicals that people, who lack sufficient scientific knowledge, are afraid of, that approach is still regarded as a readily acceptable method to prevent diseases in iodine deficiency areas (Zhang Zihe, 1156–1228).

Medical Geology in China Today

Fluoride-Related Diseases

Geology and Geochemistry of Dental and Skeletal Fluorosis

China has one of the world's most serious fluorosis issues, an issue that has afflicted this region for quite some time. In 1976, a number of Paleolithic human fossils dating 10 million years ago were unearthed in Xujiayou Village, Yang Gao County, Shanxi Province. Brown stains and spot-like depression defects were found in the five human fossil teeth showing the typical features of dental fluorosis. The discovery of ancient human teeth fluorosis is in a county where Ji Kang (AD 223–262) reported that "People living in Shanxi Province have yellow teeth" (Ji Kang, AD 223–262). This region is still a serious fluorosis area. Until the 1990s, the fluoride content of drinking water in general had been higher than 4 mg/L and reached as high as 9.5 mg/L in some individual wells. In recent years, on the basis of government funding, local residents no longer use the high-fluorine groundwater.

According to the available information endemic fluorosis is prevalent in all provinces, municipalities, and autonomous regions in China, except for Shanghai.

Excluding fluorosis from tea drinking, of all 2,221 counties in China there are 1,297 fluorosis prevalent counties with the population of nearly 650 million people, almost half of the 1.3 billion population of China.

Data released by China's Ministry of Health in 2007 indicate that the number of dental and skeletal fluorosis cases caused by drinking fluorine-rich water were 21 million and 1.3 million, respectively, representing 1.59 and 0.10% of China's population. There were more than 118,000 villages with the prevalence rates of dental fluorosis over 30% and with a total population of 84 million, accounting for 6.35% of China's population and 11.54% of the rural population in China at that time. Of these, fluorine mobilized by domestic coal combustion caused dental fluorosis in more than 16 million people and skeletal fluorosis in 1.8 million people. These two types of dental fluorosis and skeletal fluorosis patients totaled more than 37 million and 3 million, respectively, with nearly 160,000 villages having rates of dental fluorosis prevalence over 30% (Ministry of Health, P. R. China, 2009).

There are also an estimated one million people with fluorosis caused by drinking fluorine-bearing brick tea, popular in Tibet, Qinghai, Xinjiang and parts of Inner Mongolia in western China.

Relationship Between Fluorosis Rates and Levels of Fluoride Exposure

Correctly understanding the above data entails understanding the survey and statistical methods of Chinese endemic fluorosis. In medical geology, this is the bridge connecting epidemiology and the earth sciences.

A national survey was conducted in the 1970s. In the regions with fluoride content in drinking water higher than 1.0 mg/L, dental fluorosis was identified if the following symptoms occur: teeth have lost lustrous glaze and have a chalky appearance; the enamel is light yellow, yellow, brown and black in color or has stripes, light points, and plaques; and the teeth have enamel defects. Since it is impossible to use X-ray machines to inspect all the suspected population, skeletal fluorosis was determined mainly by clinical examinations. In fluorine exposure areas, skeletal fluorosis was defined as long as the patient demonstrated one or more of the following symptoms: damaged shoulder, elbow and knee, curved spine, or "humpback."

The investigation was carried out on children aged 8–12. At present, the population of a village in China is around 500–1,000, and of children aged 8–12 in a village is generally about 40–80. If the prevalence rate of dental fluorosis in children is over 30%, this village is classified as an endemic village and the rate of dental fluorosis in children represents the rate of the whole population of the village, and accordingly, the number of patients of the village is calculated. In accordance with the management approach of China's endemic disease, any county or province that has endemic villages is defined as an endemic county or province. Obviously, the population of an endemic province is greater than that of a county and much greater than that of a village.

Fluorosis from Drinking Water

This type of drinking water fluorosis is caused by excessive intake of fluorine through drinking water with high fluorine. Enrichment of fluorine in groundwater can form by leaching the fluorine from high fluorine-bearing rocks that the groundwater flows through. The more fully the water and rock interacts, the higher fluoride concentration.

Fluorosis from High Fluoride in Shallow Water

High-fluoride shallow water is widely distributed and is the source of the most serious endemic fluorosis problem in China. In this type of fluorosis, the prevalence can reach 30% (definition limit of an endemic fluorosis region) when fluoride content of drinking water exceeds 1 mg/L. In China, the highest fluoride content in these regions can be as high as 32 mg/L.

Characteristics of those regions are arid, semi-arid climate, hot summers with concentrated rainfalls, and cold and dry winters with little rainfall and intensive evaporation. Major cations in the water are potassium, sodium, and calcium released by rock weathering, forming a partial alkaline condition making fluorine more susceptible to leaching. Fluorine enters into the shallow groundwater by mixing of surface water and groundwater and is concentrated by evaporation and condensation in the shallow water by the dry climate, resulting in high-fluoride shallow water.

In China, this type of fluorosis occurs wildly from northwest to the northeast, mainly in the Inner Mongolia Autonomous Region, the central area of Heilongjiang Province, the central and western areas in Jilin Province, northern Liaoning Province, Hebei Province, Beijing, western Shandong Province, northern Henan Province, Shanxi Province, Shaanxi Province, Ningxia Autonomous Region, Gansu Province, and Xinjiang Autonomous Region. The distribution of this type of fluorosis is strictly controlled by topography.

During the past three decades, as economic development increased, the nutritional status of residents has greatly improved and the resilience of the body toward fluoride has also increased. Particularly, more and more people use bottled mineral water, purified water, and soft drinks. Many young people have given up agriculture and farm work and work seasonally in the cities or in coastal areas. Therefore, the consumption of high-fluorine groundwater has significantly declined and the dose–response relationship has also changed, for the ratio of fluorosis has declined at the same level of fluoride content of drinking water.

Since the beginning of this century, the Chinese government has invested a lot of money annually to help local residents of dental fluorosis regions improve the quality of drinking water. The main measures are looking for low-fluoride water sources including low-fluoride surface water, deep groundwater or groundwater in adjacent areas in the endemic fluorosis regions, and government-funded construction of public water supply systems. Centralized facilities for fluoride removal methods of drinking water have been constructed in a few areas because low-fluoride water could not be found and adopted.

By 2007, the drinking water of more than 55,000 villages had been changed to low-fluoride water. Of the population of 84 million that used high-fluoride drinking water in the past, more than 42 million now use low-fluoride drinking water. Because of the policy that villages with very high fluoride contents of drinking water change water sources, now most villages have water supplies with fluoride concentration of 1–2 mg/L. Therefore, we can say that China has basically eliminated the conditions resulting in skeletal fluorosis and will completely solve the drinking water fluorosis in the near future.

Fluorosis from High-Fluoride Geological Bodies

High-fluoride surface water and groundwater can form from contact with high-fluorine rocks, sediments, or ore bodies. These high-fluoride geological bodies include volcanic rocks, granites, alkaline rocks, alkaline salt lake sediments, and mineralized zones of fluorite. A wide range of fluorine-rich granites are distributed in southeastern China, and a large number of fluorite ore deposits are distributed in Zhejiang Province, Guangxi Autonomous Region, and Henan Province. High-fluorine surface water and groundwater from these areas has resulted in fluorosis in part of the local population. In the areas of Wudalianchi Volcano Group in Heilongjiang Province, Changbai Volcanoes in Liaoning Province, Arshan Volcanoes in Inner Mongolia Autonomous Region, Datong Volcanic Group in Shanxi Province, and Tengchong Volcanoes in Yunnan Province, water-soluble fluorine-rich magma and ash had been ejected during volcanic activities and formed into volcanic rocks and tuff. From contact with these rocks high-fluoride surface water and groundwater has been formed resulting in many cases of fluorosis. Fluorosis in fluorite mines in Henan Province, Zhejiang Province, and Guangxi Province has also been reported.

Fluorosis from Volcanic Hot Springs and Geothermal Regions

A large number of hot magma, volcanic ash, volcanic gases, and geothermal fluids, which contain a great amount of activated fluorine, was released by volcanic and geothermal activities. Higher temperatures in volcanoes and geothermal crustal areas caused accelerated interactions between groundwater and rocks and a higher amount of fluorine released into groundwater.

The world's first report of fluorosis was from the Vesuvius volcano area in Italy. China's initial reports of this type of fluorosis were from Xiaotangshan hot spring in the suburb of Beijing and Xiongyue and Tanggangzi hot spring areas in Liaoning Province.

In history, there had been volcanic activities in Wudalianchi area in Heilongjiang Province, Changbaishan area in Jilin Province, Arshan area in Inner Mongolia Autonomous Region, Datong area in Shanxi Province, Tengchong area in Yunnan Province, and Datun area in Taiwan, China. Now only volcanoes in Datun area in Taiwan are still active. There are high-fluoride surface water and groundwater in areas affected by all of the above volcanoes.

Fluorosis from Coal Combustion

In 1982, an investigation of fluorosis in Guizhou Province was conducted by the Institute of Geography, Chinese Academy of Sciences. They reported that a large amount of gas and dust containing fluoride was released by coal combustion and quickly combined with vapor in the air to form an aerosol or a hydrofluoric acid mist, which adhered to the surface of corn and chili. They also pointed out that fluorine concentrations in the soils in the fluorosis area were higher than those in other areas, which might potentially affect fluorine concentrations in local foodstuff. The authors demonstrated that fluorine concentrations in fresh corn from Zhijin County, Bijie County, and Guiyang City were 3.60, 2.75, and 5.35 mg/kg, respectively. Those values are higher than the world average value and may be important in that foodstuff alone can induce dental fluorosis (Li et al., 1982).

Etiological Significance of the Clay in Fluorosis from Coal Combustion

Fluorosis caused by indoor coal combustion is only prevalent in China and is a very severe health problem. China defines coal combustion-type fluorosis as an endemic disease and assigns the government health system to be charge of it. In fact, this disease is neither a traditional endemic disease nor a general "public nuisance." If a harmful element is distributed in the surface water, groundwater, foodstuff, and air of a specific area and has to be absorbed by residents during their everyday life, the disease induced by this element is endemic. If this element enters surface water, groundwater, foodstuff, and air by human activities, the disease induced by this element is a "public nuisance."

Zhou Daixing, from the Health and Epidemic Prevention Station of Qiannan Autonomous District, first noticed the relationship to clay. In the second national symposium on endemic fluorosis in 1985, he pointed out that the difference in the degrees of fluorosis of adjacent areas using the same coal is caused by the difference in fluorine concentrations in clay that was mixed with coal for combustion and suggested not to use high-fluoride clay to mix with coal (Zhou et al., 1993).

A similar conclusion was obtained from the investigation in Zhijin County, Guizhou Province (Zheng, 1992). It was pointed out that the harm from fuel made by coal mixed with clay was larger than from the coal alone and the amount of fluorine released from clay was two times more than that from the coal.

Research in collaboration with the U.S. Geology Survey on fluorine and other elements in coal and coal ash found that there was a positive correlation between the concentrations of fluorine and potassium. Because potassium is mainly distributed in clay minerals in coal and coal combustion products, it can be inferred that fluorine is also distributed in the clay minerals in the coal and coal ash and the contents of fluorine increase with increasing proportions of clay (Finkelman et al., 2002).

Recently, Wu et al. (2004) demonstrated that coal mixed with clay to form briquettes used as fuel was very common and that the clay was the main source of fluorine causing fluorosis. Based on a survey on the environment and living conditions of fluorosis areas, it was found that local residents had to use the mixture of coal powder and clay as their fuel because the coal powder was cheaper than lump coal.

In recent years, with the improvement in economic conditions, the conditions of fluorosis have greatly changed and the prevalence has gradually decreased. Now all the villages have been able to use electric energy and the usage of bio-energy sources (straw, biogas) has increased. Local residents have used government-funded stoves with chimneys for heating and cooking, and the use of stoves without chimneys has been decreasing. Now the staple food of the residents is rice, and corn is primarily used for livestock feed, which greatly reduces the fluoride intake of the residents. However, for the time being, fluorine-contaminated peppers are the main source of fluorine intake (Sun et al., 2005).

Fluoride Caused by High-Fluorine Salt

Liu Ziyue et al. (1982) reported a case of salt-inducing fluorosis in Yushan Town, Pengshui County, Sichuan Province (Liu et al., 1982). There are three sources for salt in China: sea salt, lake salt, and rock salt. Sea salt was used in the southeast coastal areas and the central region, and with a large number of salt lakes in northwest, the principal salt used by people in this area is lake salt.

Rock salt and brines from Triassic rocks are rich underground in the Zigong County in central Sichuan Province. More than 2,000 years ago, local residents began to extract and heat underground brine to make salt. This salt was called "well salt" and used in much of southwest China. Fluorine concentration is low in salt rocks of the Triassic and is only 1 mg/kg in produced salt. So there is no health risk to use this salt.

The salinity of the Cambrian strata is significantly lower than that of the Triassic. A phosphate layer with high fluorine content (1–3%) is beneath the Cambrian strata. So salt brine from the Yushan factory has lower salinity but a higher fluorine content than that of the brine in Zigong County. In the production process, in order to fully use the heat released from coal combustion, exhaust gas from coal combustion is directly contacted with brine to preheat it, therefore exhaust fluorine from coal combustion was absorbed by the brine and further increased its fluorine content. It is clear that in the pre-condensed stage, concentrations of NaCl and fluorine increase 30 and 113%, respectively, obviously due to the adsorption of fluorine by brine in contact with exhausted gas from coal combustion.

Salt production from the Yushan Factory is limited and only supplies some towns in Qianjiang County and Pengshui County, where fluorosis from coal combustion occurs. Xiaochang Village in Pengshui County is the most severe fluorosis area in China, with 100% dental fluorosis. Therefore, the majority of fluorosis in this area is the mutual results of indoor coal combustion and high-fluoride salt. In 1985, the Yushan salt factory was closed due to the high-fluoride salt products and the health risk of high-fluoride salt is no longer an issue (Zhang, 1986, 1992).

Prevention of Dental Caries by Drinking Water Fluoridation in China

Based on the data of the third epidemiological investigation on oral health in China conducted in 2005, dental caries of 12, 35–44, 65–75 years old residents of China are 28.9, 64.2, and 71.8%, respectively, with numbers of decayed teeth per person of 0.54, 2.42, and 3.22, of which 10.7, 23.3, and 8.2% has been cured, respectively. The situation of milk teeth of children is worse, with 66.0% of ratio of dental caries in 5-year-old children, with 3.5 decayed teeth per person, of which only 2.8% has been cured.

In the 1965, an experiment of tap water fluoridation was conducted in Guangzhou, China, but did not yield any reliable results, and was stopped in 1983. Since then, research on drinking water fluoridation in China has faced great challenges. Now there are no drinking water fluoridation systems in any communities in China.

In order to implement drinking water fluoridation to prevent dental caries, since 1995 Zheng Baoshan, Wang Binbin et al. have conducted research on the relationship between fluorine concentrations in drinking water and dental health in China (Wang et al., 2004a, 2005). In this project, the relationship between fluorine content in drinking water, food, and dental health of residents in 50 large cities in China was evaluated. The effects of fluorine intake by weather as well as gender were also studied.

Through this study it was found that the general concentrations of fluorine in drinking water in cities in China were low. Fluorine contents in drinking water were lower than 0.3 mg/L in 70% of the cities and lower than 0.5 mg/L in 90% of the cities. Shanghai had the highest fluorine content (0.667 mg/L) but it still not exceeded the recommended range from WHO (0.5–1.0 mg/L) (WHO, 1996a).

Fluorine concentrations in food were low and did not have much variation between cities. Fluorine contents in rice and flour were lower than the test limit, which indicates that the intake of fluorine for Chinese people was not from foodstuff generally.

Drinking water was the main source of fluorine intake for Chinese residents, in the range from 0 to 4 mg/L, indicating that the definitive factor in the difference in fluorine intake comes from tap water (Fig. 1).

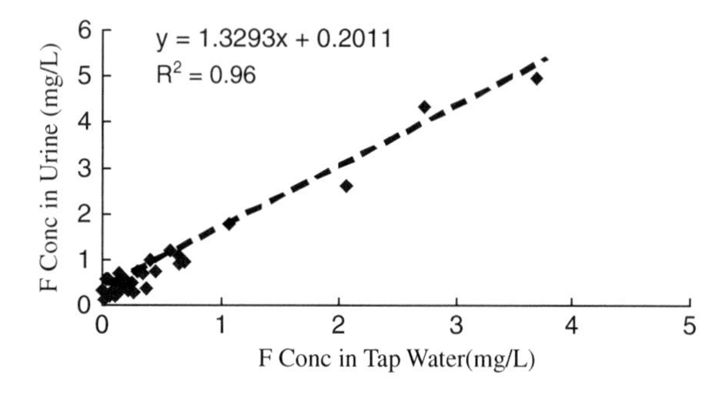

Fig. 1 The relation between fluorine concentrations in tap water and in urine of local residents of the investigated cities and villages

The relation between fluorine concentration in drinking water and dental index did not have a good linear relationship. The following two conclusions can be drawn from this study: with the increasing fluorine concentrations in drinking water in the safe range, dental caries can be limited and reduced; dental caries was not determined by fluorine alone, but eating habits, conditions of sanitation, living conditions, etc. could also affect its prevalence.

Medical Geology of Iodine in China

Iodine Deficiency Disorders (IDD) in China

Since most parts of China are far away from the sea, it is difficult for wind and rain to bring iodine from the ocean into the Chinese mainland. Also, because China is a mountainous and hilly country, rainfall not only does not add iodine but leaches some iodine from the soils. Therefore, China has the most serious problem of iodine deficiency disorders in the world. There are 375 million people living in iodine-deficient areas, accounting for 47% of the world's at-risk population and there are about 35 million people with iodine deficiency disorders in China.

In addition to Shanghai, all of the provinces and autonomous regions had endemic iodine deficiency. The most serious areas were interior mountains, areas with serious soil erosion by groundwater and surface water flowing along steep slopes, plus peat swamp areas, and limestone areas.

Endemic Cretinism in China

In iodine deficiency areas, a serious shortage of iodine intake by pregnant women will harm the development of the nervous system of the fetus and newborn infants will likely suffer from cretinism and related health problems. Patients with endemic cretinism are found in every area that has serious iodine deficiency disorders in China. During 1962–1983, an investigation on endemic cretinism in China was conducted (Institute of Epidemiological Diseases in Xinjiang Uigur Autonomous Region, 1984). The survey was carried out in 25 provinces and autonomous regions to check a population of more than 750,000 people. The survey found a goiter prevalence rate of 24.92% on average, with a range of 11.1–50.8% and a cretinism prevalence rate of 3.15% with a range of 0.04–13.6%.

The more severe the goiter problem the higher the incidence of cretinism. According to the survey the relationship between iodine deficiency endemic goiter and cretinism prevalence rates resulted in one cretinism patient in every 75.83 endemic goiter patients. Based on this ratio, there were 35 million endemic goiter patients and approximately 460,000 patients with cretinism in China prior to the implementation of salt iodization in iodine deficiency endemic areas (Institute of Epidemiological Diseases in Xinjiang Uigur Autonomous Region, 1984).

Stanbury (1996), a leading expert on iodine deficiency disorders, deemed that cretinism appeared in the region when daily iodine intake was less than 20 μg, with a prevalence threshold of 25 μg, and an epidemic threshold of endemic goiter was 50 μg. China's findings were consistent with this.

It was found from the research on the relationship between iodine concentrations of drinking water and iodine deficiency disorders in China that iodine contents in drinking water for the residents living in endemic cretinism areas are mostly less than 1.5 μg/L and that there is very little prevalence of cretinism in the areas with iodine content in drinking water is greater than 3 μg/L.

It was found that the iodine contents of salt had a significant impact on the prevalence of iodine deficiency diseases. Before 1966, there was endemic goiter but no cretinism in the south and southeast areas in Guizhou Province, but after 1966, cretinism existed with the ratio of 6.3% in children in endemic villages in this region, and the maximum ratio reached 22.8%. The only change of iodine nutritional status of residents in this region was the change in the source of edible salt. In the past, edible salt in this region was well salt from Sichuan Province with 1,600 μg/kg of iodine. In the 1950s, edible salt in this region changed into sea salt or well salt with iodine concentrations lower than 200 μg/kg (Institute of Epidemiological Diseases in Xinjiang Uigur Autonomous Region, 1984).

Endemic Goiter Induced by Iodine Exposure

From the survey on the relationship between the urinary iodine concentrations of the Chinese population and goiter prevalence, it was found that the prevalence rate of endemic goiter increased when the intake of iodine was lower or higher than the needed amount. Although the main hazard in China is iodine deficiency disorder, there are also goiter hazard problems caused by excess iodine exposure in underground water and food.

Endemic goiter was reported in Hokkaido, Japan in 1965, due to the excessive consumption of algae by the fishermen. In the 1980s, the prevalence of goiter caused by excess iodine from eating seaweed impregnated with "seaweed salt" was reported in the coastal area of Shandong Province, China.

Endemic goiter caused by iodine exposure in drinking water has been found only in China. The areas with excess iodine exposure in drinking water in China are delineated by iodine contents higher than 150 μg/L and median urinary iodine concentrations of children greater than 400 μg/L. The delineation of areas with endemic goiter induced by excess iodine exposure is iodine contents in drinking water higher than 300 μg/L and median urinary iodine concentrations of residents greater than 800 μg/L and an incidence of endemic goiter greater than 5%.

The iodine exposure region in China is located in low areas in the North China plain and inland basins in Shanxi Province and Inner Mongolia Autonomous Region. The geochemistry and formation of water with high iodine needs further research. Now the supply of iodized salt has been stopped in areas where iodine contents in drinking water are higher than 150 μg/L (Shen et al., 2007).

Prevention of IDD by Salt Iodization

In China, the supply of iodized salt was conducted in serious iodine deficiency areas since the 1950s. Confirmed by the national survey of IDD, by 1988, the total number of IDD patients had dropped from 35 million to less than 8 million (Prevention of endemic diseases for forty years in China, 1990).

In response to the World Health Organization calling for universal salt iodization, in 1991, the Chinese government formally committed to the international community to use universal salt iodization to achieve the goal of complete elimination of iodine deficiency disorders by 2000.

By the implementation of universal salt iodization, China has basically achieved the goal of eliminating iodine deficiency disorders. By 2007, iodized salt was consumed by 94.13% of the population of China and the total number of patients with endemic goiter had dropped to about 6 million people with 122,000 patients with endemic cretinism. There are no new patients of endemic cretinism in the areas covered by iodized salt. Because of the national salt iodization program, the average IQ level of Chinese newborn children increased 11–12 points (Ministry of Health of the People's Republic of China, 2009).

At present, patients with endemic goiter are mainly distributed in the Tibet Autonomous Region, Xinjiang Autonomous Region, Hainan Province, and Qinghai Province that are enriched with accessible deposits of lake salt, rock salt, well salt, or sea salt resources that the local residents use instead of the more costly commercial iodized salt. In these areas the control of iodine deficiency disorders is difficult and in some cases children are still born with cretinism.

IDD and Geochemistry of Iodine in Guizhou Province, China

Guizhou Province in southwest China is a mountainous province enriched with limestone and coal and is far from the sea, with the area of 174,700 km^2 and the population of 40 million. There are 87 counties and cities in Guizhou Province, including 36 endemic fluorosis counties with 10 million patients. IDD was prevalent in 57 counties in the southwest of the province. In 1982, there were 2.09 million patients suffered from endemic goiter, and 0.21 million cretinism patients in some endemic areas (Prevention of endemic diseases for forty years in China, 1990).

In Guizhou Province there were no IDD areas in the endemic fluorosis areas and no fluorosis areas in the endemic IDD areas. The reasons for this phenomenon are still not clear. Though one explanation may be the difference in the distribution of coal resources, coal use patterns and habits, the main food crops, and the bedrock. Communities using high-iodine coals appear to have lower incidence of IDD than communities using wood as their main energy source (Wang et al., 2004b).

Medical Geology of Arsenic in China

The Chinese people long ago discovered the arsenic-containing minerals realgar and orpiment and applied them in traditional medicine. Production of arsenic by arsenopyrite calcination method was also found so early that some historical and literary works had recorded a lot of suicide and murder cases using arsenic. The Shimen Arsenic mine in Hunan Province is the largest arsenic mine in China and has been mined for at least 1,500 years.

Arsenosis caused by drinking high-arsenic groundwater has been common in dry areas of Inner Mongolia Autonomous Region, Shanxi Province, and Xinjiang Autonomous Region in China. In Taiwan, arsenosis and blackfoot disease have been common due to drinking high-arsenic water. Indoor combustion of high-arsenic coal under poor ventilation conditions, common in some areas in Guizhou Province, has caused coal combustion-type arsenosis. Large-scale mining of non-ferrous metals and gold has caused severe arsenic contamination and arsenic poisoning of miners and residents in and around mining areas. Recently, the potential hazards of arsenic poisoning by traditional Chinese medicine methods used to treat diseases have received a lot of attention.

Studies have shown that the high ratio of lung cancer in workers of Gejiu tin mine in Yunnan Province was induced by the combined effects of arsenic and radon in the underground mine dusts. Studies on the exposure of arsenic and its relation with a variety of cancers in the blackfoot disease area in Taiwan eventually led to a worldwide correction on arsenic standards in drinking water. In the research of the world's only coal combustion-type arsenosis, coal with arsenic contents up to 3.5% with an unprecedented organic form has been found. In China, the experiment on leukemia treatment by using arsenic trioxide was a great success and significantly increased the cure rates of a kind of leukemia.

Present Situation of Endemic Arsenosis in Mainland China

In accordance with a survey co-funded by the Chinese Ministry of Science and UNICEF completed in 2003, arsenosis from drinking was prevalent in 40 counties or cities in eight provinces or municipalities and the population exposed to arsenic concentrations in drinking water higher than 0.05 mg/L is more than 500,000 people, including 7,821 people identified as arsenosis patients. Arsenosis from coal combustion was prevalent in eight counties or cities in Guizhou Province and Shaanxi Province, where arsenic contents in coal had been higher than 100 mg/kg, and coal was burned without indoor chimneys (Jin et al., 2003).

A second survey conducted during 2003–2005 (Yu et al., 2007) indicated that the actual prevalence of arsenosis from drinking water was 500,000–600,000 people who were drinking water with arsenic content higher than 0.05 mg/L, and about 10,000 arsenosis patients, including about 1,000 serious arsenosis patients.

In the above surveys, the diagnostic criterion in the investigation of arsenosis was "The diagnostic criteria of endemic arsenosis WS/T 211-2001" (Standard of diagnosis for endemic arsenism, 2009). These standards require that arsenosis patients have an excessive amount of arsenic exposure history and that there are no other reasons for the skin pigmentation and depigmentation or palmoplantar hyperkeratosis.

Arsenosis from Drinking Water in Mainland China

Arsenosis in Xinjiang Uigur Autonomous Region

Drinking water arsenosis in mainland China was first found in Xinjiang Autonomous Region and was reported by Wang et al. (1982). The drinking water arsenosis area in Xinjiang Autonomous Region is located in the alluvial fan in the southwest Junggar Basin, an arid inland basin, and in the north of the Tianshan Mountains. Along the east–west direction, there is a 250 km long zone of high-arsenic water from west of Ebi Lake to the east of the Manas River.

In the Xinjiang Autonomous Region, high-arsenic water zones also form around natural perennial or seasonal water swamps or lakes. There are similar phenomena around a number of artificial reservoirs for agricultural irrigation purposes. The high-arsenic groundwater in Xinjiang Autonomous Region is reduced groundwater, rich in humic acid, associated with methane, with high pH. High-arsenic groundwater in Xinjiang Autonomous is usually high in fluorine, but the highest arsenic groundwater does not have the highest fluoride content.

After finding arsenosis in drinking water in Xinjiang Autonomous Region, the most affected villages were moved to safe places, with low-arsenic water supply systems. However, there are still villages with arsenic concentrations in drinking water higher than 0.05 mg/L, with the maximum of 0.09 mg/L.

A number of investigations on the relationship between arsenic concentrations in drinking water and ratios of arsenosis in arsenosis areas were conducted by Wang Lianfang in Xinjiang Autonomous Region (Wang et al., 1982). Arsenosis areas in Xinjiang Autonomous Region are located in very arid regions. Under the hot, dry weather conditions, the amount of drinking water of local residents is very large, with an average of 4 L per person per day. At such a high quantity of drinking water, no clinical arsenosis exists with the arsenic concentrations in drinking water below 0.13 mg/L.

Arsenosis in Inner Mongolia Autonomous Region

There are two kinds of arsenosis in Inner Mongolia: one is induced by arsenic contamination of surface water and groundwater by high-arsenic minerals and the other is induced by high-arsenic groundwater in arid inland rivers' alluvial plains.

Chifeng City in Kosk Teng County in eastern Inner Mongolia is arid and hot in summer and cold in winter. The arsenosis village is located on the slope of a mountain with arsenopyrite deposits. Arsenic released from arsenopyrite weathering moved into groundwater and formed high-arsenic springs in some places and contaminated some wells of the village. It can be clearly observed that the closer the well, the higher level of arsenic in the well water. Years ago, the staff of a geological survey had been poisoned by mistakenly drinking from a high-arsenic fountain. Based on investigations, concentrations of arsenic in 34 well water samples in the village were in the range of 0.16–0.45 mg/L, with an average of 0.31 ± 0.21 mg/L and maximum of 0.45 mg/L. There were 22 minor arsenosis cases out of 45 people tested (Gao, 1990).

Drinking Water Arsenosis in Shanxi Province

In Shanxi Province, high-arsenic groundwater is distributed along alluvial plains in arid inland basins, including the Datong Basin in the north and the Jinzhong Basin in central Shanxi Province. The average depths of high-arsenic groundwater are in the range of 20–50 m, with the maximum of more than 100 m. Concentrations of arsenic in well water in Heigada village in Yinshan County had reached 4.35 mg/L, the highest record in mainland China, and this well has been sealed.

High-Arsenic Drinking Water Related to Blackfoot Disease in Taiwan, China

In a confined area located at the southwestern coast of Taiwan, there is a high prevalence of a unique peripheral vascular disease (Tseng, 1989; Tseng, 2002). The disease is known as blackfoot disease (BFD) after its characteristic progression from numbness or coldness of one or more extremities, to intermittent claudication, and finally to gangrene and spontaneous amputation (Tseng, 1989; Tseng, 2002). Extensive epidemiologic (Tseng, 1989) and pathological (Yeh and How, 1963) studies and water analyses (Chen et al., 1962; Kuo, 1964) have been carried out since the 1950s and the etiology of BFD has been linked to the drinking of artesian well water. Although arsenic has been suggested as the most probable cause of the disease, some investigators favored other potential etiologic factors such as humic substances (Lu et al., 1978). Despite the controversies over its real etiology, BFD has declined dramatically two to three decades after the implementation of tap water supply and termination of the use of groundwater for drinking in the endemic areas (Tseng, 2002).

Since the early 20th century, sporadic cases of a strange disease involving the lower extremities have occurred, which was characterized by progressive blackish discoloration of the skin extending from the toes gradually upward toward the ankles (Tseng, 1989; Tseng, 2002). The patients also suffered from numbness or coldness of the extremities and intermittent claudication before the development of gangrene. Ulceration might also be found in the patients. The lesions might progress to spontaneous amputation in some cases or surgical amputation was required for saving lives. In rare cases, the fingers might also be involved (Yeh and How, 1963).

Most of the patients resided in a confined area located along the southwestern coast of Taiwan.

More than 72% of the cases occurred in Hsueh-Chia Township and Pei-Men Village of Tai-Nan County and Pu-Tai Township and Yii-Chu Village of Chia-Yi County. These areas were called the "old endemic areas." Some cases were also found in the so-called new endemic areas located on the periphery of the "old endemic areas." These "new endemic areas" were confined to the counties of Yun-Ling, Chia-Yii, Tai-Nan, and Ping-Tung. A total of more than 1,600 cases had been discovered up to 1980s. In a survey carried out from 1984 to 1985, a total of 1,220 prevalent cases (690 men and 530 women) were found in the "old endemic areas."

The prognosis of BFD is relatively poor after the onset of clinical symptoms. In a series of 1,300 cases, 68% of the patients underwent spontaneous or surgical amputation. Failure of the amputation wound to heal was common and reamputation rate was as high as 23.3%. The annual death rate of the cases was 4.84 per 100 patient-years and the mean age at death was about 62 years (Lo et al., 1977).

Blackfoot disease is prevalent in alluvial plains along the southwest coast of Taiwan. The most serious four villages are located at the both sides of Bazhangxi River flowing from the Central Mountains to the southwest coast. Blackfoot disease endemic area is near the ocean, with high salinity of shallow groundwater that is not suitable for drinking, and since the beginning of the last century deep artesian water resources have been utilized. Depths of the artesian water are generally between 100 and 280 m. The artesian water is rich in organic lacustrine sediments, can ignite due to high methane content, and has a hydrogen sulfide odor. The concentrations of arsenic in the artesian water are higher than 0.35 mg/L. All of the blackfoot patients have histories of drinking artesian water. No new cases of blackfoot patients have occurred since the change of water sources in 1970s.

Extensive epidemiologic studies favor arsenic as the most probable cause (Tseng, 1989; Tseng et al., 1996, 1997). The median arsenic concentration of the artesian well water in these BFD-hyperendemic villages ranged from 0.70 to 0.93 mg/L (Kuo, 1964; Chen et al., 1962), while the shallow well water in other areas had arsenic contents between non-detectable and 0.30 mg/L with a median of 0.04 mg/L (Chen et al., 1962). A survey of the arsenic contents of 83,563 wells all over Taiwan showed that 29.1% of the wells in BFD-endemic areas had an arsenic content greater than 0.05 mg/L and 5.2% greater than 0.35 mg/L with the highest value being 2.5 mg/L. In other areas of Taiwan, only 5.7% of the wells had an arsenic content greater than 0.05 mg/L and 0.3% greater than 0.35 mg/L (Lo et al., 1977).

Because of the finding of high arsenic content in the artesian well water, Tseng (1989) evaluated the association between the arsenic concentration in well water and the prevalence of BFD in the endemic villages (Tseng et al., 1997). A dose–response relationship between them was demonstrated in different age groups. This study suggested that the prevalence of BFD increased with increasing arsenic concentrations of the well water, which could not be explained by the age difference (Tseng, 2002, 2005).

Arsenosis from Coal Combustion

Endemic arsenosis caused by indoor combustion of high-arsenic coal was confirmed in Xingren County, Guizhou Province, China in 1977 (Jin, unpublished) then more new cases in Xingyi City, Anlong County (Zhou et al., 1993), Zhijin Country (Sun et al., 1984), and Kaiyang County (Ke, 1980), all in Guizhou Province (Fig. 2), were reported (Yu et al., 2007; Zheng et al., 1996; Zeng et al., 1995). In 2004 Li Yue et al. reported similar arsenosis in south Shaanxi Province

Fig. 2 Location of endemic arsenosis area in Guizhou Province, China

where local farmers burned "stone coal" with high arsenic contents in stoves without chimneys (Li et al., 2004; Liu et al., 2002).

From a survey completed in 2003, there were 142 villages in Guizhou Province and Shaanxi Province using coal containing arsenic higher than 100 mg/kg. More than 2,400 patients (most in Guizhou Province) were found out of nearly 31,000 people tested (Jin et al., 2003).

Coal has been produced and used in Guizhou Province for the past 100 years. After the production and utilization of high-As coal, arsenosis became an epidemic disease. The earliest-known arsenosis case was found in 1953 and was called "Laizi disease," the high-As coal was called "Laizi coal," but the local residents did not know that the so-called Laizi disease was caused by arsenic exposure (Zheng et al., 1996, 1999).

All high-As coals belong to the upper layer of the Longtan Group and are located in shallow stratigraphic positions. They are commonly high in ash and sulfur and are not considered worthy for commercial mining. However, many village people collect this coal for everyday use. The majority of high-As coals were found in areas of gold mineralization, but there are a few high-As coals not located in the area of gold mineralization (Wang et al., 2006).

The practice of using high-As coals indoor without a chimney and good ventilation leads to As pollution of indoor air as well as pollution of crops and/or other foodstuff such as chili peppers which are preserved and dried indoors over the coal fires. The residents' ingestion of excessive As from polluted indoor air, food, and water was the major cause of arsenosis (Finkelman et al., 1999).

The coals related to the arsenic poisoning have low, moderate, to high ash contents, 9–56 wt% (550°C determination), but are extremely enriched in arsenic and other trace elements. The coals have As concentrations typically >100 ppm, commonly >1,000 ppm, and in a few selected cases >30,000 ppm (3 wt%) on a whole coal, as-determined basis. The coals that have 3 wt% As are exceptional, as As contents that high are usually restricted to metalliferous samples from ore deposits. The highest concentration of As in coal was 35,000 ppm with much of the arsenic in organic association (Belkin et al., 1997). The coals are also highly enriched in Sb (up to 370 ppm), Hg (up to 45 ppm), and Au (up to 570 ppb); U is also high in some samples (up to 65 ppm). An examination of the regional distribution of trace element concentrations and general geochemical relationships documents the highly variable nature of trace element enrichment (Belkin et al., 2008).

Concentrations of As in coal vary sharply, even in the same layer, for example, coal with several thousand milligrams per kilogram As may be only a few meters away from coal with only several hundred milligrams per kilogram and tens of meters away from coal with arsenic in the range of 10–20 mg/kg.

Zheng et al. (2005) have shown that the higher the As content in the coal and the longer the coal is used, the higher the As in urine and hair and the higher the incidence of the arsenosis. There is a good dose–response relationship between the As exposure and the ratio of arsenosis and but it is much more complicated than that of drinking water arsenosis.

Arsenosis Induced by Mining

During the past century, because of growing demand of non-ferrous metals and their compounds, more and more tailings and sludge have been generated by mining and smelting of these mineral resources and have become a source of serious arsenic pollution.

Almost all of the non-ferrous metal mines are associated with a large amount of arsenic. Exploitation of every ton of non-ferrous metals, the associated arsenic will be more than 1 ton. For example, exploitation of 1 ton of mercury will bring 8.36 tons arsenic into the surface environment.

By the end of 2003, 1,392 million tons of arsenic had been extracted accompanied by the exploitation of various non-ferrous metals and pyrite. Less than 30% of the arsenic has been recycled in the production process and about 100 million tons of arsenic moved into the surface environment in the forms of tailings and waste. Serious arsenic contamination has been caused by mismanagement, including 50, 25, and 10 tons of arsenic pollutions in Guangxi Autonomous Region, Yunnan Province, and Hunan Province, respectively.

According to incomplete statistics, from 1974 to 2005, due to arsenic contamination of drinking water sources by this residue, 9 cases of arsenosis events had occurred, resulting in about 2,300 arsenosis patients and 6 deaths. Arsenic contaminated water and soil caused three incidents resulting in more than 600 arsenosis patients, 2 deaths, and the death of a large number of cattle and fish. Inadvertent use of arsenic polluted sacks to load rice, 884 students and teachers of a primary school had developed acute arsenosis. Some arsenic residues can release arsine when exposed with acid and water, and sometimes even when exposed to moisture. In recent years, five arsinic-induced incidents have occurred with nearly 200 patients and 5 deaths in Yunnan Province, Guangxi Autonomous Region, and Hebei Province. In total, there had been 17 arsenic-induced incidents with more than 3,000 patients and 12 deaths from 1974 to 2005 (Xiao et al., 2008).

Medical Geology of Selenium

Selenium is essential to human health in trace amounts but harmful in excess. The safe intake range for humans is from 40 to 400 μg/day (WHO, 1996b; Yang et al., 1983). Food is the main source of selenium for humans. Geology exerts a fundamental control on the concentrations of selenium in soils, crops, and animals. Both the lowest and the highest concentrations and the flux of Se in the environment are reported in China (Wang and Gao, 2001). As a result, both endemic selenium deficiency and selenosis occurred in China.

Health Impacts of Selenium Deficiency

A low-Se belt, with Se contents in soil less than 0.125 mg/kg, stretching from Heilongjiang Province of Northeast China to Yunnan Province of southwest China was identified during the intensive geochemical surveys in 1980s in China. Lots of epidemiological data on the human diseases such as Kashin-Beck disease (KBD) and Keshan disease (KD) in this low-Se belt were obtained from the 1960s to 1970s. The data showed a close correlation between the concentration of selenium in the environment and the prevalence of the two diseases (Figs. 3 and 4) (Li et al., 2009; Tan et al., 1989, 2002). KBD and KD, especially KD, were later characterized as Se-responsive human diseases.

Selenium Deficiency and Keshan Disease (KD)

KD is a cardiovascular disease of unknown etiology and long common in northeast China. This disease manifests as an acute insufficiency of the heart function or as a chronic moderate-to-severe heart enlargement and can result in death. The disease was so severe in China that a special government office was established to coordinate the nationwide efforts against the disease in the 1950s. In the 1950s and 1960s, large-scale epidemiological investigations were launched in the provinces affected by Keshan disease, including Heilongjiang, Jilin, Liaoning, Hebei, Shandong, Henan, Inner Mongolia, Shanxi, Shaanxi, Gansu, Sichuan, Yunnan, and Tibet. About eight million people lived in the affected villages in the country

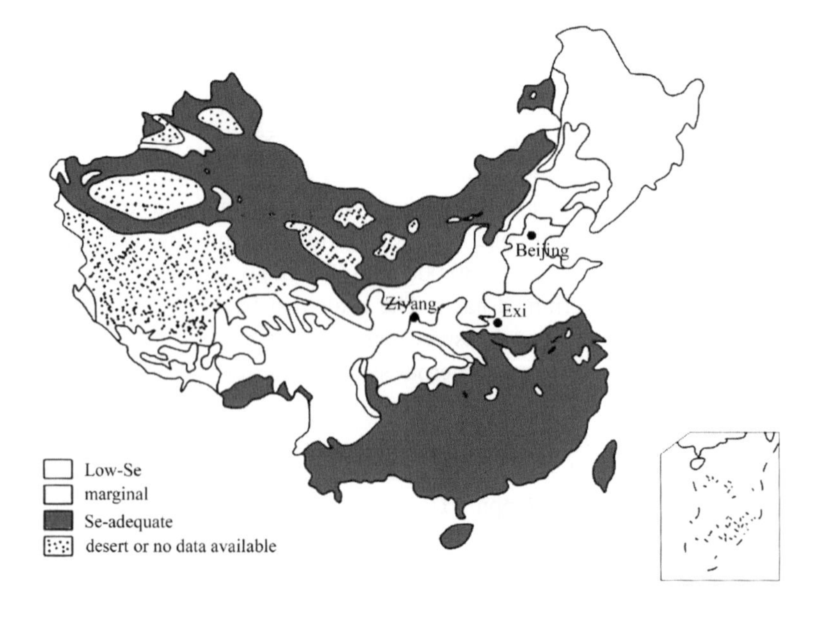

Fig. 3 Soil selenium distribution in China (low-Se: <0.125 mg/kg; marginal: 0.125–0.175 mg/kg; Se-adequate: 0.175–3.0 mg/kg. Exi and Ziyang are the Se-excessive areas where local soil Se content are higher than 3.0 mg/kg and have induced selenosis) (Li et al., 2009; Tan et al., 1989)

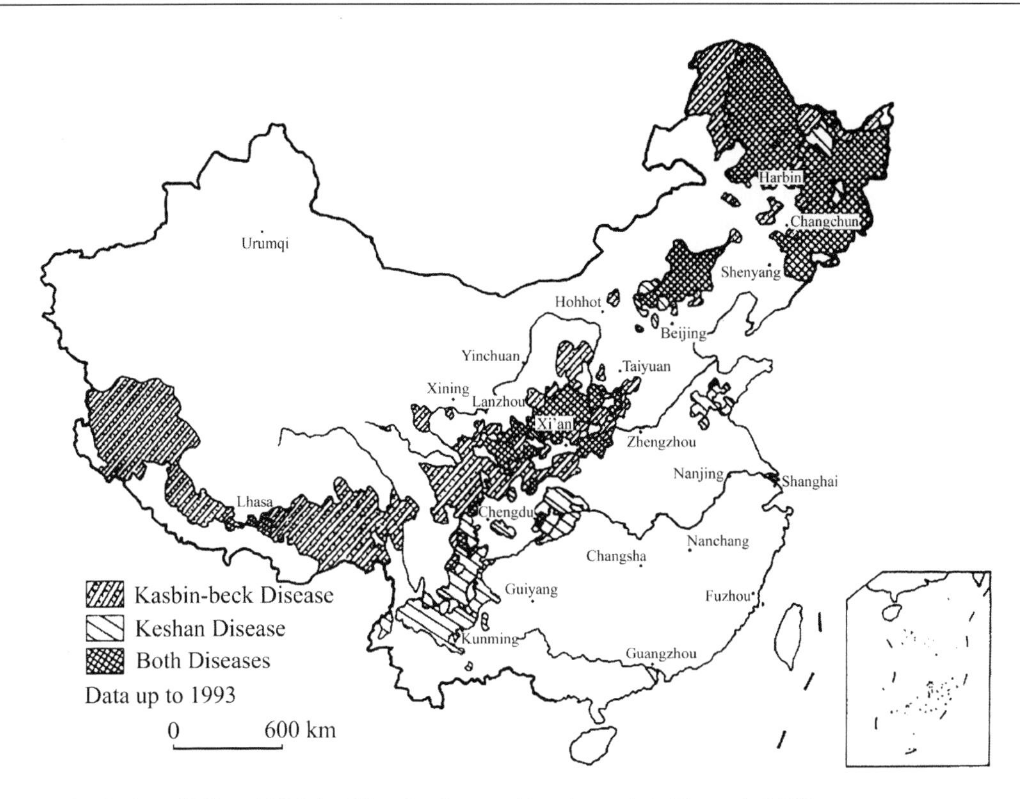

Fig. 4 Areas distribution of KD and KBD in China (by disease-affected counties) (Tan et al., 2002)

and thousands of people died of Keshan disease every year.

During the field investigations in Heilongjiang Province in 1968–1970, the interdisciplinary team visited several horse farms. It was found that all the horse farms located in the areas affected by Keshan disease suffered from an animal disease known as white muscle disease. At that time, white muscle disease had been recognized as a myopathy of livestock due to deficiency of a trace element, selenium, and supplements of Se to the livestock effectively prevented the disease (Muth et al., 1958; Muth, 1963). Given the understanding that soil Se mainly existed in the forms of selenate or selenite, two highly soluble salts, and hence could be easily leached from the soils, water samples were collected from the affected and unaffected villages in Keshan County for chemical analysis. The results indicated that Se contents in the samples from the affected areas were lower than those from the unaffected areas (<0.2 vs. >0.5 μg /L) (Tan et al., 1989). Based on the preliminary observations that (i) Keshan disease mainly occurred in the areas with strong leaching effect, (ii) Se could be leached from soils easily, and (iii) white muscle disease existed with Keshan disease in the affected areas, Se deficiency emerged as the most likely cause of Keshan disease. When this idea was presented in a workshop held in Shangzhi County in Heilongjiang in 1971, Se received great attention, especially from a group of nutritionists from the Chinese Academy of Medical Sciences. Prevention experiments were almost immediately launched to test the hypothesis in the same year.

Preventive experiments supplementing Se in the diet were conducted in most of the provinces affected by Keshan disease in the 1970s and 1980s. Selenium containing tablets or supplements were adopted to directly increase Se doses in the diets, or Se fertilizer was applied to the soil to increase the Se content in the food (Tan and Hu, 1989).

Selenium Deficiency and Kashin-Beck Disease (KDB)

Kashin-Beck disease is an endemic osteoarticular disorder. Skeletal remains indicate that the disorder goes

Fig. 5 A Kashin-Beck disease patient with a person of normal height

back at least to the 16th century and perhaps as far as the Bronze Age. In 1934, a Chinese scholar Zhang Fengshu found many patients with joint swelling living along the railway from Shenyang City to Tumenjiang City, northeast China. In 1935, Takamori, a Japanese doctor conducted research on those patients and proved that the disease was KDB.

KBD is characterized by chronic disabling degenerative osteoarthrosis affecting the peripheral joints and the spine with apoptosis of the hyaline cartilage tissues. Impairment of movement in the extremities is commonly followed by bone development disturbances such as shortened fingers and toes and in more extreme cases, dwarfism (Tan and Hu, 1989) (Fig. 5).

Etiological Research on KD and KBD

Although there is a great deal of evidence to support the theory that selenium deficiency induces KD and KBD, there are alternative theories. The evidence supporting selenium deficiency include the following: the areas in China where KD and KBD were prevalent, selenium is significantly low in soil, food crops, surface water, and groundwater; in the KD and KBD areas, white muscle disease, proved to be induced by selenium deficiency, is prevalent in animals and this disease can be prevented by using feed supplement of selenium; selenium contents in hair, serum, and urine of KD and KBD patients are significantly lower than the control group; and the incidence of KD and KBD in people can be greatly reduced by adding selenium to their diets.

Views questioning selenium deficiency inducing KD and KBD include the following: in the low-selenium areas, not all of the soil, plants, surface water, and groundwater have low contents of selenium, and not all hair, serum, and urine of the residents have lower selenium contents than those of adjacent areas; due to the relatively stable geological conditions in a rather long period, ratios of diseases induced by geological conditions should be stable in different years. However, the occurrence of both KD and KBD is much higher in some years than in others; although selenium supplementation was conducted in parts of the endemic areas, since 1987 the ratios of KD and KBD in all endemic areas have greatly decreased and the existence of selenium in geological environment has not changed during that period.

Based on the above-mentioned facts, multi-factor mechanisms for the inducement of KD and KBD have been formed: in the areas of selenium deficiency, residents cannot get enough selenium from foodstuff and drinking water, so the people are selenium deficient, with symptoms of low selenium contents in serum, urine, and hair; in the state of selenium deficiency, the subjects are invaded by different groups of biological factors, such as coxsackie virus or T-2 toxin produced by fusarium, and suffer from different diseases such as KD or KBD; cold, dampness, carbon monoxide poisoning, and poor nutritional status may be reasons for acute KD and KBD.

Past and Present Hazards of KD and KBD

KD has been prevalent in 309 counties or cities in 15 provinces in China, with a population of 76.7 million people. From 1959 to 1983, the average annual deaths due to KD were more than 1000. Based on clinical symptoms, KD is divided into acute KD, sub-acute

KD, chronic KD, and latent KD. Death is mainly caused by acute KD and sub-acute KD, and usually patients will die a half-day to 1 day after onset.

KD characteristically had outbreaks in certain years. In the years of 1959, 1964, and 1970, the incidence rates were two to three times the average. In the winter of 1943, there was an explosive KD outbreak in Long Town, Beian City, Heilongjiang Province, and about 450 people from more than 2,000 families died within 48 days, sometimes as many as 70–80 deaths a day.

There were 490 people living in Dingjiaweizi Village, Kedong County, Heilongjiang Province, where the epidemic KD occurred in 1944. By the end of 1949, the population had reduced to seven. Due to the prevalence of KD in Buteha County in Inner Mongolia in 1937, 14 out of total 24 families died. In Hebei Province, 68 out of 125 people living in Suojiayingzi Village, Guyuan County and 45 out of 51 people living Shangchang Village, Fengning County died from KD in 1945. KD risks are more serious to young women. In 1952, only a 45-year-old woman survived from KD in Xixin Village, Keshan County. Due to KD, four wives of a 60-year-old man and two wives of his brothers had died. Death from serious KD accounted for 46.6% of deaths in KD prevalent areas.

KBD has been prevalent in 31 million people in 7,840 villages of 297 counties in 14 provinces in China. There were 1.8 million KBD patients in China in 1990. The ratios of KBD are 20 and 50% for residents and adolescent of endemic areas, respectively. In some particular areas, the ratios can reach 100%. The major hazard of KBD is the loss of ability to work (Prevention of endemic diseases for forty years in China, 1990).

Since the reform and opening of China, tremendous changes have taken place in the social and economic status and nutritional and health status of the residents. Mobility of the population has greatly increased. Sources of food and drinking water have been more diverse and the dependence on local drinking water and food has been decreasing. Key factors leading to the prevalence of endemic diseases have changed and because of the government's financial support, KD and KBD in China are now basically under control. There are now 3,000 latent KD patients and 10,000 chronic KD patients in China and cases of death from KD have been very rare. According to 2008 data, there were 720,000 KBD patients in China of which 28,000 patients were under the age of 13. KBD usually occurs in the age of 7–8 years, and the symptoms may continue for life. Therefore, ratios in children under the age of 13 years old can better reflect the prevalence of the disease. During the past 20 years, KBD disease has been showing a decline and there are only a small number of new patients in the Tibet Autonomous Region, Qinghai Province, Gansu Province, and Shaanxi Province. KD and KBD may no longer be a serious public health problem in China (Sun, 2009).

Selenium Deficiencies and Cancer

Low selenium levels were also observed in biological samples such as hair, urine, blood, or serum from patients with various cancers, including liver cancer, lung caner, galactophore cancer, esophageal cancer, nasopharyngeal cancer, cervical cancer, colon cancer, gastric cancer, cardiovascular disease, diabetes mellitus, cataract and hearing loss, arsenic-associated skin lesions, and endemic goiter and cretinism. It indicates that some relationships exist between these diseases and selenium but that the low selenium levels may be a contributing factor though the exact causes are not clear. Even so, many epidemiological and clinical studies have demonstrated that dietary supplementation of selenium could reduce the prevalence of above diseases (Chen and Wong, 2009; Yu et al., 1990, 1997).

Selenosis in China

Intake of excessive selenium can cause selenosis in human and animals. Human selenosis in Western Hubei Province could be traced to 1923. A total of 477 cases of human selenosis were reported from 1923 to 1987 and 70% of the cases occurred between 1959 and 1963. Ninety percent of the patients were located in the towns of Shadi, Xintang, and Shuanghe in Enshi District and the other patients were distributed between Laiwu Town in Enshi, Nantan, and Houmen Towns in Badong County and Shatuo Town in Xuanen County (National Environmental Protection Agency, 1994).

Among them, Yutangba in Shuanghe Town suffered from the most serious endemic selenosis. Studies show that the endemic selenosis in Western Hubei Province was related to the occurrence of Se-enriched Permian carbonaceous strata (Yang et al., 1983). The selenium in soil, drinking water, crops, and vegetables all originated directly or indirectly from the seleniferous strata (Zheng et al., 1992). The average Se concentration in the local carbonaceous siliceous rocks and carbonaceous shales (known locally as "stone coal") is 143.9 mg/kg (Mou et al., 2007) and the maximum content is up to 84,123 mg/kg (Zhu et al., 2008) while the average content of selenium in world shale is only 0.06 mg/kg (Fordyce, 2005). In the Se-rich rocks, selenium mainly is present in native selenium and copper (iron) selenide assemblages including krutaite, klockmannite, mandarinoite, naumannite, Se-bearing pyrite, Se-bearing chalcopyrite, and unidentified Fe–Se minerals (Zhu et al., 2004a, 2004b).

Selenium was dispersed from the Se-enriched carbonaceous rocks through the food chain via complex biogeochemical cycling processes including weathering, rock–water interactions, and biological activities. In addition, in the endemic selenosis areas, human activities, such as stone coal conveyance by local villagers, mining of stone coal for use as a fuel or fertilizer, and discharging lime into cropland to improve soil, caused variable addition of Se to the soil and further accumulation of Se in the food chain.

Naore Village, Shuang'an Town in Daba Moutain, Ziyang County, Southern Shaanxi Province is the second region where endemic selenosis prevalence occurred (Cheng and Mei, 1980). More than ten cases of human selenosis and thousands swine selenosis cases have been reported in the region of Daba Mountain (Cheng and Mei, 1980).

Hair loss and nail loss were the prime symptoms of endemic selenosis, but disorders of the nervous system, skin, poor dental heath, garlic breath, and paralysis were also reported. Although no health investigations were carried out in the peak incidence years of 1961–1964 in Enshi District, subsequent studies in these areas carried out in the 1970s revealed very high dietary intakes of 3.2–6.8 mg/day with a range of selenium in the blood of 1.3–7.5 mg/L and hair selenium levels of 4.1–100 mg/kg (Tan and Hou, 1989; Yang et al., 1983). Due to decreasing dependence on locally grown foodstuffs in the diet, no human cases of selenium toxicity have been reported since 1987 in these areas but the local animals still frequently suffer hoof and hair loss as a result of the high environmental selenium (Fordyce, 2005).

Thallium Impacts on Human Health

Thallium is one of the most toxic metals and can cause chronic and acute poisoning (Zitko, 1975; Smith and Carson, 1977). It is widely distributed in the natural environment but is generally present in low concentrations (Fergusson, 1990; Taylor and McLennan, 1985). Thallium minerals are rare in nature, thus Tl is often excluded from the list of metals to be evaluated and the environmental impact related to the natural occurrence of Tl has received relatively little attention despite its high toxicity (Xiao et al., 2004a). However, a serious endemic thallotoxicosis incident occurred due to Tl mineralization in southwestern China's Guizhou Province.

The Tl mineralized area of Lanmuchang in Xingren County in southwest Guizhou Province, China, is known to have high Tl contents in rocks and sulfide ores. The local environment is highly enriched in Tl due to the presence of the Tl-bearing mineral lorandite ($TlAsS_2$) (Chen 1989a, b). In the 1960s–1970s, Lanmuchang, with a population of approximately 1,000 people, was affected by endemic Tl poisoning (APASSGP, 1974; Liu, 1983; Zhou and Liu, 1985). The poisoning, which affected a large portion of the population (over 400 cases) and caused symptoms such as hair loss and body-aches, was thought largely to be due to Tl contamination in drinking water and vegetables (Zhou and Liu, 1985; Long and Zhang, 1996). High concentrations of Tl in bedrocks/ores (6–35,000 mg/kg), aquatic system (0.005–1100 μg/L in groundwater and 0.07–31 μg/L in surface waters), soil layers (1.5–124 mg/kg), and food crops (1–500 mg/kg, dry weight) were observed in this area (Xiao et al., 2004b). Even though water piped in from outside of the area has been provided since the early 1990s, about 40% of the population still rely on the cheaper dug-wells or springs close to the Tl mineralized area and all rely on the natural water sources for agricultural and domestic uses. Therefore, the symptoms of chronic Tl intoxication still occurred in the area into the 2000s (Xiao et al., 2003; Xiao et al., 2007).

Previous studies demonstrated that Tl tended to accumulate in local food crops grown in Tl-polluted soils (Zhang et al., 1999; Xiao et al., 2004a). In the Lanmuchang Tl-polluted area, Tl showed the highest concentration in green cabbage, ranging from 120 to 495 mg/kg with an average of 338 mg/kg (Xiao et al.,

2004a). The high accumulation of Tl in the local food crops clearly implied that Tl in the soil can easily transfer to crops. The daily intake of 1.9 mg/day Tl from consumption of the food crops was estimated for the local adult inhabitant in Lanmuchang (Xiao et al., 2004b). This high ingestion rate of Tl is 1,000 times higher than the world average daily intake (2 μg/day) as indicated by Sabbioni et al. and also far above the element's "oral reference dose" of 0.056 mg/day (Sabbioni et al., 1984).

Medical Geology of Cancer in China

Unusually high mortality from esophageal cancer in the mountain areas in western Henan Province in Henan Province was 30.89 times that of Yunnan Province and 2.21 times that of China (Editorial Committee for the Atlas of Cancer mortality in People's Republic of China, 1979). Comparing the contents of various elements in rock, soil, water, and food in the region in Henan Province with the high ratio of esophageal cancer with those of a control region, it was found that levels of cadmium and beryllium were higher than the control area and the level of molybdenum was lower than that of the control area. However, subsequent studies confirmed that this difference was caused by different rock backgrounds and has no link to the high incidence of esophageal cancer. Esophageal cancer had been found in areas with coal containing more harmful volatile organic components, but later it was found that in the control region residents were using the same coal. Some researchers believed that high incidences of esophageal cancer were induced by the habits of eating sauerkraut and hot food, but could not confirm it.

Since the 1980s, due to the experiment of esophageal cancer prevention by a large-scale supplement of selenium and vitamins, mortality rates of esophageal cancer have dropped dramatically in the test area. But at the same time, as the reform and opening up of China and the improvement of nutrition and health status, mortality rates of esophageal cancer in other areas in China have also significantly decreased. Therefore, it cannot be proven that the deficiency of selenium and vitamins can induce high incidence of esophageal cancer.

The ratio of nasopharyngeal cancer in Guangdong Province has been 21.45 times that of Gansu Province and 3.79 times that of China. Nasopharyngeal cancer is very rare in other parts of the world, but the incidence rate in some areas in Guangdong, Guangxi, and Fujian Provinces has reached the high level of 20–50/100,000. This region happens to be a granite region rich in radioactive elements and Yangjiang City, Guangdong Province has the highest radioactive background in China. In this area, the distribution of high mortalities of nasopharyngeal cancer strongly matches the granite distribution and the distribution of high-level radioactive radon. Therefore some people thought that atmospheric radioactive particles based on the high-level radioactive background had led to high incidence of nasopharyngeal cancer. But in other areas with high-level radioactive backgrounds in the world, there is no high incidence of nasopharyngeal cancer, so this assumption could not be confirmed (Ho, 1978; Lo et al., 1996).

In accordance with the general law of the prevalence of lung cancer in the development stage of industrialization, the mortality of lung cancer in urban areas should be higher than that in rural areas and that in economically developed regions should be higher than that in undeveloped areas. But mortalities of lung cancer in tin miners in Gejiu County and farmers in Xuanwei County, Yunnan Province, has been very high. A study confirmed that mortality of lung cancer in tin miners was related with mine dusts and radon in the air of underground workplace, and high mortality of lung cancer in farmers in Xuanwei County was related with indoor heating and cooking without chimneys with coal containing high volatiles, which was also supported by the fact that the ratio of lung cancer of women was higher than men. It has been discovered recently that coal used in areas of high ratios of lung cancer contained high levels of silica particles, which were dispersed in the indoor air during the process of coal combustion, and long-term living in such indoor environment would induce high incidences of lung cancer (Large et al., 2009).

The relationship between selenium deficiency and various types of cancer has been of widespread concern in China. It is found that selenium levels in the blood of cancer patients are lower than that of control population. However, in selenium deficiency areas in China (endemic KD and KBD areas), there is no type of cancer with mortality higher than in other areas. This observation supports the view that the low intake of selenium does not cause cancer but in the development process of a cancer a large amount of selenium is consumed or the excretion of selenium is accelerated thus the lack of selenium in cancer patients is induced.

Generally, regional differences of cancer incidence, prevalence, and death rates do exist. The reasons for these differences are various, including variation in geological materials and geological process. However, determination of how geological materials and geological processes impacts the course of cancer would require a large number of detailed multi-disciplinary collaborative research projects.

Other Medical Geology Issues in China

Jiashi Disease

The main symptoms of Jiashi disease are extremely low birth rate in disease areas, diarrhea, hepatomegaly, hypoglycemia, low blood pressure, and seasonal low potassium contents in blood. The disease is prevalent in Jiashi County and Yuepuhu County in the western Tarim Basin in Xinjiang Autonomous Region, with an area of about 500 km^2. In the 1950s, there were more than 200 farmers living in the Halahuqi Farm and drinking local well water in the disease area and none of women of childbearing age had been pregnant. In 1975, they started to use drinking water from outside and by 1983 almost all of women of childbearing age had children (Wu et al., 1990).

The etiology of Jiashi disease is unclear and the two main hypotheses are "environmental geochemistry cause" and "gossypol cause." The environmental geochemistry cause is based on the observation that Jiashi disease is epidemic in a specific geochemical environment. The prevalence of Jiashi disease is strictly limited by landscapes and only occurs in the tail of the alluvial fan and the low-lying areas of Kezi River Alluvial Plain, where the land salinization is serious and drinking water is highly saline. This disease does not occur in the upper alluvial fan and high-lying areas of Kezi River Alluvial Plain, and areas of adjacent Yerqiang River flows parallel to Kezi River.

Analysis of well water in endemic areas of Jiashi disease showed that salinity, sulfate, magnesium, sodium, and calcium contents, pH, and Eh were significantly higher than non-endemic areas, but manganese and zinc contents were lower than non-endemic area. Salinity of the typical drinking water in the endemic areas was higher than 10 g/L. It was suggested that the special chemical composition of the drinking water led to the prevalence of Jiashi disease (Shi and Sun, 2008).

The gossypol cause is based on the fact that local residents had the habits of eating cottonseed oil which containing 0.42–0.97% of gossypol. Excessive intake of gossypol will poison male reproductive system and induce sperm reduction or complete disappearance. Therefore, women of childbearing age were unable to bear children. The hypothesis of gossypol cause was also supported by the fact that the ratios of Jiashi disease increased with the increase of gossypol intake.

However, no Jiashi disease occurred in other areas where people were also eating cottonseed oil in Xinjiang Autonomous Region. Since the 1980s, when the local people stopped eating cottonseed oil and drinking the local water, Jiashi disease has been despairing. But the real factor inducing this disease is still unknown (Wu et al., 1990).

Pazi Disease

Pazi disease is a chronic, multiple, and symmetrical arthropathy of bone and joint prevalent in Dayi, Guanghan, Neijiang, Zizhong, and Ziyang Counties in the western Sichuan Basin. Pazi disease and Kashin-Beck disease are similar in clinical symptoms, but different in pathological changes and population regions. Symptoms are reduced if patients emigrate from the endemic area. Healthy people can suffer from this disease if they move into the endemic area for as short as short as 1 year. Etiology of the disease is still unknown, and there are no targeted prevention and control measures (Zhu et al., 1992).

Respiratory Problems

Excess or deficiency of trace elements is just one aspect of medical geology. Exposure to mineral grains in airborne dust can also cause health problems. Exposure to silica, asbestos, pyrite, clays, and other minerals is known to cause various respiratory problems. Tian et al. (2008) and Large et al. (2008) have postulated that fine-grained silica mobilized by coal combustion may be the cause for the exceptionally high incidence of lung cancer in Xuan Wei County, Yunnan Province. Derbyshire (2005) has discussed the health impacts of silica-rich dust blowing down the Hexi corridor from northwest China to the populous east coast.

Some 440,000 coal miners in China are suffering from Coal Workers Pneumoconiosis (Black Lung Disease) and an estimated 140,000 minors have died of this problem since the 1950s (www.people.com.cn; March 18, 2005). Huang et al. (2004) have demonstrated a strong correlation between the incidence of black lung disease in the United States and the amount of pyrite in the coal. This finding indicates that reduced exposure to respirable pyrite should reduce the incidence of black lung disease.

References

Adolph WH, Ch'en Shen-Chao (1930) Iodine in nutrition in North China. Chin J Physiol 1(4):437–448.

APASSGP (Autonomous Prefecture Anti-epidemic Station of Southwest Guizhou Province, Environmental Geology Laboratory, Institute of Geochemistry, Chinese Academy of Sciences), 1974, Thallium enrichment in an ecological circulation – a case report of natural thallotoxicosis. Environ Health 2:12–15 (in Chinese).

Belkin HE, Zheng B, Zhou D, Finkelman RB (1997) Preliminary results on the Geochemistry and Mineralogy of Arsenic in Mineralized Coals from Endemic Arsenosis in Guizhou Province, P.R. China: Proceedings of the Fourteenth Annual International Pittsburgh Coal Conference and Workshop. CD-ROM p. 1–20.

Belkin HE, Zheng B, Zhou D, Finkelman RB (2008) Chronic arsenic poisoning from domestic combustion of coal in rural China: A case study of the relationship between earth materials and human health. In B de Vivo, HE Belkin, A Lima (ed) Environmental Geochemistry: Site Characterization, Data Analysis and Case Histories (Chapter 17, pp. 401–426). Amsterdam: Elsevier.

Chao Yuanfang (610) Zhu Bing Yuan Hou Lun.

Chen D (1989a) Discovery and research of lorandite in China. Acta Mineralogica Sinica 9:141–142 (in Chinese with English abstract).

Chen D (1989b) Discovery of Tl ore bodies associated with a mercury ore deposit and its mineralization mechanism. J Guizhou Inst Technol 18:1–19 (in Chinese with English abstract).

Chen KP, Wu HY, Wu TC (1962) Epidemiologic studies on blackfoot disease in Taiwan: III. Physicochemical characteristics of drinking water in endemic blackfoot disease areas. Memoirs College Med Natl Taiwan Univ 8:115–129.

Chen TF, Wong YS (2009) Selenocystine induces reactive oxygen species- mediated apoptosis in human cancer cells. Biomed Pharmacother 63:105–113.

Cheng JY Mei ZQ (1980) Initial investigation on selenosis region in Ziyang County, Shannxi Province. Shaanxi J Agri Sci 6:17–20 (in Chinese with English abstract).

Derbyshire E (2005) Natural aerosolic mineral dusts and human health. In Selinus O, Alloway BJ, Centeno JA, Finkelman RB, Fuge R, Lindh U, Smedley P (eds). (2005) Essentials of Medical Geology Impacts of the Natural Environment on Public Health, Elsevier Academic Press, 459–480.

Editorial Committee for the Atlas of Cancer mortality in People's Republic of China (1979) Atlas of Cancer mortality in the People's Republic of China. China Map Press. Shanghai, China (in Chinese with English abstract).

Fergusson JE (1990) The heavy elements: chemistry, environmental impact and health effects (614 p). Oxford: Pergamon Press.

Finkelman RB, Belkin HE, Zheng B (1999). Health impacts of domestic coal use in China. Proceedings National Academy of Science, USA. 96, p 3427–3431.

Finkelman RB, Orem W, Castranova V et al. (2002) Health impacts of coal and coal use: possible solutions. Int J Coal Geol 50(4):425–443.

Fordyce F (2005) Selenium deficiency and toxicity in the environment. In O Selinus, BJ Alloway, JA Centeno, RB Finkelman, R Fuge, U Lindh, Smedley P (eds) (2005) Essentials of Medical Geology Impacts of the Natural Environment on Public Health (pp. 373–415). Elsevier Academic Press.

Gao H (1990) First discovery of endemic arsenosis in Inner Mongolia Autonomous Region. Research on Endemic Disease Prevention in Inner Mongolia Autonomous Region 15 (4): 1(in Chinese).

Ge Hong (284–AD364) Zhou Hou Bei Ji Fang.

Ho JHC (1978) An epidemiologic and clinical study of nasopharyngeal carcinoma. Int J Radiat Oncol Biol Phys 4: 183–205.

Huang Di Nei Jing (475 BCE–AD 220) Author unknown.

Huang X, Li W, Attfield MD, Nadas A, Frenkel K, Finkelman RB (2004) Mapping and prediction of Coal Workers' Pneumoconiosis with bioavailable iron content in bituminous coals. Environ Health Perspect 113(8): 964–968.

Institute of Epidemiological Diseases in Xinjiang Uigur Autonomous Region (1984) Progress in research on endemic cretinism. The People's Press of Xinjiang Uigur Autonomous Region, Urumuchi.

Ji Kang (223–262) Yang Sheng Lun (Healthy Theory).

Jin Y, Liang C, He G, Cao J, Ma F, Wang H, Ying B, Ji R (2003) A Study on distribution of endemic arsenism in China. J Hyg Res 32(6):519–540 (in Chinese with English abstract).

Ke C (1980) Investigation report of arsenosis caused by domestic coal. J Environ Sci (in Chinese) 4:12–14.

Kuo TL (1964) Arsenic content of artesian well water in endemic area of chronic arsenic poisoning. Reports of Institute of Pathology, Taipei, Taiwan: National Taiwan University College of Medicine, 4;20:7–13.

Large, David J, Kelly, Shona, Spiro, Baruch, Tian, Linwei, Longyi, Shao, Finkelman, Robert, Zhang, Mingquan, Somerfield, Chris, Plint, Steve, Ali, Yasmin, Zhou, Yiping (2009) Silica-volatile interaction and the geological cause of the Xuan Wei lung cancer epidemic. Environ Sci Tech DOI: 10.1021/es902033j

Li R, Tan J, Wang W et al. (1982) Discussion on fluorosis induced by foodstuff in Guizhou Province. Chin J Med 62(7):425–428.

Li SJ, Li W, Hu X, Yang LS, Ruodeng X (2009) Soil selenium concentration and Kashin-Beck disease prevalence in Tibet, China. Front Environ Sci Eng China 3:62–68.

Li Y, Bai G, Zheng L et al. (2004) Investigation on the relation between coal pollution and endemic arsenosis in the south of Shaanxi Province. Chin J Endem Dis 23(6):262–265.

Liu An (180–123) BCE, Huai Nan Zi.

Liu J (1983) Report of 4 cases of neuroetinopathy by chronic thallium poisoning. Chin J Thaumat Occup Ophthal Dis 1: 22–23 (in Chinese).

Liu J, Zheng B, Aposhian HV et al. (2002) Chronic poisoning from burning high-arsenic-containing coal in Guizhou, China. Environ Health Persp 110(2):119–122.

Liu Z, Wu D, Chen K, Liu S, Li Z (1982) High fluoride salt induced endemic fluorosis. Chin J Dis Prev 16(4):204–208.

Lo KW, Cheung ST, Leung SF, et al. (1996) Hypermethylation of the *p16* gene in nasopharyngeal carcinoma. Cancer Res 56:2721–2725.

Lo MC, Hsen YC, Lin BK (1977) Arsenic content of underground water in Taiwan: Second report. Taichung, Taiwan: Taiwan Provincial Institute of Environmental Sanitation 1–21.

Long J, Zhang Z (1996) Studies on environmental effect of thallium in exploration resources in southwestern area of Guizhou. Conserv Utiliz Min Resour 3:47–49.

Lu FJ, Tsai MH, Ling KH (1978) Studies on fluorescent compounds in drinking water of areas endemic for blackfoot disease. 3. Isolation and identification of fluorescent compounds. J Formosan Med Assoc 77:68–76.

Lyth O (1946) Endemic fluorosis in Kweichow, China. The Lancet 16:233–235.

Lu Bu Wei etc. Circa 221 BCE, Lu Shi Chun Qiu.

Ministry of Health, P.R.China (2009) http://www.moh.gov.cn/publicfiles//business/htmlfiles/wsb/index.htm

Mou SH, Hu QT, Yan L (2007) Progress of researches on on endemic selenosis in Enshi District, Hubei Province. Chin J Pub Health 23(1):95–96 (in Chinese).

Muth OH (1963) White muscle disease, a selenium-responsive myopathy. J Am Vet Med Assoc 142:272–277.

Muth OH, Oldfield JE, Remmert LF et al. (1958) Effects of selenium and vitamin E on white muscle disease. Science 128:1090.

National Environmental Protection Agency, Chinese Academy of Sciences, National Education Committee (eds) (1994) The atlas of soil environmental background value in the people's Republic of China. Beijing: Chinese Environmental Science Press. (In Chinese).

Prevention of endemic diseases for forty years in China (1990). Beijing: Chinese Environmental Science Press.

Read BE How GK (1927) The iodine, arsenic, iron, calcium and sulphur content of Chinese medical algae. Chin J Physiol 1(2):99–108.

Sabbioni E, Ceotz L, Bignoli G (1984) Health and environmental implications of trace metals released from coal-fired power plants: an assessment study of the situation in the European Community. Sci Total Environ 40:141–154.

Shen H, Zhang S, Liu S et al. (2007) Study on the geographic distribution of national high water Iodine areas and the contours of water Iodine in high Iodine areas. Chin J Endemiol 26(6):658–661.

Shi T, Sun Y (2008) Analysis on the distribution of endemic diseases and the urgency of safe water supplies. West Explor Eng (5):82–84.

Smith IC, Carson BL (1977) Trace metals in the environment. Volume 1 – Thallium (p. 307). Michigan: Ann Arbor Science Publishers Inc.

Sokoloff L (1989) The history of Kashin-Beck disease. New York State J Med 89:343–351.

Stanbury JB (1996) Iodine deficiency and Iodine deficiency disorders. In EE Ziegler, LJ Filer (eds) Present knowledge in nutrition 7th edn. Washington D.C. USA. Wen Zhimei, Chen Junshi (Trans.) (1998) Modern Nutrition 7th edn. Beijing: People's Medical publishing House.

Standard of diagnosis for endemic arsenism WS/T 211-2001 http://www.moh.gov.cn/cmsresources/zwgkzt/wsbz/dfbbz/dfb/dfbwsbz/nr/149.gif

Sun B, Nie G, Zhang Y, Mai Z (1984) Investigation report of a chronic arsenosis. Collection of data of the Sanitation and Anti-epidemic Station of Guizhou Province (in Chinese) 4, 284–289.

Sun D (2009) Prospect of endemic disease control in China. Chin J Endemiol 28(1):3–6.

Sun D, Zhao X, Chen X (2005) Investigation on endemic fluorosis in China. Beijing: People's Health Press.

Sun Simiao (581–682) Qian Jin Yao Fang.

Tan JA et al. (eds) (1989) The Atlas of Endemic Diseases and Their Environments in the People's Republic of China. Beijing: Science Press.

Tan JA, Hou S (1989) Environmental selenium and health problems in China, In J Tan et al. (eds) Environmental Selenium and Health. Beijing: People Health Press.

Tan JA, Zhu WY, Wang WY et al. (2002) Selenium in soil and endemic diseases in China. Sci Total Environ 284:227–235.

Taylor SR, McLennan SM (1985) The continental crust; its composition and evolution; an examination of the geochemical record preserved in sedimentary rocks (312 p.). Oxford: Blackwell Scientific Publishing.

Tian LW, Dai S F, Wang JF, Huang YC, Ho SC, Zhou Y, Lucas D, Koshland CP (2008) Nanoquartz in Late Permian C1 coal and the high incidence of female lung cancer in the Pearl River Origin area: a retrospective cohort study. BMC Public Health, 8, doi:10.1186/1471-2458-8-398.

Tseng CH (2002) An overview on peripheral vascular disease in blackfoot disease hyperendemic villages in Taiwan. Angiology 53:529–537.

Tseng C-H (2005) Black foot disease and arsenic: A never-ending story. J Environ Sci Health 23:55–74.

Tseng CH, Chong CK, Chen CJ, Tai TY (1996) Dose-response relationship between peripheral vascular disease and ingested inorganic arsenic among residents in blackfoot disease endemic villages in Taiwan. Atherosclerosis 20:125–133.

Tseng CH, Chong CK, Chen CJ, Tai TY (1997) Lipid profile and peripheral vascular disease in arseniasis-hyperendemic villages in Taiwan. Angiology 48:321–335.

Tseng WP (1989) Blackfoot disease in Taiwan: A 30-year follow-up study. Angiology 40:547–558.

Wang B, Zheng B, Zhai C, Yu G, Liu X (2004a) Relationship between Fluorine in drinking water and dental health of residents in some large cities in China. Environ Int 30: 1067–1073.

Wang B, Finkelman RB, Belkin HE, Palmer CA (2004b) A possible health benefit of coal combustion. Abstracts of the 21st Annual Meeting of the Society for Organic Petrology, Vol. 21, p. 196–198.

Wang B, Zheng B, Wang H, Pinyakun, Tao Y (2005) Dental caries in fluorine exposure areas in China. Environ Geochem Health 27(4):285–288.

Wang L, Liu H, Xu X et al. (1982) Investigation report on endemic chronic arsenosis areas. Prev Cont Res Bull 1(1):1–7 (in Chinese).

Wang M, Zheng B, Wang B, Li S, Wu D, Hu J (2006) Arsenic concentrations in Chinese coals. Sci Total Environ 357: 96–102.

Wang ZJ, Gao YX (2001) Biogeochemical cycling of selenium in Chinese environments. Appl Geochem 16:1345–1351.

WHO (1996a) Guidelines for drinking water quality. Health Criteria and Other Supporting Information. Geneva: World Health Organization.

WHO (1996b) Trace Elements in Human Nutrition and Health. Geneva: World Health Organization.

Wu D, Zheng B, Wang A (2004) New understanding on the source of coal combustion type fluorosis in Guizhou Province. China J Endemiol 23(2):135–137.

Wu J, Deng G, Wang L et al. (1990) Discussions on etiology of Jiashi disease. J Environ Health 7(2):87–88 (in Chinese with English abstract).

Xiao T, Boyle D, Guha J, Rouleau A, Hong Y, Zheng B (2003) Groundwater-related thallium transfer processes and their impacts on the ecosystem: Southwest Guizhou Province, China. Appl Geochem 18:675–691.

Xiao T, Guha J, Boyle D, Liu CQ, Chen J (2004a) Environmental concerns related to high thallium levels in soils and thallium uptake by plants in southwest Guizhou, China. Sci Total Environ 318:223–244.

Xiao T, Guha J, Boyle D, Liu CQ, Zheng B, Wilson GC, Rouleau A, Chen J (2004b) Naturally occurring thallium: a hidden geoenvironmental health hazard? Environ Int 30:501–507.

Xiao X, Chen T, Liao, Xiaoyong, Wu B, Yan X, Zhai L, Xie H, Wang L (2008) Regional distribution of Arsenic contained minerals and Arsenic pollution in China. Geograph Res 27(1):201–212 (in Chinese).

Xiao T, Guha J, Boyle D, Liu CQ, Zheng B, Wilson GC, Rouleau A, Ning Z, He L (2007) Potential health risk in areas of high natural concentrations of thallium and importance of urine screening. Appl Geochem 22:919–929.

Yang G Wang, S Zhou R et al. (1983) Endemic selenium intoxication of humans in China. Am J Clin Nutr 37:872–881.

Yeh S, How SW (1963) A pathological study on the blackfoot disease in Taiwan. Reports, Institute of Pathology, National Taiwan University, 14:25–73.

Yu G, Sun D, Zheng Y (2007) Health effects of exposure to natural arsenic in groundwater and coal in China: An overview of occurrence. Environ Health Persp 115(4):636–642.

Yu SY, Mao BL, Xiao P et al. (1990) Intervention trial with selenium for the prevention of lung cancer among tin miners in Yunnan, China a pilot study. Biolog Trace Elem Res 24:105–108.

Yu SY, Zhu YJ, Li WG (1997) Protective role of selenium against hepatitis B virus and primary liver cancer in Qidong. Biolog Trace Elem Res 56:117–124.

Zeng W, Hou S, Ye S (1995) Skin changes by chronic arsenosis in 395 patients with black foot disease Research report from the Institute of Pathology, Taiwan National University. 16:33–40 (in Chinese).

Zhang Hua (232–300) Bo Wu Zhi.

Zhang Y (1986) Reasons for endemic fluorosis in areas supplied with salt from Yushan salt Factory. Chin J Endemiol 5(1): 72–73.

Zhang Y (1992) Research report on coal combustion type fluorosis in Fuling area. Chin J Endemiol 11(3):179–182.

Zhang Zihe circa (1156–1228) Ru Men Shi Qin.

Zhang Z, Chen G, Zheng BG, Chen YC, Hu J (1999) Contaminative markers of TI-mining district from the high contents of Tl, Hg and As of urine hair and nail in the villagers. China Environ Sci 19:481–484.

Zheng B (1992) Research on endemic fluorosis and industrial fluorine pollution (pp. 151–194). Beijing: Chinese Environmental Science Press.

Zheng B, Hong Y, Zhao W et al. (1992) The Se-rich carbonaceous siliceous rock and endemic selenosis in south-west Hubei, China. Chinese Sci Bull 37:1725–1729.

Zheng B, Yu X, Zhang J, Zhou D (1996) Environmental geochemistry of coal and endemic arsenism in Southwest Guizhou, China. Abstracts of 30th International Geological Congress Beijing, China. p. 410.

Zheng B, Zhang J, Ding Z, Yu X, Wang A, Zhou D, Mao D, Su H (1999) Issues of health and disease relating to coal use in southwestern China. Int J Coal Geol 40: 119–132.

Zheng, B, Wang B, Ding Z, Zhou D, Zhou Y, Zhou C, Chen C, Finkelman RB (2005) Endemic arsenosis caused by indoor combustion of high-As coal in Guizhou Province, P.R. China, Environ Geochem Health 27:521–528.

Zhou DX, Liu DN (1985) Chronic thallium poisoning in a rural area of Guizhou Province. China. J Environ Health 48:14–18.

Zhou D, Liu D, Zhu S, Li B, Jin D, Zhou Y, Lu X, Ha X, Zhou C (1993) Investigation of chronic arsenism caused by pollution of high arsenic coal. J Chin Prev Med (in Chinese with English abstract) 27:147–150.

Zhu JM, Li SH, Zuo W et al. (2004a) Modes of occurrence of selenium in black Se-rich rocks of Yutangba. Geochimica 33:634–640 (in Chinese with English abstract).

Zhu JM, Zuo W, Liang XB et al. (2004b) Occurrence of native selenium in Yutangba and its environmental implications. Appl Geochem 19:461–467.

Zhu JM, Wang N, Li SH et al. (2008) Distribution and transport of selenium in Yutangba, China: Impact of human activities. Sci Total Environ 392:252–261.

Zhu Yuhui et al. (1992) Endemic Disease, China Medical Encyclopedia. Shanghai: Shanghai Science and Technology Press.

Zitko V (1975) Toxicity and pollution potential of thallium. Sci Total Environ 4:185–192.

Medical Geology in Japan

Hisashi Nirei, Kunio Furuno, and Takashi Kusuda

Abstract The relationship between medical treatment and geology in Japan has an old history. However, the distribution of man-made strata deserves special mentioning in the geology of the Japanese Archipelago. The population density of Japan is high, and the man-made strata abruptly increased due to consequences of high economic growth, mass production, and mass consumption. Therefore this chapter describes the background and the different methods used in solving environmental problems linked to man-made pollution, the geological environment, and health in Japan.

Geo-pollution in Japan is an anthropogenic subterranean pollution. The geo-pollution does not mean "chemical hazards" caused by geological factors, that is, the distribution of natural elements and constituents and processes. Pollutants from geo-pollution sites are often harmful for health. Therefore geo-pollution is an integrated part of medical geology, studying the effects of our natural environment including the anthropogenic impact and processes in the natural environment.

Also hot springs are included in the concept. Heavy metals and other constituents in hot spring waters can be very useful for health but sometimes also toxic for human health. Japan's hot and cold springs have been utilized for bathing from ancient times mainly for medical and convalescent purposes. The unique methodologies for using springs have taken root as part of Japanese culture. The word *onsen* is known worldwide as a term that generally describes the culture originating from the unique utilization of hot and cold springs in Japan. The different properties of hot springs are described in this chapter.

Keywords Japan · Hot springs · Kogai · Geo-pollution · Balneotherapy · Cadmium · Mercury · Man-made strata

General Introduction

The relationship between medical treatment and geology in Japan has a surprisingly old history. Japan is an island arc and each of the geological factors that created the current chain of volcanic islands of Japan has special characteristics. Accordingly, each geological age created deposits such as metal deposits, petroleum, and coal. Mountain ranges formed around the series of volcanoes with altitudes of 2,000 m, reaching a height above sea level of 3,000 m at the center of the archipelago. Inland basins formed among the mountain ranges. Volcanic ejecta covered the volcanoes. Coastal plains spread along the coastlines during the Holocene. Around the coastal plains and inland basins, multiple terraces and terrace deposits formed as the islands were uplifted gradually from the Middle Pleistocene.

The distribution of man-made strata deserves special mentioning in the geology of the Japanese Archipelago. The population density of Japan is high, and the man-made strata abruptly increased due to consequences of high economic growth, mass production, and mass consumption. The space for factory and homes were developed into not only plateaus but land

H. Nirei (✉)
Japan Branch of the International Medical Geology Association, 1277, Kamauchiya, Motoyahagi, Katori City, Chiba Prefecture, 287-0025 Japan
e-mail: nireihisashi@msn.com

O. Selinus et al. (eds.), *Medical Geology*, International Year of Planet Earth,
DOI 10.1007/978-90-481-3430-4_12, © Springer Science+Business Media B.V. 2010

where filled into the ocean to provide waste disposal sites and surplus soil dump sites.

The lives of the Japanese people have been affected by many elements from the natural resources of the Japanese Archipelago in metal mines, oil fields, natural gas fields, hot springs, and groundwater. To discuss Japan's medical geology, one cannot ignore the diverse geological origins of the Japanese islands and nor the enormous human impact on the geological environment (Minato et al., 1977).

Two major divisions can be made when considering problems relating to geology and health in Japan. One division is the benefits and harmful effects on organisms from natural geology, starting with the health of people. The other division is geo-pollution of the geological environment caused by the pollution from human activities.

After the collapse of the Tokugawa shogunate (1867–1868) and the opening of Japan to the world, the new Meiji government adopted the strategy *fukoku kyohei* ("enrich the country, strengthen the military") to catch up on the Western European powers, including accelerated mine development and smelting. As a result, pollution increased. The development of the Ashio copper mine is an example. Modern elementary school textbooks portray the actions of the national assemblyman Shozo Tanaka who directly petitioned the Meiji emperor to protest the health afflictions of many farmers due to pollution despite being subject to absolute imperial rule. Such pollution by primary industries continued through the end of World War II. Although health afflictions also occurred prior to and during the war from the manufacture of gas bombs, no protests could be made under the imperial military regime.

After World War II, Japan's standard of living improved from 1955 to 1975 due to high economic growth and income doubling. On the other hand, chemical hazards accompanying mass production and mass consumption endangered the lives of humans. Pollution was the typical hazard. In particular, court battles among victims, defendants, and the government have continued for many years in regard to the four major diseases caused by pollution in Japan resulting from factory wastewater, mine wastewater, and toxic discharges from industrial complexes that caused many deaths and health victims. Although the government reports continuous progress toward solutions, new judicial decisions reveal unresolved issues.

Further, health afflictions due to poisons released during weapon manufacturing and industrial operations were not taken seriously prior to and during the war. Under military rule there was no consideration regarding pollution of the geological environment. Toxic substances of that period and toxic substances discharged from factories during the high economic growth period directly after the end of World War II still remain in the environment today.

Two major environmental problems appeared in the wake of efforts to implement anti-pollution measures. First, when the discharge of toxic substances into the atmosphere, ocean, and rivers by factories such as coastal industrial complexes were regulated, the toxic substances were buried in fields and mountains as industrial waste. Toxic substances reside in Japan's fields and mountains like time bombs.

Second, the mode of discharge of pollution has changed as economic growth and concentrated populations in cities have driven changes in land utilization. Stock raising, which causes nitrate–nitrogen contamination of groundwater and offensive odors, was forced by growing cities to move to remote locations. Drainage from livestock facilities was then administratively prohibited based on fears of increased water pollution in rivers, lakes, and marshes. Inevitably, subsurface disposal was carried out. Groundwater pollution by nitrate–nitrogen is occurring in many locations.

After the high economic growth period during which the surface discharge by factories producing toxic substances was deterred, the next generation of factories grouped in groundwater recharge regions such as inland plateaus and hilly regions to form leading-edge industries revolving around semiconductor manufacturing. Then, local governments began permitting subsurface disposal of factory wastewater in the competition to attract semiconductor manufacturing and other enterprises. Volatile organic compounds (VOCs) such as tetrachloroethylene (PCE) and trichloroethylene (TCE), indispensable as cleaning solutions for semiconductors, were disposed with factory wastewater by subsurface disposal in groundwater recharge regions such as plateaus and hilly regions. Volatile organic compounds such as tetrachloroethylene and trichloroethylene have now been detected in groundwater around inland industrial parks.

It has become important to investigate and clean geological strata contaminated by volatile organic compounds to provide clean groundwater resources and prevent health afflictions resulting from breathing polluted ground air created when PCB, etc., in persistent organic pollutants (POPs) and volatile organic compounds (VOCs) in the polluted strata ascend to the ground surface. Medical geology (Selinus et al., 2005) also includes Japan's radon hot springs and hydrogen sulfide hot springs formed by the chain of volcanic islands. Although both substances in their gaseous forms are harmful to human bodies, radon hot springs are popular with terminal-stage cancer patients. Such effects are also topics of medical geology.

The concept of geo-pollution science and the geo-stratigraphic unit investigation methods has been established from the necessity to scientifically elucidate pollution hazards of the geological environment and from a viewpoint of environmental resources where a polluted environment is reclaimed to add value. This scientific concept is being adapted to the investigations of the pollution of soils and groundwater, starting with the geo-pollution occurring in each age and the geo-pollution from waste disposal sites.

Although investigations of geo-pollution have contributed to the development of environmental businesses, negative contributions have also appeared because pollution problems became a major social problem when toxic substances caused serious health effects. Accordingly, analysis of the toxic substances causing health effects is regarded as important. However, analysis measurements are being relied upon excessively, and rumors are occurring relating to the extent of pollution. One aspect of such rumors is the encouragement of pernicious environmental businesses.

The determination of locations for sample collection and the methods for collecting samples during investigations fall short of clarifying the actual state of the pollution. That is, collection locations are mesh intersections of several meters. Further, methods for collecting analysis samples in the depth direction do so without regard to the behavior and state of existence of diverse contaminants. Uniform sample collection methods with such an analysis measurement mentality have been used by the central government, therefore indicating trends of pollution diffusing and moving into deeper layers at geo-pollution sites nationwide.

Hot springs represent an important benefit of natural geology related to human health deserving special mention. The Japanese have effectively utilized hot springs from ancient times for treatment, convalescence, and recreation. Therapeutic hot springs for treatment include both bathing and drinking.

Medical effects of therapeutic hot springs differ by the spring quality type determined by chemical components in the water. Therapeutic springs provide lasting bodily warmth retention effects and health maintenance effects. Generally, the temperature of the hot spring water affects whether the hot spring is used for convalescence or recreation. Simple springs having high temperatures and clarity are preferred at large leisure resorts with hot spring facilities. Men and women have used public baths from ancient ages with customarily mixed-gender bathing, and that custom continues today. Monkeys are also known to enjoy hot springs. Hot springs are utilized also for treatment of racing horses. The existence of the Japanese Archipelago as a varying entity positioned on the continental margin has provided various metal deposits from ancient geological ages. From ancient times, *daimyo* lords seeking power in each province needed military capital to assert supremacy, driving the development of metal deposits starting with gold. For a time, Japan was world renown for its gold production, and even Marco Polo describes a country of gold, Cipangu, in the travels of Marco Polo. However, health afflictions due to various heavy metals referred to as mineral pollution also have been known from ancient times as side effects of metal resource development.

Loess is carried from the Asian continent by prevailing westerlies to settle on the Japanese Archipelago lying in an arc on the eastern edge of the Asian continent. Sedimentary strata of loess are confirmed in the terraces of the Japanese Islands, and this phenomenon of loess reaching Japan from the continent has continued since geological ages. In recent years, the industrialization accompanying the economic development of east Asian countries is producing combustion soot, NO_X, SO_X, and the like. These pollutants are carried to the Japanese Archipelago to afflict human health and obstruct the growth of trees.

The geological origins and the human activities achieving abrupt economic development in the volcanic island arc forming the Japanese Archipelago described above provide many medical geological research topics centered on the harmful effects of geo-pollution and the benefits of hot springs.

Concept of Geo-pollution

Geo-pollution is an anthropogenic subterranean pollution. The geo-pollution does not mean "chemical hazards" caused by geological factors, that is, the distribution of natural elements and constituents and processes. Pollutants from geo-pollution sites are often harmful for health. Therefore geo-pollution is an integrated part of medical geology, studying the effects of our natural environment including the anthropogenic impact and processes in the natural environment. Also hot springs are included in the concept. Heavy metals and other constituents in hot spring waters can be very useful for health but sometimes also toxic for human health.

Compound Pollution from Three Different Types of Pollution in the Geological Environment

"Geo-pollution" in Japan is a compound subterranean (underground) pollution comprising three types of underground pollution: "strata pollution," e.g., top soil pollution, "groundwater pollution," and "ground air/gas pollution." These three types have a complex interaction with each other. Most geo-pollution cases are not only contained within surface strata such as man-made strata or top soil layers, but also the pollutants usually spread into deeper layers. Consequently, an examination and cleanup of a single medium is not sufficient to solve the problem (Fig. 1) (Nirei et al., 1994).

Strata Pollution

Strata pollution typically involves heavy metals (e.g., lead, cadmium, hexavalent chromium, arsenic), volatile organic compounds (VOCs, such as tetrachloroethylene (PCE), and trichloroethylene (TCE)), agricultural chemicals, and dioxins. Heavy metal pollution is often found in former sites of chemical factories or plating factories in urban areas and near waste disposal sites or old mines in suburbs. Organochlorinated substances once used in large quantity as a degreasing solvent and cleaner by semiconductor factories, mechanical manufacturing, and metal processing are found in the areas near those facilities all over Japan. Pollution with agricultural chemicals is common in golf courses as well as in the areas adjacent to farm land. These chemicals tend not to stay in the surface layers including soil layers and regularly seep into groundwater in aquifer. Dioxin pollution occurs in the strata under and around waste incineration facilities.

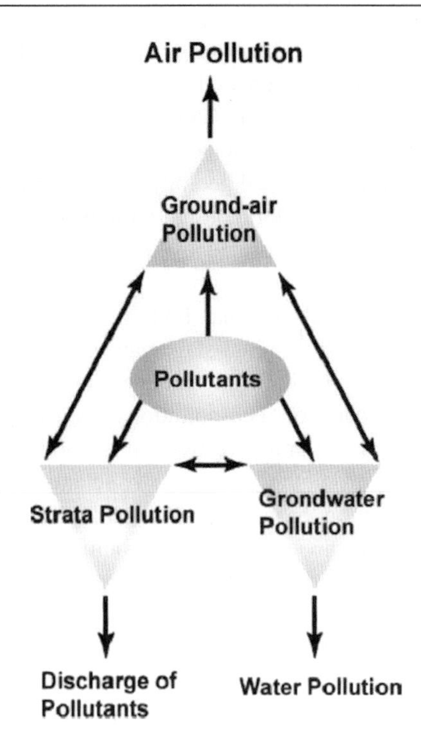

Fig. 1 Schematic model of geo-pollution

Groundwater Pollution

Groundwater pollution threatens one of the most important water sources in our country. As is the case of soil layer pollution, typical pollutants include heavy metals, organochlorinated substances, and agricultural chemicals. There are many incidents of elevated nitrate–nitrogen levels in groundwater all over Japan, caused by excessive fertilization and drainage from livestock farming. Compared to other underground mediums, groundwater flows and moves the contaminants much faster, spreading the pollution to wider areas within a short period of time. Accordingly, groundwater pollution requires prompt determination of pollution mechanisms to develop containment and cleanup plans in time.

Ground-Air Pollution

Ground-air pollution occurs when pollutants such as organochlorines, hydrocarbons, and methyl mercury are volatilized and released from polluted surface layers and groundwater, and diffuse into ground air inside soil particles or in underground openings. A severe case of ground-air pollution can even cause atmospheric pollution. In other cases, surface crop was damaged or killed due to ground air pollution with methane released from landfills and natural gas field.

Two Types of Investigation Methods in Japan

Geo-stratigraphic Unit Investigation Method

This method investigates the geo-stratigraphic unit, chronological unit, and material unit where the pollution originates (pollution source unit containing pollutants). In order to determine the pollution source unit and geo-pollution mechanism, this method investigates the physical property of the pollutants as well as the relationship between chronological unit and material unit. When taking samples for analysis, the geo-stratigraphic unit investigation method pays close attention to the unit boundaries as well as sedimentary structure within the geo-stratigraphic unit. This method elucidates the interrelation between pollutants and the geo-stratigraphic unit and can ultimately determine the pollution source and strata pollution mechanisms (Nirei et al., 2008).

A Unit-Independent Investigation Method

A unit-independent investigation method (e.g., adopted in Japanese Soil Contamination Countermeasure Act) takes analytical samples based on a predefined depth (such as 5 cm, 0.5 m, 1 m, 2 m,..., 10 m), without reference to pollution source units or unit boundaries. This type of examination cannot determine the geo-pollution mechanisms and sometimes even fails to locate the pollution source. Furthermore, the sampling and analysis methods are regulated and specified by each manual, which does not take site-specific circumstances into account. As a result, there are more and more cases of secondary pollution where boring for sample extraction causes further spreading of the pollution.

Important Role of Geo-Stratigraphic Unit Investigation Method

Careful definition and description of the sequence of deposition of man-made strata, alongside types of materials, help to establish boundaries between units that are crucial to the penetration of geo-pollutants and planning of strategies for investigation and remediation, i.e., geo-stratigraphic unit investigation method of man-made strata on geo-pollution site. The method can prevent such incidents on investigation and remediation.

The Diagnostic Standard for Cleanup on the Geo-Pollution Sites

We conduct the assessment based on our Geo-pollution Remediation Evaluation Standard. The standard rates the progress of remediation from level 0 (no examination of geo-pollution conducted) to level 6 (complete removal of pollution) (Table 1).

Each evaluation standard uses the environmental quality standards for health in Japan and the geo-stratigraphic unit investigation method.

Environmental Quality Standards for Health in Japan

From the viewpoint of geo-pollution science in medical geology, we can give an explanation about environmental quality standards for health. The environmental quality standards are enacted by the Basic Environment Act. In other words, the environmental quality standards for air, environmental quality

Table 1 Geo-pollution Remediation Evaluation Standard (defined by the geo-pollution control agency, Japan)

Eco-Level	Phase	Status	Description
Black	0	Pre-investigation stage	No investigation activities to determine if there is pollution
Red	1	Identification of pollution	Investigation is conducted to determine the existence of pollution
Orange	2	Investigation of pollution mechanism	Investigation is in progress to determine pollution mechanism
Yellow	3	Determination of pollution mechanism	Pollution mechanism is determined and verified with data
Bronze	4	Remediation in progress	Pollution mechanism is determined, but pollutants are found either in soil, groundwater, or ground air or in higher concentration than specified by the national environmental quality standards
Silver	5	Partial completion of remediation	Pollutants are still present, but their concentration is reduced to the level that meets the national environmental quality standards
Gold	6	Complete removal of pollutants	Pollutants are completely removed from the site

standards for water including groundwater, environmental quality standards for soil pollution, and environmental quality standards for noise are established as the standards to be appropriately followed to protect human health and conserve the living environment. In particular, what needs attention with respect to strata pollution that constitutes the concept of geo-pollution is that there are two sets of quality standards; one is the environmental quality standards for soil as per the Basic Environment Law versus the target substances or quality standards as per the Soil Contamination Countermeasures Act.

Each environmental quality standard may be based on drinking water quality standards, except for environmental quality standards for air quality and environmental quality standards for noise.

Drinking Water Quality Standards (DWQS)

The fundamental principle of the current DWQS is based on the standards that were revised in response to the amendment to the guidelines for drinking water quality of the World Health Organization (WHO), in May 2003, about 10 years after the previous amendment. Legally binding standards are shown in Table 2.

Table 2 Drinking water quality standards (mainly inorganic constituents) (April 2009) (by Water Supply Law)

Items	Standard values
Common bacteria	100 per 1 mL
E. coli	Not to be detected
Cadmium	0.01 mg/L
Mercury	0.0005 mg/L
Selenium	0.01 mg/L
Lead	0.01 mg/L
Arsenic	0.01 mg/L
Chromium (VI)	0.05 mg/L
Cyanide ion and cyanogens chloride	0.01 mg/L as cyanide
Nitrate and nitrite	10 mg/L as nitrogen
Fluoride	0.8 mg/L
Boron	1.0 mg/L
Zinc	1.0 mg/L
Aluminum	0.2 mg/L
Iron	0.3 mg/L
Copper	1.0 mg/L
Sodium	200 mg/L
Manganese	0.05 mg/L
Chloride ion	200 mg/L
Calcium, magnesium (hardness)	300 mg/L
Total residue	500 mg/L
Anionic surface active agent	0.2 mg/L
pH	Value 5.8 ~ 8.6
Taste	Not abnormal
Odor	Not abnormal
Color	5 degree
Turbidity	2 degree

Establishment of Environmental Quality Standards for Groundwater

Environmental quality standards (EQS) for groundwater pollution, e.g., by VOCs, were issued in 1997. EQS for groundwater are required to be maintained in order to protect human health. As a result of additions made in February 1999, EQS for ground water now regulate 26 substances including cadmium, etc. EQS for groundwater on geo-pollution science in medical geology are based on EQS for groundwater established by Ministry of the Environment Government of Japan in case of Japan (Table 3).

Environmental Quality Standards for Soil

The criteria for environmental quality of soil include the environmental quality standards for soil pollution as per the Basic Environment Law (Table 4) and the target substances and quality standards as per the Soil Contamination Countermeasures Act (Table 5). The former apply to all the elements in the ground. The latter are used for pollution investigation of former factory sites to which the Soil Contamination Countermeasures Act applies and accordingly are obligated to be investigated for pollution.

The above standards are not applicable to: (1) places where natural toxic substances exist such as the vicinities of mineral veins; (2) places designated for storage of toxic materials such as waste disposal sites by Soil Contamination Countermeasure Act of Japan.

Table 3 Establishment of environmental quality standards (mainly inorganic constituents) for groundwater (by Basic Environmental Law)

Items	Standard values
Cadmium	0.01 mg/L or less
Lead	0.01 mg/L or less
Chromium (VI)	0.05 mg/L or less
Arsenic	0.01 mg/L or less
Total mercury	0.0005 mg/L or less
Alkyl mercury	In nondetectable amounts
Selenium	0.01 mg/L or less
Nitrate–nitrogen and nitrite–nitrogen	10 mg/L or less
Fluoride	0.8 mg/L or less
Boron	1 mg/L or less

Table 4 Environmental quality standards (mainly inorganic constituents) for soil (by Basic Environment Law)

Substance	Target level of soil quality examined through leaching test and content test
Cadmium and its compounds	0.01 mg/L in sample solution and less than 1 mg/kg in rice for agricultural land
Lead and its compounds	0.01 mg/L or less in sample solution
Hexavalent chromium compounds	0.05 mg/L or less in sample solution
Arsenic and its compounds	0.01 mg/L or less in sample solution, and less than 15 mg/kg in soil for agricultural land (paddy field only)
Total mercury and its compounds	0.0005 mg/L or less in sample solution
Alkyl mercury	Not detectable in sample solution
PCBs	Not detectable in sample solution
Copper	Less than 125 mg/kg in soil for agricultural land (paddy field only)
Fluorine and its compounds	0.8 mg/L or less in sample solution
Boron and its compounds	1 mg/L or less in sample solution
Selenium and its compounds	0.01 mg/L or less in sample solution

Environmental Quality Standards for Air Quality

Japan has three kinds of environmental quality standards for air quality as follows: environmental quality standards (Table 6), environmental quality standards for benzene, trichloroethylene, tetrachloroethylene, and dichloromethane, and environmental quality standards for dioxins. Environmental quality standards for ground air pollution on geo-pollution science in medical geology are based on the environmental quality standards for air quality of Japan. The organic

Table 5 Target substances and quality standards (mainly inorganic constituents) for soil (by Soil Contamination Countermeasures Act)

Designated hazardous substances (Article 2 of the Act)		Designation standard (Article 5 of the Act): Soil Concentration Standard (risk for direct ingestion) (leaching method by 1 mol/L HCL)	Designation standard (Article 5 of the Act): Soil Leachate Standard (risk of ingestion from groundwater, etc.)
Cadmium and its compounds	Class 2 (heavy metals, etc.)	150 mg/kg	≤0.01 mg/L
Hexavalent chromium compounds		250 mg/kg	≤0.05 mg/L
Total mercury and its compounds		15 mg/kg	≤0.0005 mg/L
Alkyl mercury			Less than detection limit
Selenium and its compounds		150 mg/kg	0.01 mg/L
Lead and its compounds		150 mg/kg	0.01 mg/L
Arsenic and its compounds		150 mg/kg	0.01 mg/L
Fluorine and its compounds		4,000 mg/kg	0.8 mg/L
Boron and its compounds		4,000 mg/kg	1 mg/L
PCB	Class 3 (agrochemicals and PCBs)		Less than detection limit
Organic phosphorus compounds			Less than detection limit

Table 6 Environmental quality standards for air (by Basic Environmental Law)

Substance	Environmental conditions	Measuring method
Sulfur dioxide	The daily average for hourly values shall not exceed 0.04 ppm, and hourly values shall not exceed 0.1 ppm (notification on May 16, 1973)	Conductometric method or ultraviolet fluorescence method
Carbon monoxide	The daily average for hourly values shall not exceed 10 ppm, and average of hourly values for any consecutive 8-hour period shall not exceed 20 ppm (notification on May 8, 1973)	Nondispersive infrared analyzer method
Suspended particulate matter	The daily average for hourly values shall not exceed 0.10 mg/m^3, and hourly values shall not exceed 0.20 mg/m^3 (notification on May 8, 1973)	Weight concentration measuring methods based on filtration collection, or light scattering method; or piezoelectric microbalance method; or Beta-ray attenuation method that yields values having a linear relation with the values of the above methods
Nitrogen dioxide	The daily average for hourly values shall be within the 0.04–0.06 ppm zone or below that zone (notification on July 11, 1978)	Colorimetry employing Saltzman reagent (with Saltzman's coefficient being 0.84) or chemiluminescent method using ozone
Photochemical oxidants	Hourly values shall not exceed 0.06 ppm (notification on May 8, 1973)	Absorption spectrophotometry using a neutral potassium iodide solution; coulometry; ultraviolet absorption spectrometry; or chemiluminescent method using ethylene

pollutant standards are not covered here because they are purely anthropogenic.

Suspended particulate matter is defined as airborne particles with a diameter smaller than or equal to 10 μm. Photochemical oxidants are oxidizing substances such as ozone and peroxiacetyl nitrate produced by photochemical reactions by Basic Environmental Law.

Some Problems on Environmental Quality Standards of Japan from the View Point of Medical Geology

Japan's environmental quality standards for groundwater and for soil are based on Japan's drinking water quality standards (DWQS). DWQS is the standard that was revised in response to the amendment of the guidelines for drinking water quality of the World Health Organization (WHO). In other words, as the Japanese EQS for ground water and for soil basically use the DWQS. The EQS for groundwater and for soil are often revised according to the amendment on the guideline of WHO.

In terms of medical geology, using the uniformly revised system of quality standard has posed a limit on maintaining the lifestyle of the people who have lived in the Japanese archipelago. This may be shown in the following examples: Although the upper limit of fluorine and its compounds specified by the environmental quality standards is 0.8 mg/L or less, the typical concentration may be 1.3 mg/L or more in the seawater in some areas.

Although the upper limit of boron and its compounds specified by the environmental quality standards for groundwater and for soil pollution is 1 mg/L or less, they may exist in concentration of 3–4 mg/L in the seawater and also in the river water or groundwater.

In addition, since the Japanese archipelago is volcanic, there are many areas where the concentration of arsenic exceeds the environmental quality standards for groundwater or for soil pollution for arsenic. Those elements are major ones that enhance the medical effects of a hot spa.

On the other hand, there is no risk evaluation for composite pollution of hazardous elements. The standard values tend to be used for economic benefits such as sale of land because the health risk of a single element is emphasized in the environmental standard.

Furthermore, since the pollution investigation method specified by the Soil Contamination Countermeasures Act of Japan is based on an unit-independent investigation method [e.g., sample for analysis is taken based on specified depth (such as 5 cm, 0.5 m, 1 m, 2 m ... 10 m], without reference to pollution source units or unit boundaries, no precise evaluation of pollution can be made. Therefore, no pollution remediation evaluation can be made by the diagnostic standard for cleanup on the geo-pollution sites. Economic evaluations related to pollution remediation also turn out to be insufficiently supported by scientific facts.

As "geo-pollution" is a compound subsurface pollution comprised of three types of underground pollution: "strata pollution," i.e., top soil pollution, "groundwater pollution," and "ground air/gas pollution," the environmental quality standards for air, environmental quality standards for water including groundwater, and environmental quality standards for soil pollution should be comprehensively checked and evaluated to protect human health.

Specific Health Problems by Chemical Toxic Materials in Japan

Four Severe Health Problems in Japanese History, Kogai

Kogai

The word "kogai" is used in Japan. Although difficult to define, Japanese dictionary "Kojien" (1973) defines kogai as "a man-made disaster suffered by regional inhabitants caused by activities of private and public enterprises." According to Miyamoto (1970), "Although the word 'kogai' has been used since the middle of the Meiji era in laws and ordinances, citizens were not familiar with the word. Factories of heavy and chemical industries were constructed successively from the 1950s and the population became concentrated in major metropolitan areas of Japan's pacific coast. Kogai occurred explosively during this industrialization and urbanization." Kogai is written by the characters *koh*, which means public, and *gai*, which means harm (thus, "public pollution"). Although kogai does harm health by pollution, wrongdoers and victims always exist in all cases of kogai. On this point, the phrase "public pollution" does not fit.

The History of Kogai

After the Meiji Restoration, the Ashio Copper Mine (Tochigi Prefecture) was privatized, and mineral pollution caused harm in the Watarase River basin. Fuel during smelting caused smoke exhaust, mining pollution gases (having a main component of sulfur dioxide) were produced during the smelting, and drainage included mineral pollution (having a main component of metal ions such as copper ions). These contaminants caused considerable harm in the nearby environment. Mountains were stripped, ayu fish of the Watarase River died in large numbers (1885), rice plants began to die where they stood, and the farmers revolted. Shozo Tanaka is famous for actively leading the farmers' movement at this time. Later, this movement would be referred to as the origin of Japan's kogai movement.

The Seven Major Types of Kogai

The Basic Law for Environmental Pollution established the following seven items which are referred to as the Seven Major Types of Kogai (air pollution, water pollution, soil pollution, noise, vibrations, offensive odors and land subsidence).

The Four Big Kogai

The following are general remarks regarding the so-called Four Big Kogai.

The Minamata Disease

Minamata Bay in Kumamoto Prefecture (methyl mercury: water pollution, strata pollution, and the fish food chain).

Details of the occurrence: The Minamata disease is a neurological syndrome suffered by a human from his intake of methyl mercury originating from inorganic mercury (mercury sulfate) used by a Japanese Nitrogenous Fertilizer Company (present Chisso Corporation) as a catalyzer for acetaldehyde production. The disease is considered to be caused by liquid waste dumped into the ocean by this company. Unnatural deaths of cats and crows occurred around the Minamata Bay in Kumamoto Prefecture. Thereafter, inhabitants reported neurological symptoms and deaths occurred.

Symptoms: Methyl mercury is not readily decomposed in the human body and is deposited in inner organs and the brain. In particular its accumulation in the brain causes various central neurological disorders such as perception disorder, motor disorder, gait disorder, impaired hearing, speech disorder, restricted vision, and quadriplegia (Fig. 2). Because it is easily absorbed and not readily discharged, methyl mercury is a highly bioaccumulative substance. In other words, even though its concentration in the water is very small, it can be highly concentrated in fish and eventually toxicity will appear.

Fig. 2 An old fisherman of Minamata disease. The fingers abnormally became stiff. Dead in 1971. (Photographed by Kuwabara Shisei in 1970)

For instance, it was reported that fish lived for 10 of days in a solution of low methyl mercury concentration of 0.003–0.0003 mg/L without showing any harm but had in its body an accumulated level of methyl mercury several thousands of times higher than that in the water. Serious symptoms appeared in the fishermen and cats that ate the fish.

Countermeasures: Production of acetaldehyde started in 1932. Chisso stopped acetaldehyde production in 1963 (Fig. 3).

The Kumamoto Minamata Disease trial related to health damage by methyl mercury intake started in 1953 and still continues in 2009. The court that examined the Minamata disease case issued a ruling that Chisso pays damages of 16–18 million yen to every Minamata disease patient on March 20, 2009. But the case is not totally closed because there are groups of patients who still continue the case against the Tokyo head office of Chisso or negotiate by themselves with Chisso.

The Niigata Minamata Disease (Second Minamata Disease)

Agano River basin in Niigata Prefecture (methyl mercury: water pollution, strata pollution, and the fish food chain)

The same ailments as the Minamata disease of Kumamoto Prefecture were confirmed downstream of the Agano River in Niigata Prefecture. The disease was therefore referred to as the second Minamata disease. Methyl mercury flowing into the river accumulated in fish and shellfish, accumulated in human bodies by ingestion, and resulted in the disease.

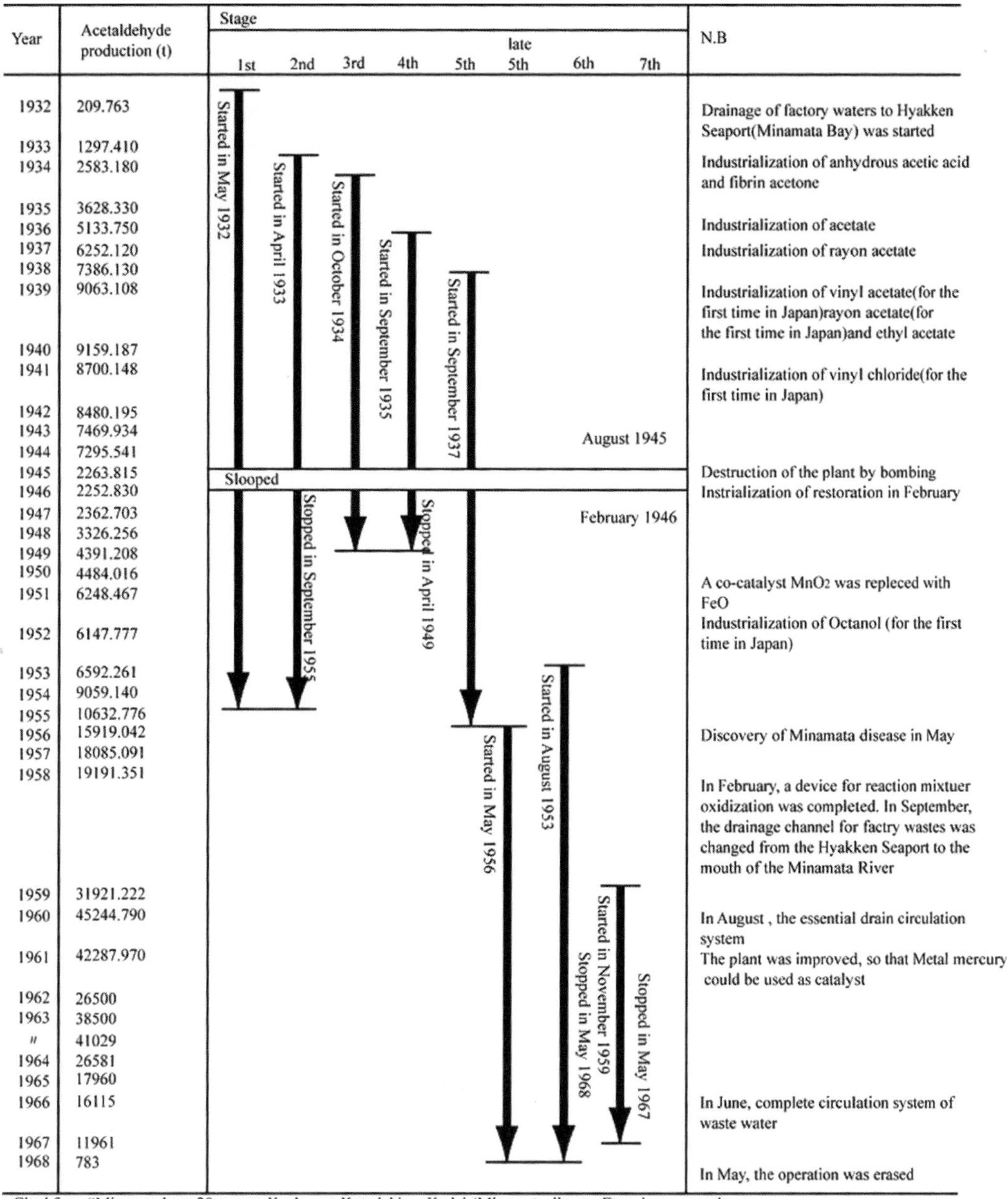

Fig. 3 Acetaldehyde production and the situation of operation of the installation in the Minamata plant. By the social scientific study group

Itai-itai Disease

Jinzu River basin in Toyama Prefecture [cadmium: water pollution (soil pollution) via rice and other food].

Details of the occurrence: The Jinzu River has an alluvial fan topography where it enters the Toyama plains. Soil and sand containing cadmium were carried downstream and settled on the river bottom. Each time a dike collapsed, rice paddies and fields were covered by the soil and sand containing cadmium, and thus the surrounding soil was polluted by cadmium. Cadmium accumulated in farm produce such as rice. Inhabitants also utilized the polluted river water as drinking water and irrigation water. Cadmium thereby accumulated in the bodies of inhabitants and resulted in the Itai-itai disease.

Symptoms: The disease mainly affects women. Complaints begin with pain in all parts of the body such as pain in the lower back, backaches, pain in the limbs, arthralgia, and ache of the pubic bone. The pain gradually worsens and finally cracks occur in bones. Then, bones break all over the body and the victim suffers, gasps, and debilitates until death, while lamenting "Itai-itai" which means, "It hurts, it hurts." Because bones lose calcium and become fragile, bone atrophy progresses and bones all over the body become fragile (Fig. 4). The body height shrinks 10–30 cm. One case is an autopsy which revealed bone breaks in 72 locations in the entire body. The disease locally drew attention as a rare disease from around 1935 (Dr. Shigejirou Hagino). In 1957, the Itai-itai disease mineral pollution theory (Noboru Hagino) was published and proposed that "the Itai-itai disease is brought about by heavy metals such as zinc and lead contained in the water of the Jinzu River." In 1961, the Itai-itai disease cadmium theory was published by Noboru Hagino and Kinichi Yoshioka.

Countermeasures: By provisions of the Farm Land Law, removal of polluted soil having more than a certain concentration, covering topsoil with soil, inspection of cadmium concentrations of rice, designation of regions, and destruction of rice having more than a certain concentration.

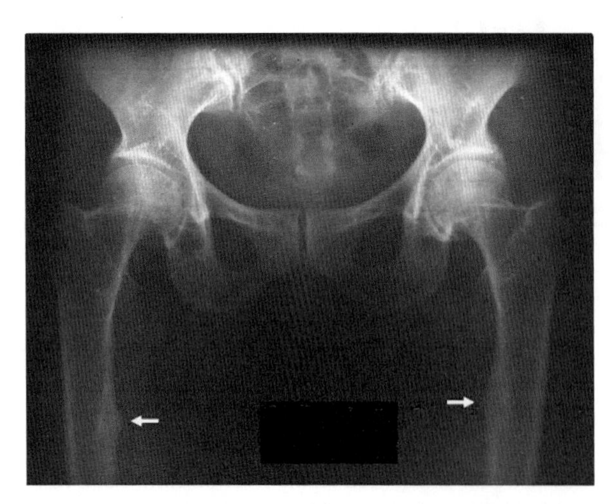

Fig. 4 69-year-old woman suffers from Itai-itai disease and had long-standing osteomalacia. This pelvic X-ray shows the symmetrical radiolucent "Looser Zones", which are characteristic osteomalacia, on the concave aspect of the femora (arrows). (Provided by Dr. Aoshima Keiko in 2010)

Each of the kogai diseases has been tried in court, and even today disputes continue among victims, enterprises, and the government regarding the scope of victims, the scope of relief, and methods thereof. A prompt solution is desired in all cases as the ages of the victims increase. In addition to the Big-Four Kogai, many other trials are being held relating to kogai.

Yokkaichi Asthma (Atmospheric Pollution Mainly Due to Sulfur Oxide)

In 1960, the Yokkaichi Kombinat industrial complex was constructed and began operations in the City of Yokkaichi in Mie Prefecture. This project was accompanied by the increase of air pollution. Complaints of difficulties in breathing and throat pain were reported, and the number of asthma patients rapidly increased. Sulfur oxide (SOx) and nitrogen oxide (NO*x*) discharged by the industrial complex were identified as causes.

The Important Role to Solve Kogai from the Viewpoint of Medical Geology

From a viewpoint of medical geology, it must be said that countermeasures encompassing the geosphere for such kogai have been insufficient.

Countermeasures are partial removal of polluted strata and covering polluted strata with unpolluted strata and are not final solutions. The substances of the pollution remain and continue to pollute the geosphere. For such regions, it is necessary to re-elucidate the mechanism of the geo-pollution, clean up the pollutants from the geosphere, and restore the geological environment.

To solve kogai problems, a medical geological viewpoint is very important to focus on substances causing kogai, exposure accumulation of the body, and effects on health.

It has recently become clear that problems involving asbestos are causing serious harm. Kogai problems, including those involving asbestos and the like, have not been concluded. Some arguments consider that kogai problems are environmental problems and generally cannot be resolved. Others point out that, contrary to recent environmental theories, that all inhabitants are both wrongdoers and victims, the victims are consistently the weak, and the essence of the environmental problem has not changed from the age of kogai (Miyamoto, 2006).

The kogai research institutes and kogai centers in various places are now changing their names to environmental research institutes and environmental research centers in national goverment and local governments.

Health Problems Due to Geo-Pollution of Chemical Toxic Material in Japan

Geo-pollution sites with volatile organic compounds (VOCs), e.g., organic chlorinated substances, once used in large quantity as a degreasing solvent and cleaner by semiconductor factories, metal processing, and laundering are found in areas close to these facilities all over Japan. VOCs used by the facilities are mainly dense non-aqueous phase liquid (DNAPL) such as tetrachloroethylene (PCE) and trichloroethylene (TCE). Geo-pollution caused by these can be seen in the polluted stratum plume, polluted groundwater plume, and the polluted ground air plume (Fig. 5).

Tetrachloroethylene (PCE) (EQS for ground water : 0.01mg/L), trichloroethylene (TCE) (EQS for ground water : 0.03 mg/L), and 1,1,1-trichloroethane (TCA) (EQS for ground water : 1mg/L) are dehalogenated and results in the occurrence of many new pollutants such

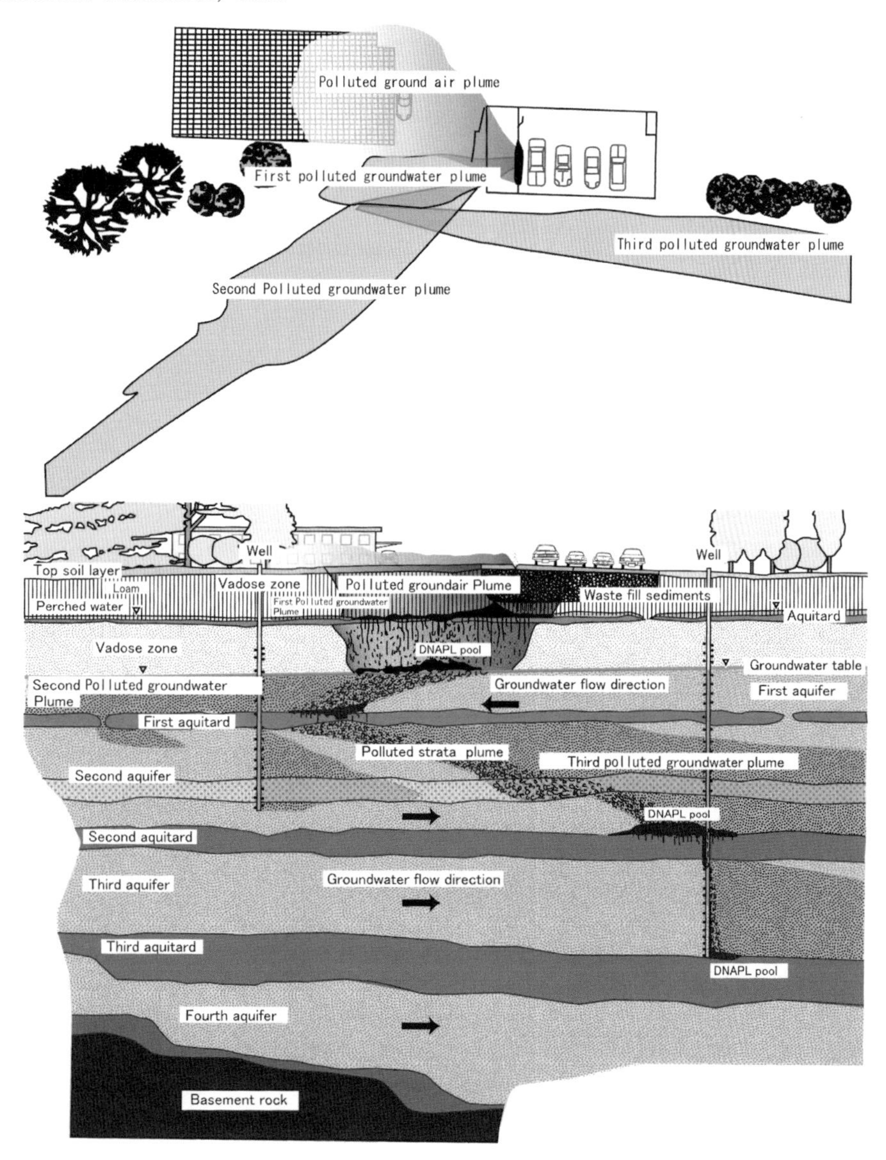

Fig. 5 Polluted strata plume, polluted groundwater plume and polluted ground air plume in VOCs (e.g., organic chlorinated substances) geo-pollution site

as dichloroethylene (DCE) and vinyl chloride (VC) at geo-pollution sites. Vogel and MaCarty (1985) proved experimentally the conversion of PCE to TCE, DCE, and VC by anaerobic biotransformation, i.e., PCE is the oldest pollutant and VC the youngest, and also conversion of TCA to 1,1-dichloroethylene (1,1-DCE) (EQS for ground water : 0.02mg/L). As a result of the transformation ranging from PCE to VC, we can say that the pollutions show four-dimensional polluting behaviors that take time and space into account, regarding to each plume at site. The same transformation is revealed in each pollution plume concerning strata and groundwater (Fig. 6a,b) and ground air (Fig. 7a,b) at the VOCs geo-pollution sites in Japan. It is important to understand that all the build up stages of the transformation of CAHs (chlorinated aliphatic hydrocarbons) and the toxic levels are different in each transformation stage.

There are two kinds of these pollution types. One type is that all the pores in sediments of strata are filled up. The other type is that the grains of sediments in the strata are coated with DNAPL resulting in remaining or residual DNAPL among the grains. Sometimes the front of the plume seeps out as spring water, causing surface water pollution, resulting in deformed carps in the pond water polluted by PCE and TCE (Fig. 8).

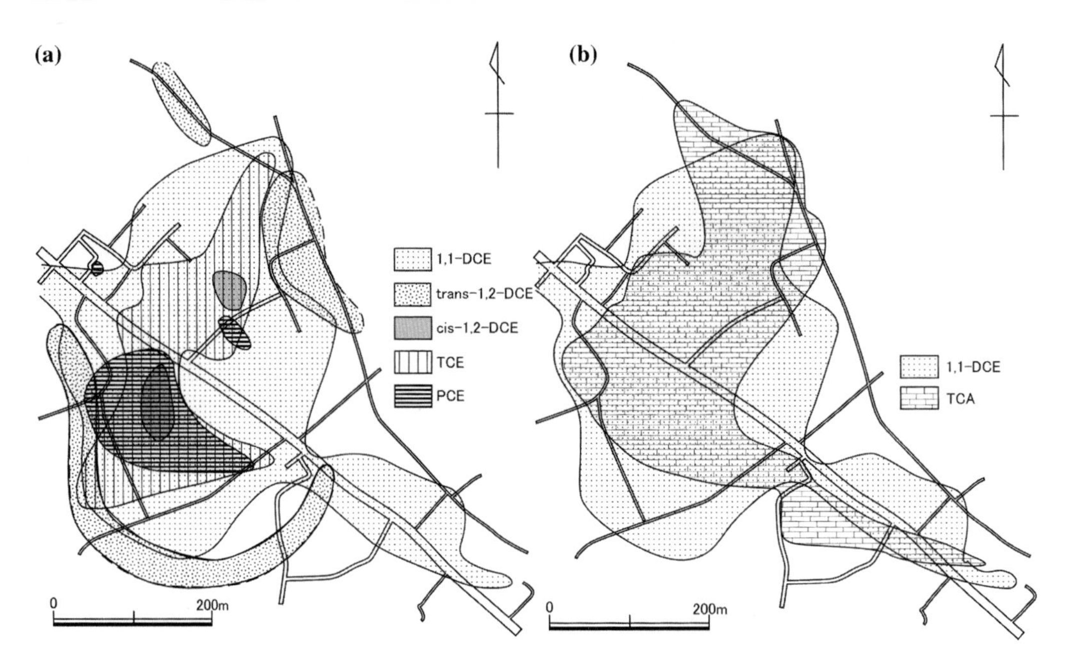

Fig. 6 (**a**) Plumes of pollutants built up from PCE and TCE in groundwater (Nirei et al., 1994); (**b**) plumes of pollutants built up from TCA in groundwater (Nirei et. al., 1994)

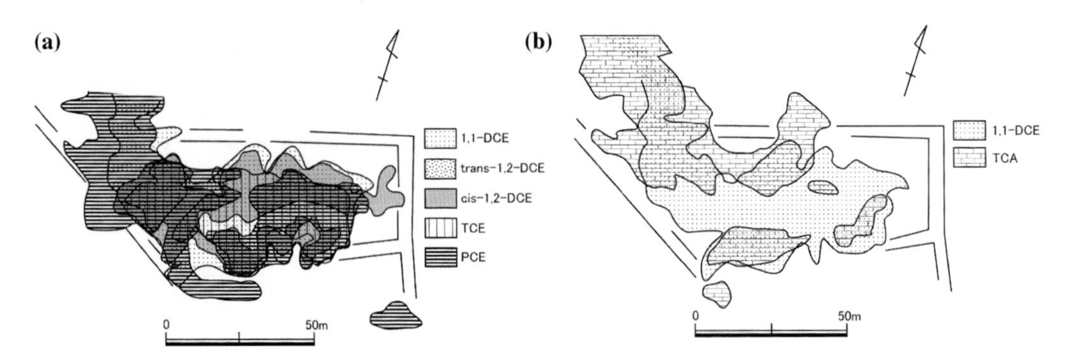

Fig. 7 (**a**) Plumes of pollutants built up from PCE and TCE in ground air (Nirei et al., 1994); (**b**) plumes of pollutants built up from TCA in ground air (Nirei et al.,1994)

Fig. 8 Deformed carps

Toxic gaseous CAHs in groundwater and on the grain surfaces in the vadose zone lead to ground air pollution (Figs. 9 and 10), the gasses flow from the geosphere to the atmosphere during low pressure, e.g., when typhoons affect geo-pollution sites. Some types of flies love to inhale TCE gas (EQS for air: $0.2 mg/m^3$) resulting in death.

Breathing of Land

Polluted ground air may move upward to the surface, pollute the air where human beings are living and cause health problems. Especially, the polluted ground air comes from geo-pollution sites such as waste disposal sites, surplus soil-dumped sites, and industrial factory sites. In order to study how to monitor the pollution of the air as well as climate variations, a monitoring system has been established near a landfill site of wastes. This is also for the purpose of managing geo-pollution hazards. As the ground air flows are controlled by the geo-stratigraphic units on solid material, we observe the solid material in the boring core samples in detail.

As a way of analyzing ground air flows, we have arranged a geo-pollution management system of observation wells for monitoring groundwater and tubes for monitoring ground air pressures, and gas concentrations as well as gas sampling. The detailed relation between the induced airflow and atmospheric pumping is not clarified, but we can show breathing with short cycles (Fig. 9) and deep breathing of land with long cycles (Fig. 10).

We can also conclude that the PCE-polluted ground air (EQS for air: $0.2 mg/m^3$), with high concentration increases from the PCE-contaminated regions to the surface as the atmospheric pressure decreases during, for example, typhoons. (Fig. 11).

Hot Springs and Mineral Springs

Introduction of Spas

Japan's hot and cold springs have been utilized for bathing from ancient times mainly for medical and convalescent purposes. The unique methodologies for using springs have taken root as part of Japanese

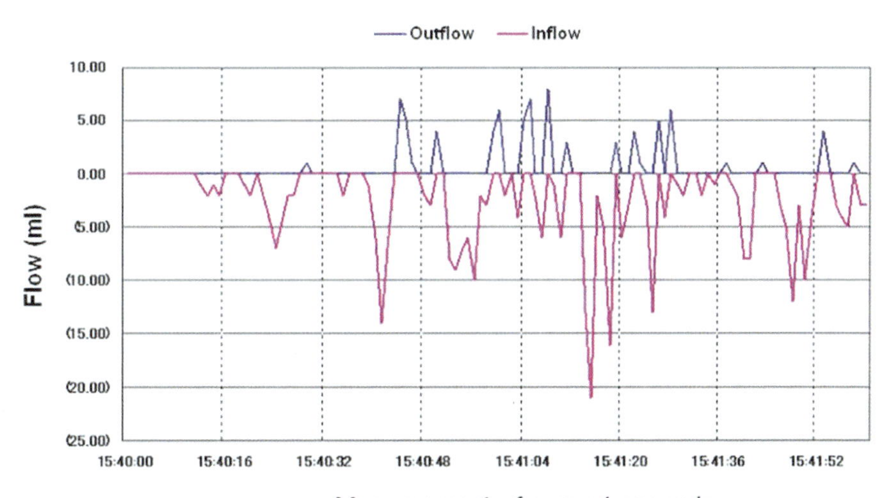

Fig. 9 The breathing of waste disposal site with short cycle (Kinjo et al., 2006)

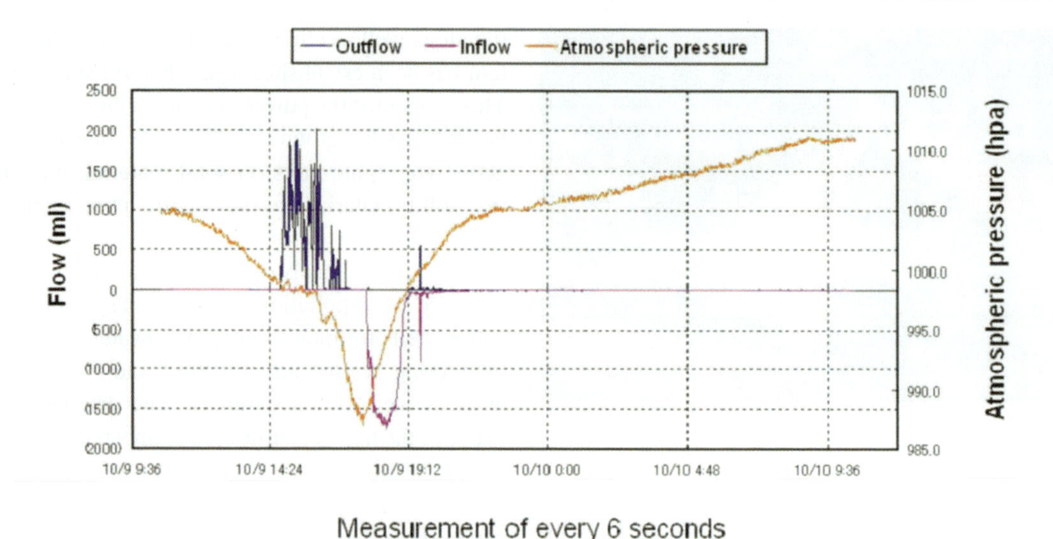

Fig. 10 The deep breathing of waste disposal site (Kinjo et al., 2006)

Fig. 11 Toxic gas blowing up from underground through a well tube at the time of a typhoon. The grass leaves bend upward by the wind force of the toxic gas (Photo by Nirei)

culture. The word *onsen* is known worldwide as a term that generally describes the culture originating from the unique utilization of hot and cold springs in Japan.

The Hot Spring Law enacted in 1948 defines a hot spring with any of the following three conditions:

(1) Water welling up from the spa has a temperature of at least 25°C.
(2) The total amount of dissolved substances in 1 kg of water is at least 1 g (excluding natural gas having hydrocarbons as the main component).
(3) The water contains at least the amount defined by law of at least 1 out of 18 types of inherent component such as free carbon dioxide.

The History of Hot Springs

The *Kojiki* (712) and the *Nihon Shoki* (720) were compiled in the eighth century and are known as Japan's oldest books. Both of these volumes indicate that hot springs were utilized as early as the seventh century. Utilization of hot springs in this period was mainly for purifying the bodies of persons with high status. However, probably people living near spas also commonly made use of hot springs.

Buddhism spread in the Kamakura and Nanbokucho periods (the twelfth to fourteenth centuries) and encouraged bathing to prevent illness and bring happiness. At Nara's Todaiji Temple, for example, water was boiled in a large caldron to make a steam bath in which bathing was performed from the viewpoint of medical treatment by the priests, poor people, sick persons, prisoners, and others to prevent illness and bring happiness. Such bathing for medical treatment, called *yusegyo* in Japanese, developed into spa treatments. It is said that spa treatments sought the medicinal effects similar to the Scandinavian sauna effects obtained by sweating in spas and stone baths (steam baths). Hot spring bathing further spread as a means for treatment from the Muromachi period (the fourteenth to fifteenth centuries). The Ikaho Onsen and the Kusatsu Onsen are two famous examples of modern resorts that began as small hot spring towns providing spa treatments.

When the Tokugawa shogunate (1603–1867) and the isolationism thereby ended, the new Meiji government was born and the country was opened to Europe and the United States. Knowledge of western medicine spread with the increased exposure to Western European culture. Hot spring treatments were rediscovered from the viewpoint of western medicine by foreign doctors and notably hot spring medical scientists from Germany. In the early Showa period (1926–1989), domestic research of hot spring treatments advanced and included the establishment of hot springs research institutes at national universities. Further, modes of traveling changed greatly due to marked developments in transportation. As hot springs began to be used for tourism as well as health treatments, the types of visitors to hot springs also changed.

The definition of hot springs was clarified with the enactment of the Hot Springs Law in 1948. The law was intended to protect hot springs, prevent disasters caused by the combustible natural gas from hot springs, realize the proper utilization of hot springs, and contribute to the promotion of public welfare. The official designation of national hot spring health resorts began in 1954 to promote recreational use. There are 91 national hot spring health resorts as of 2008.

Although the number of registered spas was 1,518 in 1962, the Furusato Sosei project from 1988 to 1989 awarded grants of 100 million yen to municipal governments. Such grants encouraged municipal governments to develop hot springs and the number of hot springs had grown to 3,139 in 2007.

Methodologies of Hot Springs

Development of Hot Spring Facilities

The names of places where hot springs are located include Deer Springs, Monkey Springs, and Crane Springs, suggesting that wild animals have utilized the medical effects of hot springs as well as humans. Accordingly, it is natural that humans also use natural pools of hot water welling up from hot springs. Even today, some spas utilize pools of hot water without permanent tub walls.

Spas of ancient times were covered outdoor baths used for bathing by inhabitants, religious uses, the convalescence of warlords, and the treatment of injuries to Daimyo (feudal lords) and men.

From the Meiji period, inns boasting indoor baths enjoyed a higher status than other inns without indoor baths. Sightseeing trips and short-term stays increased as spas developed, and bathers started to prefer inns with indoor baths. The development and enlargement of such inns led to the prosperity of modern spas (Fig. 12). During and after the period of rapid economic growth following the World War II, spas began to be utilized more for tourism than for medical treatment. As tourism increased, open-air baths (*rotenburo*) resembling natural outdoor pools increased in addition to indoor bathing areas in response to the preferences of customers. Public baths unrelated to the inn business also began installing outdoor baths. Footbaths also became popular for bathing the feet up to the ankles to treat fatigue.

Today, condominiums, villas, and other establishments are adding hot springs facilities, and the amount of hot water in the entire Japanese Archipelago is drying up due to excessive development.

Hot Water Supply Methods

Hot water supply methods include the following two types:

Fig. 12 The model indoor hot springs resort Dogo Onsen (Arima type hot spring), Ehime Prefecture

(1) Flow-through type. The flow-through type refers to feeding a constant flow rate of hot spring water into the bathing container such that the same volume overflows from the edge of the bath. In some natural bathing tubs, fresh hot spring water enters from the bottom and sides.
(2) Circulated filtration type. In the circulated filtration type, the hot spring water flows into the bathing container and is then drawn out from the bottom, passed through a filter, heater, etc., before being once again fed into the bathing container from the side walls and/or from above. It is common to add a disinfectant in this method.

Although the spring source water normally is in a reduced state directly after welling up (or being drawn up) from underground, artificial processing such as disinfection and circulation changes the water into an oxidized state. Therefore, users favor immersion in a fresh hot spring of the flow-through type to obtain the intended hot spring effects. However, sanitation problems such as microbe propagation arise in the case where the replenishment of water in the bathing container is poor due to difficulties securing the necessary amount of hot water. Recognized merits of the circulation type include solving such problems and preventing the exhaustion of hot water resources. On the other hand, microbe propagation may occur due to poor management of facilities even when using the circulation method. In one example in Hyuga City of Miyazaki Prefecture in 2002, Legionella bacteria at a hot spring facility caused the death of seven people.

Bathing Customs

Today, although men and women generally bathe in separate baths in Japan, mixed-gender bathing with strangers in unsegregated baths is called *konyoku*. This ancient custom continues in many spas. In bathing areas where hot water wells up and collects in a natural pool, little thought is given to separating baths for men and women. Mixed bathing has been practiced since ancient times where men and women, naked except for a towel, bathe together without bathing suits (Fig. 13). This natural custom seems to have been established without sexual awareness. Although men and women often can be seen enjoying conversations at spas with unseparated bathing, the custom of mixed bathing is not the norm nowadays, and many women shy away from mixed bathing. Therefore, it is common to provide bathing facilities just for women.

Fig. 13 Men and women bathing together in the Yugura Onsen (Green tuff type hot spring), Oku-Aizu district, Fukushima Prefecture

Balneotherapy

Medical effects of bathing in hot springs include physical effects of heat, buoyancy, and pressure on the human body, effects from absorbing substances contained in the water through the skin; stimulation effects from the substances absorbed into the body; and bodily modulation effects such as regulating the nervous system and adjusting internal secretion functions by stimulation from repeatedly entering the hot spring. In addition to bathing, one may visit a spa for a change of air during the vacation season. Bathing methods at hot springs may include immersing the body in water baths, being buried in warm sand baths, or relaxing in steam baths. Drinking the spring water is practiced in some locations but is not widespread.

Bathing

Bathing in Japan is carried out by immersing the unclothed body in a bathing container full of hot water at about 40°C. The bathing area is entered after washing the body outside of the bath. Mental effects of bathing include a sense of relaxation, and direct effects include thermal effects from the hot water, physical effects, effects from substances contained in the water, and modulation effects.

Bathing Methods

Bathing methods can be divided into (a) full immersion bathing, (b) partial bathing, and (c) special bathing.

Full Immersion Bathing

In this bathing method, the entire body is stretched out and immersed in a tub. Bathing is generally practiced in a hot spring of about 41°C. Alternatively, bathing may be performed for short durations at high temperatures of 43°C or more, or in baths of relatively low temperatures of about 37°C for longer periods of time. Bathing may be performed in combination with jet douches, whirlpool baths, bubble baths, and the like to apply a flow of water to the body to obtain massage effects by physical means.

Partial Bathing

In this method, part of the body is immersed in the hot spring, or water is poured over a part of the body. Partial baths include rinsing baths, waterfall baths, footbaths, and the like.

(i) Rinsing bath: The body is dowsed with hot water mainly prior to bathing. This method is performed to cleanse the body and prevent a rise in blood pressure when bathing.
(ii) Waterfall bath: Hot water falls from above to strike the shoulders, the neck, and the lower back to produce massage effects.
(iii) Footbath: Only the feet or portions of the legs below the knee are immersed in hot water. Effects include improvement of blood flow.

Special Bathing

This method includes bathing not in hot water of a hot spring but in a mud bath, sand bath, bedrock bath (Fig. 14), or a hot spring steam bath (Fig. 15).

Fig. 14 Bedrock bathing facilities used by terminal-stage cancer patients in the Tamagawa Onsen (Green tuff type hot spring), Akita Prefecture

Fig. 15 Treatment of hemorrhoids by moxibustion after treatment by steam bath at Miyanoshita, Hakone (Toto Bunso and Roka, 1811)

The Tamagawa Onsen in Akita Prefecture is famous for its bedrock bath and is utilized by terminal-stage cancer patients. Hokutolite ore of the Tamagawa Onsen contains radium. Some consider the radioactivity to be effective for the treatment of cancer. Hokutolite is designated as a special natural treasure.

Contrary to the general opinion that exposure to low concentrations of radiation should be avoided as much as possible, there is a theory (the hormesis effect) that low concentrations of radiation rather improve immunity functions. However, neither the benefits of radioactive springs on human bodies nor the hormesis effect have been proven scientifically. It may be that evaluations of low-concentration exposure may reverse views on the risks and benefits of radioactive springs. Accordingly, contributions of research from a viewpoint of medical geology are becoming important for the evaluation of low-concentration exposure in each region.

Hot spring steam bathing was practiced during the Edo period. Other treatments such as moxibustion for hemorrhoids and the like were performed in combination with hot spring steam baths (Fig. 15).

Drinking

In some cases, drinking the water of hot springs may produce effects on the stomach and intestines or on the entire body. However, drinking is not possible in some cases depending on the symptoms and the spring qualities. Prior to drinking hot spring water, it is necessary to confirm permits for drinking, ascertain indications, and contraindications for drinking, observe warnings, and drink only the appropriate amount of the appropriate hot spring.

Medical Effects of Bathing and Drinking

Indications and contraindications for hot spring water are shown in Table 7

Hot Spring Types and Geology in Japan

Hot springs are classified into the following three major divisions by geological development of the Japanese Islands Archipelago (Fig. 16) (Sakai and Oki, 1978)

The following reference numbers identify the spas in the text:

1 Arima, 2 Seikan Tunnel, 3 Moritake, 4 Isobe, 5 Yashio, 6 Nakagawa, 7 Hakone, 8 Noboribetsu, 9 Nagashima.

Arima Type Hot Springs

The Arima region (Fig. 16(1)) lies in the Rokko fault zone predominantly in the east–northeast/west–northwest directions where granodiorite of the Paleogene period (60 million years ago) intrudes into the existing rhyolitic volcanic rock. The hot spring waters can be classified into carbon dioxide cold water and neutral salt water.

(1) Carbon dioxide water: Chlorine ion concentration of 10–30 ppm with carbonic acid content reaching 2,000 ppm.
(2) Neutral salt water: Chlorine ion concentration of 5,000 ppm or more, reaching twice that of seawater.

Green Tuff Hot Springs

The Green Tuff orogenic movement occurred mainly due to volcanic activity on the ocean floor on the sea of Japan side of the Japanese Archipelago during the Neogene and early Miocene periods (24 million years ago). During the tectonic movement called the Green Tuff Geo-syncline, much volcanic ejecta, sand, and mud were thickly deposited in sedimentary regions. Green Tuff was formed by volcanic ejecta creating clay minerals such as chlorite and zeolite by diagenesis and alteration.

After the peak of volcanic activity, springs appeared on the ocean floor. Sulfide ores of useful metals such as copper, lead, zinc, gold, and silver were formed. This type of ore is called black ore or *kuroko* and is known as a metal deposit unique to the Green Tuff distribution region on the sea of Japan side. The lower portion is characteristically accompanied by large amounts of gypsum, i.e., calcium sulfate ($CaSO_4 \cdot 2H_2O$) and anhydrite ($CaSO_4$). This composition was formed when magnesium and sodium in seawater replaced the calcium in rocks, which was eluted into the seawater as calcium ions. The calcium ions formed calcium sulfate.

Table 7 Table of benefits of hot springs by spring qualities (Japanese Society of Hot Spring Sciences, 2005)

Classification of Therapeutic Spring Waters by Chemical Characteristics	Indications (○ : Bathing · ◎ : Drinking)								Contraindications (— : Bathing · × : Drinking)																		
	Cut	Scald	chronic dermatosis	Children infirmness	Chronic gynopathy	Chronic digestive organs disease	Menstrual disorder	Anemia	Chronic constipation	Constipation	Chronic gallbladder disease	Gallstone disease	Hepatopathy	Arteriosclerosis	Hypertension	Swelling	Obesity	Diabetes	Neuralgia · Myalgia · Arthralgia	Kidney disease	Hyperthyroidism	Diarrhea	Elderly xeroderma	Skin, mucous hyper-reactive especially light sensitivity	Gout	Frozen shoulder · Motor paralysis · Stiff joints · Contusion · Sprain · Hemo rrhoid · sensitivity to cold · Illness recovery · (keep health)	Acute disease · Active tuberculosis · Malignant tumor · Heavy cardiopathy · Respiratory failure · Renal insufficiency · hemorrhagic disease · pregnancy (Early and late stage)
Simple thermal waters						○													○							○	—
Carbondioxated waters	○	○				○◎			◎					○	○				○			×				○	—
Alkaline Earthy Hydrogencarbonate springs	○	○	○			○◎							◎					◎	○						◎	○	—
Sodium hydrogencarbonate springs	○	○	○			○◎							◎		×	×		◎	○	×					◎	○	—
salt water	○	○	○	○		○◎			◎						×	×			○	×						○	—
Salt water with more than 1mg/L Iodine ion	○	○	○	○		○◎			◎						×	×			○		×					○	—
Sulphate waters (Except sodium sulphate waters)	○	○	○			○			◎		◎	◎		○			◎	◎	○			×			◎	○	—
Sodium sulphate waters	○	○	○			○			◎		◎	◎		○	×	×	◎	◎	○		×				◎	○	—
Iron springs and Copper baaring Iron springs						○	○	◎											○							○	—
Sulphur waters	○		○		○	○				◎				—	—	—		○◎	○			×	—	—	◎	○	—
Simple acid sulphur thermal waters (Hydrogen sulpher type)	○		○		○	○				◎				○	○			○◎	○			×	—		◎	○	—
Acid springs and Aluminum ion bearing spring			○			○◎													○			×	—	—		○	—
Radioactive waters			○		○	○◎					○◎	○◎		○	○				○◎						○◎	○	—

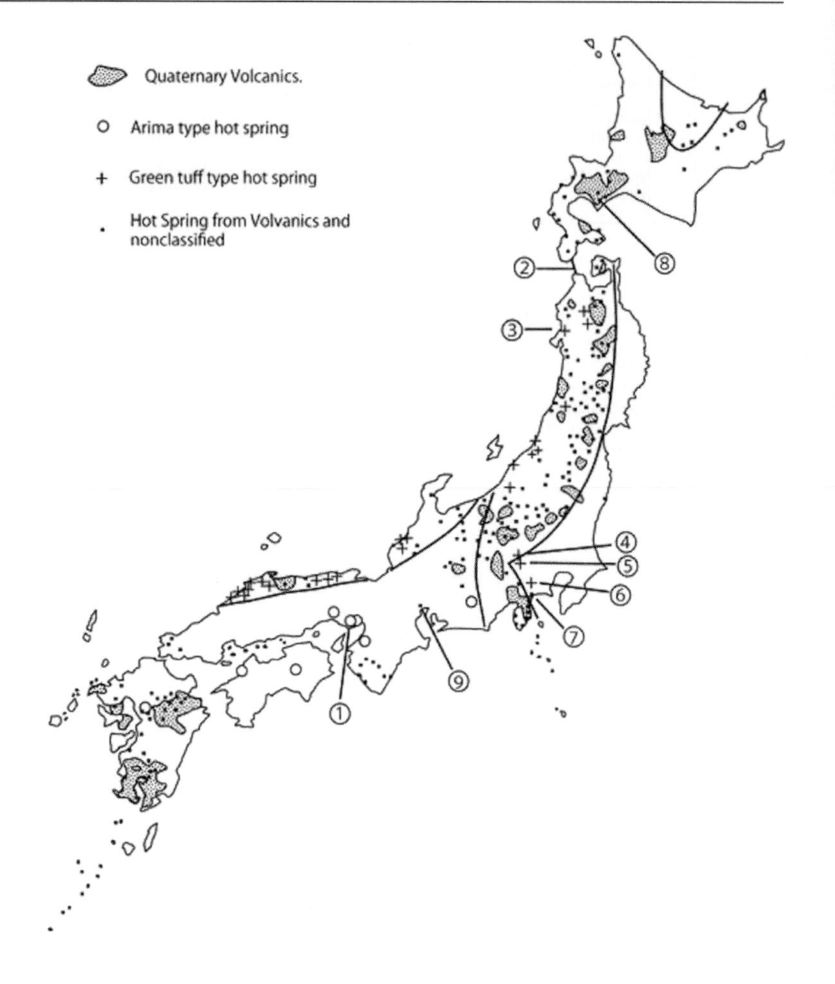

Fig. 16 Distribution map of spa and volcanoes (Sakai and Oki, 1978)

Calcium sulfate easily precipitated because its solubility in seawater decreased as the seawater was heated to high temperatures. These reactions and precipitations progressed to further precipitate gypsum, i.e., $CaSO_4$ $2H_2O$ and anhydrite. As a result, some of the seawater produced does not contain sulfate ions or magnesium ions. This phenomenon of the *kuroko* region also is common to the Green Tuff distribution region.

As the Green Tuff region underwent an uplifting movement and transformed into land, the calcium sulfate sediments and seawater existing underground in the Green Tuff region were dissolved and washed away by the flow of rainfall acting as circulation water. Hot springs flowing out from dissolved calcium sulfate sediments and seawater are referred to as Green Tuff hot springs. Measuring the isotope ratios of water shows that water inside the Seikan Tunnel (Fig. 16(2)) comes directly from the rainfall of each region. The Moritake Onsen (Fig. 16(3)) near Akita City has a high salt concentration, is poor in magnesium and sulfate, and rich in calcium. This combination is the result of the reduction of oxygen of the sulfate originally present by sulfate-reducing bacteria. Although such sulfate-reducing bacteria proliferates by combusting organic substances using oxygen from sulfate ions (SO_4^{2-}) in an environment lacking oxygen, in this case, light isotopes were consumed and the heavier isotopes of ^{34}S and ^{18}O were concentrated. Isobekosen (Fig. 16(4)) in Annaka City of Gunma Prefecture and Yashio Onsen (Fig.16(5)) in Fujioka City of Gunma Prefecture are Na–Ca–Cl-type salt water springs. This salt water was formed when all of the sulfate ions (SO_4^{2-}) in the fossil seawater were consumed to leave salt water without the sulfate. The spring qualities of the Nakagawa Onsen (Fig. 16(6)) in the central portion of the Tanzawa Mountain region is characterized by a low amount of 0.6 g/L of dissolved substances and a high pH of 10.24. The high pH is resulted from carbonate ions (CO_3^{2-}) and OH^-. It may be that the Tanzawa group is composed mainly of volcanic ejecta, little mudstone, etc., and therefore little carbon dioxide, and so the water did not become acidic and formed alkaline simple thermal waters.

Volcanic Activity of the Quaternary Period

Hot springs having thermal energy from volcanic activity as the source of heat are called volcanic hot springs. Geothermal energy is carried to the earth's surface by fluids due to the geological structure and hydrogeological structure of the volcanic belt and the surrounding region. The Hakone Onsen (Fig. 16(7)) of the Hakone volcano will be described as an example.

Thermal Structure of the Hakone Volcano

Looking at the distribution of the ground thermal structure of the Hakone volcano at a height above sea level of 0 m, the central portion of the caldera is hot and temperatures decrease outward in a concentric-circular configuration. Owakudani, Sounzan, and Yunohanasawa are located in the caldera of Hakone volcano. Hirogawara includes a high-temperature region corresponding to the Yugawara volcano. The Yumoto Onsen at the eastern edge of the Hakone caldera gushes forth from Neogenic Hayakawa volcanic breccia and the Yugashima strata forming the foundation of the Hakone volcano. The source of heat of the Yumoto Onsen depends on the volcanic activity of the Hakone volcano (Figs.17 and 18).

Spring Qualities by Sub-Zones

Many types of spring qualities exist in hot springs around a large volcano. For example, 12 types of water have been identified for the Noboribetsu Onsen (Fig. 16(8)) and 14 types have been identified for the Hakone Onsen including acidic water, alkaline water, and common salt water.

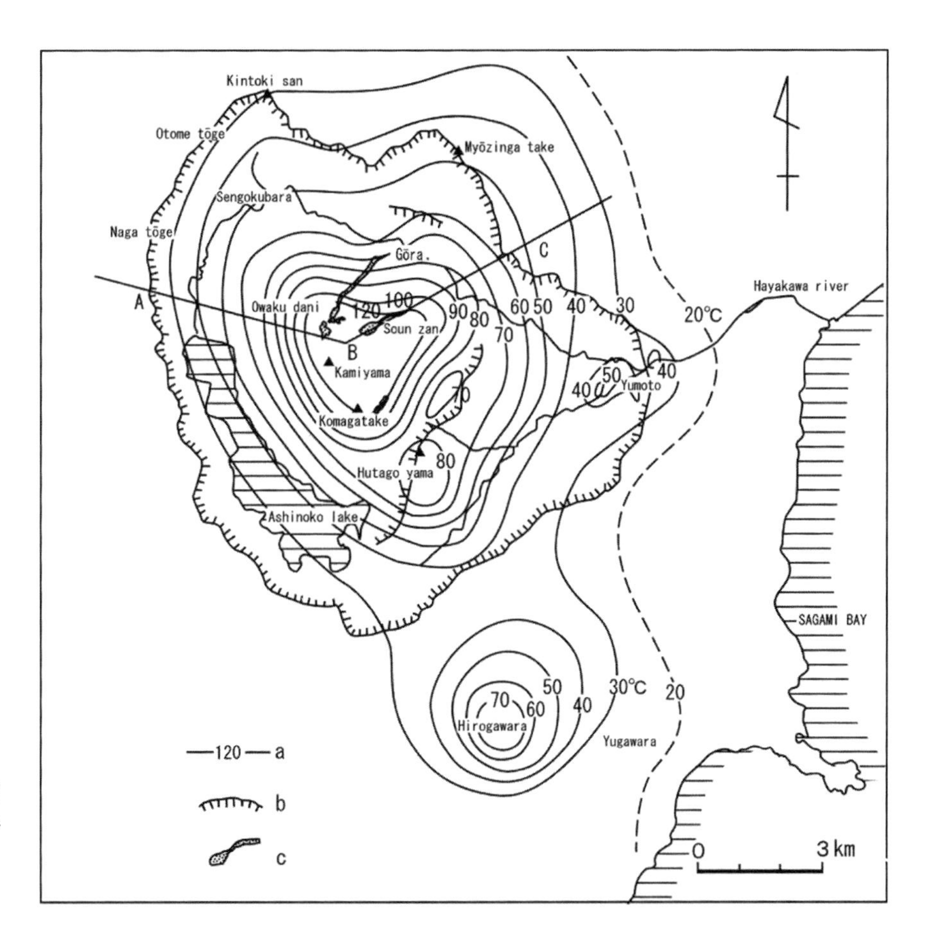

Fig. 17 Thermal structure of the Hakone volcano at the sea level (Oki and Hirano, 1970) (**a**) isotherm, (**b**) caldera, (**c**) solfatara

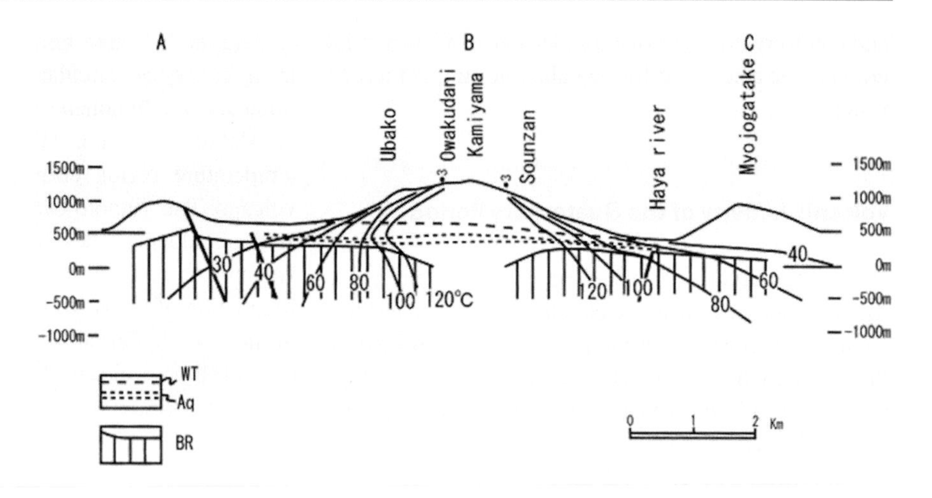

Fig. 18 East–west cross section of Hakone illustrating geological structure and isothermal profile. WT: water table, Aq: aquifer, BR: basement rocks, (Oki and Hirano, 1970)

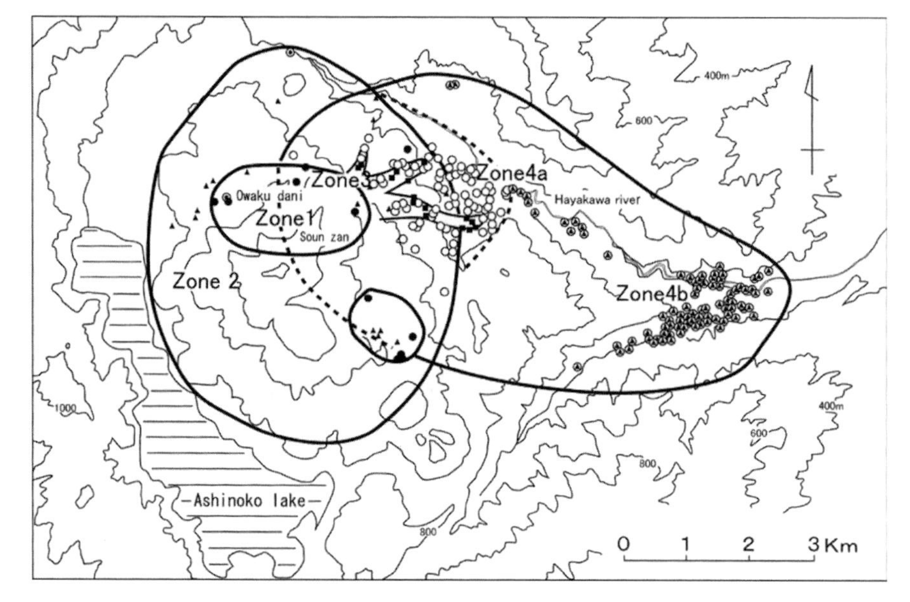

Fig. 19 Spring qualities by sub-zones (Oki and Hirano, 1970)

Hakone can be divided into four sub-zones by the solution ratios of Cl^-, SO_4^{2-},and HCO_3^- (Fig. 19).

Zone 1: Shallow hot springs with acidic sulfate water in the fumarolic area. The amount and temperature of water spring up vary with the amount of rainfall.

Zone 2: Hot springs with neutral alkaline earthy hydrocarbononate water reserved in the Hakone volcano ejecta at the bottom of the caldera. HCO_3- is supplied by decomposition of plant fossils in between volcanic deposits.

Zone 3: Hot springs with high-temperature salt water flowing downstream in three directions toward Hayakawa canyon from 300 m underground below the sulfurous gas spouts of the hillside on the eastern side of the Kamiyama central volcanic crater.

Zone 4: Hot springs with common salt/sodium hydrogencarbonate water/sulfate water formed by various proportions of the high-temperature common salt water of zone 3 mixing with the deep-layer groundwater of zone 2.

Zone 1: acidic sulfate waters

Zone 2: neutral alkaline earthy hydrocarbononate water

Zone 3: high-temperature common salt water

Zone 4: common salt/sodium hydrogencarbonate water/sulfate water formed by various proportions of the high-temperature common salt water of zone 3 mixing with the deep-layer groundwater of zone 2 (Oki et al., 1978)

Origin of Common Salt Springs

D.E. White (1957) focused on the H_2O–NaCl high-temperature high-pressure experiments of Olander and

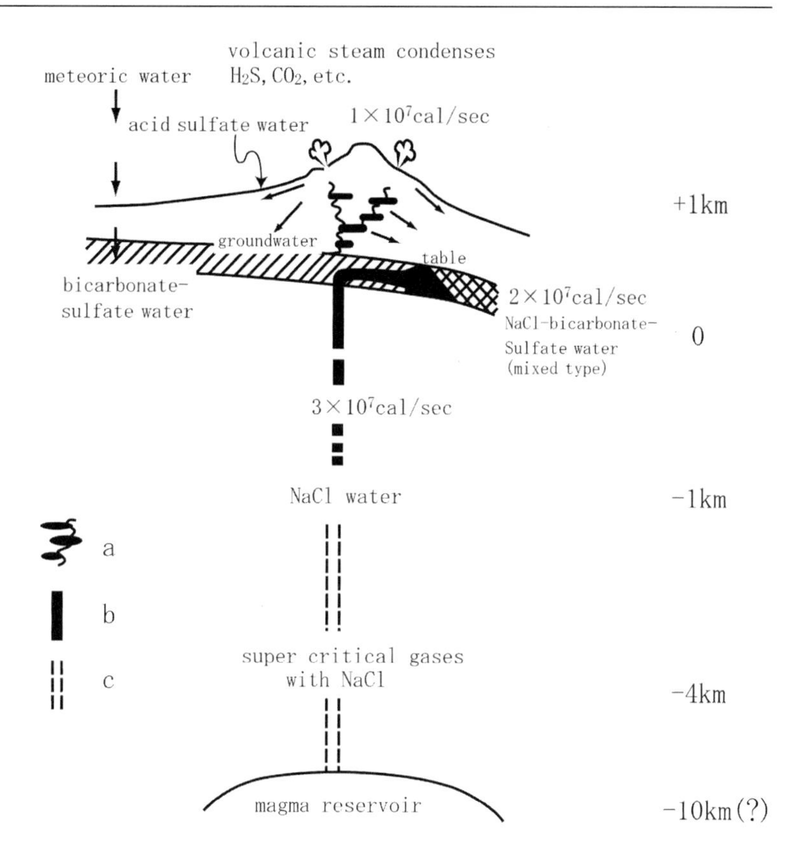

Fig. 20 Genetic model of Hakone hydrothermal system (Oki and Hirano, 1970)

Liander (1950) showing that common salt moves in water vapor and considered that common salt water in the deep portions of volcanoes is the infant stage. Figure 20 is a genetic model of the Hakone Onsen. When common salt water ascends to the ground surface (temperature and pressure decrease), common salt remains in the liquid phase and highly volatile CO_2 and H_2S are produced and spouted as gas.

(a) Repeated processes of vaporization and concentration of volcanic steam resulting in concentration of volatile components such as H_2S and CO_2
(b) Sodium chloride water (zone 3)
(c) Super critical gases (steam) with NaCl

Hydrothermal Systems and Steam-Dominant Systems

Distinctive features of hydrothermal systems include being formed where water is easily supplied. Silica sinter, calcite, zeolite minerals, etc., are formed by hydrothermal alterations of common salt water. The paths of hot water are blocked due to crystallization of alteration minerals over long periods of time, the proportion of the liquid phase decreases, and a steam-dominant system is formed. Distinctive features of steam-dominant systems include acidic sulfate water having few chlorides and a pH of 2–3. Where the hot spring gushes forth, silica sinter, carbonate, and silicate precipitates are non-existent or poor in content.

Non-Volcanic Hot Springs

The underground temperature increases about 3°C every 100 m downward. Although drilling 1,500 m should yield hot water of about 45°C, the temperature drops as the water is drawn up, and there are not many examples of springs welling up at 42°C or higher. The Nagashima Onsen (Fig. 16(9)) in Mie Prefecture obtains hot water of 60°C from a depth of 1,540 m, but generally, non-volcanic hot springs do not exceed 42°C and heat must be provided.

References

Japanese Society of Hot Spring Sciences (2005) Introduction to Hot Spring Sciences – An Invitation to Hot Springs (Vol. 48, pp. 41–52, p. 128.). Tokyo: Corona Publishing Co., Ltd.

Kinjo Y, Kusuda T, Nirei H (2006) The beginning study that will contribute to elucidate the behavior of ground air flows due to VOCs (pp. 97–102). The Proceedings of the Sixteenth Symposium on Geo-environments and Geo-technics, Japanese Society of Geo-pollution Science, Medical Geology and Urban Geology (PMUG).

Shinmuta I (ed) (1973) Kogai (p. 792). "Kojien". Tokyo: Iwanami Press.

Minato M (ed) (1977) Japan and Its Nature (p. 220). Tokyo: Heibonsya.

Miyamoto K (1970) Kogai and Resident's Campaign (p. 319). Tokyo: Jichitaikenkyusha.

Miyamoto K (2006) Toward Sustainble Society-Kogai is not Over. (p. 220). Tokyo:Iwanami press.

Nirei H, Satoh K, Suzuki Y, Furuno K (1994) Geo-pollution units application of geological concepts in environmetal studies. J Geol Soc Japan 100:425–435.

Nirei H, Maker BM, Sakunas J, Furuno K (2008) Stratigraphical classification of man-made strata and two types of geo-pollution mechanism (pp. 27–34). The Proceedings of International Symposium on Geo-pollution, Japan Branch of IUGS Commission on Geociences for Environmental Management.

Oki Y, Aramaki S, Nakamura K, Hakamata K (1978) Volcanoes of Hakone, Izu and Oshima. Bull Hot Spring Res Ins Kanagawa Pref 9(5):88.

Oki Y, Hirano T (1970) Geothermal system of Hakone volcano. U.N. symposium on the Development and Utilization of Geothermal Resources, Pisa. Geothermics Spec Issue 2(2):1.

Olander A, Liander H (1950) Acta Chem.Stand 4:1437.

Sakai H, Oki Y (1978) Hot Spring Type and Geology in Japan (Vol 48(1), pp. 41–52). "Kagaku". Tokyo: Iwanami Press.

Selinus O, Alloway B, Centeno AJ, Finkelman BR, Lindh U, Fuge R, Smedley P (2005) Essentials of Medical Geology - Impact of the Natural Environment on Public Health (p. 812). Amsterdam: Elsevier.

Toto Bunso and Roka (1811) Sawada H (trans.) (1975) Seven Hot Springs of Hakone (p. 75). Hakone Town Office.

Vogel TM, MacCarty LP (1985) Bio-transformation of tetra-chloroethylene to Trichloroethylene, Dichloroethylene,Vinyl Chloride, and Carbon Dioxide under Methanogenic Conditios. Appl Environ Microbiol 49(5):1080–1083.

White DE (1957) Thermal waters of volcanic origin. Bull Geol Soc Amer 68:1637–1658.

Medical Geology in Hellas: The Lavrion Environmental Pollution Study

Alecos Demetriades

Abstract The large volume and areal extent of metallurgical processing wastes in the Lavrion urban area, and their subsequent movement by aerial and fluvial processes, as well as by human activities resulted in the contamination of overburden materials, including residual and alluvial soil. The bioaccessibility of contaminants, such as Pb and As, is verified by the high concentrations of Pb in blood and deciduous teeth, and As in urine of children and adults alike. The present work, apart from the review of all available data and information, studied in detail every possible aspect of the Lavrion environment, i.e. overburden materials (including residual and alluvial soil), house dust, lithology, metallurgical processing wastes, and produced a multitude of geochemical distribution and other thematic maps, leading to risk assessment and the production of an environmental management plan for the rehabilitation of contaminated surficial materials. Although cost-effective solutions were proposed, no action has as yet been taken by the relevant authorities.

Keywords Geochemistry · Health · Soil · House dust · Lead · Arsenic · Blood · Teeth · Urine · Lavrion · Greece

Introduction

Lavrion (37° 42′ 38″ N, 24° 3″ 11″ E), with a population of about 10,000 people, is a town, situated in the eastern part of Lavreotiki Peninsula, and is approximately 40 km to the southeast of Athens, and 55 km by road (Fig. 1). Because of its rich mineral wealth, consisting of polymetallic sulphides and iron ore (Marinos and Petrascheck, 1956; I.G.M.E. Working Group, 1987; Vourlakos, 1992; Katerinopoulos and Zissimopoulou, 1994; Skarpelis, 2007a,b; Voudouris et al., 2008; Skarpelis and Argyraki, 2009), it was the centre of considerable mining and smelting activities from ancient to recent times (Conophagos, 1980). The first mining and smelting activities in the Lavreotiki Peninsula began early in history according to the classical Hellene historian Xenophon (c. 430–350 BC); lead isotope studies of ancient artefacts indicate that mining of argentiferous galena began between 3500 and 3000 BC (Manthos, 1990; Dermatis, 1994). Moreover, archaeological excavations at Thorikon, a settlement to the north of Lavrion, confirmed the existence of mining activities as early as 3000 BC (Gelaude et al., 1996). Mining and smelting activities continued intermittently until the 1st century AD, since Pausanias (active c. 143–176 AD), a 2nd century Hellene geographer, reported that the Athenian silver mines in the Lavreotiki Peninsula were completely forgotten by his time. It appears, however, that silver from Lavreotiki and Pangaeon (Macedonia, N.E. Hellas) was used in the 6th century AD during the building of Saint Sofia Church in Constantinople.

The peak of mining and smelting activities was between the 6th and 4th centuries BC, and especially during the 5th century BC, the Golden Age of Athens or Pericles (Conophagos, 1980). The revenue from the mines was used first to build in 483 BC the large Athenian fleet, which defeated the Persians at the naval battle of Salamis in 480 BC, and subsequently financed the construction of the renowned monuments seen

A. Demetriades (✉)
Institute of Geology and Mineral Exploration, 1 Spirou Louis Street, Entrance C, Olympic Village, Acharnae, Gr-136 77, Hellas
e-mail: ademetriades@igme.gr

O. Selinus et al. (eds.), *Medical Geology*, International Year of Planet Earth,
DOI 10.1007/978-90-481-3430-4_13,

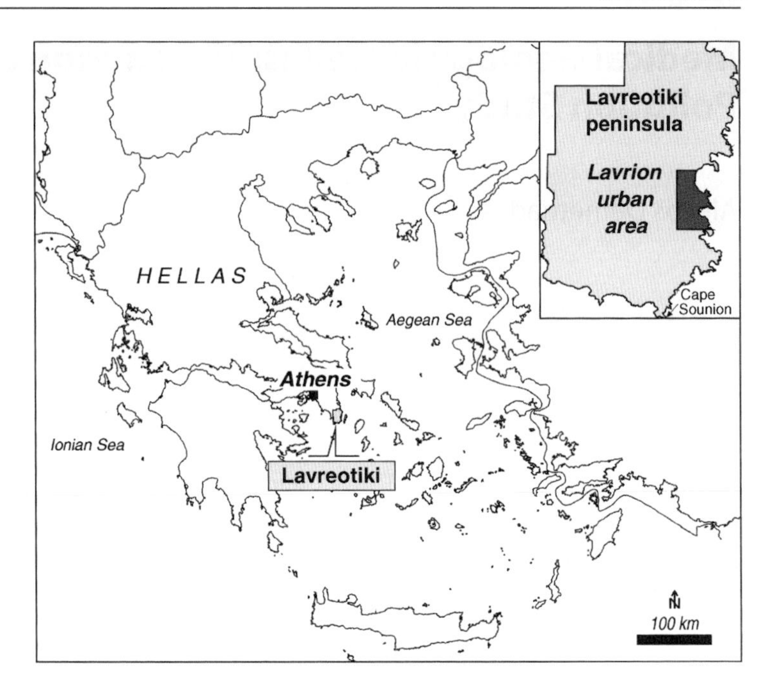

Fig. 1 Map of Hellas showing Lavreotiki Peninsula in relation to Athens. *Inset* shows the location of the Lavrion urban area

today on the Acropolis of Athens, the Parthenon, the Erechtheum, etc.

The recent mining and smelting history began in 1865 and ended in 1989 AD with the closure of the smelter. During this period Lavrion was the centre of ore beneficiation and smelting activities, which produced large volumes of hazardous metallurgical processing wastes, including pyrite, slag, flotation sand and other coarser materials. Houses, schools, parks, playgrounds, shops and roads are either situated on or are very close to these wastes. Further, the wastes are used by the local inhabitants, private companies and the Municipality for a variety of purposes, e.g. slag as hard core on roads, school yards, the new port facilities, infilling of foundations of buildings, covering of dirt roads by sand-blast material made from slag, the flotation sands as building material for houses. During their loading, transportation and unloading, fine dust is released into the ambient air and made available for transportation by wind. The strong northeasterly winds blowing in the area, almost throughout the year, transport fine- to medium-grained metallurgical processing wastes to distant places. Deposition of these materials, with high toxic element contents, and their subsequent mixing with residual and alluvial soil caused the alteration of the chemical composition of surface soil.

Because of the adverse industrial conditions in Lavrion, and the known medical effects of lead poisoning on humans, three cross-sectional epidemiological and other medical studies were carried out in the 1980s (Drossos et al., 1982; Benetou-Marantidou et al., 1985; Nakos, 1985; Hatzakis et al., 1987; Maravelias et al., 1989; Eikmann et al., 1991; Makropoulos et al., 1991, 1992a, b; Kafourou et al., 1997). Other human tissue studies included child deciduous teeth (Stavrakis et al., 1994a) and urine (Demetriades et al., 2008). The effects on plants were also investigated (Nakos, 1979a, b; Chronopoulos and Chronopoulou-Sereli, 1986a, b, 1991; Chronopoulou-Sereli and Chronopoulos, 1991a, b; Xenidis et al., 1997; NTUA, 1999), as well as on animals (Spais et al., 1987).

The Institute of Geology and Mineral Exploration with the waning down of the mineral exploration effort in Hellas began in the late 1980s environmental impact studies. Because of the mineral exploration projects, carried out in the Lavreotiki Peninsula, there was evidence of widespread contamination of surface materials (soil and stream sediment). In 1989 it was decided, therefore, to carry out, during 1990–1992, urban geochemical surveys in Lavrion and Aghios Constantinos, a village approximately 4 km to the WNW of Lavrion (Demetriades, 1992; Hadjigeorgiou-Stravrakis and Vergou-Vichou, 1992; Hadjigeorgiou-Stravrakis et al., 1993; Stavrakis et al., 1994a). Samples of garden soil (n=153), road dust (n=137) and house dust

(n=128) were collected from Lavrion and analysed by atomic absorption spectrophotometry, following a hot aqua regia extraction, for As, Cd, Cu, Fe, Pb and Zn, and Hg by a dual-beam Hg-spectrometer (Robbins, 1973; Pöppelbaum and Van den Boom, 1980a, b). Also, deciduous teeth were collected in 1991 from Lavrion primary school age children (n=82) to assess Pb absorption and to compare results with other industrial and natural background areas (Stavrakis et al., 1994a). In 1992 the project was extended to cover the whole Lavreotiki Peninsula with the collection of surface soil (n=698) and rock samples (n=136) (Stavrakis et al., 1994b; Demetriades et al., 1994a, b, c, 1996a, b, 2004). Another soil geochemical study in part of the Lavreotiki peninsula was carried out by Korre (1997) and Korre and Durucan (1995a, b).

Finally, the most comprehensive environmental impact study ever carried in an urban–sub-urban environment was co-financed by the European Community and the Hellenic State through the LIFE programme with the title *Soil Rehabilitation in the Municipality of Lavrion* (Contract No: 93/GR/A14/GR/4576). The project began in 1994 and was completed in 1999 with the submission of a massive six-volume report (Demetriades, 1999a, b, c, d, e; NTUA, 1999), as well as other site or subject-specific reports and papers (Demetriades and Stavrakis, 1995a, b; Demetriades et al., 1997, 1998). Some of these results will be described in this chapter.

Lithology

Regional and local stratigraphy and geological structure are described by Marinos and Petrascheck (1956), the I.G.M.E. Working Group (1987), Papadeas (1991), Leleu (1966, 1969), Leleu and Neumann (1969), Photiades (2003), Photiades and Carras (2001), Photiades and Saccani (2006), Photiades et al. (2004), Baziotis et al. (2008) and Skarpelis et al. (2008) and they should be consulted. For the purpose of the geochemical study it suffices to know broadly the main rock units occurring in the Lavrion urban area, i.e.

- Marble with schist intercalations,
- Schist with marble intercalations,
- Schistose-gneiss,
- Prasinite (metamorphosed mafic igneous rock), and
- Sandstone, conglomerate and alluvial deposits of Quaternary to Recent age.

Outcrops of iron and "iron calamine" mineralisation are found in the study area. Marinos and Petrascheck (1956) also referred to the occurrence of small outcrops with galena [PbS] at Kiprianos and Nichtochori hills (Fig. 2). They also mentioned the existence of an outcrop of granodiorite porphyry (eurite) at Kiprianos.

The area extent of the different rock units is tabulated in Table 1. Marble is the dominant rock type, covering 86.22% of the solid geology, followed by schist (13.24%).

Chemical Composition of Parent Rocks

One of the aims of the Lavrion urban geochemical project was to determine the background and baseline variation of elements. An indeed difficult objective, since *residual soil* in the Lavrion urban area is contaminated by toxic elements, as is described below. The only natural geological medium amenable to study natural background and baseline variation is surface rock. Rock outcrop mapping during the compilation of the 1:5,000 scale lithological map (Fig. 2) has shown that there were enough outcrops, well distributed over the whole study area, to compile lithogeochemical maps (Demetriades and Vergou-Vichou, 1999a; Demetriades et al., 1999a). Therefore, rock chip samples were collected from 140 outcrops and analysed for major and trace elements by different AAS analytical methods (SiO_2, Al_2O_3, Fe_2O_3, CaO, MgO, K_2O, Na_2O, TiO_2, P_2O_5, Ag, As, Ba, Cd, Co, Cr, Cu, Li, Mn, Mo, Ni, Pb, Sb, Sr, V, Zn and LOI) and by aqua regia and determination by ICP-AES for Ag, B, Be, Bi, Hg, La, Mo, S, Sb, Sn and U. Because of a high detection limit and poor analytical precision for some of the above determinants, analytical data from the Lavreotiki Peninsula regional rock geochemical survey were used, e.g. another 15 samples added for Ag and Mo (n=155), and only 48 rock samples for B, Be, Bi, Hg, La, S, Sb, Sn and U.

The abundance of elements occurring in rocks of the Lavrion urban area (entry All rocks, Table 2) is compared to the mean concentration in rocks of the upper continental crust (Levinson, 1980; Taylor and McLennan, 1995; Wedepohl, 1995; Reimann et al.,

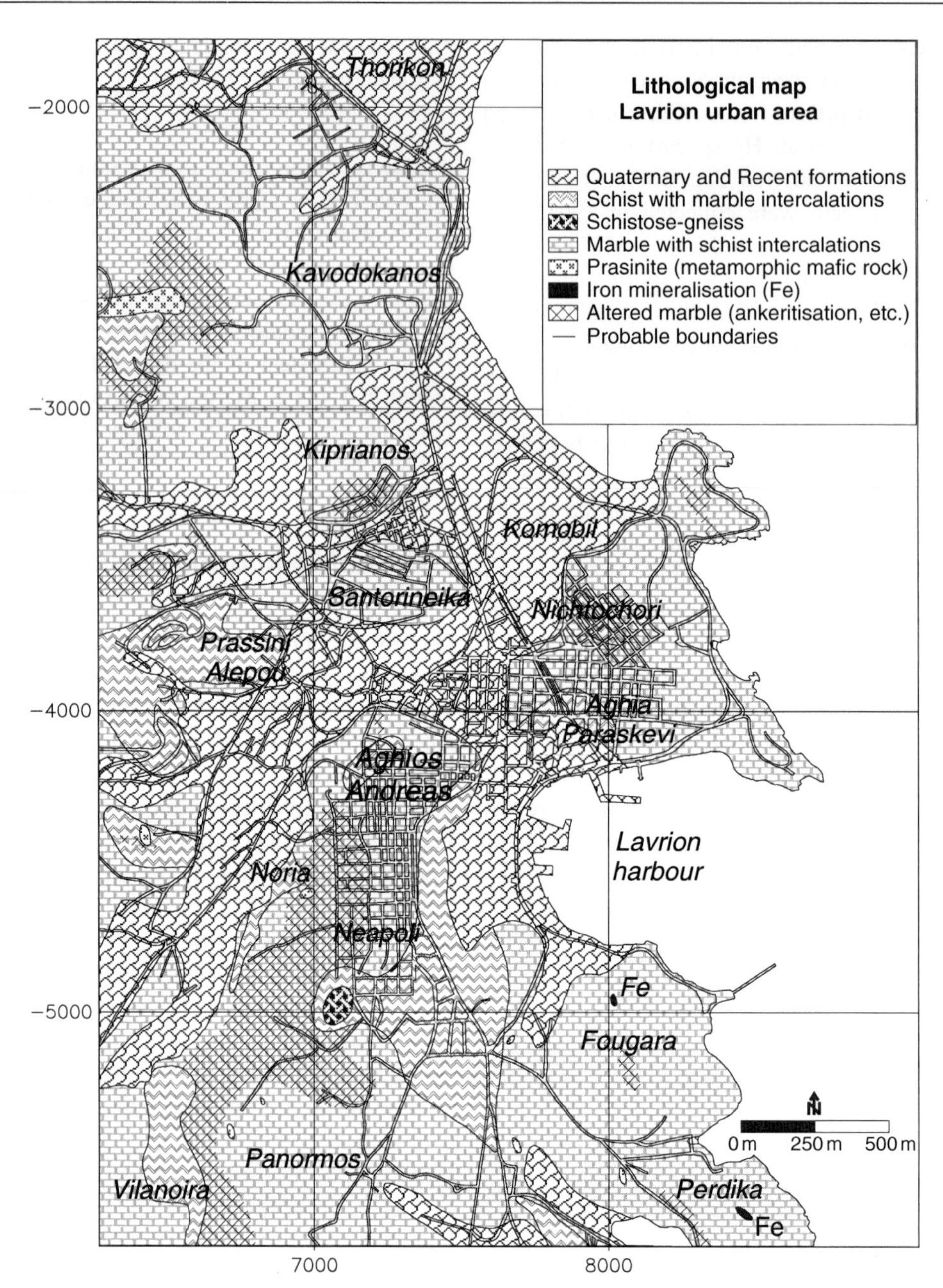

Fig. 2 Lithological map, Lavrion urban area, Hellas (from Demetriades et al., 1999c, Map 2.2; p.2.2; reduced from the original colour map at a scale of 1:5,000) (Mapped by A. Demetriades and K. Vergou-Vichou (1997))

1998). This comparison shows that the Lavrion urban area rocks are naturally

- enriched in **As**>**Sb**>**Cd**>**Ag**>**Hg**>**Ni**>Ca>**Pb**>**Zn** >P>Mn>**Cr**>Co>**Cu**>Ba>S>**U** and
- depleted in Fe>Mo>Li>Sr>La>Ti>Al>V>K>B

with respect to their abundance in rocks of the upper continental crust. Calcium enrichment is explained by the widespread occurrence of marble. Enrichment with respect to other elements, the majority of which are classified as toxic (given in bold), is either due to mineralisation or lithology. It is noted that the terms *depletion* and *enrichment* are used for mean element concentrations below and above the Global rock mean, respectively.

Since, each rock has a different chemical composition, mean element concentrations in marble, schist,

Table 1 Area covered by different rock units, Lavrion urban area

Rock unit	Area (m^2)	Area (%)
Marble with schist intercalations	6,238,189.48	86.22
Schist with marble intercalations	957,995.00	13.24
Schistose-gneiss	9,713.00	0.13
Prasinite (altered metamorphosed mafic rock)	27,301.00	0.38
Iron mineralisation	1,831.00	0.03
Total area of solid geology	*7,235,029.48*	*100.00*
Altered marble	555,173.00	8.90
Quaternary formations	2,161,342.00	29.87

Source: Data from Demetriades et al., 1999c, Table 2.1, p. 16

prasinite and sandstone of the Lavrion urban area are compared to their respective Global rock-type means tabulated by Reimann et al. (1998). This comparison is made to show geochemical differences between the different Lavrion urban area rocks and their respective Global rock averages.

All four rock types are enriched in Cd, Cu, Mn, P and Pb, which are elements associated with the polymetallic sulphide mineralisation, and depleted with respect to K and V (Table 3). Moreover, marble and schist are enriched in Ag, As, Co, Hg, Ni and Zn, elements also associated with base metal mineralisation. It is, therefore, quite apparent that mineralisation is the cause of enrichment of all rocks occurring in the Lavrion urban area, with respect to major and minor ore elements.

Prasinite, metamorphosed diabasic lava, was expected to be enriched in Fe, Ti, V, Cr, Ni and Co, because of its mafic, and locally ultramafic, mineralogy, but it is peculiarly depleted with respect to all these elements, possibly due to its alteration. The calcium enrichment is explained by its mineralogy, but the abnormal Li level is a local peculiarity, for this is an element normally enriched in granite and pegmatite, as well as in shale and schist, but not in mafic rocks.

Metallurgical Processing Wastes

Metallurgical processing of ore-grade materials from 1865 to 1989 produced a considerable amount of waste materials, which cover a large part of the Lavrion urban area (Fig. 3). Ancient wastes that may have existed are completely masked. Residual soil has been contaminated to a variable degree, and constitutes today the greatest hazard, since it covers the greatest area in Lavrion (74.91% of study area, Table 4). Since uncontaminated soil is only found in a few places, and the natural landscape has been altered considerably by deposition of metallurgical wastes, and their subsequent moving about by the local inhabitants, and importation of garden soil from other parts of Hellas, the term "overburden" is preferred for it describes the local situation perfectly well. Overburden is defined by Jackson (1997, p. 457) as the *loose soil, silt, sand, gravel, or other unconsolidated material overlying bedrock, either transported or formed in place.*

Three main categories of metallurgical processing wastes occur in the Lavrion urban area:

- beneficiation/flotation residues or tailings (they will be referred to in the text as either flotation residues or flotation tailings),
- pyritiferous tailings, and
- slag (lumpy and pelletised).

These three main types of metallurgical processing wastes have been subdivided further into other subcategories, depending on their proportion mixed with other wastes or materials, i.e.

- disseminated slag,
- sand-blast material from slag and pelletised slag,
- beneficiation/flotation sand with disseminated pyrite,
- disseminated slag and coarse-grained beneficiation/flotation residues,
- beneficiation/flotation sand and coarse-grained materials, and
- beneficiation/flotation residues and disseminated slag.

The surface area of the metallurgical processing wastes and contaminated overburden, including residual soil, are tabulated in Table 4.

Chemical Composition of Metallurgical Processing Wastes

The metallurgical processing wastes, as tabulated in Table 4, cover a large part of the Lavrion urban area, and presently they are the major primary source

Table 2 Statistical parameters of element concentrations in rock, overburden and metallurgical processing wastes in Lavrion, Global rock mean of upper continental crust (Reimann et al., 1998). Enrichment/depletion index of Lavrion rocks with respect to Global rock mean. Median values of Lavrion overburden and metallurgical wastes are given for comparison purposes

	Lavrion rocks (n=140)[1], overburden, including residual soil (n=224)[2], and metallurgical processing wastes (n=62)							Global rock mean of upper continental crust (ppm)	Enrichment/ depletion index of Lavrion rocks with respect to Global rock mean
	All rocks (n = 140) (values in ppm)					Over burden (ppm)	Metallurgical wastes (ppm)		
Element	Minimum	Maximum	Mean	Standard deviation	Median	Median	Median		
Ag*	< 0.5	41.00	0.89	3.63	0.50	12.06	18.90	0.055	16.18
Al	< 50	75,152.00	19,267.00	21,375.00	8,044.00	32,315.00	20,074.00	77,440.000	0.25
As	< 0.5	1,032.00	62.80	172.40	15.60	1,290.00	2,492.00	2.000	31.40
B**	0.25	5.24	0.52	0.87	0.25	136.00	42.97	17.000	0.03
Ba	40.00	108,000.00	1,067.00	9,110.00	210.00	479.00	243.00	668.000	1.60
Be**	< 0.5		Below detection limit			1.01	0.50	3.100	–
Bi**	< 0.5		Below detection limit			11.00	2.50	0.120	–
Ca	50.00	390,938.00	217,301.00	125,319.00	220.13	93,625.00	102,603.00	29,450.00	7.38
Cd	< 1.0	41.00	1.89	5.02	0.50	38.00	20.62	0.102	18.53
Co	< 1.0	104.00	26.90	26.10	20.50	16.00	23.83	11.600	2.32
Cr	< 1.0	610.00	100.40	145.90	20.00	183.00	73.18	35.000	2.87
Cu	3.00	225.00	32.80	31.00	25.00	186.00	630.50	14.300	2.29
Fe	979.00	107,016.00	23,790.00	20,660.00	19,515.00	44,771.00	234,500.00	30,890.000	0.77
Hg**	< 1.0	7.77	0.69	1.05	0.50	0.14	2.35	0.056	12.32
K	< 83	31,795.00	5,841.00	7,552.00	8,044.00	9,770.00	7,100.00	28,650.000	0.20
La**	< 2.0	30.69	10.21	6.47	8.86	22.70	27.32	32.300	0.32
Li	< 1.0	106.00	9.10	12.10	5.00	17.40	14.50	22.000	0.41
Mn	100.00	25,000.00	1,831.00	2,635.00	1,200.00	2,189.00	9,398.00	527.000	3.47
Mo*	< 0.5	11.00	0.83	1.16	0.50	4.90	3.57	1.400	0.59
Ni	< 1.0	1,600.00	168.30	252.20	54.50	127.00	38.52	18.600	9.05
P	44.00	8,510.00	2,381.00	1,904.00	1,855.00	992.00	1,103.00	665.000	3.58
Pb	< 1.0	1,850.00	76.85	209.06	22.00	7,305.00	20,750.00	17.000	4.52
S**	100.00	3,000.00	1,100.00	600.00	1,200.00	12,690.00	20,581.19	953.000	1.15
Sb**	< 5.0	71.04	6.94	12.05	2.50	121.00	189.00	0.310	22.39
Sn**	< 5.0		Below detection limit			18.50	27.71	2.000	–
Sr	< 1.0	800.00	121.00	129.00	98.00	118.00	178.69	316.000	0.38
Ti	< 120.0	7,014.00	920.00	1,401.00	300.00	2,162.00	737.70	3,117.000	0.30
U**	2.50	6.06	2.75	0.85	2.50	3.00	2.50	2.500	1.10
V	< 1.0	71.00	13.10	13.00	9.00	75.00	46.30	53.000	0.25
Zn	< 6.0	5,200.00	210.60	599.60	57.00	6,668.00	39,800.00	52.000	4.05

Source: Data from Demetriades and Vergou-Vichou, 1999a, Table 4.1, p. 94
[1]Rock: *Ag, Mo (n=155); **B, Be, Bi, Hg, La, S, Sb, Sn, U (n=48)
[2]Overburden: B, Bi, Hg, S, Sn, U (n=50) and Sb (n=90)

of contamination of the surface environment. It is, therefore, significant to understand their environmental impact and implications to the quality of life of the local population. Their geographical distribution and the area they cover are shown in Fig. 3. Their chemistry was studied by the collection of representative samples from the main metallurgical processing waste categories, i.e.

- slag (n=21),
- sand-blast material or wastes (n=8),
- earthy material within slag (or slag earth) (n=7),

Table 3 Comparison of mean element concentrations of Lavrion urban area rocks to Global rock-type means. Elements enriched in all rock types are displayed in bold letters and those that are depleted are italicised and underlined. (Elements below detection limit are Be, Bi, Sn and U and are not considered)

Process	Marble	Schist	Prasinite*	Sandstone
Enrichment	Ag, Al, As, Ba, **Cd**, Co, Cr, **Cu**, Fe, Hg, La, **Mn**, Mo, Ni, **P**, **Pb**, S, Ti, U, Zn	Ag, As, Ca, **Cd**, Co, Cr, **Cu**, Hg, **Mn**, Ni, **P**, **Pb**, Zn	Ca, **Cd**, **Cu**, Li, **Mn**, **P**, **Pb**	Ba, Ca, **Cd**, Co, Cr, **Cu**, Fe, **Mn**, Ni, **P**, **Pb**, Sr, Zn
Depletion	B, Ca, *K*, Li, Sr, *V*	Al, B, Ba, Fe, *K*, La, Li, Mo, S, Sr, Ti, *V*	Al, Ba, Co, Cr, Fe, *K*, Ni, Sr, Ti, *V*, Zn	Al, *K*, Li, Ti, *V*

Source: Data from Demetriades and Vergou-Vichou, 1999a, Table 4.2, p. 95

*Prasinite mineralogy: albite [$NaAlSi_3O_8$], pennine [$H_8(Mg,Fe)_5Al_2Si_3O_8$], epidote [$H_2O.4CaO.3(Al,Fe)_2O_3.6SiO_2$], clinozoisite [$HCa_2(Al,Fe)_3Si_3O_{13}$], actinolite [$Ca_2(Mg,Fe)_5(OH)_2(Si_4O_{11})_2$], hornblende {$(Ca,Na)_{2-3}(Mg,Fe^{2+},Fe^{3+},Al)_5(OH)_2.[(Si,Al)_8O_{22}]$}, glaucophane [$Na(Al,Fe,Mg)(SiO_3)_2$] and leucoxene $FeO{\cdot}TiO_2$ (Marinos and Petrascheck, 1956)

- beneficiation/flotation residues or tailings (n=8),
- pyrite tailings (n=12), and
- pyritiferous sand (n=6).

Aqua regia extractable concentrations of elements (Ag, Al, As, B, Ba, Be, Bi, Ca, Cd, Ce, Co, Cr, Cu, Fe, Ga, Ge, Hg, In, K, La, Li, Mg, Mn, Mo, Na, Nb, Ni, P, Pb, Rb, S, Sb, Sc, Se, Sn, Sr, Ta, Te, Th, Ti, U, V, W, Y, Zn and Zr) in samples of metallurgical processing wastes were determined by ICP-AES.

It must be appreciated that the metallurgical wastes in the Lavrion urban area are the products of over hundred years of ore processing. Apart from differences in the mineralogy of processed ore, changes made in the metallurgical methods resulted in the production of wastes of different chemical composition. The detailed study of their chemistry verified the existence of distinct chemical differences among the six major metallurgical waste categories, but also within each one (Demetriades and Vergou-Vichou, 1999b). The spatial chemical variation has been depicted, in general terms, by the geographical presentation of results (Demetriades et al., 1999b).

Slag heaps are distinguished into four different types in terms of their chemical composition (Fig. 3), i.e.

1. Fougara, Panormos and Neapoli are characterised by comparatively high levels in Al, (B), Ba, Be, (Cr), K, La, Li, Ni, Sb, Sn, Sr and Ti; *B is enriched in slag samples from Fougara and Panormos, and Cr from Fougara and Neapoli.*
2. Eastern part of Aghia Paraskevi and the beach section of Komobil-Kiprianos are characterised by comparatively high levels in Al, Ba, (Be), Cr, K, La, Li, Mo, Ni, Sb, Sn, Sr, Ti and V; *Be is only enriched in slag samples from Komobil-Kiprianos.*
1. South and southeast of Aghia Paraskevi are characterised by comparatively high contents in Al, Fe, K, Mo, Sn, Sr and Ti.
2. Kavodokanos and Kiprianos (western part in the metallurgical plant) are characterised by comparatively high levels in (Ag), Cu, Fe, Hg, Mn, Mo, S and Zn; *Ag is only enriched in slag samples from Kavodokanos.*

Apart from the chemical differences, slag heaps can, in fact, be distinguished in the field as well. The lumpy slag to the south of Aghia Paraskevi, behind the old iron pier of the former French Mining Company, is comparatively more weathered than slag from other sites (Fig. 3). Directly on its top store rooms of the same company have been built for the storage of pyrite, and presently they are in ruins. The lumpy slag of the heaps about Fougara hill, Panormos and Neapoli is weathered to a variable degree and should be of similar or slightly later age than that to the south of Aghia Paraskevi. The remaining slag heaps are of younger age, and not so weathered.

Sand-blast material comprises comminuted slag and fine pelletised slag. It is only found in the Kavodokanos area (Fig. 3). Although chemically similar to the Kavodokanos slag, it appears that there is a change in its chemical composition, when slag is fine- to medium-grained by either metallurgical process or crushing, i.e. it is enriched, in comparison to normal lumpy slag, with respect to most elements, e.g. Ag, As, B, Be, Ca, Cd, Cr, Cu, Fe, Hg, Mo, Ni, Pb, S, Sb, Sn, V and Zn. The earthy material within slag (slag earth) is also enriched; the Fougara slag earth is enriched with respect to Ag, Ba, Cd, Co, Cr, Cu, Hg, Ni, Pb, Sb and

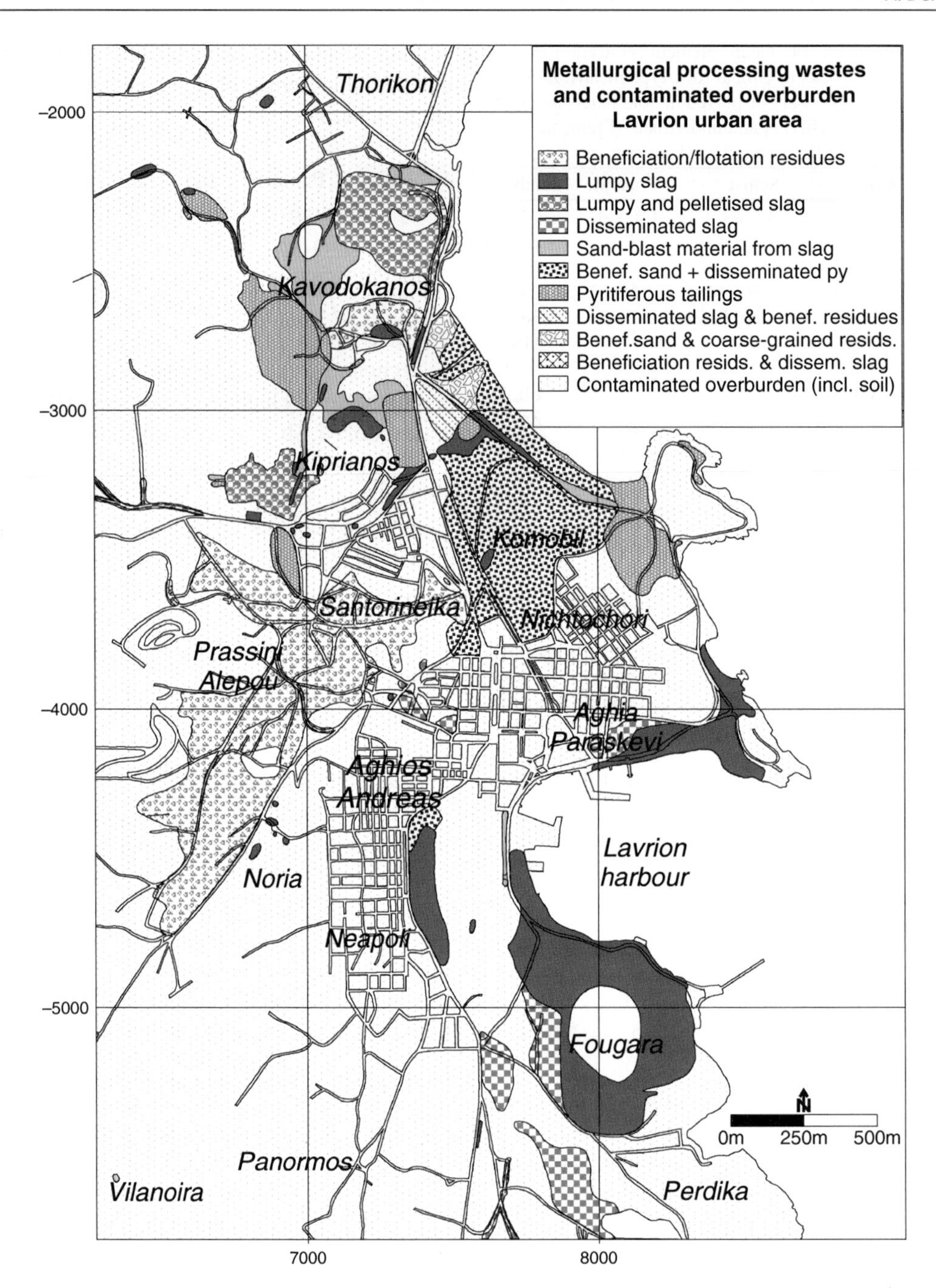

Fig. 3 Map of metallurgical processing wastes and contaminated overburden, including residual soil (from Demetriades et al., 1999f, Fig. 2.3, p.2.3; reduced from the original colour map at a scale of 1:5,000) (Mapped by A. Demetriades and K. Vergou-Vichou (1997))

V, and the Kavodokanos slag earth in Ag, As, Cd, Co, Cr, Cu, Hg, Ni, Pb, Sb and V.

Pyrite tailings can also be distinguished chemically into four broad categories:

(i) The Nichtochori-Komobil tailings, which were produced by the Mediterranean Mines company from 1953 to 1963, are characterised by high levels in Ag, As, Bi, Co, Fe, K, S, Sb, Sn and V and are weathered to a variable extent.

(ii) The Kavodokanos-Kiprianos-Santorineika pyrite-tailings are characterised by high levels in Al, Ca, Cd, (Cr), Cu, Fe, (Hg), La, Li, Mn, (Ni), P, Sr and Zn; *Cr is only enriched in pyrite samples from*

Table 4 Area covered by different types of metallurgical processing wastes and contaminated soil in the Lavrion urban area

Metallurgical processing wastes	Area (m^2)	Area (%)
Beneficiation/flotation residues or tailings	474,211.83	6.55
Lumpy slag	397,568.97	5.50
Lumpy and pelletised slag	135,764.79	1.88
Disseminated slag	145,008.07	2.00
Sand-blast material from slag and pelletised slag	105,813.29	1.46
Beneficiation/flotation sands with disseminated pyrite	305,047.01	4.22
Pyritiferous beneficiation/flotation residues	209,029.94	2.89
Disseminated slag and coarse-grained beneficiation/flotation residues	17,492.64	0.24
Beneficiation/flotation sands and coarse-grained materials	23,279.59	0.32
Beneficiation/flotation residues with disseminated slag	2,169.52	0.03
Contaminated overburden (including residual soil)	5,419,643.83	74.91
Total area	7,235,029.48	100.00

Source: Data from Demetriades et al., 1999 g, Table 2.3, p. 19

Kavodokanos and Hg and Ni are only enriched in pyrite samples from Kavodokanos and Kiprianos.

(iii) The Thorikon Gulf beach pyrite tailings are characterised by high levels in As, Bi, Co, Fe, Mn and S.

(iv) The Thorikon inland (Komobil) pyrite tailings are characterised by high levels in Ag, Al, B, Ba, Cd, Cr, Hg, K, La, Li, Ni, P, Sb, Sn, Ti, V and Zn.

The Kavodokanos pyrite tailings are the most weathered. All pyrite tailings produce acid drainage and are, therefore, potentially hazardous for polluting groundwater resources.

Flotation residues are chemically distinguished into the large mass extending from Noria to Prassini Alepou and Santorineika to the outskirts of Kiprianos and the smaller heap at Kavodokanos. The former is characterised by high levels in Al, Cr, K, Li, Mo, Ni, Ti and V and the latter in Ag, As, Ba, Bi, Cd, Cu, Fe, Hg, La, Mn, S, Sb, Sn, Sr, U and Zn. Their different chemical composition is attributed to differences in the original processed ore and its gangue.

Table 5 summarises the chemical features of the metallurgical processing wastes. In the same table are given, for comparison purposes, the median rock and overburden (including residual soil) values. This comparison definitely shows that the metallurgical wastes are the major source of contamination of overburden materials, including residual soil. For a more detailed description the report should be consulted (Demetriades and Vergou-Vichou, 1999b; Demetriades et al., 1999b).

It is finally concluded that the metallurgical processing wastes have high toxic element concentrations (As, Be, Cd, Cu, Hg, Pb, Sb, Zn). Their wide distribution in the Lavrion urban area, i.e. covering approximately 1.8 km^2 of the 7.24 km^2 studied in this project, which represents about 25% of the total area, and their transportation by aerial, fluvial and anthropogenic means, resulted in the contamination of overburden materials, including residual and alluvial soil. The first priority must, therefore, be the rehabilitation of areas covered by metallurgical processing wastes, as it will be shown below.

Land Use

The Lavrion urban area, covered by this study, extends from the residential part of the town into its outskirts, which are either agricultural or forested. The different land use types were mapped in detail by field mapping at a scale of 1:5,000. Aerial photography was used only for the exact delimitation of areas covered by forest and olive groves. Land use is an important parameter in the exposure assessment to environmental contaminants, and also the "soil" rehabilitation plan. For the same purpose other environmental management thematic maps were prepared, such as (a) property characterisation map and (b) map of urban control zones.

The original mapping of land use types had about 50 categories, which were later reduced to 30 by combining similar categories. Land use categories were classified into two broad groups (Table 6):

1. residential area, tillage, forest and open space and
2. industrial, commercial.

Both groups have 15 categories each. Areas covered by each land use category are tabulated in Table 6.

Table 5 Statistical parameters of element concentrations in metallurgical processing wastes, rock and overburden in Lavrion. Median values of rock and overburden are given for comparison purposes

	All metallurgical processing wastes (n=62) (values in ppm)						Rock (n=140)*	Overburden (n=224)**
Elements	Minimum	Maximum	Mean	Standard deviation	Coefficient of variation (%)	Median	Median (ppm)	Median (ppm)
Ag	3.20	96.00	33.80	28.22	83.49	18.90	0.50	12.06
Al	1,118.00	38,071.00	18,998.00	10,302.00	54.23	20,074.00	8,044.00	32,315.00
As	283.00	26,063.00	4,593.00	5,383.00	117.20	2492.00	15.60	1,290.00
B	< 5.00	667.07	61.78	102.80	166.40	42.97	0.25	136.00
Ba	27.70	2,059.00	368.20	419.00	113.80	243.00	210.00	479.00
Be	< 1.00	1.28	0.64	0.26	40.63	0.50	–	1.01
Bi	< 5.00	56.85	6.83	10.82	158.31	2.50	–	11.00
Ca	7,456.00	229,340.00	99,995.00	39,454.00	39.46	102,603.00	220.13	93,625.00
Cd	116.59	580.84	74.80	116.59	155.87	20.62	0.50	38.00
Co	2.98	83.95	26.35	15.96	60.57	23.83	20.50	16.00
Cr	8.11	299.15	83.47	59.98	71.86	73.18	20.00	183.00
Cu	184.00	8,700.00	1,172.90	1,362.60	116.17	630.50	25.00	186.00
Fe	35,000.00	380,000.00	217,081.00	89,238.00	41.11	234,500.00	19,515.00	44,771.00
Hg	< 1.00	10.20	2.58	1.73	67.05	2.35	0.50	0.14
K	993.00	13,408.00	6,942.00	3,911.00	56.34	7,100.00	8,044.00	9,770.00
La	< 2.00	47.25	25.13	11.64	46.32	27.32	8.86	22.70
Li	< 1.00	25.83	12.18	7.17	58.87	14.50	5.00	17.40
Mn	182.00	35,354.00	11,913.00	9,925.00	83.31	9,398.00	1,200.00	2,189.00
Mo	< 1.00	111.08	9.25	19.01	205.51	3.57	0.50	4.90
Ni	5.47	205.22	51.36	40.36	78.58	38.52	54.50	127.00
P	< 10.00	2,117.00	1,011.00	515.13	50.95	1,103.00	1,855.00	992.00
Pb	3,800.00	85,200.00	24,450.97	18,085.37	73.97	20,750.00	22.00	7,305.00
S	1,972.06	341,731.68	48,394.57	73,380.14	151.63	20,581.19	1,200.00	12,690.00
Sb	183.30	851.00	229.50	183.30	79.87	189.00	2.50	121.00
Sn	5.68	332.25	37.54	43.67	116.36	27.71	–	18.50
Sr	150.80	798.60	210.60	150.80	71.61	178.69	98.00	118.00
Ti	< 10.00	2,031.00	799.80	682.50	85.33	737.70	300.00	2,162.00
U	< 5.00	12.46	3.21	1.85	57.63	2.50	2.50	3.00
V	< 2.00	104.24	44.42	23.83	53.65	46.30	9.00	75.00
Zn	1,500.00	98,000.00	41,194.00	25,332.00	61.50	39,800.00	57.00	6,668.00

Source: from Demetriades and Vergou-Vichou, 1999b, Table 5.1, p. 128
*Rock: Ag, Mo (n=155); B, Be, Bi, Hg, La, S, Sb, Sn, U (n=48); **Overburden: B, Bi, Hg, S, Sn, U (n=50) and Sb (n=90)

Open space with or without bushes and scattered trees cover 53.86% of the study area. Together with the areas occupied by pine forest (5.15%), open space with trees (1.54%), olive groves (6.08%), vineyards (1.7%), vegetables (0.2%) and wheat fields (0.16%) make up 68.69% of the total area. All industries cover 16.82% of the total area, and only about 14.5% is utilised for residential and recreational purposes. Hence, the agricultural part dominates the study area, followed by the industrial, and the residential/recreational section of the town covers the smallest area.

A large part of the residential and recreational area of Lavrion is situated over contaminated soil, flotation residues and slag. Vineyards are mainly grown over the flotation residues, as is a large part of the olive groves. These statistics are not very encouraging, since local inhabitants live, work and play in the most hazardous areas.

Lead and Arsenic in Rock, Overburden and House Dust and Consequences to Health

The geochemical distribution of Pb in samples of rock, overburden (including residual soil) and house dust and As in overburden will be discussed, since these are the only two elements which were analysed in human tissue samples (Pb in blood and deciduous teeth; As in urine).

Table 6 Area covered by the different land use categories in the Lavrion urban area

Land use category	Area (m^2)	Area (%)
1. *Residential area, tillage, forest, open space*		
House without a garden, shop, church	481,186.45	6.65
House with a garden	416,639.21	5.76
School	44,613.16	0.62
Playground	6,724.61	0.09
Park	40,872.01	0.57
Football pitch, sports field	39,537.31	0.55
Olive grove	439,975.42	6.08
Vineyard	123,200.04	1.70
Vegetables	14,681.36	0.20
Wheat	11,222.18	0.16
Pine forest	372,498.46	5.15
Open space with trees (olive and fig trees, etc.)	111,722.65	1.54
Open space with trees (bushes)	3,896,642.50	53.86
Archaeological site	4,243.66	0.06
Cemetery	13,537.25	0.19
2. *Industrial, commercial*		
Ore treatment plant and storeroom	273,223.46	3.78
Smoke duct	23,358.94	0.32
Sand-blasting units with slag	2,436.98	0.03
Lead-acid battery factory	41,403.68	0.57
Factory (ammunition, weapons, matches), storerooms	437,996.98	6.05
Petrol station, garage and motor repairs	12,190.98	0.17
Iron converter industry and trading	119,636.64	1.65
Aluminium converter industries	3,060.38	0.04
Cotton weaving industries	119,232.05	1.65
Other industries, storage sites	9,906.59	0.14
Building materials storehouses	27,099.37	0.38
Port installations	141,574.90	1.96
Marble quarry	4,659.13	0.06
Old mining works	1,057.41	0.01
Municipal waste site	895.72	0.01
Total area	7,235,029.48	100.00

Source: Data from Demetriades et al., 1999 g, Table 2.4, p. 21

Toxic Effects of Lead

Trace amounts of Pb appear to be essential to the health of some animals, but it has never been shown as necessary in human biochemistry (Mervyn, 1985, 1986; Manousakis, 1992). Hippocrates (460–377 BC), the father of medicine, together with other Hellene philosophers and physicians, Aristotle (384–322 BC), Dioskourides (1st century BC) and Galenus (128–200 AD), has observed the effects of acute and chronic intoxication of Pb poisoning and described the classic symptoms of abdominal convulsions, constipation, anaemia and paralysis. Gaius Plinius Secundus, the classical Roman author (23–79 AD), mentioned the toxic effects of Pb and that vapours, released during smelting of Pb ore, should not be inhaled for *they are dangerous and could cause death to humans and animals*. He also warned about the danger to pregnant women in case of exposure; he advised the ones working in the treatment of "red Pb," crocoite ($PbCrO_4$), to protect themselves by wearing masks from animal organs. It appears, therefore, that ancient Hellenes and Romans knew that Pb poisoning (plumbism) was a disease contracted by workers in Pb mines and ore processing. Later medical practitioners, such as Paracelsus (1493–1541 AD), described the toxic effects of Pb at the work place. Lead apparently is the first toxic metal to receive so much attention by the medical profession from ancient to modern times.

The toxic effects of Pb are well documented, and especially for children who are more susceptible to Pb poisoning than adults (Kazantzis, 1973; Mervyn, 1985; Thornton and Culbard, 1987; Trattler, 1987; Briskin and Marcus, 1990; Ferguson, 1990; Reagan and Silbergeld, 1990; Manousakis, 1992; Wixson and Davies, 1994; US EPA, 1998). Inorganic Pb is not metabolised but is directly absorbed, distributed in different parts of the human body and excreted. The rate depends on its chemical and physical form, and on the physiological characteristics of the exposed person (e.g. nutritional status and age). Once in the blood, Pb is primarily distributed among three compartments:

- blood,
- soft tissue (kidney, bone marrow, liver and brain), and
- mineralising tissue (bones and teeth).

Absorption via the gastrointestinal track, following ingestion, is highly dependent upon the levels of calcium, iron, fats and proteins. Long-term exposure to inorganic Pb compounds is known to affect the central nervous system, peripheral nerves and kidneys. High levels of Pb in the human body have the following effects (Mervyn, 1980, 1985; Manousakis, 1992):

- *reduction of vitamin D*. If Pb is removed, however, by some means from the human body, then vitamin D is restored to normal level.

- *decrease in the amount of iron*. Lead inhibits the synthesis of haem and haemoglobin, as well as the activity of certain enzymes. In fact, anaemia is often the first sign of Pb intoxication.
- *reduction in the amount of calcium and phosphorus*. This antagonistic relationship between Pb and (Ca+P) works in the opposite way too. Calcium and phosphorus decrease the gastrointestinal absorption of Pb and aids in its elimination. Some authors suggest that people living in Pb high-risk environments they should drink additional quantities of milk. There is, however, contradictory evidence in the use of milk to counteract absorption of Pb and other toxic elements. Trattler (1987) states that milk as a source of calcium is associated with increased Pb levels and is not advised. He proposes alternatives, such as calcium orotate, calcium lactate or bone meal, and inorganic forms of phosphorus.

The US Environmental Protection Agency (US EPA, 1998, p. 5) stresses that it is important to document even exposure to low levels of Pb, because children can be permanently affected. Low-level exposure of children to Pb can cause

- nervous system and kidney damage;
- learning disabilities, attention deficit disorder and decreased intelligence;
- speech, language and behaviour problems;
- poor muscle co-ordination;
- decreased muscle and bone growth; and
- hearing damage.

While low-level exposure is most common, exposure to high levels of Pb can have devastating effects on children, including seizures, unconsciousness and, in some cases, death.

Although children are especially susceptible to Pb exposure, it can also be dangerous to adults. In adults, high Pb levels can cause

- increased chance of illness during pregnancy;
- harm to foetus, including brain damage or death;
- fertility problems (in men and women);
- high blood pressure;
- digestive problems;
- nerve disorders;
- memory and concentration problems, and
- muscle and joint pain.

Distribution of Lead (Pb) in Rocks

Parent rocks of the Lavrion urban area mainly consist of marble, schist, schistose-gneiss, prasinite, calcareous sandstone and conglomerate (Fig. 2). Fault breccia and outcrops of polymetallic and iron mineralisation are of minor significance, since they occur at a few locations. Marble and schist are the dominant rock types (Table 1). Statistical parameters of Pb concentrations in different rocks, occurring in the Lavrion study area, are tabulated in Table 7 and displayed in Fig. 4.

Marble has the greatest variation in Pb (0.5–1,850 ppm; coefficient of variation, C_V, 362.2%),

Table 7 Statistical parameters of lead (ppm Pb) concentrations in samples of rocks (marble, schist, schistose-gneiss, prasinite, sandstone and conglomerate), Lavrion urban area

Statistical parameters	All rock types	Conglomerate	Marble	Prasinite	Sandstone	Schist	Schistose-gneiss
*General rock-type mean**	17.0	–	5.0	4.0	10.0	22.0	–
Number of samples	140	3	88	4	6	33	4
Detection limit	1.0	1.0	1.0	1.0	1.0	1.0	1.0
Minimum	0.5	37	0.5	0.5	22	8	6
Maximum	1,850.00	1,040.00	1,850.00	60	58	810	54
Mean	76.85	602.33	54.61	19.63	35.67	111.82	18.75
Median	22	730	18	9	31	39	7.5
First quartile	11.5	–	10	3.25	24	16.5	6
Third quartile	54	–	39.5	36	48	72.25	31.5
Standard deviation	209.06	513.54	197.81	27.32	14.76	192.58	23.54
Coefficient of variation (%)	272.03	85.26	362.2	139.23	41.38	172.22	125.56

Source: Data from Demetriades et al., 1999 h, Table 3.1, p. 67
*Reimann et al. (1998)

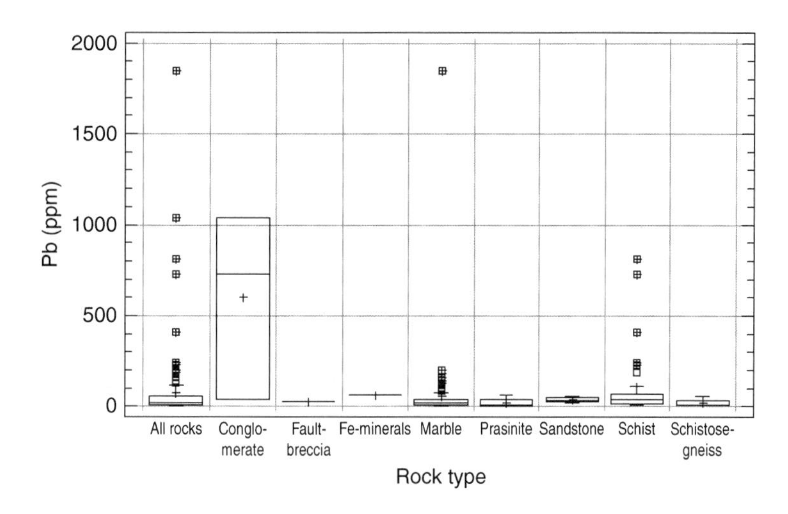

Fig. 4 Multiple *boxplot* showing the statistical distribution of lead (Pb) in different rock types occurring in the Lavrion urban area (from Demetriades et al., 1999 h, Fig. 3.1, p.67); refer to Table 7 for statistics

with a mean of 54.6 ppm and a median of 18 ppm. High Pb levels in marble occur in areas where it is altered (ankeritised) and slightly mineralised (calamine). Schist has also highly variable Pb concentrations (8–810 ppm; C_V=172.22%) and more anomalous values than marble, as indicated by the mean, median, first and third quartiles (Table 7; Fig. 4). Geochemically anomalous Pb levels in schist occur near its contact with altered marble. It is noted that base metal mineralisation occurs at the contact of marble and schist and is mainly hosted in marble. Schistose-gneiss (6–54 ppm Pb; C_V =125.56%), prasinite (0.50–60 ppm Pb; C_V =139.23%) and calcareous sandstone (22–58 ppm Pb; C_V=41.38%) do not have excessive Pb concentrations. The only deviation is by conglomerate (37–1,040 ppm Pb; C_V=85.26%), which locally contains altered mineralised marble pebbles.

From the geochemical exploration point of view, only marble and schist exhibit anomalous Pb concentrations (Fig. 5a). The outliers of the boxplot (Fig. 4) display this characteristic (Kürzl, 1988; O'Connor et al., 1988; Reimann et al., 2008), i.e. for marble there are outlying anomalous values >83.75 ppm Pb and for schist >155.88 ppm Pb. A major northeast to southwest trending geochemical anomaly extends from Aghios Andreas to Noria and a minor one in the Prassini Alepou area. Slightly elevated Pb contents occur about Fougara hill and Kiprianos. The marble in these areas is oxidised to a variable degree. Lead levels in rock samples from other parts of the Lavrion urban area are within the local geochemical background range (0.5–22 ppm Pb).

Distribution of Lead (Pb) in Metallurgical Processing Wastes

The statistical distribution of Pb in samples of metallurgical processing wastes is tabulated in Table 8 and displayed in Fig. 6. Pyrite tailings, pyritiferous sand and slag, show the greatest variation in Pb levels, because of differences in the chemical composition of the primary ore and metallurgical processing method, i.e. pyrite tailings 9,800–85,200 ppm Pb (C_V=52.65%) with a mean and median of 47,042 and 45,400 ppm Pb, respectively; pyritiferous sand with a variation from 3,800 to 41,200 ppm Pb (C_V=89.08%) and a mean of 19,130 and median of 16,765 ppm Pb and slag with values between 5,000 and 51,200 ppm Pb (C_V=72.09%). The narrowest range is shown by the flotation residues (18,500–28,100 ppm Pb; C_V=12.54%). Sand-blast material, whish is crushed slag, and the earthy material within slag have higher mean and median values with respect to slag. This is explained by the finer grain size of these materials.

Distribution of Lead (Pb) in Overburden

Figure 5b shows the spatial distribution of total Pb in overburden, including residual soil. The dominant features, governing the spatial distribution of Pb in the Lavrion surface environment, are the heaps of metallurgical processing wastes, i.e. flotation residues, pyrite

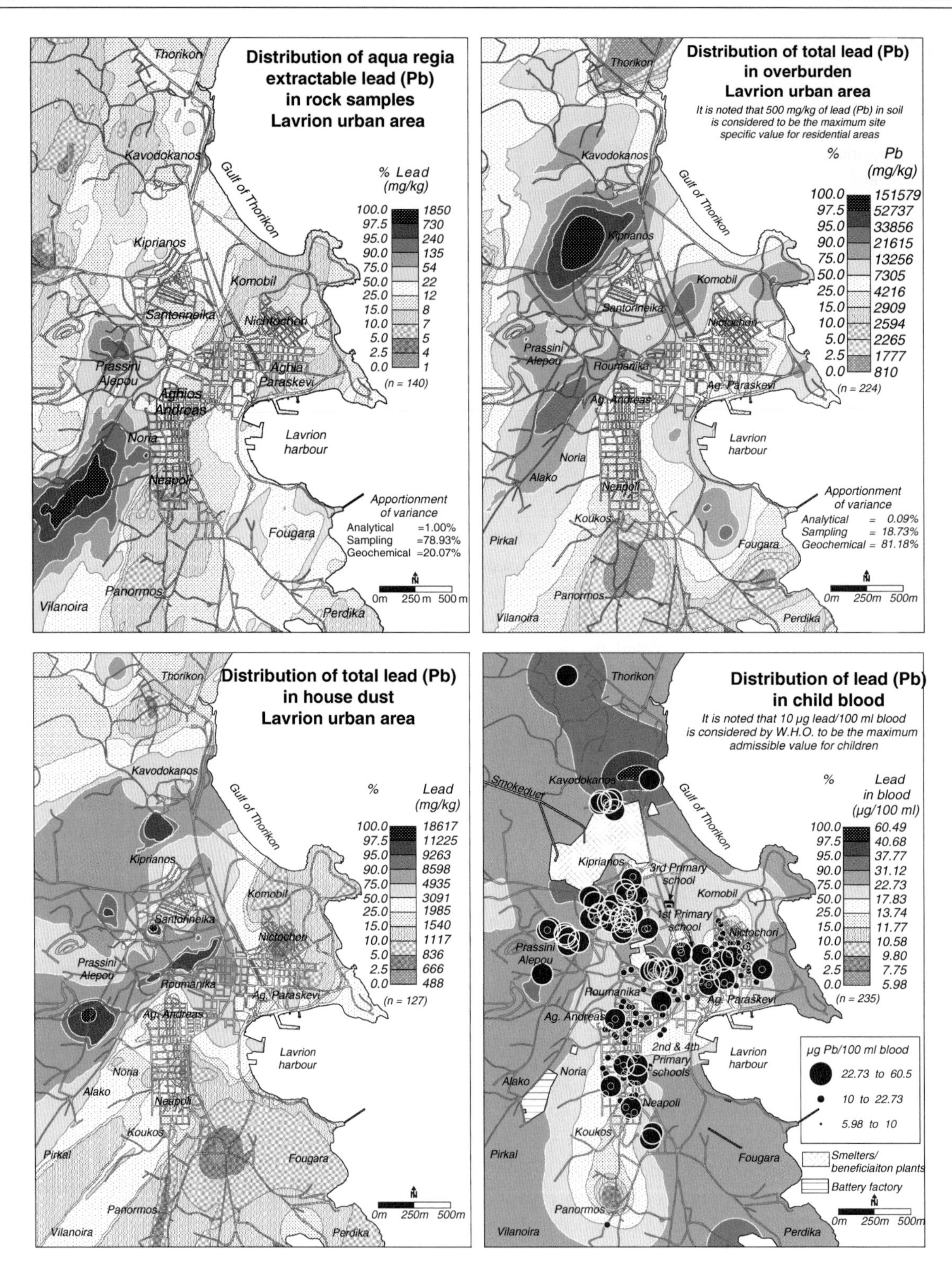

Fig. 5 Geochemical distribution maps of lead (Pb) in (**a**) rock, (**b**) overburden, including residual soil, (**c**) house dust and (**d**) child blood, Lavrion urban area (from Demetriades et al., 1999c, Map 3.1, p.3.1; Map 3.4, p.3.4; Map 3.5, p.3.5; Map 3.6, p. 3.6)

Table 8 Statistical parameters of aqua regia extractable lead (Pb) concentrations in samples of metallurgical processing wastes, Lavrion urban area (values in ppm or mg/kg)

Statistical parameters	Flotation residues	Pyritiferous sand	Pyrite tailings	Sand-blast material	Slag	Earthy material within slag
Number of samples	8	6	12	8	21	7
Detection limit	1	1	1	1	1	1
Minimum	18500	3800	9800	12300	5000	5080
Maximum	28100	41200	85200	32800	51200	30800
Mean	24537.5	19130	47041.7	19700	16495.2	19482.9
Median	24950	16765	45400	17250	11800	18500
First quartile	23050	3850	24650	14600	8575	15050
Third quartile	26850	32400	70400	24400	21550	26500
Standard deviation	3077.5	17041.8	24769.8	7741.4	11892	8757.5
Coefficient of variation (%)	12.5	89.1	52.7	39.3	72.1	44.9

Source: Data from Demetriades et al., 1999c, Map 3.2, p. 3.2

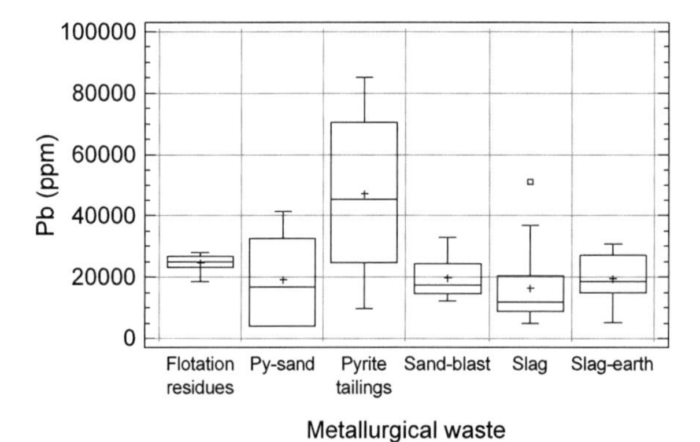

Fig. 6 Multiple *boxplot* showing the statistical distribution of lead (Pb) in samples from the different types of metallurgical processing wastes in the Lavrion urban area (from Demetriades et al., 1999 h, Fig. 3.2 , p. 68); refer to Table 8 for statistics

tailings and slag (Fig. 3). Their dominant orientations are northeast to southwest, and northwest to southeast.

Lead concentrations vary from 810 to 151,579 ppm, with a mean of 11,578 ppm, median of 7,305 ppm and a coefficient of variation of 133.79%, which indicates the great variability of Pb in overburden. It is quite apparent that high Pb contents in overburden materials, including residual soil, are mainly due to metallurgical processing activities. Emissions from smelters and the enormous amount of metallurgical wastes, deposited in the Lavrion urban area, and subsequent transportation of fine-grained fractions by aeolian processes (deflation) have contaminated even the remotest sites in the study area.

The greatest concentrations of Pb occur in areas covered by the metallurgical processing wastes. The smelter and its surrounding area (Kiprianos) have the highest Pb contents (21,615–151,579 ppm). The next area with high Pb levels occurs over the flotation residues (13,256–33,856 ppm) and extends from Alako in the south to Prassini Alepou and Kiprianos to the north (Fig. 5b). The area is inhabited and there are also agricultural activities (vegetables, olive groves, vines) and animal husbandry.

Distribution of Lead (Pb) in House Dust

Figure 5c illustrates the distribution of total Pb in house dust. The highest Pb concentrations in indoor dust occur in houses situated over metallurgical processing wastes and, especially, the flotation residues or tailings. It is indeed very disturbing to have total Pb contents in Lavrion homes varying from 488 to 18,617 ppm (mean=4,006 ppm Pb; median=3,091 ppm Pb). The notched box-and-whisker plot (Fig. 7) shows that there

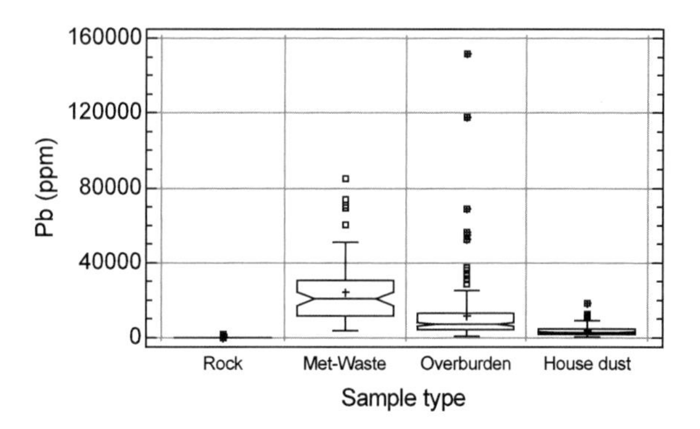

Fig. 7 Lead (Pb) contents in samples of rock (n=140), metallurgical processing wastes (n=62), overburden (n=224) and house dust (n=127), Lavrion urban area (from Demetriades et al., 1999 h, Fig. 3.11, p. 91)

is a trend of higher to lower Pb concentrations from the metallurgical wastes to overburden materials and house dust.

There appears to be a spatial relationship between total Pb in overburden (Fig. 5b) and house dust samples (Fig. 5c). Linear regression analysis of logarithmically (base 10) transformed Pb concentrations shows also that there is a strong linear correlation with a coefficient of 0.7 (Fig. 8). The coefficient of determination, $R^2\%$ statistic, indicates that the fitted model explains about 49% of the variability of total Pb in the two sample media. In Tables 9 and 10 are tabulated the calculated regression and analysis of variance parameters, respectively. The equation of the fitted model is given below:

$$\log_{10} \text{Pb in house dust} = 0.781202 + (0.711294 \times \log_{10} \text{Pb in overburden})$$

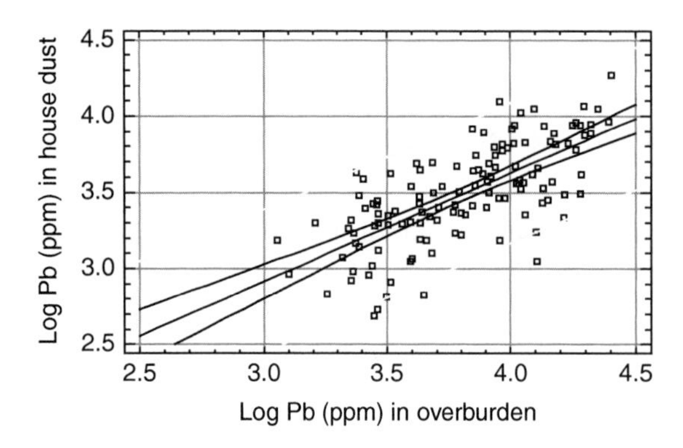

Fig. 8 *Scatter plot* of $\log_{10}$ Pb contents in house dust and overburden samples (n=126), linear regression line and 95% confidence limits (*dotted lines*) (from Demetriades et al., 1999 h, Fig. 3.3, p.70)

Table 9 Linear regression analysis of house dust and overburden total lead (Pb) contents (n=126 pairs)

Parameter	Value	Estimate	Standard error	t statistic	p value
Intercept		0.78120	0.24891	3.13851	0.0021
Slope		0.71129	0.06514	10.92020	0.0000
Correlation coefficient, R	0.700				
$R^2\%$	49.024				
Standard error of estimate	0.232				

Source: Data from Demetriades et al., 1999 h, Table 3.2, p. 70

Table 10 One-way analysis of variance of house dust and overburden total lead (Pb) concentrations (n=126 pairs)

Source	Sum of squares	Degrees of freedom	Mean square	F ratio	p value
Model	6.39155	1	6.39155	119.25	0.0000
Residual	6.64614	124	0.05360		
Total (corr.)	13.03770	125			

Source: Data from Demetriades et al., 1999 h, Table 3.3, p. 70

Since, the p value in the ANOVA table (Table 10) is less than 0.01, there is a statistically significant relationship between total Pb in house dust and total Pb in overburden materials at the 99% confidence level. Hence, the above equation may be used to predict total house dust-Pb concentrations from total overburden Pb in areas where there are no house dust samples.

Distribution of Blood-Lead in Children

Figure 5d shows the actual and interpolated blood-Pb (b-Pb) contents in 235 nursery and primary school age children in the Lavrion urban area, provided by the medical researchers of the last cross-sectional epidemiological study, which was carried out from 1 to 25 March 1988 (Makropoulos et al., 1991, 1992a, b).

Blood-Pb concentrations vary from 5.98 to 60.49 μg/100 ml, with a mean of 19.43 μg/100 ml and a median of 17.83 μg/100 ml (C_V=43.44%). An astounding 95% of the children have b-Pb levels exceeding 10 μg/100 ml, the level of "concern" (US EPA, 1986, 1998; ATSDR, 1988). This is indeed worrying, but not unexpected, when overburden in the Lavrion urban area has total Pb concentrations varying from 810 to 151,579 ppm (Fig. 5b) and house dust 488 to 18,617 ppm (Fig. 5c). Makropoulos et al. (1991) collected samples of floor dust, sand pit material and surface soil from the second and fourth primary schools, the school furthest away from the smelter (about 1,500 m); lead contents in the different materials were

- 3,780 ppm Pb in floor dust;
- 253 ppm Pb in sand pit material; and
- 10,283 ppm Pb in surface soil outside the school.

Makropoulos et al. (1991) did not realise the significance of Pb concentrations in floor dust and surface soil samples from the second and fourth primary school, which is at a distance of approximately 1,500 m from the smelter at Kiprianos, but are situated in contaminated areas. Using the regression equation above, and the concentration of 10,283.27 ppm Pb determined on soil outside the school, a value of 4,315 ppm Pb has been calculated for school dust. This value differs by approximately 14% from the analytical value of 3,780 ppm Pb, which is an acceptable difference, after considering analytical and sampling errors, and the linear correlation coefficient of 0.7. A direct relationship between surface "soil" and house dust or floor dust is thus verified, within the limits of acceptable error variation.

As it has been pointed out by many researchers, soil-Pb and house dust-Pb contribute much more to blood-Pb than does air Pb (Yankel et al., 1977; Schmitt et al., 1979; Roels et al., 1980; Brunekreef et al., 1981; Culbard et al., 1983; Duggan, 1983; Brockhaus et al., 1988; Reagan and Silbergeld, 1990; Thornton et al., 1990; Cotter-Howells and Thornton, 1991; Ferguson and Marsh, 1993; Watt et al., 1993; Ferguson et al., 1998). Hence, the geostatistical interpolation of actual b-Pb values (Fig. 5d) was based on the knowledge of Pb distribution in surficial materials, i.e. metallurgical processing wastes (Fig. 5a), overburden materials (Fig. 5b) and house dust (Fig. 5c).

High b-Pb concentrations in children are found in the areas where there are metallurgical processing wastes. Houses are either built on or very close to these wastes. Further, overburden materials, including residual soil, and house dust are highly contaminated (Figs. 5b and 5c). These confounding factors play a significant role in the high b-Pb levels of Lavrion children, found in different areas, e.g.

- in the Prassini Alepou area they are attributed to the fact that houses are built on flotation tailings;
- to the south of the disused ore washing plant (central part of the map) they are mainly ascribed to (a) aeolian transportation of fine-grained flotation tailings and (b) contaminated soil;
- to the north of the smelter in the Kavodokanos area they may be due to (a) contaminated soil, (b) smelter wastes and (c) father working at the smelter;
- the single high b-Pb value on the coastal road in the Kavodokanos area is most likely due to (a) contaminated soil and (b) proximity to metallurgical waste heaps; and
- the single high b-Pb value in the Thorikon area (north part of the map) is definitely attributed to (a) contaminated soil and (b) proximity to a metallurgical waste heap.

It is quite apparent from the above description that contaminated soil and the metallurgical processing wastes play a significant role in the b-Pb contents of Lavrion children. There is clearly a correlation between the distribution of total Pb in overburden (Fig. 5b) and b-Pb (Fig. 5d). Generally, low b-Pb levels in children are found in areas where "soil" is not intensely contaminated, e.g. (a) Panormos-Koukos-Neapoli-Aghios Andreas area, (b) Aghia Paraskevi-Nichtochori area and (c) the area directly north of Santorineika.

High Pb levels in overburden and house dust, the components in direct contact with children, are very significant to elevated b-Pb contents, when such materials are inhaled or ingested by children. Although soil pica for Lavrion children was mentioned by medical researchers (Maravelias et al., 1989), they did not realise its importance and persisted to explain observed high b-Pb levels in children to smelter emissions (Makropoulos et al., 1991). Recent studies on soil pica behaviour of children have shown that some children may ingest up to 25–60 g of soil during a single day (Calabrese et al., 1997). The US EPA on the other hand, for purposes of estimating potential health risks on children, assumed that most children ingest relatively small quantities of soil (e.g. <100 mg/day), while the upper 95th percentile are estimated to ingest 200 mg/day on average (US EPA, 1996). As Calabrese et al. (1997) pointed out, soil ingestion studies normally last for about a week or less and, therefore, it is not possible to obtain a clear understanding of intra-individual variability in soil ingestion activity. Their model, however, which is based on their long experience in soil ingestion studies on children (Stanek and Calabrese, 1995) indicates that for 1–2 days/year:

- 62% of children will ingest >1 g of soil,
- 42% of children will ingest >5 g of soil, and
- 33% of children will ingest >10 g of soil (Calabrese et al., 1997).

Their conclusions are

(a) for the majority of children, soil pica may occur only on a few days of the year, but much more frequently for others;
(b) soil pica, although highly variable, is an expected activity in a normal population of young children, rather than an unusual activity in a small subset of the population, and
(c) the implications of soil pica are significant for risk assessment.

In Lavrion, a moderate to strong northeasterly wind is blowing almost throughout the whole year. The comparatively dry climate for most months aids transportation of dust by wind. Fugitive dust, which is dust released from soil by the action of wind, with or without the assistance of mechanical disturbance (Ferguson et al., 1998) is a major hazard in Lavrion. At the head height of a young child, fugitive dust appears to be considerable, for it was experienced on the face of the sampling team during kneeling down for overburden sampling. Particles of less than 10 μm diameter are normally assumed to be respirable; larger particles are trapped in the upper respiratory tract, from where they are expectorated or swallowed (Ferguson et al., 1998). In a study of Pb exposure in young children from dust and soil in the United Kingdom, Thornton et al. (1990) estimated Pb uptake by a young child to be 36 μg Pb/day, of which 1 μg was by inhalation and 35 μg by ingestion.

It is quite apparent from the above discussion that Pb contents in "soil" and "dust" are important to exposure assessment of young children in Lavrion.

Lead (Pb) in Deciduous Teeth

Deciduous teeth were collected in 1991 from 82 Lavrion children of 6–12 years old in order to assess Pb absorption and to compare the results with non-industrial areas. The teeth were analysed for Pb at the Department of Anatomy of the University of Bergen in Norway. Lead concentrations in deciduous teeth of Lavrion children vary from 0.97 to 153.26 μg/g with a mean of 9.88 μg/g (Stavrakis et al., 1994a). The pre-industrial deciduous teeth Pb contents in children, of the same age range as Lavrion, vary from 0.20 to 4.37 μg/g with a mean 0.91 μg/g (Fosse and Wesenberg, 1981). Typical Pb levels in deciduous teeth vary from <0.1 to 50 μg/g (Ferguson, 1990). The conclusion is that Pb is in a bioaccessible form, as is shown by the comparatively high concentrations in the deciduous teeth of Lavrion children.

Toxic Effects of Arsenic

Arsenic in minute amounts is an essential element for some organisms, even humans, but in excess is highly toxic with teratogenic properties (Gough et al., 1979; Mervyn, 1985, 1986; Reimann et al., 1998; Chappell et al., 1999). Toxicity depends on the valence state of As compounds, e.g. As^{5+} compounds are less toxic than As^{3+}. Uptake of high arsenic concentrations causes gastrointestinal problems, which are accompanied by diarrhoea together with blood loss, and is followed by the destruction of capillary vessels and inflammation of intestinal walls. As a consequence of this rapid loss of liquids, circulatory insufficiency appears, which if not treated could cause death after a shock. Inflammation of lungs and dyspnoea is caused by acute inhalation of arsenical compounds.

With chronic exposure to arsenic, which could be the case in Lavrion, the more common effects include gradual loss of strength, diarrhoea or constipation, pigmentation and scaling of skin, which may undergo malignant changes, nervous manifestations marked by paralysis and confusion, degeneration of fatty tissue, anaemia and development of characteristic streaks across the fingernails.

Food is considered to be the most important source of arsenic exposure to the human population. In adults, the daily total intake of arsenic from food is estimated to be less than 200 μg. This value depends mainly on the amount of sea food consumed. The largest part of this arsenic is in an organic form. In water, arsenic predominantly exists in an inorganic form.

More information on the toxic effects of arsenic can be found in the literature, as for example in Chappell et al. (1999).

Distribution of Arsenic (As) in Rocks

Arsenic was determined on only 48 rock samples from the 140 collected. Its statistical parameters and distribution are tabulated in Table 11 and shown in Fig. 9.

Table 11 Statistical parameters of aqua regia extractable arsenic (ppm As) in all rock types, marble and schist, Lavrion urban area

Statistical parameters	All rocks	Marble	Schist
Global rock-type mean*	*2.00*	*1.50*	*13.00*
Number of samples	48	37	10
Detection limit	5.00	5.00	5.00
Minimum	2.50	2.50	5.02
Maximum	1,032.39	1032.39	42.61
Mean	62.82	76.27	18.71
Median	15.56	12.77	19.76
First quartile	6.91	5.73	10.35
Third quartile	29.01	30.06	24.13
Standard deviation	172.39	194.82	11.03
Coefficient of variation (%)	274.52	255.45	58.97

Source: Data from Demetriades and Vergou-Vichou, 1999c, Table 4.2A, p. 20
*From Reimann et al. (1998)

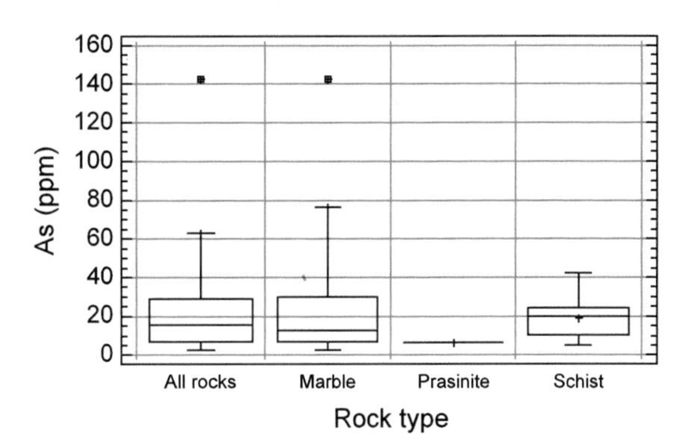

Fig. 9 Multiple *boxplot* of the distribution of aqua regia extractable arsenic (As) in different rock types occurring in the Lavrion urban area. Note that the category "All rocks" and "Marble" have four anomalous samples over 160 ppm As (i.e. 218, 377, 544 and 1032 ppm) (from Demetriades and Vergou-Vichou, 1999c, Fig. 4.2A, p. 20); refer to Table 11 for statistics

Marble has the great variation with values ranging from <5 to 1,032 ppm As, with mean and median of 76.27 and 12.77 ppm As, respectively. Schist has values varying from 5.02 to 42.61 ppm As, with a mean of 18.71 and a median of 19.76 ppm As. Anomalous As concentrations occur in marble and are found in areas where the marble is altered.

Distribution of Arsenic (As) in Metallurgical Wastes

Arsenic levels are considerably high and very variable in samples from the different types of metallurgical processing wastes, i.e. from 283 to 26,063 ppm (Table 12, Fig. 10). The widest variation of As values is found in flotation residues (2,470–26,063 ppm), followed by pyritiferous sand (3,875–14,440 ppm), pyrite tailings (2,893–14,747 ppm) and slag earth (684–15,907 ppm). Slag (283–6,236 ppm As) and sand-blast material (854–2,243 ppm As) have comparatively lower As contents.

Slag, apart from one sample high in As from Kavodokanos (Fig. 3), has the lowest values. Arsenic is enriched in slag earth and sand-blast material in comparison to samples of slag, a feature already noted for Pb, and is explained by the fact that these materials are finer grained.

Pyrite tailings of the Nichtochori-Komobil area have higher As contents than the ones at Kavodokanos, Kiprianos and Santorineika (Fig. 3). Pyritiferous sand samples from the beach, between Komobil

Table 12 Statistical parameters of aqua regia extractable arsenic (As) concentrations in samples of metallurgical processing wastes, Lavrion urban area (values in ppm or mg/kg)

Statistical parameters	Flotation residues	Pyritiferous sand	Pyrite tailings	Sand-blast material	Slag	Earthy material within slag
Number of samples	8	6	12	8	21	7
Detection limit	5.0	5.0	5.0	5.0	5.0	5.0
Minimum	2469.9	3874.9	2892.6	853.9	282.5	684.1
Maximum	26063.3	14440.1	14747.2	2243.2	6136.4	15906.8
Mean	9415.9	7225.2	7325.3	1710.3	929.3	6427.4
Median	4633.0	4740.9	5214.8	1803.5	512.6	3793.4
First quartile	2913.9	4306.5	4281.7	1387.8	399.8	2692.8
Third quartile	15850.2	11247.6	10358.5	2101.2	791.8	11610.6
Standard deviation	8853.5	4479.4	4078.5	488.5	1268.3	5955.3
Coefficient of variation (%)	94.0	62.0	55.7	28.6	136.5	92.7

Source: Data from Demetriades et al., 1999b, Map 5.3, p. 5.3

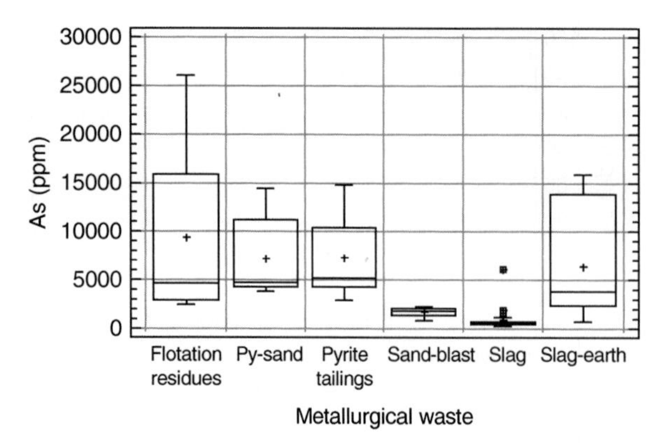

Fig. 10 Multiple *boxplot* of the statistical distribution of arsenic (As) in samples from the different types of metallurgical processing wastes in the Lavrion urban area (from Demetriades and Vergou-Vichou, 1999b, Fig. 5.3, p. 104); refer to Table 12 for statistics

and Kiprianos, have comparatively higher As levels compared to the ones further inland. Arsenic concentrations in samples of flotation residues from Kavodokanos have considerably higher than those from Noria-Prassini Alepou-Santorineika. These differences in the levels of arsenic are mainly due to the chemical composition of the processed primary ore.

Distribution of Arsenic (As) in Overburden and House Dust

Arsenic in overburden samples varies from 50 to 24,000 ppm, with a mean of 2,494 ppm ($s=\pm3,558$ ppm; $C_V=142.66\%$) and a median of 1,290 ppm ($n=224$) (Fig. 11). Whereas in house dust samples from Lavrion homes As levels vary from 130 to 3,820 ppm with a mean of 5,435 ($s=\pm507.1$ ppm; $C_V=9.33\%$) and a median of 400 ppm ($n=127$).

Major geochemical anomalies in overburden samples, with As concentrations above 1,290 ppm, have NE-SW and NW-SE trends and occur over and in the immediate neighbourhood of the flotation residues (Noria-Prassini Alepou-Santorineika), pyritiferous wastes (Komobil, Nichtochori) and slag/sandblast material (at Fougara, Aghia Paraskevi, within the smelter area at Kiprianos and Kavodokanos) (Fig. 11). The strongest geochemical anomalies >6,000–24,000 As occur over the pyrite tailings and pyritiferous sand from Nichtochori to Komobil and Kiprianos, and within the premises of the smelter. Distinct haloes extend about the geochemical anomalies down to 380 ppm As, although even lower levels, to 50 ppm As, may be attributed to aerial dispersion from the smelters and metallurgical wastes.

It is apparent from the spatial and statistical distribution of As in different sample types (Fig. 12) that the metallurgical processing wastes are the major source of contamination of overburden samples. The contribution of As from rock samples, as is also the case for all elements, is totally masked by its higher concentrations in metallurgical wastes. Both sample types, metallurgical wastes and overburden, contribute to the contamination of house dust.

If a contamination index for As in overburden and house dust samples is estimated, with respect to its median in parent rocks, this ratio varies from 3.2 to 1,538.5 in overburden samples and from 8.3 to 244.9 in house dust samples. These values show the very high contamination of overburden and house dust samples by As.

Mobility of As is generally moderate under oxidising, acid and neutral to alkaline conditions, and very low under reducing conditions (Levinson, 1980; Hoffman, 1986; Reimann et al., 1998). Ferguson (1990), however, states that an increase in pH, as for example by liming, increases the mobility of arsenite salts, presumably by bringing about a change from aluminium [$AlAsO_4$] and iron [$FeAsO_4$] arsenates to calcium arsenates [$Ca_3(AsO_4)_2$]. If this is the case, then Lavrion has a significant problem. Another important issue is the long residence time of As in soil, which according to Ferguson (1990) is 2,000 years.

It is quite apparent from the above description that As levels in Lavrion overburden materials are abnormally high and may be hazardous to biota in general. Toxicity, as in all cases, depends on the concentration of easily soluble and bioaccessible part of As, and not its total content in soil, an aspect that will be discussed below.

Distribution of Arsenic (As) in Urine

Eikmann et al. (1991) investigated in 1988 two industrial areas in Hellas, Lavrion and Elefsina-Aspropyrgos, and a background reference town, Loutraki, which is a spa town on the Bay of Corinth with no heavy or light industrial pollution. The industrial area of Elefsina-Aspropyrgos to the west of Athens is characterised by steel industry, refineries,

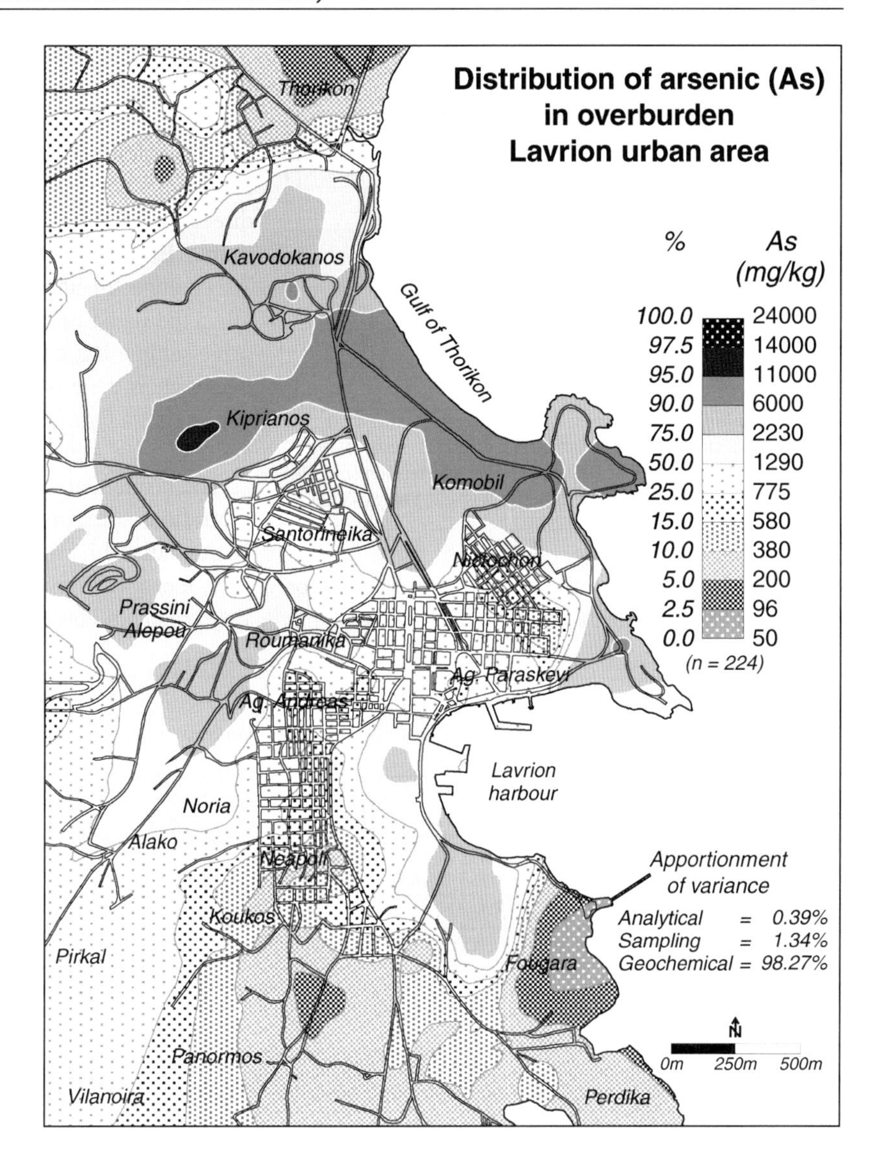

Fig. 11 Geochemical distribution map of arsenic (As) in samples of overburden, including residual soil (from Demetriades et al., 1999d, Map 6.3, p. 6.3)

a waste oil combustion plant and factories producing lead alkyls; an additional form of pollution is the emissions from traffic on the national highway between Athens and the Peloponnese.

Table 13 shows the arsenic concentrations in 24 hour urine (24 h) of workers from the Lavrion lead smelter and Alako battery factory, and the Aspropyrgos refinery. The urine-As values of Aspropyrgos workers are well below the recommended limit of 40 μg/24 h, whereas those of the Lavrion workers exceed it, i.e. 70.6% of the smelter workers and 45.2% of the battery factory workers.

In Table 14 are tabulated the urine-As concentrations in children from Lavrion, Aspropyrgos and Loutraki. The children from Aspropyrgos and Loutraki show approximately similar mean and median urine-As excretion values, whereas those from Lavrion are higher by more than two times. The frequency distribution of individual urine-As values shows once again the tendencies observed in the workers (Table 13). About 8.4% of Lavrion children have As excretion values above the recommended limit of 20 μg/24 h; the As excretion values of Aspropyrgos and Loutraki children, with only one exception in each case, are below the recommended limit.

Ten years later, on the 19th November 1998, morning urine was collected from 65 Lavrion inhabitants, and the samples were analysed at the Institute of Analytical Chemistry of the University Karl-Franzens in Graz, Austria. Evaluation of results is based on the limit of 100 μg As/l urine, which is considered to be the maximum exposure for working adults (Caroli

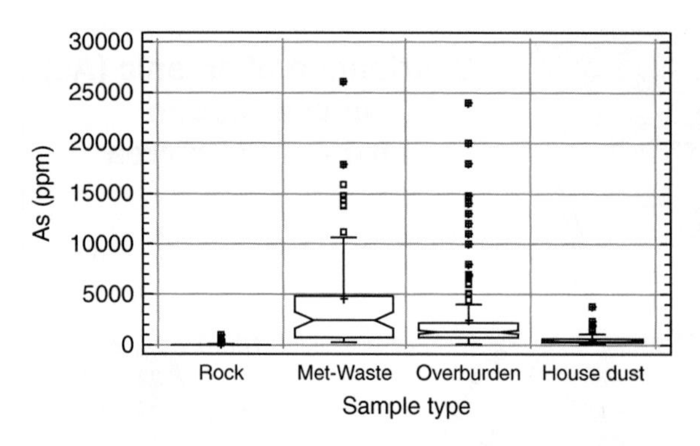

Fig. 12 Multiple *notched box*-and-*whisker plot* comparing the arsenic (As) statistical distribution in different sample types, rock (n=48), metallurgical wastes (n=62), overburden (n=224) and house dust (n=127) (from Demetriades and Vergou-Vichou, 1999d, Fig. 6.3, p. 142)

et al., 1994). It appears that 37% of the 65 Lavrion inhabitants have urine-As concentrations above this limit.

The results show that there are high concentrations of As in urine in the areas of Prassini Alepou (23.7–1,279 μg As/l, mean 229.5 and median 58.4 μg As/l) and Kavodokanos (29.7–831.5 μg As/l, mean 210.8, median 192.7 μg As/l), where the inhabitants are in direct contact with the metallurgical processing wastes (Table 15, Fig. 13). In some areas there are elevated As contents in urine in one or two samples, which affect the statistical distribution, e.g. at Roumanika one urine sample has a concentration of 1,095 μg As/l, at Nichtochori there is also one sample with 448 μg As/l and at Aghia Paraskevi there are two samples with 244 and 285.6 μg As/l.

The residential areas with the lowest urine-As concentrations are Kiprianos (10.2–28.5 μg As/l, mean 18.3, median 17.5 μg As/l) and Neapoli (19.5–59.0 μg As/l, mean 42.9, median 45.4 μg As/l), which have a very small amount of metallurgical processing wastes,

Table 13 Arsenic concentration in adult urine (μg/24 h) from workers of different factories in Lavrion and Aspropyrgos, an industrial town

Town/industry	N	Mean	Median	95th percentile	Range	Frequency <40 μg/24 h	Frequency >40 μg/24 h
Lavrion lead smelter	34	77.42	73.15	191.7	1.84–192.95	10 (29.4%)	24 (70.6%)
Lavrion battery factory	31	43.97	36.69	121.6	8.54–173.12	17 (54.8%)	14 (45.2%)
Aspropyrgos refinery	25	14.17	14.23	25	1.70–26.54	25 (100%)	0 (0.0%)

Source: Data from Eikmann et al., 1991, Tables 1 and 2, p. 464

Table 14 Arsenic concentration in child urine (μg/24 h) from Lavrion, Aspropyrgos (industrial town) and Loutraki (a spa town)

Town	N	Mean	Median	95th percentile	Range	Frequency <20 μg/24 h	Frequency >20 μg/24 h
Lavrion	261	8.59	5.76	65.9	0.53–77.23	239 (91.6%)	22 (8.4%)
Aspropyrgos	128	3.58	2.69	18.2	0.24–20.42	127 (99.2%)	1 (0.8%)
Loutraki	193	4.03	2.83	22.3	0.62–34.84	192 (99.5%)	1 (0.5%)

Source: Data from Eikmann et al., 1991, Tables 3 and 4, p. 465

Table 15 Arsenic (As) in urine samples from different residential areas in Lavrion (values in μg As/l)

Statistical parameters	Aghia Paraskevi	Prassini Alepou	Kavodokanos	Kiprianos	Neapoli	Nichtochori	Roumanika
Number of samples	7	18	17	7	6	6	4
Minimum	12.0	23.7	29.7	10.2	19.5	13.9	9.7
Maximum	285.6	1279.0	831.5	28.5	59.0	448.0	1095.0
Mean	106.8	229.5	210.8	18.3	42.9	98.0	291.0
Median	50.3	58.4	192.7	17.5	45.4	24.3	29.7
First quartile	16.3	29.2	72.4	15.7	32.8	15.6	16.6
Third quartile	212.7	161.5	289.0	20.5	55.6	62.0	565.5
Standard deviation	114.6	368.7	192.0	5.6	15.8	172.4	536.1
Coefficient of variation (%)	107.3	160.6	91.1	30.5	36.8	175.9	184.2

Source: Data from Demetriades et al., 2008, Table 6, p. 594

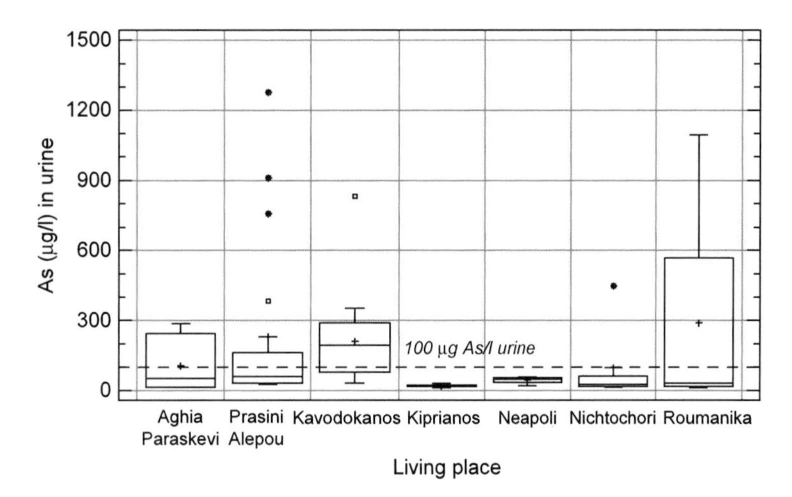

Fig. 13 Multiple *boxplot* showing the statistical distribution of urine-As in 65 people living in different residential areas of Lavrion (from Demetriades et al., 2008, Fig. 11, p. 620); refer to Table 15 for statistics

mainly slag at Kiprianos, or none at all as is the case of Neapoli.

The high As concentrations in the urine of Lavrion inhabitants show that it is still bioaccessible for uptake 9 years after the closure of the smelter complex. Therefore, the sources of the hazard still exist, and these are the metallurgical processing wastes and contaminated overburden materials, including residual and alluvial soil.

Development of an Environmental Management Plan

Besides assessing the contamination of overburden materials that are in direct contact with the Lavrion inhabitants, and especially children, the final objective of the LIFE project "Soil Rehabilitation in the Municipality of Lavrion" was to develop an *Integrated environmental management scheme* for improving the living conditions in Lavrion. After considering existing information, and data generated in the project, two aims were defined by Nikolaidis et al. (1999a):

1. The primary aim is to gradually minimise human exposure to toxic elements through
    - surgical remediation based on cost-benefit analysis,
    - lifestyle changes through education,
    - legislative action for land use changes,
    - health and safety guidelines, and
    - elimination of consumption of contaminated crops and groundwater; the latter is not used for drinking, but is utilised for irrigation purposes.
2. The secondary aim is to protect groundwater resources and biota and to rehabilitate surface soil.

Given the high degree of contamination of the Lavrion urban area and its extent, i.e. almost covering the whole study area of 7 km^2, the integrated environmental management plan focused exclusively on the primary aim. *Ten years later not even the items of this objective were implemented by the local authorities*. It was stressed that implementation of the recommendations to achieve the primary aim will be to partially protect groundwater resources and biota and the rehabilitation of "soil" to protect human receptors, and especially children. Therefore, the primary aim of the proposed action plan was related to the impacts due to "direct human exposure" to contaminants, than to the "mobility" of contaminants from soil to other receptors. This distinction was considered extremely important to comprehend, because it determined the type of analysis conducted, namely a risk-based analysis, rather than a deterministic fate and transport exposure analysis.

Development of Site-Specific Soil Action Levels

To formulate site-specific soil action levels (SALs) for Lavrion, one needs to take into consideration established risk assessment methodologies and local conditions, such as the geochemical background concentration of elements. Figure 5c presents the extent of Pb contamination of overburden materials, including residual soil, in the Lavrion urban area. Total Pb concentrations in surficial soil range from 810 to 151,579 ppm, with a mean of 11,578 ppm and a median of 7,305 ppm. These results suggest extensive soil contamination and lack of locations to collect background samples. Based on this premise, it was decided, as described above, to collect samples from the underlying bedrock to establish the local background conditions (Fig. 5a). A total of 140 bedrock samples were collected and analysed for 25 major and trace elements. In addition, 39 duplicate rock samples were analysed for 44 major and trace elements. The results of selected elements, considered to be toxic, are presented in Table 16. It is stressed that the Lavrion area bedrock, because of the different mineralising events, is enriched in heavy metals, compared to other areas without mineralisation. The median values of the elements tabulated in Table 16 were taken to represent background levels.

The European Union (Ferguson et al., 1998) and the US (US EPA, 1996) risk assessment guidelines for establishing site-specific screening action levels were reviewed. The general guidelines are similar in nature; the difference is that US EPA has developed specific equations that can be used in environmental impact assessment. It was, therefore, decided to follow the US EPA risk assessment methodology for the following reasons:

(a) it is published,
(b) it is risk based,
(c) it is similar to EU's,
(d) it has accessible data through the web page of US EPA's Integrated Risk Information System (IRIS) (http://cfpub.epa.gov/ncea/iris/),
(e) it is site-specific adaptable, and
(f) it is continuously updated, as more data become available.

The methodology differentiates between carcinogen and non-carcinogens. In general, arsenic and beryllium are the carcinogenic elements, and the rest of the heavy metals are considered to be non-carcinogenic. US EPA establishes SALs for carcinogenic elements, allowing a risk of one-in-a-million (10^{-6}, i.e. one additional cancer for every million people) for each element, and a hazard index of one for the non-carcinogenic elements. The equation accounts for ingestion and inhalation of particulates. Dermal absorption was not considered, since very limited data are available.

Soil action levels were developed for residential/recreational and industrial land uses in Lavrion and are presented in Table 17.

Using the residential or industrial SALs contamination indices can be estimated. For this purpose ten toxic elements were considered, i.e. As, Ba, Be, Cd, Cr, Cu, Ni, Pb, V and Zn. The following equation was used to calculate the contamination index for soil used for residential and recreational purposes:

$$\text{Contamination Index (CI)} = [\text{As}/25 + \text{Ba}/5{,}500 + \text{Be}/2 + \text{Cd}/40 + \text{Cr}/140 + \text{Cu}/2{,}300 + \text{Ni}/1{,}500 + \text{Pb}/500 + \text{V}/550 + \text{Zn}/20{,}000] - 10$$

For each overburden sample individual values of each element are divided by its respective soil action value; the procedure results in an element ratio in which the baseline level is represented by the number 1, i.e. when the actual concentration of an element is equal to its soil action level. Thus, element ratios >1 depict contamination. To estimate the total contamination at each sampling site, the element ratios are summed up, and from the sum is subtracted the

Table 16 Toxic elements "background concentrations" in bedrock samples from Lavrion (*n*=39)

	Toxic element concentration in ppm			
Element	Minimum	Maximum	Mean	Median
Antimony	5	71	6	5
Arsenic	2.5	540	42	11
Barium	3	4087	244	78
Beryllium	–	–	–	–
Cadmium	1	41	2	1
Chromium, total	1	610	100	20
Silver	0.5	41	1	0.5
Manganese	100	25000	1830	1200
Copper	3	225	33	25
Lead	1	1850	77	22
Mercury	1	8	1	1
Nickel	1	1600	168	55
Vanadium	1	71	13	9
Zinc	6	5200	211	57

Source: Data from Nikolaidis et al., 1999a, Table 1, p. 77

Table 17 Comparison of soil contamination levels with residential and industrial/commercial direct exposure Soil Action Levels (SALs) for Lavrion (n=50)

Element (ppm)	Lavrion Soil Samples Minimum	Maximum	Mean	Median	Residential SALs (ppm)	Industrial SALs (ppm)
Antimony	28	567	192	151	30	800
Arsenic	50	24,000	2,494	1,290	25	25
Barium	64	4,555	663	479	5,500	140,000
Beryllium	0.2	2.7	1.1	1	2	2
Cadmium	4	925	68	38	40	1,000
Chromium, total	2	1,083	264	183	140	140
Chromium, trivalent	–	–	–	–	20,000	500,000
Chromium, hexavalent	–	–	–	–	140	140
Copper	43	4,445	357	186	2,300	60,000
Lead	810	151,579	11,578	7,305	500	1,000
Mercury*	1*	728*	189*	117*	20*	600*
Nickel	40	591	141	127	1,500	40,000
Vanadium	26	325	86	75	550	14,000
Zinc	591	76,310	10,872	6,668	20,000	610,000

Source: Data from Nikolaidis et al., 1999a, Table 2, p. 77
*Values of mercury in ng/g or μg/kg or ppb

sum of baseline levels, which in this case is 10, since ten elements are considered. This is a procedure used in mineral exploration for the compilation of total geochemical anomaly maps (Sainsbury et al., 1970; Siegel, 1974). A map was subsequently plotted showing the total contamination of overburden, including residual soil (Fig. 14). Contamination indices above zero denote contamination. The contamination index map shows that the whole Lavrion urban and sub-urban area is contaminated up to 1,121 times. This is, of course, expected, because at least two elements, Pb and As, have no values below their respective SALs.

Development of Conceptual Site Model and Methodology for Remediation

A Conceptual Site Model (CSM) that is consistent with the primary aim of the integrated environmental management plan for the Lavrion urban area is presented in Fig. 15. The model conforms with the history of the area. Mining and metallurgical activities of the past 5,000 years, and especially the intensive metallurgical activities since 1865 in the Lavrion urban and sub-urban area, have left a large volume of "hazardous wastes" that constitute the primary pollution source. Anthropogenic, aeolian and fluvial processes have spread this contamination and have formed the secondary pollution source that is the present day surface soil. The adverse human health effects are attributed to direct exposure (inhalation of dust particles and ingestion of soil and dust) and consumption of locally produced agricultural and animal husbandry products (e.g. green vegetables, fruit, milk, meat, wine, olive oil).

Surgical remediation is a methodology for ranking (in a quantitative manner) contaminated areas, which need to be remediated. Typically, this prioritisation is conducted through a cost–benefit analysis. The parameters that need to be considered in such an analysis are the following:

- degree of risk (potential impact),
- sources of contamination,
- exposure route (inhalation, ingestion, etc.),
- land use type,
- population affected,
- remedial technologies and their effectiveness,
- feasibility of implementation of technology (it might be technically viable, but politically impossible), and
- social costs and benefits of remediation.

Thirty different land use categories (Table 6) and 11 with different contaminants and geochemical characteristics (Table 4) were used.

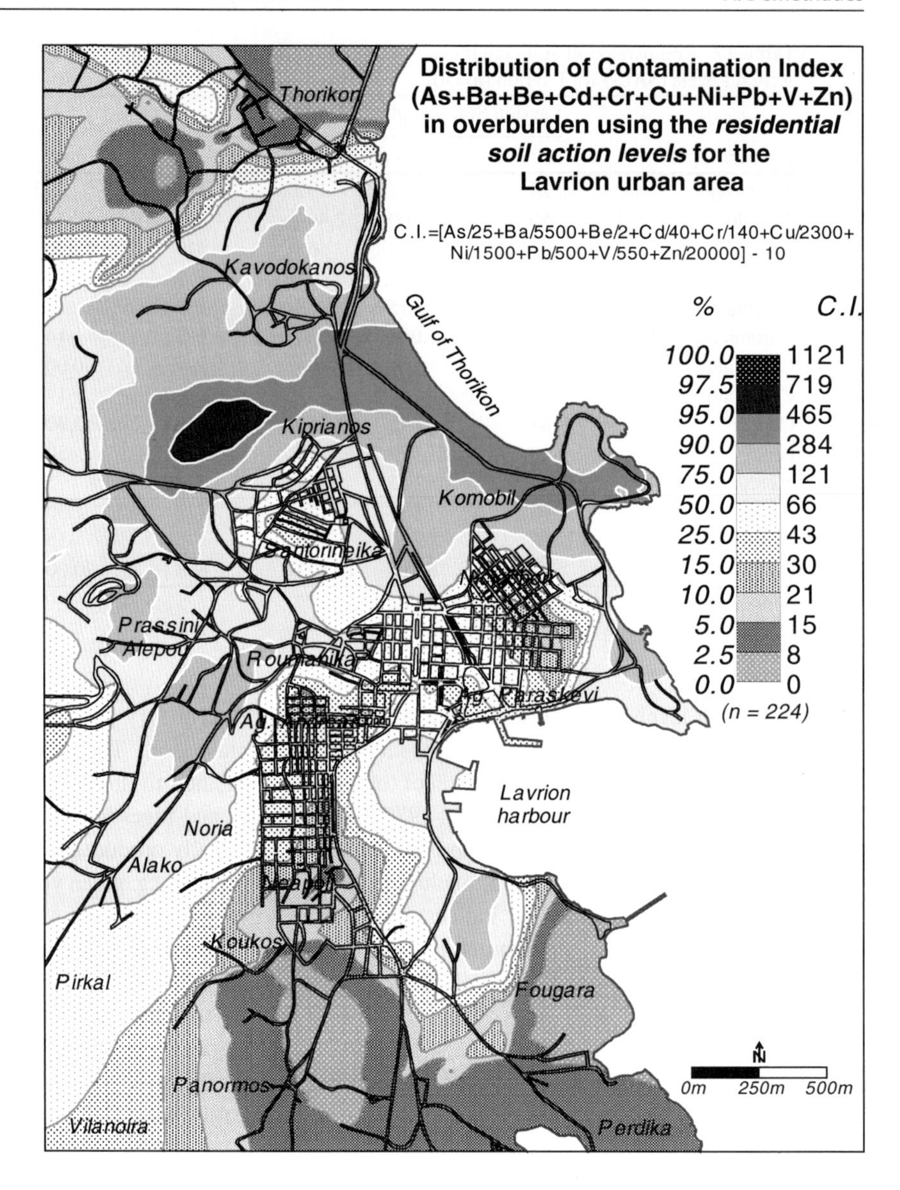

Fig. 14 Map showing the distribution of contamination index for soil using the developed residential soil action levels for Lavrion (from Demetriades et al., 1999e, Map 1.6, p. 1.6)

For each land use category, over each source of contamination, two indices were developed:

- a benefit index, BI, of benefits obtained from remediating the area, and
- a cost index, CI, of costs relating to various technologies that can remediate the area to the required clean-up level to obtain the corresponding benefits.

The ratio of cost to benefit index, CI/BI ratio, was used to rank the areas in terms of priority for remediation. The lower the ratio, the higher is the priority for remediation.

The benefit index is defined as follows:

$$\mathrm{BI}_i = \mathrm{EI}_i \times \mathrm{SBI}_i$$

where

EI_i is the exposure index of area *i* and
SBI_i is the social benefit index of area *i*.

The cost index is defined as follows:

$$\mathrm{CI}_i = (\mathrm{FI}_j / \mathrm{DE}_j) \times \mathrm{SCI}_i$$

where

FI_j is the *financial investment* of using technology *j*,
DE_j is the *degree of effectiveness* of technology *j* and
SCI_i is the *social cost index*.

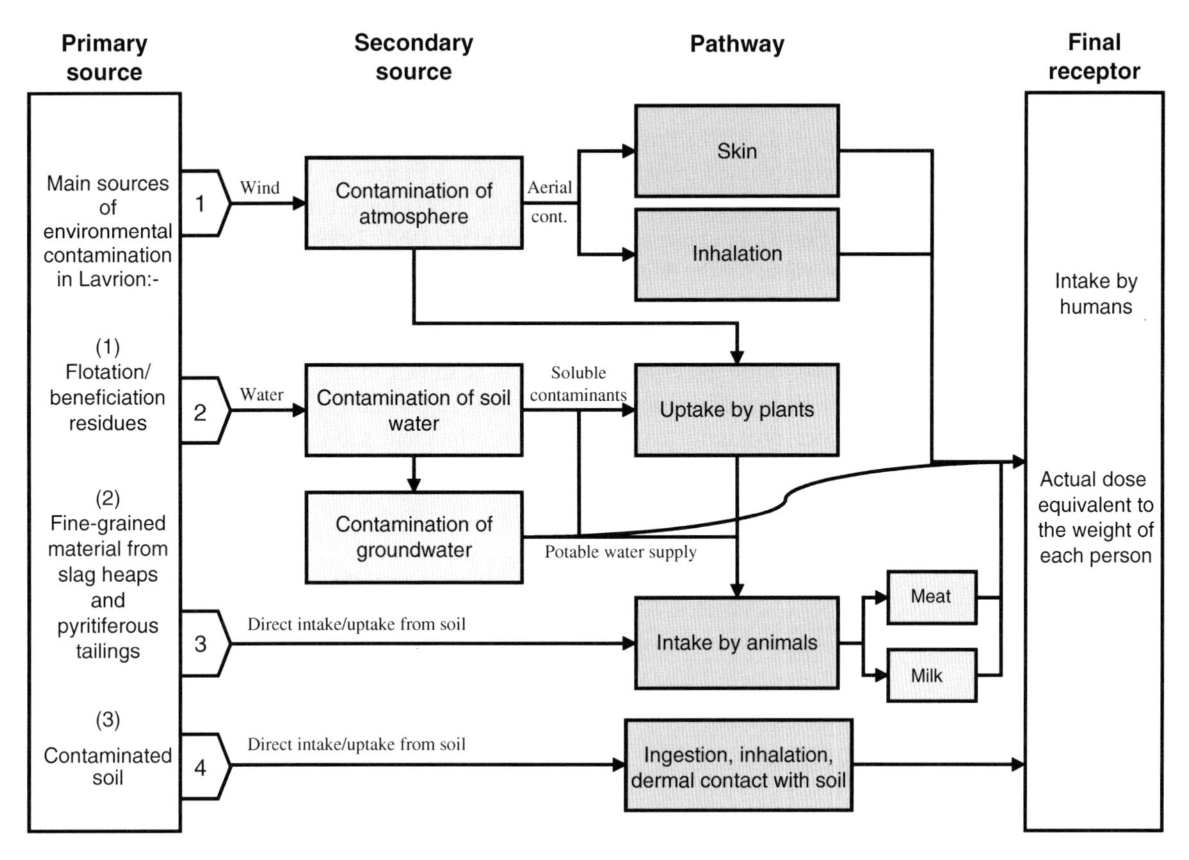

Fig. 15 Conceptual site model for the contamination of the Lavrion urban area (Compilation by A. Demetriades; from Tristán et al., 1999a, Fig. 11.1, p. 319)

Spatially resolved hazard and exposure assessments were performed as part of this EU LIFE project (Tristán et al., 1998, 1999a, b, 2000). The methodologies used to assess direct exposure of the population to environmental pollutants included both quantitative and semi-quantitative approaches. The quantitative approach employed the Integrated Exposure Uptake BioKinetic (IEUBK) (US EPA, 1994) and Human Exposure from Soil Pollutants (HESP) (Shell, 1994a, b) models to predict the spatially resolved blood-lead (b-Pb) concentrations in children and to compare them with measured epidemiological data. This approach provided an insight to the spatial and biomedical factors that affect blood-Pb concentrations, and also a measure of uncertainty by producing probabilistic hazard maps (Tristán et al., 1999a, b, 2000). The semi-quantitative approach used a multi-criteria evaluation (Eastman, 1997) for the development of a spatially resolved exposure index. In this approach, continuous criteria (factors) and Boolean constraints are standardised to a common numeric range and then combined in a weighted linear fashion to produce the exposure index. Some criteria have discrete categories – constraints (e.g. area with metal-related industry or site overlying Quaternary deposits) – and others are continuous (e.g. proximity to wastes or roads). The eight factors that were defined are

(1) lead concentration in overburden/soil;
(2) degree of dustiness of metallurgical waste;
(3) proximity to metallurgical wastes;
(4) proximity to current or previous stacks;
(5) proximity to roads;
(6) proximity to rivers;
(7) proximity to lead industry, and
(8) degree of exposure of children.

The two constraints include

(1) area with metal-related industry and
(2) area over Quaternary to Recent deposits.

Exposure index was estimated on a grid of 50×50 m, which is the grid used for the geostatistical interpolation (kriging) of total lead concentrations in overburden/soil samples (Demetriades et al., 1998; Tristán et al., 1999a, b). Other criteria were obtained from the available digital thematic maps, i.e. lithology (Fig. 2), metallurgical processing wastes (Fig. 3), land use map (Table 6), town plan and stacks. A value for the degree of dustiness was estimated for the different types of metallurgical processing residues, based on available information. Weights were assigned to the ten criteria (lead concentration in overburden, degree of dustiness of waste, proximity to smelter wastes, proximity to current or previous stacks, proximity to roads, proximity to rivers, proximity to lead industry, time of exposure of children, area with metal-related industry and over Quaternary/Recent deposits). This complex set of information was processed by the Idrisi® GIS using the Weighted Factors in Linear Combination module (Eastman, 1997) for estimation of the exposure index (Tristán et al., 1998, 1999a, b, 2000). The original arbitrary exposure index scale of 0–255 was reduced to percent (0–100) for the purposes of the environmental management scheme (Fig. 16).

As it is shown in Fig. 16 there is no 50 × 50 m block with zero child exposure to contaminants in the Lavrion urban and sub-urban area. Most blocks have exposure levels between 25 and 50% (i.e. 1,914 blocks,

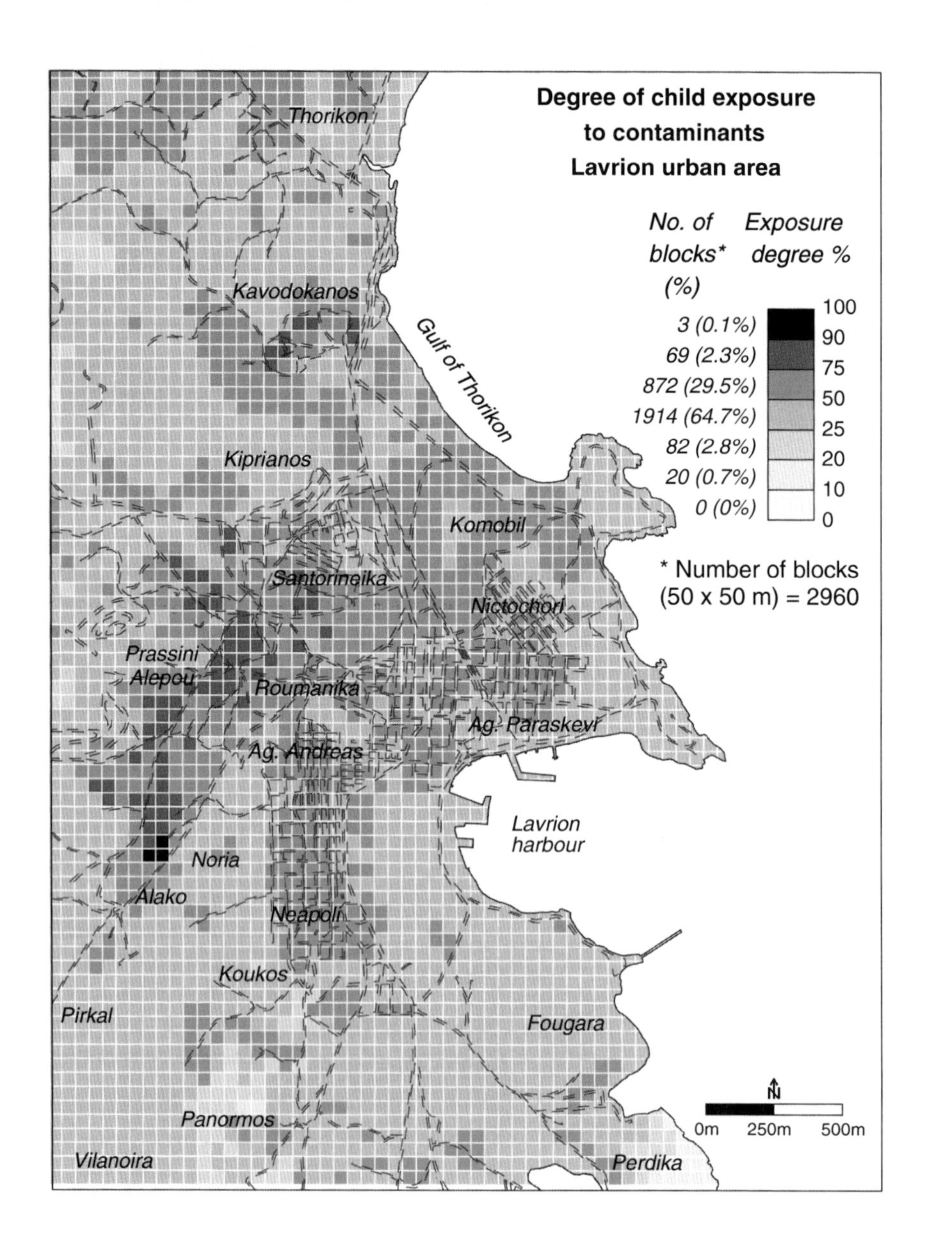

Fig. 16 Map showing the distribution of degree of child exposure to environmental contaminants, which was developed by the method of weighted linear combination, Lavrion urban area (from Tristán et al., 1999b, Map 11.17, p. 11.17)

representing 64.7% of the total number of blocks) and from 50 to 75% (i.e. 872 blocks, representing 29.5% of the total number of blocks). These are indeed very disturbing figures, since most residential areas, where children live and play are considered to be hazardous to their health.

To begin with 12 technologies were considered as viable remedial alternatives for this project. The list is not, of course, exhaustive, because other technologies do exist. The technologies evaluated are imminently applicable to the solution of the Lavrion contamination problem, by combining acceptable efficiency and low cost. Most of the technologies mentioned below were tested as part of this project (Kontopoulos et al., 1995; NTUA, 1999), i.e.

- Coverage with asphalt
- Paving
- Vegetation
- *Clean soil coverage (60 cm) and vegetation*
- Geomembrane cover including vegetation of the upper layer
- *Compacted clay cover including vegetation of the upper layer*
- *Addition of limestone and coverage with clean soil (30 cm) and vegetation*
- Chemical stabilisation with biological sludge and vegetation
- *Chemical stabilisation with biological sludge and phosphates and vegetation*
- Chemical stabilisation with biological sludge and fly ash and vegetation
- Chemical stabilisation with compost and fly ash and vegetation
- Chemical stabilisation with compost and phosphates and vegetation

The degree of effectiveness of the various remedial technologies was considered in the Lavrion project. Applicability of each technology depends on the chemical characteristics of contaminated material and on the actual or planned land use of the particular site. Application of impermeable covers and addition of limestone are appropriate for sulphide-containing wastes (pyrite and pyritiferous sands). Chemical stabilisation techniques are suitable for calcareous flotation tailings and "soil". Asphalt and paving can be applied only in urban and industrial areas.

Coverage with clean soil is in principle a technique applicable to all types of contaminated material, except the pyritiferous wastes, although even this type of waste can be covered temporarily to prevent the aerial transportation of contaminants. By covering contaminated areas with clean soil to an acceptable thickness, the risk for direct exposure of the population is minimised. The major constraint for this technique is the availability of clean soil in the Lavreotiki Peninsula. To illustrate the needs, it is simply mentioned that more than 4 million cubic metres of clean soil would be required to cover the surface of 7.2 km^2, which was examined within the framework of the present EU LIFE project. With respect to the coverage by only vegetation without prior stabilisation, it is an option with uncertain efficiency, and it largely depends on the level of contamination. It was found, however, to be totally unsuccessful on the highly contaminated flotation residues (NTUA, 1999).

Finally, four remedial technologies were considered to be cost-effective and practical, i.e.

- *Clean soil coverage (60 cm) and vegetation* for the areas with the different types of slag, contaminated residual and alluvial soil;
- *Compacted clay cover including vegetation of the upper layer* for the areas covered by pyrite tailings;
- *Addition of limestone and coverage with clean soil (30 cm) and vegetation* for the areas covered by pyritiferous sand; and
- *Chemical stabilisation with biological sludge and phosphates and vegetation* for the areas covered by the flotation/beneficiation residues.

Figure 17 shows the remedial technology to be applied at each 50×50 m block, depending on the type of waste to be remediated. The cost per square metre was estimated using 1999 prices of materials. The total cost at the end of 1999 was 58.18 million Euro. It has been estimated that about 30% of the area is covered by buildings and tar roads. Therefore, the actual cost is reduced to about 42 million Euro (1 Euro was equivalent to 330 drachmas in 1999).

Discussion and Conclusions

In this study it has been shown that overburden materials, including residual and alluvial soil, have been contaminated to an extremely high degree by metallurgical processing activities. Although mining and

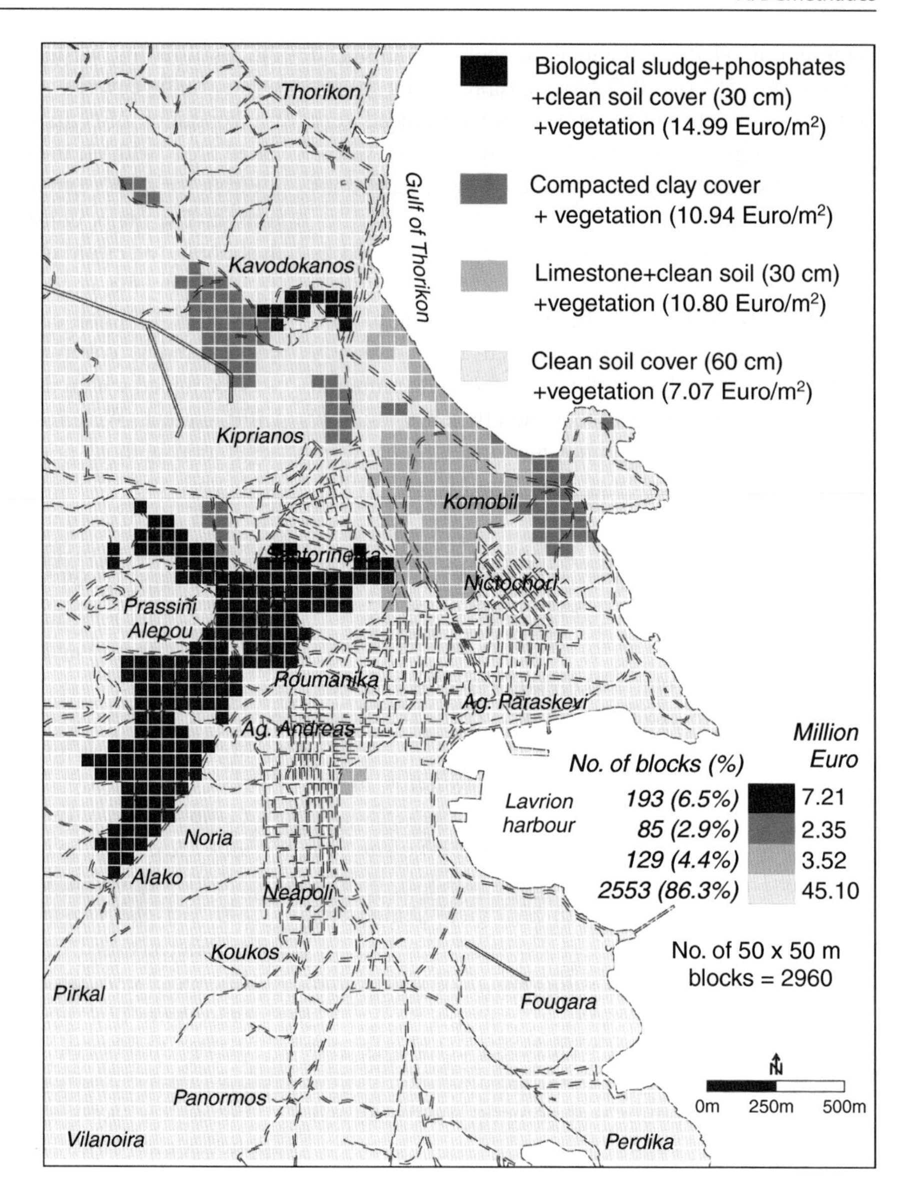

Fig. 17 Map showing the distribution of cost index of the least cost technologies for the rehabilitation of "soil" in the Lavrion urban area (from Nikolaidis et al., 1999b, Map 12.2, p. 12.2)

smelting activities in the greater Lavrion area date back to c. 3500 BC, the multi-element contamination described is largely due to the wastes produced by the intensive metallurgical processing from 1865 to 1989. These wastes cover approximately 25% of the area studied in this project (7,235 km^2). Their subsequent redistribution by aerial, fluvial and anthropogenic processes has caused the contamination of surface residual soil and house dust.

Generally, the mobility of cations in the largest part of Lavrion is comparatively low to moderate due to the alkaline pH of soil water. However, in the area of pyritiferous tailings the action of air and rainwater causes the oxidation of pyrite and generation of acid drainage (sulphuric acid, H_2SO_4) which is very corrosive. It leaches metals from the wastes, and acid drainage is produced, containing high to extreme concentrations of toxic elements, which could be harmful to humans, animals and plants. This acid drainage is also a potential hazard to groundwater supplies. Fortunately, the area covered is relatively small, i.e. the coastal area from Nichtochori-Komobil to Kiprianos and Kavodokanos.

The extreme multi-element soil contamination in Lavrion can be said that it is a permanent feature, since the residence time of heavy elements in soil is very high, i.e. As 2,000 years, Hg 920 years, Cd 280 years and Pb 400–3,000 years (Ferguson, 1990).

Since Lavrion has many peculiarities with respect to its contamination characteristics, site-specific Soil Action Levels were produced (Nikolaidis et al., 1999a) and used for the estimation of the contamination index for As, Ba, Be, Cd, Cr, Cu, Ni, Pb, V and Zn (Fig. 14). Barium, Ni and V have values below the estimated site-specific Soil Action Levels for Lavrion, whereas the concentrations of Pb and As in all Lavrion overburden samples are well above their respective SALs of 500 ppm Pb and 25 ppm As. The other elements have a smaller proportion of samples exceeding their SALs, i.e. Be (~2.5%), Cu (~2.5%), Zn (~20%), Cd (~50%) and Cr (~70%). Apart from the extreme contamination of overburden materials, including residual soil, it is disturbing to have house dust values varying from 488 to 18,617 ppm Pb and As from 130 to 3,820 ppm.

Calculation of soil lead clean-up levels requires a site-specific lead risk assessment model, which establishes a relationship between the levels of lead in soil and blood-lead. In the case of Lavrion such relationships are not necessary, since there are available epidemiological data showing that children and adults have high concentrations of (a) Pb in blood, (b) As in urine and (c) Pb in deciduous teeth. These data from human tissues show, therefore, that Pb and As are in bioaccessible forms, and remedial actions must be taken to reduce exposure to acceptable levels. Although this study was completed at the end of 1999, and the six-volume report submitted in 2000 to the Lavrion urban authorities and the Hellenic Ministry of Environment, with concrete cost-effective proposals for remediation of soil, not a single action was taken up to now. The situation remains the same with houses and schools built on metallurgical processing wastes, and children and adults alike exposed to the same environmental contaminants. It is, therefore, stressed that we are obliged to offer better and healthier living conditions to the new generation. Otherwise, it is mathematically certain that in the not too distant future we may resort to extreme solutions, because the remediation of "soil" would be more costly than moving the whole population of Lavrion and Lavreotiki Peninsula to another more suitable area.

Acknowledgements This chapter is published by permission of the General Director of the Hellenic Institute of Geology and Mineral Exploration. The results come from the LIFE programme, contract number 93/GR/A14/GR/4576, which was co-financed by the European Commission and the Hellenic State. Without the assistance of the many scientists, technical and other supporting staff this project would have not been completed. They are all sincerely thanked for their contributions.

References

ATSDR (Agency for Toxic Substances and Disease Registry) (1988) The nature and extent of lead poisoning in children in the United States: A report to Congress. US Public Health Service, Atlanta.

Baziotis I, Proyer A, Mposkos E (2008) High-pressure/low-temperature metamorphism of basalts in Lavrion (Greece): Implications for the preservation of peak metamorphic assemblages in blueschists and greenschists. Eur J Mineral 21(1):133–148.

Benetou-Marantidou A, Nalou S, Micheloyiannis I (1985) The use of a battery of tests for the estimation of neurological effects of lead in children. In TD Lekkas (ed) International Conference Heavy Metals in the Environment (Vol. 1, September, pp. 204–209). New Orleans, Edinburgh: CEP Consultants.

Briskin J, Marcus A (1990) Goals and implications of EPA's new drinking water standard for lead. In DD Hemphill (ed) Trace substances in environmental health – XXII. Supplement to volume 12 of Environmental Geochemistry and Health (pp. 33–50). Columbia: University of Missouri.

Brockhaus A, Collet W, Dolgner R, Engelke R, Ewers U, Frier I, Jermann E, Krämer U, Turfeld N, Winneke G (1988) Exposure ot lead and cadmium of children living in different areas of North-West Germany: Results of biological monitoring studies, 1982–1986. Int Arch Occup Environ Health 60:211–222.

Brunekreef B, Veenstra SJ, Bierksteker K, Boleij JS (1981) The Arnhem lead study I. Lead uptake by 1- to 3-year-old children living in the vicinity of a secondary lead smelter in Arnhem, The Netherlands. Environ Res 25:441–448.

Calabrese EJ, Stanek EJ, James RC, Roberts SM (1997) Soil ingestion: A concern for acute toxicity in children. Environ Health Persp 105:1354–1358.

Caroli S, Alimonti A, Coni E, Petrucci F, Senofonte O, Violante N (1994) The assessment of reference values for elements in human and biological tissues and fluids: A systematic review. Crit Rev Anal Chem 24(5–6):363–398.

Chappell WR, Abernathy CO, Calderon RL (eds) (1999) Arsenic Exposure and Health Effects (416 pp.). Amsterdam: Elsevier.

Chronopoulos J, Chronopoulou-Sereli C (1986a) Scwermetall-toleranz von *Crocus sieberi*, *Arisarum vulgare* und *Cyclamen graecum* in Lavrion (Attika). Verhandlungen der Gesellschaft für Ökologie (Hohenheim 1984), XIV, 357–360 (text in German with a synopsis in English).

Chronopoulos J, Chronopoulou-Sereli C (1986b) Vegetational development of halophytes to heavy metals in industrial

regions in Lavrion (Attika). Landschaft u. Stadt 18(1):42–45 (text in German with a summary in English).
Chronopoulos J, Chronopoulou-Sereli C (1991) Effects of the mining-metallurgical activity on the natural vegetation of Lavreotiki. In Abstracts of 1st Scientific Conference on Geosciences and the Environment. University of Patras, Dept. of Geology, Patras, p. 147.
Chronopoulou-Sereli C, Chronopoulos J (1991a) Untersuchungen über die Pb-belastung der vegetation in Lavreotiki (Attika). In S Riewenherm, H Lieth (eds) Verhandlungen der gesellschaft für Ökologie (Osnabrück 1989), XIX/III, 223–228 (text in German with a summary in English).
Chronopoulou-Sereli C, Chronopoulos J (1991b) Unweltbelastung der Stadt Lavrion (Attika) und Umgebung durch ein Bleinhüttenwerk. Natur und Landschaft 66(9):442–443 (text in German with a summary in English).
Conophagos EC (1980) Le Laurium antique et la technique Grecque de la production deo l'argent. National Technical University, Athens, 458 pp. (in French).
Cotter-Howells, Thornton I (1991) Sources and pathways of environmental lead to children in a Derbyshire mining village. Environ Geochem Health 13:127–135.
Culbard E, Thornton I, Watt J, Moorcroft S, Brooks K, Thompson M (1983) Metal contamination of dusts and soils in urban and rural households in the United Kingdom. In DD Hemphill (ed) Trace Substances in Environmental Health – XVII. University of Missouri, Columbia, pp. 236–241.
Demetriades A (1992) Development of integrated collaborative research programmes between the U.K. (BGS) and Greece (IGME). Environmental Geochemistry, Lavreotiki peninsula, and Multidisciplinary data interpretation, Eastern Macedonia and Thrace. Vol. 1: Text, 165 pp.; Vol. 2: Maps, diagrams and tables, 128 pp. Institute of Geology and Mineral Exploration (IGME), Athena, Hellas, Open File Report E-6700 (in English).
Demetriades A (ed) (1999a) Geochemical atlas of the Lavrion urban area for environmental protection and planning. Volume 1: Explanatory text. Project "Soil rehabilitation in the Municipality of Lavrion", EU LIFE programme Contract No.: 93/GR/A14/GR/4576. Institute of Geology and Mineral Exploration, Athena, Hellas, Open file report E-8272, 365 pp.
Demetriades A (ed) (1999b) Geochemical Atlas of the Lavrion urban area for environmental protection and planning. Volume 2: Geochemical atlas. Project "Soil rehabilitation in the Municipality of Lavrion", EU LIFE programme Contract No.: 93/GR/A14/GR/4576. Institute of Geology and Mineral Exploration, Athena, Hellas, Open file report E-8272, 210 pp.
Demetriades A (ed) (1999c) Geochemical Atlas of the Lavrion Urban Area for Environmental Protection and Planning. Volume 1A: Figures and Tables. Project "Soil rehabilitation in the Municipality of Lavrion", EU LIFE programme Contract No.: 93/GR/A14/GR/4576. Open File Report, Institute of Geology and Mineral Exploration, Athena, Hellas, 210 pp.
Demetriades A (ed) (1999d) Geochemical Atlas of the Lavrion Urban Area for Environmental Protection and Planning. Volume 1B: Appendix reports. Project "Soil rehabilitation in the Municipality of Lavrion", EU LIFE programme Contract No.: 93/GR/A14/GR/4576. Open File Report, Institute of Geology and Mineral Exploration, Athena, Hellas, 176 pp.
Demetriades A (ed) (1999e) Environmental management plan for the rehabilitation of soil in the Lavrion Urban Area. Project "Soil rehabilitation in the Municipality of Lavrion", EU LIFE programme Contract No.: 93/GR/A14/GR/4576. Open File Report, Institute of Geology and Mineral Exploration, Athena, Hellas, Volume 4, 155 pp.
Demetriades A, Stavrakis P (1995a) Risk assessment and rehabilitation of contaminated soil at the Phenikodhassos in Lavrion. Institute of Geology and Mineral Exploration, Athena, Hellas, Open file report E-7489, 38 pp. (text in Greek with a summary in English).
Demetriades A, Stavrakis P (1995b) The release of toxic elements from the Lavrion slags. Institute of Geology and Mineral Exploration, Athena, Hellas, Open file report E-7610, 46 pp. (text in Greek with a summary in English).
Demetriades A, Vergou-Vichou K (1999a) Geochemistry of parent rocks. In A Demetriades (ed) Geochemical atlas of the Lavrion urban area for environmental protection and planning. Project "Soil rehabilitation in the Municipality of Lavrion", EU LIFE programme Contract No.: 93/GR/A14/GR/4576. Institute of Geology and Mineral Exploration, Athena, Hellas, Open file report E-8272, Volume 1 – Explanatory text, Chapter 4, 92–100.
Demetriades A, Vergou-Vichou K (1999b) Chemistry of metallurgical processing wastes. In A Demetriades (ed) Geochemical atlas of the Lavrion urban area for environmental protection and planning. Project "Soil rehabilitation in the Municipality of Lavrion", EU LIFE programme Contract No.: 93/GR/A14/GR/4576. Institute of Geology and Mineral Exploration, Athena, Hellas, Open file report E-8272, Volume 1 – Explanatory text, 101–128.
Demetriades A, Vergou-Vichou K (1999c) Geochemistry of parent rocks. In A Demetriades (ed) Geochemical atlas of the Lavrion urban area for environmental protection and planning. Project "Soil rehabilitation in the Municipality of Lavrion", EU LIFE programme Contract No.: 93/GR/A14/GR/4576. Institute of Geology and Mineral Exploration, Athena, Hellas, Open file report E-8272, Volume 1A – Figures and Tables, Appendix 2A, 18–38.
Demetriades A, Vergou-Vichou K (1999d) Geochemistry of overburden. In A. Demetriades (ed) Geochemical atlas of the Lavrion urban area for environmental protection and planning. Project "Soil rehabilitation in the Municipality of Lavrion", EU LIFE programme Contract No.: 93/GR/A14/GR/4576. Institute of Geology and Mineral Exploration, Athena, Hellas, Open file report E-8272, Volume 1 – Explanatory text, Chapter 6, 129–189.
Demetriades A, Stavrakis P, Vergou-Vichou K (1994a) Maps of the environmental geochemistry study of Lavreotiki peninsula. Environmental Geochemistry Study – Lavreotiki Peninsula. Institute of Geology and Mineral Exploration, Athena, Hellas. Open File Report E-7424, Vol. 2, 36 pp. (in Greek with an English summary).
Demetriades A, Stavrakis P, Vergou-Vichou K (1994b) Environmental soil geochemical survey of Lavreotiki peninsula Attiki. Environmental Geochemistry Study – Lavreotiki Peninsula. Institute of Geology and Mineral Exploration, Athena, Hellas. Open File Report E-7424, Vol. 3, 147 pp. (in Greek with an English summary).

Demetriades A, Vergou-Vichou K, Stavrakis P (1994c) Orientation soil geochemical survey in the Lavreotiki peninsula Attiki. Environmental Geochemistry Study – Lavreotiki Peninsula. Institute of Geology and Mineral Exploration, Athena, Hellas. Open File Report E-7424, Vol. 6, 64 pp. (in Greek with an English summary).

Demetriades A, Stavrakis P, Vergou-Vichou K (1996a) Contamination of surface soil of the Lavreotiki peninsula (Attiki, Greece) by mining and smelting activities. Miner Wealth 98:7–15.

Demetriades A, Stavrakis P, Vergou-Vichou K, Makropoulos V, Vlachoyiannis N, Fosse G (1996b) Lead in the surface soil of Lavreotiki peninsula (Attiki, Greece) and its effects on human health. In Aug Anagnostopoulos, Ph Day, D Nicholls (eds) Proceedings Third International Conference on Environmental Pollution. Aristotelean University of Thessaloniki, Thessaloniki, Hellas, 143–146.

Demetriades A, Stavrakis P, Vergou-Vichou K (1997) Exploration geochemistry in environmental impact assessment: Examples from Greece. In PG Marinos, GC Koukis, GC Tsiambaos, GC Stournaras (eds) Engineering Geology and the Environment (Vol. 2, pp. 1757–1762). Rotterdam: Balkema.

Demetriades A, Vergou-Vichou K, Stavrakis P (1999a) Geochemistry of parent rocks. In A Demetriades (ed) Geochemical atlas of the Lavrion urban area for environmental protection and planning. Project "Soil rehabilitation in the Municipality of Lavrion", EU LIFE programme Contract No.: 93/GR/A14/GR/4576. Institute of Geology and Mineral Exploration, Athena, Hellas, Open file report E-8272, Volume 2 – Geochemical atlas, Chapter 4, 4.2–4.30.

Demetriades A, Vergou-Vichou K, Stavrakis P (1999b) Chemistry of metallurgical processing wastes. In A Demetriades (ed) Geochemical atlas of the Lavrion urban area for environmental protection and planning. Project "Soil rehabilitation in the Municipality of Lavrion", EU LIFE programme Contract No.: 93/GR/A14/GR/4576. Institute of Geology and Mineral Exploration, Athena, Hellas, Open file report E-8272, Volume 2 – Geochemical atlas, Chapter 5, 5.1–5.30.

Demetriades A, Vergou-Vichou K, Stavrakis P (1999c) Distribution of lead in the Lavrion urban area. In A Demetriades (ed) Geochemical atlas of the Lavrion urban area for environmental protection and planning. Project "Soil rehabilitation in the Municipality of Lavrion", EU LIFE programme Contract No.: 93/GR/A14/GR/4576. Institute of Geology and Mineral Exploration, Athena, Hellas, Open file report E-8272, Volume 2 – Geochemical atlas, Chapter 3, 3.1–3.20.

Demetriades A, Vergou-Vichou K, Stavrakis P (1999d) Geochemistry of overburden. In A Demetriades (ed) Geochemical atlas of the Lavrion urban area for environmental protection and planning. Project "Soil rehabilitation in the Municipality of Lavrion", EU LIFE programme Contract No.: 93/GR/A14/GR/4576. Institute of Geology and Mineral Exploration, Athena, Hellas, Open file report E-8272, Volume 2 – Geochemical atlas, Chapter 6, 6.1–6.30.

Demetriades A, Vergou-Vichou K, Stavrakis P (1999e) Location map, mining & regional geochemical maps of Lavreotiki peninsula and composite geochemical maps of Lavrion urban area. In A Demetriades (ed) Geochemical atlas of the Lavrion urban area for environmental protection and planning. Project "Soil rehabilitation in the Municipality of Lavrion", EU LIFE programme Contract No.: 93/GR/A14/GR/4576. Institute of Geology and Mineral Exploration, Athena, Hellas, Open file report E-8272, Volume 2 – Geochemical atlas, Chapter 1, 1.1–1.6.

Demetriades A, Vergou-Vichou K, Stavrakis P, Vassiliades E (1999f) General thematic maps of the Lavrion urban environment. In A Demetriades (ed) Geochemical atlas of the Lavrion urban area for environmental protection and planning. Project "Soil rehabilitation in the Municipality of Lavrion", EU LIFE programme Contract No.: 93/GR/A14/GR/4576. Institute of Geology and Mineral Exploration, Athena, Hellas, Open file report E-8272, Volume 2 – Geochemical atlas, 2.1–2.16.

Demetriades A, Vergou-Vichou K, Vassiliades E (1999 g) General thematic maps of the Lavrion urban area: Digital topography, Lithology, Metallurgical processing residues & contaminated soil, Soil and House dust pH. In A Demetriades (ed) Geochemical atlas of the Lavrion urban area for environmental protection and planning. Project "Soil rehabilitation in the Municipality of Lavrion", EU LIFE programme Contract No.: 93/GR/A14/GR/4576. Institute of Geology and Mineral Exploration, Athena, Hellas, Open file report E-8272, Volume 1 – Explanatory text, 14–27.

Demetriades A, Vergou-Vichou K, Vlachoyiannis N (1999 h) Distribution of lead in the Lavrion urban environment. In A Demetriades (ed) Geochemical atlas of the Lavrion urban area for environmental protection and planning. Project "Soil rehabilitation in the Municipality of Lavrion", EU LIFE programme Contract No.: 93/GR/A14/GR/4576. Institute of Geology and Mineral Exploration, Athena, Hellas, Open file report E-8272, Volume 1 – Explanatory text, Chapter 3, 63–91.

Demetriades A, Vassiliades E, Tristán E, Rosenbaum MS, Ramsey MH (1998) Child blood lead content as a basis for risk assessment of the metallurgical processing residues and contaminated soil in Lavrion Attiki. Institute of Geology and Mineral Exploration, Athena, Hellas. Open File Report E7977 (text in Greek and English).

Demetriades A, Vergou K, Tsompos P, Stefouli M (2004) The utilisation of the contamination results of Lavreotiki peninsula in land use planning. Proceedings of the 10th Scientific of S.E. Attiki, Kalivia Thorikon Attiki, 28 November – 1 December 2002. Society of Studies of South-east Attiki, Kalivia Thorikon, Hellas, 149–177 (text in Greek).

Demetriades A, Vergou K, Vlachoyiannis N (2008) The contamination of Lavreotiki peninsula and the urban environment of Lavrion from the mining-metallurgical wastes and the effects on the health of the local population. Proceedings of the 9th Scientific Meeting of S.E. Attiki, Lavrion Attiki, 13–16 April 2000. Society of Studies of South-east Attiki, Kalivia Thorikon, Hellas, 573–624 (text in Greek).

Dermatis GN (1994) Landscape and monuments of Lavreotiki. Municipality of Lavreotiki, Lavrion, 298 pp. (text in Greek).

Drossos Ch, Papadopoulou-Ntafioti Z, Mavroidis K, Michalodimitriadis D, Salamalikis L, Gounaris A, Varonos D (1982) Environmental contamination by lead in Greece. Paediatrics, Athena, Hellas, 45, 114–124 (text in Greek).

Duggan MJ (1983) Contribution of lead in dust to children's blood lead level. Environ Health Perspect 50:371–381.

Eastman JR (1997) Idrisi for Windows (version 2.0) (1997). Clark University, Massachusetts.

Eikmann Th, Michels S, Makropoulos V, Krieger Th, Einbrodt HJ, Tsomi K (1991) Cross sectional epidemiological study on arsenic excretion in urine of children and workers in Greece. Gordon and Breach Science Publ. Toxicol Environ Chem 31–32:461–466.

Ferguson C, Marsh J (1993) Assessing human health risks from ingestion of contaminated soil. Land Contam Reclamat 1:177–185.

Ferguson C, Darmendrail D, Freier K, Jensen BK, Jensen J, Kasamas H, Urzelai A, Vegter J (eds) (1998) Risk Assessment for Contaminated Sites in Europe, Vol. 1, Scientific Basis (165 pp.). Nottingham, UK: LQM Press.

Ferguson JE (1990) The heavy elements: Chemistry, Environmental Impact and Health Effects (614 pp.). Oxford: Pergamon Press.

Fosse G, Wesenberg GBR (1981) Lead, cadmium, zinc and copper in deciduous teeth of Norwegian children in the pre-industrial age. Int J Environ Stud 16:163–170.

Gelaude P, Kalmthout PV, Rewitzer C (1996) Laurion – The minerals in the ancient slags (195 pp.). Nijmegen, The Netherlands: Janssen Print.

Gough LP, Shacklette HT, Case AA (1979) Element concentrations toxic to plants, animals and man. US Geological Survey Bulletin, 1466, 80 pp.

Hadjigeorgiou-Stravrakis P, Vergou-Vichou K (1992) Environmental geochemistry study of the Lavrion and Ayios Constantinos (Kamariza) area in Attica. Institute of Geology and Mineral Exploration, Athena, Hellas, Open File Report E6778, 33 pp. (in Greek with an English summary).

Hadjigeorgiou-Stravrakis P, Vergou-Vichou K, Demetriades A (1993) The contribution of geochemical exploration in the study of interior and exterior quality of the Lavrion and Ayios Constantinos (Kamariza) urban areas Attiki. In Proceedings Heleco'93, First International Exhibition and Conference of Environmental Technology for the Mediterranean Region. Technical Chamber of Greece, Athena, Hellas, Vol. II, 301–313 (in Greek with an English abstract).

Hatzakis A, Kokkevi A, Katsouyanni K, Maravelias C, Salaminios F, Kalandidi A, Koutselinis A, Stefanis K, Trichopoulos D (1987) Psychometric intelligence and attentional performance deficits in lead-exposed children. In SE Lindberg, TC Hutchinson (eds) International Conference Heavy Metals in the Environment, New Orleans, September. CEP Consultants, Edinburgh, Vol. 1, 204–209.

Hoffman SJ (1986) Soil sampling. In WK Fletcher, SJ Hoffman, MB Mehrtens, AJ Sinclair, I Thompson (eds) Exploration geochemistry: Design and interpretation of soil surveys. Reviews in Economic Geology, Vol. 3. Society of Economic Geologists, Univ. of Texas, El Paso, Texas, 39–77.

IGME Working Group (1987) Exploration results in Lavrion for the location of mixed sulphide mineralisation (geology-ore deposits-prefeasibility study). Institute of Geology and Mineral Exploration, Athena, Hellas, Open File Report E6411, 124 pp. (in Greek).

Jackson JA (ed) (1997) Glossary of geology. American Geological Institute, Alexandria, Va., 769 pp.

Kafourou A, Touloumi G, Makropoulos V, Loutradi A, Papanagiotou A, Hatzakis A (1997) Effects of lead on the somatic growth of children. Arch Environ Health 52(5): 377–383.

Katerinopoulos A, Zissimopoulou E (1994) Minerals of the Lavrion mines. The Greek Association of Mineral and Fossil Collectors, Athena, Hellas, 304 pp.

Kazantzis G (1973) Metal contaminants in the environment. Practitioner 210:482–489.

Kontopoulos K, Komnitsas A, Xenidis A, Papassiopi N (1995) Environmental characterisation of the sulphidic tailings in Lavrion. Miner Eng 8(10):1209–1919.

Korre A, Durucan S (1995a) The application of geographic information systems to the analysis and mapping of heavy metal contamination around Lavrio mine workings. APCOM XXV Conference, Brisbane, 579–585.

Korre A, Durucan S (1995b) Assessment of soil contamination. In Environmental Management in the Minerals Industries (ENVIRO-MIN), Med-Campus C035, 5.1–5.19.

Korre A (1997) A Methodology for the Statistical and Spatial Analysis of Soil Contamination in GIS. Unpublished Ph.D. thesis. University of London (Imperial College of Science, Technology and Medicine), 205 pp.

Kürzl H (1988) Exploratory data analysis: Recent advances for the interpretation of geochemical data. J Geochem Explor 30(3):309–322.

Leleu M (1966) Données nouvelles sur la paléogéographie et les rapports des séries métallifères du Laurium (Attique, Grèce). Comptes Rendus de l'Académie des sciences, Paris, 262:2008–2011.

Leleu M (1969) Essai d'interprétation thermodynamique en métallogénie: les minéralisations karstiques du Laurium (Grece). Bulletin Bureau de Recherche Géologiques et Minières (BRGM), 2e s., 4, 1–66.

Leleu M, Neumann M (1969) L'âge des formations d'Attique (Grèce): du Paléozoïque au Mésozoïque. Comptes Rendus de l'Académie des sciences, Paris, 268: 1361–1363.

Levinson AA (1980) Introduction to Exploration Geochemistry (924 pp.). Wilmette, Illinois, USA: Applied Publishing Ltd.

Makropoulos V, Konteye C, Eikmann Th, Einbrodt HJ, Hatzakis A, Papanagiotou G (1991) Cross-sectional epidemiological study on the lead burden of children and workers in Greece. Gordon and Breach Science Publ., U.K. Toxicol Environ Chem 31–32:467–477.

Makropoulos W, Stilianakis N, Eikmann Th, Einbrodt HJ, Hatzakis A, Nikolau-Papanagiotou A (1992a) Cross-sectional epidemiological study of the effect of various pollutants on the health of children in Greece. Fresenius Environ Bull 1:117–122.

Makropoulos W, Jakobi K, Stilianakis N, Vlachogiannis N, Pesch T, Tambakis S (1992b) Blood and cadmium burden in pregnant women, newborns and schoolage children in Lavrion (Greece). Wissenschaft und Umweit 3:221–224 (in German with an abstract in English).

Manousakis G (1992) Trace Elements in Human Health (204 pp.). Thessaloniki, Greece: Kiriakidis Brothers. (text in Greek).

Manthos GK (1990) Mining and Metallurgical Lavrion (168 pp.). Lavrion, Hellas: Municipality of Lavrion. (text in Greek).

Maravelias C, Hatzakis A, Katsouyanni K, Trichopoulos D, Koutselinis A, Ewers U, Brockhaus A (1989) Exposure to lead and cadmium of children living near a lead smelter at Lavrion, Greece. Sci Total Environ 84:61–70.

Marinos GP, Petrascheck WE (1956) Laurium. (In Greek with an extended abstract in English). Geological and Geophysical Research, IV, no. 1. Institute for Geology and Subsurface Research, Athena, Hellas, 246 pp.

Mervyn L (1985) The dictionary of minerals. The Complete Guide to Minerals and Mineral Therapy (224 pp.). Wellingborough, UK: Thorsons Publishing Group.

Mervyn L (1986) Thorsons Complete Guide to Vitamins and Minerals (336 pp.). Wellingborough, UK: Thorsons Publishing Group.

Nakos G (1979a) Environmental contamination by lead: The fate of lead in soil and its effects on *Pinus halepensis*. Ministry of Agriculture, Hellenic Forestry Research Institute, Report No. 105, Athena, Hellas, 34 pp. (in Greek with an English summary).

Nakos G (1979b) Lead pollution: Fate of lead in the soil and its effects on *Pinus halepensis*. Plant Soil 53:427–443.

Nakos S (1985) Blood Lead Levels and Renal Tubular Function in Children Living in an Area Polluted with Lead. Unpublished Ph.D. thesis. University of Ioannina, Hellas, 97 pp. (in Greek with English summary).

Nikolaidis N, Demetriades A, Vergou-Vichou K, Vassiliades E, Papassiopi N, Theodoratos P, Varelidis N, Zamani A (1999a) Environmental management plan for the rehabilitation of soil in the Lavrion urban area. In A Demetriades (ed) Environmental management plan for the rehabilitation of soil in the Lavrion urban area. Project "Soil rehabilitation in the Municipality of Lavrion", EU LIFE programme Contract No.: 93/GR/A14/GR/4576. Institute of Geology and Mineral Exploration, Athena, Hellas, Open file report E-8272, Volume 4, 72–121.

Nikolaidis N, Demetriades A, Vergou-Vichou K, Vassiliades E, Papassiopi N, Theodoratos P, Varelidis N, Zamani A (1999b) Environmental management and planning. In A Demetriades (ed) Environmental management plan for the rehabilitation of soil in the Lavrion urban area. Project "Soil rehabilitation in the Municipality of Lavrion", EU LIFE programme Contract No.: 93/GR/A14/GR/4576. Institute of Geology and Mineral Exploration, Athena, Hellas, Open file report E-8272, Volume 2 – Geochemical Atlas, 12.1–12.6.

NTUA (1999) Environmental Characterisation of Lavrion Site – Development of remediation techniques. Volume 3. Project "Soil rehabilitation in the Municipality of Lavrion", EU LIFE programme Contract No.: 93/GR/A14/GR/4576. National Technical University of Athens, Athena, Hellas. Open File Report, 174 pp.

O'Connor PJ, Reimann C, Kürzl H (1988) A Geochemical Survey of Inishowen, Co. Donegal. Geological Survey of Ireland, Dublin.

Papadeas GD (1991) Recent considerations for the geological-tectonic evolution of the metamorphic rocks in Attiki and Variskia mineralization. Proc Acad Athens 66: 331–370.

Photiades A (2003) Geological structure of the Lavreotiki area (Attica, Greece). In A Demetriades (ed) Lavreotiki Excursion Guide, 29th August 2003, Excursion leaders A Demetriades, A Photiades and C Panagopoulos, 7th Biennial S.G.A. Meeting "Mineral Exploration and Sustainable Development, August 24–28, 2003, Athena, Hellas, 13–18.

Photiades A, Carras N (2001) Stratigraphy and geological structure of the Lavrion area (Attica, Greece). Bull Geol Soc Greece 34(1):103–109.

Photiades A, Carras N, Mavridou F (2004) Geological Map of Greece in Scale 1:50.000 "Lavrion-Makronisos" Sheet. Athena, Hellas: Institute of Geology and Mineral Exploration.

Photiades A, Saccani M (2006) Geochemistry and tectonomagmatic significance of HP/LT metaophiolites of the Attic-Cycladic zone in the Lavrion area (Attica, Greece). Ofioliti 31(2):89–102.

Pöppelbaum M, Van den Boom G (1980a) Prospecting for hidden ore deposits. A method for determining mercury (Hg) concentrations in hard rock. In W Ernst, JH Hohnholz, HR Lang, KH Jacob, SV Wahl (eds) Natural Resources and Development, a Biannual Collection of Recent German Contributions Concerning the Exploration and Exploitation of Natural Resources (Vol. 11, pp. 68–77). Tübingen, Germany: Institute of Scientific Co-operation.

Pöppelbaum M, Van den Boom G (1980b) Eine Methode zur Bestimmung der Quecksilbergehalter in Festgesteinen. Geol Jb Hannover D37:5–14.

Reagan PL, Silbergeld EK (1990) Establishing a health based standard for lead in residential soils. In DD Hemphill (ed) Trace Substances in Environmental Health – XXII. Supplement to Volume 12 of Environmental Geochemistry and Health (pp. 199–238). Columbia: University of Missouri.

Reimann C, Äyräs M, Chekushin V, Bogatyrev I, Boyd R, Caritat P de, Dutter R, Finne TE, Halleraker JH, Jæger Ø, Kashulina G, Lehto O, Niskavaara H, Pavlov V, Räisänen ML, Strand T, Volden T (1998) Environmental Geochemical Atlas of the Central Barents Region. Geological Survey of Norway, Trondheim, 745 pp.

Reimann C, Filzmoser P, Garrett RG, Dutter R (2008) Statistical Data Analysis Explained – Applied Environmental Statistics with R (343 pp.). Chichester, UK: J. Wiley & Sons, Ltd.

Robbins JC (1973) Zeeman spectrometer for measurement of atmospheric mercury vapour. In MJ Jones (ed) Geochemical Exploration 1972 (pp. 315–323). London: Institution of Mining & Metallurgy.

Roels HA, Buchet JP, Lauwerys RR, Bruaux P, Claeys-Thoreau F, Lafontaine A, Verduyn G (1980) Exposure to lead by the oral and the pulmonary routes of children living in the vicinity of a primary lead smelter. Environ Res 22: 81–99.

Sainsbury CL, Hudson T, Kachadoorian R, Richards D (1970) Geology, mineral deposits, and geochemical and radiometric anomalies, Serpentine Hot Springs Area, Seward Peninsula, Alaska. U.S. Geological Survey Bulletin, 1312-H, H1-H19.

Schmitt N, Philion JJ, Larsen AA, Harnadek M, Lynch AJ (1979) Surface soil as a potential source of lead exposure for young children. Can Med J 121:1474–1478.

Siegel FR (1974) Applied Geochemistry (353 pp.). New York: J. Wiley & Sons.

Shell (1994a) Human exposure to soil pollutants, version 2.10a (HESP). Shell Internationale Petroleum Maatschappij B.V. The Hague, The Netherlands.

Shell (1994b) Human exposure to soil pollutants, version 2.10a (HESP). The concepts of HESP-Reference manual. Shell Internationale Petroleum Maatschappij B.V. The Hague, The Netherlands.

Skarpelis N (2007a) The Lavrion deposit (SE Attica, Greece): Geology, mineralogy and minor elements chemistry. Neues Jahrbuch für Mineralogie Abhandlungen 183: 227–249.

Skarpelis N (2007b) Lavrion: Ancient silver mines in a young extensional structural domain. In G Lister, M Forster, U Ring (eds) Inside the Aegean Metamorphic Core Complexes. J Virt Explor, Electronic Edition, ISSN 1441-8142, 27 pp, paper 8.

Skarpelis N, Argyraki A (2009) Geology and Origin of Supergene Ore at the Lavrion Pb-Ag-Zn Deposit, Attica, Greece. Resour Geol 59(1):1–14.

Skarpelis N, Tsikouras B, Pe-Piper G (2008) The Miocene igneous rocks in the Basal Unit of Lavrion (SE Attica, Greece): petrology and geodynamic implications. Geol Mag 145:1–15.

Stavrakis P, Vergou-Vichou K, Fosse G, Makropoulos V, Demetriades A, Vlachoyiannis N (1994a) A multidisciplinary study on the effects of environmental contamination on the human population of the Lavrion urban area, Hellas. In SP Varnavas (ed) Environmental Contamination. 6th International Conference, Delphi, Greece, CEP Consultants, Edinburgh, 20–22.

Stavrakis P, Demetriades A, Vergou-Vichou K (1994b) Environmental impact assessment in the Lavreotiki peninsula Attiki. Institute of Geology and Mineral Exploration, Athena, Hellas. Open File Report E7424, Vol. 1, 44 pp. (in Greek with an English summary).

Spais AG, Omirou A, Argiroudis S, Roubies N (1987) Lead poisoning in thoroughbred foals and mares of a stud established in the mining area of Lavrion. 4th Hellenic Veterinary Congress. Book of abstracts (in Greek). Hellenic Veterinary Medical Society, Athena, Hellas.

Stanek EJ, III, Calabrese EJ (1995) Daily estimates of soil ingestion in children. Environ Health Persp 103: 276–285.

Taylor SR, McLennan SM (1995) The geochemical evolution of the continental crust. Rev Geophys 33:241–265.

Thornton I, Culbard E (eds) (1987) Lead in the Home Environment (224 pp.). Northwood, UK: Science Reviews Ltd.

Thornton I, Davies DJA, Watt JM, Quinn MJ (1990) Lead exposure in young children from dust and soil in the United Kingdom. Environ Health Perspect 89:55–60.

Trattler R (1987) Better Health Through Natural Healing (624 pp.). Wellingborough, Northamptonsire, UK: Thorsons Publishing Group.

Tristán E, Rosenbaum MS, Ramsey MH (1998) Evaluation of child exposure to lead in Lavrion as a basis for risk assessment, Part II. In A Demetriades, E Vassiliades, E Tristán, MS Rosenbaum, MH Ramsey (eds) Technical Report, Child blood lead contents as a basis for risk assessment of the metallurgical processing residues and contaminated soil in Lavrion Attiki. Institute of Geology and Mineral Exploration, Athena, Hellas, Open File Report 7977-E, 69 pp.

Tristán E, Ramsey MH, Thornton I, Kazantzis G, Rosenbaum MS, Demetriades A, Vassiliades E, Vergou-Vichou K (1999a) Spatially resolved hazard and exposure assessments. In A Demetriades (ed) Geochemical atlas of the Lavrion urban area for environmental protection and planning. Project "Soil rehabilitation in the Municipality of Lavrion", EU LIFE programme Contract No.: 93/GR/A14/GR/4576. Institute of Geology and Mineral Exploration, Athena, Hellas, Open file report E-8272, Volume 1 – Explanatory text, Chapter 11, 311–349.

Tristán E, Ramsey MH, Thornton I, Kazantzis G, Rosenbaum MS, Demetriades A, Vassiliades E (1999b) Risk assessment. In A Demetriades (ed) Geochemical atlas of the Lavrion urban area for environmental protection and planning. Project "Soil rehabilitation in the Municipality of Lavrion", EU LIFE programme Contract No.: 93/GR/A14/GR/4576. Institute of Geology and Mineral Exploration, Athena, Hellas, Open file report E-8272, Volume 2 – Geochemical Atlas, Chapter 11, 11.1–11.18.

Tristán E, Demetriades A, Ramsey MH, Rosenbaum MS, Stavrakis P, Thornton I, Vassiliades E, Vergou-Vichou K (2000) Spatially resolved hazard and exposure assessments: An example of lead in soil at Lavrion, Greece. Environ Res Section A, 82:33–45.

US EPA (United States Environmental Protection Agency) (1994) Integrated Exposure Uptake Biokinetic Model for Lead in Children (version 0.99d). Washington, DC: Environmental Protection Agency.

US EPA (United States Environmental Protection Agency) (1986) Carcinogen Risk Assessment Guidelines. 51 FR 33992. Washington, DC: Environmental Protection Agency.

US EPA (United States Environmental Protection Agency) (1996) Exposure Factors Handbook. Washington, DC: Environmental Protection Agency.

US EPA (United States Environmental Protection Agency) (1998) Lead in Your Home: A Parent's Reference Guide. EPA 747-B-98-002, (70 pp.). Washington, DC: Environmental Protection Agency.

Voudouris P, Melfos V, Spry PG, Bonsall TA, Tarkian M, Solomos Ch (2008) Carbonate-replacement Pb–Zn–Ag±Au mineralization in the Kamariza area, Lavrion, Greece: Mineralogy and thermochemical conditions of formation. Mineral Petrol 94(1–2):85–106.

Vourlakos N (1992) The minerals of Lavreotiki and the mineral components of its rocks. Library of the Society of Lavreotiki Studies, Lavrion, Publication No. 5, 31 pp. (text in Greek).

Watt JM, Thornton I, Cotter-Howells J (1993) Physical evidence suggesting the transfer of soil Pb into young children via hand to mouth activity. Appl Geochem Suppl. Issue 2:269–272.

Wedepohl KH (1995) The composition of the continental crust. Geochim Cosmochim Acta 59:1217–1232.

Wixson BG, Davies BE (1994) Lead in soil – Recommended guidelines. Science and Technology Letters, Northwood, Middlesex, UK, 132 pp.

Xenidis A, Komnitsas K, Papassiopi N, Kontopoulos A (1997) Environmental implications of mining activities in Lavrion. In PG Marinos, GC Koukis, GC Tsiambaos, GC Stournaras (eds) Engineering Geology and the Environment (Vol. 3, pp. 2575–2580). Rotterdam: A.A. Balkema.

Yankel AJ, Von Linden IH, Walter SD (1977) The Silver Valley study: The relationship between childhood blood levels and environmental exposure. J Air Pollut Cont Assoc 27: 763–767.

Index

O. Selinus et al. (eds.), *Medical Geology,* International Year of Planet Earth,
DOI 10.1007/978-90-481-3430-4, © Springer Science+Business Media B.V. 2010